AGRICULTURAL ECOLOGY

AGRICULTURAL ECOLOGY

AN ANALYSIS OF WORLD FOOD PRODUCTION SYSTEMS

George W. Cox
Michael D. Atkins

SAN DIEGO STATE UNIVERSITY

W. H. FREEMAN AND COMPANY
San Francisco

The cover drawing is from "The Chinampas of Mexico" by
Michael D. Coe. Copyright © 1964 by Scientific American, Inc.
All rights reserved.

Sponsoring Editor: Gunder Hefta
Project Editor: Nancy Flight
Manuscript Editor: Kate Maloy
Designer: Robert Ishi
Production Coordinator: M. Y. Mim
Illustration Coordinator: Batyah Janowski
Artist: Eric G. Hieber, E H Technical Associates
Compositor: Advanced Typesetting Services of California
Printer and Binder: Kingsport Press.

Library of Congress Cataloging in Publication Data

Cox, George W 1935–
    Agricultural ecology.

    Includes bibliographies and index.
    1. Agricultural ecology. 2. Agriculture.
I. Atkins, Michael D., joint author. II. Title.
S589.7.C69   630   78–25745
ISBN 0–7167–1046–3

9 8 7 6 5 4 3 2 1

To

Ira and Edna Cox
William and Dorothy Atkins

for their many years of help.

CONTENTS

PREFACE

The heritage of agriculture is a heritage of oneness with nature. About twelve thousand years ago, human populations in scattered regions of the earth began gradually to engage more in husbandry than in simple hunting and gathering. Their techniques were tested rigorously by the laws of survival, and the ones that met the test were sustained by a strong ecological rationale. For thousands of years these lines of agricultural evolution were pursued. Each growing season, established practices were subjected to new tests, and new modifications were screened by the challenges of nature. The results of the developmental pathways, the systems of crop and animal agriculture that we see around us today, are not only of practical importance but also of immense theoretical interest. They constitute one of the most intricate examples of cultural and biological evolution acting in concert, a natural process unique to humans and their biological associates. They also demonstrate the actual unity of applied and theoretical ecology.

This unity has been mistreated by the artificially branching growth of modern science. Although no absolute barriers to communication among scientists have been erected, some significant discontinuities have developed. In liberal arts schools, natural scientists have tended to pursue studies that concentrate upon theoretical topics relating to the physical and biological world and have generally avoided applied topics. In the biological sciences, botanists, vertebrate zoologists, entomologists, geneticists, and ecologists have fragmented science still further. In the agricultural schools, real-world problems have been attacked with strongly applied programs of research and education, and ivory tower approaches have been scorned. In the social sciences, archeology and social anthropology have been pursued in even greater isolation and have often been regarded by scientists in the other areas as a quaint set of activities of neither theoretical nor applied interest.

The unity of human ecology has been lost in these patterns of separation. It is our conviction that the time is ripe for an organized, interdisciplinary approach to agricultural science. Each of the major areas that touch upon agricultural science is demonstrating a growing interest in the activities and developments in neighboring areas. What was theoretical yesterday is being applied today. Programs of applied research are increasingly the source of new theoretical insights. Examination of human cultures of both the past and the present reveals striking ecological rationales behind what were once thought to be arbitrary patterns of human activity.

We have undertaken the writing of this textbook in order to accelerate the integration of these areas. Our efforts have been supported strongly by the interest of students in this interdisciplinary topic. To them it is an approach that is meaningful and satisfying, and they are quick to explain their reasons for this feeling. Agricultural ecology has a strong theoretical basis, yet is eminently practical; it deals with processes in nature, yet relates these processes to man in intimate fashion; it concentrates on scientific phenomena, yet recognizes the ties that exist to economics, politics, and other fields of human activity. This response from students has motivated us to carry this project to completion.

We must acknowledge the assistance and contribution of many persons. Here at San Diego State University, several colleagues participated with us in developing a formal course in agricultural ecology, and we have drawn liberally from their ideas and information. We are therefore grateful to James Alexander, Charles F. Cooper, Donald Eidemiller, Albert W. Johnson, Calvert Norland, Ron Monroe, and Joy and Paul Zedler. Other colleagues here have provided helpful ideas and information and have offered criticism on early drafts of chapters; they are Adela S. Baer, Boyd D. Collier, David Farris, Warren Johnson, Philip Miller, and several of those listed above as course participants. Major portions of the manuscript were read and criticized by James W. Avault of Louisiana State University, Harold Breimyer of the University of Missouri, Gary W. Fick of Cornell University, and James D. McQuigg of Columbia, Missouri; we are exceedingly grateful for their valuable suggestions. Still other colleagues who have contributed in various ways are Solomon Bekure, of the International Livestock Center for Africa, Addis Ababa, Ethiopia; Michael Evenari, of the Botany Department, Hebrew University, Jerusalem, Israel; Roger and Cecile Hay, of UNICEF and the United Nations Development Program, Addis Ababa, Ethiopia; Ernst Hajek, of the Universidad Católica, Santiago, Chile; Zev Naveh, of the Technion, Haifa, Israel; Dwain Parrack, of Johns Hopkins University, Baltimore; Robert Ricklefs, of the University of Pennsylvania, Philadelphia; and William G. Wellington, of the Institute of Animal Resource Ecology, University of British Columbia. We especially wish to thank Harvey C. McCaleb, former editor at W. H. Freeman and Company, for encouragement, good advice, and effective editorial assistance. We are grateful to our wives, Darla G. Cox and Elinor Atkins, for their patience and assistance during the gathering of materials and the preparation of the manuscript. To our typists, Donnie Bocko, Kristine Creed, Jackie McClanahan, and Ann Smith, we extend sincere appreciation for accurate and efficient work.

We especially acknowledge the enthusiasm and encouragement provided by the students in Agricultural Ecology at San Diego State University and in Ecología Agraria at the Universidad Católica, Santiago, Chile.

George W. Cox
Michael D. Atkins

February 1979

INTRODUCTION

Man's agricultural activities have become the dominant ecological force over nearly one-third of the land area of the earth. The rest of the terrestrial environment is being subjected to more and more mineral exploitation, intensive forestry, recreational use, and watershed manipulation, with the result that man is rapidly advancing upon direct management of two-thirds of the earth's land area. In aquatic environments, a major portion of what many scientists believe is the maximum sustainable fisheries harvest is already being taken, and these systems, particularly in the oceans, are already being looked to as a future source of large amounts of energy and minerals. In rising to this position as a dominating ecological influence, man has become a significant geochemical agent. He is mobilizing mineral elements, converting substances from one chemical form to another, increasing the flux of chemical substances through the environment, modifying the composition of the atmosphere and hydrosphere, and stimulating the geological erosion of the continents—all on a scale rivaling that of natural processes. Both the intentional and the inadvertent consequences of these actions benefit and damage the food production activities upon which man depends absolutely.

As man's activities increase in scale, his dependence upon their successful pursuit becomes greater and more complex. The human population, now over four billion, is growing at a rate of about 2 percent per year. Continued growth at this rate means that by the year 2000 another three billion people will be added, a number roughly equal to the total population of the earth in 1960. The short-term success and long-term capability of the food supply systems that are developed to feed this population will determine much of the course of humanity's future.

Human agricultural systems are diverse in their characteristics, perhaps as diverse now as at any time in human history. Permanent systems of crop cultivation range from those that depend almost completely on human and animal labor to intensive, highly mechanized systems that depend on large inputs of fuels and man-made agricultural chemicals. Throughout the tropics a variety of nonpermanent systems, including shifting cultivation and pastoralism, are pursued by millions of people. All these systems function within distinctive cultural and economic contexts, an understanding of which is essential to any attempt to improve or modify them.

The pattern of human agriculture is also a

changing one, and the rate of change is destined to increase. Future trends will reflect to a great extent the pattern of expansion of mechanized agriculture. We estimate that roughly 40 percent of the world's crop cultivation is now carried out by intensive, relatively mechanized techniques. The remaining 60 percent seems to be divided about equally between nonmechanized, permanent cultivation and shifting cultivation. Continued expansion of intensive, mechanized cultivation in the immediate future is a certainty; this expansion will be at the expense of less intensive systems and arable lands now occupied by permanent pastures or natural plant communities. The natural communities likely to be affected most are tropical forests and savannas, temperate woodlands and prairies, and, through irrigation, deserts. As crop cultivation encroaches further upon pasture land and natural communities, intensive ranching operations will be forced to expand further into natural grassland areas, including the vast areas of dry tropical savanna now occupied by native pastoralists. This "domino effect" will not only magnify the local and global ecological perturbations related to intensive agriculture but will also generate severe cultural upheaval. The lives of people engaged in subsistence agriculture are intimately tied to their agricultural practices. Mechanized agriculture may both displace these people and disrupt the long-established family and tribal patterns of social organization through which they have successfully adjusted to their environment.

All these patterns of agriculture operate within the context of the global ecosystem. Agricultural systems, often spatially discontinuous, are interspersed among natural ecosystems with which they are intimately associated. Not only do they operate according to the same natural laws and ecological principles that govern surrounding natural systems, they also exert a substantial biological and physical impact upon both bordering and remote ecosystems.

Man's agricultural systems are the product of cultural evolution operating within an ecological framework. When man first walked across the face of the earth his impact on his environment was no greater or smaller than that of any other omnivorous species. His impact increased with his acquisition of fire and improved hunting technology. Through the constant refinement and use of his innate intelligence, man continued to expand his ability to exploit the environment. The biotic and physical resources of this environment nevertheless influenced the direction and extent of his cultural evolution. Changes occurred at different rates in different situations. In certain environments, man progressed little beyond hunting and gathering; in other regions he evolved from hunter-gatherer to semisedentary herder or settled farmer. Great civilizations developed where the climatic and biotic resources were conducive to settlement and permanent crop cultivation.

The early agricultural systems created by man stimulated further cultural evolution. The cultivation of plants and the domestication of animals expanded the need for simple tools. The response to this need signaled the beginning of a revolution that changed the nature of man's interaction with his environment over the entire earth. At the outset, when man relied on simple tools and his own labor, he lived in close association with his crops and herds. Agricultural products were harvested and consumed, and wastes were returned to the environment in the same locality. Later, as his technology improved and he harnessed the energies of animals, water, and wind for agricultural work, the spatial constraints on his activities were eased. It became possible for people to gather in villages and towns. At first this was largely for protection, with villagers moving out daily to cultivate surrounding fields. Subsequently, the capability of individual farmers to produce beyond a subsistence level released segments of the human population to other activities such as specialized professions pursued largely in expanding urban centers. This initial separation of centers of production and consumption was perpetuated under various feudal systems, and it later formed the basis for the establishment of plantation systems throughout the world.

The foregoing developments formed the beginning of what has become one of the dominant features of modern agriculture: the exportation of farm products from the site of production, their consumption elsewhere, and the minimum return of wastes to the land. This pattern reached full development with the industrial revolution. Industrialization fostered rapid urbanization in areas with access to major sources of energy and raw materials. In addition to isolating large human populations from areas of food production, the industrial revolution also gave birth to the powerful agricultural tools that enabled an increasingly smaller fraction of the human population to meet the food and fiber needs of growing urban populations. The pattern of modern society, characterized by food production in rural areas, consumption in remote urban centers, and the disposal of wastes into sewer systems far from the site of production, was thus established.

The patterns created by urbanization and industrialization have major economic and political implications for agriculture. The livelihood of the farmer has come to depend on patterns of supply and demand established largely in urban centers. Public policy with respect to agriculture strongly reflects the attitudes of populations so removed from agriculture that they know little about how it operates. Indeed, even the farmer's perception of what constitutes sound agricultural practice frequently has been influenced by industrial and commercial interests centered in urban areas.

Not all societies have evolved this extreme separation of production and consumption, however. All the cultural stages that lead to intensive, mechanized agriculture are in existence today. Some groups of Australian aborigines and African Bushmen continue to be hunters and gatherers; various African and Asian peoples pursue relatively undisturbed pastoralism; swidden (slash-and-burn) cultivation remains a major form of subsistence agriculture throughout the tropics. In Africa, Asia, and South America a variety of permanent cultivation systems that are based largely on human and animal labor still operate. In areas such as the Valley of Mexico and parts of the Middle East, vestiges of specialized agricultural systems of the past still survive. Ecologists and agriculturalists are thus provided with an opportunity to study and evaluate a wide variety of food supply and production systems.

A number of studies already conducted have shown that some of these nonmechanized systems possess a broad ecological rationale. It has also become apparent that the disappearance of some of the specialized agricultural systems of the past was not the result of ecological failure. Nevertheless, there is lasting evidence that environmental destruction has been caused by ill-conceived agricultural practices in both the past and the present. Careful examination of the ecological basis for the long-term success of some systems and the failure of others should help us to develop ecologically sound approaches to agriculture.

These questions relate directly to the imbalance between human populations and food resources. Despite the establishment of a number of national family planning programs, the global trend is still one of rapid population growth. Unless food production is substantially increased, this trend will simply add to the millions already suffering from malnutrition and, in some areas, facing starvation.

Attempts to achieve a revolutionary increase in food production through the expansion of mechanized agriculture without a full awareness of ecological and social repercussions are premature at best. Modern agricultural technology is not time-tested, even in areas where it has led to greatly improved yields with few apparent ill effects. We can state with reasonable certainty that intensive, mechanized agriculture is causing some degree of long-term ecological damage. We should not look blindly to current agricultural technology for a quick solution to the problems of overpopulation and underproduction. Our objective must be to achieve an ecologically sound approach with the potential for meeting the long-range needs of a stabilized human population.

We can readily recognize some of the problems that have been generated by mechanized agriculture in countries such as the United States. Intensive cultivation of the soil, disruption of decomposer food chains, depletion of organic matter in the soil, exhaustion of native nutrient pools, and the exposure of the soil surface to sun and precipitation have led to severe deterioration of soil structure and fertility. The addition of large quantities of highly soluble fertilizers, combined in many locations with irrigation, have increased the outflow of nutrients from cultivated areas and caused the eutrophication of adjacent aquatic ecosystems. The synchronized cultivation of single, genetically homogeneous crop species over large areas has increased the vulnerability of crops to insect and disease outbreaks that in turn have led to the massive use of pesticides on certain crops. In many cases, this strategy of crop protection has backfired. By killing beneficial as well as pest species, chemical pesticides have frequently encouraged the rapid resurgence of pests and outbreaks of secondary pests. This increased instability within crop systems has been accompanied by pesticide contamination of remote ecosystems.

Mechanized, exporting agricultural systems further rely on constant inputs of energy, nutrients, and water, thereby contributing to problems of resource use and supply. In some systems the total input of fossil fuel calories (used in manufacturing and operating farm machinery, producing and distributing synthetic fertilizers and pesticides, controlling water for irrigation and processing and marketing crops)

far exceeds the caloric value of the harvest. The costs of many of these activities will increase. The demand for phosphorus to replenish depleted nutrient pools is making serious inroads on the world's supply of high-phosphate rock. Production of nitrogen fertilizers, which currently relies on natural gas both as a raw material and an energy source, is becoming more expensive. Extensive irrigation in arid regions is contributing to regional deficiencies in water supply that can be met only by major interbasin transfers of water, again at an increased energy cost.

The human race is committed to preventing starvation and reducing malnutrition. The past achievements of modern agriculture have contributed substantially to this goal; they have, at the least, enabled us to maintain existing standards of nutrition in the face of growing populations. To continue to do even this is an enormous task. In 1969 an FAO report concluded that underdeveloped countries must increase food production 80 percent by 1985 simply to maintain present nutritional standards, and that the increase must be 140 percent if these countries are to meet the greater food demands related to the greater affluence and purchasing power that have been projected for them. Industrialization, accompanied by an exponential increase in resource exploitation throughout the world, has made greater affluence a major factor in the food demand equation of underdeveloped as well as developed countries. The newfound wealth of producers of raw materials is accompanied by a general demand for better nutrition that creates more competition for commodities in world markets and drives prices higher. Furthermore, this increased ability to compete for food in world markets has diminished concern about overpopulation in some parts of the underdeveloped world.

The challenge is clearly enormous. It involves not only agriculture but the total relationship between man and the systems from which he extracts his food and fiber. To meet this challenge we must greatly reorganize our thinking in several areas. We must recognize that the earth possesses a human carrying capacity and that we must take deliberate action to keep populations from exceeding this capacity. We must also recognize agricultural systems as ecosystems and must apply an ecologically sound agricultural technology to increase production without, in the long run, destroying agricultural lands or damaging global ecology. Furthermore, we must be aware of the cultural and economic contexts within which change must occur. The weaknesses of single-crop economies in parts of the underdeveloped world, and the recent socio-political setbacks of the Green Revolution, attest to the dangers of ignoring such relationships.

The human strategy for agriculture must be designed to avoid a variety of predictable disasters that could seriously reduce the earth's productive capacity. The developed countries are in a position to lead in this regard. The strategy that these nations offer must not only be ecologically sound but must consider the economic and sociological implications for the underdeveloped countries.

In the following chapters we shall explore all these considerations. We hope that what follows will not only stimulate an interest in agricultural ecology but will also reveal the ecological fitness of past and present agricultural systems as a basis for developing an ecologically sound approach to agriculture in the future.

Part One

THE ECOLOGICAL AND HISTORICAL CONTEXT OF AGRICULTURE

1

THE WORLD FOOD BALANCE

The world's most influential nations have changed during this century from societies dominated by rural attitudes and values into urban-industrial societies with very different value systems. This transformation has been due primarily to a technological revolution that has affected all areas of human activity: production, transportation, communication, and health. More specifically, this change has been effected by the mobilization of new and more effective sources of energy, which has increased the productivity of human labor; that is, the output of goods and services per worker.

Through business and industry, energy-intensive technology has created a high material standard of living for some of the world's population and has increased the opportunity for creative human potential to be realized in other areas of endeavor. These benefits, which are most readily available in urban areas, have acted as magnets for rural dwellers, many of whom have been permitted or even forced to migrate because of increased labor productivity in rural areas that has come about through the application of mechanized techniques, controlled irrigation, concentrated fertilizers, and many other technological achievements.

These changes have been viewed as steps toward a human independence from nature. In reality, just the opposite is true. Man's dependencies on the natural world are greater and more varied than ever; his impacts upon his environment are also more intense and diverse. Changes in the cost of materials unimportant a century ago can now throw national economies into recession. Pollutants produced in one location enter the global systems of air and water circulation to reappear in unexpected fashion.

This interaction with nature is even more obvious in food supply systems. Farming, although it requires good business sense and the use of varied products of industry, remains basically an ecological enterprise. It is an activity in which natural ecosystems, open to the influences of climate, substrate, and wild biota, are modified to increase yields of desired food and fiber products. The greater the changes in the basic patterns of structure and function that prevail in the natural system, the greater is the human effort necessary to maintain the agricultural system (Cox and Atkins 1976).

The influences of the technological revolution, intense in the developed nations but distributed very unevenly throughout other areas of the world, have not all been beneficial. Rapid depletion of nonrenewable resources, accelerated human population growth, and the disruptive impacts of industrial societies on the cultures and environments of preindustrial peoples have prevented most of the world's population from enjoying the full benefits of the revolution. We shall begin our discussion of agricultural ecology by examining this discrepancy in detail. It is the primary motivating force, not only for this book, but for any ecological approach to the topic of human food production.

HUMAN POPULATIONS AND FOOD DEMAND

The world food balance is the relation between the need for food and the supply of food available to meet that need. The need itself is a function of the size of the human population and the per capita demand of its individuals.

The world population, which had grown to 4.2 billion by the beginning of 1976, is still growing explosively. Table 1-1 indicates the extent to which this growth deviates from the pattern it has exhibited over most of human history. Before the agricultural revolution, about 8000 B.C., the rate of population growth was at most one-tenth of one percent each

century, despite the fact that during this period man spread from Africa to nearly all parts of the globe. After man began to practice agriculture, which revolutionized his relationship to his food resources, growth of the world population accelerated, but only to a rate of 5 to 6 percent per century. It was not until the industrial revolution that populations began to grow explosively. From 1750 to 1975, the population grew at an average rate of 110 percent per century. At the rate of growth that existed in 1975, however, the increase per century would be over 800 percent.

The size, age structure, and growth rate of regional populations vary greatly (Table 1-2). North America, Europe, and the USSR presently display population growth rates of 1.0 percent or less per year, which are high by standards of total human history but low compared to current rates in other regions. These populations are also characterized by both low birth rates and low death rates, and the low birth rates have prevailed long enough to create a population age structure with the smallest percentage of individuals under 15 years of age. Latin America, in contrast, shows the highest growth rate, 2.9 percent per year. This figure reflects relatively high birth rates, low death rates, and the fact that most of the population are young, with 43 percent under 15 years of age. This last fact means that there is a population growth momentum in this region; that is, even if families elect to have fewer children, the sheer percentage of the population at or approaching active reproductive age will force growth to continue for a long time. Much the same can be said for Asia, where about 60 percent of the world's population now live.

The pattern in Africa is somewhat different. Because it has the highest birth rates in the world, the greatest percentage of population under 15 years of age, and an annual growth rate close to that of Latin America, it has strong momentum for future growth. However, Africa also shows higher death rates than any other

TABLE 1-1
Human population size and growth rate per century over human history.

Period	Estimated population	Percent increase per century
400,000 B.C.	50,000 (?)	0.1
8000 B.C.	5,000,000	5.3
1 A.D.	300,000,000	5.7
1750 A.D.	791,000,000	110.1
1975 A.D.	4,147,000,000	

TABLE 1-2

11
Estimated world population statistics, 1975.

Region	Population (millions)	Annual growth rate (%)	Rates/1000/year Birth	Rates/1000/year Death	Percent urban	Percent under 15
North America	242.2	1.0	15.0	9.1	76.5	27
Latin America + Caribbean	327.6	2.9	38.0	11.0	60.4	43
Europe	474.2	0.8	15.4	10.4	67.2	26
USSR	254.3	0.9	18.2	8.7	60.5	29
Africa	420.1	2.8	47.0	21.0	24.5	44
Asia	2407.4	2.5	39.0	14.0	—	—
Oceania	20.9	2.1	23.0	9.7	71.6	33
World	4146.9	2.2	35.0	13.0	39.3	37

continental region. The death rates will probably decline before the birth rates do, and the result is likely to be the highest rates of population growth in human history.

It is clear from this discussion that the highest rates of population growth exist in the parts of the world where food and population imbalances are already greatest: Latin America, Asia, and Africa. At the current rates, the annual food production in these areas has to increase 2.5 to 3.0 percent simply to maintain existing levels of nutrition.

Rapid growth, however, is not the only major trend in human populations. The population explosion is accompanied by a **population implosion,** the rapid increase in the percentage of the population living in cities (Table 1-3). Although the first cities appeared about 5500

TABLE 1-3
Past and estimated future trends of urbanization of the world population.

	Year 1850 Number	%	Year 1950 Number	%	Year 1970 Number	%	Year 2000 Number	%
World population (millions)	1262		2502		3628		6335	
Rural	1181	93.6	1796	71.8	2229	61.4	2797	44.2
Urban	81	6.4	706	28.2	1399	38.6	3538	55.8
In cities (>100,000)	29	2.3	406	16.2	864	23.8	2399	37.9
In cities (>1 million)	13	1.0	182	7.3	448	12.4	1497	23.6

Source: Davis 1972.

TABLE 1-4
Land use, 1970.

Region	Total area	Arable land	Percent of world total	Grazing land	Forest land	Other	Potential arable land	Percent land arable
			Land area = hectares $\times$ 10^6					
Europe	493	148	10.4	93	139	113	174	85.1
USSR	2,240	233	16.4	375	910	722	356	65.4
North America	2,241	253	17.7	374	815	799	465	54.4
South America	1,784	90	6.3	408	927	359	680	13.2
Asia	2,753	449	16.3	449	571	1,284	627	71.6
Africa	3,030	204	14.3	842	647	1,337	733	27.8
Oceania	851	47	3.3	460	82	262	154	30.5
World total	13,392	1,424	10.6	3,001	4,091	4,876	3,189	44.6
Percent	100.00	10.63	—	22.41	30.55	36.41	23.81	—

Sources: UN Production Yearbook 24, 1970; President's Science Advisory Committee, *The World Food Problem*, vol. 2, 1967.

years ago in the Tigris and Euphrates valleys of the Middle East (Sjoberg 1965), the fraction of the human race living in cities remained small until the nineteenth century A.D. As recently as 1850, for example, only 6.4 percent of the population lived in cities with more than 5000 residents (Davis 1972). This means that, until quite recently, most people worked in rural occupations and relied on their own efforts for the bulk of their food supply.

Urbanization began in Western Europe and North America with the Industrial Revolution of the 1800s (Davis 1965). England was the first urbanized nation. In 1800 about 19 percent of the English population lived in cities of over 100,000; by 1900, primarily because of rural-urban migration, this had risen to about 40 percent, with over 70 percent of the population living in cities of at least 5000 persons. At this time, cities were unhealthy places to live, and mortality rates generally exceeded birth rates. This migration was so widespread in England that the rural population declined in absolute numbers, that is, in actual numbers as well as in the percent of total population, a situation that has occurred in all industrialized nations, including the United States.

In the developing countries urbanization is a much more recent phenomenon, dating from the period 1920 to 1940, and it has followed a somewhat different pattern. For example, health conditions in the cities of developing areas are frequently better than those in the countryside, and high birth rates prevail in both urban and rural areas. As indicated in Table 1-4 for Costa Rica, urbanization is the combined result of migration and population growth. Between 1927 and 1963, about 80 percent of the increase in Costa Rica's urban population was attributable to population growth and 20 percent to rural-urban migration. This means that urbanization in areas that show this pattern does not create a decline in the rural population.

Projections to the year 2000 A.D. (Table 1-3) indicate that this trend will continue and that by that date the majority of the world's population will live in cities. The significance of this

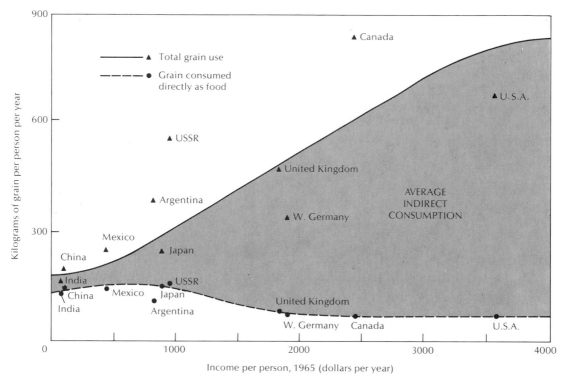

FIGURE 1-1.
Direct and indirect grain consumption per capita in relation to per capita income for selected countries. (Modified from Brown 1973.)

for food production is twofold: First, the bulk of the world population is becoming dependent upon food produced in one area and transported to another; second, the majority that influences political decisions relating to food production is becoming increasingly urban.

The demand for food by the human population is influenced by still another factor, however. The level of affluence, or per capita income, is closely related to the total agricultural yield that is consumed directly or indirectly by individuals (Brown 1973). Direct consumption refers to harvested foods that are eaten as such; indirect consumption refers to harvested foods that are fed to livestock or intensively refined to produce smaller amounts of high-quality food for human use. Figure 1-1 illustrates this in terms of per capita grain consumption. In the developing countries, such as China, India, and Mexico, per capita grain consumption lies between 100 and slightly over 200 kilograms annually. Almost all of this is consumed directly. In the developed countries, such as Canada, the United States, and the USSR, per capita consumption ranges between 450 and 900 kilograms per year. The greatest portion of this is not consumed directly, of course, but is fed to animals whose meat is consumed (Figure 1-2). Affluence, the ability to buy high-quality meats and other expensive foods, means that

FIGURE 1-2.
Most grain produced in the United States is used in intensive livestock feeding operations. (Photograph by E. E. Hertzog, courtesy of the U.S. Bureau of Reclamation.)

the average American is diverting for his own use 3 to 5 times as much grain as an individual in the developing areas mentioned.

Although the consumption permitted by affluence has been carried beyond the limits of good health in some countries, it is obviously beneficial up to a point. As we shall see, protein deficiency is a major problem in developing areas, and it is to be hoped that the amount of high-quality protein available to people in these countries can be increased. This will probably require increased per capita grain consumption. Furthermore, when incomes rise for any reason, the populations benefiting seek foods of better quality. Thus increased per capita food demands are now being created in the nations belonging to the Organization of Petroleum Exporting Countries (OPEC) because of incomes from oil exports (Brown 1975).

HUMAN FOOD PRODUCTION SYSTEMS

The systems that function to meet the food needs of the human population are diverse, but they can generally be grouped under six major headings:

1. Crop cultivation
2. Livestock and domestic animal production
3. Marine and freshwater fisheries
4. Hunting and gathering of wild foods
5. Aquaculture and mariculture
6. Production of synthetic food materials

While agriculture, in the traditional sense, has been considered to comprise only the first two of these systems, it is apparent that all are closely interrelated to the acquisition of food. The collapse of the Peruvian anchoveta fishery in 1972, for example, led to a sudden shift, in intensive U.S. beef and poultry production, from fish meal to soybean meal as a protein-rich feed component. This in turn reduced the availability and increased the price of soybeans for export, a product that Japan relied upon heavily for human consumption. Likewise, in China, where population density demands that virtually all grain production be used directly for human consumption, the culture of freshwater fish has developed much more than in other countries. Intensified efforts by nations throughout the world to control fishing activities in their coastal waters have been sparked by concern for the potential contribution of marine fisheries to the world food supply.

For these reasons we will consider all of the systems listed in our examination of agricultural ecology. Furthermore, if we examine systems of organized agriculture—crop and livestock production—on a worldwide basis, we find that these too are highly diversified. These systems have undergone both biological and cultural evolution over the past few thousand years, and the course of these developments has been influenced by characteristics of the natural environment. Man himself has been influenced by these evolutionary events. For example, his genetic capacity for lactose (milk sugar) metabolism seems to have been influenced by the pathway of agricultural development. European and North American societies, with a long history of lactose tolerance, sometimes forget that a large fraction of the world's adult population are not only unable to assimilate lactose, but even suffer digestive upset if they ingest it.

The history of biological and cultural evolution has led to an array of strikingly diverse human food production systems, many of which are unfamiliar to both ecologists and agricultural scientists. Systems of organized agriculture range from pastoralism, which itself involves many kinds of animals in a wide range of both moderate and hostile environments, through nonmechanized systems of both shifting and permanent crop cultivation, to the mechanized farming and livestock production systems of the industrialized nations. The diversity of nonmechanized systems of cultivation is astonishing; some emphasize grains, others root crops, and in some cases their efficiencies and levels of productivity are competitive with those of so-called modern systems of food production.

The world now possesses, in fact, a greater diversity of human food production systems than ever. Some of these systems face extinction, but all are more accessible to examination and scientific study now than they have ever been. Major portions of the human population still depend primarily on nonmechanized systems of food production. All of these systems are products of long developmental adjustment to local conditions, and many have been practiced for hundreds or thousands of years in essentially the same form and location. All possess strong ecological rationales and have been proven over time. Failure to examine these systems in detail and to evaluate their rationales can only be described as foolhardy. Nevertheless, the current global trend has been toward the rapid replacement of these systems by energy-intensive, mechanized techniques of recent origin. Any approach to agricultural ecology implies a serious, ecological examination of these longstanding patterns of food production before they are modified or destroyed by programs intended to bring about quick increases in food production (Cox and Atkins 1975).

AGRICULTURAL LAND

Agricultural lands are those used for crop cultivation or grazing by domestic animals. **Arable land** is that portion of agricultural land used for growing crops, including grains and vegetables, fiber plants such as cotton and sisal, specialty

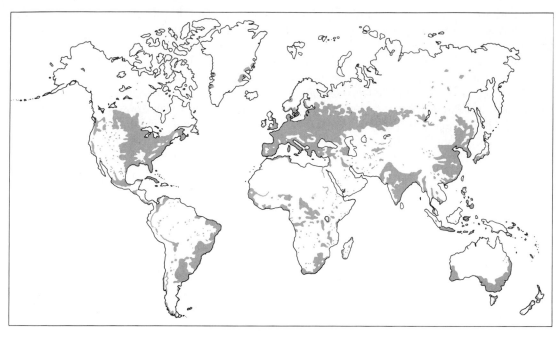

FIGURE 1-3.
Approximate world distribution of arable land. Because of the inadequacy of available information, and because of the scale of the map, unshaded areas include small amounts of cropland and shaded areas include small amounts of land not used for crops. (Photograph courtesy of the USDA.)

crops such as tobacco, harvested forages such as alfalfa, and permanent tree crops such as fruits, nuts, and rubber. Arable land also includes regularly cultivated land that may be fallow at certain times to allow moisture or nutrient build-up, or for other reasons. Land that might be converted to cultivation but that is not now in a regular cultivation cycle is called **potentially arable land.** Worldwide, **grazing land** includes areas that range from irrigated or naturally well-watered pastures to dry rangelands utilized by domestic species.

The world's arable land now constitutes slightly over 10 percent of its total area (Table 1-4). This percentage varies significantly for different continents, ranging from 3.3 percent in Oceania (largely Australia and New Zealand) to 17.7 percent in North America. The value for Africa, 14.3 percent, is deceptively high since it includes a large, but difficult to estimate, fraction of bush land that lies fallow under systems of shifting cultivation (see Chapter 5). The fairly low value of 6.3 percent for South America is correlated with the fact that much of the southern part of the continent is open, relatively dry grazing land while the Amazon Basin and other humid tropical areas are largely forest. Of course arable land is not uniformly distributed throughout these continental regions, and although exact mapping of its distribution is probably still impossible an approximation is given in Figure 1-3.

Relating arable land to population density creates a somewhat different picture (Table 1-5). Worldwide, about one-third of a hectare of arable land exists per person. In Asia, where almost 60 percent of the world's population is centered, arable land per capita equals only

TABLE 1-5
Per capita arable and potentially arable land in relation to 1975
regional world populations.

| Region | 1975 population (millions) | Land per capita (hectares) | |
		Arable	Potentially arable
Europe	474.2	0.31	0.37
USSR	254.3	0.92	1.40
North America	300.0	0.84	1.55
South America	269.0	0.33	2.53
Asia	2407.4	0.19	0.26
Africa	420.1	0.49	1.75
Oceania	20.9	2.25	7.37
World total	4146.9	0.34	0.77

0.19 hectares. Oceania, with the smallest population and smallest arable land area, possesses over two hectares per person and exports food. North America and the USSR both have nearly a hectare of arable land per person. In the USSR, however, much of this land is either at high latitudes or borders the arid central-Asian deserts and is therefore subject to unfavorable temperature and moisture conditions. North America, with its abundant arable land more favorably situated, is the world's largest exporter of basic food materials. Africa and South America possess half a hectare or less of arable land per capita.

The area of arable land changes, of course, as human population density and activities change. Tracing the growth of the arable land area of the world is difficult, and for most of South America and Africa it has been virtually impossible until quite recent times. Grigg (1974) has summarized much of the information available, for most of the world, over the period 1870 to 1970 (Table 1-6). These data cover about 70 percent of the world's total arable land that existed in 1970, and they reveal a number of interesting patterns. For example, by 1870 nearly all of the arable land of present-day Europe was under cultivation. Between 1870 and 1930, most of the increase in arable land occurred in North America, portions of South America, Russia, and Australia, all largely Temperate Zone regions undergoing active colonization. Much less spectacular increases occurred in China and the Indian region. By 1930, the major growth of arable land in several of these regions had been accomplished; little subsequent growth has occurred in North America and temperate South America. Since 1930, major increases in arable lands have occurred only in certain areas: in the 1950s, for example, the USSR achieved a large, if not spectacularly successful, increase in arable land through its virgin lands development program; beyond this, most recent increases in arable land have been in developing areas of the tropics and subtropics and are closely correlated with population growth.

Projecting future possibilities for increases in arable land is also difficult. Several estimates of the world's potentially arable land have been prepared, the most authoritative of which is the report by the President's Science Advisory Committee (PSAC 1967) on the world food problem. In this analysis, potentially arable land

TABLE 1-6
Change in the arable land area in various world regions, 1870–1970.

Region	Hectares $\times 10^6$					
	1870	1910	1930	1950	1960	1970
Europe	141	147	150	148	151	147
Russia	102	160	160	175	220	232
China	81	91	98	91	113	110
India and Pakistan	68	86	110	157	151	193
North America	80	154	196	220	183	218
Argentina and Uruguay	0.4	29	33	35	28	32
Australia	0.4	4	12	11	12	19
Japan	3.2	5.4	6.0	5.7	6.0	5.5
Java	5.6	7	8.4	–	8.8	–
Southeast Asia	3	9	12	13	15	17
Total	485	693	775	864	887	982

Source: Grigg 1974.

is considered to be land presently cultivated plus that which, on the basis of topography, soil depth, and climate, might reasonably be cultivated by means of agricultural technology equivalent to that used in the United States (Figure 1-4); it is therefore clear that a number of significant assumptions were made in this estimate. In any case, the areas of potentially arable land in various continents are indicated in Table 1-4. Nearly 3.2 million hectares of potentially arable land are judged to exist, of which about 45 percent are now being used. At face value, this suggests that the arable land of the world might be increased twofold.

We feel that the foregoing analysis is deceptive and dangerous. First, all arable land areas are not equally productive. The most productive of the earth's potentially arable lands are now in use; even if put into production by the same technology the remaining areas are unlikely to yield at comparable levels. Furthermore, major capital investments and technolog-

ical advances must be made before much of the remaining potentially arable land can be used for permanent cultivation. Irrigation is necessary, for example, before a single crop can be grown on any of the remaining potentially arable land in Asia. Likewise, in the humid tropical areas of Africa, South America, and other continents, major advances in agricultural technology will be necessary to permit profitable sustained-yield farming. Finally, as Table 1-4 clearly illustrates, the area of arable land can be increased only by decreasing the areas of other important ecosystems, such as grazing lands and forests. It will be instructive for the reader to determine how much potentially arable land exists on his continent and to visualize the changes that would occur in the landscape if it were put under cultivation.

Grazing lands, mostly rangelands with native vegetation supported by natural rainfall (Figure 1-5), cover about 22.4 percent of the earth's surface, or roughly twice the area covered by

FIGURE 1-4.
Arable land consists of land used for the production of food and fiber crops, including both fallow land and land used for tree and bush crops. It is exemplified by the land under irrigation in this photograph of the Coachella Valley in California. Potentially arable land also includes areas, like those in the righthand portion of the photograph, that might be cultivated by means of modern agricultural technology. (Photograph by E. E. Hertzog, courtesy of the U.S. Bureau of Reclamation.)

FIGURE 1-5.
Most of the world's grazing land, like this area of mountain rangeland in the western United States, consists of areas of native vegetation that depend on natural rainfall. (Photograph courtesy of the USDA.)

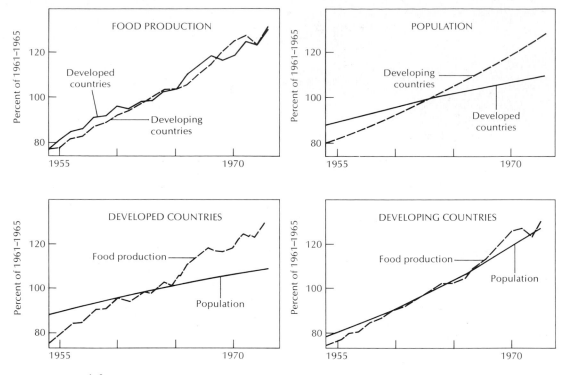

FIGURE 1-6.
Over the past two decades overall growth of food production has been nearly parallel in the developed and developing countries. In the developed countries, this growth has markedly exceeded that of population; but in the developing countries food production has little more than kept pace with population growth. Data do not include Communist Asia. (From the USDA 1974.)

land under cultivation (Table 1-4). Relative to total continental area, these lands are most extensive in Oceania, Africa, and South America. Altogether, crop cultivation and animal grazing dominate one-third of the earth's surface.

TRENDS IN WORLD FOOD PRODUCTION

Over the past 20 years or so, gross production of foods has increased at almost exactly the same rate in both the developed and developing countries (Figure 1-6); this rate averages about 2.8 percent per year (Walters 1975). In the developed countries, this increase in food production has been accompanied by a rate of population growth that has declined from 1.3 percent per year in the 1950s to about 0.9 percent per year in the mid-1970s. In the developing world, population growth rates have actually increased, rising from about 1.9 percent per year in 1950 to 2.5 percent per year in the mid-1970s.

Thus if we examine per capita trends in food production, we see very different relationships for the developed and developing countries. In developed areas (Figure 1-7) real per capita increases have averaged about 1.5 percent per

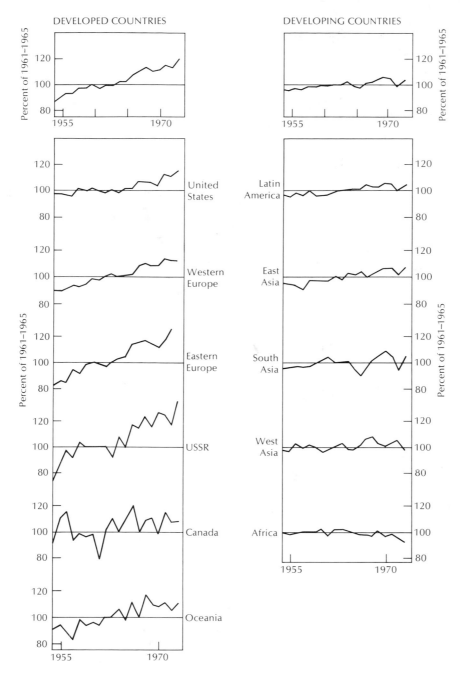

FIGURE 1-7.
In sharp contrast to patterns in most areas of the developed world, per capita food production in parts of the developing world has remained nearly constant over the past two decades or has even declined, as in Africa. (From the USDA 1974.)

FIGURE 1-8.
Pickers at work on tea plantations in the Toro District of Uganda illustrate the emphasis that is placed on cash crop production in many developing countries. (Photograph by Yutaka Nagata, courtesy of the United Nations and the World Bank.)

year. In many areas this has shown up as an increase in the diet of the amounts of animal protein and other highly processed foods; in other cases part of the increased production has been exported, instead of being consumed in the developed countries themselves. In developing countries, growth of food production has very nearly been cancelled by population growth (Figure 1-7). Overall, there has been an annual increase of 0.4 percent per year in per capita food production. If we examine patterns of increased food production in different regions of the developing world, however, we see that even this small rate of improvement has not been realized everywhere. Western Asia, for example, shows no clear trend for increase, and the developing areas of Africa show an apparent decline in per capita production since about 1961.

The slow growth of production in developing areas is further complicated by a heavy em-phasis on cash crops. Many of the countries involved, especially in Africa, have dualistic agricultural economies. One sector of the economy is engaged in cash crop production and uses modern machinery, productive inputs, and efficient transportation systems (Lofchie 1975). Most of the coffee, cocoa, tea, bananas, ground nuts, and nonfood products such as cotton, sisal, and tobacco are exported (Figure 1-8). Much of the modernization effort in developing countries has benefited this sector. The second sector of the economy is that which produces food for local consumption. Relegated to less productive lands, and benefiting only weakly from agricultural modernization, this sector often operates by subsistence methods that have changed little over the past century. In short, the small per capita improvement in agricultural production in the developing countries implies an even smaller increase in the food available to the mass of the people.

Much has been said and written in recent years about the so-called **Green Revolution.** Based on the development of short-stemmed varieties of irrigated rice and wheat that have high-yield responses to fertilizers and careful management practice, the Green Revolution was expected to produce a rapid increase in per-acre grain production in many developing tropical areas. Where these varieties have been introduced and grown under controlled irrigation and full management practices, yields have increased significantly. However, the small fraction of arable lands that are subject to fully controlled irrigation, and the difficulties and rising costs of supplying fertilizer and other requirements, have made the actual gains quite modest. In some cases they have been further reduced by unexpected pest and disease problems. Moreover, in the Philippines, where high-yielding rice varieties have been grown for more than ten years, their yields on rain-fed lands are no better than those of traditional varieties and are only 16 percent better on irrigated lands (Herdt and Wickham 1975). In India, where high-yielding varieties of both rice and wheat have been introduced, the trend line of annual increase in grain production fails to show a significant change in slope after the introduction of Green Revolution technology (Figure 1-9). No revolutionary increase in overall food production seems to have occurred, and the net effect of the grain varieties involved may have been, at best, to maintain the same rate of increase in production that already existed (Sen 1974).

Inadequacy of food production has not just been quantitative. The quality of food, particularly of protein, has become seriously deficient in many areas. The production of protein-rich plant and animal foods has not grown parallel with overall food production. In some cases, for example, grain production has increased by cultivating grain on lands that were once used for growing protein-rich plants such as legumes. In addition, world fisheries harvests, which

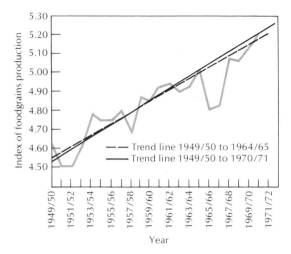

FIGURE 1-9.
The rate of growth in overall cereal grain production in India, as indicated by the difference in slope of the lines in the graph, has changed little as a result of the Green Revolution. The period 1949/50 to 1964/65 is before the Green Revolution; the period 1949/50 to 1970/71 includes the early years of the Green Revolution. The index of production is the natural logarithm of cereal grain production in millions of 1000-pound lots. (Modified from Sen 1974, courtesy of the author and Wiley Eastern Limited.)

peaked at 70 million tons in 1970, experienced consecutive years of decline from 1971 through 1973, which suggests that we may be approaching the limits to which wild fisheries can be exploited by traditional methods. We shall examine the implications of protein insufficiency more closely in later discussions of human nutrition.

INTERNATIONAL TRADE IN FOOD COMMODITIES

Since World War II, a major change has occurred in the degree to which national populations depend on imported food. Before the war, the only nations that regularly imported basic food commodities for a major portion of their

FIGURE 1-10.
Wheat is loaded aboard a ship at Newport News, Virginia. A railroad car (*upper left*) is tipped on its side and emptied into the ship's hold in about three minutes. (Photograph by Jack Schneider, courtesy of the USDA.)

national food supply were the most secure imperial powers, such as England at the height of its empire (Rothschild 1976). Since World War II, an increasing number of nations, including many developing countries, have become totally dependent on food imports along trade routes that reach halfway around the globe (Figure 1-10). This dependence, the product of population growth, insufficient attention to the development of internal agriculture, and international market influences, culminated in disorderly and dangerous market conditions in the mid-1970s.

Net grain imports by various sectors of the world economy show these trends in the 1960s and early 1970s (Figure 1-11). Although imports by developed countries have shown little average change over this period, those of developing countries have steadily increased. Imports have also increased for Communist countries, in which purchases have been irregular in occurrence and amount.

As imports continued to increase in the 1970s world grain stocks declined. United States stocks, over 100 million tons in the early 1960s (corresponding to about three years' export volume), fell to under one-third of that amount by 1974/75; what is more, the reserve production capacity of the United States, in the form of lands withheld from production under the soil bank program, dropped to zero.

In 1972, the United States suddenly rose to greater importance in world grain trade, with exports amounting to half or more of the world

FIGURE 1-11.
World grain stocks, gross exports, and net imports by various world regions from 1960/61 through 1974/75. (From Poleman 1975. Copyright © 1975 by the American Association for the Advancement of Science.)

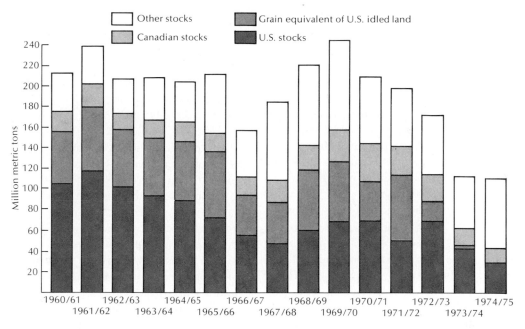

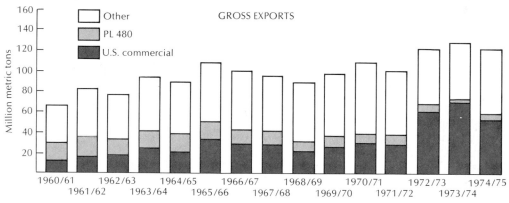

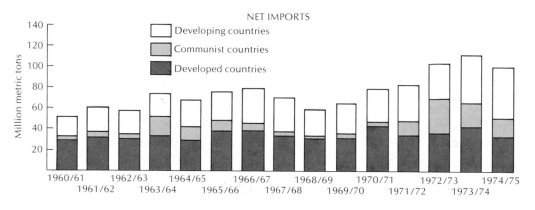

total. In addition, American exports changed from being largely subsidized to being almost entirely commercial. Before the early 1970s, about half of American exports were sold under some sort of foreign aid arrangement, much of them under Public Law 480 ("Food for Peace"). By the mid-1970s these arrangements applied to less than 15 percent of exported shipments (Rothschild 1976). Current American policy is to establish firm, long-term arrangements with major importers, particularly with those, such as the USSR, whose buying has tended to be erratic. Strong arrangements of this sort, involving the USSR, Japan, and various European countries, may cause problems for developing nations that need to import food and that may increasingly find themselves in competition for the leftovers from the American market.

This disorderly set of conditions in the world grain markets owes at least some of its origin to the developmental policies followed by some underdeveloped countries, and to American food aid policy during the surplus years of the 1950s and 1960s. Most countries saw the road to development and modernization as one that required emphasis on industrialization. This policy was supported by international organizations and by the U.S. Agency for International Development. The approach was essentially a shortcut: the development of a strong agricultural sector was to be skipped in favor of efforts to stimulate urban-industrial development. The possibility of obtaining cheap imports of U.S. grain, much of it subsidized under PL480, made it possible to embark upon this effort. In effect, however, what was occurring, deliberately or inadvertently, was the cultivation of grain markets by the United States (Poleman 1975; Rothschild 1976).

Under these circumstances—increased demand due to population growth and rising affluence, decreasing opportunities for expansion of agricultural land, and disorderly market conditions—a variety of problems relating to food adequacy are rising to a new prominence.

CHRONIC MALNUTRITION

Malnutrition can be caused by a dietary deficiency in any of a large number of required food components: vitamins, minerals and other essential elements, protein or its specific amino acid constituents, and total calories. Whereas all forms of malnutrition contribute to the global food imbalance, two of them—protein and calorie deficiencies—are difficult to remedy except by readjusting the basic relationship between human populations and the food productive capacity of their environment. When this relationship is bad it leads to an interlocked syndrome termed **protein-calorie malnutrition** (PCM). This syndrome is the dominant nutritional problem in the underdeveloped world (Jelliffe and Jelliffe 1975). PCM constitutes a sort of nutritional "iceberg" (Figure 1-12); severe forms are conspicuous and easily diagnosed, but they are indicative of much greater incidence of moderate to mild forms of similar deficiency.

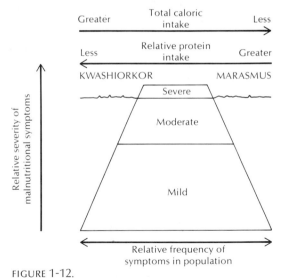

FIGURE 1-12.
Severe expressions of kwashiorkor and marasmus exhibit an "iceberg" effect. These expressions are indicative of much greater frequencies of moderate and mild forms of malnutrition in the population. (Modified from Jelliffe and Jelliffe 1975.)

TABLE 1-7
Essential and nonessential amino acids for man.

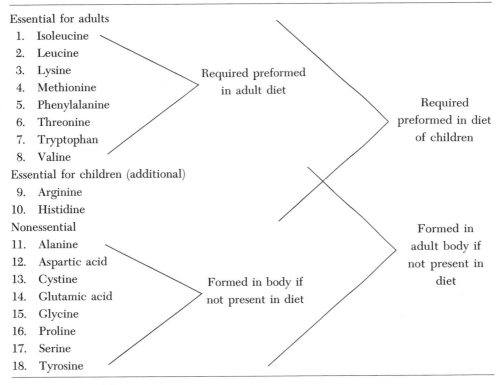

Essential for adults
1. Isoleucine
2. Leucine
3. Lysine
4. Methionine
5. Phenylalanine
6. Threonine
7. Tryptophan
8. Valine

Essential for children (additional)
9. Arginine
10. Histidine

Nonessential
11. Alanine
12. Aspartic acid
13. Cystine
14. Glutamic acid
15. Glycine
16. Proline
17. Serine
18. Tyrosine

Required preformed
in adult diet

Required
preformed in diet
of children

Formed in body if
not present in diet

Formed in
adult body if
not present in
diet

Source: Barrass 1974.

Two forms of severe PCM have been recognized: **kwashiorkor** and **marasmus.** Kwashiorkor results from a serious dietary deficiency of protein, even though total caloric intake is adequate. Proteins, composed of amino acid units, are basic constituents of practically all body systems and are required for both body maintenance and growth. Some 18 different amino acids are involved in the structure of human body proteins, of which eight (in adults) to ten (in children) must be ingested preformed in the diet (Table 1-7). These are termed the **essential amino acids.** Thus not only the quantity of protein but its quality (i.e., its quantitative spectrum of essential amino acids) is of dietary significance.

The amount of protein required in the diet varies with age, sex, body weight, and the demands of pregnancy and lactation (Table 1-8). Safe dietary levels, or intake levels that meet the needs of most persons, have been estimated to be about 37 grams and 29 grams per day respectively for men and women of average size and activity. For individuals engaged in strenuous activity, intake needs would be somewhat higher. These estimates, furthermore, are based on protein of optimal amino acid composition, such as that present in milk or eggs. In practice, the mixed diet contributes protein of less optimal composition. Diets of persons in developed countries contain protein averaging only 80 percent of the quality of

TABLE 1-8
Estimated safe levels of protein intake.

| | Average body weight (kg) | Requirement for egg or milk quality protein | | Corrected for protein quality | | |
		Per kg of body weight per day (g)	Per individual per day (g)	80%	70%	60%
Children						
1–3 years	13.4	1.19	15.9	20	23	27
4–6 years	20.2	1.01	20.4	26	29	34
7–9 years	28.1	0.88	24.7	31	35	41
Male adolescents						
10–12 years	36.9	0.81	29.9	37	43	50
13–15 years	51.3	0.72	36.9	46	53	62
16–19 years	62.9	0.60	37.7	47	54	63
Female adolescents						
10–12 years	38.0	0.76	28.9	36	41	48
13–15 years	49.9	0.63	31.4	39	45	52
16–19 years	54.4	0.35	29.9	37	43	50
Adults						
Men	65.0	0.57	37.1	46	53	62
Women	55.0	0.52	28.6	36	41	48
Extra for women						
Pregnancy (2nd half)	—	—	+8.5	+11	+13	+15
Lactation (1st 6 months)	—	—	+17.0	+21	+24	+28

Source: From FAO/WHO (1973).

milk or egg protein. In developing countries this value usually averages about 70 percent, and in areas of severe imbalance it equals about 60 percent. This means simply that, to make up for the poor amino acid composition of the particular proteins present, as indicated in Table 1-8, total food protein intake must be higher than the figures given earlier.

Kwashiorkor appears when there is a chronic insufficiency of protein nutrition. The symptoms of this disease include wasting of muscle tissues, hair and skin disorders, fluid accumulation in body tissues, and a distended abdomen (Figure 1-13). Because of their high protein requirements, children are especially susceptible. In early infancy, while brain growth and increase in cell number are still occurring, protein deficiency can limit brain cell numbers, an irreversible effect (Winick and Rosso 1975). Whether protein deficiency that occurs at later stages of development can be reversed is the subject of some controversy, but it seems likely that any major developmental impairment at any period of childhood will carry over to adulthood to some degree.

Kwashiorkor is prevalent where major reliance is placed on starchy foods such as cassava and maize. Cassava, a tropical tuber, contains

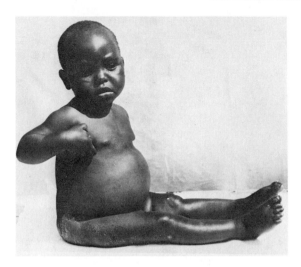

FIGURE 1-13.
Typical symptoms of kwashiorkor, as seen in this child from Kenya, include thinning of the hair, enlargement of the abdomen, edema of the legs, and dermatosis along the thighs. (Photograph courtesy of the Government of Kenya and the FAO, Rome.)

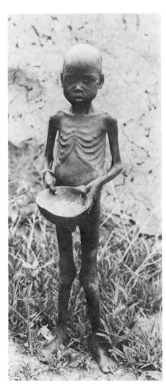

FIGURE 1-14.
Marasmus is evidenced by the loss of body hair and extreme emaciation, as seen in this child from Upper Volta, whose daily diet consisted of boiled millet or sorghum flour mixed with water. (Photograph courtesy of the FAO, Rome.)

only 1.2 grams of protein per 100 grams. The energy needs of a 1-to-3-year-old child would be met by about 350 to 400 grams of this food, which would provide only 4.2 to 4.8 grams of protein, an amount far below the need shown in Table 1-8.

Marasmus is the condition resulting from caloric inadequacy of the diet. It occurs as a chronic nutritional problem primarily in children (Figure 1-14). If it occurs before weaning, it is usually caused by inadequate lactation by the mother; after weaning, or in bottle-fed children, it is the result of food that is inadequate in quantity or else is too dilute (Bailey 1975). Actually, both marasmus and kwashiorkor have increased in many areas where there has been a shift away from breast feeding. In place of human milk containing adequate amounts of high-quality protein, bottle-fed infants often receive starchy synthetic formulas deficient in protein.

Describing the quantitative extent of PCM is difficult because severe cases grade gradually into moderate and mild forms of malnutrition. The 1970 Rome World Food Conference estimated that 25 percent of the population of the underdeveloped world (434 million people) had inadequate protein-calorie diets (Table 1-9). Although this estimate certainly includes a large number of people whose dietary deficiencies are only moderate, human populations exist in a number of areas of Asia, Africa, and Latin America in which up to 7 percent of the children exhibit severe PCM and the majority show some degree of debilitation. Overall it has been estimated that at a given time 10 to 20 million children exist with symptoms of PCM

TABLE 1-9
Numbers of people estimated by the FAO to have had an
insufficient protein-energy supply in 1970, by region.

Region	Population (millions)	Portion with insufficient supply	
		Percent	Number
Developed	1074	3	28
Developing[a]	1751	25	434
Latin America	283	13	36
Far East[a]	1020	30	301
Near East	171	18	30
Africa	273	25	67
World[a]	2825	16	462

[a]Excluding China.
Source: Poleman (1975).

severe enough that death is probable unless they are treated (Bailey 1975; Jelliffe and Jelliffe 1975; Okeahialam 1975).

Many of the debilitating effects of PCM lead to increased susceptibility of the individual to infection, which may lead in turn to further deterioration of nutritional condition. This type of interaction is called **synergism,** and essentially means that the two problems combined lead to more than their additive effects. The normal defense mechanisms of the body, including the immune system and the phagocytic white cell system, are weakened by nutritional deficiency (Latham 1975), and infectious agents can more easily enter the body and become established. Poor nutrition also weakens the ability of the body to combat disease when it flares up. Diseases such as measles that are mild and easily overcome in well-nourished populations can become fatal.

Disease also leads to nutritional deterioration. Infection and fever increase the loss of protein from the body and thus raise the dietary intake requirements. More seriously, intestinal infections, particularly bacterial diarrheas, can impair the ability of the intestine to absorb nutrients, creating a digestive inefficiency.

FAMINE

Famine can be defined as the failure of regional food production systems, which leads to increased mortality due to starvation and associated disease. Famine differs from chronic malnutrition in that its most significant effect is not the physical and mental crippling of individuals by long-continued poor nutrition, but the sharp rise in mortality caused by a relatively sudden lack of food. Two ecological features characterize regions prone to famine: seasonality of climate, and reliance by human populations on subsistence food production systems (Cox 1976). Virtually all famines, except those caused by wars and civil disruption, have occurred in areas where patterns of temperature or rainfall are both strongly seasonal and highly variable from year to year. Combined with subsistence food production, seasonality means that what is produced during a favorable season must be stored locally to meet the food needs of a subsequent unfavorable season. Any factor causing food production to fail or even to be severely reduced during the favorable season can thus lead to famine during the unfavorable season.

Hundreds of famines have been recorded throughout human history. In some regions, such as China and India, it is probably safe to say that before the development of modern transportation systems, famine somewhere within the country was a nearly annual occurrence. The immediate causes of food production failures are well known: unfavorable weather, plant and livestock diseases, war and civil disruption, and natural catastrophes such as earthquakes. Unfavorable weather conditions including droughts, cold and wet growing seasons, floods, and cyclones, constitute the most frequent cause of famine. In the Old World, two famine belts can be recognized. One, stretching from the British Isles across northern Europe and Asia to Japan, is an area in which famines have been triggered largely by cold, wet growing seasons, or by crop diseases that flourish under such conditions. Further south, from Africa eastward through the Middle East to India and China, lies a belt of drought-induced famine. In the northern belt, the risk of famine has largely been eliminated by modernized systems of transportation and effective government administration. In the southern belt, these systems remain much less fully developed, and averting incipient famine is more difficult.

The world may now be emerging from a half-century respite from famines triggered primarily by climatic factors. Following the 1928/29 famine in China, in which an estimated three million people died, most famines have been due to war and civil disruption until recently. The almost 50 years following the famine in China have represented one of the longest intervals of favorable food supply conditions in the last two thousand years. Climatically, the start of this half-century coincides with an increase in the mean temperature of higher latitudes of the northern hemisphere and with an increase in the reliability of monsoons in India (Bryson 1974). This period has also seen the most rapid growth of technological ability

and economic activity in human history and has provided man with new tools of production and improved means of transport. Both these factors have helped to alleviate problems of food production failure where they have tended to occur.

This period of favorable conditions and freedom from famine may be coming to an end, however. The climates that we have come to think of as normal, and under which national boundaries have grown rigid, may actually have been more abnormally favorable during this period than at any time during the past thousand years (Bryson 1974). A return to less favorable conditions, and to climatic problems such as the Sahelian drought (Figure 1-15) and weather-related crop failures in the USSR, may now be under way, and it may be complicated by the possibility that population growth has eroded the margin of safety that technology and economic development have provided, making the danger of climate-triggered famine a real and present one. Although chronic malnutrition and potential famine may be alleviated by short-term food relief, it is clear that deeper problems exist and that eliminating the threat of these problems requires recognition of the underlying causes of imbalance between populations and their food supply systems.

UNDERLYING CAUSES OF MALNUTRITION AND FAMINE

Prior to the worldwide spread of Western influences and the urban-industrial culture that grew out of the Industrial Revolution, many human societies possessed a stable relation with the environments in which they existed. The food supply systems that supported these societies were mainly those of subsistence and local market food production. It is certainly true that by modern standards most of these societies suffered many undesirable qualities: for example, low material standard of living, primitive

FIGURE 1-15.
During the height of the Sahelian drought, pastoral peoples, especially the women and children, were forced to congregate at relief locations to which emergency food was air-lifted. This photo was taken in Chad in August 1973. (Photograph by T. Page, courtesy of the FAO, Rome.)

and inadequate medical services, and limited opportunity for intellectual growth. Nevertheless, many of these groups had something of equal or greater importance: an ecologically stabilized relationship between the human population and the productive capacity of the environment.

These **ecologically stabilized societies** tended first of all to possess strong mechanisms of population limitation (Wilkinson 1974). Some of these, such as taboos on sexual intercourse under various circumstances, acted to reduce birth rates to well below the biological maximum; others, such as infanticide, reduced the effective recruitment of individuals into the population; and still others, such as ritualized warfare, raised death rates. Moreover, cultural beliefs and fears, such as those related to sorcery, helped to maintain wide dispersion of the human population in certain environments of low

productive potential. We regard many of these practices as cruel and inhuman, but we must also recognize that they fulfilled a critical ecological role in maintaining a stable population density.

The food supply systems of many of these groups were also characterized by a sound ecological rationale. The fact that these systems are more complex and diversified than most members of developed societies recognize and that they have survived for hundreds or thousands of years attests to the fact that they are ecologically adjusted to their environments. In areas subject to periodic climatic problems, there are also emergency food procurement systems of considerable capacity.

The spread of Western civilization and the industrial and economic activities growing out of the Industrial Revolution have transformed many of these indigenous cultures; and although there is good reason to believe that Western technology is capable of establishing ecologically stabilized societies at a new level, with improved material, medical, and intellectual conditions, the current trends in many of these indigenous cultures do not appear to be moving in this direction. The impact of Western civilization has been highly disruptive, and there is considerable doubt that the developed countries can themselves achieve a new, ecologically stabilized state.

Perhaps the most serious aspect of this problem is the fact that the impact of modern civilization upon indigenous cultures is unbalanced and that trends promoted in developing nations remain unbalanced. Culture is an integrated whole, a set of practices and beliefs that forms an integrated functional complex. This fact has two corollaries: (1) successful modification of one component requires accomodating change in other components, and (2) change in one cultural component can lead to second- and third-order changes in other components, some of which are not easy to anticipate (Foster 1962).

The imbalance in impacts and trends that relate to populations and food supply are evident in many forms: death rate reduction without birth rate reduction, cash crop development without attention to the food supply of cash crop laborers, improvement of water supply for livestock without regulation of grazing intensity, and emphasis on urban-industrial development without parallel development of a strong agricultural sector. Earlier we suggested that this last trend has been permitted by the surplus of production and the patterns of aid coming from the developed countries following World War II. Now we see, in a number of developing areas of the world, what amounts to the breakdown of internal food production systems that have been neglected for several decades while population growth and urbanization have continued.

PROBLEMS OF AGRICULTURAL INTENSIFICATION

Worldwide modernization of agriculture by the adoption of intensive mechanized agriculture patterned after that of Europe and North America has been widely believed to be the solution to world food production problems. Intensive mechanized agriculture has both weaknesses and strengths, however, that must be identified and evaluated in planning future patterns of agricultural development.

Mechanized agriculture has amply demonstrated its ability to generate distinctive environmental problems. The traditional problems of declining fertility and erosion are now accompanied by pollution with agricultural chemicals; chemical residues in harvested foods; secondary pest outbreaks triggered by pesticide treatments; and vulnerability of large monocultures to new pest and disease varieties. Powerful agricultural chemicals, ranging from hormones to pesticides, have been found to trigger cancer in experimental animals, and more and more forms of cancer are thought to be environmentally produced or aggravated. We tend to have strong faith in our technological ability to solve such problems when they arise, but this ability may not be infallible.

Intensive mechanized agriculture is typically regarded as being efficient, with efficiency measured in terms of production achieved per worker on the farm. The impression of efficiency suffers, however, when the system is viewed in terms of the entire labor force engaged in manufacturing the products and providing the services required by farm workers; and it appears still less efficient when one considers individuals employed in processing, transporting, and marketing food products. Furthermore, when production is compared to other investments—capital, land, synthetic chemicals, and energy—intensive mechanized agriculture fares poorly compared to many

FIGURE 1-16.
Agricultural lands have increasingly become sites for assembling inputs supplied through technology. Crop production on this land in the Imperial Valley of California is totally dependent on irrigation (see foreground) and on the use of synthetic nitrogen fertilizers (produced in the plant in the background). (Photograph by G. W. Cox.)

other systems (Figure 1-16). Intensive use of energy and chemicals is part of the price paid for labor efficiency on the farm, but it becomes increasingly clear that we must view agricultural activities in terms of efficiencies other than those of labor. The growing concern about adequacy of future energy supplies makes energy efficiency one of the central concerns of agricultural ecology (see Chapter 24).

THE EDUCATIONAL GAP

Andre and Jean Mayer (1974) have recently described the gap that has developed in science and education between agricultural science and other areas of natural science, a gap for which both groups of scientists are responsible. They point out that traditional biologists have regarded agricultural science as an applied subject that is important only to its practitioners. Research in applied aspects of biology related

to agriculture has been actively discouraged. Likewise, agricultural scientists have tended to regard biology and ecology as abstract, academic fields in which little of the activity pursued has meaning in the real world. The system of higher education in the United States, which separates agriculture from other natural sciences in different schools and universities, has formalized these differences. As a result, scientists in the two areas talk only among themselves, read their own journals, attend their own meetings, and develop their own courses and curricula.

The unfortunate consequence of this academic isolation, according to the Mayers, is that in the curricula of many major universities agricultural science is not represented at all. Basic ecology has similarly been ignored in many agricultural schools and universities. Where both areas are represented, they are generally separated by administrative and curricular barriers that are rarely crossed by students or faculty. This isolation, unfortunately, is not limited to universities, but exists in government as well.

We agree that this long-maintained separation has been detrimental to both fields. Agricultural research has been allowed to proceed without adequate regard for genetic and nutritional factors or for the environmental impact of agricultural practices within and outside agricultural ecosystems. Ecologists have been discouraged from pursuing basic studies in agroecosystems, where there may arise not only important theoretical advances, but discoveries with important practical significance.

This separation is now beginning to break down. Farming is being recognized for what it really is: an ecological enterprise that deserves the concerted attention of natural scientists in both applied and theoretical areas. We suggest that the study of agricultural ecology is a unifying approach to the pursuit of this goal.

SUMMARY

Widespread concern has emerged, both in science and in society at large, about the balance between human populations and their food supply systems. Explosive population growth continues, with world population increasing at 2.2 percent per year. In certain areas, the food need created by this increase in numbers is compounded by increased per capita food consumption resulting from greater affluence. Worldwide, human populations are becoming increasingly concentrated in cities, creating urban majorities that have little firsthand contact with food production activities.

The food production systems serving this population are diverse and must be considered in a coordinated fashion; for example, factors that affect marine fisheries also have an impact upon crop agriculture. On a global scale, dealing with human food production also requires evaluating subsistence and nonmechanized agricultural systems as well as intensive mechanized systems.

Although some increase in the world's arable land seems possible, the potentially arable land that exists is less productive and more difficult to bring into use than that presently farmed. Thus most of the current and future growth in production must be through increased yields per acre. In recent years, annual increase in food production has barely kept pace with population growth in developing areas. In some developing countries, the per capita production of food has actually declined. Severe problems of chronic malnutrition and the danger of famine exist in a number of areas.

Much of this crisis in population and food supply reflects the breakdown of internal production systems in the developing countries, and this breakdown in turn reflects the uneven impacts of developed economies on those of developing areas—impacts that tend to destabilize what in many instances were ecologically stabilized societies. Patterns of world food trade since World War II have also encouraged inattention to agricultural development on the part of many developing nations.

In education, a serious gap has existed between agricultural science and other areas of natural science. The world food imbalance is a problem of agricultural ecosystems and their management and cannot be dealt with except through the coordinated efforts of both applied and theoretical natural scientists with a wide range of interests. Agricultural ecology is a combined discipline designed to foster this approach.

Literature Cited

Bailey, K. V. 1975. Malnutrition in the African region. *WHO Chronicle* 29:354–364.

Barrass, R. 1974. *Biology: Food and People.* New York: St. Martin's Press.

Brown, L. R. 1973. Population and affluence: growing pressures on world food resources. *Population Bull.* 29(2):1–31.

———. 1975. The world food prospect. *Science* 190:1053–1059.

Bryson, R. A. 1974. *World Climate and World Food Systems. III. The Lessons of Climatic History.* Univ. Wisc. Env. Stud., Report #27.

Cox, G. W. 1976. *The Ecology of Famine. Summary and Discussion.* AAAS Meeting, Boston. 20 Feb.

Cox, G. W., and M. D. Atkins. 1975. Agricultural ecology. *Bull. Ecol. Soc. Amer.* 56(3):2–6.

———. 1976. *Overfarming, Underfarming, and Agricultural Ecology.* (Unpublished manuscript.)

Davis, K. 1965. The urbanization of the human population. *Sci. Amer.* 213(3): 40–53.

———. 1972. The role of urbanization in the development process. *Int. Tech. Coop. Ctr. Rev.* (Tel Aviv) 1:1–13.

FAO/WHO. 1973. *Energy and Protein Requirements.* WHO Technical Report Series, No. 522.

Foster, G. M. 1962. *Traditional Cultures and the Impact of Technological Change.* New York: Harper and Row.

Grigg, D. B. 1974. The growth and distribution of the world's arable land, 1870–1970. *Geography* 59(2):104–110.

Herdt, R. W., and T. M. Wickham. 1975. Exploring the gap between potential and actual rice yield in the Philippines. *Food Res. Inst. Studies* 14(2):163–181.

Jelliffe, D. B., and E. F. P. Jelliffe. 1975. Human milk, nutrition, and the world resource crisis. *Science* 188:557–561.

Latham, M. C. 1975. Nutrition and infection in national development. *Science* 188:561–565.

Lofchie, M. F. 1975. Political and economic origins of African hunger. *J. Mod. Afr. Stud.* 13(4):551–567.

Mayer, A., and J. Mayer. 1974. Agriculture, the island empire. *Daedalus* 103:83–95.

Okeahialam, T. C. 1975. Non-nutritional aetiological factors of protein-calorie malnutrition (PCM) in Africa. *J. Trop. Pediatrics and Env. Child Health* 21:20–25.

Poleman, T. T. 1975. World food: A perspective. *Science* 188:510–518.

President's Science Advisory Committee. 1967. *The World Food Problem*, vol. 1, 2, 3. Washington, D.C.: U.S. Government Printing Office.

Rothschild, E. 1976. Food politics. *Foreign Affairs* 54(2):285–307.

Sen, B. 1974. *The Green Revolution in India.* New York: Halsted Press.

Sjoberg, G. 1965. The origin and evolution of cities. *Sci. Amer.* 213(3):54–63.

USDA 1974. *The World Food Situation and Prospects to 1985.* Econ. Res. Serv., Foreign Agricultural Economic Report No. 98.

University of California Food Task Force. 1974. *A Hungry World: The Challenge to Agriculture.* Div. of Agr. Sciences, University of California.

Walters, H. 1975. Difficult issues underlying food problems. *Science* 188:524–530.

Wilkinson, R. G. 1974. Reproductive constraints in ecologically stabilized societies. In *Population and Its Problems: A Plain Man's Guide.* H. B. Parry (ed.). pp. 294–299. Oxford: Clarendon Press.

Winick, M., and P. Rosso. 1975. Brain DNA synthesis in protein-calorie malnutrition. In *Protein-Calorie Malnutrition.* R. E. Olson (ed.). pp. 94–102. New York: Academic Press.

2

THE ECOSYSTEM CONCEPT

Now that we have surveyed the world food situation let us examine the food producing systems upon which human beings depend. Long ago man obtained most of his food from natural assemblages of plants and animals; now most of his food comes from manipulated systems frequently referred to as agroecosystems. Agroecosystems are often highly productive, but their productivity sometimes fails through instabilities inherent in their ecological design. In order to understand both the potential productivity and the frailty of agricultural systems, we must be aware of the fundamental ecological relationships and processes that are common to all ecosystems. The objective of this chapter, therefore, is to explain the fundamentals of the ecosystem concept and thus provide a background for the discussions that follow. This chapter will serve as a review for some students; for others who wish to learn more about basic ecology we recommend a modern text such as Collier et al. (1973) or Ricklefs (1973).

NATURAL ECOSYSTEMS

Any collection of organisms that interact or have the potential to interact form, along with the physical environment in which they live, an ecological system or **ecosystem.** Although in the terrestrial environment we tend to distinguish one ecosystem from another according to the most apparent features of its biological components, we are actually recognizing the limitations placed upon it by its nonliving components. By observing the life-forms of the dominant organisms in an ecosystem we can make some fairly accurate inferences about the physical environment. For example, if the vegetation is sparse and the most common plants are woody with small leathery leaves, we could infer that the climate is characterized by an inadequacy of available moisture. Many other inferences, however, would be impossible without a more detailed study.

The abiotic components of all ecosystems consist of energy and the solid, liquid, and gaseous elements of a relatively thin layer of the earth's crust and its atmospheric envelope. Each of these components has its own specific properties, and each contributes to spatial and temporal variations that affect the distribution and abundance of living organisms. Thus the earth is covered by a mosaic of ecosystems, each different from the next yet functioning in accordance with the same set of physical and biological laws.

Each ecosystem is said to have a unique structure, characterized by its species composition, **biomass** (total weight of constituent organisms), and the spatial organization of its biotic components. This structure is, of course, the result of the evolutionary process by which the component species have adapted to the constraints of the physical environment and have developed the ability to interact successfully with the other species in the system. Thus when we speak of ecosystem structure, we refer not only to the physiognomy of the system but also to the way that nutrients and energy flow through it (Odum 1969).

Each ecosystem, however, is continuous to some degree with adjacent systems. Sometimes the boundary between two ecosystems, such as between a terrestrial and an aquatic system, is abrupt and well defined; more often the systems merge gradually, creating a transitional zone such as the scattered trees and grass that usually occur between a forest and open grassland.

Regardless of whether the boundary between adjacent ecosystems is abrupt or gradual, there is always some flow of energy and materials from one system to the other. For example, organic matter containing chemical energy moves by gravity from ecosystems at high elevations to lower ones; and in a similar manner dissolved nutrients flow in and out of ecosystems in runoff and groundwater. Animals, too, move back and forth, exploiting the resources of adjacent ecosystems and thus connecting the grazing and detritus food chains of the two systems. In this respect ecosystems are open and capable of responding not only to changes in their own environment, but also to the movement of energy and nutrients to and from adjacent systems.

In addition, biological communities are initially limited by the existing physical properties of the site on which they occur. A newly formed lava bed, for example, can at first support only a few species of lichens, whereas a newly plowed field may be successfully invaded by a host of annual plants. But the living community soon modifies the nonliving environment. Even the most simple layer of vegetation changes the pattern of energy exchange between the sun, the earth's surface, and the adjacent atmosphere: vegetation lowers the reflective power (**albedo**) of the surface and more solar radiation is therefore absorbed. Plants, by drawing water from subsurface layers of the soil and by transpiring water vapor, change the moisture characteristics of the adjacent layer of air and thereby influence patterns of energy exchange still further. The plant roots and vegetative litter change the structure and composition of the soil. These changes in the physical environment often induce biotic changes, which in turn continue to modify the physical environment. Thus the living and nonliving components of every ecosystem continually adjust to one another through a series of feedback mechanisms (Figure 2-1). Clearly, then, ecosystems are not static entities; they are dynamic systems with characteristic patterns of **energy flow, nutrient cycling,** and **structural change.**

Energy Flow

The ultimate source of energy for almost all the work done in an ecosystem is the sun. The amount of **solar energy** that reaches the outer surface of the earth's atmosphere is relatively constant at about 2 calories per square centimeter per minute, but not all of this energy reaches the ground (Oort 1970); in fact, the

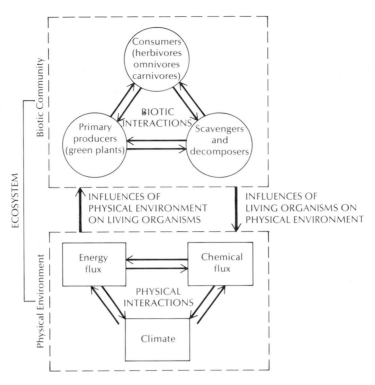

FIGURE 2-1.
A diagram representing the interactions between the biotic and abiotic components of a typical terrestrial ecosystem.

amount that does reach the earth's surface varies greatly from place to place (see Chapter 7). Much of it is reflected back into space from atmospheric dust particles and clouds; some is absorbed in the atmosphere and converted into chemical energy or heat or is reradiated in the form of electromagnetic waves. Of the solar energy that completely penetrates the atmosphere and reaches the earth's surface, still more is reflected, absorbed by the land or water, or reradiated into space. Consequently, only a very small portion of the sun's energy is available to sustain life (Figure 2-2). Extremely efficient biological communities convert about 3 percent of the total insolation into biochemical energy; most communities convert closer to 1 percent (Woodwell 1970).

Living organisms use energy in either a radiant form such as light, or in a fixed form bound up in organic molecules. Solar energy is converted into chemical energy by **photosynthesis** in green plants (**autotrophs**), a process by which carbon dioxide and water are converted to sugar within plant cells that contain chlorophyll. The total amount of solar energy thus converted into chemical energy is called **gross primary production.** Some of the sugars that constitute this production are used by the plant for cellular respiration. The chemical energy that is left after the respiratory requirements have been met (**net primary production**) may be stored as starch or related compounds in existing plant tissues or may be combined into specialized carbohydrates such as cellulose. The

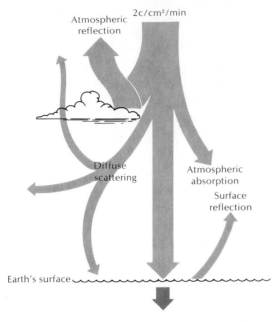

2c/cm²/min

Atmospheric reflection

Diffuse scattering

Atmospheric absorption

Surface reflection

Earth's surface

FIGURE 2-2.
The disposition of solar radiation once it has entered the earth's atmospheric envelope. Only 1 to 3 percent of the radiation that enters the top of the atmosphere is used in photosynthesis.

products of photosynthesis are also converted into other complex substances, such as proteins and lipids, that are incorporated into new plant structures. The net primary production thus forms the total converted solar energy available for use by the nonphotosynthetic organisms (**heterotrophs**) in the ecosystem.

The level of net primary production of an ecosystem depends on a variety of factors, including total solar radiation, temperature, rainfall, and the availability of essential nutrients such as nitrogen, phosphorus, and sulfur. Generally speaking, ecosystems in humid climatic zones are more productive than those in dry regions, and those in warm areas are more productive than those in cool ones (see Chapter 7). Furthermore, ecosystems with a substantial inflow of water, nutrients, and chemical energy are more productive than

systems with low inflows. For example, the net primary production of a tropical rain forest is roughly 80 times that of a warm desert, and that of a temperate coniferous forest is over 100 times that of tundra (Deevey 1971). The relationship between the temperature and moisture characteristics of climate and the primary productivity of several types of soil-vegetation formations is represented by the data of Rodin et al. (1975) given in Table 2-1.

The chemical energy contained in the tissues of green plants is the primary source of energy used by all other organisms in the ecosystem. We can, therefore, organize the biotic components of any ecosystem according to the feeding or **trophic** relationships that exist among them. Green plants form the base of the trophic structure because as we have said they are the **primary producers.** The plant tissue is fed upon by various kinds of herbivores collectively called **primary consumers;** thus some of the compounds manufactured by plants are converted to animal tissue that is in turn eaten by carnivores. Many carnivores are in turn eaten by higher-level carnivores. Consequently each species can be assigned to a trophic level, that relates to others in the system like a link in a chain (Figure 2-3), but this **food chain concept** is an oversimplification that applies only to ecosystems with very few species. In most ecosystems there are a number of species on each trophic level. Some herbivores are specialists, while others feed on a variety of plant species. Similarly, few carnivores are specific enough in their food habits to be assigned to a single trophic level: they may eat both herbivores and other carnivores. Some organisms are omnivorous, feeding as herbivores some of the time and carnivores the rest. In more complex ecosystems, therefore, the trophic levels are abstract and energy does not merely flow along a simple species chain. Instead the pathways for energy flow become intricately interwoven in a netlike arrangement referred to as a **food web.**

The trophic relationships described above

TABLE 2-1

The relationship between the temperature-moisture characteristics of climate and the annual primary productivity (tons per hectare) for some of the world's major soil-vegetation formations.

Thermal zone	Hydrothermal character	Soil-vegetation formation	Annual primary production
Polar	Humid	Polar desert	1.0
	Humid	Tundra on tundra soils	2.5
Boreal	Humid	Needle forest of taiga on podzolic soils	7.0
	Humid	Needle forest of southern taiga on turf-podzolic soils	7.5
	Humid	Broadleaf forest on gray forest soils	8.0
Subboreal	Humid	Broadleaf forest on brown forest soils	13.0
	Arid	Semishrub desert on gray-brown desert soils	1.5
Subtropical	Humid	Broadleaf forest on red and yellow soils	20.0
	Arid	Desert on subtropical desert soils	1.0
Tropical	Humid	Humid tropical forest on red ferralitic soils	30.0
	Arid	Desert on tropical soils	1.0

Source: Rodin et al. 1975.

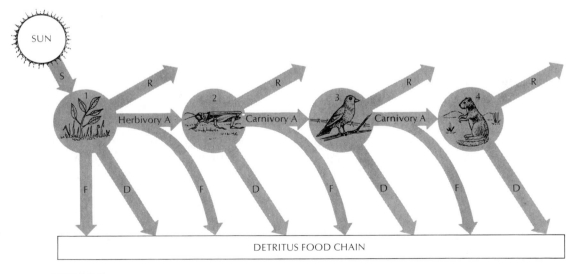

FIGURE 2-3.
Diagram of a typical grazing food chain showing the disposition of energy as it flows through an ecosystem.
S—Solar energy entering the system.
A—The energy available for assimilation by the members of the next trophic level.
D—Energy lost through death and decomposition.
F—Energy lost in the form of feces and other waste products.
R—Energy lost through respiration.

form what is commonly called a **grazing food chain** or **web**. Although this is the most visible aspect of an ecosystem, it rarely accounts for more than about half the total energy that passes through the entire system. Not all plants and animals are removed by being eaten; some dead plant and animal matter accumulates and is gradually broken down by detritus-feeding animals and decomposers. The same is true for the waste products produced by all organisms that comprise a grazing food chain. These wastes, along with the uneaten dead, usually constitute half or more of the energy that passes through the ecosystem. The organisms that consume and break down this material constitute a **detritus food chain** or **web**. This less visible mélange of organisms is not as easily characterized as the species complex of a grazing food chain because decomposers cannot be assigned to specific trophic levels. Many of these organisms have overlapping and opportunistic feeding habits, and energy flow cannot be described simply as a step-by-step transfer between discrete trophic groups.

Herbivore and detritus food chains are not independently functioning subsystems. The two pathways of energy flow are closely interrelated, first by the continual contribution of material to the detritus food chain by components of the grazing food chain and subsequently by a return of energy to the grazing chain. Many of the larger species of scavengers that contribute to a breakdown of detritus form at least part of the diet of many carnivores. Insectivorus birds, for example, consume both earthworms and detritus-feeding insects as well as foliage-feeding species. Furthermore, detritus forms the main source of food for the larval stages of many insects, the adults of which feed on the plant or animal members of a grazing food chain. Consequently, in addition to a flow of energy through the detritus food chain that provides for the gradual breakdown and recycling of vital nutrients to green plants,

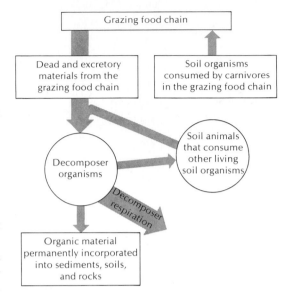

FIGURE 2-4.
Diagram of the energy transfers between a detritus food chain and a grazing food chain and of the typical routes of energy loss (Redrawn with permission from W. B. Clapham, Jr., 1973, *Natural Ecosystems*, Macmillan Publishing Company, Inc., New York. Copyright © 1973 by W. B. Clapham, Jr.)

there is a constant short-circuiting of energy to a variety of carnivores (Figure 2-4).

The total flow of energy through an ecosystem is governed by the first and second **laws of thermodynamics.** The first law states that energy can be converted from one form to another, as green plants convert solar energy into the stored chemical energy of plant tissue, which is in turn converted into heat energy by herbivore metabolism. The second law states that no conversion of energy from one form to another is complete: part of the energy is always lost in the conversion process. Thus the second law imposes an upper limit on the number of possible energy transfers from one trophic level to the next. Much of the energy taken in by an organism is used to build

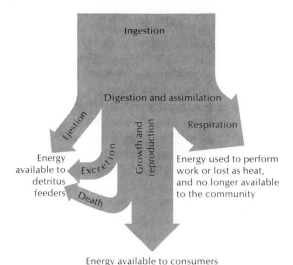

FIGURE 2-5.
The disposition of energy within a link of a grazing food chain. (Redrawn with permission from R. E. Ricklefs, 1973.)

or repair body tissues and to do useful work, but much of it is ultimately dissipated as heat or is passed from the body in the form of excretory products and feces (Figure 2-5). The efficiency of energy transfer varies from about 5 percent to 30 percent; that is to say, usually less than one-fourth of the energy consumed by an animal is converted into its own tissue. As a result, most ecosystems have herbivore food chains that consist of only three or four trophic levels.

The efficiency with which energy flows through an ecosystem depends on its biological composition as well as on the state of the physical environment. We are reasonably aware of the influence that light and temperature have upon photosynthesis and upon the activity of the animals within an herbivore food chain, but we are much less aware of the importance of the physical environment to the functioning of detritus food chains. The rate at which detritus is broken down depends greatly

on temperature, moisture, and the availability of oxygen. When it is warm, moist, and well aerated, detritus decomposes rapidly because the respiratory metabolism of decomposer organisms is high and the activity of earthworms and arthropods continually churns the soil, thereby increasing the availability of oxygen. When the temperature drops, respiratory metabolism and decomposition slow. Similarly, when there is inadequate oxygen organic matter can only be broken down very slowly by a few highly specialized anaerobic microorganisms.

The speed with which detritus is broken down is important to the whole ecosystem because this breakdown releases nutrients essential to plant growth and to continued primary productivity. If the release of nutrients is too rapid, they may be flushed from the root zone before they can be taken up by the plants; conversely, if the detritus breaks down too slowly, the nutrients remain locked in undegraded materials where they are unavailable for plant use.

Each unit of solar energy assimilated in photosynthesis passes through an ecosystem only once. How long this passage takes varies widely according to how much energy is stored in the ecosystem as biomass. In some aquatic systems, for example, the turnover rate of phytoplankton is rapid and the energy assimilated by them during photosynthesis passes through the system in less than three weeks. In terrestrial ecosystems, on the other hand, much of the energy assimilated is allocated to supportive tissue, such as the woody parts of trees, and its turnover may take hundreds of years.

Biogeochemical Cycles

The basic chemical elements from which all organic matter is synthesized are constant in form and finite in supply. Unlike energy that enters an ecosystem from without and flows

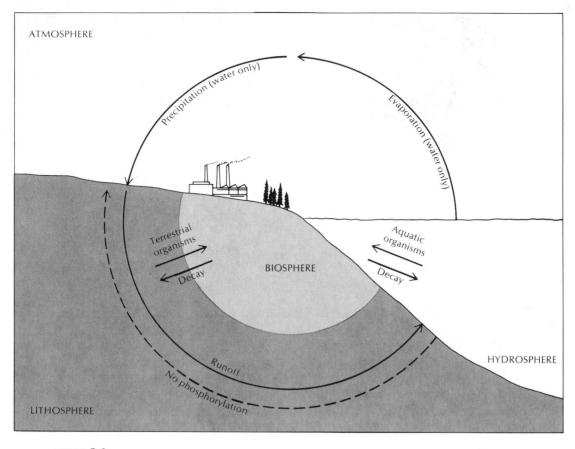

FIGURE 2-6.
The typical pattern of soluble-element cycles followed by elements that are not volatile but dissolve readily in water. (From "The Mineral Cycles" by Edward S. Deevey, Jr. Copyright © 1970 by Scientific American, Inc. All rights reserved.)

through it in one direction until it is completely dissipated as heat, basic nutrients must be used over and over again. Their passage is therefore cyclical, and, although each chemical element follows a somewhat different route, all are eventually restored to their inorganic form suitable for reuse by other organisms.

We refer to this pattern of movement of basic nutrients as a **biogeochemical cycle** because it consists of an organic phase involving living organisms (food chains) and a geologic or abiotic phase governed by the basic chemical properties of each element. Although some nutrients pass through herbivore and detritus food chains in essentially complete organic subcycles, the main reservoirs of most of these materials are in abiotic components of the ecosystem. The main abiotic reservoirs of mineral nutrients, such as phosphorus, that are water soluble but not volatile are sedimentary deposits to which living organisms have rather limited access (Figure 2-6); on the other hand,

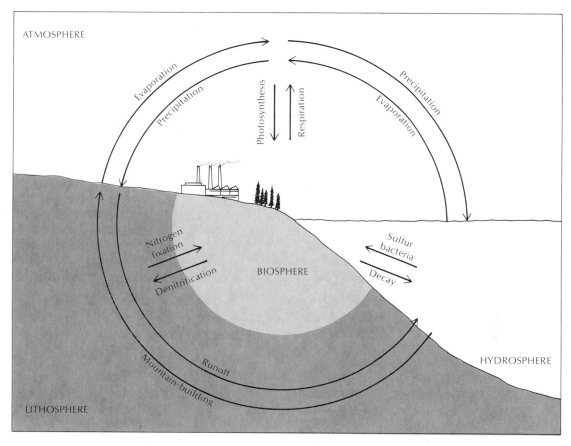

FIGURE 2-7.
The typical pattern of the carboxylation cycle that circulates essential elements, such as oxygen, carbon, and nitrogen, that readily enter a gaseous phase. (From "The Mineral Cycles" by Edward S. Deevey. Copyright © 1970 by Scientific American, Inc. All rights reserved.)

the major abiotic reservoirs of carbon, oxygen, hydrogen, and nitrogen occur in the atmosphere and therefore are readily accessible to living things (Figure 2-7). Deevey (1970) characterized these two cycles as **soluble element cycles** and **carboxylation cycles,** respectively.

Biogeochemical cycles are described in terms of the amount of basic nutrient that exists in an ecosystem component (**nutrient pool**) and the amount of that nutrient that is transferred from one pool to another per unit of time (**nutrient flux rate**). In the biosphere as a whole, or a large ecosystem in its mature stage of development (a **climax**), the major biogeochemical cycles are essentially in a steady-state condition; that is, the inflow and outflow of nutrients to and from their major pools are about evenly balanced. In small ecosystems, however, and in those undergoing rapid change, such as that which can occur on a site recently disturbed by a landslide, these inflows and outflows of nutrients may be substantially out of balance. Although nutrients cycle at the same time that energy flows unidirectionally through an ecosystem, the relationship between available energy and decomposition creates a similarity in the movement of both energy and nutrients through biological communities. This is particularly well illustrated by the carbon cycle; in fact, radioactive carbon is used to trace patterns of energy flow through ecosystems.

THE CARBON CYCLE   The carbon cycle (Figure 2-8), although similar in many respects to other biogeochemical cycles, is unique in that its organic phase does not exist as a complete subcycle. Instead, the biotic and abiotic phases are closely intertwined. Carbon enters the biotic

portion of the ecosystem through the uptake of atmospheric carbon dioxide by the tissues of green plants, which then incorporate this carbon into organic molecules by photosynthesis. Energy is thus stored by living organisms in the form of fixed carbon. Whenever these organic carbon-containing compounds are used in cell respiration, energy is released, and carbon in the form of carbon dioxide is returned to the atmosphere. However, not all carbon-containing molecules are used as they pass through the grazing and detritus food chains. Some are incorporated into carbon-rich deposits such as coal, oil, and gas, where they are more or less permanently stored. Large quantities of carbon dioxide react with water to form bicarbonate and carbonate, which organisms such as clams combine with dissolved calcium to produce the calcium carbonate in their shells. When these animals die the calcium carbonate either enters into solution or forms sediments that are later uncovered and oxidized during upheaval and weathering.

Regardless of how carbon compounds are degraded, however, they ultimately yield carbon dioxide, the raw material from which they were initially made. The turnover rate of carbon is so rapid that although it is present in the atmosphere in a relatively low concentration, there is almost always an adequate—although not necessarily optimal—supply for living organisms.

THE NITROGEN CYCLE   The pattern of movement of nitrogen through an ecosystem (Figure 2-9) differs in several respects from that of carbon. First, like carbon, the main reservoir of nitrogen is the atmosphere, but although the concentration is high, atmospheric nitrogen is not available to most living organisms. Second, there is a distinct biotic subcycle for nitrogen that involves the biological breakdown, in several steps, of organic nitrogenous compounds to inorganic forms.

Nitrogen, in gaseous form, constitutes approximately 79 percent of the earth's atmo-

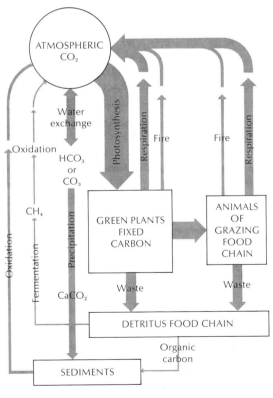

FIGURE 2-8.
The carbon cycle.

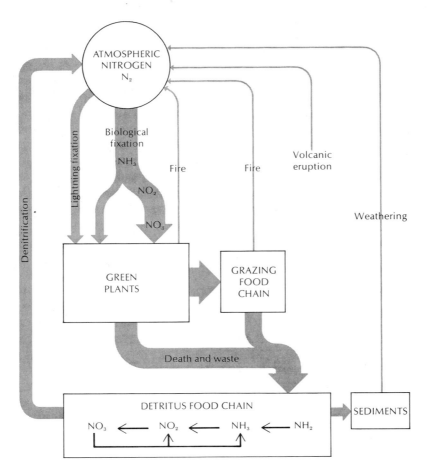

FIGURE 2-9.
The nitrogen cycle.

sphere. Some gaseous nitrogen enters the atmosphere through volcanic action, but this is balanced by a roughly equivalent amount that is lost from the atmosphere through its incorporation into deep oceanic sediments. Although the atmospheric pool of nitrogen is virtually inexhaustible, it consists mainly of molecular nitrogen (N_2), which can be utilized directly by only a few specialized microorganisms. Most organisms obtain nitrogen in the form of soluble nitrates (NO_3^{2-}) or ammonium ions (NH_4^-).

Some atmospheric nitrogen is converted by lightning-induced oxidation to nitrogen oxides that react with water to form nitrate. However, most of the nitrogen used by green plants, and subsequently by animals and decomposers, is made available through the process of **biological fixation**, which is carried out by certain bacteria, actinomycetes, and blue-green algae that extract molecular nitrogen from the atmosphere and combine it with hydrogen to form ammonia (NH_3). In terrestrial ecosystems **nitrogen-fixing bacteria** that live symbiotically in root nodules of legumes and some tree species are extremely important components of the nitrogen cycle (see Chapter 13).

In a typical grazing food chain, green plants take up nitrates through their roots and incorporate the nitrogen into organic molecules such as amino acids, proteins, nucleic acids, and vitamins. Except for a small portion that is returned to the atmosphere by fire and biological processes as gaseous nitrogen or nitrogen oxides, the nitrogen-containing organic molecules are recycled through the interaction of the herbivore and detritus food chains. Nitrogenous wastes and dead organisms are degraded by detritus feeders into the amino form (unless they are already in such a form). The amino groups (NH_2) liberated from various organic molecules are then converted to ammonia (NH_3) by a process called **deamination.** Certain bacteria then oxidize the ammonia to form nitrate by the process of nitrification. These water soluble nitrates are subsequently taken up again by green plants to complete the organic phase of the cycle.

Sometimes nitrate is degraded to nitrite before it can be utilized by green plants. This degradation is carried out by **denitrifying bacteria** that utilize the energy released by the conversion of the nitrate back to nitrite, ammonia, and molecular nitrogen. The molecular nitrogen thus produced either reenters the atmosphere, is refixed by nitrogen-fixing bacteria, or becomes incorporated into sediments. Nitrate is usually highly soluble and is normally washed away by groundwater.

THE PHOSPHORUS CYCLE   The carbon and nitrogen cycles contain most of the basic features of nutrient cycles, but phosphorus is an important and well-known element, the cycling of which characterizes that of all the important elements that have a strictly sedimentary, as opposed to atmospheric, reservoir (Figure 2-10).

Phosphorus occurs naturally in the environment as a form of phosphate (PO_4^{3-}, HPO_4^{2-} or $H_2PO_4^{-}$). It may occur as soluble inorganic phosphate ions or as a part of other organic or inorganic molecules of varying solubility. But the ultimate source of phosphate, as with all sediment-reservoir nutrients, is some form of rock; hence these nutrients are first made available to living organisms through the weathering of these rocks.

Plants obtain soluble phosphate ions through their root systems and incorporate them into living tissue. These tissues are ultimately consumed by herbivores or decomposers, and the phosphorus is thus passed along food chains. Excess phosphate in the diet of members of a grazing food chain is excreted by them and can form massive local phosphorus-rich deposits such as the guano beds created by sea birds along the west coast of Peru and Chile.

As detritus is degraded by decomposers, the phosphate component of organic molecules that enter the detritus food chain is liberated in the form of inorganic ions. These ions may then either be taken up again by plants or incorporated into sediments. Because some phosphate compounds are not very soluble, phosphorus is not readily removed from sediments once incorporated into them. Furthermore, phosphates in the soil may react with other chemicals to form insoluble compounds or become physically incorporated into clay minerals, thereby reducing their availability to living organisms. The amount of phosphorus trapped by these processes depends on a variety of factors, such as the mineral composition and pH of the soil. Consequently, the phosphate cycle is imperfect and only a variable portion recycles through the organic or biotic phase.

THE HYDROLOGIC CYCLE   Water is the most abundant chemical compound in the biosphere. Although it is of vital importance to a variety of chemical reactions, water differs from the

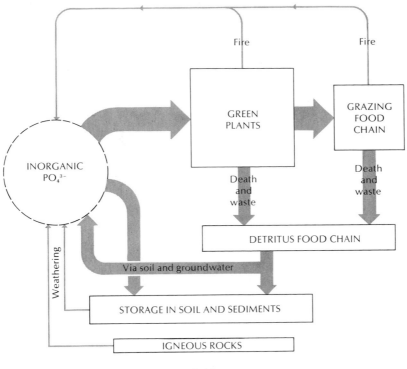

FIGURE 2-10.
The phosphorus cycle.

basic nutrients in that most of it retains its basic chemical state throughout the biotic and abiotic phases of its cycle. Water is an essential component of all living tissues and is their source of hydrogen; it is also the prime mover of soluble nutrients from one place to another and is the medium in which these nutrients enter plants (Penman 1970).

The cyclical movement of water in the biosphere is driven by a combination of solar energy and gravity. A great deal of the total solar energy that reaches the earth is consumed by the conversion of water to water vapor, which is then readily transported by atmospheric circulation. When water vapor in the atmosphere condenses into water droplets, it falls as precipitation under the influence of gravity.

Approximately 95 percent of the water on earth is bound chemically into the crystalline structure of rocks and therefore does not cycle (Nace 1967). Most of the remainder is in the ocean reservoirs with only a very small portion existing as water in cells, as atmospheric water vapor, or as free fresh water. Although there is enough water in the atmosphere to cover the entire earth to a depth of about 3 centimeters, the average annual precipitation over the earth is about 81 centimeters (Furon 1967). This means that all the water in the atmosphere is turned over approximately every 13 to 14 days. However, the worldwide distribution of evaporation and precipitation is far from even. There is generally more evaporation over the oceans and more precipitation over the land. Over the oceans the hydrologic cycle is exceed-

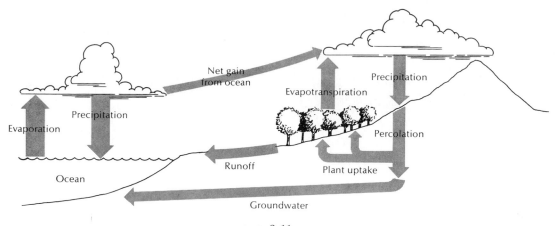

FIGURE 2-11.
The hydrologic cycle.

ingly simple: water leaves by evaporation and returns by precipitation. The water that falls on the land, however, may return almost immediately to the atmosphere by evaporation or may be retained temporarily by terrestrial ecosystems. Some of the water returns quickly to various reservoirs in the form of surface runoff; the rest penetrates the soil and subsequently follows one of two possible pathways: it is either taken up by living organisms and eventually returned to the atmosphere by transpiration or perspiration, or it percolates slowly through the soil as groundwater (Figure 2-11).

The considerable volume of water that enters plants through their root systems is almost entirely lost by evaporation from the surface of the foliage (transpiration). This part of the hydrologic cycle is particularly important to the function of the biogeochemical cycles discussed earlier, because all mineral nutrients that enter the roots of plants do so in an aqueous solution. Similarly, the gross movement of nutrients within and among ecosystems is accomplished largely by runoff and by the flow of water through the soil. Sometimes this process removes vast quantities of material from an ecosystem, but the total depletion of nutrients

rarely occurs because they are also carried in by water from elsewhere. The movement of water is also an important weathering force that helps to uncover nutrient-bearing rocks and to recycle chemicals that are trapped in sediments. The gravitational flow of water then continually moves these nutrients from high- to low-lying ecosystems and eventually into the deep ocean basins where they are again trapped in sedimentary deposits.

Succession and Stability

Ecosystems, like the organisms that are members of them, change with time; they originate, develop, and mature (and sometimes decline and disappear). This change, called **ecological succession,** is a result of the dynamic interaction that takes place between the biotic and abiotic components of each community that develops. These interactions form a series of feedback mechanisms that drive the succession toward a steady-state system called a climax (Figure 2-12). For example, wherever the physical environment is sufficiently hospitable, a bare plot of ground will eventually be invaded by

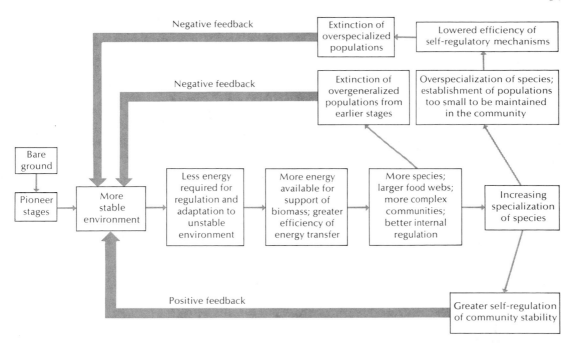

FIGURE 2-12.
Representation of feedback loops involved in the progressive changes that take place in ecosystems undergoing succession. During early stages the dominance of the positive feedback loop leads to a self-sustained increase in diversity and stability. As succession proceeds, the establishment of more specialized species is counteracted by the elimination of those that are overgeneralized or overspecialized. The climax community is maintained by the dynamic equilibrium that develops between the positive and negative feedback loops. (Redrawn with permission from W. B. Clapham, Jr., 1973, Natural Ecosystems, Macmillan Publishing Company, Inc., New York. Copyright © 1973 by W. B. Clapham, Jr.)

pioneer species able to withstand the unmodified physical conditions that prevail there. These few species will exert a modifying influence on the environment, and this slight modification, plus a gradual increase in the organic content of the soil and the influx of nutrients from adjacent systems, increases the suitability of the site for the establishment of additional species. As new species enter the system some of those established earlier are forced out. In the early stages of succession the community structure may change rapidly, but the rate of change slows as the more permanent climax community becomes fully developed.

The early stages of succession are characterized by the presence of relatively few, often ephemeral, species. Both primary production and accumulated biomass are at a low level, but the ratio of production to biomass is higher than at any subsequent stage. Usually there is little accumulated organic litter, so that the detritus food chain is poorly developed, as is the organic phase of nutrient cycling within the system. These characteristics gradually change, however, until the climax stage is attained. Under climax conditions, primary production and accumulated biomass are high, but the ratio between them is at its lowest level.

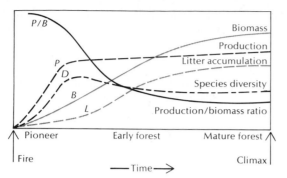

FIGURE 2-13.

Changes in ecosystem parameters that accompany succession following fire in a generalized terrestrial environment. (Modified from R. H. Whittaker, 1970, *Communities and Ecosystems*, Macmillan Publishing Company, Inc., New York.)

Species diversity has increased, although it may have been higher in some intermediate stage of succession. The detritus food chain is well developed and as a result the organic phase of nutrient cycles is complete. These successional changes are illustrated graphically in Figure 2-13.

Succession is the process by which ecosystems evolve toward the most stable state possible within the constraints of the environment. The greater the number of alternative pathways along which energy and nutrients can move through an ecosystem, the more stable the system seems to be. In general, as species diversity increases so do the number of trophic links and alternative pathways for energy flow (Southwood and Way 1970). It is probably for this reason that many ecologists have equated diversity and stability. This correlation can be quite variable, however, and there may be no simple relationship (Goodman 1975). It may change according to the balance of species that play generalized as opposed to specialized roles, as well as in response to the constraints imposed by the physical environment. For example, major fluctuations in the weather are often accompanied by broad fluctuations in primary production and in the food available to consumers. Under such conditions food generalists (species that eat a wide range of foods) that can respond readily to changes in the abundance of specific foods contribute to the long-term stability of the system by creating an effective buffer against major changes in energy and nutrient flux rates.

The level of ecosystem stability that can develop varies from place to place in relation to regional climatic and edaphic (soil) conditions. Continually warm, moist climates generally favor the development of complex and stable ecosystems. A wide variety of species can survive under such physical conditions, but many tend to be specialists that have evolved strategies for reducing the impact of competition from other species. Although specialist species may be vulnerable to rather minor environmental fluctuations, the large number of interspecific interactions that are possible among them contributes to the overall stability of the system. Conversely, in harsher, more variable environments only a few species prevail, and although many of them may be trophic generalists, their populations often fluctuate markedly in response to the alternating intensification and relaxation of the physical environmental constraints. Consequently, these systems tend to be more unstable.

An increase in species diversity usually alters their spatial distribution in such a way that the populations of interacting species tend to fluctuate less markedly. However, spatial heterogeneity in more simple systems can also have a stabilizing influence (Murdoch and Oaten 1975).

AGRICULTURAL ECOSYSTEMS

Approximately 1.4 billion hectares, or slightly over 10 percent of the ice-free land in the world, are now used for some form of crop culture. About 3 billion additional hectares are

TABLE 2-2
The total area of the world's major natural ecosystems and crop agriculture.

Ecosystem	Area (billion hectares)
Tropical forest	2.0
Temperate forest	1.8
Desert scrub	1.8
Savanna	1.5
Crop Agriculture	1.4
Boreal forest	1.2
Temperate grassland	.9
Arctic and alpine tundra	.8
Woodland and scrub	.7

used regularly for hay production or as pasture and range for domestic animals. Agriculture in the broad sense is therefore currently practiced on about 30 percent of the earth's land area. Land used for crop culture alone now roughly equals that taken up by any one of the world's largest natural ecosystems (Table 2-2). Obviously, all the acreage used for human food production has been claimed from natural ecosystems, especially from grasslands and deciduous forests. The land thus claimed has generally been the best available in terms of soil fertility, topography, sunlight, rainfall, and other aspects of the climate, and, since only about 7 percent of the earth's land area has a combination of features ideally suited for crop culture, it is safe to assume that most of it has already been put to such use. Any further expansion of agriculture will undoubtedly take the best of what is left and reduce the extent of natural ecosystems still further.

Agriculture in one form or another has become such an integral part of our environment that the casual observer views farming and grazing as activities that blend harmoniously into the local ecology. When we look more closely, however, we find that ecosystems managed by man are often fundamentally different

from natural ecosystems in several important respects. In fact, in recent years many of the environmental problems that have developed in agricultural regions have been blamed on agricultural practices. There is no doubt that some forms of agriculture are particularly disruptive, but it is both unfair and unwise to tar all agroecosystems with the same brush. As Smith and Hill (1975) suggest, natural and agricultural ecosystems form a continuum; they are not discrete systems. Using species diversity and the intensity of human intervention (management) as two major means of distinguishing natural and manipulated ecosystems, Smith and Hill prepared a chart (Figure 2-14) that shows the relationships among several different ecosystems.

Although such an analysis does suggest a continuum it is still possible to identify some important differences between natural and even modestly managed systems. Let us compare, for example, the basic structure and function of natural grassland and managed pasture. The fundamental differences revealed by this comparison will form a useful basis for examin-

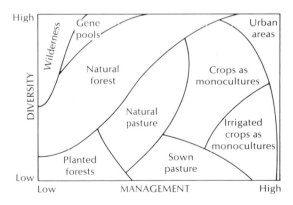

FIGURE 2-14.
The relationships among a variety of ecosystems, based on their biological diversity and the level of managerial input. (Redrawn from Smith and Hill, 1975, *Journal of Environmental Quality*, vol. 4, 1975, No. 2, by permission of the American Society of Agronomy, Crop Science Society of America, and Soil Science Society of America.)

ing the ecological features of intensive mechanized agriculture in Chapter 6.

In a natural grassland community the primary producers consist of a variety of plants dominated by perennial grasses. The vegetation is fed upon by a variety of herbivores ranging from insects to rodents and large hoofed mammals. Some of the herbivores are eaten by carnivores and they in turn by other carnivores. The wastes from and deaths of members of each trophic level contribute material to a well-developed detritus food chain. As a result, the system contains a high level of internal nutrient cycling and a reasonably good balance in the fluxes of nutrients in and out of its major components. Through these same relationships the solar energy fixed by the green plants flows in a complex fashion through the food web and is gradually dissipated by the life activities of a diverse group of organisms.

In a managed pasture ecosystem there are several ecological differences that are the results of man's attempt to concentrate energy flow through a particular species of herbivore. Intense management may involve efforts to eliminate nonpreferred species of plants and to increase the abundance of those preferred by the domestic animals being raised; this can be accomplished mechanically or with selective herbicides. In addition, predator control is implemented, fences are constructed, and pesticides applied to reduce the number and variety of competing herbivores and to eliminate the larger species of carnivores. In other words, the complex food web of the natural grassland ecosystem is reduced to a simple food chain that consists mainly of only a few plant species, a single species of domesticated herbivore, and man as the primary carnivore.

These changes may seem innocent, but let us consider their ecological impact. The overall simplification of trophic structure reduces the inherent stability of the system and increases the likelihood of outbreaks of pest species such as plant-feeding insects and weeds.

More specifically, changing the composition of the vegetation alters the pattern of nutrient uptake and use as well as the seasonal pattern of primary production. More intensive grazing by a single species reduces the amount of dead plant tissue that is returned to the soil. The fact that the food chain is shortened and few animals ever die within the system greatly reduces the accumulation of detritus and weakens the detritus food chain. The internal recycling of nutrients is reduced even more by the fact that the flesh (or milk) of the domesticated herbivores is consumed outside the system.

The degree to which these disruptions are manifested depends on the form of management that man implements. In a natural grazing system such as that of the Serengeti Plain in Tanzania (Bell 1971) there is a minimum of human disruption. In this case a complex of mammalian herbivores capably and effectively utilizes different plants and different stages of plant growth in a grazing succession (Figure 2-15). Moreover, there are seasonal migrations of the large herbivores that enable them to exploit regional differences in the seasonal pattern of plant growth.

People of long-standing pastoralist societies may display a high level of ecological awareness. The animals, often several species, that are used for food and other products are moved from place to place according to seasonal changes in range conditions. They are protected to some extent and only on rare occasion are they fed food brought in from elsewhere. Likewise, few of the animals are exported and nearly all waste products are returned to the system.

In more intensively managed systems the number of plant species is reduced to make room for the forage considered to be best for a single species of herbivore. The food supply is more heavily supplemented from outside the system, and the entire new animal biomass is exported for consumption elsewhere. Steps must be taken to avoid overgrazing, and the productivity of the system must be maintained

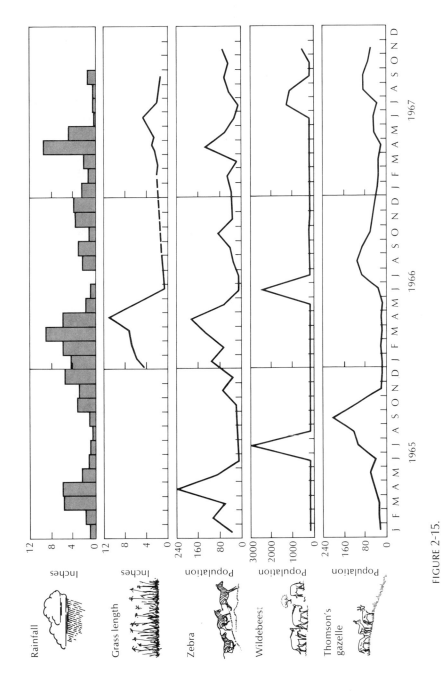

FIGURE 2-15.
The relationships among rainfall, grass length, and populations of different species of large herbivore on the western Serengeti plains of Africa. (From "A Grazing Ecosystem in the Serengeti" by Richard H. V. Bell. Copyright © 1971 by Scientific American, Inc. All rights reserved.)

artificially through energy and nutrient supplements, including human and animal labor, fossil fuels, pesticides and chemical fertilizers, and irrigation.

Most ecosystems exploited by primitive pastoralists and even by modern ranchers closely resemble the natural ecosystems from which they are derived. Other types of agricultural systems depart more markedly from natural systems, and the most pronounced differences appear in greenhouses and systems that cultivate a single species of annual crop. The largest portion of the 1.4 billion hectares devoted to crop production is used for the intensive culture of annual grains and starchy root-crops. The agroecosystems in which these staple commodities are produced are characterized by high productivity (most of which is exported) and by huge energy and material subsidies.

Although some agricultural societies still emphasize the planting of multiple-species crops, the trend over the past half century in the world's major food producing areas has been toward **monoculture**, or the cultivation of a single species. The reason for this trend is obvious: Each cultural step from planting to harvest can be synchronized and undertaken uniformly. This leads to increased labor efficiency, which is often further enhanced by the development of new crop varieties and cultural practices that lend themselves to mechanization. Although some intensive, non-mechanized agricultural systems still exist, such as those in China (Wortman 1975) and southeast Asia, the trend is toward the greater use of specialized machines, and the result is often a highly mechanized agroecosystem that bears little similarity to a natural ecosystem in either structure or function (see Chapter 6).

In terms of food production these intensive, mechanized agroecosystems have been a major accomplishment. We must view this accomplishment in perspective, however, and recognize that the supplies of agricultural land,

nutrients, water, and fossil fuels are finite. We are in fact facing some vital ecological question: How much land can be converted from natural ecosystems to agroecosystems without upsetting the overall balance in the ecosphere? How much water can we divert to crop irrigation, especially in arid areas, before we significantly affect the global distribution of precipitation and the rate of turnover in the hydrologic cycle? How long can we continue to tamper with the natural pattern of nutrient cycling by industrially fixing nitrogen and by mining phosphates? How long can we continue to supplement agroecosystems with energy derived from declining fossil fuel supplies?

Agriculture has already expanded beyond the point at which its ecological impact is primarily local. We must now view agriculture in terms of its global impact. We have taken the world's most productive acres and converted them to highly modified types of ecosystems. These acres are managed by a strategy that varies considerably from nature's balancing of fertility and productivity and the resulting progress toward stability. The world's natural ecosystems act as a buffer against the ecological disturbances created by man; and the more acreage we convert to agriculture the more we weaken that buffer. Some regions already face the threat of water shortages, and our supplies of fossil fuels and certain nutrients are declining steadily.

Furthermore, the large input of energy, nutrients, and pesticides needed to maintain the productivity of some mechanized monocultures can only be supported by affluent industrialized societies that can afford the required materials and implements. Yet developing countries that lack the means to support this form of agriculture are tempted, by the spectacular levels of productivity they promise, to develop mechanized agroecosystems; the results can be disastrous.

Clearly the long-term future of agriculture rests in the balance, and the need to develop alternative strategies for food production is upon us. Our knowledge of ecosystem structure and function has grown rapidly in recent years, and we are now able to introduce more ecological rationale into our own food production systems and to provide ecologically sound advice to emerging nations. Nevertheless, we are not likely to make these desirable changes unless politicians, industrialists, farmers, scientists, and the general public realize that some of our most highly productive systems cannot be supported indefinitely in their present form. Too many people have been led to believe that machines and the development of miracle grains will solve the world's food problems. But as the Mayers (1974) so aptly stated, " . . . miracle grains are not simply new seeds which can replace older ones and automatically give much larger crops. They are fertilizer-responding varieties which can indeed increase yields but only if cultivated under optimum conditions." These optimum conditions, which involve the application of fertilizer, water, and pesticides along with a considerable input of energy cannot be provided forever. The long-term success of our food producing systems will ultimately depend on a more complete appreciation of the ecosystem concept as it applies to both natural and agricultural systems.

SUMMARY

An ecosystem is an assemblage of organisms that interact, or have the potential to interact, with each each other and with their physical environment. All ecosystems, whether they are untouched by man's influence or intensely managed, operate according to the same basic, natural laws.

The ultimate source of energy of all ecosystems is the sun. Solar energy is transformed into chemical energy by the process of photosynthesis in green plants. The chemical energy of the plant tissue is then utilized by organisms and their tissues in turn by others until all the energy is finally dissipated as heat. About half of the sun's energy converted by plants flows through a chain-like sequence of animal groups called an herbivore food chain, and the remainder flows through a less structured mélange of scavengers and decomposers that collectively form the detritus food chain.

Whereas the flow of energy through ecosystems is unidirectional, the flow of essential nutrients, the supplies of which are finite, is cyclical. Several key substances, such as carbon, nitrogen, and oxygen, have large atmospheric pools, whereas others, like phosphorus and sulfur, have less accessible sedimentary reservoirs. All of them have a biotic phase and an abiotic phase in their cycles and the efficiency with which these cycles interact greatly influences the structure and functioning of ecosystems.

The structure and function of different types of ecosystems form a continuum in time and space that reflects the adaptability of the species of which they are composed. Ecosystems thus reflect both long-term evolutionary trends and shorter-term interactions in which the physical environment and the assemblage of organisms influence each other through a continuous system of feedback controls. Whenever human beings intercede with even minimal management activities they generate basic changes in the functioning of the system. The more intensive man's management the more pronounced these changes become and the more likely it is that further management and subsidization will be necessary. Understanding the ecosystem concept, and applying this understanding to the development of ecologically sound management practices, is the primary objective of agricultural ecology.

Literature Cited

Bell, R. H. V. 1971. A grazing ecosystem in the Serengeti. *Sci. Amer.* 225(1):86–93.

Collier, B. D., G. W. Cox, A. W. Johnson, and P. C. Miller. 1973. *Dynamic Ecology.* Englewood Cliffs, N.J.: Prentice-Hall, Inc.

Deevey, E. S. 1970. Mineral cycles. In *The Biosphere.* pp. 81–92. San Francisco: W. H. Freeman and Company.

Furon, R. 1967. *The Problem of Water: A World Study.* New York: American Elsevier Publishing Co., Inc.

Goodman, D. 1975. The theory of diversity-stability relationships in ecology. *Quart. Rev. Biol.* 50(3):237–266.

Mayer, A., and J. Mayer. 1974. Agriculture, the island empire. *Daedalus* 103(3):83–95.

Murdoch, W. W., and A. Oaten. 1975. Predation and population stability. *Adv. Ecol. Res.* 9:1–131.

Nace, R. L. 1967. Are we running out of water? *U.S. Geol. Surv. Circ.* 536:1–7.

Odum, E. P. 1969. The strategy of ecosystem development. *Science* 164:262–270.

Oort, A. H. 1970. The energy cycle of the earth. In *The Biosphere.* pp. 13–24. San Francisco: W. H. Freeman and Company.

Penman, H. L. 1970. The water cycle. In *The Biosphere.* pp. 37–46. San Francisco: W. H. Freeman and Company.

Ricklefs, R. E. 1953. *Ecology.* Newton, Mass.: Chiron Press.

Rodin, L. E., N. I. Bazilevich, and N. N. Rozov. 1975. Productivity of the world's main ecosystems. In *Productivity of World Ecosystems.* pp. 13–26. Washington, D.C.: National Academy of Sciences.

Smith, D. F., and D. M. Hill. 1975. Natural and agricultural ecosystems. *J. Env. Qual.* 4(2):143–145.

Southwood, T. R. E., and M. J. Way. 1970. Ecological Background to pest management. In *Concepts of Pest Management.* R. L. Rabb and E. E. Guthrie (eds.). pp. 6–29. Raleigh: North Carolina State University.

Woodwell, G. M. 1970. The energy cycle of the biosphere. In *The Biosphere.* pp. 25–36. San Francisco: W. H. Freeman and Company.

Wortman, S. 1975. Agriculture in China. *Sci. Amer.* 232(6):13–21.

3

THE EVOLUTION OF AGRICULTURAL SYSTEMS

In ecological terms, agriculture represents a symbiotic relationship between humans and domesticated plants and animals. This relationship is, of course, unusually complex in the number of species and the diversity of specific interactions, but it is basically one that benefits both man, through the harvest of food items, and his domestic species, which depend on him for their propagation. Examples of this general type of symbiosis abound in nature: Various species of ants tend populations of aphids, protecting them from predators and harvesting honeydew from them in return. Other types of ants cultivate varieties of fungus in underground gallery "farms," thus practicing a form of symbiosis much like that of crop agriculture. The honeybee carries on a loose symbiotic association with a wide variety of flowering plants, pollinating the flowers and harvesting nectar and pollen.

Many agricultural symbioses have evolved and they all seem to be characterized by factors that are important to the survival of the participants. This suggests that we might understand the orgins of human agriculture in similar evolutionary terms; that is, we may view the cultural and biological evolution of human food production systems in terms of specific ecological advantages to the participants in this system. Moreover, if we adopt an ecological and evolutionary perspective, we may assume that agricultural systems evolved by gradual, step-by-step modification of systems of food procurement.

No organism, including man, has achieved such a perfect balance with its environment that it is free from intraspecific competition and other processes that lead to continuing biological or cultural adaptation within its population. Thus we can assume that, at any stage in the development of agriculture, modifications either of human practice or of organismic characteristics could improve this mutual relationship and therefore be favored in the process of evolution. Although unique historical factors, such as human migrations and sudden climatic shifts, certainly influence this process, there is no reason to believe that they are essential to its occurrence.

The environmental setting clearly exerts a strong control over the direction and rate of change in relationships such as those that exist in food procurement: It not only generates a need for improvements in food procuring techniques, but also offers opportunities and resources for these improvements. Certain environments—for example, the high arctic— offer few opportunities for the development of new food production technologies, although the need for them may be great. Others, such as certain aseasonal tropical environments, may offer many opportunities for which there is little need. One of our principal objectives will be to identify the environmental settings in which agriculture arose and diversified.

The diversity of existing agricultural systems is very great; it is characterized by a variety of crop species, domestic animals, and management techniques. We know that this diversity has been achieved within a short period of human history, roughly the last 12,000 years (Ucko and Dimbleby 1969). We also know that early accomplishments occurred in several widely separated world regions. In attempting to trace the history of agriculture, however, we must keep in mind that particular features of an agricultural system in a given locality may be a result either of their independent origin in that locality or of their introduction through cultural diffusion from other areas. There is no reason to assume that either of these processes predominated in general or in a particular situation. Specifically, there is no reason to assume that agriculture itself originated in one region and spread to all others or that it originated independently in all the regions where it now possesses distinctive characteristics.

THE ENVIRONMENTAL SETTING OF AGRICULTURAL ORIGIN

The transition from preagricultural systems of hunting and gathering to formal agriculture seems to have occurred in only about six or seven areas that differ greatly in extent, climate, and geography. From these centers, agriculture spread rapidly to most other parts of the world. In certain climatic zones, such as the Arctic, and in a few other regions, it neither evolved internally nor arrived by cultural diffusion until colonization in historical times by Western man.

The most extensive area of the latter type is Australia. Archeological evidence indicates that early man was present in Australia at least 16,000 years ago (Mulvaney 1966) and probably much earlier. The arrival of man in the New World probably occurred more than 30,000 years ago (Dragoo 1976, Haynes 1970), but the main peopling of North and South America may have been the result of a migration that began only about 13,000 years ago (Martin 1973). In both instances man arrived as a hunter and gatherer, and just the fact that he colonized the areas involved indicates that he possessed a relatively aggressive, adaptable culture. In the New World, this culture diversified in an amazing fashion, producing some of the world's most remarkable civilizations. The transition from hunting and gathering to agriculture occurred in different ways in different areas. In Australia, however, despite long population, the material culture remained one of the simplest in the world and the transition to agriculture never occurred.

The fact that the native inhabitants of Australia never developed agriculture is strong evidence that the environmental setting is greatly important to agricultural origins. In examining the features of the Australian environment, we may note first that Australia is geologically, geographically, climatically, and biotically one of the least diverse continental areas of the world (Mulvaney 1966). The principal mountain and river systems are restricted to a narrow eastern and southeastern strip along the edge of the continent, and over 90 percent of the land surface lies below an elevation of 610 meters. The bulk of the continent is arid or semiarid. Biotically, it possesses relictual fauna and flora that show very little major variation from region to region. Various species of kangaroos and eucalyptus are ubiquitous, with closely related

FIGURE 3-1.
An Aborigine woman and child foraging for grubs in a sandy area of the Western Desert, near Mt. Madley, Western Australia (photograph taken in 1970). The wooden bowl is one of the few utilitarian artifacts that are used in daily life. Despite their simple material culture and lack of agriculture, however, these people are able to meet their food needs quite efficiently through diversified gathering and hunting activities. Their nonmaterial culture, moreover, is rich and diversified. (Photograph courtesy of Richard A. Gould, University of Hawaii, Honolulu.)

forms replacing one another from one region to another.

The Australian environment presents few opportunities for gradual development of complex food procurement systems by man. We can cite several specific examples of this environmental impoverishment. First, the continent is poorly endowed with the raw materials for tool making. Hard, fine-grained rocks such as flint are rare, and less easily worked materials such as quartz and quartzite must be used; even these are unavailable in many areas. There are also no animals with horns, antlers, or tusks, the other major source of materials for primitive tools. Not surprisingly, most tools and weapons used by Australian aborigines are made of wood (Figure 3-1). Hafted stone hammers and adzes, together with hand-held stone scapers, constitute the only important tools fashioned out of other materials. The environmental restrictions against the development of more powerful cutting, chopping, and digging tools may thus have restrained the development of agriculture as well.

Specific biotic features unsuitable to the domestication of plants and animals may also have played important roles. Marsupial mammals, although they nurse their young on milk as placental mammals do, are poorly suited for artificial milking by man; and eucalypts, which dominate the woody plant flora almost everywhere, do not produce abundant edible fruits.

The dominant perennial grasses do not produce abundant seed as do the annual grasses that are abundant in many other areas of the world.

This negative evidence from Australia forms a contrast with much positive evidence from areas that did develop agriculture. Environmentally, these regions are characterized by diversity of physiography, climate, and ecology, and by the availability of suitable materials for extensive tool making. All these features exist in the centers of origin of the major systems of agriculture based on the cultivation of grains; these systems, which we shall collectively term **granoculture**, are the ones for which our understanding of evolutionary origins is best. A second major group of agricultural systems, centered on starchy root crops and called **vegeculture**, also appears to have originated in environmentally diverse regions, although we know less about the details of this origin. These systems seem to be of more recent origin and they appear to have developed in more tropical environments than granoculture systems.

Granocultural systems originated in three independent areas of the Northern Hemisphere. These areas, termed **centers** or **nuclear areas**, are not, of course, specific localities. They are, however, relatively well-defined areas no more than a few hundred kilometers in extent. The most important of these centers, in terms of the diversity of forms of domesticated plants and animals that it contributed to modern agriculture, was located in the Middle East, and extended from present-day Israel north and east to southern Turkey and thence southward to the Iran-Iraq boundary. A second Old World center, as yet very poorly known, was located in northwestern China; and the third center, in the New World, occupied the southern portion of the Mexican Plateau. We shall examine each of these centers in detail and attempt to trace the ecological and evolutionary patterns of agricultural development associated with them.

THE ORIGINS OF AGRICULTURE IN THE MIDDLE EAST

The Middle Eastern center of origin occupies an elliptical region that extends from the highlands of Israel and Jordan in the west through Syria and southeastern Turkey to the flanks of the Zagros Mountains just east of the border of Iraq and Iran (Figure 3-2). This region is rich in archeological sites, and has been investigated more thoroughly than the other centers. More than 20 early farming villages have been excavated and studied in detail. Several recent summaries exist that outline our knowledge about agricultural origins in this region (Flannery 1965, 1969; Wright 1968, 1976; Zohary and Spiegel-Roy 1975).

The region is semiarid to arid in overall climate, but possesses wide physiographic variations that are reflected in extensive climatic and ecological diversification. This environmental diversity is illustrated by the region in and around the Zagros Mountains, where many of the most important archeological sites are located (Figure 3-3). This area can be divided into four major ecological zones: high plateau, intermontane valley, piedmont steppe, and alluvial desert. To the east of the Zagros Mountains lies the high, central plateau of Iran, an area that is actually an enclosed basin with an average elevation of 900 to 1500 meters. Most of this plateau is cool desert, with an annual precipitation averaging less than 230 millimeters and vegetation that is dominated by low, shrubby forms of sagebrush (*Artemisia*) and saltbush (*Altriplex*). Herds of gazelles and wild asses live throughout the region, but even today the area has little potential for crop agriculture. Mineral deposits, however, particularly copper and turquoise, exist in the area and have been exploited by man from very early times.

The Zagros Mountains, reaching elevations over 3000 meters, separate the Iranian Plateau from the lowlands of Mesopotamia to the west. Within these mountains, at elevations between

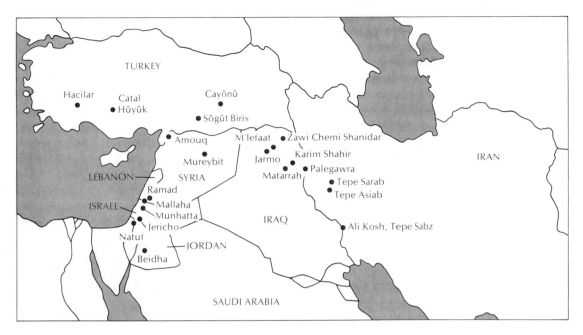

FIGURE 3-2.
Sites of early farming villages in the Near East. Dates for these sites range from about 9500 B.P. to about 8500 B.P. (From Harlan 1971. Copyright © 1971 by the American Association for the Advancement of Science.)

600 and 1350 meters, there are a series of well-watered, fertile valleys with a much richer and more diverse flora and fauna than are found elsewhere in the region. Although precipitation ranges only from about 250 to 1000 millimeters per year, the geological structure of the mountains creates major aquifer strata that receive water as rain and snow at high elevations and discharge it into valley streams. The climate is similar to that of the Mediterranean region in general, with warm, relatively dry summers and cool, damp winters. The vegetation of the hilly slopes is an open oak-pistachio woodland in which the wild ancestors of barley, wheat, and oats still grow along with several fruit-bearing woody plants, including hawthorne and wild pear. Furthermore, the ancestors of four of our most important domestic animals—sheep, goats, cattle, and pigs—once occupied these valleys.

Westward from the Zagros Mountains lies the piedmont steppe, which ranges in altitude from 150 to 300 meters. Here, precipitation drops off rapidly, with the annual total being 250 to 380 millimeters, most of which falls during the winter. The region, is a winter grassland with well-drained, rich soils deposited by streams flowing westward out of the mountains. Herds of gazelle, wild ass, and cattle once roamed here, and deposits of asphalt that occur in several locations within this zone were exploited by early inhabitants of the area, principally as a cementing material for tool construction.

Still further to the west this steppe grades into the alluvial desert region adjacent to the

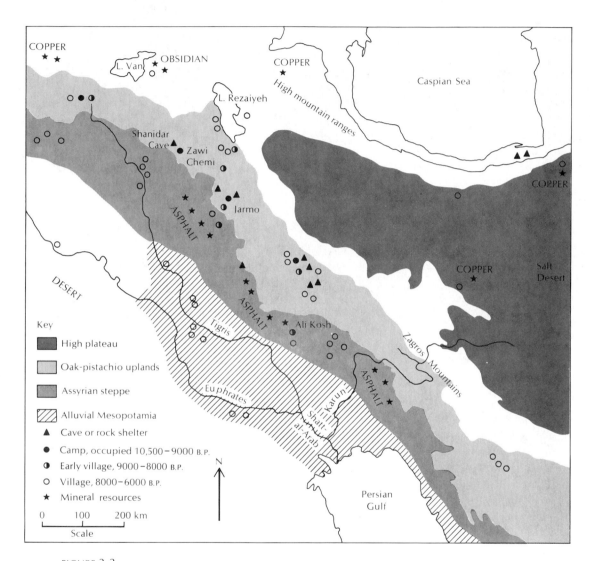

FIGURE 3-3.
The Mesopotamian lowlands and the Zagros Mountain region of Iraq and Iran, showing features important in the early history of agriculture. (Modified from Flannery 1965. Copyright © 1965 by the American Association for the Advancement of Science.)

Tigris, Euphrates, and Karun rivers. Lying at elevations below 150 meters, this region is hot and dry with an annual precipitation again below 250 millimeters. The large rivers contain water throughout the year and possess unique riverine flora and fauna.

This eastern arm of the Middle Eastern center is thus diverse in climate, soils, vegetation, and animal life, yet one can pass through all these zones in a trip of only about 300 kilometers. It is a region with a variety of mineral deposits, including gems, metals, and asphalt, all of which were useful to primitive man. The area has been occupied by man from very early times: Archeological sites establish his presence as early as 42,000 B.P.[*] (Flannery 1965) and show that he was harvesting and hunting many of the plants and animals that he later domesticated. One could hardly characterize this region as an ideal area for the pursuit of agriculture as we now know it; but its basic environmental characteristics seem to have been ideal for the origin of agriculture.

The early, major steps of domestication took place in this region, between about 11,000 and 8000 years ago (Higgs and Jarman 1972; Protsch and Berger 1973). Both animal domestication and domestication of grain species occurred very early. Remains of sheep and goats, together with varieties of einkorn and emmer wheat, are radiocarbon dated to between 11,000 and 9000 years ago in several different localities. Cattle, pigs, and barley appear in the dated record between 9000 and 8000 years ago.

What was the nature of the transition from hunting and gathering activities to organized agriculture, and what stimulated it? Several investigators have been attracted by the possibility that climatic change was a major influence. Childe (1951), one of the earliest to formulate a general theory on the subject,

termed the transition "The Neolithic Revolution." He postulated that increasingly dry conditions that developed at the end of the Pleistocene forced a close association between man and animals in the few areas that had water, and he suggested that his association led first to familiarity and then to domestication.

Wright (1968, 1976) also has asserted the importance of climatic change in the Middle East at the end of the Pleistocene, but in a manner much different than Childe's. Analyses of pollen and of the remains of small aquatic crustaceans in lake sediments of the region testify to significant climatic changes between about 14,000 and 11,000 years ago. Evidence from the same sources suggests that, before these changes took place, a cool-temperate climate prevailed throughout the region, perhaps as cool as that of southern Scandinavia at present, but relatively dry as well. Under such conditions the terrestrial vegetation would have been that of cool desert, with predominant forms being similar to those that now occur over the high plateau areas of the Middle East.

Between 14,000 and 11,000 years ago, the period during which continental glaciers retreated rapidly in more northern regions, a change toward warmer, drier conditions took place. The effects of this change on the vegetation of the Zagros Mountains, for example, may have been great, especially for those species with potential for domestication. Both wheat and barley, in fact, may have spread into the region only after this climatic change. Thus the origin of agriculture may have been triggered or encouraged by the effects of climatic shifts at the end of the Pleistocene. New biotic material, suitable for domestication, may have been introduced just when older patterns of hunting and gathering may have become harder to pursue. Climatic change may in fact have been an essential to the creation of conditions optimal for plant and animal domestication (Wright 1976).

---

[*]B.P. = before present

Other workers, especially Flannery (1965, 1969), note that climatic changes of this sort have occurred many times and in many places without leading to the development of agriculture. Furthermore, sheep and goat remains from as early as 42,000 B.P. are present in cave deposits of the Zagros Mountain area, suggesting that environmental conditions at that time were not totally dissimilar to those of the later epoch during which agriculture did originate. Cereal pollen is also present in cave deposits dated to 16,000 B.P., indicating that the biotic raw material was present well before the transition to agriculture occurred.

Flannery argues that agriculture arose slowly and gradually over several thousand years and that the driving force was the interaction between man and the diversified natural environment of this Middle Eastern center. The key feature of this interaction was the development of a "broad spectrum" system of hunting and gathering that was affected by the seasonality of various plant and animal foods and that probably involved seasonal migration between some or all of the environmental zones of the region. The seasonality of some wild foods would give rise to food procurement systems that were scheduled to permit maximum effectiveness of exploitation of the various materials available.

The sequence leading from diversified hunting and gathering to agriculture can be considered in three stages, although this is of course arbitrary. The earliest stage—preagrarian hunting and gathering—is presumed to have spanned the period from before 42,000 B.P. to about 12,000 B.P. During this period the inhabitants of the region perfected a complex, closely scheduled system of wild plant and animal food procurement. During the summer, for example, various fruits and nuts were available in the mountain valleys; wild grains also matured there, large game could be hunted, and forays could be made into the high plateau region for hunting and mineral collection. During

fall and spring, hunting and gathering parties may have moved to the lower, warmer piedmont steppe. When other food sources in these areas were least abundant, populations of fish, turtles, crabs, clams, and other riparian foods could be harvested in the alluvial desert region near major rivers. Diversified hunting and gathering activities of this sort would have fostered the development of a variety of tools for hunting, food collection and preparation, and travel. Toward the end of this period, the various methods of food procurement probably began more closely to resemble the actual husbandry of wild plants and animals. The main food species, while still reproducing in the wild, experienced important shifts in the evolutionary forces affecting them. More and more, human food procurement came to have a dominant effect on their evolution.

The second stage, involving approximately a 4000-year period from 12,000 B.P. to 8000 B.P., was marked by a transition from the husbandry of wild populations to the management of domesticated species. Domestication is considered to have occurred only when man began exercising intentional, purposeful selective breeding of the species involved, and it is likely that this began in association with three factors: the establishment of more permanent or regularly used settlements, the increase in the capacity of human technology to modify the physical and biotic environment, and an increase in human population density.

In this transition to domestication, environmental disturbance may also have played a very important role. For example, disturbance of soil and natural vegetation favors annual plants characterized by rapid vegetative growth and abundant seed production. These are characteristics of weeds, but also of the ancestors of many crop plants, including wheat, barley, and especially oats. Species of this sort are also adapted to environments that are intensively grazed and in which fire is frequent (Lewis 1972). Thus environmental disturbance near human settlements probably produced conditions that selectively favored plant and animal species with a high potential for domestication. These disturbed environments may have been colonized naturally, or, in the case of certain plant species, the seeds may have been transported to settlement areas along with collected plant material.

The behavior of these "protodomesticates" in disturbed environments may have contributed still further to the process of domestication. Disturbed conditions, and the removal of competition from other native plants and animals, may have permitted the survival of genotypes, or even of interspecific hybrids, that would otherwise have been quickly eliminated in nature but that might have had characteristics useful to humans. Furthermore, environmental disturbance combined with the disposal of organic wastes or deposition of ash following fires may have especially favored species that responded well to fertilizer.

Thus it is likely that gradual transitions occurred from simple collection to the husbandry of wild populations and to unconscious and conscious control of reproduction. Protodomesticates, though in many cases still in reproductive contact with wild populations, gradually became subjected to new evolutionary pressures and to accelerated evolution. Certain forces of natural selection, such as natural grazing pressure and competition from wild forms, were replaced by new pressures, including those of new soils and climates in areas to which man transported the species involved.

Identifying the point at which evolutionary control shifted to man, however, is not easy; it requires the accumulation in archeological material of the morphological changes that distinguish husbanded forms from wild forms. In animal species, it may also be recognized through differences in the age or sex structure of the population indicated by the bone material accumulated in archeological sites.

Dating the time of change is also difficult. Radio-carbon dating of bones and plant materials is the best method, but for grains and other remains of early plants this is rarely possible. Dates based on radiocarbon analysis of charcoal associated with the remains of early domesticates can occasionally give false information because of the redeposition together of materials of different ages. Nevertheless, detectible changes occurred in many species at a very early stage in domestication. Some of these appear to be the results of unconscious selection; that is, changes that automatically accompany human influence without being chosen deliberately (Darlington 1970). One must be careful in concluding that particular characteristics were the result of unconscious or conscious selection, because our understanding of what was important to early man is of course incomplete.

It is very possible that unconscious selection was responsible for several characteristic and parallel changes in early crop plants (Harlan et al. 1973, Harlan 1975, Evans 1976). For example, the early ancestors of wheat, barley, and many other grains had seed heads with a brittle or "shattering" central stalk. In wild populations this characteristic is strongly adaptive because, when the seed is mature, the action of the wind or a passing animal will cause the seeds to break away and be carried from the parent plant. Once the planting of the seed is under human influence, however, this trait causes most seeds to be lost; those that remain attached are perpetuated by human cultivation. In most grains the characteristic of shattering or nonshattering is under simple genetic control and easily modified by selection.

Other changes also increase the fraction of the seed recovered by the farmer: These include growth patterns that reduce the number but increase the size of inflorescences and also synchronize their maturation. The best example of this is the cultivated sunflower, the single giant head of which has replaced many smaller

FIGURE 3-4.
Spikelet of wild oats, *Avena fatua (left)*, contrasted with two spikelets of cultivated oats, *A. sativa*. In wild oats, the lemmas (outer of the two bracts enclosing the floret) are thickly covered with stiff brown hairs and have a long, twisted awn. This awn untwists when moistened, creating a turning motion that tends to force the seed into damp soil after a rain. The dense covering of hairs is protective. In cultivated oats, however, the functions of protection and planting have been taken over by man, and both these characteristics have therefore been lost. The cultivated grain is also larger and has less protein (19 percent instead of 30 percent as in the wild oats) because of the relative increase in amount of stored carbohydrate. (Photograph courtesy of the USDA.)

heads that, in the wild form, tend to flower and mature over a longer period. Human harvest and planting also tend to increase the percent of seed set, and to convert sterile into fertile florets. This occurs simply because plants with these characteristics contribute more to the next season's seed and thus perpetuate their qualities.

Still another set of automatic selection pressures affect seedling growth in dense stands; these often produce, for example, greater seed size, usually by increasing the amount of carbohydrate relative to protein, and a reduction in germination inhibitors, as well as in the glumes and other appendages that often contain such substances (Figure 3-4).

Many other trends can be discovered in the evolution of early domesticates, some of which are clearly the product of conscious selection. Among them are the loss of defensive structures such as spines or bitter and toxic chemicals, the increase in size of the food organ, and many other characteristics of appearance and taste.

Domestication of a number of other species, particularly of legumes, seems to have occurred along with that of wheat and barley (Zohary and Hopf (1973). Peas (*Pisum sativum*) and lentils (*Lens culinaris*) are recorded from a number of early farming villages by 9000 to 8000 B.P. Bitter vetch (*Vicia ervilia*), chick pea (*Cicer arientinum*), and broad bean (*Vicia faba*) were added to the list of domestic species shortly thereafter.

At this point, roughly 8000 B.P., we may consider that Middle Eastern agriculture had reached the state of formal, organized practice. From then on, the inhabitants expanded and diversified agricultural activity and spread it to new regions. We shall discuss certain aspects of this stage shortly, but first let us examine the centers of origin of other granocultural systems.

THE MIDDLE AMERICAN CENTER OF ORIGIN

In Middle America, the pattern of agricultural evolution shows interesting similarities to that of the Middle East (Flannery 1968). The highlands of southern Mexico seem to have been the center of origin in the Northern Hemisphere, although, as in the case of the Old World, this center was certainly not restricted to a single locality. Much of the evidence bearing on agricultural origins again comes from cave deposits, several of which are located in the Tehuacan Valley, in the state of Puebla, south and east of Mexico City (Figures 3-5 and 3-6). These sites, studied by MacNeish (1964a, 1964b) and later by several other workers, together with several sites further south in Oaxaca, attest to the fact that agricultural origin was a regional phenomenon.

The physiography of the southern Mexican highlands is diverse; although the interior of Mexico here is generally semiarid, mountains, intermontane valleys, and interior basins combine to create a variety of climates, soil types, and natural communities. The highlands fall away to coastal plains on the Pacific and Atlantic slopes, the latter of which has a distinctly humid, tropical climate. Within this environment, early man developed a diversified hunting-gathering system of food procurement and eventually evolved a distinctive form of

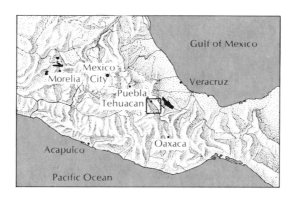

FIGURE 3-5.
The Tehuacan Valley is located in a semiarid highland area of southern Mexico, on the border of the states of Puebla and Oaxaca. The dry climate favors the preservation of plant materials in archeological sites. (From "The Origins of New World Civilization," by Richard S. MacNeish. Copyright © 1964 by Scientific American, Inc. All rights reserved.)

FIGURE 3-6.
Excavations at Coxcatlan Cave in the Tehuacan Valley, Puebla, Mexico. These excavations revealed a record of human occupancy from before 9000 B.P. through about 1500 A.D. and provided much information about the origin of agriculture and the domestication of corn in the New World. (Photograph courtesy of the Robert S. Peabody Foundation for Archeology.)

agriculture (Flannery 1968). A great deal can be inferred about these early hunting and gathering patterns, since many of these activities are still practiced by Indians of the region. Although agriculture is now the main food procurement system, the various methods of hunting and gathering are still practiced on a more limited scale.

Over the period from about 13,000 to 7000 years ago, the inhabitants of this region evolved a broad range of hunting and gathering methods closely keyed to the seasonality of various food resources. Each type of food required a particular type of technology. In the Tehuacan Valley, six major systems of food procurement seem to have been predominant: The first cen-

tered on the maguey, a succulent-leaved rosette plant of the genus *Agave*, commonly known as the century plant. Typical of many arid and semiarid regions of tropical and subtropical America, these plants grow without flowering for a number of years; eventually they send up a tall flower stalk and then bloom and die. The heart, or interior, of the stem base at the time this flower stalk begins to grow, is edible if it is roasted for a long time to make it tender and to eliminate bitter juices. The chewed, fibrous remains of this material are a common component of food remains in Middle American archeological sites.

A second food procurement system centered on fruits of various cacti, including prickly pears and columnar forms. The fruits of these plants are abundant at particular times of year, generally toward the end of the dry season. They must be harvested quickly before they are eaten by animals, and then used within a short time before they rot.

A third system involved collection of the seed pods of several species of leguminous trees, including members of the genus *Prosopis*, to which the mesquite belongs. The young, fresh pods of several of these trees are tender and can be eaten raw; some can also be boiled to produce a thick syrup.

Inhabitants of the area also hunted white-tailed deer and the cottontail rabbit. White-tailed deer range widely through different habitats of the region and they were the most important large game animal. They were hunted principally with spears, thrown by hand or with a mechanical spear-thrower from points along animal trails. Cottontail rabbits were captured mainly in traps and snares.

Finally, another system, initially of relatively minor importance, involved the procurement of seeds of wild grasses, including species of foxtail grasses of the genus *Setaria*, and, more important, the ancestral forms of corn. This system eventually became central to the origin of New World agriculture.

Certain of these foods were available only during particular seasons, especially the cactus fruits, legume pods, and grass seeds. Food procurement thus had to be scheduled so that harvesting these foods took precedence over hunting, which was possible year round. By this scheduling, and by local or regional migrations, hunting-gathering groups were able to maintain a diversified food exploitation system.

Between 7000 and 4000 years ago, the transition to formal agriculture took place, apparently through gradual processes similar to those in the Middle East. Some workers feel that special stimuli may have precipitated the transition: climatic change, demographic pressure of an unusual degree, or accidental

genetic changes in the ancestral form of corn that converted this plant food to a more important food source and initiated a sequence that led to domestication.

Whatever the details of this transition, the fact remains that in Middle America agriculture arose in an environment that had many of the same characteristics as the Middle Eastern center. Here we may note that a secondary, or perhaps independent, center of origin existed in the Andean region of South America, again a physiographically and ecologically diversified region. The agricultural system that evolved here, and of which we know few details, is unique in its reliance upon both a grain—corn—and a root crop—the potato.

THE NORTHERN CHINESE CENTER

The third and least well-known center of granoculture origin is located in northwestern China (Chang 1970; Ho 1969). This is the northernmost of the centers of granocultural origin; it is situated between the fertile North China Plain in the east and the deserts of western China and Mongolia (Figure 3-7). Like the previous two centers, however, it occupies a region with a generally semiarid climate and diverse physiography. The archeological sites from which this center is known lie on terraces above the Wei Ho and Yellow Rivers, where elevations are 600 to 700 meters. To the south, in a northwest-southeast band, extends the Tsinling Shan Mountain Range, the highest range east of the Plateau of Tibet. The highest peak of this range is slightly over 4100 meters. To the north and west extends the loess plateau, a unique and awesome area. Wind-deposited loess lies 50 to 250 meters deep over the plateau, but erosion has cut ravines 90 to 150 or more meters deep over more than half of it. The surface of the plateau lies at elevations of 1300 to 1600 meters, a landscape that has apparently been barren of trees in post-

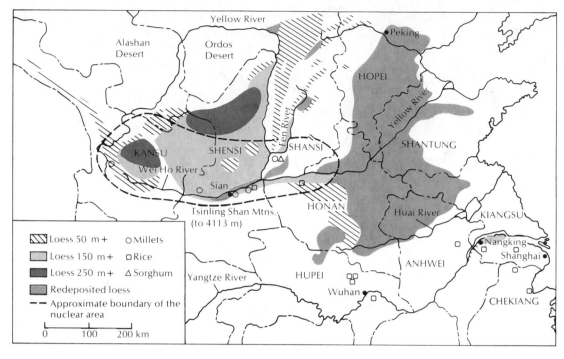

FIGURE 3-7.
The northern Chinese center of agricultural origin.
(Modified from Ho 1969.)

Pleistocene times, and that seems to have supported, at best, a desert steppe vegetation of sagebrush and similar species.

Information on the life of preagricultural peoples in this region is scarce, but archeological sites indicate that the first cultivated plants were grains distinct from those cultivated in other areas of the world. These early domesticates were millets of the genera *Setaria* and *Panicum*, and their cultivation began at least 7000 years ago (Ho 1969). Historical records attest to the fact that a great many varieties of these millets existed. Later on, rice, wheat, and barley were added to the inventory of grains, and various vegetable species were domesticated.

At this point all we know is that a fairly discrete center of agricultural origin existed in northern China, and that it shares with the other centers we have described a semiarid climate and certain features of physiographic diversity.

GRANOCULTURAL ORIGINS AND ANIMAL DOMESTICATION

The three centers of granoculture origin differ widely in terms of the endemic animal species that were domesticated. The earliest and most important animal domestications occurred in the Middle Eastern center, where the four

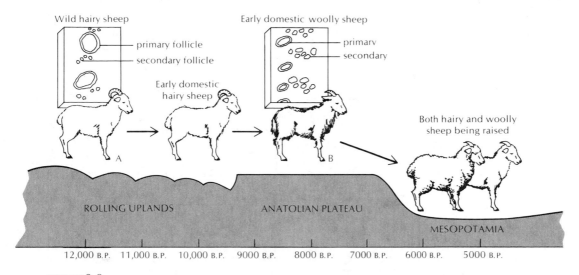

FIGURE 3-8.

Diagrammatic representation of the changes in hair follicle characteristics of sheep during the early history of domestication in the Near East. In the wild ancestral forms (A), the coat was hairy because it had fewer and smaller secondary follicles. Woolly sheep (B) originated about 8000 B.P. through selection for more and larger secondary (wool) follicles. Both hairy and woolly forms are still raised in the Near East. (Modified from Flannery 1965. Copyright © 1965 by the American Association for the Advancement of Science.)

most important livestock species of modern agriculture—sheep, goats, pigs, and cattle—first appear as domesticates. Interestingly, if the original geographical ranges of the ancestors of these species are mapped, the Middle East appears as the only region of Eurasia in which all four existed together.

Sheep may have been the first species ever domesticated (Higgs and Jarman 1972). Remains from one of the northern Zagros Mountain sites, dated at 11,000 B.P. are mostly of young animals; this suggests that husbandry practices involved selective harvest of the young. One of the earliest morphological changes associated with domestication was the loss of horns in female animals; the remains of a hornless female have been radiocarbon dated to 9500 B.P. (Flannery 1969). At the time of domestication, the sheep had a hairy coat like a deer's (Flannery 1965). Woolly sheep

were the product of selection that modified the location and emphasized the importance of the hair follicles that produce dense underfur. They probably did not appear until about 8000 B.P. (Figure 3-8).

The goat was also domesticated very early, apparently between 10,500 and 10,000 B.P. as indicated by sites in which bone remains suggest selective harvest of particular age classes (Flannery 1965; Protsch and Berger 1973). After domestication, the shape of the goat's horn changed from that of a scimitar to the familiar twisted form, a change that could be the result of unconscious selection leading to a nonadaptive character.

Both cattle and pigs were apparently domesticated somewhat later, perhaps around 9000 B.P. (Protsch and Berger 1973). For both of these species, one of the early changes following domestication was an appreciable reduction in size. Interestingly, domestic pigs appeared in more northern, humid sites in the Middle Eastern center, and in situations that clearly represent permanent settlements. Further south, in more arid regions, where animals had to be herded between lowland winter ranges and mountain or upland summer ranges, goats, sheep, and cattle rather than domestic pigs appear in the archeological record. Thus early livestock systems seem to have been divided, on ecological grounds, into those with and without pigs. Religious and dietary laws relating to the pig were probably the result rather than the cause of this differentiation.

One of the more interesting points of speculation about the origins of agriculture concerns the relationship between plant and animal domestication. The Middle Eastern center of origin appears to have been both the earliest and the richest in animal domesticates. There is little evidence that either form of domestication preceded the other and considerable reason to believe that the two processes were closely associated. One of the early stages of animal domestication probably involved capturing young animals born in wild populations and confining or incorporating them into herds associated with human setttlements. At times these animals may have required artificial feed, and the disturbance of vegetation near settlements by these herds, together with seeds introduced in forage brought to such herds, may have been one of the ways in which man came into close association with potential crop species such as wheat and barley. It could be that agriculture first appeared in the Middle East because it was only in this area that a strong animal-plant interaction promoted domestication.

Very few animals were domesticated early in either the Middle American or Chinese centers. In Middle America only the turkey and the dog, which was used partly as a food animal, were domesticated. The failure of New World peoples to domesticate major livestock species is correlated, however, with differences that still exist in the diversity of the large animal fauna of North America. North America had an abundant and diverse fauna until the late Pleistocene, but it disappeared at roughly the same time that skilled large animal hunters appeared on the continent. Although little-known hunting and gathering cultures apparently existed much earlier, in the New World there is good evidence that many skilled hunters of mammoths and other large game came out of Alaska about 13,000 years ago (Haynes 1970). When the continental and mountain glaciers receded in the northwest, these hunters were able to move from Alaska southeast along the Rocky Mountains. Upon reaching west-central Canada, they encountered an extensive fauna poorly adapted to human predation and thus extremely vulnerable to hunting. This contrasted to the situation in the Old World, where wild animals were able to adjust gradually as man developed into a skilled hunter over a period of many thousands of years. The result it that these hunters may have expanded rapidly and suddenly over North and South America, reaching Tierra del Fuego in little more than a thousand years, replacing existing human cultures and hunting most large fauna to extinction (Martin 1973). This may have selectively eliminated many species with potential for domestication, thus accounting for the less extensive domestication of animals in the Middle American area.

In northwestern China, there is no evidence for endemic domestication of any animals. Cattle and pigs were present in this region, but sheep and goats, the first species domesticated in the Middle East, were not. Perhaps the greater ability of sheep and goats to tolerate extreme, semiarid environments was an added important feature in the first stages of animal domestication, an advantage absent in the northern Chinese center.

DIVERSIFICATION AND SPREAD OF GRANOCULTURE

As early agriculture spread outward from the Middle Eastern center, new species of plants and animals suitable for domestication were encountered. This, along with continued elaboration within the center itself, eventually led to the domestication of a number of long-lived woody plants and vines. Earliest among these were the date palm, *Phoenix dactylifera*, which appeared in lower Mesopotamia about 6000 B.P., and the olive, *Olea europea*, which first appeared about 5700 to 5500 B.P. in Palestine (Zohary and Spiegel-Roy 1975). The grape, *Vitis vinifera*, was domesticated by 4900 B.P. and the fig, *Ficus carica*, by 3900 B.P., both in areas bordering the eastern Mediterranean.

It is not surprising that tree and vine crops were added only after the earlier domestications. The wild forms of all of these species are long-lived, and the time required to convert them from marginally important sources of wild fruit to highly productive domesticates must have been longer than for annuals such as wheat and barley. It is important to note that, although all these species commonly reproduce by seed, they are all also particularly amenable to vegetative propagation (by cuttings and transplanted shoots), which, under domestication, is the exclusive means of reproduction for all four species. The susceptibility of these forms to vegetative propagation undoubtedly played an important role in their early domestication.

As agriculture expanded, interaction between domesticated and wild plant and animal species probably contributed in an important way to the further evolution of domestic varieties. New patterns of hybridization emerged as domesticates were transported to areas with different climates and soils, where they came into contact with different crop varieties or species of crop relatives. The importance of such interactions in crop evolution is examined more closely in Chapter 4.

The spread of agricultural technology from the Middle East into Europe via Greece occurred by 8500 to 8000 B.P. (Rodden 1965). From there, further expansion apparently followed two general routes, one westward along the coastal areas of the Mediterranean and eventually the Atlantic, the other north and west through the Balkans and Hungarian Plain into northern Europe (Figure 3-9). The rate of expansion has been estimated at only about one kilometer per year, and in all probability it involved the migration of the peoples who actually practiced agriculture, rather than simple technological diffusion (Cavalli-Sforza 1974). This spread brought early agriculture, with its four major animal domesticates and

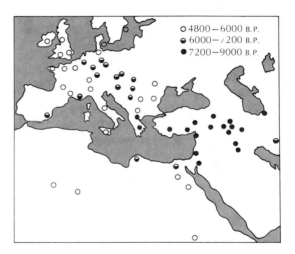

FIGURE 3-9.
The location and approximate age of early farming settlements in Europe and North Africa. These seem to indicate two routes of agricultural dispersal, one along the Mediterranean littoral and the other northward through the Balkans and the Hungarian Plain into central Europe. (Reproduced with modification from Harlan, *Crops and Man*, 1975, Chapter 8, page 186, by permission of the American Society of Agronomy and Crop Science Society of America.)

several grains, to Scandanavia and England by approximately 5000 B.P. (Waterbolk 1968).

The technology of agriculture also spread eastward, carrying the basic Middle Eastern domesticates, including wheat, barley, sheep, goats, cattle, and pigs, into northern China between 4000 and 3000 B.P. (Ho 1969; Watson 1969).

THE ORIGINS OF VEGECULTURE

Systems centered on grain production may not be the only forms of early agriculture that evolved independently. A growing amount of evidence exists for the origin in several parts of the tropics of systems of vegeculture that were partially or completely independent in their early stages (Harris 1972). These systems were based on the cultivation of starchy tubers such as manioc, taro, yams, and sweet potato, but they also include many vegetables and fruits and a few domestic animals.

Unfortunately, the archeological record of vegecultural origins is much less complete than that for granoculture. Humid tropical conditions favor rapid decomposition of organic remains, and the basic foods, primarily fleshy tubers, are especially vulnerable to such decomposition. Furthermore, fewer specialized tools of stone and other durable materials seem to have been used in vegeculture. Many of our ideas about vegecultural origins must therefore be inferred from present-day patterns of human culture and plant distribution. The few radiocarbon dates that are available, however, indicate that these systems are fairly old. In southeastern Asia the remains of several vegetable and legume species that may have been domesticated have been recovered from Spirit Cave in northwestern Thailand. These materials have been radiocarbon dated to between 10,000 and 8000 B.P. (Gorman 1971). In the Peruvian Andes of South America, domesticated common and lima beans have been dated as early as

7680 B.P. (Kaplan et al. 1973). The llama and its relatives may have been domesticated between 4500 and 3750 B.P. (Pires-Ferreira et al. 1976).

Vegecultural systems have evolved in three parts of the tropics: the New World tropics, Africa, and southeastern Asia. Like granoculture, these systems appear to have originated under characteristic ecological conditions and then dispersed into areas with a broader range of environments (Harris 1972). Unlike granocultural systems, however, the areas of origin do not appear to be well-defined centers of relatively small size, but broad regions of much greater extent. The characteristics of these regions can be inferred from the nature of the principal cultivated crops. The original biological function of starchy tubers permits the survival of a plant over an unfavorable season, which in the tropics almost always means a dry season. With the return of moist conditions, the stored food energy of the tuber permits rapid vegetative growth of the plant. Thus, the staple crop species of vegeculture systems are perennials rather than annuals. Several, such as yams and sweet potatoes, are vines; others, including manioc, have a shrublike form. In both cases, it appears that these plants, now widely cultivated in many tropical environments, are derived from ancestors that lived on the fringes of the humid tropics in regions with dry seasons that range from two and one-half to five months long. The viny species probably grew at forest edges where light was available in abundance, but where woody plants were also available to support their climbing foliage. The shrublike forms probably inhabited more open, dry areas, or areas with longer dry seasons.

Thus it is likely that vegecultural systems originated in tropical areas that had pronounced dry seasons and an intermixture of forest and nonforest vegetation, probably areas in which riparian forests give way to savannas or grasslands, or areas in which the interactions of soils, animal grazing, fire, and vegetation led to a mosaic of different plant communities. In South America, these conditions are met in large areas of the Orinoco Basin and coastal lowlands of Venezuela and Columbia. In Africa, environments of this general type extend from West Africa eastward across the northern edge of the Congo Basin. In Southeast Asia, similar conditions prevail from the Ganges lowlands of India eastward across the upper Indo-chinese region.

Vegecultural systems have spread widely through the tropics, however, and in particular have expanded into humid tropical environments. These systems appear ecologically well adapted to the humid tropics. The vegecultural crop complex includes many species besides the principal tuber producers. More important, these include a variety of plant forms: herbaceous annuals, climbing perennial vines, shrubby forms, and large long-lived herbaceous species such as banana and sugar cane. Various grains, such as rice in Southeast Asia, and grain amaranths in highland parts of South America, were also domesticated in these regions of vegeculture origin. A vegeculture crop ecosystem may thus be quite diverse in structure and composition. This not only makes efficient use of soil nutrients and light energy, but also creates a relatively permanent cover for the soil that protects it from the direct action of sun and rain. The greater degree of biotic diversity may also make the crop system as a whole less vulnerable to destruction by crop pests. Finally, the fact that the major crop is a starchy tuber means that the soil is not as rapidly depleted of nutrients such as nitrogen and phosphorus as it would be under grain cultivation.

Nevertheless, shifting cultivation is necessary in nearly all vegecultural systems because of the eventual decline of fertility and the increased losses from weeds and animal pests. The fact that the principal crop is primarily starch also means that vegeculture must be supplemented to a greater extent than granoculture by other sources of protein. Fishing,

hunting, and the gathering of wild fruits are important parts of the total food economy of people who practice vegeculture.

Despite these disadvantages, it appears that granoculture tends to infiltrate or even to displace vegeculture in many areas of the tropics (Harris 1972). Grain is a better food, nutritionally; it is also easier to store and transport. People who practice granoculture are consequently less dependent on accessory food resources such as fish and game. Not surprisingly, such groups seem more prone to colonize new regions. Within the tropics, the more detrimental impact of granoculture on the environment may, in fact, force such groups to be more mobile.

Early vegecultural systems seem to have been replaced by granoculture in parts of northern South America. Archeological sites in northern Colombia and northwestern Venezuela show a transition from implements used to prepare manioc to those used to grind corn and other seed crops at approximately 2500 B.P. (Harris 1972). In southeastern Asia, as well, rice cultivation has shown a progressive expansion from a continental source area through the Indonesian archipelago, where it has displaced vegecultural systems based on yams and taro (Figure 3-10). This trend has intensified even more of course as a result of the worldwide dissemination of modern mechanized grain cultivation.

INTERACTIONS BETWEEN GRANOCULTURE AND VEGECULTURE

Granocultural systems lack several of the adaptive features of vegeculture, especially when considered in relation to humid tropical environments. Monocultural stands of grains are highly susceptible to pests and diseases, and because the crop plants are annuals, the soil is exposed more to sun and rain during early crop growth and after harvest. This depletes

FIGURE 3-10.
In parts of Southeast Asia, vegecultural systems based on root staples such as taro, *Colocasia esculenta*, have been replaced by granocultural systems based on rice. Taro, seen here in a planting in Hawaii, is cultivated in wet paddies in a manner similar to paddy rice culture. (Photograph by Herbert Y. T. Loo, courtesy of the USDA Soil Conservation Service.)

the soil of nutrients more rapidly and means that cultivated plots must be abandoned sooner than in vegeculture.

SUMMARY

The transition from hunting and gathering to organized agriculture occurred independently in several centers or world regions. Granocultural systems, centered on the cultivation of grains and seed crops, have three major centers of origin in subtropical and temperate regions. Vegecultural systems, which cultivate starchy tubers, originated in the three main areas of the world tropics. In a few areas, agriculture neither evolved nor entered by cultural diffusion.

Centers of agricultural origin are characterized by a high degree of ecological diversity expressed in local or regional differences in climate, soils, and natural communities. These conditions permitted the development of diversified hunting and gathering systems that led gradually to the intensive manipulation by human beings of wild plant and animal populations. As the density and permanence of human populations increased, plant and animal species preadapted to domestication tended to concentrate in the disturbed environments surrounding settlements. Manipulation of these populations gradually reduced natural selection and increased artificial selection. These species eventually came under the reproductive control of man, thus creating early, diversified agricultural systems.

Both granoculture and vegeculture spread widely and continued to diversify. Vegeculture, which originated on the drier margins of the humid tropics, tended to invade more humid tropical areas, where its characteristics appear ecologically compatible with basic ecosystem processes. Granoculture, tends to be a more aggressive colonizer, mainly because of the greater nutritive value of its principal food and the independence from other procurement systems that results from this fact. In at least some humid tropical areas, early vegecultural systems were later displaced by granoculture.

Literature Cited

Cavalli-Sforza, L. L. 1974. The genetics of human populations. *Sci. Amer.* 231(3):80–89.

Chang, K. 1970. The beginnings of agriculture in the Far East. *Antiq.* 44:175–185.

Childe, V. G. 1951. *Man Makes Himself.* London: C. A. Watts and Co.

Darlington, D. C. 1970. The origins of agriculture. *Nat. Hist.* 79(5):47–57.

Dragoo, D. W. 1976. Some aspects of eastern North American prehistory: A review 1975. *Amer. Antiq.* 41:3–27.

Evans, L. T. 1976. Physiological adaptation to performance as crop plants. *Phil. Trans. Roy. Soc. London.* B275:71–83.

Flannery, K. V. 1965. The ecology of early food production in Mesopotamia. *Science* 147: 1247–1256.

————. 1968. Archeological systems theory and early Mesoamerica. In *Anthropological Archeology in the Americas.* B. J. Meggers (ed.). pp. 67–87. Washington, D.C.: Anthropological Society of Washington.

————. 1969. Origins and ecological effects of early domestication in Iran and the Near East. In *The Domestication and Exploitation of Plants and Animals.* P. J. Ucko and G. W. Dimbleby (eds.). pp. 75–100. Chicago: Aldine Publishing Co.

Gorman, C. F. 1971. The Hoabinhian and after: subsistence patterns in Southeast Asia during the late Pleistocene and early Recent periods. *World Arch.* 2:300–320.

Harlan, J. R. 1971. Agricultural origins: centers and noncenters. *Science* 174:468–474.

————. 1975. *Crops and Man.* Madison, Wis.: Amer. Soc. Agron.

Harlan, J. R., J. M. J. de Wet, and E. G. Price. 1973. Comparative evolution of cereals. *Evolution* 27:311–325.

Harris, D. R. 1972. The origins of agriculture in the tropics. *Amer. Scientist* 60:180–193.

Haynes, C. V. 1970. Geochronology of man-mammoth sites and their bearing on the origin of the Llano complex. In *Pleistocene and Recent Environments of the Central Great Plains.* W. Dort, Jr. and J. K. Jones, Jr. (eds.). pp. 77–92. Univ. of Kansas, Dept. of Geol., Spec. Pub. 3.

Higgs, E. S., and M. R. Jarman. 1972. The origins of animal and plant husbandry. In *Papers in Economic Prehistory.* E. S. Higgs (ed.). pp. 3–13. Cambridge: University Press.

Ho, P. 1969. The loess and the origin of Chinese agriculture. *Amer. Hist. Rev.* 75(1):1–36.

Kaplan, L., T. F. Lynch, and C. E. Smith, Jr. 1973. Early cultivated beans (*Phaseolus vulgaris*) from an intermontane Peruvian valley. *Science* 179:76–77.

Lewis, H. T. 1972. The role of fire in the domestication of plants and animals in Southwest Asia: A hypothesis. *Man* 7:195–222.

Martin, P. S. 1973. The discovery of America. *Science* 179:969–974.

MacNeish, R. S. 1964a. The food gathering and incipient agriculture stage of prehistoric Middle America. In *Natural Environments and Early Cultures*. R. C. West (ed.). *Handbook of Middle American Indians*, vol. 1, pp. 413–426. Austin: University of Texas Press.

———. 1964b. The origins of New World civilization. *Sci. Amer.* 211(5):29–37.

Mulvaney, D. J. 1966. The prehistory of the Australian Aborigine. *Sci. Amer.* 214(3):84–93.

Pires-Ferreira, J. W., E. Pires-Ferreira, and P. Kaulioke. 1976. Preceramic animal utilization in the central Peruvian Andes. *Science* 194:483–490.

Protsch, R., and R. Berger. 1973. Earliest radiocarbon dates for domesticated animals. *Science* 179:235–239.

Rodden, R. J. 1965. An early Neolithic village in Greece. *Sci. Amer.* 212(4):83–92.

Struever, S. (ed.). 1971. *Prehistoric Agriculture*. Garden City, N.Y.: Natural History Press.

Ucko, P. J., and G. W. Dimbleby (eds.). 1969. *The Domestication and Exploitation of Plants and Animals*. Chicago: Aldine Publishing Co.

Waterbolk, H. T. 1968. Food production in prehistoric Europe. *Science* 162:1093–1102.

Watson, W. 1969. Early animal domestication in China. In *The Domestication and Exploitation of Plants and Animals*. P. J. Ucko and G. W. Dimbleby (eds.). pp. 393–395. Chicago: Aldine Publishing Co.

Wright, H. E., Jr. 1968. Natural environment of early food production north of Mesopotamia. *Science* 161:334–339.

———. 1976. The environmental setting for plant domestication in the Near East. *Science* 194:385–389.

Zohary, D., and M. Hopf. 1973. Domestication of pulses in the Old World. *Science* 182:887–894.

Zohary, D., M. Hopf and P. Spiegel-Roy. 1975. Beginnings of fruit growing in the Old World. *Science* 187:319–327.

4

THE ECOLOGY OF DOMESTICATION

We have seen that the earliest systems of agriculture developed gradually in several widely separated areas of the world that had seasonal climates and a diversity of natural ecosystems in close proximity. Now we can focus more closely on where various plant and animal species were domesticated and how, in ecological and evolutionary terms, this process occurred. Because crop and domestic animal origins are broad topics, that have been thoroughly dealt with elsewhere (Heiser 1973; Zeuner 1965), we must, even in our closer examination, confine our attention to the general patterns and processes of domestication. We shall begin with the major geographical patterns of crop and animal domestication, after which we shall consider the basic evolutionary mechanisms that operated during the domestication process, and we shall illustrate these by detailed examination of several plant and animal examples.

We have already discussed some of the geography of domestication, having described the world regions in which the species that form the basis of the earliest agricultural systems were domesticated. These species, however, are only a few of the plant and animal domesticates that are important to modern agriculture. We must therefore concern ourselves with the areas that provided the biotic raw material for the enrichment of early agriculture and thus created the systems of diversified plant and animal agriculture that now exist.

We have also mentioned some of the ecological relationships and evolutionary processes important to domestication, including, for example, the significance of environmental disturbance around human settlements. The domestication process, however, involves much more than gradual change growing out of the interaction of man and a protodomesticate under particular environmental conditions. The entire process of evolutionary change is an accelerated one, due to the imposition of strong, new selective pressures by man and sudden changes in basic characteristics of the plant or animal species are often important. Many such changes occur by hybridization, which obviously depends on the kinds of the species that exist in and around agricultural areas. Domestication is thus an intricate evolutionary process that typically involves not only man and one or more potential domesticates, but also one or more nondomesticated forms.

GEOGRAPHY OF CROP AND DOMESTIC ANIMAL ORIGINS

The first comprehensive studies of the geographic origins of crop plants were carried out by N. I. Vavilov, a Russian plant geneticist. Vavilov was Director of the All-Union Institute of Plant Industry in Leningrad from 1920 until his arrest in 1940 for his opposition to the government-backed genetic views of Lysenko. Vavilov died in a Siberian labor camp in 1943 (Caspari and Marshak 1965). Between 1921 and 1934, however, he founded more than 400 research institutes and experiment stations with a combined staff of about 20,000 persons. Between 1923 and 1931 he led expeditions to various parts of Asia, Africa, and Central and South America to collect cultivated plants, and he directed still other expeditions to various parts of the world to collect plants of economic importance. From 1931 to 1939 he concentrated on similar studies of domesticated animals, especially cattle, horses, and reindeer. His influence, however, is felt most strongly in his ideas about the geography of crop origins (Leppik 1969).

Examination of the enormous collections of crop species and their varieties often revealed wide differences in the number of varieties of a particular crop species that were cultivated in different geographical areas. For example, Vavilov accumulated some 26,000 strains of wheat from various parts of the world. The greatest number of bread wheats, *Triticum aestivum,* occurred in Afganistan, which suggested that bread wheats had been under cultivation, and selective modification by man, in this region longer than anywhere else. Vavilov thus inferred that such areas of maximum intraspecific diversity corresponded to the area in which the species was first domesticated.

In some cases, several areas of high intraspecific diversity could be found. For the durum wheats, *T. durum,* large numbers of varieties exist in Ethiopia, in the Middle East, and in the area around the Mediterranean Sea. No wild relatives of wheat are known in Ethiopia, which forced Vavilov to recognize that there were sometimes subcenters of diversification in areas to which the domesticated form was transported from the primary centers in which the original domestication took place.

Vavilov also noted that the centers of diversity, or origin, of different crop species frequently coincided or overlapped. In 1926 he proposed a number of **world centers of origin of crop plants** based on these observations, and in 1935 he refined and expanded his concept of centers of origin. Many of Vavilov's publications were in Russian, but his selected writings on this subject have been published in English (Vavilov 1951). Vavilov identified eight world centers of origin of cultivated plants, two of which had associated subcenters (Figure 4-1). These centers are listed below, along with a general indication of the types of plants involved:

1. The Chinese Center. Several species of legumes and fruits, together with tea.
2. The Indian Center. Rice, several legumes and vegetables, and black pepper.
   2a. The Indo-Malayan Subcenter. Yams and tropical fruits.
3. The Central Asiatic Center. A diverse assortment of grains, legumes and other vegetables, fruits, and nuts.
4. The Near Eastern Center. Many grains, several legumes, vegetables, fruits and nuts.
5. The Mediterranean Center. A few grains, vegetables and legumes, and the olive.
6. The Abyssinian Center. Several grains and legumes.
7. The Middle American Center. Corn, beans, peppers, squash, and a few others.
8. The South American Center. White and sweet potatoes, tomato, papaya, tobacco.

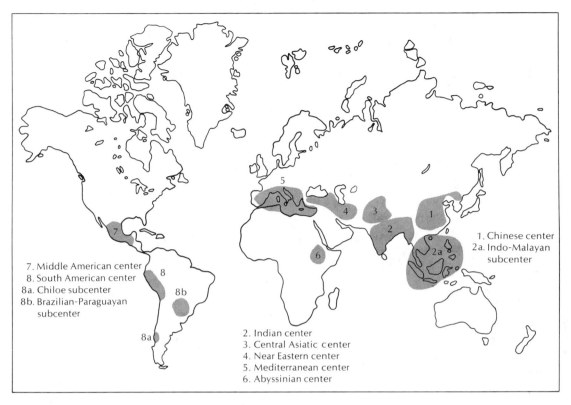

FIGURE 4-1.
Vavilov's world centers of origin of crop plants. (Modified from Harlan 1971. Copyright © 1971 by the American Association for the Advancement of Science.)

8a. The Chiloe Subcenter. White potato.
8b. The Brazilian-Paraguayan Subcenter. Manioc, peanut, cacao, and pineapple.

Vavilov's theories of crop origins, backed by data from his extensive worldwide collections, strongly influenced scientific thought on this subject. Nevertheless, as more information on crop origins has accumulated it has become necessary to modify the idea that almost all domestication occurred in discrete centers of limited extent. One of Vavilov's associates, P. M. Zhukovsky (1968), has modified, enlarged, and added to the system of centers proposed by Vavilov, to the point that in several instances the concept of a center seems unrealistic (Harlan 1971).

The principal weakness of Vavilov's approach, however, lies in his central assumption: that centers of diversity correspond to centers of origin. Historical records of the results of transporting plant species to new geographical areas show that many turn out to be more successful in new areas than in their native region. For certain crop species, such as the white potato introduced into Europe, this has proved beneficial to agriculture. On the other hand, in-

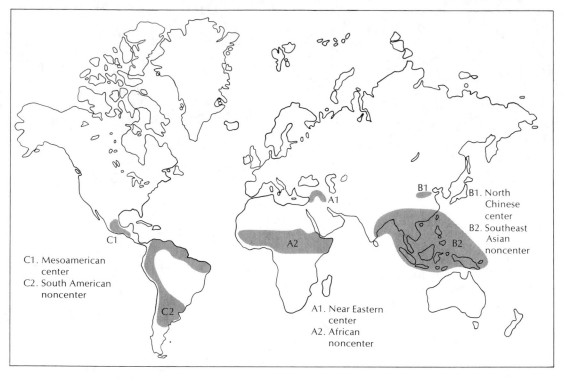

FIGURE 4-2.
Centers and noncenters of domestication of crops and agricultural animals, designated by Harlan. (Modified from Harlan 1971. Copyright © 1971 by the American Association for the Advancement of Science.)

troduced weeds that were innocuous in their original habitat can become serious pests in a new environment. Many species may thrive better in new geographical areas, because they are freed from the constraints of predation, parasitism, disease, or competition from the other organisms that coevolved with them. When a new crop plant is introduced into a new region with diversified ecological and cultural contexts, rapid evolutionary diversification of the species is highly probable. Vavilov found China to be an area of high diversity for the common bean, *Phaseolus vulgaris,* a species introduced to the Far East from the New World within historical time.

Harlan (1971) has suggested a simple, ecologically realistic scheme to describe the geography of early domestication (Figure 4-2). He suggests that certain regions of the world fit the idea of centers of origin for domesticated plants and animals (i.e., distinct areas within which domestication occurred and from which agricultural forms were dispersed), while others do not. Harlan terms these **centers** and **noncenters,** respectively. He recognizes three centers of domestication that correspond to the regions of the Middle East, North China, and Middle America in which agriculture first evolved.

Harlan's three noncenters lie in Africa, Southeast Asia and the East Indies, and South

America. A rich variety of plant domesticates exists in Africa, many species of which are little known outside the continent (Figure 4-3). Although the evidence from archeological studies is still quite fragmentary, the probable areas of domestication for which we have the best evidence from archeological studies lie in a sub-saharan belt stretching 7000 kilometers from Senegal in the west to Ethiopia in the east (Fig-

1. *Brachiaria deflexa*
2. *Digitaria exilis* and *D. iburua*
3. *Oriza glaberrima*
4. *Dioscorea rotundata*
5. *Musa ensete* and *Guizotia abyssinica*
6. *Eragrostis tef*
7. *Voandzeia* and *Kerstingiella*
8. *Sorghum bicolor*
9. *Pennisetum americanum*
10. *Eleusine coracana*

FIGURE 4-4.
Probable areas of domestication of selected African crop plants. (From Harlan 1971. Copyright © 1971 by the American Association for the Advancement of Science.)

FIGURE 4-3.
In Sidamo Province in the southern highlands of Ethiopia, the Guraghe rely upon enset, *Musa ensete*, as their food staple. Enset, a relative of the banana, is unique to this region. The plant is not cultivated for its fruit, but for a carbohydrate-rich material deposited in the base of the trunk. Enset is harvested by cutting off the leafy trunk a short distance above the ground surface. The material in the center of the trunk is then pounded to a pulp, which is dipped out. (Photograph courtesy of Dr. Solomon Bekure.)

ure 4-4). In southeastern Asia, crop domestication seems to have occurred over an area ranging from India through the Indonesian Islands to New Guinea and perhaps to the Solomon Islands and New Caledonia, a region about 10,000 kilometers wide. In South America, probable areas of domestication of various crops form a curved band that extends from Argentina northward through the Andes to Colombia and eastward to the northeastern tip of the continent, another region that is many thousands of kilometers in extent.

It is interesting that subtropical or temperate centers of very early origin seem to be paired with noncenters of possibly later origin that are more tropical. Even though the earliest stages of agricultural development may have been independent, it is likely that interaction and interchange took place between the center and the noncenter in each of these pairs. In particular, it may well be that granoculture was introduced into the more southern noncenters from the more northern centers, thus promoting the development of agricultural systems with greater crop diversification in tropical regions. In the Far East, influences from the northern granocultural center may have fostered the domestication of rice, which apparently occurred in southern Asia. The northward dispersal of this crop may in turn have stimulated further development of granoculture in northern China. Early elaboration of agriculture in southern areas may have formed the foundation of the early tropical civilizations, many of which, including the Mayan of Central America, the Khmer Empire centered on Angkor Wat in Cambodia, and the Nok, Ife, Benin, and Sudanic Kingdoms of Africa, are now represented only by ruins.

Harlan's scheme of centers and noncenters does not pretend to represent the areas of origin of all agricultural plants and animals. Table 4-1 summarizes current thought about the area of origin of major crop plants, and Table 4-2 does the same for important domestic animals.

TABLE 4-1
Regions of domestication of major crop plants (compiled from various sources).

	Middle East and Europe	
Southwestern Asia	Mediterranean Region	Europe
Triticum dicoccum, Emmer wheat	*Vicia faba,* Broad bean	*Avena sativa,* Oats
T. durum, Macaroni wheat	*Brassica oleracea,* Cabbage, etc.	*Secale cereale,* Rye
T. aestivum, Bread wheat	*Brassica napus,* Turnip	*Ribes* spp., Currants
Hordeum vulgare, Barley	*Olea europaea,* Olive	*Rubus* spp., Raspberries
Lens esculenta, Lentil	*Ceratonia siliqua,* Carob	*Prunus domestica,* Plum
Pisum sativum, Pea	*Allium sativum,* Garlic	*Cydonia oblonga,* Quince
Cicer arietinum, Chick-pea	*A. cepa,* Onion	
Brassica campestris, Rape	*A. porrum,* Leek	
Papaver somniferum, Opium poppy	*Lactuca sativa,* Lettuce	
Cucumis melo, Melon	*Beta vulgaris,* Beet	
Daucus carota, Carrot	*Asparagus officinalis,* Asparagus	
Ficus carica, Fig	*Pastinaca sativa,* Parsnip	
Punica granatum, Pomegranate	*Rheum officinale,* Rhubarb	
Prunus avium, Cherry	*Humulus lupulus,* Hop	
P. amygdalus, Almond	*Raphanus sativus,* Radish	
P. armeniaca, Apricot		
Pyrus communis, Pear		

continued

TABLE 4-1 (*continued*)
Regions of domestication of major crop plants (compiled from various sources).

Middle East and Europe		
Southwestern Asia	Mediterranean Region	Europe
P. malus, Apple		
Vitis vinifera, Grape		
Pistacia vera, Pistachio nut		
Diospyros lotos, Persimmon		
Phoenix dactilifera, Date palm		
Crocus sativus, Saffron		
Carthamus tinctorius, Safflower		
Linum usitatissimum, Flax (Linseed)		
Juglans regia, English walnut		

Africa		
West Africa	Central Africa	Ethiopian Region
Oryza glaberrima, African rice	*Sorghum bicolor*, Sorghum	*Eragrostis tef*, Tef
Digitaria exilis, Fonio	*Pennisetum americanum*, Pearl millet	*Eleusine coracana*, Finger millet
Brachiaria deflexa, Guinea millet	*Cola acuminata*, Cola nut	*Dolichos lablab*, Lablab bean
Elaeis guineensis, Oil palm	*Sesamum indicum*, Sesame	*Guizotia abyssinica*, Niger seed
Ceiba pentandra, Kapok	*Colocynthis citrullus*, Watermelon	*Ricinus communis*, Castor-oil bean
Blighia sapida, Akee fruit		*Coffea arabica*, Coffee
Dioscorea cayenensis, Yam		*Hibiscus esculentus*, Okra
Vigna unguiculata, Cowpea		*Musa ensete*, Enset
Voandzeia subterranea, Groundnut		*Catha edulis*, Chat

Asia		
Indo-Burmese Region	Southeast Asia	China
Oryza sativa, Rice	*Coix lachryma-jobi*, Job's tears	*Aleurites moluccana*, Tung oil
Cajanus cajan, Pigeon pea	*Dendrocalamus asper*, Giant bamboo	*Avena nuda*, Naked oat
Phaseolus aconitifolius, Math bean	*Dioscorea esculenta*, Yam	*Glycine max*, Soybean
Phaseolus calcaratus, Rice bean	*Zingiber* spp., Ginger	*Stizolobium hasjoo*, Velvet bean
Dolichos biflorus, Horse gram	*Citrus* spp., Citrus fruits	*Phaseolus angularis*, Adzuki bean
Vigna sinensis, Asparagus bean	*Musa* spp., Bananas and plantains	*Phyllostachys* spp., Small bamboos
V. radiata, Mung bean	*Cocos nucifera*, Coconut	
Amaranthus paniculatus, Amaranth		

continued

TABLE 4-1 (*continued*)
Regions of domestication of major crop plants (compiled from various sources).

Asia		
Indo-Burmese Region	Southeast Asia	China
Solanum melongena, Eggplant	*Elettria cardamomum*, Cardamoms	*Zizania latifolia*, Manchurian rice
Raphanus caudatus, Rat's tail radish	*Myristica fragrans*, Nutmeg	*Brassica chinensis*, etc., Pak-choy
Colocasia esculenta, Taro	*Curcuma longa*, Tumeric	*Allium fistulosum*, etc., Welsh onion
Cucumis sativus, Cucumber	*Cyamopsis tetragonolobus*, Guar	*Raphanus sativus*, Radish
Mangifera indica, Mango	*Artocarpus communis*, Breadfruit	*Prunus armeniaca*, Apricot
Gossypium arboreum, Tree cotton	*A. integrifolia*, Jackfruit	*Prunus persica*, Peach
Corchorus olitorius, Jute	*Saccharum officinarum*, Sugarcane	*Citrus* spp., Citrus fruits
Piper nigrum, Pepper	*Alocasia macrorhiza*, Elephant ear	*Broussonetia* sp., Paper mulberry
Acacia arabica, Gum arabic	*Tacca leontopetaloides*, Arrowroot	*Morus alba*, White mulberry
Indigofera tinctoria, Indigo		*Camellia (Thea) sinensis*, China tea
Panicum miliare, Slender millet		*Setaria italica*, Foxtail millet
Tamarindus indica, Tamarind		*Panicum miliaceum*, Broom-corn millet
		Echinochloa frumentacea, Japanese millet
		Fagopyrum esculentum, Buckwheat
		Castanea henryi, Chinese chestnut
		Litchi chinensis, Litchi nut
		Trapa natans, Water chestnut
		Wasabia japonica, Horseradish

New World		
North and Central America	Peruvian Region	Brazil-Paraguay
Zea mays, Maize (corn)	*Solanum tuberosum*, Potato	*Manihot esculenta*, Manioc
Ipomoea batatas, Sweet potato	*Chenopodium quinoa*, Quinoa	*Arachis hypogaea*, Peanut
Phaseolus vulgaris, Kidney bean	*Phaseolus lunatus*, Lima bean	*Phaseolus caracalla*, Caracol
Maranta arundinacea, Arrowroot	*Canna edulis*, Canna	*Theobroma cacao*, Cocoa
Capsicum annuum, etc., Red pepper	*Gossypium barbadense*, Sea island cotton	*Ananas comosus*, Pineapple
Gossypium hirsutum, Upland cotton	*Carica papaya*, etc., Papaya	*Bertholletia excelsa*, Brazil nut
	Nicotiana tabacum, Tobacco	*Anacardium occidentale*, Cashew nut
	Cinchona calisaya, Quinine	

continued

TABLE 4-1 (*continued*)
Regions of domestication of major crop plants (compiled from various sources).

New World		
North and Central America	Peruvian Region	Brazil-Paraguay
Agave sisalana, Sisal hemp	*Amaranthus* spp., Grain amaranths	*Passiflora edulis*, Passion fruit
Psidium guayava, Guava	*Oxalis tuberosa*, Oca	*Hevea brasiliensis*, Para rubber
Helianthus annuus, Sunflower	*Tropaeolum tuberosum*, Anu	*Dioscorea trifida*, Yam
Helianthus tuberosus, Jerusalem artichoke		*Xanthosoma sagittifolium*, Tannia
Cucurbita spp., Squash, pumpkin, gourd, etc.		
Amaranthus spp., Grain amaranths		
Pachyrrhizus erosus, Jicama		
Persea americana, Avocado		
Sechium edule, Chayote		
Vanilla planifolia, Vanilla		
Lycopersicon esculentum, Tomato		
Physalis ixocarpa, Tomatillo		

TABLE 4-2
Regions of domestication of important agricultural animals (compiled from various sources).

Southwestern Asia
  Ovis aries, Sheep
  Capra hircus, Goat
  Sus domesticus, Swine
  Bos taurus, Cattle
  Canis familiaris, Dog
  Camelus dromedarius, Dromedary camel
  Equus hemionus, Onager
  Anas platyrhyncha, Duck

Western and Central Asia
  Equus caballus, Horse
  Camelus bactrianus, Bactrian camel
  Bos grunniens, Yak

China
  Phasianus colchicus, Pheasant
  Cyganopsis cygnoides, Chinese goose
  Cyprinus carpio, Carp
  Carassius auratus, Goldfish
  Bombyx mori, Silk moth

India and Southeast Asia
  Bos indicus, Humped cattle
  Bubalus bubalus, Water buffalo
  Bos frontalis, Mithan
  Bos javanicus, Bali cattle
  Gallus gallus, Fowl

Arctic Eurasia
  Rangifer tarandus, Reindeer

Europe
  Oryctolagus cuniculus, Rabbit
  Coturnix coturnix, Quail

Mediterranean Region
  Anser anser, Goose
  Columba livia, Pigeon
  Muraena muraena, Roman eel

Africa
  Equus asinus, Ass
  Numida meleagris, Guinea fowl
  Chenalopex aegyptiacus, Egyptian goose
  Apus mellifera, Honeybee

Central America
  Meleagris gallopavo, Turkey

South America
  Lama glama, Llama
  Lama pacos, Alpaca
  Cavia porcellus, Guinea pig
  Cairina moschata, Muscovy duck

MAJOR PROCESSES IN THE EVOLUTION OF DOMESTICATED FORMS

Before examining specific examples of the evolution of domesticated plants and animals, we shall look at two major processes that have often played important roles in domestication. These are (1) hybridization between domesticates and wild forms, and (2) the occurrence of polyploidy in plant evolution.

The hybridization of various domestic varieties has always been an important technique in plant and animal improvement; but it is only one-half of a cyclical differentiation-hybridization sequence (Harlan 1966, 1970). In other words, hybridization between two varieties of a domesticated plant or animal presupposes a period of selection that led to the differentiation and divergence of the two varieties.

Differentiation-hybridization cycles can be repeated many times, but the length of the cycle units varies greatly from one species to another. In a sense, cycle length is simply an index of the degree of divergence that can take place without preventing hybridization between the forms (Figure 4-5). This depends upon the extent to which the genotype of the organism is buffered to permit reasonably normal gene function in a sharply modified genetic context. For example, self-fertilizing plant species with diploid chromosome numbers tend to be weakly buffered; the normal reproductive process rarely creates new genotypes that differ markedly from those of the parent, and mechanisms that permit normal gene function in novel genetic contexts are unnecessary and

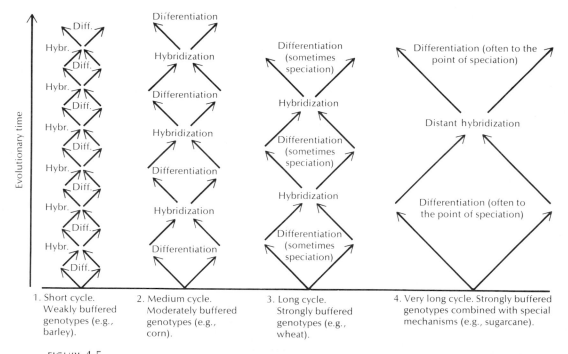

FIGURE 4-5.
Schematic representation of differentiation and hybridization cycles in crop plants with different degrees of genotypic buffering. (Reprinted, with modification, by permission from *Plant Breeding* by K. J. Frey. Copyright © 1966 by Iowa State University Press, Ames, Iowa 50010.)

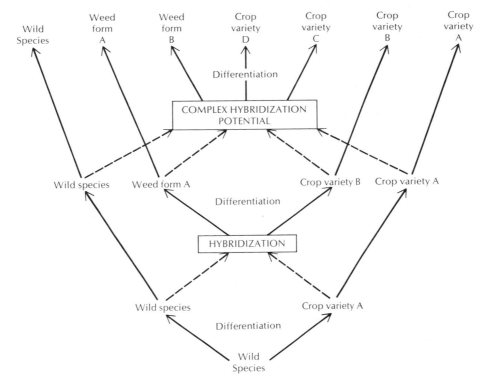

FIGURE 4-6.
Schematic representation of the increase in complexity of the relationships among wild, weed, and crop varieties through evolutionary time in rapidly evolving crop species complexes.

usually absent. Crosses are therefore limited to varieties that have not diverged greatly, and the results of such crosses, due to the lack of buffering, are likely to be major. As indicated in Figure 4-6, barley is an example of this sort of plant.

Maize, a cross-fertilizing diploid species, has more genetic buffering than the self-fertilizing species. Each reproduction involves two parent individuals with at least slight differences in genetic constitution, and the constitution of the offspring is still different. For the offspring to survive, its genetic system must have some buffering mechanism that permits reasonably normal gene function in the new context. This means, in effect, that hybridization can occur in maize between strains that have diverged to a greater degree.

Other species, such as the wheats and sugarcane, are even more strongly buffered, mainly because they are polyploid. The polyploid condition means simply that they possess more than two sets of chromosomes and that their genetic constitution has a major element of redundancy in it, much as a carefully designed spaceship possesses alternative control systems for each critical operation. This redundancy tends to strengthen the degree of genetic buffering, and it permits hybridization between still more divergent forms.

In the environments in which most evolution under domestication has actually gone on, however, hybridization is often spontaneous and often involves wild as well as domesticated forms. In most centers of origin or differentiation of crop plants, wild and weed races exist alongside cultivated races of a particular crop. The occurrence of weed races—forms adapted to disturbed habitats created by man—is very common and has been demonstrated for wheat, barley, rice, sorghum, maize, rye, oats, carrot, radish, lettuce, potato, tomato, sugarcane, and many other plants (Harlan 1970; de Wet and Harlan 1975). Feral populations of domesticated animals are analogous to the weed races of plant species. Hybridization remains infrequent because most weed races are partially isolated, reproductively, from cultivated races. Nevertheless, it does occur and has played a significant role in the differentiation-hybridization cycle and the overall evolution of domesticated plants and, perhaps, animals.

Weed races, adapted to man-created habitats, must have evolved along with cultivated races, and the sequence of differentiation-hybridization cycles may have approximated that shown in Figure 4-6, in which the diversity of cultivated and weed races, and thus of potential hybridization combinations, increases over time. Extinctions of weed races and replacement of early cultivated forms by later ones limit this diversity, of course, but it is clear that the process of evolution under domestication is complex. We should emphasize that the differences among wild, weed, and cultivated forms are genetically minor; domestication processes increase variation, but do not tend to lead to speciation (Harlan et al. 1973).

This complexity is often reflected at the chromosome level, through differences in chromosome numbers among wild, weed, and cultivated forms. Even without changes at the gene level, changes in chromosome number may conspicuously modify hereditary characteristics. Polyploidy, or an increased number of complete sets of chromosomes, is one of the types of chromosomal changes most important in crop evolution (Darlington 1973; Schwanitz 1966). Normally, the nonreproductive cells of a flowering plant or of an animal are diploid in chromosome number. At some point in the reproductive process, corresponding chromosomes of these sets pair and duplicate, after which two successive cell divisions occur, reducing the chromosome number to a haploid state. This process, **meiosis,** leads to the formation of gametes that contain one member of each chromosome pair. The fusion of gametes in sexual reproduction returns the chromosome number to the diploid state.

Polyploidy (triploidy, tetraploidy, etc.) can originate in several ways, all basically the result of unusual events such as interspecific hybridization or failure of meiotic reduction in chromosome number. **Autopolyploidy** refers to polyploidy that originates by intraspecific processes. During vegetative growth, for example, the diploid chromosome set may duplicate without subsequent cell division and thus produce a cell with a tetraploid number. If this cell subsequently divides normally, it may produce tissues that are generally tetraploid and that perpetuate this tetraploid number even through sexual reproduction, producing diploid gametes that, in fusing, restore the tetraploid condition. Autopolyploidy can also arise out of accidents of meiosis that produce gametes with unreduced chromosome numbers. Thus a diploid female cell can be fertilized by a haploid male cell, creating a triploid cell, or by a diploid male cell, creating a tetraploid cell.

Allopolyploidy, in contrast, results from hybridization. For example, female gametes of one species can be fertilized by male gametes of another, as shown schematically in Figure 4-7, to produce a diploid cell with one set of chromosomes from each parental species. These two sets may not even include the same number of chromosomes, and may, in general, be so different that they are unable to pair in the preliminary stages of meiosis. As a result the offspring of such unions are generally sterile, and unless something further occurs, or unless vegetative reproduction enables this genotype to be perpetuated indefinitely, this kind of hybridization creates an evolutionary dead end. However, a

AUTOPOLYPLOIDY

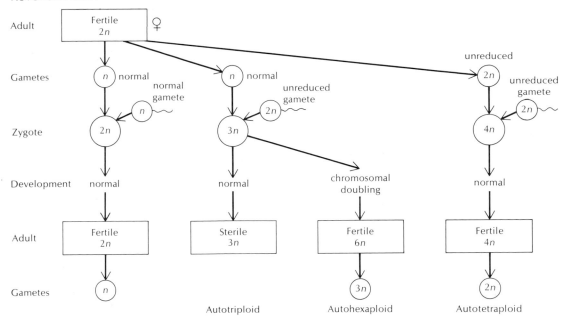

ALLOPOLYPLOIDY

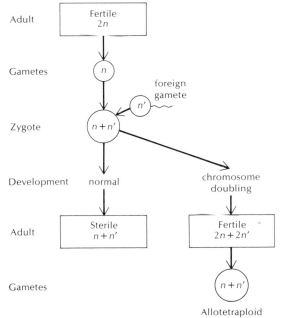

FIGURE 4-7.
Common patterns of origin of autopolyploid and allopolyploid forms. Polyploids in which meiotic pairing is possible between even-numbered sets of similar chromosomes tend to be fertile; those with odd numbers of chromosome sets, or sets that are dissimilar, tend to be sterile.

second accident, such as those that produce autopolyploids can lead to a doubling of these chromosome sets, and because each set of chromosomes may then pair and undergo normal meiosis, the result may be a restoration of fertility.

Chromosomal aberrations can be much more complicated than these, but most polyploid forms arise in one of the ways described above. Actually, between 30 and 35 percent of all flowering plant species have polyploid chromosome numbers.

The specific effects of polyploidy are not predictable, but they often involve quantitative changes of considerable significance to the plant breeder. Polyploid forms often have larger, thicker leaves, sturdier stems, fleshier roots, larger flowers and fruits, and higher photosynthetic rates. Such plants are termed "gigas" varieties by breeders, suggesting the gigantism expressed in one or more features (Schwanitz 1966).

Autopolyploidy or allopolyploidy followed by fertility restoration essentially results in the sudden emergence of a new variety that is quite reproductively isolated from its parental form or forms. Hybrids between a tetraploid and diploid will possess three sets of chromosomes, creating pairing problems at meiosis. Polyploids with uneven numbers of chromosome sets are usually sterile. Nevertheless, the possibility of some genetic interchange still remains through rare events of the type that gave rise to the polyploid forms to begin with.

To illustrate the patterns we have mentioned, we shall now turn to several examples of the evolution of crops and domestic animals under domestication.

THE EVOLUTION OF WHEATS

As an evolutionary group, the wheats comprise 22 wild and 13 cultivated species belonging to the genera *Triticum* and *Aegilops* (Kuckuck

1970; Riley 1965; Zohary 1970). These species show a polyploid sequence based on a diploid chromosome number of 14 and include species with tetraploid (28) and hexaploid (42) chromosome complements. Most of these wild and cultivated species have been subjected to careful cytogenetic study, and the evolutionary sequence culminating in the polyploid domestic forms has been accurately worked out in all but one case.

The wild members of the wheat group are distributed in various parts of the Middle East and central Asia and are diploid in chromosome number. The diploid, tetraploid, and hexaploid chromosome complements in the cultivated wheats appear to be derived from three wild species, which thus form the ancestors of modern wheats. Figure 4-8 shows the presumed relationships among these wild forms and their cultivated descendents. The first of the wild species, einkorn, *Triticum boeoticum*, has a chromosome complement that we shall designate A. This species occurs from Greece and the Balkans eastward through Turkey to the Caspian Sea and western Iran (Harlan and Zohary 1966). The second species, a wheat grass, *Aegilops speltoides*, with a chromosome complement designated B, is less widely distributed, occurring from northern Syria and southern Turkey eastward to northern Iraq. The third species, also a wheat grass, *Ae. squarrosa*, with a chromosome complement D, occurs from eastern Turkey and western Iraq eastward through the Caspian region to Soviet Central Asia, Pakistan, and Kashmir.

Wild einkorn was domesticated early, producing what we shall call cultivated einkorn, *Triticum monococcum*. The name *einkorn* refers to the fact that the individual spikelets of the fruiting head tend to have a single grain, or occasionally two. This form of wheat is not as abundant as one of the tetraploid forms in the earliest archeological sites of the Middle East, but was apparently under domestication by about 7500 to 6750 B.C. (Renfrew 1973). Cul-

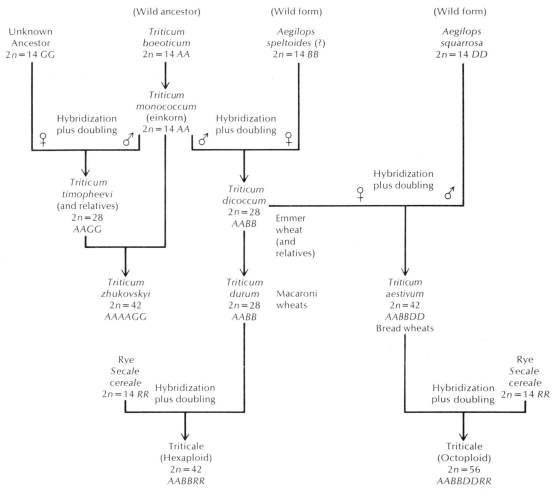

FIGURE 4-8.
Tentative scheme of evolutionary relationships involving hybridization and polyploidy in the wheats and their relatives. Letters (A, B, D, etc.) identify chromosome sets from respective wild ancestors.

tivated einkorn is diploid in chromosome number, and its domestication seems to have involved simple selection by early farmers. Einkorn is little used today; in parts of Turkey and Yugoslavia it is grown for fodder and is occasionally used in pilaf, but not for breadmaking.

Tetraploid wheats appear to have resulted from hybridization between wild or cultivated einkorn and the wheat grass *Aegilops speltoides*. However, it is not certain that this species is the source of the *B* chromosome complement, despite analyses of protein polypeptides, from various possible sources, that support *Ae. speltoides* as the source (Chen et al. 1975). The result of such a cross, and the doubling of chromosome sets from the two parents, is a wheat with a chromosome complement *AABB*. One species of this sort, *T. dicoccoides,* grows wild over much of the Middle East and is known as wild emmer. A closely related form, *T. dicoccum,* or cultivated emmer, is the oldest cultivated wheat and the form that appears most commonly in early archeological sites (Kuckuck 1970). The cultivated form of emmer is very similar to the wild form and was apparently derived by selection. Wild emmer may, of course, be of natural hybrid origin and may represent the form from which cultivated emmer was derived by selection. Emmer wheat is the form from which durum wheat, *T. durum,* and several related forms have been derived by selection and by crossing tetraploid varieties. Durum wheats are used widely at present for making macaroni and other forms of pasta. One other member of the tetraploid wheat group, *T. timopheevi,* cultivated in the Russian Republic of Georgia, has a somewhat uncertain origin. Its chromosomal complement contains the *AA* contribution of a *Triticum* parent, but a *GG* complement from another source. This *GG* set may be nothing more than a *BB* set considerably modified by evolution of isolated forms of *T. dicoccoides,* but in any case it seems likely that this last species represents an independent domestication of a tetraploid form from wild populations.

The hexaploid wheats include the bread wheats, and the most important of them are apparently derived from a cross between emmer, either wild or cultivated, and the wheat grass *Ae. squarrosa.* The resulting chromosome complement can be symbolized *AABBDD.* Among these hexaploid forms are several minor cultivated varieties of the Middle East and Europe that possess a hull around the seed that is somewhat difficult to remove. The free-threshing bread wheats, *T. aestivum* and its relatives, are forms in which the grain readily separates from the surrounding hull; these forms are apparently of the most recent origin, but they were nevertheless under cultivation by about 5000 B.C. (Renfrew 1973).

Still other hybridizations have occurred and are being encouraged among wheats. One hexaploid wheat, *T. zhukovskyi,* a form also grown in the Russian Republic of Georgia, apparently has a different origin from that of other hexaploids. It appears to have only the *A* and *G* chromosome complements and therefore to be the product of a cross between *T. timopheevi* and *T. monococcum* (Johnson 1968). The triticales, derived in recent years by crosses between wheats and rye, are discussed in detail in Chapter 22.

The evolution of the various forms of wheat thus exemplifies both the importance of interspecific hybridization and the major role of polyploidy in the origin of modern cultivated forms. Furthermore, we can see in this plant group the coexistence of several wild, weed, and cultivated forms in many parts of the Middle East. Later in this chapter we shall comment on the significance that these different forms have for the future.

THE EVOLUTION OF MAIZE

Probably no aspect of crop origins has attracted more controversy than has the origin of corn, or maize* (Galinat 1971, 1975; Mangelsdorf 1974). Much of this controversy results from the fact that cultivated maize, *Zea mays,* unlike wheats and most other cultivated plants, is strikingly different in appearance from even its closest relatives. The diploid chromosome number of maize is 20, and this number is the same as that of its closest relative, teosinte, *Zea mexicana,* a weedy annual grass that grows wild from Mexico southward to Honduras and can often be found in and around maize fields.

---

*We use the term *maize* in this historical account because of its nearly universal usage in this context, but we use the term *corn* in reference to modern cultivation systems.

Both teosinte and maize possess separate male and female inflorescences, the staminate being terminal and the pistillate lateral, but in teosinte, the pistillate spikelets are solitary, instead of occurring in pairs as they do in maize. They bear 6 to 12 kernels, each of which is enclosed in a "shell" formed by a segment of the rachis of the spike and by one of the glumes or bracts that occur at the base of flower spikelets in grasses. Several other differences occur, involving genes on almost all chromosome pairs, but teosinte nevertheless crosses readily with maize to produce fertile offspring that are capable of backcrossing to either parent.

A second genus, *Tripsacum*, with similarities to maize, also exists. This genus includes eleven species distributed from the northeastern United States to Paraguay (de Wet et al. 1976). All are perennials, all have male and female spikelets located in different sections of the same spike, and all produce kernels enclosed in a shell much like that of teosinte. The diploid chromosome number is 36, but several of the species are tetraploid ($2n = 72$), or have polyploid races ($2n = 54$ or 72). One, with a $2n$ number of 64, seems to possess a set of *Zea* chromosomes (54 from *Tripsacum*, 10 from *Zea*).

Maize was widely distributed in North and South America in Precolumbian times, with about 100 well-defined varieties in existence (Mangelsdorf 1974). Maize and its relatives are native only to the New World, and despite occasional suggestions of an origin in southeastern Asia maize is almost certainly of New World origin.

Four principal theories of maize origin exist (Mangelsdorf 1974). One holds that maize evolved directly, and without a great deal of hybridization with other species, from pod corn, an unusual variant of maize with kernels enclosed in individual bracts. This characteristic, which shows up only rarely in modern corn, is much like that which generally prevails for other grasses and thus suggests that the kernels, now naked and crowded together on the cob,

might originally have been enclosed and distributed along a thinner, elongate rachis. The second theory suggests that maize originated from teosinte by selection and by occasional hybridization between varieties of teosinte and early maize. The third theory is that maize, teosinte, and *Tripsacum* evolved independently from some common ancestor, and that the original wild forms of maize have been completely displaced by domesticated varieties. The fourth, and most elaborate, theory is that of Mangelsdorf and Reeves (1939) and is known as the **tripartite theory.** This theory suggests (1) that the earliest ancestor of maize is a form of pod corn, (2) that this form hybridized with a species of *Tripsacum* to give rise to teosinte as a weed product, and (3) that modern corn strains are the product of selection of the variations formed by crossing maize with both teosinte and *Tripsacum* (Figure 4-9). Recently, Mangelsdorf (1974) has abandoned the idea that teosinte originated by hybridization between maize and *Tripsacum*, and instead considers it to be "essentially a mutant form of maize." Mangelsdorf has marshalled much data to support the rest of the tripartite hypothesis, and in view of our earlier comments about the importance of hybridization in crop evolution and the example offered by evolution of wheats, this sequence of evolution is not impossible.

A certain amount of fossil and archeological evidence exists that bears on the origin of maize. Pollen that has been identified as maize pollen has been recovered from depth of 69 to 72 meters in sediments of the Mexico City Basin, actually in a drill core below the Palacio de Bellas Artes in the City itself. The age of these sediments is estimated at 80,000 years. If these pollen grains are not contaminants, and if they represent maize and not teosinte, this is strong evidence for the preagricultural existence of an ancestral form of maize. The presumed age of this material predates not only agriculture but also, by more than 50,000 years, the existence of man in the New World. Independent verification of presence and age of maize pollen in this situation would be very useful.

Indisputable remains of maize have been recovered from caves of the Tehuacan Valley of southern Mexico. These tiny popcornlike cobs are 19 to 25 millimeters long, possess a terminal spike of male flowers, and are between 7200 and 5400 years old (Figure 4-10). They are considered by Mangelsdorf to represent the wild

maize ancestors that existed before domestication and before any *Tripsacum* hybridizations that, among other things, might have given rise to teosinte.

De Wet and Harlan (1972), however, note the recent finding of clear teosinte remains that have been radiocarbon dated to 7040 B.P. They further argue that relatively few changes may actually be required to derive, from teosinte, a

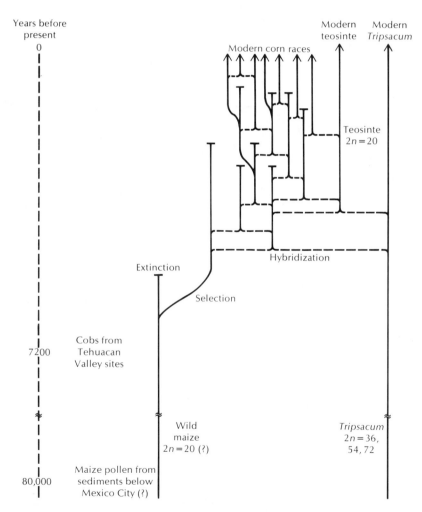

FIGURE 4-9.
Schematic representation of the original version of the tripartite hypothesis for the origin of cultivated corn.

Early maize

Teosinte

Tripsacum

FIGURE 4-10.
Inflorescences of *Tripsacum*, teosinte, and the form of early maize recovered from the oldest levels of excavation in Coxcatlan Cave (about 7200 B.P.), Tehuacan Valley, Puebla, Mexico. This form of maize, which had male flowers as a terminal spike on the cob (female inflorescence), was believed by supporters of the tripartite hypothesis to be the wild ancestor of cultivated maize or corn. Now it is believed by most workers to be an early variety of domesticated maize. (From "The Origins of New World Civilization" by Richard S. MacNeish. Copyright © 1964 by Scientific American, Inc. All rights reserved.)

cob structure like that found in the Tehuacan Valley caves; for example, the cupule in which the kernel of maize rests can seemingly be derived simply from the fruit case of teosinte. Furthermore, the changes required to derive maize from teosinte appear simpler than those required to do the opposite (Galinat 1975).

One of the weakest features of the tripartite hypothesis, however, relates to the differences in chromosome numbers of maize and *Tripsa-*cum. It is difficult to imagine, and so far impossible to demonstrate experimentally, the derivation, from such a cross, of a chromosomal arrangement like that of teosinte. At the same time, it is easy to derive the chromosome arrangement of maize from that of teosinte, since the numbers are identical and since the two species cross easily with apparently complete chromosome pairing and fertility in the hybrid offspring.

The following sequence thus seems reasonable for the origin of maize (Galinat 1975; P. H. Zedler, personal communication): Early hunter-gatherers discovered that they could use teosinte kernels, either by roasting them and popping them like popcorn or by cracking and grinding them on a metate, or stone mortar. Certain strains of teosinte gradually emerged as "habitation weeds" around campsites. Cultural selection then led slowly to various modifications such as the development of a nonshattering rachis, the shortening and thickening of the rachis, the occurrence of paired spikelets, naked grains, an enclosing husk, and so on. With the development of full-scale seed agriculture, maize was transported to new regions and strain differences arose. Evolution of new forms was thus favored by increased possibilities of hybridization between various maize strains and between maize and teosinte.

The exact origin of maize remains somewhat in doubt, however, and future studies may still provide surprises. Whatever the exact story, the origins of this crop have clearly been influenced by hybridization; only the exact nature and the specific participants of the relationships remain to be clarified.

In the case of maize, it is interesting to note the reciprocal influence of the crop's evolution upon the culture of the human groups that guided it. Katz, Hediger, and Valleroy (1974) have recently summarized data on the techniques of maize preparation by 51 cultural groups from the northern United States to southern Chile. Compared to other grains, maize tends to be deficient in niacin and certain essential amino acids, particularly lysine and tryptophan. These deficiencies are great enough that a significant degree of malnutrition would be expected in people who relied exclusively on maize for their principal carbohydrate food. However, almost all groups that rely heavily on maize also practice some form of alkali treatment in the preparation or cooking of maize. This usually involves heating the maize kernels in a solution of lye, lime, or ashes. Analysis of the nutritional change produced indicates that the absolute quantities of certain vitamins and amino acids are reduced by this treatment, but that the availability of niacin and niacin precursors (niacin is a member of the Vitamin B complex) is increased. In addition, the relative abundance of essential amino acids is modified to a more balanced state, with that of lysine being improved the most. Cultural groups that do not practice alkali treatment of maize possess, in almost all cases, some important source of essential amino acids in meat obtained by hunting and fishing. Thus it can be seen that evolutionary relationships between man and his domesticates are reciprocal: man guided the biological evolution of maize, the nutritional characteristics of which channeled the cultural evolution of the groups that depended upon the species.

THE EVOLUTION OF THE POTATO

No crop species illustrates more clearly than the potato the importance, both in a historical and a modern context, of hybridization and polyploidy in crop evolution. The potato belongs to one taxonomic section of the very large genus, *Solanum*, the type genus of the family Solanaceae. Authorities differ widely in the number of species involved, but there are between 90 and 150 wild species, all in the New World, from Nebraska and Colorado in North America southward through Mexico and Central America to central Chile in South America. There are two areas in which the wild forms are very diverse (Hawkes 1970). The most important lies in South America, specifically in the Andes, from Peru southward through Bolivia to northwestern Argentina; the second area is in the highlands of Mexico.

Authorities also disagree on the number of cultivated species, with opinions ranging from 1

FIGURE 4-11.
An Indian farmer examines his potato crop in the Altiplano of Bolivia. Fields like this tend to have many strains of wild, weed, and crop potatoes growing in close association and frequently hybridizing among themselves. (Photograph by H. G. Dion, courtesy of the FAO, Rome.)

to about 20. There is an amazing variety of cultivated forms in the Andean region of South America, with over 400 named varieties being found in Peru and Bolivia alone. These forms fall into four chromosome groups: diploids with 24 chromosomes, triploids with 36, tetraploids with 48, and pentaploids with 60. All the cultivated forms were apparently domesticated in South America, although the crop was introduced to Central America and Mexico soon after the Spanish conquest.

The environment in which domestication took place was probably not unlike that now existing in the Altiplano of Bolivia and the high Andes of adjacent Peru (Ugent 1970). Here, the fields prepared and planted by Andean farmers tend to be invaded by many of the noncultivated wild or weed forms of potato that flourish around cultivated areas (Figure 4-11). Their in-

vasion of cultivated fields, however, provides a situation ideal for hybridization between cultivated and noncultivated forms, and there is considerable evidence that such hybridization is an ongoing process.

Ugent (1970) suggests that domestication of the potato involved a process much like that which we outlined in Chapter 3 for the origins of granoculture. Hunter-gatherer groups probably gathered wild potatoes, which gradually became associated, in weed form, with human campsites or settlements. Here, various forms mingled, hybridized, and were eventually subjected to stronger evolutionary selection by man. The capability that desirable strains had for vegetative propagation probably played an important role in this process. In fact, vegetative reproduction, combined with frequent hybridization that formed new genotypes from

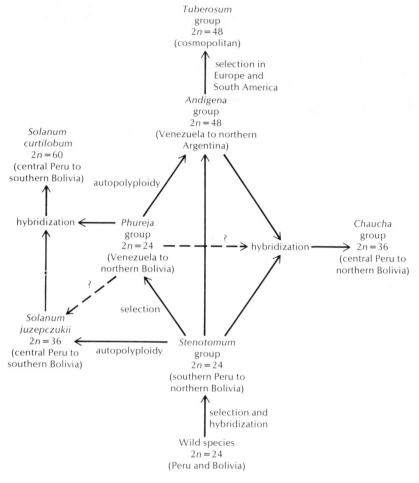

FIGURE 4-12.
Schematic representation of the relationships thought to exist among the various groups of potatoes cultivated in the Andean region of South America. (Based on data from Ugent 1970.)

which desirable strains could be chosen and propagated, is certainly responsible for the great diversity of cultivated forms of potatoes.

The cultivated varieties and species of potato fall into several major groups, the evolutionary relationships among which can be surmised (Figure 4-12). Wild species possess the diploid chromosome number 24. The *Stenotomum* group of varieties, distributed from southern Peru to northern Bolivia, also has the diploid number, and was probably derived simply by hybridization and selection among diploid forms. The *Phureja* group, extending from Bolivia northward through the Andes to Venezuela, likewise has 24 chromosomes and is probably derived by selection. This group is characterized by having short tuber dormancy, a feature correlated with the less seasonal climates close to the equator. The group *Andigena* is tetraploid in chromosome number, however, and probably originated by autopolyploidy from members of the *Stenotomum* or *Phureja* groups (or both!). This group is widely distributed through the Andes and is the ancestral stock from which the *Tuberosum* group of modern, cosmopolitan varieties was derived by selection.

Several other remarkable evolutionary events have occurred within this group of plants, however. A group of triploid varieties, the *Chaucha* group, occurring in Peru and Bolivia, appears to have originated by hybridization between members of the tetraploid *Andigena* group and diploid *Stenotomum* (or possibly *Phureja*) group. Two other forms in central Peru and Bolivia also seem to have had either a hybrid origin or one involving polyploidy. The species *Solanum juzepczukii*, a triploid, seems to have originated by autopolyploidy from a form of the *Stenotomum* (or possibly *Phureja*) group, but then to have hybridized with a form of the *Phureja* group to give rise to *S. curtilobum*, a form with the pentaploid chromosome number 60.

Not surprisingly, tracing such relationships with exactitude soon becomes very difficult. However, as we shall describe shortly, the capacity for genetic and evolutionary interactions of this sort is important to future agriculture.

THE EVOLUTION OF CATTLE

So far we have confined our attention to crop origins, but the origins of domestic animals follow the same principles except that the occurrence of polyploidy is generally infrequent in animals. The origin of modern cattle, and of other domesticated cattle-like animals, is perhaps the most complex and interesting example

to consider. The cattle-like animals are those of the subfamily Bovinae of the family Bovidae. The family itself includes several other subfamilies containing goats, sheep, and antelopes. The Bovinae include some eight genera and approximately 15 species, at least six of which have been domesticated and two others partially domesticated (Figure 4-13).

We shall not concern ourselves with most of these species, except to point out that they are generally similar enough that interspecific crosses are possible between many of them, and to suggest that the potential for domestication has not by any means been exhausted in the group. Instead, we shall concentrate on the origins of cattle belonging to the genus *Bos*, including the humped cattle, *B. indicus*, and the humpless forms, *B. taurus*.

The species ancestral to western cattle was the aurochs, *B. primigenius*, a form widely distributed in Europe, northern Africa, the Middle East, and northern Asia (Epstein 1971; Isaac 1970). The characteristics of this animal are well known, since it survived well into modern history; the last individual died in a Polish park in 1627. The bulls had horns that curved first outward and then forward, with a slight upward turn at the tips. Overall body coloration of bulls was black, at least in Europe, with a light streak along the backbone. A characteristic patch of curly hair occupied the forehead. Cows were apparently brownish-red in color, although some color variants existed, especially in southern and southwestern Europe.

Domestication seems to have occurred first about 9000 to 8000 years B.P. in areas of Greece and Turkey (Protsch and Berger 1973). The first clear evidence of domestication in the archeological record is the appearance of a distinctive variety of cattle, the *longifrons* form (*B. taurus longifrons*°). This type had shorter horns than

---

°Also known as *B. t. brachyceros* in some classifications.

GENUS   SUBGENUS   WILD SPECIES

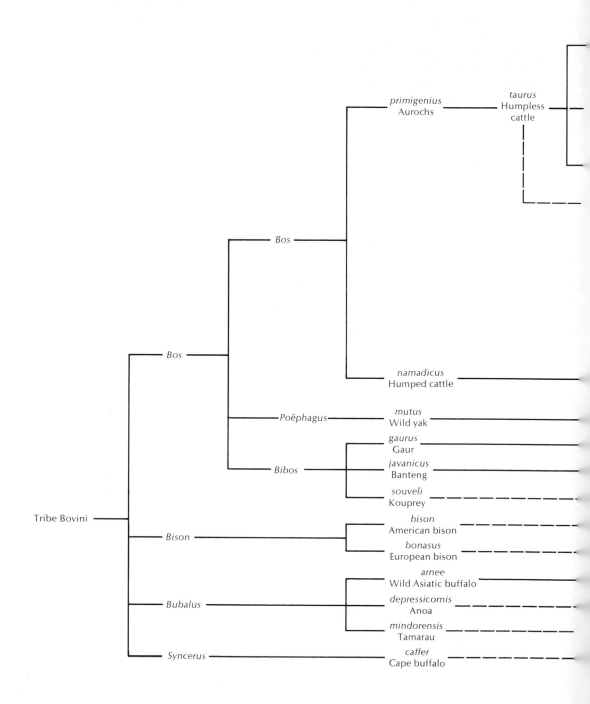

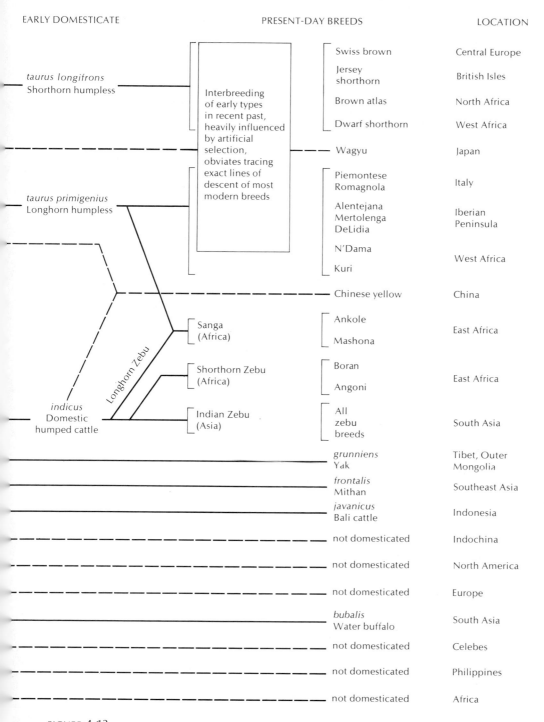

FIGURE 4-13.
Relationships among cattle and their near relatives, showing the role of hybridization in the origin of modern breeds. (Modified from Rouse 1970.)

the wild form, which seems to have been the result of artificial selection, and the lighter weight of the horns permitted the frontal and parietal bones of the skull to grow in a domed fashion, producing a narrower, higher facial structure. Other domestications, apparently in western areas of Eurasia, resulted in a second domestic form, the *primigenius* variety (*B. taurus primigenius*), that still retained heavy horns and a normal skull structure.

These early domesticated humpless cattle were the ancestors of most western breeds. Many varieties have been produced by selection, and hybridization among these forms has been very complex. The existing form that may be closest to these early domesticates is the Japanese Wagyu, apparently a very early introduction to the islands and one that has been somewhat protected from subsequent modification by interbreeding (Rouse 1970).

Still another early domestication gave rise to domestic humped cattle. The wild ancestor of this form is not known for sure, although a wild species, *B. namadicus,* once existed throughout much of southern and southeastern Asia. This animal may simply have been a variety of *B. primigenius* (Isaac 1970), but the skull structure of this fossil species shows characteristics of facial slenderness that suggest it may have been the ancestor of humped cattle. Furthermore, the degree to which humped cattle have adapted physiologically to tropical conditions argues strongly that their ancestors were distinctly different from those that gave rise to humpless domesticates (Zeuner 1963). Domestication of this humped form occurred in the middle East as early as 4500 B.C.

Hybridization between humped and humpless cattle has happened profitably several times throughout the history of domesticated cattle, and of course is now being practiced with greater intensity than ever. The Chinese Yellow breed is perhaps one of the earliest products of this cross; it is used primarily as a draft animal in central and southern China (Rouse 1970).

It was in Africa, however, that some of the most interesting and complex patterns of cattle evolution developed. The earliest cattle, recorded in Egyptian tomb drawings as early as 4000 B.C., were *primigenius* forms with long, upsweeping horns. The N'Dama breed of West Africa may be the purest modern derivative of this form. The N'Dama are remarkable for their resistance to various African pests and diseases, including trypanosomiasis, which suggests a long history of evolutionary adjustment to African conditions (Epstein 1971). Shorthorn humpless cattle, the *longifrons* type, arrived later, and for a period both were present. An Egyptian temple mural dated 1482 B.C. shows both forms. Several present-day breeds, principally of North and West Africa, may be nearly direct derivatives of this form. Humped cattle arrived still later, possibly entering the continent by boat in what is now Somalia at about 2000 B.C. Hybridization of longhorned humpless forms with humped cattle in the area of Ethiopia and Somalia apparently gave rise to a base stock that has subsequently produced many of the unique breeds of East African cattle. This basic type, the Sanga, has long, sweepingly upcurved horns (Figure 4-14). Still later introductions, about 700 A.D., brought shorthorn zebu cattle from India. Africa has thus become a sort of melting pot for many ancestral breeds of cattle.

Hybridization involving cattle is not limited, however, to crosses between *B. taurus* and *B. indicus* forms. Crosses of these two species have been carried out with banteng, gaur, mithan, yak, water buffalo, and American and European bison; but these crosses generally show some degree of infertility among the offspring (Simoons 1968). Often the female offspring are fertile and male offspring sterile, but in other cases the offspring of both sexes are sterile. Some of these crosses, such as be-

FIGURE 4-14.
The Ankole cattle of Uganda are one of the distinctive Sanga breeds derived from crosses between longhorn zebu and longhorn humpless cattle in East Africa. (Photograph courtesy of the FAO, Rome.)

tween cattle and yak, are deliberately practiced (Rouse 1970), but none has yet had important consequences in terms of deriving new breeds of domestic animals.

The motive for the domestication of cattle has been much argued (see Zeuner 1963 and Simoons 1968). The bull, of course, has been a symbol of fertility, masculinity, and power to many peoples throughout much of history. Some students (Simoons 1968) have argued that its symbolic significance was the most powerful driving force for early domestication. Others (Zeuner 1963) suggest that the domestication was an economic enterprise, the result of the animal's many uses: as a source of food in the form of meat and milk, of other products such as hides and dung, and of traction power.

It seems ecologically unlikely that a symbiotic relationship as important as that of man and cattle requires more of an explanation than mutual ecological benefit.

THE CONTINUING EVOLUTION OF DOMESTICATED PLANTS AND ANIMALS

The foregoing examples demonstrate much more than just the complexities inherent in the origin of domesticated crops and animals. They illustrate not only the genetic complexities that have permitted the development of man's most useful and productive agricultural species, but also the fact that this process depends as well on plant and animal species that have little direct productive importance. We shall examine this aspect of domestication more closely in a later chapter, but two major points should be made while these examples are still fresh: the importance of wild species as a source of macrovariation for artificial selection, and the importance of local varieties and weed forms as sources of microvariability for breeding.

Macrovariability (i.e., dramatic new patterns of form or function) often appears when distinct species are crossed. The examples of wheat, the potato, and to a lesser extent maize and cattle, demonstrate that major evolutionary advances can often result from crosses between distinct species or even members of distinct genera. In plants, we find that several of our important domesticates, such as the tetraploid and hexaploid wheats, are the products of allopolyploidy. With modern advances in our understanding of cellular physiology and genetics, especially of antibody production and the immune reaction, the possibility of distant crosses is increasing.

Plant and animal breeders are in constant need of sources of microvariability, principally new alleles for existing genes. With crop plants, one of the continuing struggles in breeding crops is to create patterns of genetic resistance to pests and agents of diseases, which are constantly evolving and finding ways to crack existing patterns of genetic resistance. We do find, however, that many minor wheat varieties carry distinctive patterns of resistance to a variety of rusts, smuts, and mildews, as well as pronounced differences in their resistance to drought and lodging (Kuckuck 1970). The potato is especially vulnerable to fungal blights, and some wild and weed forms possess remarkably variable blight resistance systems (Ugent 1970). In fact, the seemingly ill-kept potato fields of the Andes may have been important to the survival of potato farming and Inca culture over thousands of years.

In the modernization of world agriculture, we must be careful not to seriously deplete the reservoirs of genetic variability that exist in the wild and weedy relatives of crop plants. The current trend, however, is to do just that. Widespread dissemination of so-called miracle wheats, modern potato varieties, hybrid corn, European and North American cattle, and clean cultivation techniques is rapidly eliminating these wild and weedy forms as well as local races of domestic forms.

SUMMARY

Early studies of the geographical origins of crops and domestic animals suggested that these occurred in several highly discrete centers scattered widely over the world. These studies were based primarily on the assumption that centers of varietal diversity corresponded to centers of evolutionary origin, an assumption that was true in some but certainly not all cases. It now appears that domestication sometimes originated in localized centers and other times over broad geographical areas called noncenters. Centers of origin developed in the Middle East, northern China, and Mesoamerica; noncenters arose in South America, in Africa just south of the Sahara, and in the Indo-Pacific area. Interaction and exchange between the neighboring centers and noncenters may have played an important role in the early diversification of agriculture.

Two major processes have greatly influenced the evolution of crops and animals under domestication: hybridization between different varieties and species, and the occurrence of polyploidy in plants. Hybridization is part of a hybridization-differentiation cycle, the length of which depends on the degree of genetic buffering that an organism possesses; the cycle is short when buffering is low, longer when it is high. The hybridization-differentiation cycle tends to generate a diverse assemblage of crop varieties, weed forms, and wild ancestral forms that may continue to coexist. Polyploidy, an increase in the number of chromosome sets, may be caused by accidents of cell division or sexual reproduction within a species, or by interspecific hybridization. Polyploidy often produces quantitative changes in species characteristics that are useful to humans.

Several examples illustrate these processes. Some 13 species of cultivated wheats have evolved out of a sequence including interspecific hybridization among three ancestral grass species and the resulting formation of diploid, tetraploid, and hexaploid forms. The origins of maize are still somewhat uncertain, but nevertheless involve hybridization with teosinte and perhaps a third grass, *Tripsacum*. The potato, domesticated in the Andean region of South America, also has a complicated evolutionary history involving autopolyploidy, hybridization between forms of different polyploid levels, and the appearance of weed races as well as of wild and cultivated forms. The evolution of present-day cattle is similarly complicated; domestication of humpless and humped cattle apparently occurred in different areas, with the origin of basic modern breeds resulting from several early differentiations and subsequent crosses.

These examples show the importance of different varieties, wild and weed forms, and related species to the evolutionary genetics of domesticated species, and they argue strongly for the preservation of such forms to be used in future genetic improvement of domestic plants and animals.

Literature Cited

Caspari, E. W., and R. E. Marshak. 1965. The rise and fall of Lysenko. *Science* 149:275–278.

Chen, K., J. C. Gray, and S. J. Wildman. 1975. Fraction 1 protein and the origin of polyploid wheats. *Science* 190:1304–1306.

Darlington, D. C. 1973. *Chromosome Botany and the Origins of Cultivated Plants.* New York: Hafner Press.

De Wet, J. M. J., and J. R. Harlan. 1972. Origin of maize: the tripartite hypothesis. *Euphytica* 21:271–279.

———. 1975. Weeds and domesticates: Evolution in the man-made habitat. *Econ. Bot.* 29:99–107.

Epstein, H. 1971. *The Origin of the Domestic Animals of Africa,* vol. 1. New York: Africana Pub. Corp.

Galinat, W. C. 1971. The origin of maize. *Ann. Rev. Genet.* 5:447–478.

———. 1975. The evolutionary emergence of maize. *Bull. Torrey Bot. Club* 102:313–324.

Harlan, J. R. 1966. Plant introduction and systematics. In *Plant Breeding.* Ames: Iowa State University Press.

———. 1970. Evolution of cultivated plants. In *Genetic Resources in Plants—Their Exploitation and Conservation.* O. H. Frankel and E. Bennett (eds.). pp. 19–32. Philadelphia: F.A. Davis Co.

———. 1971. Agricultural origins: centers and noncenters. *Science* 174:468–474.

Harlan, J. R., J. M. J. de Wet, and E. G. Price. 1973. Comparative evolution of cereals. *Evolution* 27:311–325.

Harlan, J. R., and D. Zohary. 1966. Distribution of wild wheat and barley. *Science* 153: 1074–1080.

Hawkes, J. G. 1970. Potatoes. In *Genetic Resources in Plants—Their Exploitation and Conservation.* O. H. Frankel and E. Bennett (eds.). pp. 311–319. Philadelphia: F. A. Davis Co.

Heiser, C. B., Jr. 1973. *Seed to Civilization.* San Francisco: W. H. Freeman and Company.

Isaac, E. 1970. *Geography of Domestication.* Englewood Cliffs, N.J.: Prentice-Hall.

Johnson, B. L. 1968. Electrophoretic evidence on the origin of *Triticum zhukovskyi.* In *Proceedings of the Third International Wheat Genetics Symposium.* K. W. Findlay and K. W. Shepherd (eds.). pp. 105–110. New York: Plenum Press.

Katz, S. H., M. L. Hediger, and L. A. Valleroy. 1974. Traditional maize processing techniques in the New World. *Science* 184:765–773.

Kuckuck, H. 1970. Primitive wheats. In *Genetic Resources in Plants—Their Exploitation and Conservation.* O. H. Frankel and E. Bennett (eds.). pp. 249–266. Philadelphia: F. A. Davis Co.

Leppik, E. E. 1969. The life and world of N. I. Vavilov. *Econ. Bot.* 23:128–132.

Mangelsdorf, P. C. 1974. *Corn: Its Origin, Evolution, and Improvement.* Cambridge: Belknap Press. Harvard University.

Mangelsdorf, P. C., and R. G. Reeves. 1939. *The Origin of Indian Corn and Its Relatives.* Texas Agr. Expt. Sta. Bull. 574:1–315.

Protsch, R., and R. Berger. 1973. Earliest radiocarbon dates for domesticated animals. *Science* 179:235–239.

Renfrew, J. M. 1973. *Palaeoethnobotany.* New York: Columbia University Press.

Riley, R. 1965. Cytogenetics and the evolution of wheat. In *Essays on Crop Plant Evolution.* J. B. Hutchinson (ed.). Cambridge: Cambridge University Press.

Rouse, J. E. 1970. *World Cattle,* vols. 1, 2. Norman: University of Oklahoma Press.

Schwanitz, F. 1966. *The Origin of Cultivated Plants.* Cambridge: Harvard University Press.

Simoons, F. J. 1968. *A Ceremonial Ox of India.* Madison: University of Wisconsin Press.

Ugent, D. 1970. The potato. *Science* 170:1161–1166.

Vavilov, N. I. 1951. The origin, variation, immunity and breeding of cultivated plants. *Chronica Botanica* 13:1–366.

Zedler, P. H. Personal communication. October, 1974.

Zeuner, F. E. 1963. *A History of Domesticated Animals.* New York: Harper and Row.

Zhukovsky, P. M. 1968. New centres of the origin and new gene centres of cultivated plants including specifically endemic micro-centres of species closely allied to cultivated species. *Bot. Zh.* (Leningrad) 53:430–460.

Zohary, D. 1970. Wild wheats. In *Genetic Resources in Plants—Their Exploitation and Conservation.* O. H. Frankel and F. Bennett (eds.). pp. 239–247. Philadelphia: F. A. Davis Co.

5

SUBSISTENCE AGRICULTURE

Several thousand years of cultural evolution have led not only to the development of modern systems of intensive, mechanized agriculture, but also to an amazing variety of systems geared to produce food either for individual use or for sale and barter through local markets. Collectively, we shall term these latter systems **subsistence agriculture** because their central emphasis is on self-sufficiency in food production. Our objective in this chapter is to understand the nature, extent, and significance of these systems.

Subsistence agriculture deserves serious examination for several reasons. First, despite the prominence of intensive agriculture as it is practiced in the developed countries, large portions of the world's population still depend on subsistence systems, and we estimate that about 60 percent of the world's cultivated cropland is still farmed by these methods. Obviously, a realistic examination of world food production cannot ignore these systems. Second, these systems reflect a high degree of ecological awareness that has come from hundreds or even thousands of years of continuous practice, often under challenging environmental conditions; for ex-

ample, the Ingorot of northern Luzon, Philippine Islands, have successfully grown paddy rice in remarkable mountainside terraces for over 3000 years (Figure 5-1). These food production technologies, combined with the wider culture of their practitioners, deserve study because they represent steady-state solutions to problems of human life in major varied world environments.

Studying these systems, finally, may reveal important clues to the development of ecologically sound systems of modern agriculture for some of the most intractable portions of the underdeveloped world. Fortunately, food production systems that represent almost all the stages of agricultural evolution and diversification still survive, and modern transportation has made these systems more accessible to scientific study than at any time in history. At the same time, however, the explosive growth of the human population and the spread of influences from industrialized societies is disrupting and displacing them at an increasing rate. Opportunities for studying undisturbed patterns of subsistence agriculture are beginning to diminish rapidly.

FIGURE 5-1.
Parts of the mountainous area of northern Luzon, Philippine Islands, have been farmed
continuously for over 3000 years by means of remarkable paddy rice terraces. (Photograph by P. A. Pittet, courtesy of the FAO, Rome.)

In our examining these systems of agriculture we must be especially careful to avoid biased conclusions, either favorable or unfavorable. On the one hand, we may tend automatically to interpret differences from the systems with which we are familiar as indications of inefficiency and backwardness in food production; this psychological trap is known as **ethnocentrism.** On the other hand, we may erroneously conclude that such systems, lacking the troublesome technology-induced problems of mechanized agriculture, offer a trouble-free route to "new" patterns of food production. The truth, of course, is that all agricultural systems possess their own particular advantages, problems, and limitations. Our goal is to examine these systems objectively and learn what we can from them.

We shall begin by looking closely at several major, present-day systems of subsistence agriculture, including pastoralism, shifting cultivation, and nonmechanized permanent cultivation. We shall focus on several specific examples instead of trying to survey world farming systems; then we shall consider two ecologically interesting systems of specialized subsistence farming in special environments—the *chinampa* system of swampland farming practiced in the Valley of Mexico and the ancient systems of runoff farming practiced in the Negev Desert of southern Israel and in portions of the Arabian Peninsula.

PASTORALISM

Pastoralism can be defined as a food production system in which herding peoples depend for subsistence entirely, or almost entirely, on live-stock products such as milk, meat, blood, and hides (Brown 1971). Pastoralist systems are now limited almost entirely to the arid and semiarid parts of Africa and Asia and parts of arctic Asia. The major animal domesticates involved in these systems are cattle, sheep, goats, camels, yak, and reindeer. Pastoralist systems merge with those of subsistence cultivation, and, as we shall see, many peoples can be classified as semipastoralists.

It is difficult to determine the total number of pastoralists that exist because they live in some of the most remote areas of the world, and they merge culturally with other subsistence agricul-turalists. Brown (1971) estimates that there are about 50 million pastoralists in Africa south of the Sahara; in Kenya, Tanzania, and Uganda

alone there are some 894,000 pastoralists and 629,000 semipastoralists (Morgan 1972). Al-though pastoralism, as a system of food produc-tion, is probably more important per unit popu-lation in the area of Africa south of the Sahara, it is also practiced across the northern Saharan fringe, through the Middle East, and into the semiarid and high mountain areas of east-central Asia. Thus we estimate that close to 100 million pastoralists live in Africa and Asia.

A number of interesting studies of pastoralist food production systems have been carried out. Lustig-Arecco (1975) gives capsule summaries of the original pastoral systems of the Reindeer Chukchi in northeastern arctic Siberia, and of Tibetan Plateau pastoralists whose system was based on the yak. The best known systems are those practiced in Africa, however. These vary considerably, depending on the history of the peoples involved and the characteristics of the environment (Ruthenberg 1971). Table 5-1 shows the general manner in which pastoral and

TABLE 5-1

Relationship of pastoral and semipastoral grazing systems to climate in Africa.

Rainfall (mm/year)	Predominant system	Principal animals
Under 50	Occasional use by complete pastoralists	Camels
50–200	Complete pastoralism with long migrations	Camels
200–400	Complete pastoralism and semi-pastoralism with supplemental crop cultivation	Cattle, goats, sheep
400–600	Semipastoralism with greater emphasis on crop cultivation	Cattle, goats, sheep
600–1000	Semipastoralism primarily as result of ethnic tradition	Cattle
Over 1000	Permanent stock-keeping with occa-sional semipastoralism as result of ethnic tradition	Cattle

Source: Modified from Ruthenberg 1971.

semipastoral systems vary with rainfall in Africa. In areas with less than 200 millimeters of annual rainfall, complete pastoralism, usually based on camels, is the rule. In areas with 200 to 600 millimeters of rainfall, pastoral and semi-pastoral systems involving cattle, sheep, and goats flourish. Where rainfall exceeds 600 milli-meters per year, complete pastoralism is rare, and semipastoralism is maintained more as the result of strong cultural tradition than of clear economic advantage.

The Dyson-Hudsons (1969, 1970) have de-scribed the semipastoral system of the Karimo-jong Tribe of northeastern Uganda in ecological terms. This system is similar to those of many of the other East African tribes, such as the Masai and Samburu, but differs principally in the fact that the Karimojong pursue limited crop cultivation.

The Karimojong are a tribe of about 60,000 persons occupying a 4000-square-mile area of northeastern Uganda, an area of high plains savanna and grassland. Rainfall varies some-what with altitude and greatly from year to year, but it generally ranges between 400 and 1400 millimeters and comes primarily from March to September. Lying in the semiarid region of Africa just south of the Sahara, this region has been affected to some extent by the recent Sahelian drought.

The Karimojong women, girls, and infants live in permanent settlements of a few hundred people. The women practice agriculture in an area of roughly one square mile around the set-tlement, growing sorghum, millet, and some corn. The success of their efforts varies greatly, depending on rainfall conditions. The men and boys spend most of their life in shifting livestock camps tending the herds of shorthorn zebu cat-tle, together with small numbers of sheep and goats. These herds number about 2 to 3 cattle per person in the total population. The loca-tions of these camps change seasonally. During the rainy season, the herds graze at lower eleva-tion, in drier areas, and generally close to the settlements; during the dry season they must be moved to higher elevations and areas with fav-orable dry-season grazing conditions.

The food economy of the Karimojong is thus based partly on agriculture and partly on live-stock. In the shifting camps, the men and boys subsist almost entirely on milk and blood. Their daily diet generally consists of about two and one-half pints of milk mixed with a portion of blood. To provide this, the cattle, about 12 percent of which are lactating at a given time, are milked twice a day. Depending upon range conditions, each cow yields from one and one-half to five pints of milk per day. Individual animals are bled at three- to five-month in-tervals, with four to eight pints taken on each occasion. A thong is placed around the neck of the animal and tightened to make the jugular vein stand out. The vein is then lanced with an arrow, and the blood drained into a bowl. When the thong is released, bleeding ceases.

The diet in the settlements is quite different. Here, the principal food is some form of por-ridge, made with the available grains and gar-nished with other materials such as milk (when the cattle herds are nearby) and a variety of gathered wild foods, including mushrooms, honey, and various herbs. When the herders are near the settlements, the men and boys may also eat porridge brought to them by the women. A few other foods are also utilized: Wild fruits are collected and eaten; meat is eaten on ceremon-ial occasions and when animals die; and a form of sorghum beer is prepared by fermentation.

The food production system of the Karimo-jong thus responds, in a complex fashion, to the ecological realities of a harsh environment char-acterized not only by a highly unpredictable pattern of rainfall, but also by geographical variability that creates a variety of seasonal grazing situations. The tropical conditions and lack of modern technology prohibit extensive food storage and do not allow forage to be trans-ported to the livestock as it is in many modern ranching operations. Moreover, disease and drought characterize the region and effectively prevent the people from depending entirely on either livestock or crop cultivation.

There are many variations on the system of Karimojong. The Masai and Samburu of Kenya are pure pastoralists, for example, and they live entirely in shifting camps. They are completely dependent on their herds and on other food obtained by gathering and trading. The Gabra, and several other tribal groups in northern Kenya, Somalia, Ethiopia, and the Sudan, are pastoralists who rely on camels, as well as cattle, sheep, and goats (Torry 1974).

Pastoralist systems have their advantages and disadvantages. It should be pointed out that the livestock practices of these peoples are not at all market-oriented. Raising livestock for market involves the association of small numbers of people with large numbers of animals in order to obtain a maximum, occasional yield of meat. Pastoralists, however, associate large numbers of people with large numbers of animals under a system designed to provide regular daily food harvests for subsistence. This need for a regular food yield has, of course, been the factor responsible for dietary emphasis on milk and blood, although for most pastoralist groups blood is somewhat less important, and meat more important, than for the Karimojong. Thus these systems involve the harvest of all basic animal food products.

The reliance on milk is a highly unpredictable, semiarid environment creates problems, however. It means, first of all, that most of the herd must be female. In fact, mature females alone regularly make up about 50 percent of the entire herd (Brown 1971). The milk taken by man is often needed by calves as well, leading to poor growth and high mortality of calves. On the other hand, when good range conditions prevail, the large mature female component means that herds increase rapidly, roughly doubling in the time it takes an animal to grow to reproductive maturity. Although this is good from the standpoint of the individual herder,

over the short run, it means that when dry conditions return, livestock numbers may exceed range carrying capacity, leading to range deterioration and to starvation among both livestock and people.

Supporting a pastoralist way of life requires a daily caloric intake of about 2300 calories per adult, or about 15,000 calories for an average family of 6.5 persons (Brown 1971). This in turn necessitates an average daily yield from herd animals of 16 liters of milk and 2.4 kilograms of meat, assuming a diet that is three-fourths milk and one-fourth meat. To provide this, a family must maintain a herd of about 35 to 40 cattle, about half of which must be mature females, and the rest bulls and replacement stock (2 to 3 bulls, 15 to 18 heifers, and several young males). The area of grazing land needed for such a herd depends on precipitation, of course. With an annual precipitation of 750 millimeters, such a herd could be supported on about 40 to 60 hectares; with only 250 millimeters, this increases to about 400 hectares. If we use an intermediate figure corresponding to a grazing area of about 17 hectares per person, our earlier estimate of 100 million pastoralists means that 17 million square kilometers, or roughly 11.4 percent of the earth's land surface, may be under pastoralist systems.

SHIFTING CULTIVATION

Shifting cultivation is a general agricultural pattern in which an area of land is cleared of vegetation, prepared for planting, and cultivated for a relatively short period; then it is abandoned, and the farmer moves to a new area where the process is repeated. This technique is known under the names of slash-and-burn, swiddening, *milpa, ladang, kaingin,* and more than 25 other local and regional terms (Russell 1968). Today, shifting cultivation is practiced mainly in the tropics, where it remains as one of the most important general systems of subsistence agriculture. In 1957, the United Nations estimated that 200 million people depended on this form of agriculture; today, because of population growth, the number has probably increased to almost 400 million.

Historically, shifting cultivation was not limited to the tropics. The earliest systems of agriculture in Europe were apparently of this type (Russell 1968). Forest plots were cut, the woody material was burned, and wheat and barley were then cultivated for perhaps 10 to 25 years before the land was abandoned. In northern Europe, in areas covered by spruce-pine forests growing on poor, sandy soils, the technique has been used until recent times. There, the system involved clearing the forest during the first year, burning the dead material the second year, and cultivating the area for 4 to 6 years. In parts of northern Russia and Scandanavia, this technique may still be in use.

Shifting agriculture is still practiced extensively, however, in all major tropical and subtropical regions of the world. The particular systems in use seem to have evolved independently in several regions, and contrary to some statements the practice shows many ecologically significant variations. Shifting cultivation is not a single technique practiced in a standard manner, but a general class of techniques that vary greatly in detail. To understand this type of agriculture, we shall look at a particular example and then consider some of the variations.

Rappaport (1967, 1971) has analyzed in detail the cultural ecology of the Tsembaga, a clan group of Maring-speaking Melanesian natives in the mountains of east-central New Guinea. He has emphasized the agricultural economy of these peoples, which is a little-modified form of shifting cultivation.

The Tsembaga clan group numbers 204 people who occupy an area of about 8.3 square kilometers of tropical lowland and montane forest lying between elevations of 670 and 1525 meters. The Tsembaga practice shifting cultivation over an area of approximately 405

hectares, but at any one time only about 36 to 40 hectares of this total are actually under cultivation. The rest, about 90 percent, is in "bush fallow," or, in ecological terms, is in the process of secondary succession back to forest. The cultivated land thus amounts to about 0.2 hectare per person. The population density of the Tsembaga is thus 25 people per square kilometer of total occupied land or about 40 per square kilometer of cultivated land.

The Tsembaga's technique of shifting agriculture involves clearing secondary forest with machetes. Most of the land used has been farmed before, and only rarely are the Tsembaga faced with clearing primary forest. They cut brush first, then trees. Some trees, too large to cut completely, are simply trimmed of branches and left standing (Figure 5-2). The plots are then fenced to keep out swine, which are kept in a semi-wild state. The dried plant material is then burned, a process that may have to be repeated two or more times because of the relatively humid conditions that prevail.

After burning, crops are planted. The soil is not cultivated in the normal sense; instead, holes are punched in the surface and cuttings of various species are placed in them. The total crop plant inventory of the Tsembaga is remarkably large, consisting of about 36 species, many of which have several different varieties. The Tsembaga's staples are root crops, of which taro and sweet potato are of primary importance and yams and cassava of secondary importance. Other crops include banana, beans, corn, sugarcane, cucumber, pumpkin, and a number of leafy vegetables. These species are interplanted to form a complex crop ecosystem (Figure 5-3) in which the foliage of different crop species grows to different heights and the root systems show considerable differences as well.

The main activity during the period of crop growth is weeding, during which the Tsembaga distinguish between weeds and tree sprouts

FIGURE 5-2.
A Tsembaga farm plot recently cleared for shifting cultivation in the mountains of New Guinea. Piles of dried branches and brush are being burned. Many of the larger stacked poles will be used in making fences to exclude swine, which range freely through the area. (Photograph courtesy of Roy A. Rappaport, University of Michigan.)

FIGURE 5-3.
Tsembaga woman harvesting materials from a young garden in the mountains of New Guinea. The most conspicuous plants in this graden are taro (large arrow-shaped leaves), sugarcane (narrow grasslike leaves), corn (broader grasslike leaves), and banana (broad-leaved plants on hillside), but yams and cassava are also present. (Photograph courtesy of Roy A. Rappaport, University of Michigan.)

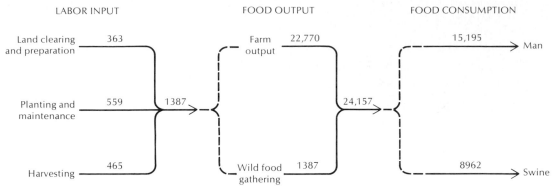

LABOR INPUT · FOOD OUTPUT · FOOD CONSUMPTION

Land clearing and preparation — 363 ·
Planting and maintenance — 559 · 1387 ·
Harvesting — 465 ·
Farm output — 22,770 ·
Wild food gathering — 1387 ·
24,157 ·
Man — 15,195 ·
Swine — 8962

kcal × 10³ per hectare

FIGURE 5-4.
Energy input (as kilocalories expended in human labor) and output (as food calories) for shifting cultivation as practiced by the Tsembaga in the mountains of New Guinea. (Based on data from Rappaport 1971.)

and seedlings, which are left undisturbed. These gardens are cultivated for only two years, after which they are allowed to regrow in brush and forest.

Rappaport, in addition to describing this system qualitatively, analyzed the energetic efficiency of the Tsembaga food production operation. By observing the time necessary for various activities and estimating the energy then required, he calculated that the total input to gardening was slightly over 1.3 million kilocalories per hectare (Figure 5-4). Food harvests from this activity, however, were over 22 million kilocalories per hectare, giving an efficiency, calculated as an output-input relationship, of 16.4 kilocalories per kilocalorie invested. The food yield for this system, in terms of kilocalories per hectare, is slightly greater than that obtained for modern, intensive corn farming in the Midwestern United States (see Chapter 24). This yield is supplemented by a smaller amount of food gathered in the surrounding forest, giving a total supply, per hectare of cultivated land, of about 24.1 million kilocalories per year (Figure 5-5).

Of this total food supply, about two-thirds is consumed by humans and one-third by swine. A swine herd is maintained by the Tsembaga, the function of which is ceremonial rather than agricultural. These animals are important in a ritualized warfare cycle of about 10 years' duration. Over this period the swine herd is built up; at its height the animals are slaughtered and consumed in a ceremonial feast that is frequently followed by warfare against neighboring native groups. As we shall see, however, this seemingly wasteful use of food energy may play an important role in the tribe's overall adjustment to its tropical forest environment.

A great many modifications of the technique of shifting agriculture actually exist (Miracle 1967; Ruthenberg 1971). Techniques of land preparation and planting vary as indicated by the examples listed and described in Table 5-2. The length of the fallow period also varies with environment and with human population density. In humid tropical forest environments, such as New Guinea, the fallow period tends to be two, three, or more decades. Here, the land under cultivation is usually less than 10 percent of the total land subject to shifting cultivation. In drier regions, such as the

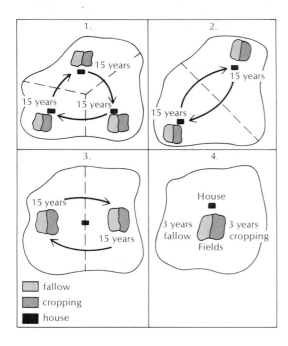

FIGURE 5-5.
Changes in the pattern of shifting cultivation in the Kilombero Valley, Tanzania, as population increases from low (1) to high (4) density. Increasing population density restricts the ability of farmers to move to new sites in a progressive fashion. Details of the patterns are: (1) short rotation of 3 years of cropping and 3 years fallow, long rotation of 15 years of occupation and 30 years of nonoccupation, total duration of rotation 45 years with frequent changing of house sites; (2) total duration of rotation 30 years, with semistationary house site; (3) total duration of rotation 30 years, with permanent house site; (4) short rotation only, with 3 years cropping and 3 years fallow, house site permanent. (From Ruthenberg 1977: *Farming Systems in the Tropics,* Oxford University Press, Oxford, England. Based on E. Baum 1968, "Land Use in the Kilombero Valley." In *Smallholding Farmer and Smallholder Development in Tanzania,* H. Ruthenberg, ed., p. 33. Afrika-Studien, No. 24, IFO Institut, München.)

TABLE 5-2
Techniques used in clearing and preparing farm plots in shifting agricultural systems.

System	Characteristics
1. Burn and plant	Thick, dry secondary vegetation is burned; crop is planted without extensive soil cultivation.
2. Burn, hoe and cut, plant	Dry savanna vegetation is burned; remaining trees and bushes are cut; the soil is hoed and the crop planted.
3. Cut, burn, plant	The most common system, in which vegetation is cut and allowed to dry during the dry season; it is burned at the end of the dry season, and the crop is planted at the start of the rainy season.
4. Cut, plant, burn	Crops are planted while the vegetation, generally forest, is being cut; after the crops are established, the dead material is burned in a manner that causes little crop damage.
5. Cut, bury refuse in mounds, plant	Cut vegetation is composted (with or without previous or subsequent burning) and the crops planted at a later time.
6. Cut, add extra wood, burn, plant, hoe	Cut vegetation is augmented with wood and brush from other areas (5 to 20 times the farm plot area); the plot is burned, crops planted, and the soil cultivated by hoeing.
7. Cut, wait one season, plant	Forest vegetation is partially cut, bananas planted; clearing is completed the following year, and other crops are interplanted with the first.
8. Kill trees by ringing, ridge, plant	Trees are girdled but not cut; when defoliation has occurred, the soil is hoed into ridges and crops are planted.

Source: From Ruthenberg 1971.

moist borders of the African savanna zone, fallow periods may be less than ten years, and the fraction of potentially arable land under cultivation may increase to 20 to 30 percent. Figure 5-5 shows the general pattern by which land tenure varies in relation to human population density in the Kilombero Valley, Tanzania.

Shifting cultivation represents a response to two major features of the tropical environment: rapid depletion of soil nutrients in newly cleared land, and rapid growth of pest populations in areas that lack an unfavorable season. For reasons that we shall examine more closely in Chapter 11, the nutrient holding capacity of many tropical soils is low; decomposition, under warm, moist climatic conditions, is rapid; and, where rainfall is high, the leaching rates for nutrients are high. Thus a rapid decline in fertility takes place after land clearing, and this decline (Figure 5-6) is soon reflected in the yields of crops (Nye and Greenland 1961).

Weeds and crop pests, particularly herbivorous insects, also proliferate (Janzen 1970).

In ecological terms, shifting cultivation represents the periodic exploitation of the nutrient capital of natural ecosystems that has accumulated over a period of ecological succession. Success of shifting cultivation thus depends on a fallow period of adequate length, which is possible only below some critical human population density. Systems of shifting cultivation have existed for several thousand years in various parts of the tropics, and they apparently represent a successful way of dealing with ecological realities in these areas. Nevertheless, in many of these areas the systems are now breaking down. Why?

Meggers (1971) has examined this question in conjunction with a comparative analysis of the cultural ecology of various Indian tribes in the Amazon Basin. She notes that, without exception, tribes that live in upland areas above the floodplains of major streams practice shift-

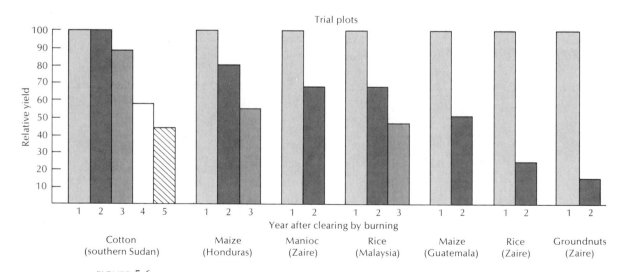

FIGURE 5-6.
Decline in relative yields during successive years of cropping in shifting cultivation systems in humid tropical areas (without fertilizer use). (From Ruthenberg 1977: *Farming Systems in the Tropics*, Oxford University Press, Oxford, England. Based on Nye and Greenland 1961 and B. Andreae 1972, *Landwirtschaftliche Betriebsformen in den Tropen*. Parey, Hamburg and Berlin.)

ing cultivation; they also carry on various hunting and gathering activities, which apparently contribute more to their overall food economy than is the case for the New Guinea Tsembaga. What is particularly interesting, however, is the existence of other cultural mechanisms that seem to be important to the total adjustment of Indian groups to their tropical forest environment. Each tribe has a variety of customs and practices, some of which we would be likely to consider crude and cruel, that influence birth and death rates. Contraception and abortion are employed, and taboos against sexual intercourse exist for various reasons. Infanticide is practiced under certain circumstances. Among adults, warfare, blood revenge, punishment by death, and sorcery occur in certain tribes. No single tribe has all of these mechanisms, but each has some. The net result of such customs and practices seems to be zero population growth and a wide dispersion of the population as a whole. Durham (1976) has analyzed this relationship more rigorously, concentrating on the role of aggression in relation to resource competition among primitive human groups. He concludes that existing evidence supports the idea that aggression in many of the societies we have discussed is indeed adaptive in the context of scarce resources. He cautions, however, against interpreting such cultural practices as conscious means of population control. Their function is to guarantee access to resources and to maximize the survival and reproductive success of the participants. Their overall consequence, however, may be population regulation.

Thus we must recognize that all aspects of human cultures are closely integrated and that the success of systems like shifting cultivation can depend on the integrity of other cultural mechanisms that may not appear to be involved. It is tempting to conclude that the Tsembaga's investment of food energy in swine functions in this manner through the involve-

ment of these animals in the ritualized warfare cycle. In a sense, the caloric value of this food is a measure of the cost of zero population growth!

Modern civilization, through medicine, public health, and cultural encroachment, tends to modify many of the mechanisms essential to systems of shifting cultivation. In the Philippine Islands, Kowal (1966) has documented the breakdown of the system of shifting cultivation, *kaingin*, that originally existed there. Where kaingin is practiced on relatively level land, and with a long fallow period, it appears not to be destructive. However, when population grows to the point where it must be practiced on steeper slopes, a process of destructive ecological change is set in motion. Erosion occurs on steeper slopes, damaging the soil surface and severely depleting its fertility. This favors invasion by grasses and pine trees, which creates a community that tends to maintain itself, or even to spread, through its susceptibility to and its tolerance of fire. This process has led to the very extensive spread of pine forests in the mountains of central Luzon (Figure 5-7). These highly flammable pine forests allow higher rates of soil erosion than do the original broadleaf forests. Their productivity is lower, and their action in rebuilding soil fertility is weaker.

As in the case of pastoralist systems, calculating the portion of the globe that is subject periodically to shifting agriculture gives surprising results. The Tsembaga's cultivated land was equal to about 0.2 hectares per capita. If this value holds more or less true for shifting cultivation in general, and if land under cultivation is about one-sixth of the total land in the rotation system, then about 4.8 million square kilometers of the world's arable land are subject to shifting agriculture. This is 34.3 percent of the 14 million square kilometers frequently cited as the total cultivated land for the world (Whittaker 1975).

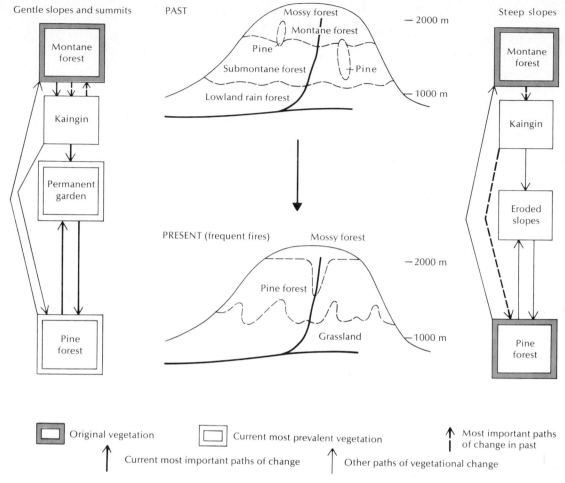

FIGURE 5-7.
Landscape degradation caused by shifting cultivation (*kaingin*) in mountain areas of Luzon, Philippine Islands. As kaingin is extended to steeper hillsides, and as cultivation becomes more permanent, erosion and depletion of soil fertility tend to favor the expansion of pine forests. These forests, highly subject to fire, are low in productivity and poor in restoring soil fertility. (Modified from Kowal 1966.)

PERMANENT NONMECHANIZED AGRICULTURAL SYSTEMS

Systems of permanent subsistence agriculture vary in their characteristics as much or more than those of shifting agriculture (Ruthenberg 1971). Some of these systems depend on rainfall for moisture; in others irrigation is practiced. Where labor is abundant and inexpensive there may develop quite complex systems of mixed-species cropping, staggered planting for different crops, or crop rotation. Fertilization is used in almost all these systems, although it is usually inadequate to maintain the original fertility of the soil, and productivity is thus lower than in shifting systems.

Some of these systems are monocultural, however, and some have dealt successfully with the problem of fertilization. We shall examine in detail two systems of rice cultivation employed in Thailand (Hanks 1972). In parts of southeastern Asia, rice is cultivated under shifting agricultural systems. In other areas, permanent cultivation that relies on either flood irrigation or canal (paddy) irrigation is practiced. Hanks (1972) compared all of these systems, but we shall consider only the last two.

Rice cultivation with flood irrigation is, of course, carried out only on flood plain areas of major rivers. The flood plain land is plowed before the rainy season, typically with the aid of oxen. The rice seeds are scattered over the plowed land and lightly covered with soil. The variety of rice that is planted depends on the depth to which the soil will be covered when the floods come. Taller, slow maturing, "floating" varieties are sown in areas that will be covered most deeply (Figure 5-8). These varieties must also be characterized by rapid ger-

mination; they must also grow fast enough to avoid being completely drowned by the rising water. At higher elevations slower germinating, shorter, more rapidly maturing varieties must be used, since they will be the last to be covered by the rising water and the first to be uncovered when the water recedes.

Flood irrigation is one of the most efficient agricultural systems known, in terms of energy and nutrients required. Both water and nutrients, contained in the silt carried by the river, are supplied by nature; in a sense these systems receive a natural subsidy of the sort that man must provide at considerable cost in other situations. A general estimate of the efficiency of this system is given in Table 5-3. The average yield of 1332 kilograms per hectare is obtained at a cost of almost 71 man-days of labor per crop season. Hanks (1972) estimates the food cost of

FIGURE 5-8.
A planting of "floating" rice in Thailand. The height of the plant indicates the depth of water that this variety can tolerate. (Photograph by G. De Sabatino, courtesy of the FAO, Rome.)

TABLE 5-3
Yield per hectare, labor input, and output/input efficiency in systems of flood- and paddy-irrigated rice production in southeastern Asia.

System	Rice yield (kg/ha)	Labor inputs (man-days)	Labor cost (kg rice/man-day)	Output/input
Flood irrigation	1332	70.74	0.50	37.67
Paddy irrigation	2186	193.13	0.50	22.68

Source: From Hanks 1972.

his labor at 0.50 kilogram of rice per day. Thus the return is equal to 37.67 kilograms of rice for each kilogram invested.

Canal irrigation requires a greater investment of energy. Diked paddies and canals must be constructed and maintained, and artificial fertilization must be supplied. In southeastern Asia paddies are fertilized variously with green manures, addition of straw stubble, and the use of animal manure and night soil. In addition, to maximize production, the rice plants are germinated in separate nurseries and transplanted into the paddies at the appropriate stage of growth (Figures 5-9, 5-10, 5-11).

Under careful management the crop yield per season is greater than for flood irrigation. Hanks (1972) gives the average value of 2186 kilograms per hectare. The labor investment is also greater, however: just over 193 man-days per crop season. The output/input ratio, again

FIGURE 5-9.
Paddy rice culture in Indonesia. Here oxen are being used to prepare terraced paddies for the planting of rice seedlings. (Photograph by F. Botts, courtesy of the FAO, Rome.)

Throughout the country there are about 175 million of these animals, which are regarded as sacred by Hindus. It has been suggested by various individuals, Indians as well as outsiders, and ecologists as well as agriculturalists, that these animals are a major liability to food production in India.

Harris (1966) has argued convincingly, however, that in an ecological sense these animals are very useful. They provide a variety of services, especially in plowing and transport. The highly seasonal activities of agriculture require that bullocks be available simultaneously to many farmers at critical periods, such as when land is being prepared just before the rainy sea-

FIGURE 5-10.
Paddy rice culture in Indonesia. Rice seedlings, germinated in nurseries, are being transplanted to newly prepared paddies. The marked stick is used to assure straight rows and optimal spacing. (Photograph by F. Botts, courtesy of the FAO, Rome.)

FIGURE 5-11.
Paddy rice culture in Indonesia. Mature rice is being harvested by hand. (Photograph by F. Botts, courtesy of the FAO, Rome.)

based on 0.50 kilogram of rice per man-day, is 22.68. Where a reliable year-round supply of irrigation water exists, of course, more than one crop per year can be obtained from paddy rice.

In many systems of permanent subsistence agriculture, domesticated animals play an important role, the ecological rationale of which is often not apparent to the casual observer and is sometimes not clear to the agricultural expert. In India, for example, where much of the crop production is carried out by techniques like those just described, zebu cattle are a conspicuous component of the agricultural system.

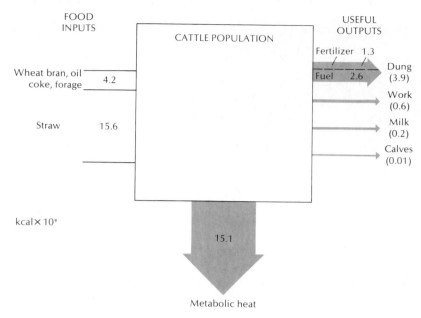

FIGURE 5-12.
Energy input and output for a population of zebu cattle in a rural area of West Bengal, India. Numbers indicate the total annual energy flow through the animal population in a 14.9-square-kilometer study area. The energy output as dung is divided according to whether it is used for fuel or for fertilizer. (Modified from Odend'hal 1972.)

son. In addition, they provide many useful products including milk, meat, hides, bones for fertilizer production, and dung. It should be noted that millions of Indians are not prohibited from consuming beef or else do not respect this traditional prohibition. The hides of the more than 25 million animals that die each year contribute to the world's largest leather industry, and dung is the principal fuel in many areas, as well as a fertilizer in some. Throughout much of the country, no suitable, inexpensive substitute for this fuel exists.

Recently, Odend'hal (1972) has analyzed the importance of Indian cattle in terms of energy. He carried out his study in a 14.9-square-kilometer area in rural West Bengal, not far from Calcutta. Within this area, a census revealed populations of 1100 people and 252 cattle per square kilometer. Odend'hal was able to obtain a record of reproductive performance for part of the cattle population over the period of a year. He recorded the types and quantitites of food consumed by cattle and measured the caloric value of these materials; he also collected data on milk and dung production (also in terms of energy), on production of calves, and on the use of the animals for traction work.

From these data, Odend'hal was able to construct an energy balance sheet (Figure 5-12) for the entire population. Here, we should note that the food inputs to the cattle population are rice straw, the largest single item, together with banana trunks, sugarcane tops, rice hulls, wheat bran, and mustard oil cake, this last being the only commercially obtained food item. In addition, the animals graze on grasses and forbs. The main foods for the cattle population are thus materials that man cannot use directly except as fertilizers. The cattle population turns these materials into useful work and products, the most important of which are dung, milk, and calves.

Viewed in terms of meat- and milk-production efficiency, these cattle perform very poorly: The products equal only 0.96 percent of the food energy input, whereas, on western U.S. rangeland, beef production efficiency alone averages about 4.1 percent (Odend'hal 1972). However, when the energy values of all the products that are useful in the rural Indian agricultural system, including work and dung, are considered, this efficiency rises to 17.02 percent of the food energy input.

We must recognize, of course, that the existence of an ecological rationale for a situation such as that examined by Odend'hal is not the same as optimal ecological organization. Instead, it indicates that the role of each component of a system such as that of rural India must be understood and taken into account in programs designed to improve agricultural and social conditions in underdeveloped regions.

We shall now turn to an examination of two specialized systems of subsistence agriculture that give further insight into the integration of successful agricultural systems into the ecosystems in which they exist.

CHINAMPA CULTIVATION IN THE VALLEY OF MEXICO

When the Spanish conquistadors reached the Valley of Mexico, they found that the capital of the Aztec empire was a city, Tenochtitlan, located on an island in a shallow lake (Figure 5-13). Over much of the swampland bordering the lake system they also encountered one of the most original and productive systems of agriculture in the western hemisphere. This system,

known as **chinampa agriculture,** was at the height of its development at the time of the conquest; now it survives in the "floating gardens" area of the suburb of Xochimilco (Coe 1964, Armillas 1971).

The extent of the lake system in the Valley of Mexico, which is a closed basin, has varied over the past few thousand years. During periods of moist climate, a single lake covering 180 to 200 square kilometers (essentially the entire valley floor) has existed. In drier times this lake system retreats, forming five connected shallow lakes. During these periods the Indian inhabitants of the valley learned to construct elevated farming plots in the swampy and shallow parts of the lake. These plots, or *chinampas*, were from 2.5 to 10 meters wide and up to 100 meters long. They were constructed out of mud scooped from the intervening areas and piled on the plot; this also created a series of canals that separated the plots. The chinampas were built up to a height of 0.5 to 0.7 meter above water level. Their sides were reinforced by posts interwoven with branches and by willow trees planted along their edges (Figure 5-14).

During the Aztec period, the chinampas were organized into farm units of 6 to 8 plots each (Figure 5-15). A series of large canals served as transportation routes through the system, and moisture for crop growth was provided largely by horizontal seepage of water from the canals, which maintained a saturated zone at a relatively shallow depth in the soil. When there were unusually low water levels, however, no part of any plot was far from the water in a canal. Fertility of the chinampas was maintained by regular mucking and composting. Today the farmers, or *chinamperos*, use canvas bags on long poles to dip muck from the canals and place it on the plots, and a similiar procedure was presumably used in Aztec times. In addition, aquatic weeds are harvested, spread over the chinampa surface, and covered with muck. The plots were also manured, probably

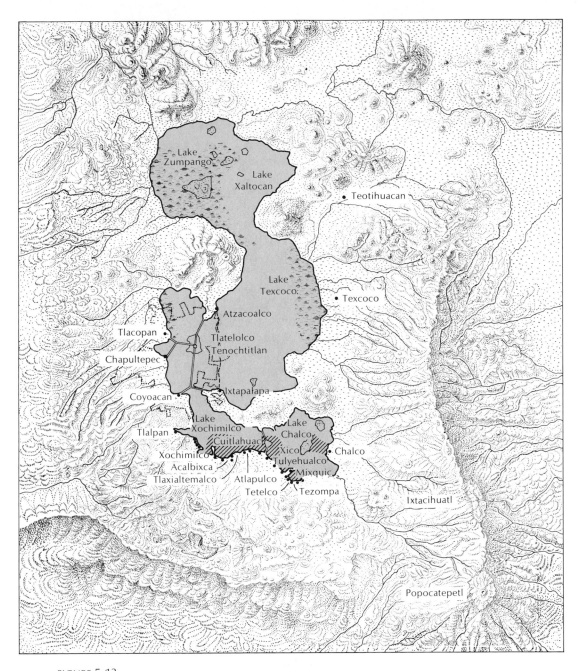

FIGURE 5-13.

Chinampa areas (*shaded*) and the Valley of Mexico are shown as they appeared in summer at the time of the Spanish conquest in 1521. In the rainy season the five lakes coalesced into one large lake: the Lake of the Moon. Tenochtitlan-Tlatelolco was the Aztec captial. The dotted line marks the limits of modern Mexico City. The broken line between Atzacoalco and Iztapalapa shows the part of the great Aztec dike that sealed off and protected the chinampas from the salty waters of Lake Texcoco. Causeways and aquaducts leading to the Aztec capital are also shown, and the names of the nine chinampa towns that remain are in bold type. The large black dots without names are the sites of the freshwater springs that fed the chinampa zones. (From "The Chinampas of Mexico" by Michael D. Coe. Copyright © 1964 by Scientific American, Inc. All rights reserved.)

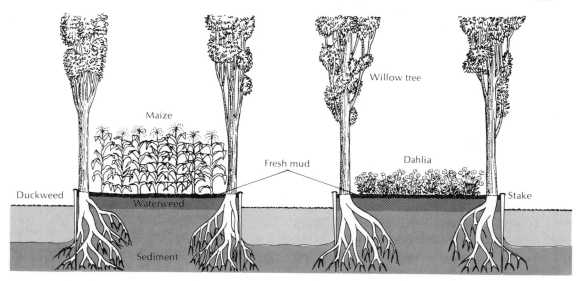

FIGURE 5-14.
Structure of the chinampas of the Valley of Mexico, as seen in cross section. (From "The Chinampas of Mexico" by Michael D. Coe. Copyright © 1964 by Scientific American, Inc. All rights reserved.)

with night soil during Aztec times, although animal manure is used now.

The chinampas are cultivated in a manner designed to maximize the number of annual crops that can be obtained. Today, special seedling nurseries are used, which are constructed on one end of a chinampa, close to the water. A thick layer of muck is spread over a bed of waterweeds, allowed to dry somewhat, and planted with the seeds or cuttings of crop plants. When these seedlings have reached proper size, the muck bed is simply cut into blocks containing individual seedlings, and these are transferred to the chinampa proper. These nurseries were originally built on floating mats of vegetation 6 to 10 meters long and narrow enough to pass through the smaller canals; they could then be towed from place to place. In fact, it was apparently the existence of these nurseries that gave the Spaniards the idea that the agricultural plots themselves were floating.

The crops grown on the chinampas originally included corn, beans, peppers, and species of grain amaranths. Today, quite a variety of truck crops are grown.

Archeological investigations in the Valley of Mexico show that island village sites in the lake were first occupied several centuries B.C. It is presumed that chinampa farming was begun around such villages. However, from the time of Christ until about 1200 A.D. the lake level rose to the point where extensive chinampa cultivation was probably impossible. It is interesting that this corresponded to the period during which the Valley of Mexico was dominated by inhabitants of Teotihuacan, which may reflect the fact that land-based agriculture was more profitable because of a wetter climate.

The lake level fell and chinampa agriculture began to expand again at about 1200 A.D.; it reached its maximum development about 1400 to 1600 A.D. At its height, the Aztecs had achieved control of the water system of the entire basin. By means of a system of causeways with control gates, they had isolated the saline waters of the lowest portion of the lake system,

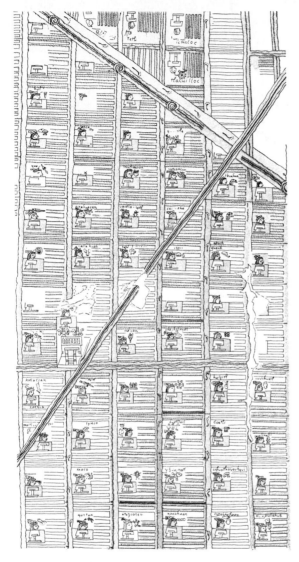

FIGURE 5-15.
Ancient Aztec map of a portion of Tenochtitlan-Tlatelolco shows that it was a chinampa city. Six to eight plots are associated with each house. Profile of the householder and his name in both hieroglyphs and Spanish script appear above each house. Footprints indicate a path between plots or beside a canal. This is a copy of a small part of the damaged map, which is in the National Museum of Anthropology in Mexico City. (From "The Chinampas of Mexico" by Michael D. Coe. Copyright © 1964 by Scientific American, Inc. All rights reserved.)

Lake Texcoco, so that water could enter, but not leave this area. They also had aquaducts to bring fresh water to the city of Tenochtitlan itself. The system of chinampas, and its success, thus depended on the strong political organization of the valley as a whole.

After the conquest, alternative systems of agriculture were introduced by the Spanish and the lake system was eventually drained as the new Mexico City grew. In 1900, under the administration of Porfirio Diaz, a drainage tunnel was constructed through the mountains on the northern edge of the valley. This project, together with heavy withdrawals of water from wells for urban use, has eliminated most of the lake system except for traces in the Xochimilco area. Aerial photographs, however, still reveal the outlines of chinampa plots over much of the valley. Reconstruction of this system on maps indicates that, at its maximum extent, the chinampa system covered 120 square kilometers, with the actual cultivated land equaling about 90 square kilometers. It is estimated that the productive capacity of this land was able to feed 100,000 people. Only a few small remnants survive.

Although the chinampa system in Mexico is the best-known system of swampland agriculture, it is not the only one. In central Burma, the Intha, an ethnic group of native Burmese numbering about 70,000, practice a similar type of agriculture (Jeffrey and Garrett 1974). The Intha live in some 200 small villages in and around Lake Inle, which in the wet season has an area of about 155 square kilometers. Their farm plots, narrower than chinampas, are worked almost entirely from boats. The plots are first constructed as floating structures consisting of mats of vegetation covered with muck. After a year or so, these structures become waterlogged and sink, but may eventually be converted into permanent agricultural islands like chinampas. These swampland plots are highly productive, enabling the Intha to meet their own food needs and also to export eggplant, cabbage, cauliflower, cucumbers, peas, and beans to neighboring areas of Burma.

Similarly, in the Valley of Kashmir, the Mirbahri practice a system of swamp agriculture in Dal Lake (Lawrence 1895). Two major types of plots are used: One is an actual floating garden, built, like the Intha's plots, on a heavy mat of reeds and other buoyant lake vegetation. These mats, about 2 meters wide, are covered with conical mounds of muck and aquatic plants in which 2 to 4 seedlings of melons, cucumbers, or tomatoes are planted. The second type of plot, near the shore, is an elongate chinampalike plot constructed by techniques almost identical to those used in the Valley of Mexico. A great variety of crops are likewise grown successfully on these plots.

These remarkably similar systems of swampland agriculture appear to be successful because they simultaneously solve problems of both moisture and fertility by a relatively simple system of technological manipulation of the landscape. Western agriculturalists, in emphasizing machinery-intensive systems of crop cultivation, have given little or no attention to this unique and highly productive approach to agriculture.

DESERT RUNOFF AGRICULTURE

From 200 B.C. to about 630 A.D., parts of the Negev Desert of what is now southern Israel and the northern part of the Arabian Peninsula were occupied by a prosperous group known as the Nabateans. In the Negev Desert these people occupied six major cities, all now abandoned, and enjoyed a long period of political stability as a part of the Roman and Byzantine Empires (Figure 5-16). The prosperity of the Nabatean Kingdom was not the consequence of a wetter climate at that time. Instead, it appears to have been due to the perfection of a unique

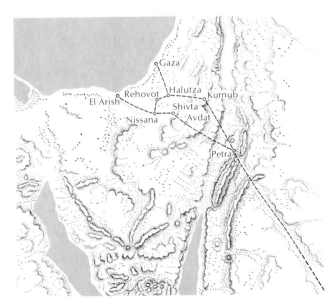

FIGURE 5-16.
The Nabatean Kingdom included the now-abandoned cities of Petra, the capital (located in what is now Jordan), and six cities in the Negev Desert of present-day Israel (Avdat, Shivta, Nissana, Rehovot, Halutza, and Kurnub). The dotted lines indicate trade routes through the region. (From "Ancient Masters of the Desert" by Michael Evenari and Dov Koller. Copyright © 1956 by Scientific American, Inc. All rights reserved.)

FIGURE 5-17.
Small catchment runoff farms of the Nabateans consisted of terraced wadi channels, where crops were planted, and an associated watershed about 17 to 30 times larger in area. This diagram shows an area of farm plots, separated by low rock dikes, together with a conduit that controlled water flow within the farm plot and an inclined collector that shunted runoff from a portion of the watershed to the desired part of the farm. (From "Ancient Masters of the Desert" by Michael Evenari and Dov Koller. Copyright © 1956 by Scientific American, Inc. All rights reserved.)

system of agricultural exploitation of the desert environment (Evenari and Koller 1956, Evenari et al. 1961).

The Negev Desert receives an annual rainfall of about 100 millimeters, most of which comes in short rain showers of less than 10 millimeters each. Despite this, surface runoff tends to be high. The soils of the Negev are largely wind-deposited loess, and, when wetted, they tend to form a relatively impervious crust. From small watersheds, runoff frequently amounts to 20 to 40 percent. From larger watersheds, because of greater infiltration through the coarser materials accumulated in streambeds, runoff is less, averaging 3 to 6 percent. The Nabateans had learned how to take advantage of these runoff patterns to support corps with relatively high water demands. The design of their farm systems varied somewhat with the size of the watershed and its drainage channels, or *wadis*, as they are termed in the Middle East.

Small catchment **runoff farms** (Figure 5-17)

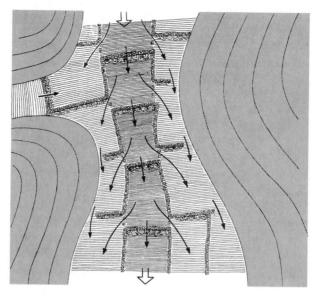

FIGURE 5-18.
Design of rock barriers used to spread the water in large,
shallow wadis onto lateral portions of the floodplain in runoff
farms of the Nabateans. (From "Ancient Masters of the
Desert" by Michael Evenari and Dov Koller. Copyright
© 1956 by Scientific American, Inc. All rights reserved.)

were created in areas with watersheds 10 to 100 hectares in area. These farm units consisted of catchment areas located on the watershed slope and cultivated plots located in drainage bottoms below these catchments. Cultivated plots varied in size from less than a hectare to about 5 hectares, so that a ratio of catchment to farmed plot was about 17:1 to 30:1. The farm plots were actually created by the construction of rock dikes across the wadis, thus accumulating soil suitable for cultivation.

The catchment slopes were modified to maximize runoff. Scattered rocks were collected into piles, a practice that increased the proportion of the surface that would seal when it rained and thus increased runoff. Stone conduits were also built to carry water to various parts of the bottomland farm plots in equitable amounts. This runoff, of course, was absorbed by the accumulated soils of the farm plots, providing adequate moisture for a variety of crops. Approximately 3000 to 4000 cubic meters of water per hectare were required for good crop production.

Along the bottoms of larger wadis, more elaborate systems of water control had to be used. In broad, relatively shallow wadis, a series of rock barricades were used that were structured to spread water laterally from the main channel onto the adjacent floodplain (Figure 5–18). In the larger, deeper wadis, in which the water flowed too fast for the channel itself to be cultivated, a complicated system of barrier dams, diversion channels, and rock dikes was necessary to raise the water and divert it into floodplain farm plots (Figure 5-19).

The few written records that survive from the period show that a variety of crops were

134

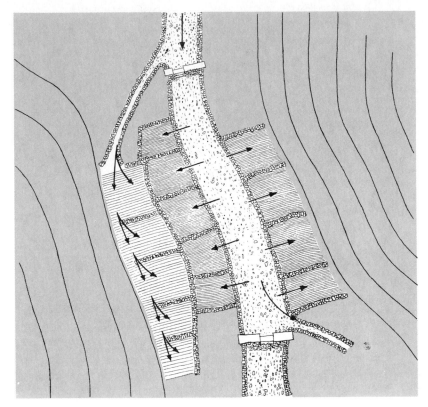

FIGURE 5-19.
Design of Nabatean water-spreading dams in large, deep wadis. The dams raised the
water level in the unfarmed wadi channel so that water flowed laterally into terraced
plots on the floodplain. (From "Ancient Masters of the Desert" by Michael Evenari
and Dov Koller. Copyright © 1965 by Scientific American, Inc. All rights reserved.)

grown, including barley, wheat, legumes, grapes, figs, and dates. In fact, the yields obtained were quite respectable. For barley, records show that yields were about 8× the quantity of seed used; for wheat, the yields were about 7×. Today, in the moister northern Negev region, with modern practices, barley yields are only 9 to 11× and wheat 8× the quantities of seed required.

The success of this system again seems to be the result of relatively simple technological manipulation of the natural landscape in a manner that deals simultaneously with the problems of both water and nutrients. The porous rock

dikes of these runoff farms impede the flow of water so that the nutrient-containing particulates are deposited to form the soil and the water soaks in deeply to provide moisture for crop growth. No attempt is made to create large reservoirs from the surface of which major losses by evaporation would occur and at the bottom of which nutrient-rich sediments would accumulate without benefit. Instead, these materials are directed immediately to the areas of cultivated soil and, in a sense, stored on location.

Like the chinampa system of the Valley of Mexico, however, the runoff farms of the Negev depended on a strong and stable government. The coordination of activities over whole watershed systems was essential to the success of individual parts of the system. Thus the demise of the Nabatean system was brought about by political rather than ecological factors in about 700 A.D., when the Byzantine Empire was overthrown by invading Moslem Arabs.

GENERAL PATTERNS AND TRENDS IN SUBSISTENCE AGRICULTURE

Two major ecological strategies are incorporated in the various systems of subsistence agriculture that we have examined. First, practitioners of many systems, such as pastoral grazing and shifting cultivation, take advantage of accumulated nutrients or harvestable food energy by being highly mobile. In a sense, these systems operate by periodically "mining" this accumulated capital; thus they depend on population densities low enough to allow adequate resources to accumulate before the same area is used again. The second strategy, seen in systems such as flood irrigation, chinampa farming, and Nabatean runoff agriculture, takes advantage of natural through-flows of water and nutrients by simple technological manipulations that simultaneously solve moisture and fertility problems. These systems depend greatly on special environmental circumstances and require a degree of regional political stability that protects the flowing systems on which agriculture depends.

Under optimal conditions these systems may, as in the case of the Tsembaga of New Guinea, achieve high levels of productivity and can support population densities comparable to those supported by modern mechanized agriculture. These systems are labor intensive, but their yields, considered in terms of invested energy, are highly efficient. Furthermore, as Odend'hal (1972) has observed, they employ what can be termed the "principle of increased product utilization by low-energy cultures." Materials that, in industrialized countries, are often considered wastes are used and recycled. Therefore, although these systems may not themselves be solutions to problems of world food production, they can provide partial answers and clues to the design of improved agricultural systems for areas such as the tropics.

Population growth, however, is forcing rapid changes in many of these systems. Ruthenberg (1971) has summarized the major patterns of change that are tending to occur in farming systems in the tropics (Figure 5-20). In this figure, the size of the arrows indicates, in a very general way, the principal patterns of change. Most of these can be seen to undermine the ecological basis on which systems such as shifting cultivation and pastoralism succeed. Under these circumstances, four major possibilities exist, which should be clearly stated:

1. Productivity per unit area of agricultural land will be maintained by increasing human and animal labor to compensate for the increased nutrients taken from the land in the form of food and for other kinds of deterioration caused by more intensive manipulation.
2. Productivity per unit area of agricultural land will be maintained by increasing fertilizers, irrigation, and other energy-intensive, high technology factors that compensate for the deterioration caused by more intense exploitation.
3. Productivity per unit area of agricultural land will be maintained by changes in the structure and ecology of the agricultural system that will reduce deterioration, allow greater yields to be taken more often, and involve little increase in labor or other inputs.
4. Productivity per unit area of agricultural land will decline.

Of course, any combination of these is also

HUMID CLIMATES

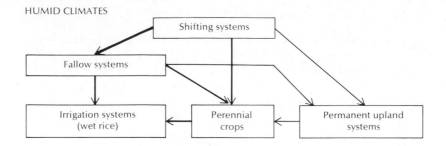

SEMIHUMID CLIMATES

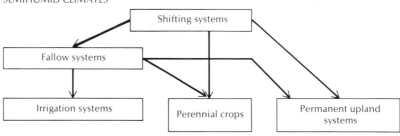

SEMIARID CLIMATES

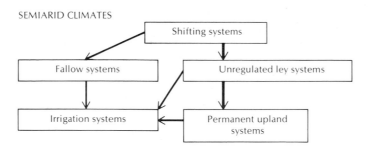

HIGH ALTITUDES

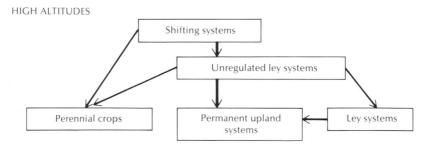

FIGURE 5-20.
General tendencies for change in farming patterns accompanying population growth and development in tropical areas. These trends apply best to small landholders. Ley farming refers to the cultivation of an area with irregular (unregulated) or regular (regulated) fallow periods. (From Ruthenberg 1977, *Farming Systems in the Tropics,* Oxford University Press, Oxford, England.)

possible. The point is that greater yields are not possible without greater effort.

SUMMARY

Many important systems of subsistence and local market agriculture exist in various parts of the world. These systems, normally thought of as "primitive," are based on intrinsic ecological rationales and in some cases show high levels of crop productivity. It is necessary to understand the rationale and the limitations of these systems in order to deal properly with problems of food production in the underdeveloped parts of the world.

Pastoralist systems of subsistence agriculture, practiced by about 100 million people, are characterized by the association of large numbers of people with large herds of domestic animals. The objective of these systems is to obtain regular yields of foods such as milk, meat, and blood. Pastoralist systems are pursued on about one-third of the earth's grassland and savanna environments.

Shifting cultivation, now restricted largely to the tropics, is a farming system in which plots of land are cleared, cultivated for a few seasons or years, and then abandoned because of a decline in productivity that is caused by the exhaustion of soil nutrient capital and the buildup of weeds and insect pest populations. There are many forms of shifting agriculture, differing in the details of land preparation, duration of tenure and fallow periods, and other characteristics.

Permanent nonmechanized systems of agri-culture exist where problems of moisture and fertility can be dealt with by labor-intensive technology. Perhaps the most efficient are those that are flood irrigated. Flooding, along with its deposition of water-carried silt, simultaneously provides both moisture and nutrients at no cost to the farmer. In other situations, exemplified by paddy rice culture, great investments of human labor must be made, although the levels of productivity that result may be higher.

Specialized agricultural systems of swamp-land and desert areas demonstrate the manner in which natural through-flows of water and nutrients can be taken advantage of to develop productive, permanent agricultural systems. Farm plots that are elevated a few feet above the water level of shallow lakes and fertilized by mucking and composting aquatic weeds have been developed in several parts of the world; these plots generally exhibit high levels of productivity. In the Negev Desert, a unique farming system was evolved by the Nabateans between 200 B.C. and 650 A.D. Runoff, carrying dissolved and suspended nutrients from large watersheds, was concentrated in bottomland farm plots of much smaller area, enabling traditional crops to be raised in an area with only 100 millimeters of rainfall per year.

All these systems are being modified and displaced by forces resulting from rapid population growth and the spread of industrial civilization. It seems, however, that the intensification of food production in many of the areas can occur only if greater investments of energy, materials, and skill in the ecological design of agricultural systems are made.

Literature Cited

Armillas, P. 1971. Gardens on swamps. *Science* 174:653–661.

Brown, L. H. 1971. The biology of pastoral man as a factor in conservation. *Biol. Conserv.* 3:93–100.

Coe, M. D. 1964. The chinampas of Mexico. *Sci. Amer.* 211(1):90–98.

Durham, W. H. 1976. Resource competition and human aggression. I. A review of human warfare. *Quart. Rev. Biol.* 51(3):385–415.

Dyson-Hudson, R., and N. Dyson-Hudson. 1969. Subsistence herding in Uganda. *Sci. Amer.* 220(2):76–89.

——. 1970. The food production system of a seminomadic society: The Karimojong, Uganda. In *African Food Production Systems.* P. F. M. McLoughlin (ed.). pp. 91–123. Baltimore: John Hopkins Press.

Evenari, M., and D. Koller. 1956. Ancient masters of the desert. *Sci. Amer.* 194(4):39–45.

Evenari, M., D. Koller, L. Shanan, N. Tadmor, and Y. Aharoni. 1961. Ancient agriculture in the Negev. *Science.* 133:979–996.

Hanks, L. M. 1972. *Rice and Man.* Chicago: Aldine-Atherton.

Harris, M. 1966. The cultural ecology of India's sacred cattle. *Cur. Anth.* 7:51–59.

Janzen, D. 1970. The unexploited tropics. *Bull. Ecol. Soc. Amer.* 51:4–7.

Jeffrey, D., and W. E. Garrett. 1974. Burma's leg rowers and floating farms. *Nat. Geog.* 146(6): 826–845.

Kowal, N. E. 1966. Shifting cultivation, fire, and pine forest in the Cordillera Central, Luzon, Philippines. *Ecol. Monog.* 36:389–419.

Lawrence, W. R. 1895. *Valley of Kashmir.* Oxford: Oxford University Press.

Lustig-Arecco, V. 1975. *Technology: Strategies for Survival.* New York: Holt, Rinehart, and Winston.

Meggers, B. J. 1971. *Amazonia: Man and Culture in a Counterfeit Paradise.* Chicago: Aldine-Atherton.

Miracle, M. P. 1967. *Agriculture in the Congo Basin.* Madison: University of Wisconsin Press.

Morgan, W. T. W. (ed.). 1972. *East Africa: Its Peoples and Resources.* Nairobi: Oxford University Press.

Nye, P. H., and D. J. Greenland. 1961. *The Soil under Shifting Cultivation.* Reading, England: Commonwealth Agriculture Bureau.

Odend'hal, S. 1972. Energetics of Indian cattle in their environment. *Human Ecol.* 1:3–22.

Rappaport, R. A. 1967. *Pigs for the Ancestors.* New Haven: Yale University Press.

——. 1971. The flow of energy in an agricultural society. *Sci. Amer.* 225(3):116–132.

Russell, W. M. S. 1968. The slash-and-burn technique. *Nat. Hist.* 78(3):58–65.

Ruthenberg, H. 1971. *Farming Systems in the Tropics.* Oxford: Clarendon Press.

Torry, W. 1974. Life in the camel's shadow. *Nat. Hist.* 83(5):60–69.

Whittaker, R. H. 1975. *Communities and Ecosystems,* 2nd ed. New York: Macmillan Publishing Company, Inc.

6

ECOLOGICAL FEATURES
OF INTENSIVE AGRICULTURE

Now that we have reviewed some fundamentals of ecology and examined the productivity, ecological rationale, and environmental impact of several types of subsistence food production systems, we can examine some aspects of intensive agriculture.

Although subsistence agriculture is of considerable local importance, most of the world's food supply is produced by more intensively managed agroecosystems. As populations grow and become more highly urbanized in countries with low food production, the need increases for changes that will raise the productivity of local agriculture. The productive, mechanized agricultural systems of countries such as the United States often become the model that leaders of less developed countries adopt. This approach can be dangerous, because subsistence systems often based on built-in ecological rationales are replaced by more intensive systems that require, at the very least, technological resources that may not be in adequate supply.

A number of authors, including Hooper (1976), Revelle (1976), and Wortman (1976), have suggested that the intensification of agriculture through the application of modern technology is the best route for the developing world to follow. From a socio-economic point of view this may be logical, but we must not lose sight of the need for a sound, ecological approach to development. We must realize that intensive agroecosystems are often strikingly different in structure and function from the natural ecosystems that previously occupied a given area. Our objective in this chapter is to examine the ecological features of these intensive agricultural systems, and it is to be hoped that recognizing the ecological strengths and weaknesses of such systems will guide the formulation of methods for increasing agricultural production. New systems must maximize the role of ecological mechanisms and emphasize the use of renewable resources.

The intensification of agriculture does not necessarily involve a high level of mechanization, although in industrialized countries it usually does. Agriculture can also be intensified biologically, by reducing the variety of crop species cultivated, increasing the stocking level (density) of either plants or animals, and decreasing the time that fields are allowed to be fallow or pastures to go ungrazed. The increased effort that these activities require may

FIGURE 6-1.
Aerial view of a portion of California's San Joaquin Valley, showing the mosaic of agricultural plots separated by an irrigation canal, roadways, and fence lines. (Photograph by J. C. Dahilig, courtesy of the U.S. Bureau of Reclamation.)

be provided by human and draft animal labor rather than by machines. Regardless of whether the intensification involves mechanization, we can identify differences between natural and intensive agricultural ecosystems in features such as ecosystem continuity, the evolutionary adaptation of the various biotic components, specific and intraspecific diversity, the degree to which outside subsidies are necessary, and the nature of biotic regulation.

CONTINUITY

One of the most obvious features of intensive agricultural systems is their well-defined spatial arrangement. Natural ecosystems, except where terrestrial and aquatic ecosystems come together, tend to blend one into the next through zones of transition called **ecotones**. They are interrelated in a rather delicate manner by the back-and-forth movement of species and the exchange across their boundary zones of energy, nutrients, and water. Agroecosystems usually have much more clearly defined boundaries, which are quite noticeable when farmland is viewed from the air (Figure 6-1). The boundary between cultivated land and the natural ecosystem from which it was carved is also typically well delineated. Within the converted area there are plots clearly separated by fences, ditches, canals, windbreaks, and access roads. Sometimes a single crop extends over many square miles, but usually a large agricultural area like California's San Joaquin Valley is broken up into a mosaic of plots that in many cases match specific crops with local edaphic, topographic, or climatic conditions.

Of major significance, however, is the fact

that the original flora and fauna are completely replaced over vast areas. Where patches of natural vegetation persist they often occur on sites unsuitable for agriculture and contribute only minimally to the ecological stability of the area. On the other hand, high-quality agricultural land is frequently taken over by agriculture-related activities that may disrupt local ecology more than actual cultivation does. For example, commercial developments spring up to provide goods and services to the surrounding agricultural system; these are followed by housing tracts to accommodate a labor force. As a result, an extensive area that once supported natural, subtly integrated ecosystems is replaced by a patchwork of crops, commercial developments, and residential areas that generate a variety of environmental conflicts.

Natural ecosystems are self-regulating and self-perpetuating, and they display considerable temporal continuity, especially during the late stages of succession. Most agroecosystems, particularly intensively managed monocultures, are neither self-regulating nor self-perpetuating. Some, like orchards and vineyards, may persist for 50 years or more, but only when they are adequately cared for; if they are not, native species soon take over. Most agricultural crops are planted and harvested only months apart. The seeds are usually planted in bare soil that has been prepared by machines, the crop is managed intensively for a short time and then almost completely removed at harvest. The soil is subsequently tilled in preparation for the next planting. When this type of cultural practice is followed, the soil is disturbed repeatedly, and the soil surface is exposed periodically to the baking effect of direct sun as well as to the erosive and compacting actions of wind and rain (Figure 6-2).

FIGURE 6-2.
Soil destruction caused by repeated improper cultivation. (Photograph courtesy of the USDA, Soil Conservation Service.)

Under natural circumstances landslides, fires, and other disturbances create scattered openings in otherwise continuous, well established communities. The patches of more or less bare ground thus created are occupied by a sequence of successional stages that gradually restores the site to the biotic conditions that prevailed before the disturbance. Under intensive agriculture, however, the site is disturbed year after year. The planted crop replaces the pioneer community of ecological succession, and subsequent successional stages are never allowed to develop.

As outlined in Chapter 2, ecological succession involves progressive patterns of change in both the structure and the function of a system. Structurally, there is an increase in the number of species, the accumulated biomass, and the ratio of the animal-to-plant components of that biomass. Functionally, there is a gradual rise in productivity and a decline in the year-to-year variability of net production as the system progresses toward a steady-state climax condition. The productivity and stability of this steady-state system reflect the action of regulatory controls that buffer the effects of environmental fluctuations, maintain favorable levels of nutrient flux, and stabilize interspecies relationships. As succession proceeds these benefits are paid for with more and more of the system's photosynthetic production. In the climax ecosystem, most or all of the system's primary productivity is needed to maintain its structure and organization, so that no further increase in the standing biomass is possible. The system persists in this climax state until some natural disturbance occurs that allows succession to begin again.

Intensive agriculture prevents this normal course of events from taking place. Every newly planted crop represents the first stage of succession that is neither persistent nor steady-state. The objective of growing a crop is to obtain the greatest possible harvest. The best means for this is to establish a system for which the ratio between primary production and biomass is at its highest level; that is, one in which little primary production is required for maintenance. To maintain a system of this type it is necessary for man to assume responsibility for the costs of maintenance and regulation normally taken care of by the natural processes that lead to the establishment of a climax ecosystem. In some of the agricultural systems described in Chapter 5, ingenious elements of design keep effective natural regulatory mechanisms in operation. In most intensive agricultural systems, mechanized ones in particular, these natural mechanisms are replaced by technology and large inputs of energy.

ADAPTABILITY

The later stages of natural succession are characterized by assemblages of species that have evolved a high degree of coadjustment. In fact, in early stages, the lack of coadjustment of many of the biotic components helps to drive succession (Horn 1976). Agroecosystems consist of unnatural assemblages of selected, domesticated species and an assortment of native or imported opportunistic species that manage to invade the site. These two groups have not been integrated into a steady-state system by the process of coevolution, and the opportunistic species frequently constitute weed, insect, and disease pests that must be dealt with by the farmer.

Some natural ecosystems—a northern coniferous forest, for example—at least superficially resemble a monoculture; but close examination reveals that they consist of a fair number of species that have evolved a high level of interdependence. In these as in all natural systems the assemblage of organisms is the result of natural selection. The species adapt to withstand the normal fluctuations of the physical environment and to survive the sometimes intense pressures of intra- and interspecific competition. Each species exhibits a degree of genetic variability that buffers the impact of many environmental forces and thereby assures its own persistence as well as that of the system.

Crop plants and domestic animals, on the other hand, are species chosen by man and cared for by him to produce a usable commodity. Although such domesticates are often grown or raised under climates similar to those under which they evolved, they are also frequently cultivated under conditions to which they are not well adapted. Near the limits of climatic tolerance, the productivity of crop plants may decline and their susceptibility to weeds, insects, and pathogens may increase. In other words, crop communities are poorly buffered, easily disrupted systems.

INTERSPECIFIC DIVERSITY

Obviously, the goal of intensive agriculture is to maximize the portion of net productivity that can be harvested as a usable commodity. Efforts to maximize crop production often involve management plans that become self-reinforcing feedback loops. It is easy to see how certain interventions in the way a system functions can create a need for additional interventions. For example, broad spectrum insecticides can kill both harmful and beneficial species, so that, after spraying, the crop environment is highly vulnerable to invasion by another pest or else to a rapid regeneration of the original pest in the absence of its natural enemies. The result is often a program of intensive chemical control that is characterized by more and more frequent interventions on the part of the farmer and by a greatly simplified biotic community.

In natural ecosystems, most of the chemical energy stored by green plants is dissipated as heat as it passes along the four or five links of the grazing food chains, and along the asso-

ciated detritus food chains. Man learned early in his history as an agriculturalist that he could divert more food energy to himself simply by shortening the grazing food chain. Consequently, wherever intensive agriculture is practiced man is the primary consumer in plant agroecosystems and the principal carnivore in food chains that consist of forage plants, a domesticated herbivore, and man. In addition to shortening grazing food chains, man systematically eliminates all undesirable species, his competitors, from the trophic levels that remain; the complexity of the food web is thus greatly simplified.

As we pointed out in Chapter 2, the solar energy that is chemically fixed by green plants fuels all the work done within a natural ecosystem. When most of this energy is diverted directly to man or to a domestic animal, the normal network through which energy flows is disrupted and the processes that normally occur are eliminated. Because these processes usually contribute significantly to the overall maintenance and stability of the system, their disruption can have a major ecological impact.

A monoculture is by design an exceedingly simple system. Ideally, at least from man's point of view, the crop plant and perhaps some pollinators should constitute the entire system. In such a case there are many ecological vacancies that can be filled by potential invaders. Crop environments are therefore highly vulnerable to weeds, pathogens, insects, and other herbivores. Even large crop acreages cannot be isolated from those components of natural ecosystems that can and do interfere with man's attempt to maximize the crop output. In fact, large areas have a greater perimeter for invasion, and create a larger target for dispersing species. Once such an ideal environment has been invaded, the population growth and dispersal of pest organisms within the system may proceed rapidly. The new arrivals often benefit at first from the absence of natural enemies and thus proliferate unchecked. The resulting high populations may then represent a substantial energy loss as far as man is concerned, and so steps are often taken to control them.

For these reasons, pest control is one of the major areas of concern in intensive agriculture. Many high-yielding crops have been found to be particularly susceptible to insects and pathogens; but the uniformity of all monocultures provides an unnatural abundance of host material, and the optimal environment created by irrigation and fertilization promotes the rapid growth of weeds. Weeds left unchecked can completely choke out patches of the crop (Figure 6-3) and substantially reduce yields by effectively competing with the crop for sunlight, water, and nutrients. Pathogens weaken the ability of the crop to compete with weeds and generally reduce yields by directly weakening or killing the plants. Insects and other herbivores form another unwanted trophic link that can result in a major decline in

yield. A common method of eliminating these potentially costly energy leaks is to use various poisons, which can have a number of undesirable ecological consequences.

Insecticides are often nonselective and kill useful species as well as pests. If large numbers of pollinators are killed, for example, fruit crop yields can be markedly depressed; destroying the natural enemies of insect herbivores and disease vectors can lead to a rapid resurgence of a pest population or to secondary pest problems that involve previously innocuous species (see Chapter 15). When either pest resurgence or secondary pest problems develop the common response is often to apply another pesticide. Each intervention has an associated energy and monetary cost, but more important it further simplifies the system and thus damages its stability.

Instability, however, is not the only functional change brought about by the community simplification that characterizes many agro-

FIGURE 6-3.
A small truck farm showing the impact of weeds on part of a cabbage patch. Note the larger cabbage plants in the weed-free area to the right. (Photograph by M. D. Atkins.)

ecosystems. The fertility of the system is also substantially modified. In Chapter 2 we noted that the natural interplay between the grazing and detritus food chains was important to the overall pattern of energy flow and nutrient cycling. At first it would seem that the substantial amount of energy flowing through the detritus food chain of a natural ecosystem would become a major energy leak in a system managed for maximum productivity. We must remember, however, that this energy is used for vital kinds of work associated with nutrient cycling. It is used largely by the decomposer organisms that break down dead organic matter and make the constituent nutrients available to plants. If these nutrients are not provided by biochemical processes that operate within the system, soil nutrients are depleted and must be added from outside the system.

Intensive agriculture clearly places its emphasis on a short, efficient grazing food chain that is maintained at the expense of the detritus food chain. The net primary production comes mainly from a single plant species and then passes to man, either directly or through one of his domesticated herbivores. In addition, many crop varieties (such as short-stemmed wheat) have been selected to provide a high ratio between their usable and nonusable parts, and these crops contribute very little to the organic debris in the soil.

INTRASPECIFIC DIVERSITY

The species that make up natural ecosystems usually display a wide range of genetic variability. Within their genetic composition are characteristics that have been tested and proven beneficial over thousands of years. Often there are recently acquired genetic traits being tested in the arena of natural selection for the first time. This high degree of heterogeneity reduces the impact of a variable environment. Natural ecosystems are certainly not immune to devas-

tation merely because they contain a variety of genotypes, but their more diverse gene pools play an important role as a fine tuning mechanism that continually adjusts to changing conditions.

Crop plants and domestic animals, on the other hand, are genetically much more uniform as a result of intensive selection and breeding. The gene pool is often narrowed by intensive breeding for yield, flavor, and conformation characteristics that may have little direct ecological significance. But increased genetic homogeneity also increases the potential impact of a variable environment and can increase the damage caused by insects and pathogens. The rice improvement program of the International Rice Research Institute (IRRI) experienced some of these problems. When the dwarf, high-yielding IR-8 variety was released in 1966, it was heralded as a major breakthrough and the beginning of the Green Revolution. However, IR-8 proved to be susceptible to several pathogens and an insect pest that, in combination, caused serious economic and sociological problems (Franke 1974, Harris 1973). In all fairness to the IRRI, their continued efforts resulted in the 1974 release of IR-28, which is almost completely resistant to the pests and pathogens that plagued IR-8 (Jennings 1976).

The cultivation of a genetically homogeneous crop is important in terms of efficiency, because it fosters the standardization of cultural practices over a large acreage. Such a crop can be sown, weeded, irrigated, and fertilized uniformly. As a result, the growth and development of the crop plants are highly synchronized: Throughout the system, individual plants germinate, leaf, flower, and set seed at the same time. Although this is desirable in terms of crop management and marketing, this synchronization introduces still other ecological problems that can be corrected only by human intervention. Bare seed beds are vulnerable to invasion by weeds that must be removed either mechanically or with herbicides. As the crop develops, all of the plants draw upon the same resources at the same time, sometimes making repeated applications of fertilizer or irrigation water necessary. The uniform development throughout the crop planting increases its susceptibility to invasion by pests that often must be controlled with chemicals. Even the synchronization of flowering can increase the level of competition for natural pollinators, thereby creating a need to bring domesticated bees into the crop area.

Despite these dangers, the breeding of more productive crop varieties with a more uniform genetic constitution has been a major goal in the biological intensification of agriculture in recent years. The breeding of crops adapted to intensive cultivation must go on, but the facts remain that genetic variability does buffer the impact of natural calamities, and new crop varities with little genetic heterogeneity are vulnerable to destruction by new pest genotypes. It is to be hoped that future crop development programs will not only emphasize varieties that make efficient use of the costly resources often supplied under intensive agriculture, but also varieties that can be grown in marginal areas. Chapter 22 focuses more closely on the subject of breeding for productivity.

NUTRIENT AND WATER SUBSIDIES

The principal aim of intensive agriculture is to produce a harvest that greatly exceeds the food needs of the farmer or the farm community. After harvest, most of the commodity produced is transported to another location, often an urban center, for processing and consumption. Intensive agroecosystems are therefore exporting systems, and this can have a substantial impact on soil structure and fertility. The amount of plant material left in the soil varies according to the type of crop and the harvesting technique employed, but it is frequently small. When some root crops, such as onions, are harvested, virtu-

ally no plant material is left behind. For others, such as sugar beets, a large quantity of fiber in the form of branch roots remains in the soil after harvesting. Pest control practices also influence the quantity of plant material left in the soil: For example, after sweet corn is harvested the stalks and leaves are often disked into the soil to cut down the overwintering populations of corn borer; in contrast, the straw and stubble of certain grains are often burned to eliminate disease agents (Figure 6-4).

The failure to return plant debris to the soil after harvest, and the fact that there is not a normal complement of herbivores and carnivores to contribute their waste products and remains, leads to a reduction in organic matter in the soil. The detritus food chain of such systems is greatly reduced in importance, and the tight internal biotic phase of various nutrient cycles is nearly eliminated. The balanced nutrient flux rates that characterize established natural ecosystems are replaced by a large outflow of nutrients, which leads to the gradual depletion of internal nutrient pools. As these nutrient pools are depleted, the crop yield declines. High yields can then be maintained only if nutrients in the form of chemical fertilizers are brought into the system (see Chapter 12). These fertilizers usually consist of mined phosphates and industrially fixed nitrogen, the production of which requires a substantial expenditure of energy outside the system. In other words, external sources of energy are used to compensate for the fact that the energy normally expended by decomposers in the process of maintaining fertility has been sidetracked.

Finally, fertility is also affected adversely by the repeated disturbance of the soil that is characteristic of many crop systems. Mechanical seed-bed preparation, weed removal, and other cultural practices curtail succession below the soil's surface as well as above it. As indicated in Chapter 2, the soil environment changes along with the system of which it is a part. By the

FIGURE 6-4.
Cloud of smoke rising from fields where grain stubble is being burned in Northern California. (Photograph by M. D. Atkins.)

time natural succession has led to the formation of a climax community, the soil environment has become highly organized and plays a major role in the storage and movement of water and nutrients. When it is disturbed repeatedly, the soil becomes unable to sustain its unique functions and can become little more than a support medium for a crop grown in what resembles a hydroponic system.

Ideally soil consists of a mixture of living organisms and a variety of nonliving components. This mixture is not random, but is highly integrated and has a precise architecture (see Chapter 11). The architecture of a particular soil governs its ability to hold or release vital nutrients. Soils that are composed of a good mixture of sand, clay, and organic matter consist of aggregations of small particles known as **crumbs.** The particles that make up these crumbs are bound together primarily by complex organic substances. The size and shape of the spaces between the crumbs determine the amount of air in the soil, its water holding capacity, and the ease with which it can be penetrated by the growing roots of plants.

The repeated cultivation of the soil, combined with a reduction in the input of humus, alters both the structure of the crumbs and the way they are organized with regard to one another. When the organic content of the soil is reduced, the crumb structure of the soil may be weakened, and cultivation then tends to destroy the crumbs. Fine inorganic particles no longer bound together in crumbs are easily eroded away by wind and water. Soils with a low organic content are also easily compacted by rain and machines. Once compacted, the soil is more difficult to cultivate, and irrigation water tends to run off, carrying soluble nutrients with it (Figure 6-5). It may eventually be necessary to add synthetic conditioners to the soil—another external cost.

FIGURE 6-5.
Erosion of topsoil by water flowing down a slope. Note accumulation of silt in the foreground. (Photograph by R. L. Kent, courtesy of the USDA, Soil Conservation Service.)

The plants of natural ecosystems are usually well adapted to the edaphic and climatic character of their region. The seasonal availability of water, a combined effect of precipitation and soil type, often plays a major role in the growth cycles and productivity of wild species. It is evident from our examination of subsistence systems in Chapter 5 that man's initial selection and cultivation of crops usually exploited these relationships. However, the intensification of agriculture has included both multiple cropping (growing several crops in a single season), and the cultivation of crops in areas with less available moisture than they require. In both cases a water subsidy in the form of irrigation is required (Figure 6-6). Extensive irrigation can cause a related increase in the energy subsidy, a change in soil structure and increased erosion, the silting of natural and manmade reservoirs, and a basic change in the hydrologic cycle (see Chapter 12).

ECOSYSTEM REGULATION AND SUBSIDIZATION

Clearly, then, intensive agricultural systems are artificial assemblages of plants, and sometimes of animals, that are different in both structure and function from permanent natural ecosystems. As we have already stated, the former are specifically designed to maximize the production of a specific commodity and, compared with natural systems, are most like the ephemeral pioneer communities that invade disturbed sites. In our intensive manipulation of these systems little emphasis has been placed on ecologically sound management. Regardless of whether the farmland involved has been created by converting complex natural ecosystems, such as a deciduous forest, or by irrigating otherwise unproductive arid land, we too often fail to apply basic ecological principles.

There is no question that our approach to in-

FIGURE 6-6.
The huge amount of water needed for sloped furrow irrigation of grain sorghum in Texas has contributed to a drop in the area's water table. (Photograph by F. S. Witte, courtesy of the USDA.)

tensifying agriculture has permitted fantastic gains in crop production; but we must realize that these levels of production can be obtained only as long as man is willing to provide substantial material and energy inputs. These systems are not self-perpetuating. Think of what soon happens to abandoned farm land: Natural communities of a more permanent and stable kind soon become reestablished if the site has not been too severely damaged. This is the paradox of agriculture: To maximize our production we must often provide the energy and nutrients that create an environment that favors succession. Then we must continue to expend energy to prevent that succession from occurring.

In pursuing this course of action we are clearly putting the resiliency of nature to a hard test (Figure 6-7). It is not yet clear to what extent we can control the flow of energy through ecosystems without triggering long-term degradation of basic productivity. There must be some point at which further interference will cause a major collapse in the mechanisms that control fertility and productivity. Holling (1973) postulated that the degree of variation in basic ecosystem properties is related directly to the level of system disturbance. If we extend this thinking to agriculture, we can expect that at some point intensification will increase the variability of productivity. We cannot yet determine how serious this variability may be or how close we are to the critical points at which long-term deterioration of our more intensive agroecosystems will begin.

Some of the oldest (and thus time-tested) forms of intensive agriculture, such as the several-thousand-year-old terraced rice fields of the Philippines, demonstrate an ecological soundness that probably developed gradually over centuries of trial and error. The same is true for the ancient practice of rice culture

FIGURE 6-7.
How much disturbance can natural ecosystems tolerate? The removal of natural vegeta-
tion can cause pronounced changes such as those seen in this Kentucky landscape.
(Photograph by E. W. Cole, courtesy of the USDA, Soil Conservation Service.)

on the flood plains of major rivers. The flood
waters provide a natural subsidy drawing on
the energy of the hydrologic cycle; they also
provide the water needed by the crop and
deliver nutrients from areas upstream. But
not all ancient forms of intensive agriculture
have survived. There is considerable evidence
that some of the natural ecosystems converted
to agricultural production have been too fragile
to withstand the altered conditions. Vast areas
of China that once supported deciduous forests
are now barren because of soil erosion induced
by agriculture. The intensive crop cultivation
of the Sumerians in the Tigris-Euphrates Valley
failed because of salinization caused by improper irrigation. Overgrazing in many parts
of the world has also converted once productive
areas to virtually barren land (Figure 6-8).

Mechanization is often seen as one of the
most ecologically detrimental features of intensive agriculture because it promotes monocultural practices, leads to compaction and
other destructive soil changes, and requires
a substantial input of energy. But the labor-
intensive systems that are often held up as an
example of a better approach to food production are not without problems. The increases
in agricultural productivity achieved in China
(Wortman 1975) and in India (Mellor 1976),
where major portions of the population are
engaged in farm labor, have resulted primarily
from the adaption of high yielding strains of
wheat and rice and from intensified multiple
cropping (Sprague 1975). But it must be remembered that these crop varieties require
more fertilizers and pesticides than lower yielding ones; and very little consideration has been
given to either the cost or the environmental
impacts of these materials.

In the industrialized western nations the

FIGURE 6-8.
This area near Pearsall, Texas, shows impact of overgrazing. (Photograph by John McConnell, courtesy of the USDA, Soil Conservation Service.)

trend toward mechanized agriculture began about 200 years ago, but it has been most noticeable in the last few decades. It is not just the use of machines, however, that has changed agriculture in countries like the United States. There has been a growing trend for agricultural production to be subsumed under a recently emerged agribusiness sector of the economy that manipulates land, water, crops, and animals as production factors in the same way that it does labor, capital, and technology (Jansen 1974).

Before World War II over 20 percent of Americans lived and worked on farms. Now only 4 to 5 percent work farms, and this small fraction is able to realize the production of huge surpluses of food (Kastelic 1974). Over this same period the size of the average farm has increased, as one might expect (Figure 6-9). How far these trends will continue will depend upon a variety of socio-political and economic influences, as well as environmental considera-

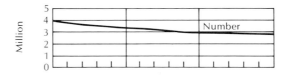

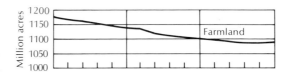

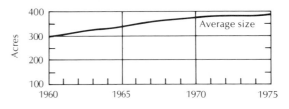

FIGURE 6-9.
Changes in the number of farms, acreage of farmland, and average farm size in the United States between 1960 and 1975. (USDA.)

tions. So far the extremely large agricultural units that have been established in the Soviet Union have not appeared in the United States. One- or two-family farms (units of variable size but usually less than 800 hectares) are still favored in this country. A study by Raup (1970) suggests that these smaller units are not only economically sound but may not be as damaging to the environment as larger farming operations.

Large agricultural companies obtain volume discounts on bulk purchases of fertilizers, pesticides, soil conditioners, machinery, and even fuel; but these lower costs may lead to ecologically harmful misuse. According to Pearlman and Shea (1972), in 1964 American farms with gross annual incomes over $40,000 contained only 28 percent of productive agricultural land but used more than 40 percent of all the pesticides sold. Small farms (less than $5,000 gross annual income) accounted for 9 percent of the total United States farm acreage and used 11 percent of the pesticides. On the other hand, medium-sized farms accounted for 63 percent of the acreage and used only 49 percent of the pesticides. A similar trend was noted for fertilizer use. According to Raup (1970), the seriousness of agricultural pollution due to fossil fuel consumption and to the use of fertilizers and pesticides, together with the cost of environmental clean-up, increase with farm size. In regions where agricultural production is diffused over the landscape, agricultural pollution has been avoided to a great extent. Preserving natural ecosystems in agricultural areas apparently provides a buffer that reduces the need for expensive environmental protection.

Environmental problems arise not only out of an increase in farm size, but also out of the intensification of agriculture. Recent changes in the approach to livestock production clearly reveal this. Not many years ago livestock were dispersed widely over large range acreages and among many small farms where they were maintained in pastures at a low density. Under such conditions wastes were distributed over the land much the way they are in pastoral ecosystems. Today large numbers of animals are confined on small acreages and are fed special diets of high-protein grains and fodder. Feed lots capable of handling 30,000 to 50,000 cattle annually, but only about 10 hectares in size, are now common (Figure 6-10). Poultry and egg "factories" occupying about one hectare house several hundred thousand birds and produce millions of eggs annually.

The intensification of animal agriculture has created some major ecological problems related to the disposal of a large volume of solid and liquid wastes. Livestock raised in the United States produce about two billion tons of solid and liquid waste per year, an amount roughly equivalent to that from two billion human beings. Where feed lots are adjacent to productive land the waste can be used to fertilize crops, such as alfalfa or corn, that can be fed back to the livestock. This reduces the transportation costs for both the feed and the manure, but often the cost of this form of waste removal exceeds the value of the manure as fertilizer (Kastelic 1974). Furthermore, the manure can only be spread on the land at certain times, and so it must be stockpiled for long periods during which pollution problems can arise (Loehr 1970). Large accumulations of manure give off a strong odor, and the seepage of concentrated liquids into groundwater supplies can cause serious problems. Problems created by other agricultural wastes such as pea vines, oil seed pulps, and nutshells have been partially solved by processing them into useful products. Unfortunately this is not yet true for animal wastes.

It is clear that agricultural efficiency cannot be measured simply in terms of yield or cost per unit of production. It must also be evaluated in terms of its ecological impact, and the costs of environmental protection must be added into the cost of each unit of production. From

FIGURE 6-10.
A large cattle feed lot in the San Joaquin Valley of California. Note the density of animals
in the background. (Photograph by M. D. Atkins.)

an ecological point of view, agriculture must also be evaluated in terms of energy; that is, the energy value of the food produced compared to the total energy input made by man. Although the yield of labor-intensive systems such as Asian wet-land rice culture may not be exceptional, the energy ratio of cost to benefit, of about 1:30, is outstanding. The reverse is usually the case for highly mechanized systems, in which the energy input often approaches or exceeds the energy value of the return. For national food production systems it has been estimated that the energy expended to produce one food calorie is 1.14 calories in the United States (Hirst 1974), 2.50 calories in England (Blaxter 1975), and 3.85 calories in Israel (Stanhill 1975). The energy cost of intensive livestock production greatly exceeds the caloric value of the meat produced.

QUANTIFYING THE INTENSITY OF AGRICULTURE

Recent experience has shown that high levels of agricultural production can be achieved even in difficult environments by increasing the input of energy either directly or indirectly. The fossil fuels used by intensive agroecosystems to drive tractors, pump water, and produce fertilizers and pesticides represent a form of capital provided by nature. In a very real sense, portions of this capital liquidated now will not be available for use later, and vice versa. The maximization of gross production now, by wasteful use of this capital, thus represents a small short-term gain at the cost of a greater long-term loss. Obviously, we should attempt to design agroecosystems from which reasonable yields can be extracted without unnecessary depletion of our natural resources. In attempting to do this we can use two measures of performance: (1) the ratio of energy output (food) to energy input (solar and fossil); (2) the magnitude of energy transfer rates per unit area (MacKinnon 1975).

Earlier, we discussed the costs of maintenance and regulation in natural systems. The greater the degree of modification we implement, the higher are the costs of maintenance and regulation. If, for example, we manage an area of deciduous forest to encourage and selectively exploit a few of the many species present, the extra input costs will not be great because the natural system will still be reinvesting much of its own productivity in its own maintenance. If, however, we replace the forest with a monoculture of corn, the costs that we must bear increase greatly. We must attempt to prevent natural succession; we must maintain fertility artificially; and we must take on the task of pest control. If, in addition, we wish to raise the level of net primary production (*NPP*) above that of the deciduous forest, we can do so only by applying larger or more effective inputs that more than compensate for the loss of the mechanisms that operated in the diverse forest system. The further the agroecosystem deviates from natural structure and function, the greater is the cost of maintaining this artificial condition.

Thus it is ecologically sound to gear production input policy to the *NPP* level of the natural ecosystems replaced by agroecosystems. To do this we must develop some quantitative measure of energy intensiveness. There are many ways in which this might be done, but to illustrate the fact that it can let us examine a particular example: the eastern deciduous forest and the intensive corn farming that has replaced much of it. The average *NPP* of the deciduous forest is

4700 kcal m²yr (Leith 1972). The *NPP* of corn in the United States has changed dramatically over the past 30 years as agricultural technology has intensified. Estimates of the yield of grain and total corn *NPP* (including grain and vegetative parts) are given in Table 6-1, together with fossil fuel and total energy inputs to their production. We can see that since 1945 grain yield has increased about two and one-half times, while fossil fuel inputs have more than trebled.

Mathematical functions can be constructed to integrate these relationships. One of the simplest is:

$$\frac{(FF + D)(T + D)}{NPP_{agr}}$$

where FF is the fossil fuel input, T the total energy input, and D the *NPP* of the natural ecosystem (NPP_{nat}) minus the *NPP* of the agroecosystem (NPP_{agr}).

As applied to policy and planning, the objective would be to minimize the value of this expression, which is essentially a cost (numerator) to benefit (denominator) ratio. The

TABLE 6-1
Net energy relationships of midwestern United States corn production from 1945 through 1970.[a]

Year	Corn production (kcal/m²/yr)		Energy inputs (kcal/m²/yr)		Agroecosystem production deficit[b]	Efficiency expression value[c]
	Grain	Total *NPP*	Fossil fuel	Total		
1945	847	2668	217	229	2032	1906
1950	946	2982	286	298	1718	1263
1954	1021	3217	376	382	1483	1078
1959	1345	4237	456	469	463	202
1964	1694	5335	545	554	0	56
1970	2018	6355	699	716	0	79

[a]Data on production and energy input from Pimentel et al. 1973.
[b]*NPP* of deciduous forest (4700 kcal/m²/yr) minus total *NPP* of corn ecosystem.
[c]See p. 154 for description.

numerator is constructed to do two things: (1) encourage input to the point at which the agroecosystem achieves an *NPP* comparable to that of the natural system; (2) favor efficiency in use of fossil fuel inputs. The presence of the deficit term *D* makes the numerator large when NPP_{agr} is well below NPP_{nat}. In effect this is saying that investment in such cases is desirable in order to reestablish the "normal" levels of fertility and productivity that can prevail. At any level of NPP_{agr}, but more easily when NPP_{agr} is below NPP_{nat}, increased investment can reduce the value of the expression if it produces a great enough increase in NPP_{agr}. This is easier to do, however, with investments of non-fossil fuel energy than with fossil fuels, since the fossil fuel investment is squared in the numerator (as *FF* and as part of *T*). This also means that any change that replaces fossil fuel inputs by other energy inputs of equal, or even somewhat larger, amounts will reduce the value of the expression. Finally, once NPP_{agr} reaches NPP_{nat}, *D* falls to zero and additional increases in NPP_{agr} are possible only if the additional inputs are highly effective.

Using the data for corn from 1945 to 1970, and the value of deciduous forest *NPP* given in Table 6-1, we can calculate values of this expression. These are also included in the table. Expression values from 1945 through 1964 suggest that the improved yields represent reasonable applications of improved technology. The increase from 1959 to 1964, raising NPP_{agr} above NPP_{nat}, reduces the value of the expression largely because the additional inputs are associated with an increase in NPP_{agr} more than 12 times as great. However, the continued increase in inputs through 1970 had a much smaller effect, and the value of the expression increased.

This last observation suggests that in Midwestern corn farming we have arrived at the point where additional inputs are becoming much less effective in increasing productivity. Further deviation in the pattern of ecosystem function from that which represents a steady-state adjustment of climate, soils, and natural biota may be very costly. Instead of working toward further gains by increasing inputs, it seems more appropriate to work for an increase in net productivity of the corn system by reducing inputs, particularly fossil fuels, while maintaining present levels of yield. The resources freed by such a strategy can then be used to raise production in other ecosystems with NPP_{agr} levels well below those of natural systems.

This approach applies the net energy concept to food production in ecosystems in which the natural climate, soils, and living organisms are still powerful forces. Obviously, agricultural production that is low because of poor methods, ineffective inputs, and inadequate equipment, can be improved only by an increase in the input of energy and other resources. However, the exhorbitant use of nonrenewable inputs and the mining of agroecosystem capital for short-term gain must be avoided. The development of productive agroecosystems that maximize the use of natural mechanisms and renewable resources must become a major goal of agricultural ecology.

SUMMARY

Most of the world's food supply is produced by agricultural systems that are either biologically or technologically more intensive than the subsistence farming systems engaged in by some of the world's people. Highly managed agroecosystems, although they operate according to the same natural laws, as natural ecosystems, are strikingly different in structure and function.

Intensive agroecosystems usually have sharp, distinct boundaries and are shorter lived than mature natural communities. Because they lack both spatial and temporal continuity, agroecosystems, particularly crop monocultures, are neither self-regulating nor self-perpetuating.

Succession is prevented and the detritus food chains rarely have an opportunity to develop fully. The plants and animals of intensive agriculture have experienced centuries of selection for characteristics of little ecological importance; they are the products of intensive breeding programs with specific objectives. The biotic components of agroecosystems, therefore, lack the adaptive flexibility that buffers the flora and fauna of natural ecosystems against the vicissitudes of nature. Often, the genetic homogeneity of crop species increases their susceptibility to physical and biological challenges.

Structurally, agroecosystems are shortened grazing food chains that consist essentially of a crop plant and man or a crop plant, a domestic animal, and man. In addition to maximizing the flow of energy and materials to man, these intensive agroecosystems are exporters to urban centers. The depletion of nutrients and the alteration of soil structure that result are often compensated for by large subsidies of fertilizer, irrigation water, and sometimes soil conditioners. To these subsidies must be added those of pesticides needed to compensate for a lack of self-regulation. Together these subsidies require large amounts of energy from outside the system to accomplish the work done by solar energy in natural systems.

The intensification of agricultural practices is a crucial test of the resiliency of nature. We do not know how much further man can proceed with the control of energy flow through ecosystems without setting in motion a major decline in future food production potential. Before we discover this critical point through unfortunate experience we should endeavor to design agroecosystems that compare in productivity to natural ecosystems, and in doing so we should make maximum use of natural ecosystem processes and renewable rather than nonrenewable resources.

Literature Cited

Blaxter, K. L. 1975. The energetics of British agriculture. *J. Sci. Fd. Agric.* 26:1055–1064.

Franke, R. W. 1974. Miracle seeds and shattered dreams. *Nat. Hist.* 83(1):10.

Harris, M. 1973. The human strategy: The withering Green Revolution. *Nat. Hist.* 82(3):20.

Hirst, E. 1974. Food-related energy requirements. *Science* 184:135–138.

Holling, C. S. 1973. Resilience and stability of ecological systems. *Annu. Rev. Ecol. Syst.* 4:1–23.

Hooper, W. D. 1976. *The development of agriculture in developing countries. Sci. Amer.* 235(3): 196–205.

Horn, H. S. 1976. Succession. In *Theoretical Ecology.* R. M. May (ed.). pp. 187–204. Oxford: Blackwell Scientific Publications.

Jansen, A. J. 1974. Agroecosystems in future society. *Agro-Ecosystems* 1:69–80.

Jennings, P. R. 1976. The amplification of agricultural production. *Sci. Amer.* 235(3):180–194.

Kastelic, J. 1974. Symbiotic relations between plants and animals and the pollution problem. In *Animal Agriculture.* H. H. Cole and M. Ronning (eds.). pp. 18–28. San Francisco: W. H. Freeman and Company.

Leith, H. 1972. Modelling the primary production of the world. *Nat. and Res.* 8(2):5–10.

Loehr, R. C. 1970. Changing practices in agriculture and their effect on the environment. *CRC Crit. Rev. Env. Cont.* 1:69–99.

MacKinnon, J. C. 1975. Design and management of farms as agricultural ecosystems. *Agro-Ecosystems* 2:227–291.

Mellor, J. W. 1976. The agriculture of India, *Sci. Amer.* 235(3):154–163.

Perelman, M., and K. P. Shea. 1972. The big farm. *Environment* 14(10):10–15.

Pimentel, D., L. E. Hurd, A. C. Bellotti, M. J. Forster, I. N. Oka, O. D. Sholes, and R. J. Whitman. 1973. Food production and the energy crisis. *Science* 182:443–449.

Raup, P. M. 1970. Economies and diseconomies of large scale agriculture. *Amer. J. Agr. Econ.* 51:1274–1283.

Revelle, R. 1976. The resources available for agriculture. *Sci. Amer.* 235(3):164–178.

Smith, R. F. 1971. *The Impact of the Green Revolution on Plant Production in Tropical and Subtropical Areas.* 1970 Founders Memorial Lecture. Presented to the Entomol. Soc. Am., Los Angeles, Nov. (Mimeographed.)

Sprague, C. F. 1975. Agriculture in China. *Science* 188:549–555.

Stanhill, G. 1975. Energy and agriculture: a national case study. In *Heat and Mass Transfer in the Biosphere,* part 1. *Transfer Processes in the Plant Environment.* D. A. de Vries and N. H. Afgan (eds.). pp. 513–533. Washington, D.C.: Hemisphere Publishing Corporation.

Wortman, S. 1975. Agriculture in China. *Sci. Amer.* 232(6):13–21.

——. 1976. Food and agriculture. *Sci. Amer.* 235(3):30–39.

Part Two

THE DYNAMICS OF AGROECOSYSTEMS

7

CLIMATE AND AGRICULTURE

Most life on earth exists in a narrow zone where the earth and its gaseous envelope, the atmosphere, meet. The character of the atmosphere therefore greatly affects the kinds of organisms that can flourish at any particular place. We refer to the overall condition of the atmosphere over a short span of time, as described by a combination of its temperature, pressure, moisture content, and movement as the weather. **Weather** is, therefore, the momentary state of the atmosphere at a particular place, and the study of the physical aspects of this state and related events is called **meteorology.**

Climatology, on the other hand, describes weather patterns over time and space; thus **climate** is a composite of day-to-day weather conditions described in terms of both averages and variability. It is mainly climate that shapes the basic patterns of ecosystem structure and function and thus limits the types of agroecosystems that man can establish to replace natural systems.

On the broadest scale, meterologists and climatologists are concerned with the processes that determine seasonal patterns of temperature, precipitation, and other conditions over large geographical areas. We will examine basic meteorological processes on this broad scale only as preparation for a discussion of regional climates and the agricultural opportunities they govern. Several good texts, such as Trewartha (1954), provide more information about broadscale meteorological processes, and a detailed discussion of bioclimatology at the level of specific agroecosystems appears in Chapter 8.

PRINCIPAL METEOROLOGICAL PROCESSES

The principal meteorological processes that determine the distribution pattern of climates over the earth are **solar radiation, atmospheric circulation,** and the **mechanisms of precipitation.** Although they will be discussed separately, these processes are closely interrelated and act in concert to produce regional climates.

Solar Radiation

The sun is the primary source of energy for almost all the work done in the ecosphere. Solar radiation is spread over a broad spectrum of

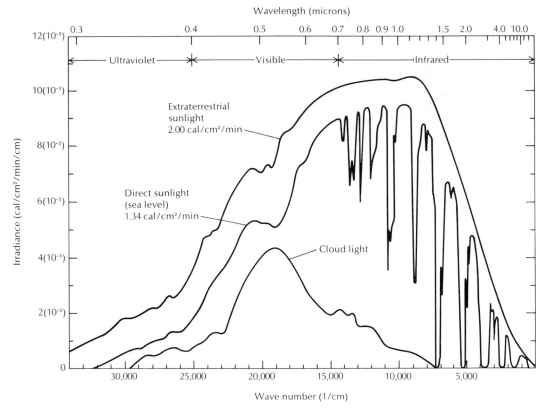

Wavelength (microns)

FIGURE 7-1.
Spectral distribution of solar radiation at the earth's surface under clear and cloudy conditions and on a horizontal surface outside the earth's atmosphere. (Modified from Gates 1965.)

ultraviolet, visible, and near infrared wavelengths (Figure 7-1). The total value of this input, measured outside the atmosphere in a plane perpendicular to the path of radiation, is approximately 2.00 cal/cm²/min.° This quantity varies little and is termed the **solar constant**. Most of this radiation occurs in wavelengths between 0.2 and 3.0 microns. Outside the atmosphere, about 7 percent of this energy

lies in the ultraviolet range (below 0.4 microns), about 50 percent is concentrated in the visible range (0.4 to 0.7 microns), and the rest falls in the near infrared portion of the spectrum.

As solar radiation passes through the atmosphere, certain portions of the spectrum are depleted through absorption by atmospheric constituents (Figure 7-1). Most of the ultraviolet radiation, for example, is absorbed by the ozone layer of the stratosphere and different wavelengths in the near infrared portion of the spectrum are selectively absorbed by water and molecular oxygen. Particulates such as dust and

°International usage now tends to favor the expression of energy flow as Watts (W) per square meter (1 Watt = 1 joule per second). W/m² = 698 × (cal/cm²/min).

smoke absorb both visible and infrared radiation and thus significantly reduce the total incident solar radiation that reaches the earth's surface. Under clear sky conditions at sea level, the solar radiation incident on a surface perpendicular to the incoming path averages no more than about 1.4 cal/cm^2/min—a loss of about 30 percent (Collier et al. 1973).

The amount of solar radiation incident on an actual land surface per unit time, termed **irradiance,** also depends on the angle of incidence of the radiation and the length of its path through the atmosphere. As the angle of incidence decreases, the incoming radiation is spread over a larger area of surface and less energy is delivered per unit area (Figure 7-2). Oblique rays also pass through a thicker layer of atmosphere and as a result are subject to more depletion by reflection and more scattering by atmospheric particles and water droplets. Because both day length variation and angle of incidence are the same for all places with the same latitude, they all potentially receive the same amount of solar energy. Insolation decreases as latitude increases, however, so that places at higher latitudes receive less solar energy than those at lower latitudes.

Seasonal changes in the distribution of solar energy over the earth's surface are caused by the tilt of the earth on its axis and by the earth's revolution around the sun (Figure 7-3). Throughout the year, the band of maximum insolation thus swings back and forth across the equator in relation to the angle of the sun's rays and changes in day length. This generates a series of global insolation curves that characterize the seasonal progression of insolation at different latitudes (Figure 7-4). The zone between 23.5°N and 23.5°S, referred to as the tropics, receives a nearly constant, high level of solar radiation with very little seasonal variation. However, even in the tropics there are two periods of maximum insolation that coincide with the passage of the sun directly overhead. In the middle latitudes (23.5° to 66.5°) there is a period of maximum insolation at the summer solstice (June 21 in the Northern Hemisphere) and a period of minimum insolation at the winter solstice (December 22). The greatest variation in insolation occurs at latitudes above 66.5° where insolation varies from maximum at the summer solstice to zero at the winter solstice.

Of the insolation that reaches the outer surface of the atmosphere, approximately 35 percent is reflected back into space by clouds and other components of the atmosphere and by the earth's surface. About 14 percent is absorbed by the atmosphere (mostly by water vapor), and the remaining 51 percent is absorbed by the earth's land and water surfaces. Most of the latter returns to the atmosphere either through shortwave radiation or as the latent energy of evaporated water.

In fact, more heat passes to the atmosphere from these secondary processes than from direct incoming solar radiation. However, the quantity of insolation that reaches the earth and becomes available to heat the bottom of the atmosphere is much more variable than the quantity that reaches the outer atmosphere. Clouds, for example, markedly reduce the

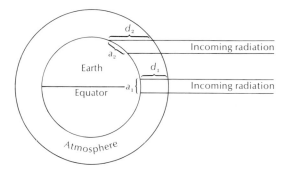

FIGURE 7-2.
Diagramatic representation of difference between the length of the radiation path $(d_1:d_2)$ and spread of radiation at the earth's surface $(a_1:a_2)$ in relation to latitude.

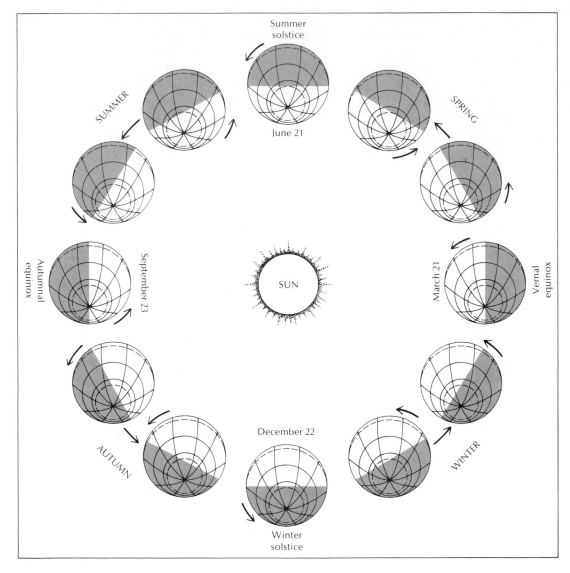

FIGURE 7-3.
Seasonal changes in the distribution of solar radiation over the earth's surface, caused by the tilt of the earth on its axis and its revolution around the sun. (Modified from Trewartha 1954.)

amount of solar energy that actually reaches the earth's surface (Figure 7-1). This variability in the insolation that strikes the surface exerts a major effect on climate. During a clear period, for example, the ground warms rapidly and forms a heat reservoir that transfers heat to the lower part of the atmosphere during a subsequent cloudy period.

The surface of the earth is thus immensely important to the overall scheme of energy exchange. The degree to which any area of the earth is effective in capturing and then reradiat-

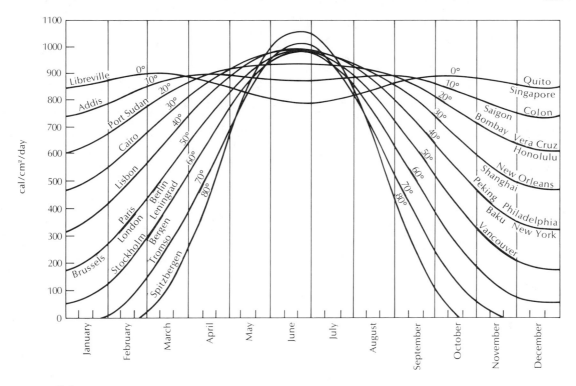

FIGURE 7-4.
Seasonal curves of daily solar radiation on horizontal surfaces at different latitudes. (Modified from Gates 1962.)

ing solar energy depends on the nature of the earth's surface in that area. For example, snow will reflect up to 90 percent of the insolation that strikes it, whereas black soil will reflect only about 10 percent. Water usually reflects more solar radiation than land, and land areas blanketed with vegetation heat and cool more slowly than barren areas do.

Atmospheric Circulation

The pressure of a gas is affected by temperature. Simply stated, warm air is less dense than colder air; warm air therefore rises and is replaced by descending colder air. Consequently, even if the surface of the earth were uniform, seasonal and latitudinal differences in solar in-

solation would create enough differences in the heating of the atmosphere to generate some large-scale patterns of atmospheric circulation. However, the earth's surface is divided unequally between water and land, the physical properties of which result in still more heating differences and thus constitute another influence on atmospheric movement. The specific heat of water (the energy required to raise one cubic centimeter 1°C) is about 2.5 times that of dry soil. Furthermore, the transfer of heat in water is largely by convection and turbulence, whereas in soil it occurs by molecular conduction. Consequently, both surface heating and cooling are much slower for water than for land, and both diurnal and seasonal fluctuations in temperature are less pronounced over large bodies of water than over land.

A large fraction of the solar energy that reaches the earth is consumed by the evaporation of water from the soil, from bodies of water, and from the foliage of plants. This energy is stored as **latent heat,** however, and is returned to the atmosphere whenever water vapor condenses. This produces a transfer of energy from ocean areas, where evaporation is high, to continental areas, where condensation is high.

Although the atmosphere itself tends to cool as its heat is radiated back into space, this heat loss is counteracted by a constant upward transfer of heat from the earth's surface and by the condensation of water vapor at lower levels. This upward transfer of energy is the major source of fuel for the atmospheric heat engine (Oort 1970). The schematic diagram of the average atmospheric energy cycle presented in Figure 7-5 reveals the importance of the latent heat of evaporation and the transfer of heat from the earth's surface.

When we consider the basic heat exchange properties of the earth and its atmosphere in terms of the latitudinal gradient of solar insolation, it becomes clear that nature's tendency toward equilibrium would create major energy transfers between equatorial and polar regions; this in turn creates large-scale atmospheric movements. The high level of heating around the equator causes the air there to expand and rise, thus forming a zone of low atmospheric pressure. The warm air that rises is replaced by cooler surface air that flows in from a polar direction, while the upper air flows toward the poles. On the other hand, because of the low level of solar insolation over the poles the air in those regions cools markedly by radiation back into space (back radiation). This cold air descends and flows toward the equator near the surface. Between equatorial and polar regions is another circulation system that includes a zone of descending cool air at around the thirtieth latitude and another of rising warm air at

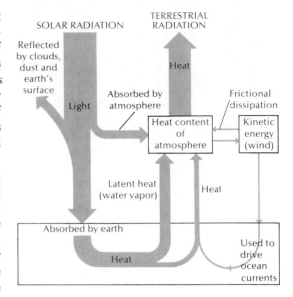

FIGURE 7-5.
Diagramatic representation of the relative importance of various energy flows involved in the transfer of heat at the earth's surface. The thickness of the arrows indicates the relative magnitude of the various energy flows and conversion rates. (From "The Energy Cycle of the Earth" by Abraham H. Oort. Copyright © 1970 by Scientific American, Inc. All rights reserved.)

around the sixtieth latitude. A series of atmospheric circulation cells is thus created (Figure 7-6).

These atmospheric cells create a series of latitudinal bands of vertical circulation that can be identified at the surface as regions of different atmospheric pressure. Normal atmospheric pressure, or the force applied to the earth's surface by a column of air that extends from sea level to the top of the atmosphere, is sufficient to support a column of mercury 29.92 inches or 760 millimeters high (1013 millibars). When the atmosphere expands and contracts as it heats and cools respectively, the pressure recorded at sea level varies from about 982 to 1033 millibars. The globe can therefore be divided into a series of horizontal bands of alternating high

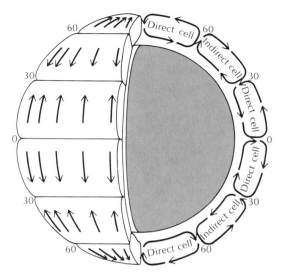

FIGURE 7-6.
The latitudinal arrangement of atmospheric circulation cells created by differences in the solar radiation at different latitudes. (From "The Energy Cycle of the Earth" by Abraham H. Oort. Copyright © 1970 by Scientific American, Inc. All rights reserved.)

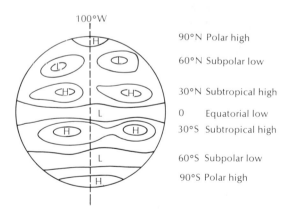

90°N	Polar high
60°N	Subpolar low
30°N	Subtropical high
0	Equatorial low
30°S	Subtropical high
60°S	Subpolar low
90°S	Polar high

FIGURE 7-7.
The distribution pattern of atmospheric pressure at the earth's surface, showing the influence of the large land masses of the Northern Hemisphere.

and low pressure (Figure 7-7). These bands in turn become broken up by pressure changes that correspond to the differences in heating of the air over the oceans and the continents. The global pattern of atmospheric pressure is thus more accurately described as a series of alternating high- and low-pressure cells, especially in the Northern Hemisphere, where the topography of the major land masses is a major influence.

Just as energy is transferred between bodies of unequal temperature, air moves between areas of unequal pressure: we refer to this as wind. The steeper the gradient between pressure cells, the more rapidly the air moves from the area of high pressure to the adjacent area of lower pressure. If the pressure gradient were the only determining factor, winds would always blow at right angles to imaginary lines (**isobars**) connecting areas of like pressure. However, the rotation of the earth creates a second force (**Coriolis force**), which deflects the flow and causes the wind to blow obliquely across the isobars. Thus winds are deflected to the right of a pressure gradient in the Northern Hemisphere and to the left in the Southern Hemisphere. North of the equator the air that flows out of a high pressure cell is deflected in a clockwise direction to form a divergent system called an **anticyclone.** Air flowing into a low pressure cell is deflected counterclockwise to form a convergent system called a **cyclone** (Figure 7-8).

The effect of the Coriolis force on atmospheric movement on a global scale is a series of latitudinal zones of **prevailing winds** (Figure 7-9). Although Figure 7-9 oversimplifies global circulation, this general pattern dominates the worldwide distribution of climatic conditions. Apart from the exchange of energy between air masses, the friction between moving air and the ocean surfaces creates major currents that redistribute the large quantity of energy absorbed by the tropical oceans. All interrelationships

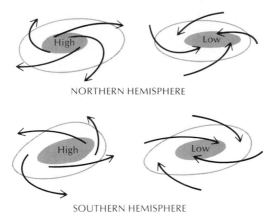

FIGURE 7-8.
The basic pattern of circulation associated with cells of
high and low pressure in the Northern and Southern
hemispheres.

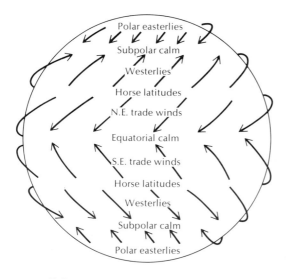

FIGURE 7-9.
The pattern of prevailing winds created by a combination
of the Coriolis force and global atmospheric circulation.

among temperature, pressure, and the movement of air masses affect not only the distribution of different temperature regimes and
wind systems, but also the global pattern of
water movements that forms the basis of the
hydrologic cycle (see Chapter 2).

Principal Mechanisms of Precipitation

The capacity of air to hold water vapor depends on its temperature and pressure. The
amount of water vapor in the air relative to the
total amount that the air can hold at a given
temperature is called the **relative humidity**
(100 percent relative humidity equals saturation). Temperature and relative humidity therefore determine the drying capacity or evaporative power of an air mass and greatly affect the
rates at which plants and animals lose moisture
and heat. As an air mass cools, its capacity to
hold moisture declines and relative humidity increases. Once the air becomes saturated it cannot cool further without losing some of its moisture. The temperature at which water vapor in
the air starts to condense is called the **dew
point.** The fine droplets of water that condense
on surfaces colder than the overlying saturated
air mass are called **dew** and can be an important
form of precipitation.

Clearly, then, water taken into unsaturated
air masses by evaporation is subsequently deposited as some form of precipitation (rain,
snow, sleet, dew, etc.) when the air mass cools
sufficiently. The cooling results primarily from
contact with a cold underlying surface, by contact or mixing with a colder air mass, or by what
is termed **adiabatic cooling,** that is, the change
in temperature that occurs when air rises and
expands. The rate of adiabatic cooling of unsaturated air is approximately 10°C for each
1000-meter increase in altitude. An air mass
that is warmer and lighter than the air surrounding it will tend to rise; as it rises it gradually
expands and cools. This will continue until it
reaches the same temperature and pressure as
the surrounding air. The cooling that results can
reduce the temperature of an extensive air mass
below its dew point, causing water vapor to
condense. Nearly all the precipitation that falls
on the earth results from the cooling of rising
air. There are three main mechanisms that
cause air masses to rise, and although there is

often an interplay among these mechanisms, we tend to label precipitation according to the dominant lift mechanism involved.

1. **Convectional precipitation.** Surface air heated by the underlying ocean or land mass expands and begins to rise through the cooler, heavier air around it. Because the rising air cools at almost twice the rate of the normal temperature drop associated with altitude (normal lapse rate), it can reach the same temperature as the surrounding air fairly quickly and stop rising. However, if the initial rise is great enough to depress the temperature below the dew point, the resulting condensation warms the air mass and causes it to rise still further. This form of convectional activity is responsible for thunderstorms. In some parts of the world convectional precipitation is the main source of moisture. It is often highly beneficial because it is usually associated with warm weather and so it coincides with the growing season of many crop plants. However, convectional precipitation often takes the form of heavy showers that can damage both the vegetation and the soil.

2. **Orographic precipitation.** Air masses are also forced to rise by topographic features, such as mountain ranges, that lie across their path. As the air is forced upward it cools and condensation occurs. Wherever relatively high mountains flank the western margin of a continent, as they do in North and South America, the moisture-laden air coming off the ocean produces heavy precipitation on the western slopes. On the eastern, or leeward, side of these mountains, however, there is often a low rainfall area, called a **rain shadow,** where the air descends and warms above its dew point (Figure 7-10). The precipitation associated with topographic features is often not purely orographic but results as well

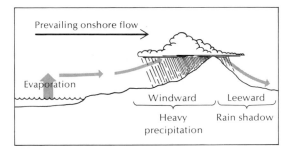

FIGURE 7-10.
Diagram of the precipitation pattern associated with an onshore flow of moisture laden maritime air intercepted by a mountain range.

from a complex interplay of convectional currents set in motion by the topography and the convergence of horizontal air currents.

3. **Frontal precipitation.** Whenever two large air masses converge, some upward movement is inevitable, and the associated cooling can cause precipitation. When the converging air masses are of similar temperature, as is often the case in the equatorial convergence zone, the leading edges of both masses are rapidly deflected upward and the precipitation is often heavy and of short duration. In the middle latitudes the temperatures of the converging air masses may be substantially different. When this occurs the warm, moister air rides up over the heavier cold air. Unlike convectional uplifting, the warm air rises obliquely and more slowly over the wedgelike leading edge of the cold air. This slower ascent and cooling often results in more gentle precipitation of longer duration (Figure 7-11).

The distribution of precipitation over the earth depends to a large extent on global atmospheric circulation. Rainfall in the tropics, for example, is mainly convectional and is caused by the convergence of the tradewinds in a zone known as the Doldrums, and the convection

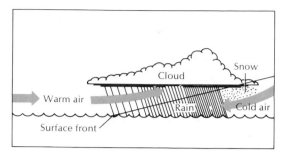

FIGURE 7-11.
Diagram of the mechanisms involved in frontal precipitation.

currents generated by the high level of solar heating. As the sun moves north or south the zones of intertropical convergence (ITC) shift as well, so that many areas near the equator have two wet seasons when the sun is directly overhead, and two drier seasons when the sun is inclined to the north or south. A little farther from the equator there tends to be a single wet season when the sun is overhead and a dry season when the sun is on the other side of the equator. However, warm continental land masses also tend to produce thermal low-pressure cells, and the flow of moisture-laden maritime air onto the land can produce a year-round pattern of adequate to heavy rainfall.

In the vicinity of 30° north and south the subsiding air associated with subtropical high-pressure cells is warm and dry; the world's major arid regions are therefore located in these zones. The land in this latitude is little influenced by the precipitation-producing intertropical convergence and the temperate frontal zones that lie to either side. The margin toward the equator is influenced by occasional tropical storms during the summer, and its other flank may receive some precipitation during winter from frontal storms that move farther south than usual. Most of the moisture falls along the coastal plains, however, and the inland areas, which may be extremely dry, are the great warm deserts of the world (the Sahara and

Kalahari in Africa, the Atacama in Chile, the Sonoran in Mexico, and the deserts of the Arabian Peninsula, the Middle East, and the outback of Australia).

The middle latitudes (35° to 60°) are greatly influenced by the westerly winds and the alternating high- and low-pressure cells that move from west to east around the globe. Cyclonic storms associated with low-pressure cells develop over the oceans and move onto the continents. The contrasting air masses of adjacent high- and low-pressure cells generate a considerable exchange of energy and precipitation along fronts that are swept eastward by an undulating circulation in the Northern Hemisphere or by the more regular zonal circulation in the Southern Hemisphere.

The global atmospheric circulation and its attendant zones of ascending and descending air masses produces a series of generalized precipitation zones illustrated in Figure 7-12. The interrelationships of the oceans and continents, however, cause a variety of nonzonal precipitation patterns. For example, the extreme low pressures that develop over the northern land masses with cooler oceans on their southern flanks produce convectional systems of a gigantic scale. These systems, called monsoons, produce a regional reversal of the prevailing winds and generate heavy rains during the summer months.

EFFECTS OF MAJOR CLIMATE TYPES ON ECOSYSTEM STRUCTURE AND FUNCTION

Having examined the principal elements of climate (i.e., the basic distribution of solar radiation, global patterns of atmospheric circulation, and the main causes of precipitation), we can now consider briefly the geographical distribution of the world's major climatic regimes and their effects on ecosystems. No simple classification of world climates can be suited

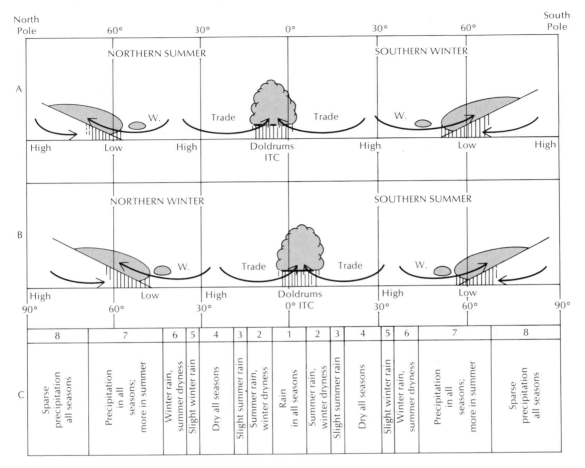

FIGURE 7-12.
Schematic cross section through the atmosphere showing the major zones of ascending and descending air at different seasons: (A) pattern during the northern summer; (B) pattern during the northern winter; (C) zones of precipitation. (From Petterssen 1941, *Introduction to Meteorology*. Copyright © 1941. Used with permission of McGraw-Hill Book Company.)

ideally to all purposes. Since climate is a combination of weather types, and weather is composed of many individual elements, no two places can have truly identical climates that share certain major characteristics that control the composition of vegetation and therefore the structure of ecosystems. In fact, some classifications use vegetational transitions to establish the boundaries between different climates. As one would expect, however, climates are characterized mainly by mean monthly temperatures and the quantity and distribution of precipitation. One such classification and the principal characteristics of each climate type are presented in Table 7-1; the distribution of these climatic types over the globe is shown in Figure 7-13. Obviously, general kinds of plants and animals that have become adapted to specific physical environments will prevail in specific climates. Nevertheless each area under the influence of

TABLE 7-1
Classification and characteristics of climate.

Groups of climate	Types of climate	Pressure system and wind belt		Precipitation
		Summer	Winter	
A. Tropical rainy	Af (Am), tropical wet	ITC, doldrums, equatorial westerlies	ITC, doldrums, equatorial westerlies	No dry season
	Aw, tropical wet and dry	ITC, doldrums, equatorial westerlies	Trades	High-sun wet (zenithal rains) low-sun dry
B. Dry	BS, semiarid (steppe)			
	BSh, tropical and subtropical	Subtropical high and dry trades	Subtropical high and dry trades	Short moist season
	BSk, middle latitude		Continental winter anticyclone	Meager rainfall, most in summer
	BW, arid (desert)			
	BWh, tropical and subtropical	Subtropical high and dry trades	Subtropical high and dry trades	Constantly dry
	BWk, middle latitude		Continental winter anticyclone	Constantly dry
C. Humid mesothermal	Cs, dry summer subtropical	Subtropical high (stable east side)	Westerlies	Summer drought, winter rain
	Ca, humid subtropical	Subtropical high (unstable west side)	Westerlies	Rain in all seasons

Climate type			
Cb, Cc, marine climate	Westerlies	Westerlies	Rain in all seasons, accent on winter
D. Humid microthermal — Da, humid continental, warm summer	Westerlies	Westerlies	Rain in all seasons, accent on summer; winter snow cover
Db, humid continental, cool summer	Westerlies	Westerlies and winter anticyclone	Rain in all seasons, accent on summer; long winter snow cover
Dc (Dd) subarctic	Westerlies	Winter anticyclone and polar winds	Meager precipitation throughout year
E. Polar — ET, tundra	Polar winds	Polar winds	Meager precipitation throughout year
EF, ice cap	Polar winds	Polar winds	Meager precipitation throughout year
H. Undifferentiated highlands			

Key:

A = temperature of coolest month over 18°C (64.4°F)
B = evaporation exceeds precipitation
C = coldest month between 18°C (64.4°F) and 0°C (32°F)
D = temperature of coldest month under 32°F (0°C); warmest month over 10°C (50°F)
E = temperature of warmest month under 10°C (50°F)
a = warmest month over 22°C (71.6°F)
b = warmest month below 22°C (71.6°F)
c = warmest month below 22°C (71.6°F); less than four months above 10°C (50°F)
d = coldest month below −38°C (−36.4°F)

With A climates:

f = no dry season; driest month over 6 cm (2.4 in.)
s = dry period at high sun or summer; rare in A climates
w = dry period at low sun or winter; driest month under 6 cm (2.4 in.)

With C and D climates:

f = no dry season; difference between rainiest and driest months less than in s and w; driest month of summer over 3 cm (1.2 in.)
s = summer dry; at least 3 times as much rain in wettest month of winter as in driest month of summer; driest month less than 3 cm (1.2 in.)
w = winter dry; at least 10 times as much rain in wettest month of summer as in driest month of winter

Source: From S. T. Trewartha 1954. *An Introduction to Climate.* Copyright © 1954 by McGraw-Hill, Inc. Used with permission of McGraw-Hill Book Company.

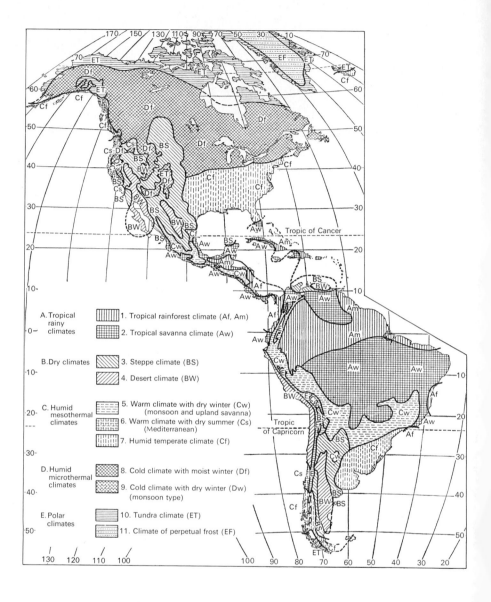

A. Tropical rainy climates
1. Tropical rainforest climate (Af, Am)
2. Tropical savanna climate (Aw)

B. Dry climates
3. Steppe climate (BS)
4. Desert climate (BW)

C. Humid mesothermal climates
5. Warm climate with dry winter (Cw) (monsoon and upland savanna)
6. Warm climate with dry summer (Cs) (Mediterranean)
7. Humid temperate climate (Cf)

D. Humid microthermal climates
8. Cold climate with moist winter (Df)
9. Cold climate with dry winter (Dw) (monsoon type)

E. Polar climates
10. Tundra climate (ET)
11. Climate of perpetual frost (EF)

a particular climate will have its own particular assemblage of organisms based on its biogeographical history. This too is an oversimplification, of course, because climate is superimposed upon a mosaic of soil types with different fertilities. Although it is true that climate has a profound effect on the development of soils and on soil fertility, all soils begin with parent materials the distribution of which is independent of climate.

Even though great differences in the distribution of solar energy occur from the tropics to polar regions and insolation varies considerably in relation to atmospheric conditions and local topography, it is normally adequate for photosynthesis and plant growth. Lack of solar radia-

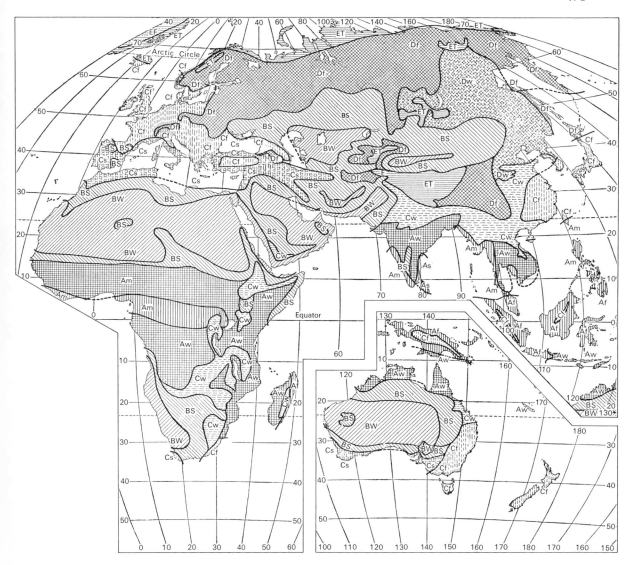

FIGURE 7-13.
The major climates of the earth. (From S. T. Trewartha 1954, *An Introduction to Climate.* Copyright © 1954 by McGraw-Hill, Inc. Used with permission of McGraw-Hill Book Company.)

tion seldom limits the productivity of terrestrial ecosystems. Productivity depends more on the efficiency with which solar energy is captured, and this is a function of the growth form of the vegetation, which in turn is controlled largely by temperature and moisture. Light usually only becomes deficient when an area becomes overstocked with plants. In natural ecosystems plant structure has evolved to maximize the capture of solar energy in relation to the growth and

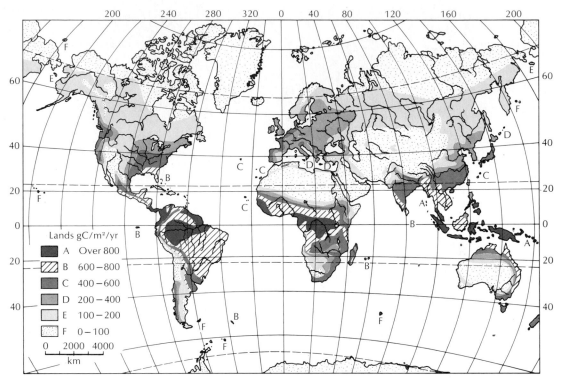

FIGURE 7-14.
Worldwide pattern of net primary production in grams of carbon per square meter per year based on approximate values. (Modified from Reichle 1970.)

productivity that can occur under the local constraints of temperature and moisture. Because photosynthesis is a chemical process that fixes or incorporates carbon into organic compounds, the amount of carbon fixed per unit area each year serves as a measure of net production. Figure 7-14 shows the global distribution of annual net primary production. When this distribution is compared with the distribution of global climates some relationships appear. If we consider the climates under which various ecosystems have developed, we can develop some generalizations about climatic effects on vegetation; for example, moist areas are generally more productive than dry areas. The net primary production for major terrestrial ecosystems is presented in Figure 7-15.

Water is of course essential for life itself, but it is also important to plant growth as the carrier of nutrients through the conduction system into the plant tissue. Many desert areas have soils rich in nutrients and temperatures highly favorable to plant growth, but the lack of water isolates the nutrients from the plant tissue. On the other hand, too much water at one time tends to flush vital nutrients from the upper soil so that they become inaccessible to the roots. Most low-lying wetlands are highly productive because they are often catch-basins for nutrients carried in from surrounding uplands; in this sense they are subsidized systems.

Temperature governs the rate of all biochemical processes, with growth proceeding more quickly at high temperatures than at low ones.

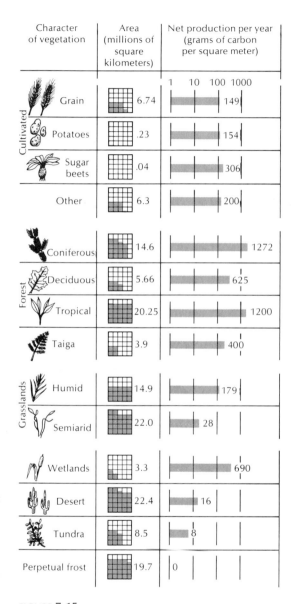

Character of vegetation		Area (millions of square kilometers)	Net production per year (grams of carbon per square meter)
Cultivated	Grain	6.74	149
	Potatoes	.23	154
	Sugar beets	.04	306
	Other	6.3	200
Forest	Coniferous	14.6	1272
	Deciduous	5.66	625
	Tropical	20.25	1200
	Taiga	3.9	400
Grasslands	Humid	14.9	179
	Semiarid	22.0	28
	Wetlands	3.3	690
	Desert	22.4	16
	Tundra	8.5	8
	Perpetual frost	19.7	0

FIGURE 7-15.
The net primary production of major terrestrial ecosystems in grams of carbon per square meter per year. (From "The Human Population" by Edward S. Deevey, Jr. Copyright © 1960 by Scientific American, Inc. All rights reserved.)

Thus the length of the growing season alone does not determine productivity. In the subarctic, for example, the period of time suitable for plant growth is short, but during that time temperatures can be quite high and the photoperiod is long, so that growth occurs rapidly. Net primary production per unit of time under these conditions can be fairly high.

Temperature also affects the rate of decomposition and thus the nutrient flux rate. In cool climates, such as those of northern climax coniferous forests, there may be a substantial build-up of organic debris because it accumulates more rapidly than it decomposes. Like the high-standing biomass of these systems, the undecomposed litter forms a nutrient bank for use by later successional communities that develop after some form of disturbance. Although the foregoing examples indicate the importance of precipitation or temperature alone to some basic ecosystem functions, their influence cannot be separated. Moisture is also important to the process of decomposition and nutrient cycling. Decomposition proceeds more rapidly in damp than in dry soil, but saturated soil creates anaerobic conditions that retard decomposition. Similarly, temperature influences evaporation, so that what might be an adequate moisture for plant growth in a cool environment may be inadequate in a warm one.

CLIMATE AND AGRICULTURE

In agriculture there are many criteria by which we can characterize climate, including length of the growing season, number of frost-free days, and effective versus absolute precipitation. All these features, however, relate to the fundamental climatic conditions imposed on a region by insolation, atmospheric circulation, and precipitation. Climate clearly imposes some very definite limitations not only on natural ecosystems but on the structure and function of agro-ecosystems as well.

TABLE 7-2
Summary of the major climates of the world suitable for agriculture, and the predominant forms of native vegetation supported.

Climate group	Climate type	Temperature	Precipitation	Typical vegetation
Tropical rainy	Tropical wet	Warm year-round	No dry season	Tropical rain forest
	Tropical wet and dry	Warm year-round	High sun wet, low sun dry	Savanna and dry forest
Dry	Midlatitude Steppe	Warm summers, cold winters	Meager, moist in summer	Short grass and bunch grass
Humid mesothermal	Dry summer Subtropical	Hot summers, warm winters	Summer drought, winter rain	Sclerophyllic woody shrubs
	Humid subtropical	Warm summers, mild winters	Rain year-round	Mixed forest, tall grass
	Marine	Cool summers, mild winters	Rain all seasons but most in winter	Coniferous to deciduous forest
Humid microthermal	Humid continental, warm summer	Warm to hot summers with mild periods, cold winters	Summer rains prevail, some regions sub-humid	Deciduous hardwood forest
	Humid continental, cool summer	Warm to hot summers, cool to cold winters	Total rainfall variable, summer usually wet	Forest and tall grass

As we pointed out in Chapter 2, agricultural land has been claimed from a variety of natural ecosystems. Likewise, domesticated plants and animals have been selected for cultivation from the vast assemblage of wild species, and like their wild relatives crop plants, and to a lesser extent domestic animals, have specific climatic requirements. In fact, the origin of the cultivation of some crops can readily be traced to areas with particular climatic characteristics. In other words, man recognized early in his history as an agriculturist the relationship between climate and vegetation and selected species for domestication accordingly. Although modern agriculture has introduced some crops into areas that have been cleared of very dissimilar ecosystems (e.g., corn in areas cleared of deciduous forest or rice in areas cleared of tropical forest), we can still see relationships between crop type, native vegetation, and regional climate throughout the world.

Of the major climates summarized in Table 7-1 the only climates suitable for agriculture without a high level of irrigation are the tropical rainy, mid-latitude dry, humid mesothermal, and warmer humid microthermal. Table 7-2 shows the major chracteristics of these climates and the form of the associated dominant natural vegetation associated with them.

Tropical Rainy Climates

The tropical wet climate is characterized by heavy precipitation distributed throughout the year and by uniformly high temperatures. Such conditions foster the growth of the world's most luxuriant vegetation known as tropical rain forests. As we have noted, rapid plant growth and high levels of leaching by the heavy tropical rains produce generally infertile forest soils, and shifting native agriculture within these forest areas reflects the problems of maintaining soil fertility. The principal centers of crop production are in areas where there are young unleached soils, often on well-drained uplands or on deltas and floodplains where the soils are rejuvenated by nutrients carried in from upstream by normal river flow or periodic flooding. The forest soils are best suited to crops, such as banana and rubber, that have growth characteristics similar to those of the native vegetation. Rice is the most important crop grown in the more fertile soils, and it is cultivated on terraced slopes as well as on the deltas and floodplains.

The tropical wet and dry climate is quite different from the tropical wet climate in that it usually has less precipitation and the rainfall is distributed unevenly throughout the year so that there is a distinct dry season. This type of climate occupies a zone of transition between constantly wet and constantly dry climates so that there is considerable variation between the amount of rainfall on its two margins; thus the native vegetation ranges from tropical deciduous forest on the wet side, through moist tree savanna with tall grasses in the middle, to a dry tree savanna with short grasses on the dry side. Under a tropical wet and dry climate the soil often forms a hard surface crust that impedes tillage. As in tropical wet climates, floodplains and deltas are the most attractive sites for cultivation. In the drier regions there is little natural hay provided by the brittle dead grass, making some of the short-grass savanna poor even for grazing during the dry season. Nevertheless, a number of important crop plants have been selected from among the natural vegetation that evolved under tropical wet and dry climates. Coffee (Figure 7-16) is an important cash crop that needs a warm wet season for berry growth and a cooler dry season for ripening. Many perennials form large tuberous roots as an adaptation to the dry season, and some of these, such as yams, sweet potato, taro, and cassava, are cultivated as annual crops. Sugarcane is also an important crop in areas with this climate: luxuriant growth occurs during the wet season while the dry season induces a buildup in the sugar content and accommodates harvest. Corn, another tall grass, can also be highly productive under a tropical wet and dry climate regime, and its cultivation in the tropics is spreading rapidly.

Dry Climates

The most important characteristic of dry climates is that the potential loss of moisture through evaporation from the soil surface and vegetation exceeds the average annual precipitation. Because this moisture loss is much greater in warm regions than in cool ones, climates cannot be defined by the amount of precipitation they receive. The steppes of the middle latitudes are the only important, unirrigated agricultural regions dominated by a dry climate. Middle-latitude steppes occupy transitional zones between deserts and humid climates. Precipitation is low, and although it is distributed throughout the year it is heaviest during the summer. Even so, rainfall is unreliable, and the strong winds and higher temperatures that also occur during the summer create a high level of evapotranspiration.

The natural vegetation of the steppes is dominated by short, shallow-rooted grasses mixed with widely spaced bunch grasses. The Great Plains region of the United States is a good ex-

FIGURE 7-16.
A coffee plantation in El Salvador showing the shade provided by a closed canopy of
deciduous trees. The overall appearance of the plantation is that of a tropical deciduous
forest. (Photograph courtesy of the USDA.)

ample of a high-quality, short-grass ecosystem, and it forms one of the best and most extensive natural grazing areas in the world. The short grasses return a reasonable amount of humus to the soil and reduce the extent to which small clay and silt particles are removed by the wind. In fact, areas covered by short grasses often receive fine soil particles blown in from less protected desert areas. The soil is therefore highly productive whenever there is adequate rainfall.

The most important crops in these areas are a variety of small grains, some of the ancestors of which can be found among the wild grasses of steppes. Spring wheat is by far the most widely cultivated of these grains; planted in the spring, it grows slowly until the warm summer temperatures coincide with the summer rains, when it fills out rapidly. It ripens in the dry warmth of the early fall and is ready for harvest in the nor-

mally dry autumn. Where water is available for irrigation the climate and soil of the middle-latitude steppes will produce a bountiful harvest of other crops such as sugar beets and corn.

Humid Mesothermal Climates

The humid mesothermal climates are characterized by seasonal changes in temperature that make temperature as great an influence as rainfall on the nature of the vegetation. Mesothermal climates have relatively mild winters and are thus restricted to the lower latitude, continental areas or on coastal strips along the western sides of the continents. Three distinct subclimates are recognized within this climatic type: the dry-summer subtropical or Mediterranean climate, the humid subtropical climate, and the marine climate.

FIGURE 7-17.
Groves of citrus and avocado carved from the typical Mediterranean scrub vegetation that is common to southern California.

Mediterranean climates receive a modest amount of precipitation in the winter and have warm-to-hot, dry summers. The native vegetation consists mainly of stunted trees and woody shrubs. Because the periods of rainfall do not coincide with high temperatures, plants, such as grasses, with superficial root systems do not do well; most wild grass growth follows the winter rains. It is therefore not surprising that these climates are well suited to the cultivation of winter grains such as barley. Some deep-rooted summer perennials, such as olive and grape, also do well, and with some irrigation, the Mediterranean climate supports the cultivation of citrus and avocado (Figure 7-17). Young alluvial soils that have collected in coastal basins and mountain valleys, as well as on floodplains and deltas, are highly suitable for the cultivation of a variety of truck crops.

The humid subtropical climate, typified by abundant precipitation either distributed throughout the year or concentrated during the warm season, is ideal for agriculture. Native vegetation is generally well developed; forests predominate, but in areas of moderate rainfall, or where moisture is retained near the surface, tall grass prairies may predominate instead. The grassland soils are generally the most productive because they are subject to less leaching, but areas cleared of deciduous forest produce high yields when fertilizers are added. Extensive areas within regions of subtropical climate are devoted to specialty nonfood crops such as tobacco and cotton. However, the ample summer precipitation, the hot summer days, and the fairly long growing season are ideal for the cultivation of corn, sorghum, soybeans, peanuts, and a variety of other fruits, nuts, and vegetables.

Areas where a humid marine climate prevails generally lie poleward of the Mediterranean climate. The influence of the ocean keeps the summers cool and the winters mild. Precipitation occurs throughout the year but is often more abundant in winter than summer. The cooler summer temperatures and moist maritime air reduce evapotranspiration making drought uncommon. Depending on the specific location, the natural vegetation is dominated by deciduous forest, coniferous forest, heaths, or moor.

Marine climate areas are well suited to permanent pasture and the cultivation of hay and forage crops to support a dairy industry. The most widely grown stable crop of this climate type is probably the potato, but a variety of vegetables, legumes, soft fruits, and orchard crops also do very well.

Humid Microthermal Climates

The humid microthermal climates include the humid continental, warm-summer and cool-summer climates, each of which is important from an agricultural standpoint, plus the subarctic climate, which is agriculturally unimportant. Each of the humid continental climates has a distinct winter with snow. Temperature influences plant growth more than moisture does. As their names suggest, these climates are distinguished by the difference in their summer temperatures.

The natural vegetation of humid microthermal climates ranges from forest in the more humid areas to tall-grass prairie in the subhumid areas. In the virgin state, these prairies provide some of the world's best natural rangeland; they usually have fertile dark soils and are generally too valuable as cropland to be used for grazing. The forests in these climates range from coniferous at higher latitudes where the summers are cool to mainly deciduous at lower latitudes where the summers are warm. Similarly, there is a north-south gradient of soil fertility that ranges from poor podzols (see page 238) at higher latitudes to considerably more fertile soils where the deciduous forests grade into tall-grass prairie.

The warm moist summers and fertile soil of the tall-grass prairies and deciduous forests make them highly suitable for corn and soybean production. Sugar beets are a valuable crop in the transition zone between the warm-summer and cool-summer humid climates. Spring wheat dominates agriculture in the subhumid areas; forage crops such as clover and alfalfa, and small grains such as barley, oats, and rye are best suited to the cool-summer regions with poorer soils.

Although potentially highly productive agroecosystems have been developed under humid microthermal climates, they are the systems most subject to climatic variability. These climates more than any other involve constant conflicts between dissimilar air masses, particularly in North America where the major mountain ranges run north and south and thus permit the free movement of cold dry air masses from the north and warm moist air masses from the south. Violent weather systems develop along the frontal contact between these air masses and often cause severe crop damage.

An Overview

By examining the pattern of distribution of major climates we can predict the types of agroecosystems that can be supported best and with the least intervention in a specific area. When we look at agriculture worldwide, however, the most logical predictions are not confirmed: Crops are not matched up with climates the way natural vegetation and climate are matched. This departure is most pronounced in the developed nations, as one might expect. Crops for which there is a great demand are cultivated in areas where they can be grown productively only with subsidies of energy, nutrients, and water. Furthermore, plant breeders have developed new crop varieties that can be cultivated under climates very different from those that molded the evolution of the ancestral stock. Various strains of wheat, for example, are now cultivated from the tropics to the edge of the subarctic.

Many of the crop varieties grown today lack some of the traits, common to wild species, that tend to buffer the impact of climatic variability on productivity over a long period of time. The yields of some species, like corn, fluctuate widely in response to changes from year to year. Although such crops are generally suited to cultivation over a large area with a similar climate, local weather variations have different effects on yield. For example, higher-than average temperatures in August favor higher corn yields in Oklahoma, Kansas, and Nebraska, whereas rain in June is more important in Texas; in Oklahoma and Kansas July and August rains, respectively, are most beneficial to increased yield (Thompson 1964). Second, despite what would appear to be ample opportunity to match highly productive crop varieties with suitable climatic characteristics, there has been a tendency to expand the area of cultivation of a relatively few popular (and profitable) food commodities. As a result some crops are now grown in areas with unsuitable natural climates, and new agricultural areas are being carved from natural ecosystems that are better suited to the conditions there. Two areas being invaded by agriculture are the tropics and warm deserts where the temperature and long growing season offer possibilities for high productivity.

Because large areas within the tropics are the most productive on earth (see Figure 7-15), they are often thought to be able to support highly productive agroecosystems. But tropical forests are characterized by a rapid turnover of nutrients; organic matter decomposes rapidly, and the indigenous plants are adapted to use the released nutrients quickly before they are washed from the soil by the heavy rains. A few crop plants like coffee, rubber, and bananas, which are related to the local vegetation, do quite well; but these crops are not useful as staple foods, and removing the local vegetation and introducing a field crop can have disastrous results (see Chapter 11).

Warm deserts can become highly productive if they are irrigated; but modifying the climate in this way can be both costly and ecologically destructive. To grow one ton of corn under irrigation requires about 1000 tons of water;

one ton of cotton fiber requires approximately 10,000 tons of water (Revelle 1963). Such large-scale applications of irrigation water can profoundly affect the hydrologic cycle. Much of the irrigation water leaves the crop by evaporation from the leaves and soil, and although it eventually returns to the earth as precipitation it may fall on a totally different area.

We cannot leave this general discussion of climate and agriculture without mentioning, at least briefly, the effects of climate on the raising of livestock. Obviously the climate exercises a major influence on the quantity and quality of livestock feed, expecially on pasture and range vegetation. But we will restrict this discussion to the direct effects of climate on the animals themselves.

Because domestic animals and poultry are homoiotherms (warm-blooded animals), they are most sensitive to climatic effects on their body temperature. For most species the optimal ratio between input (food) and output (weight gain, milk, or eggs) is achieved when the air temperature is between 13° and 18°C, the relative humidity between 55 and 65 percent, the wind velocity between 5 to 8 kilometers per hour, and when there is a medium amount of sunshine (Bayley 1974). Photoperiod may be important as a regulatory mechanism in breeding cycles, but temperature exerts an overriding effect on most physiological functions. Between 10° and 20°C, generally called the comfort zone, there is usually no discernible change in body processes. Outside the comfort zone temperature effects can become significant, and those of other components of weather can increase. Relative humidity, precipitation, wind velocity, and solar radiation can all be either beneficial or detrimental, depending on whether the ambient temperature drops below or rises above the comfort zone. High humidity can inhibit an animal's ability to cool by the evaporation of sweat, whereas rain during a hot period will increase evaporative cooling. At low temperatures wind increases the chill factor by further dissipating body heat, whereas at high temperatures it helps an animal to stay cool. Solar radiation warms both an animal and its environment, so it can help to maintain thermal balance when the air is cold; but it can substantially increase the heat load when the air is hot.

Animals, like plants, are adapted to the climatic conditions of their native range, but in the course of evolution priority has been placed upon reproduction, survival, and comfort instead of productivity. Man has therefore put great emphasis on selecting the most productive varieties for specific types of climate. For example, animals with pigmented skin, which prevents sunburn, and lightcolored coats, which reflect solar radiation, are less likely to be put under stress by high levels of insolation in warm areas. Productivity generally declines as temperatures move outside of the comfort zone. Milk production drops measurably when the temperature falls below 6°C or rises above 26°C; these figures vary somewhat with the breed of cattle. Growth and weight gain are also affected by temperatures outside the comfort zone. Most animals require more feed for maintenance under cold temperatures, and they eat less and gain less weight at high temperatures. Thus climate is important to animal production because it influences both the efficiency of food utilization with respect to milk and egg production and the time required to attain desired weight levels.

Climatic Variability

The idea of a climatic average enables us to describe the temperature, sunlight, moisture, wind, and other features of a region's physical environment in a rather general way. We expect therefore that different elements of the climate will vary within and between seasons. Most of these variations have little effect on the persistence of local native flora and fauna, although even in natural communities we sometimes see catastrophic effects of weather. But the relationshiop of natural vegetation and the animals that rely on it for food and shelter is one of long-term homeostasis; that is, a decline in the productivity of an ecosystem due to weather in one year will be compensated for by increased productivity in other years. The natural variability in climate also affects the yield of crops. But man's agricultural strategy depends on maximum yield each and every year, and when climate reduces productivity, the consequences are more serious in agroecosystems than in natural ecosystems. Although the dependence of crop growth and development on weather is generally recognized, it has been difficult to determine the influence of specific factors because of the complex interactions among them. Furthermore, the weather pattern of the previous year can affect yield as much as current weather. Generally speaking, specific weather elements are more dependable towards the middle of a climatic zone than they are along the periphery, and for some crops the dependability of a specific factor may be very important. For example, the dependable August rainfall in the Soybean Belt has been suggested as the main reason that the United States produces two-thirds of the world's soybean crop (Thompson 1963). In recent years different grain-growing areas of the world have experienced marked weather-induced fluctuations in production; as a result the large grain surpluses that existed a few years ago have been consumed.

The degree to which the yields of specific crops vary with variations in climate depends partly on where they are grown compared to where they originated and on the climatic tolerance they have developed. For example, many tropical crops, such as bananas, do not suffer greatly from weather variations because they simply cannot be grown successfully outside the fairly dependable climates under which they evolved. Other crops, such as tomatoes, that were originally adapted to the warm days, cool nights, and adequate moisture of high elevation tropical areas, are now grown in a variety of temperate climates and they suffer substantially reduced yields whenever the climate is suboptimal at a critical time.

In order to increase yield and to expand the acreage on which important or high value crops are grown, man often attempts to modify the climate, particularly by making its most important elements more dependable. The most common example of course is irrigation, since water shortages limit productivity more often than adverse temperatures or a lack or excess of sunlight. Attempts are also made to modify temperature, however, by mulching with either dark or reflective materials and by using wind breaks, smudge pots and livestock shelters. The ultimate climate modification, of course, occurs in greenhouses and other enclosed agricultural systems where crops are grown in rigorously controlled environments.

Went (1967) describes detailed studies of the climatic requirements of specific crop plants, an understanding of which is essential if we are to base our cultivation of more plant species on knowledge instead of trial and error. Such studies will also stimulate even greater efforts to modify weather by large-scale means such as cloud seeding. However, it is necessary to realize that weather is a continuous phenomenon that cannot be changed in one place without affecting the climate in adjacent areas. In desert areas large-scale irrigation increases the relative humidity of the air, making adjacent areas less comfortable for people and livestock. This is particularly noticeable in the Imperial Valley of California and near Phoenix, Arizona. Increased atmospheric moisture can also increase cloudiness and induce atmospheric instability and harsh convectional precipitation. There have already been claims of flood damage due to cloud seeding (Halacy 1968).

Over the past few years meteorologists have noted a gradual deterioration of the climate, at least in the Northern Hemisphere. After about 75 years of moderate climate there now seems

to be a cooling trend, and indications are that we are about to enter a period of greater climatic variability. Reports indicate that between 1945 and 1968 there was a one-half-degree drop in the temperature of the lowest atmospheric layers in the Northern Hemisphere. This may seem like an insignificant decrease, but it may have shifted the major patterns of atmospheric circulation enough to increase local weather extremes. Recent prolonged droughts, extended cold spells, flooding, and delayed summer precipitation may all be results of such a shift, and they have a pronounced impact on agricultural productivity. If we bear in mind that many of our more productive crop varieties were developed and established during a period of climatic stability, the impact of greater variability in weather could be significant.

There is some controversy over what has caused the observed decline in temperature (Kellogg and Schneider 1974). Some meteorologists see it as part of the normal long-term pattern of warming and cooling that has characterized interglacial periods. Some environmentalists contend that the change is related to the effects of industrialization on the chemical composition of the atmosphere. Others attach considerable significance to the fact that man has drastically changed 20 percent of the earth's land surface—an amount that may have changed the earth's albedo (Kukla and Kukla 1974). Probably all these factors are involved, but over a long term atmospheric pollution should have a warming effect and therefore temper any interglacial cooling (Thompson 1975).

Most important to our discussion here is the impact that the current cooling trend will have on the climate of major agricultural regions and on agricultural productivity at a time when there is a growing demand for staples. As we mentioned earlier, a relatively small decline in the average temperature changes the earth's major patterns of atmospheric circulation.

When a band of circulation that includes pressure cells and their frontal systems shifts, the areas influenced by the old and new positions of what is often called the storm track experience more variable weather. In recent years we have seen a protracted drought in the sub-Sahara region of Africa, a failure of the monsoons in India and Southeast Asia, and a significant increase in the variability of weather in the middle latitudes. In 1976 alone there was a considerable increase in cloudiness and precipitation in the Pacific Northwest and in southwestern British Columbia, a drought in California, a continuation of the drought in the upper Great Plains, a serious drought in European Common Market countries, and drier than normal conditions in the grain-growing regions of the Soviet Union. The impact of these climatic conditions on agricultural production and world food supply can be significant. Crop failures in Europe have been widespread. Rangelands in South Dakota have been so seriously affected by drought that it will take several years of normal precipitation and reduced grazing to restore their normal carrying capacity. In 1976 California suffered agricultural losses amounting to over 450 million dollars. Decreased grain yields in the Soviet Union prior to the bumper crop of 1976 forced that nation back into the world grain market as a major buyer.

The extent to which continued climatic variability will affect the world's agricultural productivity is extremely difficult to forecast. Productivity is influenced by the extent of the crop area, the varieties grown, and cultural technology. Improvements in yield attributable to these factors would partially offset declines caused by climatic viability. The Institute of Ecology (1976) evaluated the impact of climatic variation on wheat production in Canda and the United States and on corn, sorghum, and soybean production in the United States. The evaluation was based on scenarios assuming the 1975 crop area and on 1973 technology (and presumably crop varieties) and used climatic data for the periods 1933–1936, 1953–1955, 1961–1963, and 1971–1975. The results indicate that crop production would have varied substantially. For example, if 1936 weather recurred, the United States could lose 71 million metric tons, or 27 percent, of the 1975 production of corn, wheat, sorghum, and soybeans. The graphs generated by the studies for wheat in Canada and corn in the United States are presented in Figure 7-18 as further examples.

Given a prospect of several decades of climatic instability, many methods have been suggested for favorably modifying the world's climate (Figure 7-19); but there are a number of national and international consequences that need to be weighed very carefully (Kellogg and Schneider 1974). Certainly there are less hazardous alternatives that could be explored first. Newman and Pickett (1974) have pointed out that many of the areas most subject to climatic variations (i.e., high risk areas) usually border areas that are generally favorable to agriculture. As one might expect, monocultures often extend into these areas, where they are subject to large variations in yield. Newman and Pickett suggest that it should be possible to develop an agricultural technology that would allow rapid shifts in crop production patterns to improve productivity in favorable areas and to stabilize productivity in high risk areas. Among other aspects of such a program would be to develop crop varieties that give good yields in the risky transition zones between wet and dry areas. Most of the arable land that could be brought into production lies in areas subject to considerable climatic variation and to drought, and the development of crop varieties suited to such conditions would not only permit an expansion of agriculture in some regions, but would help to soften the impact of climatic deterioration.

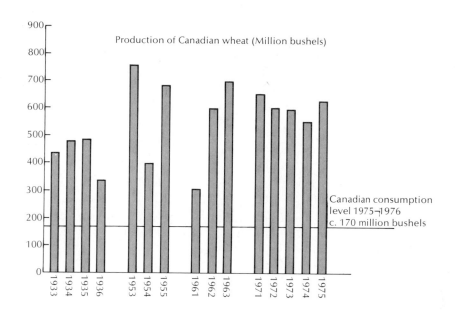

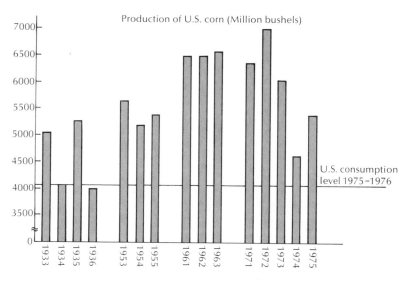

FIGURE 7-18.
Scenario production of (A) Canadian wheat and (B) United States corn showing the possible impact of climatic variability on food production. See text for details. (From Institute of Ecology 1976.)

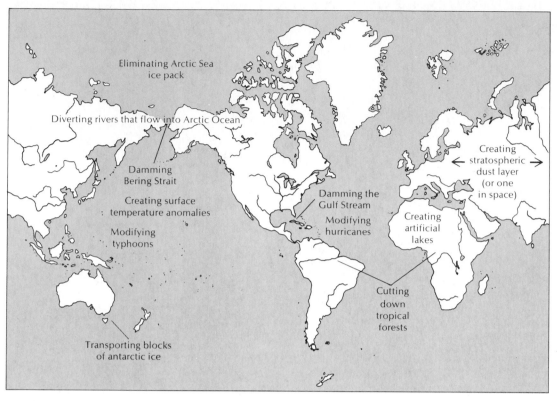

FIGURE 7-19.
Schematic illustration of the engineering schemes that could be proposed to modify or control climate. (From Kellogg and Schneider 1974. Copyright © 1974 by the American Association for the Advancement of Science.)

SUMMARY

The distribution of the major types of climate over the surface of the earth is determined by regional differences in insolation, by large-scale patterns of atmospheric circulation, and by the mechanisms that produce precipitation. Climate is a major determinant of the structure and functioning of natural ecosystems and of the agroecosystems that man establishes in their place.

Because the basic agricultural plants are descendants of wild species adapted to rather specific climatic conditions, there are broad zones that are better suited to the culture of some crops than to that of others. When crops are well matched to a particular climate they can be grown under its influence with minimal subsidization; however, the high demand for some crops has led to their cultivation in areas where large-scale subsidies become necessary.

Many modern crop varieties have been developed and proven during a period of relative climatic stability. The world now seems to be entering a period of greater climatic variability. How this will effect food production in the years ahead remains to be seen; but most experts predict a decline in agricultural production, and studies of some North American food crops support their view. Some of the production decline could be made up through the development of new crop varieties, improved technology, and an increase in crop area. Indications are that plans to do these things should be made now.

Literature Cited

Bayley, N. D. 1974. The environment versus man and his environment. In *Animal Agriculture.* H. H. Cole and M. Ronning (eds.). pp. 455–469. San Francisco: W. H. Freeman and Company.

Collier, B. D., G. W. Cox, A. W. Johnson, and P. C. Miller. 1973. *Dynamic Ecology.* Englewood Cliffs, N.J.: Prentice-Hall, Inc.

Gates, D. M. 1962. *Energy Exchange in the Biosphere.* New York: Harper and Row.

Gates, D. M. 1965. Energy, plants and ecology. *Ecology* 46:1–24.

Halacy, D. S. 1968. *The Weather Changers.* New York: Harper and Row.

Heiser, C. B., Jr. 1973. *Seed to Civilization.* San Francisco: W. H. Freeman and Company.

Institute of Ecology. 1976. *Impact of Climatic Fluctuation on Major North American Food Crops.* Stant Litho.

Kellogg, W. W., and S. H. Schneider. 1974. Climate stabilization: For better or for worse? *Science* 186:1163–1173.

Kukla, G. J., and H. J. Kukla. 1974. Increased surface albedo in the Northern Hemisphere. *Science* 183:709–714.

Newman, J. E., and R. C. Pickett. 1974. World climates and food supply variations. *Science* 186:877–881.

Oort, A. H. 1970. The energy cycle of the earth. In *The Biosphere.* pp. 13–23. San Francisco: W. H. Freeman and Company.

Revelle, R. 1963. Water. In *Plant Agriculture*, 2nd ed. J. Janick, R. W. Schery, F. W. Woods, and V. W. Ruttan (eds.). pp. 96–107. San Francisco: W. H. Freeman and Company. 1970.

Thompson, L. M. 1963. *Weather and Technology in the Production of Corn and Soybeans.* CAED Report 17, Iowa State University.

Thompson, L. M. 1964. *Weather Influences on Corn Yield.* Proc. Agr. Res. Inst. Washington, D.C.: National Academy of Sciences.

Thompson, L. M. 1975. World weather patterns and food supply. *J. Soil and Water Cons.* 30:44–47.

Trewartha, G. T. 1954. *An Introduction to Climate,* New York: McGraw-Hill Book Company, Inc.

Went, F. W. 1957. Climate and agriculture. In *Plant Agriculture*, 2nd ed. J. Janick, R. W. Schery, F. W. Woods, and V. W. Ruttan (eds.). pp. 108–118. San Francisco: W. H. Freeman and Company.

8

BIOCLIMATOLOGY OF AGROECOSYSTEMS

In Chapter 7 we examined the large-scale meteorological processes that define regional climates and set the basic limits on the global distribution of different agroecosystems. We can also examine in detail the patterns of weather and climate in more restricted areas, such as San Diego County, or the Imperial Valley, in southern California. Studies on this scale are frequently termed **mesometeorology** and **mesoclimatology,** and they are obviously important to agriculture also, especially where complex topographic features interact with regional meteorology to produce sharp differences in weather and climate over short distances.

The conditions that affect plants and animals exist on a still smaller scale, one that is **micrometeorological** or **microclimatological** (Rosenberg 1974). The precise conditions to which organisms respond are those that prevail at the immediate surface of the plant or animal body, or in a very narrow zone just above the surface, termed the **boundary layer** (Gates 1962). These conditions are variations, often major ones, on those that prevail some distance above the surface and that are usually taken to characterize the local weather and climate. Near the ground or vegetation surface, and especially within a stand of vegetation, the conditions are quite different, and the degree of difference may be increased by the microtopography of the site; that is, direction and steepness of the slope, surface conformation, and many other characteristics. Microclimatology thus concerns itself with conditions in the lower atmosphere to some height above the zone in which plants and animals actually exist. Generally speaking, microclimatology deals with conditions from the surface to a height about four times that of the objects of interest. To understand the origin of microclimates in a grassland area, for example, one must consider processes from the surface to a height of perhaps 6 meters; in a forest, microclimatic investigations extend to a much greater height.

MICROCLIMATOLOGY

Given the general temperature and humidity of the air mass brought to a region by global atmospheric circulation, specific microclimates arise out of the interaction of local meteorological processes, topography, and vegetation. The principal meteorological processes involve solar and far infrared radiation, precipitation, and wind. Topography determines the degree to which a particular surface area is exposed to these weather conditions and can create other conditions that influence the local movement of air masses. Vegetation interacts with these conditions too, but it also influences the CO_2, water vapor, and heat content of the lower atmosphere by gas-exchange processes.

Because temperature influences or is affected by many of these relationships, we can examine their basic interplay by summarizing their influence on energy-exchange processes. If we concern ourselves with the major energy-exchange processes that create microclimatic conditions on a plot of agricultural land, we can summarize these in an **energy budget equation:**

$$S(1 - \alpha) + L_d - L_u$$

$$\pm H_{air} \pm H_{ev} \pm H_{soil} = 0$$

where S = the downward flux of shortwave solar radiation, α = the albedo of the surface, L = flux of longwave infrared radiation to (L_d) or away from (L_u) the surface, and H = gain or loss by heating or cooling of the air and soil, or by evaporation or condensation of water. In this equation, positive values represent a flow of heat to the surface, negative values a flow of heat away from the surface. The **albedo** of the surface, expressed as a decimal fraction, is simply the proportion of the solar radiation that is reflected back into space.

In Chapter 7, we noted that incoming solar radiation, equal to 2.00 cal/cm²/min outside the earth's atmosphere, is modified by a number of factors, including latitude and characteristics of the atmosphere. On a more local scale, the direction and steepness of sloping land surfaces also influence the irradiance that accumulates during daylight (Figure 8-1). For example, in a locality in Colorado (39°N) the total daily input of solar radiation is lowest on a steep northfac-

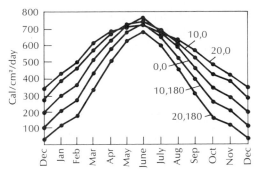

FIGURE 8-1.
Seasonal curves of total daily irradiance for slopes of different degree and direction at the latitude of central Colorado (39° N). The first number gives slope (degrees from horizontal); the second, direction (0° = southfacing, 180° = northfacing). (Modified from Collier et al. 1973.)

ing slope at all seasons. Seasonal change in daily irradiance is more than tenfold for this slope, whereas for a southfacing slope of equal steepness it is only about twofold.

An area of land surface receives, in addition to solar radiation, a downward flux of longwave infrared radiation or thermal radiation. The thermal radiation spectrum is generally regarded as extending from 3 to 100 microns (Monteith 1973), but the greatest energy flux is in the 9 to-11 micron range. The downward flux from the sky consists of energy absorbed from both incoming solar and outgoing thermal sources and reradiated downward by various atmospheric components, especially water vapor, carbon dioxide, ozone, and particulates. The downward flux of thermal radiation can be approximated by a relationship that includes the air temperature near the surface and the moisture content of the air. Thermal radiation from the sky is thus not influenced directly by solar radiation, and a downward flux occurs both day and night. This downward flux can range from about 0.35 to about 0.60 cal/cm²/min.

Returning to our energy budget equation, we see that there is also an outward flux of thermal radiation. This loss is related to the surface temperature and to emittance. Emittance, expressed as a decimal fraction, is high, equalling 0.95 to 0.98 for most vegetated surfaces and somewhat less for surfaces such as sand and rock. Heat can also be lost from the immediate surface by convection to the air above it, by conduction to deep layers of the soil, and by the evaporation of water. Convectional loss from the surface depends principally on the temperature gradient from the surface to the air, and is also increased by wind, which increases the exchange of air between locations near the surface and those away from it. Conduction through the soil depends on the temperature gradient and the thermal conductivity of the materials involved. Evaporation removes approximately 580 calories per gram of water lost. Heat is transferred from the hotter to the colder object, so if the surface is colder than the air above it (or the soil beneath it) the heat will flow in the opposite direction. When the surface is colder than an overlying air mass that is nearly saturated with water, the condensation of water (as dew) can also add heat to the surface.

An example of how these processes can interact to create distinctive microclimates is given in Table 8-1. These data, collected by Aizenstadt in central Asia and summarized by Miller (1965), compare the daytime energy flux for an open desert area with that for a nearby irrigated cotton field. The data pertain to a warm, clear-sky summer day when total downward radiation (solar and far infrared) exceeds slightly the value of the solar constant. For both systems, the downward flux is the same. Reflected solar radiation is less in the cotton field, due to a lower albedo, and thermal radiation loss is

TABLE 8-1

Radiation budgets, heat fluxes, and temperature profiles for open desert and an irrigated cotton field in central Asia.

	Radiation exchange ($cal/cm^2/min$)							
	Open desert				Irrigated cotton field			
	Solar				Solar			
	Direct	Diffuse	Thermal	Total spectrum	Direct	Diffuse	Thermal	Total spectrum
Downward	1.26	0.17	0.65	2.08	1.30	0.13	0.65	2.08
Upward	0.37		0.95	1.32	0.27		0.70	0.97
Net gain (+) or loss (−)	1.06		−0.30	0.76	1.16		−0.05	1.11

Heat flux ($cal/cm^2/min$)				
	to air	0.60	to air	0.05
	by evaporation	0	by evaporation	1.00
	to soil	0.16	to soil	0.06
		0.76		1.11

Temperature profile (°C)				
	150 cm	36°	150 cm	31°
	20 cm	38°	20 cm	31°
	0 cm	59°	0 cm	32°
	−5 cm	40°	−5 cm	35°

Source: Aizenstadt, as presented in Miller 1965.

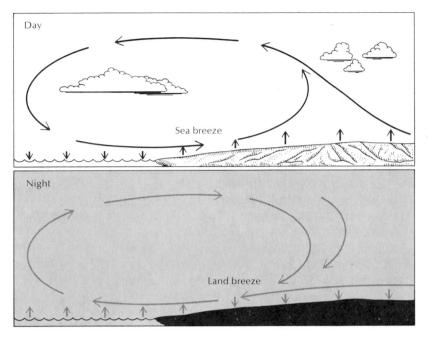

FIGURE 8-2.
Near large lakes or the ocean, sea breezes develop during the day as a result of differential heating of the land; this creates a pressure gradient that brings marine air landward. Differential cooling of the land at night can reverse this process, creating a land breeze. (Modified from "Micrometeorology" by Sir Graham Sutton. Copyright © 1964 by Scientific American, Inc. All rights reserved.)

lower because of the lower surface temperature. In the cotton field, much of the radiation that reaches the surface is expended in the evaporation of water, which is nil in the desert area. Evaporation, in fact, is of such magnitude that the transfer of heat to air and to deeper soil is small, and temperatures of the air, surface, and deeper soil are all lower in the irrigated field than on the desert.

These processes of energy exchange can interact with local physiographic conditions to produce a variety of distinctive microclimatological phenomena including sea and lake breezes, slope winds, and inversions. **Sea and lake breezes** are perhaps the best example of the local circulation patterns that can develop between adjacent areas that have different energy-exchange characteristics. Where land and water areas are in contact, the lower specific heat and thermal conductivity of the land surface will cause more of the energy received to be transferred to the air above it. As it heats, this air will expand and tend to rise, and as it rises it will tend to be replaced by cooler, heavier air from areas lying over water (Figure 8-2. Thus during the day a sea breeze or lake breeze tends to develop. The process begins in the morning near the coastline, and the breeze produced reaches further inland during the day. The entire process reverses at night, because the land surface and the air above it tend to cool faster than the surface of the water due to heat loss by thermal radiation.

Slope winds are another expression of energy exchange by land surfaces (Sutton 1964). As the land surface loses heat by thermal radiation at

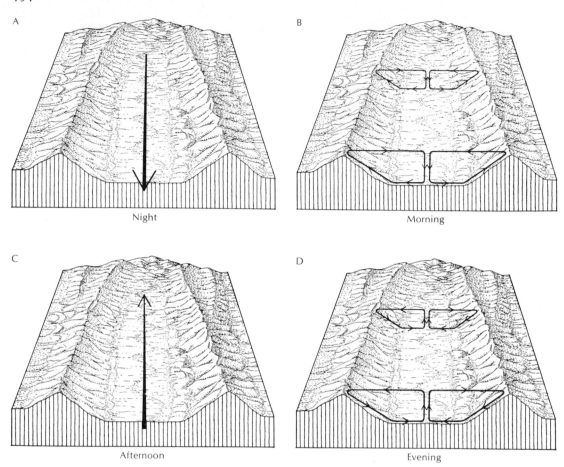

FIGURE 8-3.
Slope winds follow a complex diurnal pattern that is based on the rapid heating and cooling of the land surface on clear days and nights. At night and through the predawn hours (A) cooled air flows downslope by gravity, creating a mountain wind. During the forenoon (B) the warming valley floor induces a cross-valley circulation of air that concentrates into a general upslope flow, or valley wind, in the late afternoon (C). An opposite cross-valley flow begins after sunset (D). (From "Micrometeorology" by Sir Graham Sutton. Copyright © 1964 by Scientific American, Inc. All rights reserved.)

night, the air adjacent to the surface also cools and, being heavier, tends to flow downhill if a significant slope exists. These movements are local at first, but if they begin in an extensive valley area, they can intensify and consolidate to form a **mountain wind,** that is, a cool nightly air flow down the valley in general. During the day an opposite process occurs, creating a **valley wind** that moves into the hills surrounding a valley area (Figure 8-3).

Where the land surface is level, the cool air layer created at night near the ground may remain in place and gradually become colder and thicker. Where a basin or gentle valley bottom exists, a trap can form for cold air from the slopes above, and a thick layer of unusually cold air can develop at night. Such cold layers, or **inversions** (the normal condition is one of warm air near the surface that cools toward higher levels) are of particular significance to farmers because they can cause frosts that damage plants. Several techniques are used to weaken or destroy such inversions, including smudge pots (Figure 8-4), which lay down a blanket of particulates and thus reduce the net loss of heat from the surface by radiation, and fan or propeller devices that mix the air and break up the development of extremely cold layers immediately above the surface (Figure 8-5). When temperatures have dropped to freezing or below, water can also be sprayed on plant surfaces (Figure 8-6); as this water freezes heat of fusion is released, and as long as freezing continues the temperature of the surfaces themselves will not

drop below 0°C. Spraying must be continued until temperatures have warmed enough to melt the ice that has formed (Rosenberg 1974).

The vegetation of an area, whether it is natural or agricultural, also modifies microclimatic conditions, both by influencing energy-exchange processes and by imposing an additional set of gas-exchange processes on the atmosphere near the surface. Vegetated surfaces, first of all, have a lower albedo than most nonvegetated land surfaces, and so a higher proportion of the incident solar radiation is absorbed. Absorption and emittance in the far infrared range are also high, and so thermal energy exchange is high. Furthermore, as we noted in our earlier example, the presence of a well-developed vegetative layer implies the presence of moisture; thus in vegetated areas evaporation and transpira-

FIGURE 8-4.
Smudge pots operating in cherry orchards near Boise, Idaho during freezing spring weather. (Photograph by J. D. Roderick, courtesy of the U.S. Bureau of Reclamation.)

FIGURE 8-5.
Citrus groves near Indio, California, equipped with propeller units to break up surface temperature inversions during freezing weather. (Photograph by E. E. Hertzog, courtesy of the U.S. Bureau of Reclamation.)

tion also play important roles in energy exchange. Where the vegetation cover is dense or forms a solid canopy it has still another influence on energy exchange and microclimate. Under these conditions the canopy surface becomes the energy exchange surface, and the area beneath the canopy is insulated against the extremes of temperature that occur at the exchange surface. This effect, which might be termed the lath house principle, is used in a variety of ways to control the microclimate of agricultural situations. At this point we should also note that although photosynthesis and respiration of the organisms present in an area also involve energy exchange, they do so in a very limited way and do not play a significant role in creating microclimates.

Gas-exchange processes, however, do contribute to the creation of local microclimates, particularly of those that involve carbon dioxide. Because carbon dioxide is produced by decomposition processes as well as by the respiratory metabolism of all tissues, the soil tends always to be a carbon dioxide source. The above-ground biotic system tends to be a carbon dioxide source at night (when respiration occurs but photosynthesis does not), when a high concentration of CO_2 is likely to develop near the ground surface and within the vegetation. Still air conditions and temperature inversions keep this CO_2 accumulation from being dispersed, and it can in fact be used to estimate the total respiration of the ecosystem. Concentrations of CO_2 can reach 400 parts per million

FIGURE 8-6.
Sprinkler system combined with propeller units to help protect a Florida citrus grove
from frost. This photo was taken at 9:15 A.M. on 13 December 1962, when the sprinkler
system was still operating. The weight of heavy ice accumulations under extreme conditions can obviously damage trees. (Photograph by B. C. Beville, courtesy of USDA Soil
Conservation Service.)

(ppm) or more when conditions favor high respiration rates. During the day, respiratory
production of CO_2 continues, but uptake by
photosynthesis also occurs, and this can cause
the CO_2 concentration near actively photo-
synthesizing tissues to drop below normal. In
the canopy zone of crop stands, levels of CO_2
may be 270 ppm or less.

Transpiration of water by plants can also
modify the amount of moisture in the air, especially during daytime hours when the highest
transpiration rates occur. At night, in crop
stands with well-developed canopies, temperatures tend to be lowest at the canopy surface,
where they may fall below the dew point. The
condensation of dew on plant surfaces can lower
the moisture content of the air at this level to a
point somewhat below that of the air farther
away from the foliage surface.

The result of these processes is the creation of
vertical microclimates, or microclimatic profiles, within a crop ecosystem (Figure 8-7).
These microclimatic conditions both influence
and are influenced by the activities of the organisms present. We shall now examine how
they influence the activities of organisms.

MICROCLIMATIC INFLUENCES
ON ORGANISMIC ACTIVITIES

Organisms are linked, or **coupled,** to conditions
of their physical environment to the point
where a change in a particular environmental

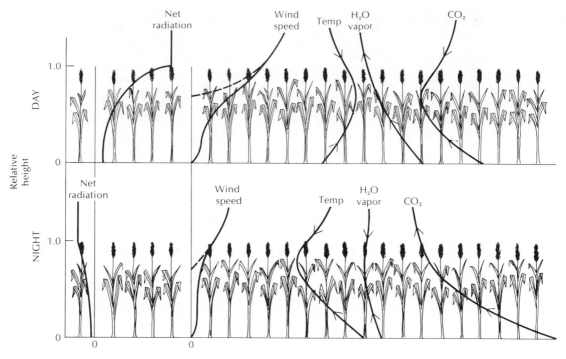

FIGURE 8-7.
Schematic representation of profiles of net radiation (i.e., incoming solar and thermal radiation minus outgoing thermal radiation), wind speed, air temperature, vapor pressure of water, and CO_2 concentration within the stand of a grain crop. (Modified from Monteith 1973.)

factor induces a response in the organism. If the response to the environmental change is relatively large the organism is said to be tightly coupled; loosely coupled if the response is relatively small. The characteristics of the organism that determine the nature and extent of its responses can be termed **coupling factors.** The principal factors of the physical environment, the coupling relationships, and the organismic responses are diagramed in Figure 8-8. In this figure, air temperature is coupled to the body temperature of the organism; the coupling factor in this case would be the one that determines the rate at which heat is transferred between the air and the organism,

namely the organism's surface conductivity. An organism with a highly insulated body surface would be loosely coupled to air temperature; one with a poorly insulated surface would be tightly coupled.

The organism's initial response to changes in most of the physical environmental factors with which we are concerned is a change in surface temperature. Thus we can integrate an organism's response to complex changes in its physical environment by considering its overall energy balance, as we did for our consideration of microclimate origins. Here, however, we shall find that certain relationships that we did not consider before take on major importance.

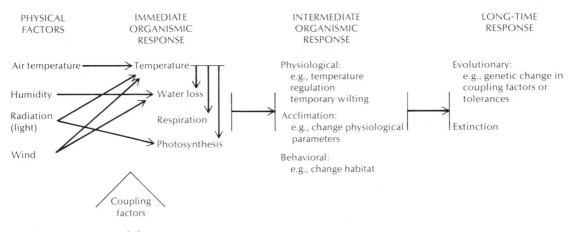

FIGURE 8-8.
Physical factors of the environment, coupling relationships, and organismic responses.
(Modified from Collier et al. 1973.)

We can write an energy budget equation that summarizes the exchanges of energy that occur between a plant or animal and its environment:

$$S(1 - \alpha) + L_d - L_u + M \\ \pm H_{ev} \pm CD \pm CV = \text{Storage}$$

where M = metabolic heat production, and CD and CV = the gain or loss of heat by conduction and convection, respectively; the other symbols are the same as those in the earlier energy budget equation. Storage refers to the net gain (or, if negative, loss) of heat by the organism, which will be reflected in a change in temperature. This term is insignificant for organisms with small mass, or for thin structures such as leaves, but it is important for larger organisms, particularly birds and mammals. As in the earlier equation, positive values indicate a heat gain by the organism, negative values a heat loss.

Solar radiation reaches the surface of a plant or animal as three somewhat different components: the direct solar beam from the sun; diffuse downward solar radiation (i.e., that which is scattered by particulates and condensed water droplets); and the solar radiation that is reflected upward to the surfaces of organisms from points below. The total radiation that is absorbed by the organism depends on several relationships, including the reflectance of the surface, the angle at which the direct solar beam strikes the surface, and the intensity of the radiation beam. We have already discussed the importance of reflectivity (albedo) and angle of incidence; radiation intensity requires further examination.

The intensity of the radiation beam is influenced by absorption and reflection by the atmosphere. In addition, in a vegetation canopy, the intensity that reaches a given leaf depends on how much is absorbed and reflected by leaves higher in the canopy. In general, in a well-developed, homogeneous vegetation canopy, the decline in intensity is exponential and can be described by the following equation:

$$S_F = S_0 e^{-KF}$$

where S_F = the radiation intensity at a point within the canopy, S_0 = that above the canopy, K = the extinction coefficient, and F = the leaf area index above the point at which radiation intensity is being calculated. The **leaf area index** is the total area of leaf surface in a vegetational canopy per unit area of ground. A leaf area index of 3.0 means simply that the total area of foliage above 1 square foot of gound is 3 square feet. The **extinction coefficient** describes the rate of light depletion per unit leaf area index. Thus the foregoing equation says simply that the radiation intensity at a particular level equals the radiation intensity above the canopy multiplied by an exponential depletion factor that takes into account both the depletion per unit leaf area index and the increment of leaf area index through which light must pass.

The infrared radiation that reaches the organism comes from both above and below. The downward flux of thermal radiation that reaches the upper surface of a vetetational canopy or an organism in the open depends, as we noted earlier, on the temperature of the air and on its moisture content. At some height above the surface, an organism or a structure such as a leaf also receives thermal radiation from below and from the sides. The intensity of this radiation is a function of the temperature of the radiating objects. A flat object such as a leaf receives both a downward and an upward flux of thermal radiation, and an animal such as a cow receives thermal radiation from all sides in a complex pattern that varies according to the nature of the surroundings.

Loss of thermal radiation by the organism is a function of the surface temperature. This loss, like the emission of thermal radiation by all objects, is proportional to the fourth power of the absolute temperature ($°C + 273$).

Three additional physical factors, H_{ev}, CD, and CV, can act as routes either for heat gain or heat loss. Because the change of water from liquid to vapor requires energy (580 calories per gram), and because the vapor produced is carried away in the air, this process serves as a route of heat loss. Condensation on the body surface of an organism is a means of heat gain.

Any factor such as wind that increases the rate of evaporation, increases heat loss. Here, the boundary layer phenomenon becomes important: evaporation in still air tends to saturate the air in the thin layer in close contact with an organism's surface, and further evaporation can occur (assuming no temperature differential) only when water molecules move out of this zone by diffusion. Wind, however, increases the rate at which saturated air is replaced with unsaturated air in the boundary layer.

Conduction (i.e., the transfer of heat by the exchange of energy between molecules in contact with each other) and convection (i.e., the transfer of heat from place to place by currents in air or water) can also be routes of heat gain or loss. For an organism, the rate of heat exchange by both of these processes is affected by the temperature difference between the organism and its surroundings and by the effectiveness of any insulating layer that might exist. Convection is further affected by a boundary layer phenomenon. In still air, the molecules of air in a thin layer adjacent to an organism's surface may be heated by conduction if the surface is warmer than the air. This heated boundary layer expands and small currents develop that tend to carry this heated air upward and away from the organism. Cooler air then moves into the boundary layer. Wind may enhance this exchange, however, and thus increase the potential for the loss of heat through convection. When the air is warmer than the organism, wind will encourage a heat gain by the organism for similar reasons.

The metabolism of an organism can also play an important role in energy exchange, particularly in the case of higher animals. In confined situations, plant metabolism can generate significant amounts of heat, and in a number of situations (e.g., wet hay piles and damp grain stores) microorganisms can produce large quantities of metabolic heat through decomposition. For birds and mammals, however, metabolic heat is a major component of the total heat budget.

TABLE 8-2
Summary of daytime and nighttime energy-exchange relationships for oak and aspen leaves near the canopy top. Measurements taken in central Colorado during August.

cal/cm^2/min[a]	Gambell's Oak		Quaking Aspen	
	Day	Night	Day	Night
	0600–1800 hr	1800–0600 hr	0600–1800 hr	1800–0600 hr
Energy gain				
S (solar radiation)				
Direct	0.37	—	0.46	—
Diffuse	0.17	—	0.24	—
Reflected	0.22	—	0.22	—
Total	0.76	—	0.92	—
Absorption	0.50		0.50	
S absorbed	0.38	—	0.46	—
L (thermal radiation)				
Downward	0.61	0.40	0.45	0.37
Upward	0.69	0.53	0.59	0.47
Total	1.30	0.93	1.04	0.84
Absorption	0.97	0.97	0.97	0.97
L absorbed	1.26	0.90	1.01	0.82
Total absorbed	1.64	0.90	1.47	0.82
Energy loss				
L (thermal radiation)	−1.24	−1.02	−1.10	−0.92
Convection	−0.08	+0.14	0.00	+0.08
Transpiration	−0.28	−0.04	−0.36	+0.02
Total lost	−1.60	−0.92	−1.46	−0.82
Leaf temperature				
(°C)	26.0°	11.5°	17.0°	4.5°

[a]Both sides of a leaf section 1 cm × 1 cm in size.
Source: Miller 1967.

To see how a number of these energy-exchange processes are interrelated in actual environmental situations, let us consider an example (Table 8-2). Miller (1967) measured energy-exchange processes for leaves of two tree species, Gambell's Oak and Trembling Aspen, in relatively dense, pure stands of each species in Colorado during August. Located within a few miles of each other, the oak stand was at nearly the same elevation (2286 meters) as the aspen stand (2621 meters). In each of these stands, instruments were placed at various heights to measure physical conditions and leaf responses. The effects of metabolism and conduction in this case were considered to be negligible and were ignored. As noted in Table 8–2, the albedo of the leaves of both species, covering wavelengths from the ultraviolet through the near infrared, was assumed to be 0.50 (absorption = 0.50), and the absorption of both species in the far infrared, about 0.97. The measurements obtained allowed Miller to compute the overall energy budget for oak and aspen leaves near the top of the canopy in each stand.

The energy budgets for these two situations demonstrate several very interesting points. First, the solar radiation input, although obviously essential for processes such as photosynthesis, was less than one-half that of thermal radiation. Second, even in a region where clear sky conditions occurred much of the time, diffuse and reflected solar radiation make up a large part (about one-half) of the solar radiation received by the leaves. Since these leaves were near the top of a canopy, much of this reflected energy came from other leaves, of course.

Under the conditions that prevailed, leaf temperatures rose to several degrees above air temperatures during the day and fell several degrees below air temperature at night. Over the general range of temperatures that were involved, however, heat loss by radiation was the principal mechanism of energy loss, being somewhat more important in oak than in aspen because of the higher temperatures of oak leaves. Part of this difference may also have been related to the drier habitat of the oak stand; less energy was dissipated by transpiration for oak than for aspen, which may have been one reason for higher leaf temperatures in the oak and thus for higher losses through radiation and convection.

Several instances can also be noted in which convection or change of water between liquid and vapor states acted as routes of heat loss under one set of conditions and heat gain under another. For the oak convection was a heat gain route at night, when leaf temperatures fell below air temperature; but at the same time humidities that were still below saturation allowed transpiration to continue as a route of heat loss. In aspen, under cooler nighttime conditions, convection not only became a route of heat gain, but dew also formed, making the change of water from vapor to liquid a route of heat gain.

MICROCLIMATE AND CROP ECOSYSTEM PRODUCTIVITY

Microclimatic profiles, coupled with the quantity of photosynthetically active radiation that enters the ecosystem, determine the potential rate of primary production. In crop ecosystems, of course, this is the basis of agricultural productivity. **Photosynthetically active radiation** (PAR) equals about 50 percent of the total incoming solar radiation in wavelengths between 0.3 and 4.0 microns (Monteith 1969). This percentage seems to be almost independent of atmospheric conditions and the height of the sun in the sky (Szeicz 1974) and represents specifically the energy in the 0.4-to-0.7 micron wavelength range (visible light).

Several research groups are currently investigating in detail the relation between crop microclimate and production (Biscoe et al. 1975; Lemon et al. 1971; Loomis et al. 1971). The most comprehensive analysis carried out for a single crop species may be that conducted on barley, *Hordeum vulgare* (variety Proctor), in England by a group of scientists at the University of Nottingham (Biscoe et al. 1975). In this study, an intensively instrumented barley stand was observed over an entire growing season; microclimatic measurements were continuously recorded and combined with regular measurements of the biomass and metabolic activity of various plant components. The instrumentation complex was designed to describe profiles of radiation, wet and dry bulb temperature, wind, and CO_2 concentration, as well as to allow a complete breakdown of the radiation balance for the crop system (Figure 8-9). The data obtained were automatically stored and summarized by a computer (Gregson and Biscoe 1975).

The barley crop studied was sown on 18 March and harvested on 21 August, with intensive study beginning in late April and continuing through late July. The principal objective

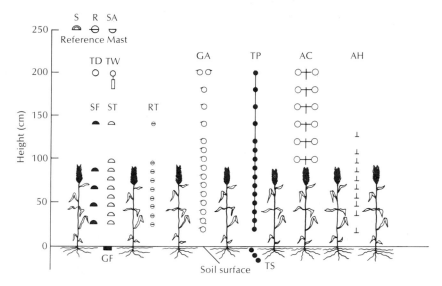

FIGURE 8-9.
Instrumentation complex used in studying the relationships between production and microclimate within a barley stand in England. The instruments are: AC, cup anemometer; AH, hot bulb anemometer; GA, gas sample intake; GF, flux plates for estimating the ground heat flux; R, reference net radiometer; RT, tube net radiometer; S, Kipp solarimeter; SA, albedo instrument; SF, filtered tube solarimeter to measure infrared radiation; ST, tube solarimeters to measure total solar radiation; TD, dry bulb thermometer and TW, wet bulb thermometer on the reference mast; TS, soil thermometer; TP, thermocouple array for measuring temperature differences in the profile. (From Biscoe et al. 1975.)

was to understand the pattern and control of carbon fixation by the plant, especially the storage of fixed carbon in the harvestable seed.

Both the soil and the atmosphere constitute sources of CO_2 for photosynthetic uptake by the plant (Figure 8-10). The CO_2 released from the soil consisted essentially of two components: that produced by the respiration of decomposer organisms (R_s) and that due to respiration of the barley roots (R_r). Decomposer production of CO_2 remained relatively constant over the principal growing period, at about 0.11 g/m²/hr, but root respiration varied with the stage of crop growth. Root biomass reached a maximum during the first week of June, declining gradually thereafter. Over the growing season about half the CO_2 released from the soil came from each source.

Aboveground, CO_2 was released by respiration of plant tissues (R_c) during both the day and night. During the day, however, photosynthetic uptake (P_c) also occurred to the extent that, under all normal daytime conditions, it exceeded the combined release of CO_2 from all soil and aboveground sources. Thus during the day there was a net flux of CO_2 from the air above into the crop canopy. At night, there was a net flux from the crop canopy to the air above. Net primary production (NPP) could therefore be estimated by the equation:

$$NPP = \Sigma_d(P_a + R_s) - \Sigma_n(R_a - R_s)$$

or

$$\Sigma(P_a - R_a) + \Sigma R_s$$

where d represents daylight hours and n night-

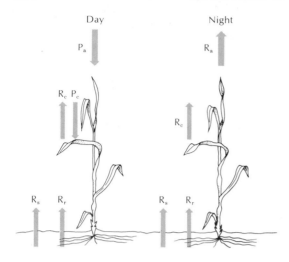

P_a = Net downward CO_2 flux
P_c = Photosynthetic CO_2 uptake
R_c = CO_2 release by shoot respiration
R_s = CO_2 release by soil decomposers
R_r = CO_2 release by root respiration
R_a = Net upward CO_2 flux

FIGURE 8-10.
Diagramatic representation of the direction of major fluxes of CO_2 in a barley stand during the day and night. (From Biscoe et al. 1975.)

pattern of carbon fixation could be followed throughout the daily and seasonal cycle and correlated with microclimatic factors such as the intensity of solar radiation (Figure 8-11). Furthermore, detailed measurements on the rate of photosynthesis by individual plant parts, including leaves, leaf sheaths, and the grain head itself, permitted the influence of microclimate and radiation regime to be related to specific processes within the plant (Figure 8-12).

As the process of carbon fixation was followed from day to day, the dramatic impact of short-term weather variation became evident (Figure 8-13). During the growing season, most of the weather variables, such as temperature and cloudiness, affected the input of solar radiation to the crop system; and so this factor showed a very close relation to the trend in daily net production. Over most of the growing season (Figure 8–14) several interesting patterns were discernible: during the first five weeks, aside from fluctuations caused by weather, the relationship of net production to respiration remained essentially the same. During the week

time hours. Gross primary production (GPP), or total photosynthetic fixation of carbon before it is reduced by plant respiration, would thus equal:

$$GPP = \Sigma_d(P_a + R_s) + \Sigma_d(R_r + R_c)$$

However, since R_c cannot be determined during the day because it is offset by CO_2 uptake in photosynthesis, GPP is calculated in practice as:

$$GPP = \Sigma_d(P_a + R_s) + \Sigma_n(R_a - R_s)\ (TC)$$

where TC = temperature correction factor.
The fluxes of CO_2 from the soil could be measured fairly simply and those in the atmosphere calculated from profile data with the aid of mathematical theory relating to heat flow and turbulent transfer in the air. As a result, the

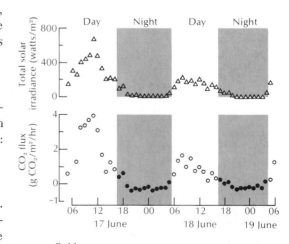

FIGURE 8-11.
Solar irradiance and net CO_2 uptake in a barley stand in England over a two-day period in June. (Modified from Biscoe et al. 1975.)

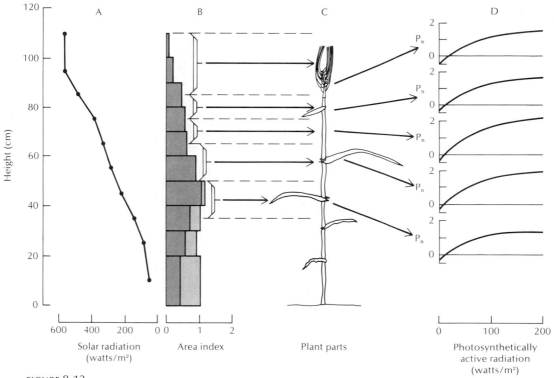

FIGURE 8-12.
Diagrammatic representation of the manner in which radiation profile was related to the area and productivity of different organs of the barley plant. Measurements of irradiance taken at different levels (A), together with measures of surface area index at the same levels (B), were assigned to the plant organs existing at those levels (C). Curves relating net photosynthesis (P_n) and irradiance were then developed for each organ. (Modified from Biscoe et al. 1975.)

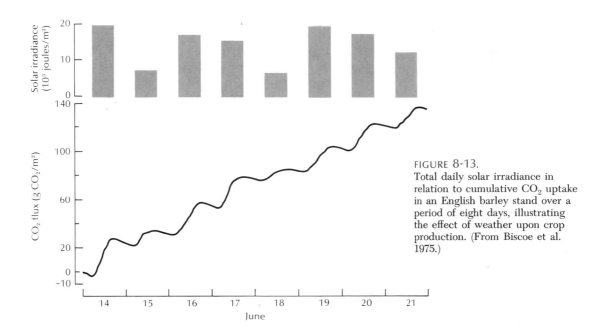

FIGURE 8-13.
Total daily solar irradiance in relation to cumulative CO_2 uptake in an English barley stand over a period of eight days, illustrating the effect of weather upon crop production. (From Biscoe et al. 1975.)

206

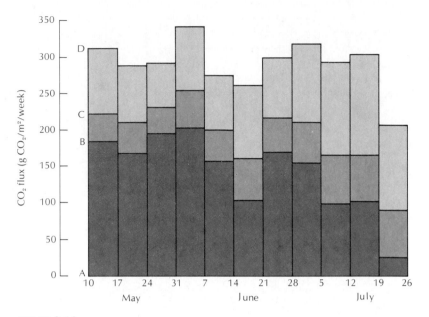

FIGURE 8-14.
Photosynthesis and respiration in a barley stand in England over an eleven-week period.
AD measures gross photosynthesis; CD, daytime plant respiration; BC, nighttime plant
respiration; and AB, net photosynthesis. (Modified from Biscoe et al. 1975.)

of June 14 to 21, when the developing barley
head was nearly ready to emerge, both diurnal
and nocturnal respiration rates increased. These
rates dropped back a bit after the head
emerged, but during July marked increases
in respiration rates accompanied senescence
and the final storage of material in the
seed head.

The final process of grain head development
is particularly interesting. During most of the
development of the head, much of the carbon
fixation required to produce the organic matter
in the seeds actually occurs by photosynthesis
in the head. However, the final maturation and
storage of matter require translocation of material from other parts of the plant (Figure
8-15). The stem tissues decline in weight by
about one-third as the head undergoes its final
development, and it is presumably the trans-

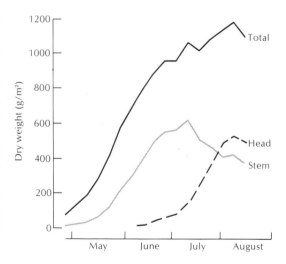

FIGURE 8-15.
Changes in weight of head and stem components of a barley stand as the head matures. (From Biscoe et al. 1975.)

location of materials from other plant parts and their deposition in the seed that also account for much of the respiratory increase in the head toward the end of the growing season.

The data available were also used to calculate production efficiencies for different periods during the growing season. Comparing gross and net photosynthesis to photosynthetically active radiation, during the early part of the growing season, gave efficiencies of 5.5 and 2.8 percent, respectively. On individual days maximum efficiencies of about 10 percent occurred, but these were under cloudy conditions when the total radiation intensity was quite low. Although efficiency was high under such conditions, the total rate of carbon fixation was less than on clear days when efficiencies were lower. Near the end of the crop season, of course, the efficiency of photosynthetic production dropped markedly.

Studies similar to those conducted by Biscoe and his co-workers for barley have also been carried out for corn (maize) by a group of workers at Cornell University (Lemon et al..1971; Shawcroft et al. 1974). One of the objectives of these studies has been to develop a predictive systems model of crop physiology and microclimate. This model, termed the *soil-plant-atmosphere model* (SPAM), attempts to describe in specific mathematical terms the major interactions that occur within the crop system (Figure 8-16). Based on inputs at two boundaries—the soil surface and an imaginary plane in the atmosphere above the crop canopy—the model predicts physiological responses of the crop and of the profiles of microclimatic conditions within the crop stand. To make such predictions, a number of features of the crop stand structure, such as the density, height distribution, and angle of leaves, must also be specified. The various atmospheric inputs are those listed in association with the climate boundary in Figure 8–16; input values at the soil surface consist of the water potential of the

soil surface and the rate of CO_2 output from this surface. Each of the submodels, represented in simple graphical terms, is actually a complex set of mathematical statements that describe the behavior of a particular process in relation to those variables that influence it. The output, or prediction, of the model consists of the set of relationships shown in the central "crop prediction" box in Figure 8–16. Organismic processes, such as transpiration, photosynthesis, and respiration, can be integrated over the entire height profile to give estimates of their rates for the stand as a whole.

Obviously it is very difficult to describe such a complex system in simple, predictive terms. This model, for example, gives reasonable predictions only for crop stands that are simple and uniform in structure, free from horizontal variation in conditions, and in steady-state or only slowly changing conditions. Figure 8-17 compares some of the predicted microclimatic profiles with those actually observed. Although the predictions obtained from such models are not perfect, the models may nevertheless have considerable value. One of the most useful aspects of a model of this type is the identification of particular relationships that have major impacts on organismic processes. Soil moisture conditions, CO_2 input rates, foliage density, height distribution, and geometry, as well as many other relationships, can all be manipulated mathematically and their influence upon model predictions determined. Using a model in this way, which is relatively inexpensive compared to field experimentation, can help identify and limit the range of experiments that it is worthwhile to actually carry out. The results of such experiments, of course, can be used to modify and expand the model so that it is more realistic and useful.

For example, where there is more than enough solar radiation for maximum rates of photosynthesis by a crop, but where efficient use of irrigation water is necessary, one thing

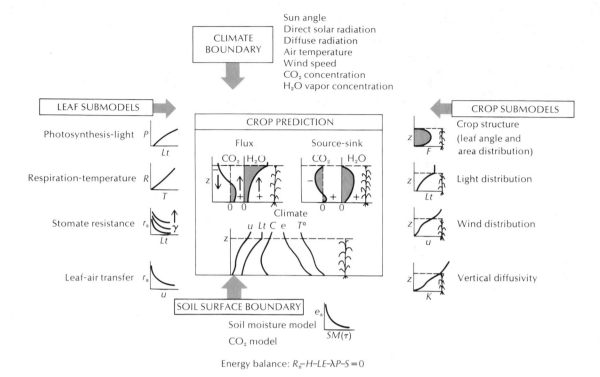

FIGURE 8-16.

Schematic summary of a mathematical soil-plant-atmosphere model (SPAM) giving required inputs, submodels, and representative daytime predictions of climate and community activity (that is, water vapor and carbon dioxide exchange). Abbreviations: z, height; u, wind; Lt, light; C, concentration of carbon dioxide; e, water vapor; $T°$, air temperature; e_s, surface vapor pressure; $[SM(\tau)]$, surface soil moisture or water potential; P, photosynthesis; R, respiration; T, leaf temperature; r_s, stomate resistance; γ, minimum stomate resistance at high light intensities; r_a, gas diffusion resistance; F, leaf surface area; K, vertical diffusivity; R_n, net radiation; H, sensible heat; LE, latent heat; λP, photochemical energy equivalent; and S, soil heat storage. (From Lemon et al. 1971. Copyright © 1971 by the American Association for the Advancement of Science.)

that such a model (or the energy budget equation itself) can predict is that an increase in reflectance will reduce transpiration by reducing the net input of heat energy to the crop system. Doraiswamy and Rosenberg (1974) have tested this prediction in a field experiment in which the whitish clay kaolinite was sprayed onto soybean foliage. In Nebraska, soybeans reach maximum photosynthetic rates at about one-third to two-thirds full sunlight. Increasing the reflection of incoming solar radiation would thus reduce photosynthesis only during early morning and late afternoon. Reducing the total energy input through reducing transpiration and thus water losses and irrigation requirements might be valuable enough to justify the slight decline in production. Detailed energy balance studies of treated and untreated fields showed that treatment resulted in an increase of about 20 percent in reflection of shortwave radiation and about an 8-percent decrease of total radiation input. Full evaluation of the

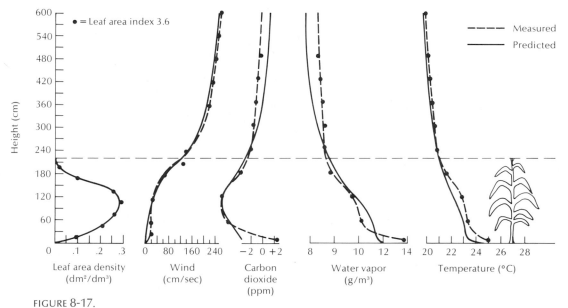

FIGURE 8-17.

Measured (*dashed line*) and predicted (*solid line*) profiles of microclimate during midday (11:45 A.M. to 12:15 P.M.) in a New York corn field, using the soil-plant-atmosphere model (SPAM). (From Lemon et al. 1971. Copyright © 1971 by the American Association for the Advancement of Science.)

usefulness of this technique is yet to be completed, but it is clear that experimental modification of crop microclimate and energy-exchange processes has considerable unexplored potential.

Obviously, the interrelationships between crop species and microclimate depend greatly on the morphology, physiology, and developmental stage of the crop plant itself (Figure 8-18). Plant species vary in many specific details of morphology and physiology that influence their relations with the microclimate. Leaf width, for example, is an important variable that influences the rate of convectional heat exchange between the leaf and the air; the rate of exchange is greater for narrower leaves. The physiological status of the stomata—the tiny openings on the surface of leaves through which exchanges of CO_2, O_2, and water vapor occur—rigidly controls the rates of transpiration and photosynthesis of a plant. Opening and closing of the stomata are influenced in a very complex fashion by microclimatic influences and plant metabolism.

There are also major differences in systems of respiration and photosynthesis among different plant groups. These differences concern the initial end products of photosynthesis and the extent to which they are broken down by respiration during the day. These characteristics, both minor and major, adapt to particular microclimates and are subject to manipulation through breeding programs designed to improve productivity in crop species. In Chapter 22, we consider the details of these relationships and examine the potential for crop improvement.

MICROCLIMATOLOGY AND DOMESTIC ANIMAL PRODUCTION

The physiological relationship between animals and their physical environment can also be

FIGURE 8-18.
In this field in the Sacramento Valley of California, recently transplanted tomato plants
are covered by paper cones to prevent damage from wind, rain, and sun until they
become established. This procedure in effect provides a special microclimate for these
plants during a sensitive stage of development. (Photograph by D. C. Schuhart, courtesy
of the USDA Soil Conservation Service.)

evaluated in terms of energy-exchange processes and the energy budget equation (Porter and Gates 1969; Spotila and Gates 1975; Wang 1972). In approaching the ecology of domestic animals from this point of view, we should recognize first the basic differences in the energy-exchange processes of plants and animals. First, because most domestic animals are birds and mammals, metabolism is a major source of body heat. These animals, termed **homoiotherms,** maintain a high body temperature that fluctuates between rather narrow limits. For mammals, body temperatures tend generally to fall in the range from 34 ° to 40°C. Birds show consistently higher body temperatures, usually between 38° and 45°C. Maintaining these high temperatures is the purpose

of a complex homeostatic mechanism for temperature regulation, one major component of which is the release of metabolic heat by respiratory combustion of food materials. For many domestic animals, such as dairy cattle, we should note that metabolic heat is generated, not only by maintenance metabolism and body growth processes, but also by the production of materials harvested by man. A cow that yields about 50 kilograms of milk per day, for example, generates more than twice the total metabolic heat production of one that yields only 10 kilograms per day (Bianca 1976).

Water balance is also more complicated for animals than for plants. Animals take in water by drinking, but it is also available to them in their food. Under normal conditions, all food materials contain a certain quantity of **preformed water,** or moisture. For example, the grasses or leaves eaten by an herbivore in a humid region contain 80 percent or more moisture; even in desert regions such materials contain about 60 percent moisture. As food materials are consumed in cellular respiration, however, one of the end products is **metabolic water,** which is produced by all animals and is just as useful in maintaining the body water balance as any other water. Different food materials produce different amounts of metabolic water: about 0.556 grams H_2O per gram of starch; 0.396 grams H_2O per gram of protein; and 1.071 grams H_2O per gram of fat. Loss of water from the body in the urine and feces does not directly influence energy exchange; but losses of water through **insensible** and **heat regulatory evaporation** are both important factors in energy exchange. Insensible losses are those from the skin and lungs and are essentially unavoidable due to the nature of the surfaces involved. Heat regulatory evaporation includes the extra losses from the body surface (by sweating) and from the respiratory passages (due to panting or other accelerated ventilation). Finally, we should note that work may appear as a component of the energy budget of draft animals, but we will not consider this component in detail.

We can now restate the energy budget equation given earlier, thus subdividing the evaporation term:

$$S(1 - \alpha) + L_d - L_u + M - H_{ev(i)} - H_{ev(hr)}$$
$$\pm CD \pm CV = \text{Storage}$$

where $H_{ev\,(i)}$ = insensible evaporative heat loss and $H_{ev\,(hr)}$ = heat regulatory evaporative heat loss. When body temperature remains constant, storage equals zero; when the right hand portion of the equation is positive, body temperature rises; when negative, it falls. This version of the equation, however, neglects the possi-

bility that water condensation will add heat to the animal's body, and does not include work.

In the case of homiotherms we must recognize the importance of the surface layers of fat, fur, feathers, or skin in impeding the flow of heat between the outer exchange surface (e.g., outer surface of fur) and the body interior (beneath the skin and surface fat layer). Insulating layers of this type determine the temperature gradient that can be maintained between outer and inner surfaces for a given level of heat production (internally) or for heat transfer to the surface from the air (externally) as shown in Figure 8-19. More simply, the temperature difference across an insulating layer, ΔT, will be given by:

$$\Delta T = d_b/k_b \, (M - H_{ev})$$

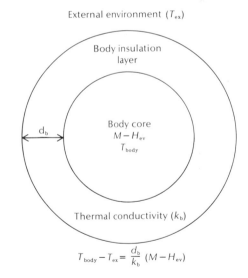

External environment (T_{ex})

Body insulation layer

d_b

Body core
$M - H_{ev}$
T_{body}

Thermal conductivity (k_b)

$$T_{body} - T_{ex} = \frac{d_b}{k_b} \, (M - H_{ev})$$

FIGURE 8-19.
Cylinder model of the body of a warm-blooded animal, illustrating the factors that determine the temperature difference ($T°C$) between the body core (T_{body}) and the external environment (T_{ex}). M = metabolic heat production (cal/min); H_{ev} = heat dissipation by evaporation of water (cal/min); d_b = thickness of insulation layer (cm); and k_b = thermal conductivity of insulation layer (cal/cm/min/°C).

where d_b = the thickness of the insulating layer (cm); k_b = the thermal conductivity of the insulating layer (cal/cm/min/°C); M = metabolic heat production (cal/min); and H_{ev} = the portion of M dissipated by evaporation on or beneath the layer in question (cal/min).

Based on their general characteristics, such as body size, basal metabolism, insulation, and general morphology and physiology, homoiothermic animals show a **zone of thermoneutrality** (Figure 8-20), or range of external temperatures (in still air) over which body temperature can be maintained at an optimal level without significant extra expenditure of energy (Bianca 1976). The dilation or constriction of surface blood vessels, or a change in the angle of erection of hairs or feathers that thus modifies the effective thickness of the insulating coat, are examples of regulatory mechanisms operating in the zone of thermoneutrality. The thermoneutrality zones of several domestic animals are given in Figure 8–21. The thermoneutral zones for smaller animals tend to be in a higher range than those for large animals; and the thermoneutral zone for young animals is both smaller and at a higher temperature range than for adults (Bianca 1976). Below the zone of thermoneutrality, the animal must increase its metabolic heat production. Metabolic heat is generated primarily by respiratory activity in muscle tissues; shivering for example, is simply an intense form of muscle activity that releases heat when there is a great need for it.

Above the thermoneutrality zone, energy must be expended to lose extra heat through

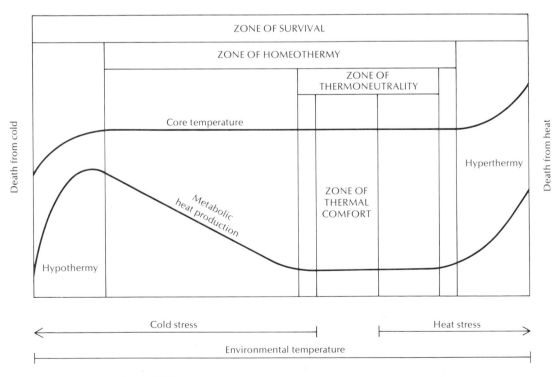

FIGURE 8-20.
Body temperature relationships and terminology for homoiothermic animals.

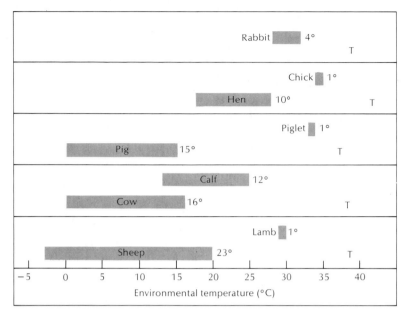

FIGURE 8-21.
Thermoneutral zones and normal body temperatures (*T*) for several adult and infant domestic animals.

mechanisms such as sweating or panting. Although these activities actually generate some heat through increased metabolism, more heat is lost through evaporation and body temperature is thus lowered.

Many factors modify this pattern, of course. Animals may physically leave an unfavorable situation and move into a favorable one, or they may show behavioral responses. Other environmental factors, particularly wind and atmospheric vapor pressure, can also influence heat exchange processes.

The most important environmental conditions and organismal responses can be combined in a **climate space diagram** (Porter and Gates 1969) that shows the range of conditions under which the animal can exist. Such a diagram is shown in Figure 8-22 for the domestic pig. In this figure air temperature is plotted against total radiation absorbed, per unit of body surface, from the external environment. Within the diagram is an enclosed area that represents the extremes of temperature and radiation absorption that the species can tolerate given its metabolic capacities, evaporative heat loss capability, insulation thickness and conductivity, and internal body temperature limits. The overall space in the enclosed area is actually the composite of climate spaces defined for three wind speeds that cover the range usually found outdoors. A diagram of this type thus summarizes a great deal of information.

We should note, however, that the optimal conditions for the species constitutes relatively small proportion of this climate space. Outside this optimal area, the animal must devote more and more of its food energy intake to regulatory activity, usually at the expense of growth and reproduction. Maintaining a favorable microclimate is thus as important in animal agriculture as it is in plant agriculture (McDowell 1974). Figure 8-23 gives an example of how

214

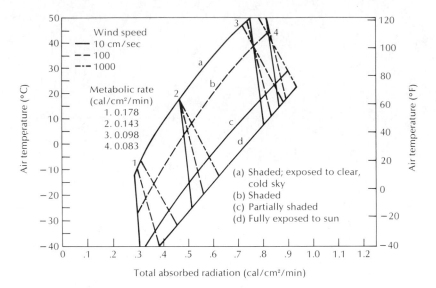

FIGURE 8-22.
Climate space diagram for the domestic pig, showing relations between air temperature, absorbed radiation, and wind speed under which body temperature can be regulated successfully. (Modified from Porter and Gates 1969.)

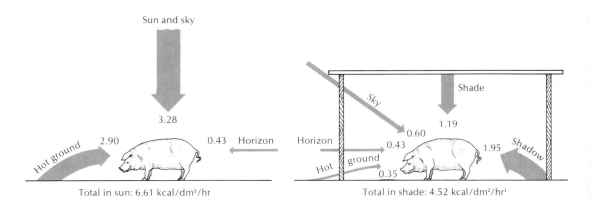

FIGURE 8-23.
Modification of the heat load on a domestic pig by a roofed shelter under conditions of a clear summer sky in the Imperial Valley of California. (Modified from Bianca 1976.)

FIGURE 8-24.
The eland, an East African ungulate, is a species that is adapted to semiarid tropical regions and has a strong potential for domestication. (Photograph by G. W. Cox.)

shelter can be used to reduce the radiation heat load on domestic animals in a feedlot.

Analyzing the physiological capabilities of domestic animals and of potential domesticates is an important part of developing ecologically sound agricultural systems for arid zones and parts of the tropics. We know, for example, that domestic cattle are somewhat ill-adapted for many arid tropical and subtropical regions because of their high water requirements. Taylor (1969) has studied the environmental physiology of two African ungulates, the eland (*Taurotragus oryx*) and the oryx (*Oryx beisa*). Both of these species occur in drier areas of Africa, and the eland has also been successfully domesticated, at least on a limited scale (Figure 8-24).

Taylor found that both species possess a number of striking adaptations that enable them to reduce water loss by evaporation and to maximize the water taken in through feeding; as a result, both species essentially do not depend on free water for drinking, even though their environment is hot and dry. They "store" heat during the day; that is, their body temperature rises to the upper edge of their tolerance range before extensive evaporative cooling begins. In this way they maximize the losses that occur by radiation and convection. By letting its body temperature rise 7.3°C, an eland weighing 500 kilograms avoids expending about 5 liters of water for evaporative cooling. When dehydrated, both species have other mechanisms that reduce evaporative loss. The body temperature of the oryx, in particular, can rise to 45°C, a level much higher than what is normal for most mammals. Brain damage is avoided with the aid of a mechanism that supplies cooler blood to the brain than to the rest of the body. Both species also pant when they are dehydrated, instead of sweating extensively. Panting removes heat from the body core, leaving surface temperatures high, and thus encourages radiation and convection losses that sweating, by cooling the skin surface, inhibits. Both species also feed on plant food

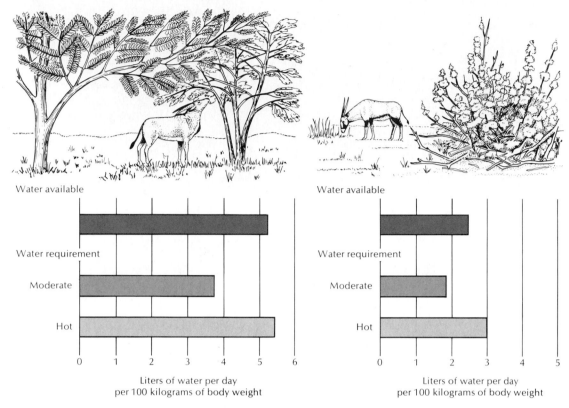

FIGURE 8-25.
Water requirements of eland and oryx under moderate and hot conditions, relative to moisture intake in their food. (From "The Eland and the Oryx" by C. R. Taylor. Copyright © 1968 by Scientific American, Inc. All rights reserved.)

that is high in moisture, the oryx apparently doing much of its feeding at night, when leaves of the various grasses on which it depends have absorbed maximum moisture from the air. Thus, even under hot conditions, both the eland and the oryx can meet their water needs with moisture from food (Figure 8-25).

SUMMARY

The climatic conditions to which an organism responds are those that immediately surround and impinge upon it. These conditions, which form the microclimate, result from the inter-

action of regional weather and climate with local physiographic and vegetational relationships at the land surface. The general nature of this interaction can be described by the energy budget equation, which summarizes the gain and loss of short- and longwave radiation at the surface, together with the change in heat content of soil and air. These interactions, along with gas exchange processes that take place in the vegetation, produce distinctive microclimate profiles of conditions within and immediately above the vegetation.

The influence of microclimatic factors on the organism can also be evaluated to a great extent by analyzing energy budget relationships. The net gain or loss of energy by radiational exchange, convection, and evaporation or condensation determines the pattern of change in the temperature of an 'organism; under some circumstances conduction can also be an important influence on an organism's temperature. For plants, the quantity of photosynthetically active radiation, and the manner in which it is depleted as it passes downward through the vegetational canopy, determine the potential rate of photosynthetic production.

Detailed studies of microclimatic relations during crop growing seasons are now providing information on the precise ways in which microclimate influences various plant processes and developmental stages. In some cases these influences may permit the development of a predictive systems model of crop production processes. The study of such models can help to identify crop-microclimate relations of critical importance, or to suggest new approaches to the manipulation of crop environments.

Domestic animals also exhibit close relationships to their microclimate. Since most are homoiotherms, metabolic heat production and body insulation are important factors in energy exchange; and routes of intake and loss of water are particularly important, especially in hot, arid regions.

Literature Cited

Bianca, W. 1976. The significance of meteorology in animal production. *Int. J. Biometeor.* 20:139–156.

Biscoe, P. V., J. A. Clark, K. Gregson, M. McGowan, J. L. Monteith, and R. K. Scott. 1975. Barley and its environment. I. Theory and practice. *J. Appl. Ecol.* 12:227–247.

Biscoe, P. V., J. N. Gallagher, E. J. Littleton, J. L. Monteith, and R. K. Scott. 1975. Barley and its environment. IV. Sources of assimilate for the grain. *J. Apple. Ecol.* 12:295–318.

Biscoe, P. V., R. K. Scott, and J. L. Monteith. 1975. Barley and its environment. III. Carbon budget of the stand. *J. Appl. Ecol.* 12:269–293.

Bond, T. E., C. F. Kelly, and N. R. Ittner. 1954. Radiation studies of painted shade materials. *Agri. Eng.* 35:389–392.

Doraiswamy, P. C., and N. J. Rosenberg. 1974. Reflectant induced modification of soybean canopy radiation balance. I. Preliminary tests with a kaolinite reflectant. *Agron. J.* 66: 224–228.

Gates, D. M. 1962. *Energy Exchange in the Biosphere.* New York: Harper and Row.

———. 1965. Energy, plants, and ecology. *Ecology* 46:1–24.

Gregson, K., and P. V. Biscoe. 1975. Barley and its environment. II. Strategy for computing. *J. Appl. Ecol.* 12:259–267.

Lemon, E., D. W. Stewart, and R. W. Shawcroft. 1971. The sun's work in a cornfield. *Science* 174:371–378.

Loomis, R. S., W. A. Williams, and A. E. Hall. 1971. Agricultural productivity. *Ann. Rev. Plant Physiol.* 22:431–484.

McDowell, R. E. 1974. The environment versus man and his animals. In *Animal Agriculture.* H. H. Cole and M. Ronning (eds.). pp. 455–469. San Francisco: W. H. Freeman and Company.

Miller, D. H. 1965. The heat and water budget of the earth's surface. *Adv. Geophys.* 11:175–302.

Miller, P. C. 1967. Leaf temperatures, leaf orientation and energy exchange in quaking aspen (*Populus tremuloides*) and Gambell's oak (*Quercus gambellii*) in central Colorado. *Oecol. Plant.* 2:241–270.

Monteith, J. L. 1969. Light interception and radiative exchange in crop stands. In *Physiological Aspects of Crop Yield*. J. D. Eastin, F. A. Haskins, C. Y. Sullivan, and C. H. M. Van Bavel (eds.). pp. 89–110. Madison, Wisc.: American Society of Agronomy.

——. 1973. *Principles of Environmental Physics*. London: Edward Arnold, Ltd.

Porter, W. P., and D. M. Gates. 1969. Thermodynamic equilibria of animals with environment. *Ecol. Mono.* 39(3):227–244.

Rosenberg, N. J. 1974. *Microclimate: The Biological Environment*. New York: John Wiley and Sons.

Shawcroft, R. W., E. R. Lemon, L. H. Allen, Jr., D. W. Stewart, and S. E. Jensen. 1974. The soil-plant-atmosphere model and some of its predictions. *Agri. Meteor.* 14:287–307.

Spotila, J. R. and D. M. Gates. 1975. Body size, insulation, and optimum body temperatures of homeotherms. In *Perspectives of Biophysical Ecology*, D. M. Gates and R. B. Schmerl (eds.). *Ecological Studies*, vol. 12. pp. 291–301. New York: Springer-Verlag.

Sutton, Sir G. 1964. Micrometeorology. *Sci. Amer.* 211(4):62–76.

Szeicz, G. 1974. Solar radiation for plant growth. *J. Appl. Ecol.* 11:617–636.

Taylor, C. R. 1969. The eland and the oryx. *Sci. Amer.* 220(1):89–95.

Wang, J. Y. 1972. *Agricultural Meteorology*. 3rd ed. San Jose, Calif.: Milieu Information Service.

9

SOIL FORMATION AND STRUCTURE

In ecological terms, soil is that portion of a terrestrial ecosystem whose matrix consists mainly of solid-phase materials, and in this sense it is a subsystem of natural or man-dominated ecosystems. In another sense it is an ecosystem in its own right. Soil consists of both living and nonliving components that interact by exchanging energy and specific chemical substances. The living components of the soil ecosystem include not only the roots of higher plants but also populations of many kinds of autotrophs, consumers, and decomposers. The members of these populations vary in size from macroscopic to microscopic and in presence from temporary to permanent. The nature and extent of their energy- and nutrient-exchange activities make them a dominant force in the overall ecosystem of which the soil is a part. Nonliving components of the soil include fragments of parental rocks and minerals, secondary minerals produced in the soil itself, organic matter, water, atmospheric gases, and of course small amounts of various organic and inorganic materials in solution. The complexity and variability of this ecosystem are extraordinary, and because many of its important functions take place among microscopic organisms within a dense, opaque matrix, it is one of the most difficult of ecosystems to study. Our knowledge of the structure and dynamics of this system is therefore far from perfect.

Soil ecosystems, like others, show complicated patterns of development in response to controlling features of the environment. The characteristics of a given soil are a function of climate, biota, topography, parent material, and the length of time through which these factors have interacted. In recent history, of course, man has become one of the most important biotic forces in the development of many soil systems: His activities have converted some lands from deserts to rich agroecosystems, others from productive lands into wastelands.

The concept of soil as a complex ecosystem that develops in response to complex environmental conditions implies several important things: first, that soil characteristics change over a continuum from place to place in response to corresponding changes in the environmental complex. It is inaccurate to think of soils as distinct types scattered in patchwork fashion over the landscape, as they are often shown on soils maps. For practical purposes we distinguish and name soil types, but we must remember that this is an intellectual compromise with the reality of continuous variation.

A second implication of the soil ecosystem concept is that the pattern of development in one location may, because of topographic influences, be affected by what occurs elsewhere. Given otherwise identical conditions, the soil that develops on a ridge will differ from the soils that develop on the slopes and valley floor below. This has led to the concept of the **soil catena,** or the sequence of soil types that tend to develop from the same parent material but in different topographic situations (Hunt 1972). This concept emphasizes the fact that soil ecosystems, as they exist in nature, are three-dimensional; and it raises the question of what represents the spatial unit of integrated and unified function under natural conditions. Huggett (1975) has suggested that this unit, which he terms the **soil landscape system,** corresponds to the valley basin or erosional drainage basin. It is the three-dimensional system bounded by the soil surface, the watershed boundaries, and the depth to which weathering has taken place in the soil. It is the natural unit because it represents the spatial unit through which materials move within the soil, both vertically (under the influence of gravity) and horizontally (as they are carried by water).

Finally, the concept of the soil ecosystem implies that the behavior of the system as a whole cannot be understood entirely by extrapolating or summing the independent properties of its individual components. At higher levels of organizational complexity, including those of soil ecosystems, properties emerge that are unique to the level of organization. Our objectives, therefore, are to examine the nature and behavior of various components of the soil ecosystem, to understand how these are integrated into overall patterns of soil ecosystem function, and to determine the significance of these relationships to agricultural activities.

We shall begin this examination of the soil ecosystem by looking at the physical structure of the soil and how this structure develops over time.

PEDOGENESIS

The **parent material** of a soil is the surface geological deposit that, through exposure to climatic and biotic influences, changes and contributes to the solid phase of the soil ecosystem. This parent material may be rock, unconsolidated sediments, organic matter, or a combination of these three materials. In mountain areas, for example, landslides can expose new, unweathered rock surfaces from time to time. Elsewhere, the action of wind or water can create new deposits of sands or still finer mineral particles, sometimes with a certain amount of mixed-in organic matter. Deposits of peat can be reexposed to the action of the environment, constituting an almost purely organic parent material for soil formation. Glaciers can scrape up, transport, and deposit mixtures of rock, unconsolidated sediments, and organic matter in areas where they are eventually exposed to processes of soil formation, or **pedogenesis.**

In general, we can distinguish two major groups of soils in terms of their parent material relationships: **residual soils** and **transported soils.** Residual soils are those that develop on consolidated bedrock parent material, regardless of whether the material is igneous, metamorphic, or sedimentary. Transported soils are those that develop on materials, generally unconsolidated, that have been transported to the particular location by wind, water, ice, or gravity in relatively recent geological times. Wind-transported parent materials include both sand and finer particles, the former deposited as dunes, the latter as **loess.** In several major areas in North America, Europe, and Asia, extensive wind deposition of silt and clay

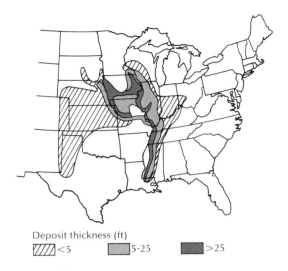

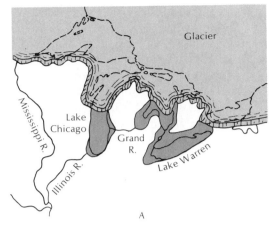

Deposit thickness (ft)

▨ <5   ▨ 5-25   ■ >25

FIGURE 9-1.
Areas of wind-deposited soil in the central United States.
Most of the deeper deposits consist of material exposed by
or washed out from melting continental glaciers at the end
of the Pleistocene. (From *Geology of Soils* by Charles B.
Hunt. W. H. Freeman and Company. Copyright © 1972.)

particles occurred during the retreat of Pleistocene continental glaciers; these loess deposits
have subsequently become important agricultural soils (Figure 9-1). Water-deposited parental materials, including those of floodplains,
deltas, and lake bottoms, have also become
important agricultural soils. During the cool,
moist period of Pleistocene glaciation, extensive freshwater lakes occupied basins that, in
many parts of the world, are now dry or contain
only small bodies of water. Likewise, large
temporary lakes were formed as continental
glaciers retreated and the ice blocked major
river drainage systems (Figure 9-2). During the
Pleistocene, continental and mountain glaciers
deposited material called **till** widely throughout higher latitudes of the Northern Hemisphere and in mountain areas throughout the
world. Transport of materials by gravity (e.g.,
through landslides, rock falls, and soil slippage)
is of more local importance and occurs primarily in mountainous regions.

FIGURE 9-2.
Post-glacial lakes in central North America: (A) lakes that
existed while the St. Lawrence River remained blocked by
ice and the Great Lakes region was drained via the Mississippi River; (B) lakes that existed later in glacial retreat
while drainages into Hudson Bay remained blocked by ice.

Weathering processes begin with the exposure of the parent material. These processes fall into two major categories: **mechanical weathering** and **chemical weathering.** Mechanical weathering includes processes that cause consolidated parent materials, such as rocks, boulders, and bedrock, to disintegrate. Expansion and contraction caused by temperature changes, for example, tend to produce fracture lines and small crevices in rock surfaces. If water penetrates these crevices and later freezes, it can exert even greater forces of disintegration. The roots of vascular plants and the rhizoids of lower plants can grow into cracks and force them open, and abrasive materials carried by wind, water, and moving ice can erode rocks and other solid parent materials.

Chemical weathering, on the other hand, includes any natural chemical process that tends to break down parent materials, to convert materials chemically from one form to another within the soil, or to transport materials from one place to another within the soil system. The principal chemical processes involved in breaking down parent materials are hydration, hydrolysis, solution, and oxidation (FitzPatrick 1972). **Hydration,** the addition of water molecules to the chemical structure of a mineral, can cause the crystalline structure of certain minerals to swell and split open. Generally, however, the hydration of mineral compounds occurs later in the weathering process. In **hydrolysis** various cations in the original crystalline structure of silicate minerals are replaced by hydrogen ions, which leads to the decomposition of those minerals. The lower the soil pH, the greater is the concentration of hydrogen ions and thus the greater is the intensity of hydrolysis of primary minerals. This form of chemical weathering may thus be fostered by living organisms, which release organic acids as a result of their metabolism, or by the decomposition of dead organic matter. **Solution** can also act as a process of chemical weathering where certain minerals with high solubility in water or in weak acids are present. Many chlorides, nitrates, and sulfates are soluble in water, although these materials tend to be minor constituents of most parent rock. Calcite and dolomite, the principal minerals of limestones, are quite soluble in weak carbonic acid solutions; limestone caves and caverns are created by the erosion of these materials in subterranean channels of water flow. **Oxidation** is still another process of chemical weathering; it is of particular importance for iron compounds, which in primary rocks exist mainly in a reduced state.

Of perhaps greater significance, however, are the chemical processes that modify the structure of minerals once they are released from consolidated parent material and that create new secondary minerals within the soil. We shall concern ourselves only with processes that relate to the origin of various clay minerals, which constitute one of the most important structural and functional components of agricultural soils. Clay minerals fall into two major groups: the silicate clays, and the hydroxide clays. Clay minerals form the finest particulate group in soils, the particles being generally less than 0.002 millimeters in diameter. Clay minerals are extremely variable in structure, and are very difficult to study. Because clay mineralogy is extremely complex we will confine our attention to rather general aspects of this topic.

The **silicate clays** are primarily hydrated aluminum silicates, most of which possess a crystalline structure visible only with the aid of the electron microscope (Figure 9-3). This structure consists of a series of laminated plates or crystalline sheets of aluminum and silicon atoms (Figure 9-4). The structure of various clay minerals differs according to the arrangement of these two types of crystalline layers and the presence or absence of other elements, particularly iron and magnesium. Of the many forms and varieties of clay minerals distin-

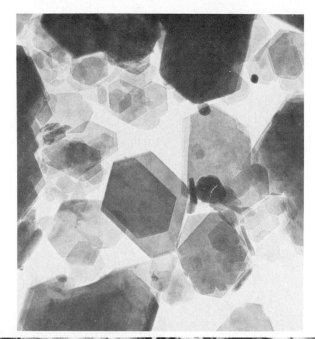

FIGURE 9-3.
Electron photomicrographs of the fine structure of representative crystalline clays: (A) transmission electron micrograph of kaolinite from Georgia, U.S.A. (magnification about 45,000×); (B) transmission electron micrograph of platinum-carbon replicas of montmorillonite from Montmorillon, France (magnification about 12,500×). (Photographs courtesy of Kenneth M. Towe, Smithsonian Institution, Washington, D.C.)

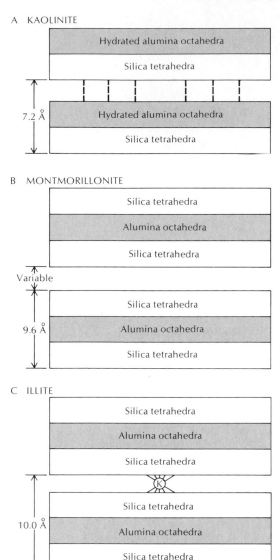

FIGURE 9-4.
Microcrystalline structure of kaolinite, montmorillonite, and illite. (A) Kaolinite, a 1:1 clay mineral, possesses basic units made up of adjoining sheets of aluminum and silicon atoms. These units are rigidly bound together and do not separate when the clay crystal is wetted. (B) Montmorillonite, a 2:1 mineral, has units consisting of a sheet of aluminum atoms sandwiched between sheets of silicon atoms. When the crystal is wetted, these units separate, admitting water and solutes to the interior of the crystal and causing it to swell. (C) Illite, also a 2:1 mineral, has units that are held together more strongly by bonds involving potassium atoms.

guished by mineralogists, five basic types are both common and of major importance: kaolinite, illite, montmorillonite, vermiculite, and allophane.

Kaolinite is characterized by a 1:1 crystalline structure formed by alternating sheets of aluminum and silicon atoms. **Illite, montmorillonite,** and **vermiculite,** on the other hand, have a 2:1 structure formed by a sheet of aluminum atoms sandwiched between two sheets of silicon atoms; this structure is more complex and variable than that of kaolinite, and the composition of these three minerals includes atoms of potassium (in the case of illite), iron and magnesium (in the case of montmorillonite and vermiculite), and calcium (in the case of montmorillonite). The crystals of these clay minerals consist of repeated sets of these 1:1 or 2:1 basic units, which are held together with varying strength by hydrogen bonds between adjacent crystalline units. For kaolinite, these hydrogen bonds are strong and the overall crystal is a relatively stable structure. For illite and vermiculite, and to an even greater extent for montmorillonite, this interunit bonding is weak, so that, when clays of these types are wetted, the individual 2:1 units tend to become hydrated. In the process these sheetlike units separate, water molecules move into position and form weak chemical attachments between them, and the crystal as a whole swells. As masses of these clays dry out, the clay crystals shrink. This tendency to swell and shrink is greatest for montmorillonite, and so clayey soils high in montmorillonite are often quite easily identified by their tendency to form large cracks as they dry (Figure 9-5). This alternate swelling and shrinking can create serious engineering problems in highway and building construction. **Allophane** is a clay mineral common in soils derived from volcanic ash, and although it possesses aluminum and silicon atoms in a ratio of 1:2, an extensive sheetlike crystalline structure is apparently absent. Particles of allophane usually appear

FIGURE 9-5.
Shrinking and cracking of a clayey soil high
in montmorillonite, near Las Cruces,
New Mexico. (Photograph by R. E. Neher,
courtesy of the USDA Soil Conservation
Service.)

as irregular spherules, and this clay mineral is usually described as amorphous.

More important for agriculture, however, is the fact that silicate clays are able to hold cations in adsorbed form on their surfaces. This is because their crystals or particles carry a net negative electrical charge, and thus attract and hold positively charged ions, many of which are important plant nutrients. These cations are available to plants by means of processes examined in detail in Chapter 10. Here, however, we shall only mention several general points that relate to our discussion of the structure of these materials. The ability of various clay minerals to hold and exchange cations, termed **cation exchange capacity,** varies with the mineral. For kaolinite, in which the basic crystalline units tend not to separate

when wetted, cation adsorbtion takes place only on the external surfaces of the crystal. For illite, montmorillonite, and vermiculite the internal "surfaces" of the sheets can also hold adsorbed cations, since these sheets tend to separate when wetted, thus admitting cations to the interior of the crystal. Cation exchange capacity is therefore great for montmorillonite and vermiculite, less for illite, and quite small for kaolinite (Figure 9-6). Allophane also has a high cation exchange capacity correlated with the very large total surface area of its combined particles. Clearly, this ability can determine to a great extent the fertility potential of a given soil.

The **hydroxide clays** also lack a clear crystalline structure. These materials are no more than hydrated iron and aluminum oxides with the

226

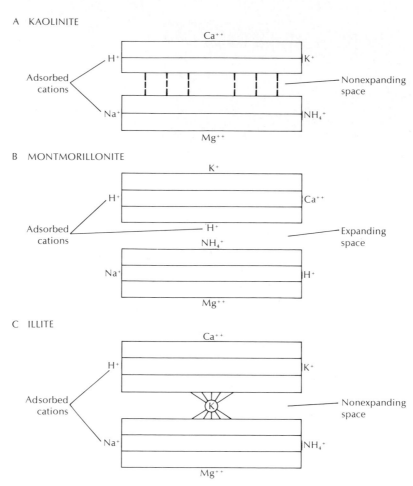

FIGURE 9-6.
Microcrystalline structure of kaolinite, montmorillonite, and illite showing surfaces on which cations are held by electrostatic forces (adsorbed). (A) For kaolinite exchangeable ions are held on external surfaces only. (B) For montmorillonite, separation of units upon wetting allows ions to be held on internal as well as external surfaces. (C) For illite, the adsorbtion surfaces are external only.

generalized chemical structures $Fe_2O_3 \cdot xH_2O$ and $Al_2O_3 \cdot xH_2O$, where x indicates a variable number of hydrated water molecules attached to the iron or aluminum oxide. The cation exchange capacity of these clays is lower than that of any of the silicate clays.

These clay minerals are formed in the soil by complex processes by which silicate minerals in parental rock are chemically modified and reorganized. It is important to note, however, that the resulting clay component of any soil is a mixture of many secondary clay minerals, together with their precursors and intermediates. Under particular climatic conditions, however, particular types of groups of these minerals tend to predominate (Gradusov 1974).

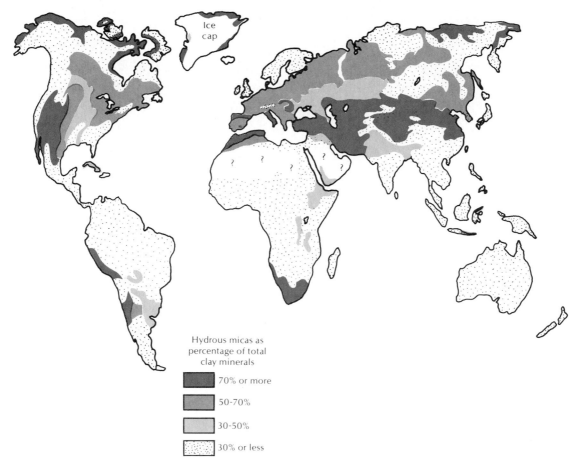

FIGURE 9-7.
World distribution of hydrous mica clay minerals (mostly illite and vermiculite) in the A horizons of zonal soils. (Modified from Gradusov 1974.)

In particular, the hydroxide clays predominate in most humid tropical areas, a feature with important implications for the agricultural potential of these regions. Kaolinite, in the United States, is most abundant in the soils of the southeastern states, where serious problems of low fertility are common.

As shown in Figures 9-7 and 9-8, the silicate clays that have higher cation exchange capacities, such as illite, montmorillonite, and allophane, are widespread in the Northern Hemisphere and in the southern portions of South America, Africa, and Australia. In general, this group of clay minerals is characteristic of relatively young soils, and so these minerals tend to predominate where soils have been derived from glacial till in relatively recent geological times. With time, and with continued intensive leaching under warm climate and acidic soil conditions, montmorillonite is apparently converted first to kaolinite, as in the southeastern United States, and even-

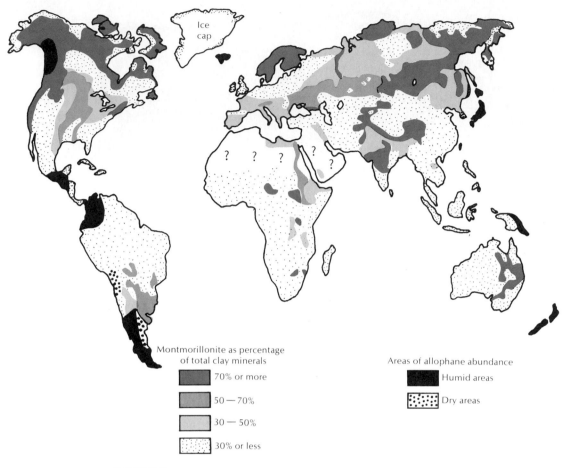

FIGURE 9-8.
World distribution of montmorillonite and allophane in the A horizons of zonal soils. (Modified from Gradusov 1974.)

tually to hydroxide clays, as in the humid tropics. The general nature of this process is suggested in Figures 9-9 and 9-10.

These features of mechanical and chemical weathering create a particular textural composition for the mineral portion of the soil. Soil texture is described in terms of the percentage, by weight, of the total mineral soil that falls into various particle size classes. The major classes, in order of decreasing particle size, are sand, silt, and clay, although the critical limits for these classes vary somewhat with the classification system used (Table 9-1). Soil texture is analyzed by a technique that separates individual soil particles and suspends them in water in a sedimentation tube. After the soil sample is thoroughly shaken to suspend soil particles, hydrometer readings are taken at various times as the particles gradually settle to determine the quantity of soil still in sus-

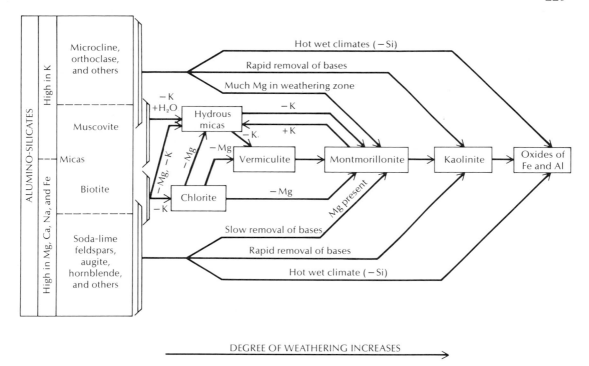

FIGURE 9-9.

Approximate sequence of weathering conversions of primary and secondary clay minerals. (Reprinted with permission of Macmillan Publishing Co., Inc., from *The Nature and Properties of Soils,* 7th ed., by H. O. Buckman and N. C. Brady. Copyright © Macmillan Publishing Co., Inc., 1969.)

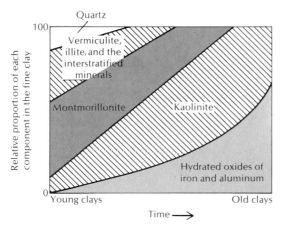

FIGURE 9-10.

Change in the proportions of various secondary clay minerals as a function of the length of weathering (schematic). (From L. M. Thompson, *Soils and Soil Fertility.* Copyright © 1957 by the McGraw-Hill Book Company. Used with permission of McGraw-Hill Book Company.)

TABLE 9-1
Soil texture classification according to the United States Department of Agriculture System and the International System.

	USDA	International
Category	Diameter limits (mm)	Diameter limits (mm)
Very coarse sand	2.00–1.00	—
Coarse sand	1.00–0.50	2.00–0.20
Medium sand	0.50–0.25	—
Fine sand	0.25–0.10	0.20–0.02
Very fine sand	0.10–0.05	—
Silt	0.05–0.002	0.02–0.002
Clay	<0.002	<0.002

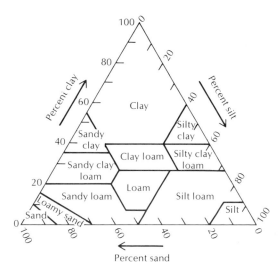

FIGURE 9-11.
Textural soil names based on the percentage composition of sand, silt, and clay. (After USDA drawing.)

pension. The larger particles settle out first; the clay fraction remains in suspension for a number of hours or longer.

Based on the percentage of sand, silt, and clay, a name based on its texture may be given to a soil (Figure 9-11). Soil texture is related to several characteristics that influence a soil's suitability for agriculture. In general, a **loam,** consisting of less than 30 percent clay and about equal portions of sand and silt, is most suitable. Loamy soils tend to be easy to cultivate, possess a high degree of aeration and water-holding capacity, and are well able to retain nutrients.

The composition and structure of the soil are affected not only by mechanical and chemical weathering, but also by biotic processes, which are important in terms of what they contribute physically to the soil system—organic matter—and what they contribute to its organization.

Except in soils that develop on fossil deposits of organic matter, such as peat, the organic matter present in them is derived primarily from organisms that live within or on the surface of the soil in question. All the organic matter involved comes ultimately either from green plants or from other specialized autotrophs of minor overall importance. The root systems of higher plants penetrate throughout the soil; the soil can, in fact, be defined as extending only the the maximum depth of root penetration, where it usually changes to parent material. Death of these root tissues introduces bulk organic matter into deeper layers of the soil, and in some crop systems, where most aboveground plant parts are harvested, this is the principal route of introducing organic matter into the soil. Even during the life of the plant, however, organic substances can pass from roots into the surrounding soil (Borner 1960).

Most of the organic matter in most natural ecosystems enters from the surface, however. In some instances, especially in forests, a surface layer of dead leaves, branches, and other plant materials can accumulate to considerable depth, and animal carcasses are also deposited on the surface from time to time. All of these materials, termed detritus, are fed upon by vertebrate and invertebrate consumers, which fragment them and convert the organic substances into their own tissues and organic wastes. This action, together with that of microscopic decomposer organisms, ultimately mixes much of the detritus into the surface layers of mineral soil as a formless, chemically complex organic substance termed **humus.** At this stage the original source of the material may have been obscured by its passage through a complex network of detritus-feeding and decomposer organisms; for example, it may be possible to identify much of the organic matter present only as having most recently been part of the body tissues of decomposer bacteria.

Humus comprises a number of distinct groups of organic compounds, each of which itself contains many specific substances (Kononova 1966). Some of these compounds are residues of plant and animal tissues that have undergone extensive decomposition; others are residues of the tissues of bacterial and fungal decomposer organisms; that is, organic molecules secondarily synthesized by these organisms from the dead matter on which they are dependent for nutrition. **Polysaccharides** constitute one of the most important components of humus; they are polymers, or long-chain molecules that consist of repeating units of simple sugars. Cellulose and lignin, constituents of the cell walls of higher plants, are among the decomposition residues present. Two other groups, **muco-polysaccharides** and **polyuronide gums,** are derived largely from bacterial cell walls. Muco-polysaccharides are polymers that consist of simple sugars, some of which possess attached amino groups; these molecules can reach molecular weights of several million. Polyuronide gums are polymers composed of units that are derived from simple sugars and that possess attached amino groups. In general, they are smaller in size than muco-polysaccharides, with molecular weights up to 100,000 or so. Both of these groups, as we shall see, can exert an important cementing action in the soil.

Still another diverse and, as yet, poorly known group of organic substances are the **humic acids.** In general, these compounds are smaller than the others, with molecular weights under 50,000, and are more nearly spherical than the elongate polysaccharides. Humic acids appear to be products of decomposition of plant tissues (Baver 1968). **Lipids,** which are products of the decomposition of both plant and animal tissues, are also present in soils and, because of their water resistance, they can play a special role in soil structure.

Organic matter is important for a number of reasons: Through its decomposition, it can release nutrients that can be taken up by growing plants; and, like silicate clays, it holds cations by adsorption, thus contributing to the overall cation exchange capacity of the soil. These two aspects of the function of soil organic matter are described in Chapter 10, but here, we shall consider another function—the cementation of mineral particles into larger aggregates.

In addition to the textural composition described earlier, soils possess a macrostructure formed by the aggregation of soil particles into units that vary in size from small to large (Figure 9-12). The more massive soil aggregates typically occur in finer-textured soils as a consequence of various chemical and physical processes, some the result of improper farming practices, and they are generally unfavorable to crop growth. Macrostructure of the "crumb" or granular type, however, is an important benefit to agricultural soils.

Soil crumbs are aggregates that range in diameter from about 1 to 3 millimeters (Buckman and Brady 1974), and they vary in stability. Some, described as being air stable, retain their structure as long as they are dry but lose it when wetted. Water stable aggregates, on the other hand, maintain their form even when wetted. Good crumb structure in a soil is detectible in the field; it is, in fact, what gives a soil the quality of **tilth** (porosity and ease of tillage) desired by the farmer (Figure 9-13). When

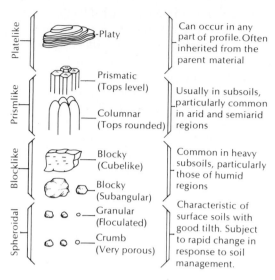

Platelike	Platy	Can occur in any part of profile. Often inherited from the parent material
Prismlike	Prismatic (Tops level) / Columnar (Tops rounded)	Usually in subsoils, particularly common in arid and semiarid regions
Blocklike	Blocky (Cubelike) / Blocky (Subangular)	Common in heavy subsoils, particularly those of humid regions
Spheroidal	Granular (Floculated) / Crumb (Very porous)	Characteristic of surface soils with good tilth. Subject to rapid change in response to soil management.

FIGURE 9-12.
Diagrammatic representation of different patterns of soil aggregation and macrostructure. (Modified from Buckman and Brady 1974.)

good tilth or crumb structure is present, a lump of soil may be crushed easily by hand, yielding a mass of fine aggregates in the size range noted above.

The significance of good crumb structure is manifold. First, it maintains a low bulk density; that is, a low weight of solids per unit volume. This is equivalent to maintaining a high percentage of pore space within the soil, providing for effective aeration and allowing water to percolate readily into the soil and be retained in a portion of the available pore space. Low bulk density, and the absence of massive, hard aggregates means that the soil is easy to till. It also allows the roots of growing crop plants to penetrate the soil easily, for essentially the same reason. Stability of these aggregates in the face of exposure to wind and water means also that the soil is more resistant to erosion by these agents. Good tilth,

FIGURE 9-13.
The soil on the left shows a dense, blocky macrostructure unfavorable to tillage and root penetration. That on the right possesses good crumb structure, allowing easy penetration by roots and facilitating tillage. (Photograph courtesy of the USDA.)

FIGURE 9-14.
Massive aggregates formed by plowing, under wet conditions, a central Ohio soil that lacks adequate crumb structure. (Photograph by G. W. Cox.)

finally, allows the farmer greater freedom in timing his plowing and cultivating. As crumb structure deteriorates, the moisture range in which such activities can be carried out narrows and the danger increases that massive, blocky aggregates will be produced due to overly moist soil conditions during cultivation (Figure 9-14).

The origin of crumb structure is a complex process (Russell 1971) that consists of two steps: the formation of units of mineral soil from 1 to 3 millimeters in diameter, and the cementation of these soil particles into a unit with some degree of stability. Mechanical processes such as freezing and thawing, wetting and drying, and the compressional forces created by growing roots are all important in forming the soil into crumb-sized units. In areas of rela-

tively moist climatic conditions, especially in the Temperate Zone, earthworms are also important in creating such units. The soil ingested by these organisms is organized into crumb-like units during its passage through the earthworm intestine (Figure 9-15).

Cementation of these units is largely the function of organic matter (Harris et al. 1966). Although weak chemical forces between clay particles can contribute, interparticle linkages that result from long-chain organic molecules seem to be especially important. Lipids and related substances can also give crumb units a degree of "waterproofing," thus contributing to their water stability. Evidence of the importance of many of the naturally occurring polysaccharides as cementing agents is supplied by the fact that long-chain synthetic polymers

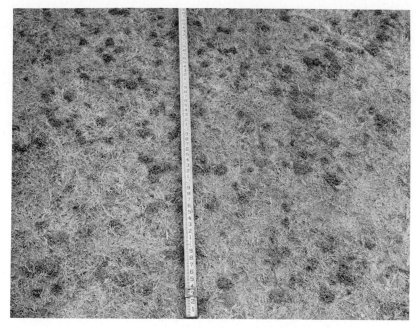

FIGURE 9-15.
Earthworms pass large quantities of soil through their digestive tracts, molding and
cementing the mineral particles into soil crumbs. The intensity of this action is evident in
this photo of earthworm casts in a Bermudagrass pasture in Texas. (Photograph by J. W.
Huckabee, Jr., courtesy of the USDA Soil Conservation Service.)

of various types can be used to create aggregates of the crumb type experimentally (Harris et al. 1966; Stefanson 1973).

In addition to acting as agents that organize soil particles into crumb-sized units, earthworms exert a cementing action on these same units (Barley 1961). The soil ingested by earthworms is formed into mucus-covered pellets as it passes through the digestive tract; during the course of a year up to 9 or more kilograms dry weight of soil per square meter can be converted into stabilized soil crumbs in this manner.

Clay soils may develop a granular structure primarily through electrostatic attraction of clay particles, a phenomenon known as **flocculation** (Taylor 1972). In the case of colloidal clay particles or crystals, aggregation represents the state of thermodynamic stability. Aggre-

gation can be prevented, however, by the nature of the surface-related charge on the individual particles, which can repel instead of attract. This charge, in turn, is influenced by the ionic composition of the soil solution surrounding the clay particles. This relationship is quite complex and relates to the thickness of the charge-bearing zone that surrounds the clay particles. When sodium ions predominate in the soil solution, a thick, repulsive charged zone exists, and the individual clay crystals remain dispersed; this condition is described as **puddling** of the clay and is characterized by soil that is sticky when wet and becomes hard and impermeable when it dries. A predominance of calcium ions in the soil solution has the opposite effect and allows the individual clay crystals to aggregate into granular units that give a more favorable texture to the soil as a whole. This, of course, is one of the principle reasons for using gypsum (calcium sulphate) on heavy, clayey soils.

Most of the pedogenic processes that we have discussed so far operate differently from one another at different locations in a soil landscape system or at different soil depths. Adding to the complexity of the process of pedogenesis are factors that transport materials, either from one depth to another or from a location topographically higher in the landscape system to a lower location. The action of water in translocating materials is termed **leaching.** Considering first the vertical movement of material in a particular location within a soil system, we should note that the type and quantity of material leached depends on total precipitation, the acidity of the water passing through the soil, temperature, and many other factors. Materials can be dissolved from one layer and deposited in another, or they can be carried in "suspension" as water filters to greater depths in the soil. Where precipitation is great, some materials can be carried completely through the soil and into the groundwater system, thus effectively flushing them out of the ecosystem of which the soil is part. Where rainfall is less, materials are carried to some consistent depth, such as the average maximum depth of moisture penetration, and deposited there. In some cases, where the groundwater table is close to the surface, materials can move upward in the soil with capillary movements of water. Evaporation of the water then leaves these transported solutes at or near the surface.

Water moves not only vertically but laterally through the soil, parallel to the surface, in a downslope direction. In general, once pedogenesis has been carried on to a significant stage, a profile of the soil's permeability to water tends to develop. Permeability tends to decrease with depth, which means that, at any given depth, the resistance to lateral water movement is almost always less than to vertical

downward movement. Thus a strong tendency exists for lateral movements of water wherever there is some degree of slope.

We should also note that other factors can produce vertical and horizontal movements of soil materials. Freezing and thawing, especially in arctic and alpine areas, can sort soil components and produce a vertical "mixing" of the soil profile (Hunt 1972). The burrowing activities of small vertebrates and invertebrates can transport considerable quantities of material from deeper soil layers to the surface (Abaturov 1972), and, of course, deeply rooted plants are continually engaged in extracting nutrients from the soil and in translocating them to their above-ground parts, where many of them end up deposited on the soil surface as dead tissues. Thomas (1969), for example, showed that in forests of the southeastern United States, the flowering dogwood, *Cornus florida*, acts as a sort of calcium "pump" in the forest ecosystem. This species, a small understory tree, is a minor constituent of the vegetation, constituting only 0.2 percent of the total dry weight biomass of the organic matter in the above-ground vegetation and undecomposed litter. Flowering dogwood tissues, however, contained nearly 1.8 percent of the total calcium in the organic components of the forest, making the species nine times as important in calcium content as in biomass.

SOIL PROFILES

All these processes of pedogenesis operate within the soil ecosystem, and one of the results of their concerted action (in a sense one of the unique, or emergent, properties of the ecosystem level of organization) is the development of a characteristic soil profile, or spectrum of profiles, that represent the range of conditions throughout a given soil landscape system. Because it reflects most of the major pedogenic processes that we have been considering, the

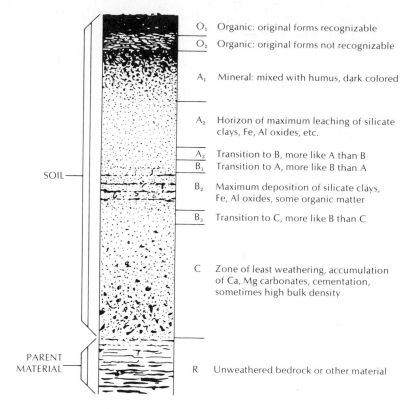

FIGURE 9-16.
A representative soil profile, illustrating the various horizons that might be present.

soil profile has become one of the basic features used to describe and classify soils.

Soil profiles consist of several more or less recognizable layers, or **horizons.** These horizons are designated by a set of standardized symbols, diagramed in Figure 9-16. The O horizon, which is nearest the surface, consists of organic matter, as yet unmixed with the underlying mineral soil. This horizon is often well developed in forest ecosystems, but it may be virtually absent in other environments. The O_1 subdivision of this horizon designates the portion of the layer in which the original forms of the materials can still be distinguished; in the O_2 subdivision, decomposition is advanced

enough that the sources of the organic residues are unidentifiable.

The A horizon is the topmost layer of mineral soil and is usually the layer that has the highest content of organic matter. It is also the layer in which the leaching of various materials is most intense. The A_1 subdivision is characterized by dark coloration, indicating a high organic content; the A_2 portion of this horizon is the layer in which leaching is greatest. The types of materials removed by leaching, as we shall see, vary greatly with climate and other conditions. The A_3 subdivision is a transitional layer to the next horizon that is nevertheless similar to the main part of the A horizon.

The B horizon is the soil layer in which many of the leached materials are deposited. The B_1 portion, like the A_3 directly above it, is a transitional layer. The B_2 subdivision is the layer in which the greatest deposition takes place, and it is typically darker in color and higher in organic matter than the most strongly leached portions of the A horizon. The B_3 horizon is another transitional layer.

The C horizon consists of weathered parent material. In certain soils it is also the layer in which highly soluble materials, such as calcium and magnesium carbonates, and some other materials are deposited. Deposition, in these cases, indicates that the soil is a closed system, and precipitation does not generally penetrate to the water table. The deposited materials might form a hardpan or cemented layer that can be rocklike and impervious to water and plant roots. The underlying, consolidated rock layer, whether or not it is of the same material as the parent material of the soil, is often designated R.

Not all these horizons and their subdivisions are necessarily present in a given profile. In addition, various horizons can be characterized by special conditions that should be noted in a profile description. A number of the most important of these are indicated in Table 9-2. We might, for example, characterize a particular alkaline grassland soil as having the following horizonal sequence: A_p, B_2, B_{3ca}, C_{ca}, C_{sa}, C (Hunt 1972). This would indicate a surface layer, rich in organic matter but disturbed by plowing, below which is a layer rich in deposited materials leached from the A horizon, followed by transitional B and C horizon layers in which there is a zone of calcium carbonate deposition. Still deeper lies a zone in which other soluble salts have been

TABLE 9-2
Symbols commonly subscripted to soil horizon designations to indicate special characteristics.

Subscript	Characteristic
b	Indicates a soil layer buried by a surface deposit; for example, a leached layer buried under a sand dune would be designated A_{2b}.
ca	An accumulation of calcium carbonate.
cn	An accumulation of concretions, usually of iron, manganese and iron, or phosphate and iron.
cs	An accumulation of gypsum, calcium sulfate.
f	A permanently frozen layer.
g	A waterlogged or gley layer.
h	An accumulation of organic matter.
ir	An accumulation of iron.
m	A cemented layer or hardpan, due to deposition of silica or carbonate.
p	A layer disturbed by plowing.
sa	An accumulation of soluble salts.
t	An accumulation of clay.

Source: Adapted from *Geology of Soils* by Charles B. Hunt. W. H. Freeman and Company. Copyright © 1972.

deposited, and finally, at the bottom of the profile, a layer of weathered parent material with no significant deposition of materials from above.

The process of pedogenesis, and the profile it creates, vary with the regional climate and with a great many local relationships. Regional patterns are strong, however, and we can distinguish four major pedogenic processes that are closely related to climate: gleization, podzolization, calcification, and laterization. We shall examine each of these in detail, and then consider briefly the resulting patterns of geographical distribution of soils.

MAJOR SOIL FORMATION PATTERNS

Gleization is the characteristic soil formation process of arctic and subarctic regions with cold, humid climates. The humid conditions that prevail in these areas result not so much from high precipitation as from the facts that evaporation and transpiration are low, due to the prevailing low temperatures, and that drainage is often impaired by the presence of permafrost. **Permafrost** is the zone of permanently frozen substrate that extends from the maximum depth of summer thaw downward to great depth in the underlying material. Under these conditions, organic matter decomposes very slowly, and for much of the summer period the thawed zone of surface soil tends to remain waterlogged, due to the poor drainage. The low decomposition rate causes large quantities of organic matter to accumulate in the surface layers of the soil, and waterlogging creates anoxic conditions in the lower, mineral soil layers, that tend to maintain iron components in a reduced, or ferrous, state. In this state they have a blue-gray color, which gives this horizon a distinctive appearance.

These conditions lead to a shallow soil with a profile of only a few layers (Figure 9-17).

A surface O horizon of undecomposed peaty material typically occurs, overlying a deeper A_1 horizon that is dark in color and that usually has an organic matter content over 50 percent. In some soils this is underlain by a lighter-colored, sandier A_2 horizon. Below these lies a B horizon that is high in clay and that varies in color from solid blue-gray to blue-gray mottled and streaked with yellow and red from the oxidation of iron compounds along root channels and aggregate boundaries.

Obviously, many problems must be overcome before soils of this type can be cultivated: Artificial drainage is often required; plowing and liming are generally necessary to raise the pH of the soil, mix the organic surface layers with the mineral soil, and encourage decomposition; and where there is permafrost, serious problems can arise with the melting of massive ice deposits in the subsoil after the land has been cleared and plowed.

Gley soils show a transition to those created by a second major soil formation process, podzolization (Karavayeva 1974). **Podzolization** occurs characteristically in regions with cool, humid climates where precipitation is high and soil pH is acid; it is largely the result of acidic decomposition products from plant material. **Podzol soils,** the most characteristic product of this process, usually occur in areas that were originally covered by coniferous forest. **Podzolic soils,** mainly the result of podzolization but influenced by other processes as well, characterize many of the areas originally covered by temperate deciduous forests.

The characteristic features of podzolization are the leaching of organic matter and clay minerals from the A horizon and their deposition in the B horizon (Figure 9–17). This tends to produce a so-called classic profile in which most of the basic horizons described earlier are clearly represented. In undisturbed situations, surface accumulations of forest litter create a well-developed O horizon, beneath which lies

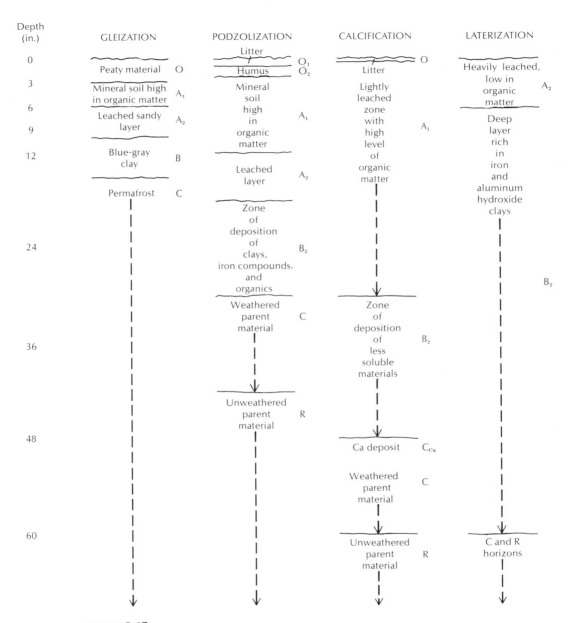

FIGURE 9-17.
Representative soil profiles resulting from typical sequences of gleization, podzolization, calcification, and laterization.

a relatively thin, dark A_1 horizon with a high content of organic matter mixed with mineral soil. The next layer is a thick, intensely developed A_2 horizon, characteristically lighter in color, that in well-developed podzols might consist of little more than quartz sand and might have an ashy-white color. Transitional layers to the B horizon might be recognizable, and the B_2 horizon is typically a deep layer with a reddish or reddish-brown color due to higher concentrations of clay minerals, iron oxides, and organic matter.

Podzols and podzolic soils are "open" soil systems, in the sense that they receive more water by precipitation than they lose by evapotranspiration (i.e., the combined loss of water by evaporation from the soil and from plant surfaces). As a result, quantities of water pass through these systems, either vertically into groundwater systems or laterally into streams. These flows carry materials in solution and occasionally in suspension. This throughflowing water has an acid pH, which significantly affects the fertility of the soils in question. We noted earlier that the cation exchange capacity of a soil is its ability to hold positively-charged ions in adsorbed form on clay minerals and organic matter. These ions, many of which are important plant nutrients, are actually held with different electrostatic force; that is, some ions are held more strongly than others. The hydrogen ion has a very strong affinity for these adsorption sites, and, as a result, when hydrogen ions are present in abundance in the soil solution, they tend to displace other cations from adsorption sites, leaving them free in the soil solution and thus prone to be carried away by leaching. (These relationships are discussed in more detail in Chapter 10.) Podzols, then, tend to be low in fertility. When converted to agriculture they require heavy liming to raise the pH, and regular, heavy fertilization with all major plant nutrients. With podzolic soils of temperate deciduous forest regions, these problems are less severe,

although still of major importance.

The third major soil formation process, **calcification,** takes place in dry climates with cool to hot temperature regimes. This process, in its broadest nature, is responsible for soil development in grasslands, steppes, and deserts. We can consider the typical process to be that which occurs in perennial grassland areas, where relatively dry conditions prevail but where a continuous cover of natural vegetation is present. Soils of this type are "closed" systems, in which evapotranspiration is greater than precipitation, and leaching of materials completely out of the system is infrequent. Although some of the more soluble components of the surface layers are translocated to deeper layers, the clay fraction is not leached and texture often changes little with depth.

Typically, O horizons are absent or only weakly developed (Figure 9-7), but the A_1 horizon can be thick and dark, reflecting the incorporation of organic matter to a considerable depth by the death and decomposition of the extensive fibrous root systems of grassland plants. Where calcium or magnesium carbonates, or other soluble materials, are present, these are leached and deposited at a depth determined by the average depth of water percolation. In progressively drier environments these deposits occur at shallower depths, and where the quantities of mineral are large, a hardpan can form that is impervious to water and plant roots.

The natural fertility of soils formed by calcification processes tends to be high, and except where unusual conditions exist, such as the presence of a hardpan or high salinity, these soils are very well suited to agriculture, given adequate moisture.

The last major soil formation process, **laterization,** is typical of the humid tropics, although it is by no means the only important pedogenic process in tropical latitudes. **Lateritic soils** tend to develop under conditions of high precipitation and year-round high temperatures

(McNeil 1964), and so they predominate in areas of the lowland tropics that have humid climates, such as the Amazon Basin, the Congo Basin, and much of southeastern Asia and the East Indies. However, soils showing the influence of this process also occur in subtropical areas such as the southeastern United States and parts of Australia.

Laterization is characterized by the rapid decomposition of organic matter under favorable conditions of moisture and temperature and by intense leaching due to high precipitation. Furthermore, the soils of the tropics are old soils that have been worked and reworked by surface processes of erosion, alluvial deposition, and soil formation for long periods of geological time. Under the conditions of warmth and humidity, leaching has tended to remove silicates, leaving high concentrations of iron and aluminum hydroxide clays as the primary constituents of surface soil (Figure 9-17). The abundance of aluminum, in some areas, is toxic to certain crops. In addition, the cation exchange capacity of these soils tends to be low, and this, combined with the acidity of the materials produced by decomposition, means that nutrient cations are subject to rapid leaching. Although areas of tropical rain forest that occur on lateritic soils have the highest productivity of any natural vegetation type, most of the nutrients are bound up in the organic biomass of the system at any one time and are well conserved by ecosystem processes. When this natural vegetation is removed, nutrient dumping tends to occur; that is, the accumulated nutrient capital of the system is leached away.

We should note that there has been an unfortunate tendency among some people to equate tropical soils with lateritic soils, and these in turn with a particular type of lateritic soil (Sanchez and Buol 1975) known commonly as a **ground water laterite,** which dries, upon exposure to the sun and air, to form a material of bricklike hardness (Kellogg 1950). Most soil scientists, as a result, prefer to restrict the use of the term **laterite** to the material thus formed, rather than to lateritic soils. Although highly leached soils formed by the general process of laterization occupy perhaps 51 percent of the land area of the tropics, no more than about 7 percent of the land possesses soil that is subject to lateritic rock formation upon exposure (Sanchez and Buol 1975). Moreover, it is usually only after a significant degree of surface erosion that a layer prone to lateritic rock formation is exposed. Nevertheless, the area of such soils is not inconsequential, and their potential for permanent, traditional sorts of agriculture is almost nil.

SOIL CLASSIFICATION AND DISTRIBUTION

Several major systems for naming and classifying soils have been proposed, and a number of these are now simultaneously in use (Buol et al. 1973). Unfortunately, in recent years, there has been a tendency among soil scientists to use systems of naming and description whose synthetic terminology is almost unintelligible to the nonspecialist. Here we shall use a traditional terminology and consider only the major soil types derived by the processes just described.

Soils can be classified, first, according to whether they are azonal, intrazonal, or zonal. **Azonal soils** lack the development of a profile, chiefly because they are of very recent origin, or sometimes because they are subject to more or less constant reworking and redeposition; these soils include beach deposits, dunes, and talus accumulations. **Intrazonal soils** are formed as a result of local, specialized conditions of parent material, topography, or habitat; bog soils, saline and alkaline soils, and limy soils are examples. **Zonal soils** are those with major characteristics related to regional soil formation processes.

Climate			Pedogenic process	Charac-teristic plant formation class	Zonal soil group	Main soil class
Moisture	Temperature	Climate type				
High	Cold	ET	GLEIZATION	Tundra	Tundra	PEDALFER
	Cool	D	PODZOLIZATION	Needle-leaf forest	Podzol	
	Warm	C		Deciduous forest	Podzolic	
	Hot	A	LATERIZATION	Rain forest	Lateritic	
Low	Cool-warm	BS	CALCIFICATION	Prairie	Chernozem	PEDOCAL
				Steppe	Chestnut and brown	
	Hot	BW		Desert	Desert and red desert	

FIGURE 9-18.
The principal soil formation processes in relation to climate, vegetation, and resulting soil type.

Zonal soils can be subdivided into two major groupings: pedalfers and pedocals (Figure 9-18). **Pedalfers** are soils that tend to be acidic, open-system soils in which calcium and magnesium salts tend to be leached out of the system. **Pedocals** tend to be basic in pH and are closed soil systems in which calcium and magnesium salts are deposited at some depth within the soil profile. Zonal soils, as we have suggested in our discussions of pedogenesis, show strong relationships with climate and natural vegetation (Figure 9-19). Gleization is the pedogenic process that predominates in arctic tundra and in certain alpine tundra situations creating tundra and gley soils. The geographical distribution of these and other types is shown in Figure 9-19.

Podzolization becomes the predominant process as one moves south into forested regions. True podzols correspond very closely in distribution to the northern coniferous forest, and podzolic soils, which show characteristics influenced by the process of laterization, occupy regions of temperate hardwood forest and extend into certain forested subtropical and even tropical areas. In the temperate zone, where parent materials are low in carbonate, gray-brown podzolic soils with well-defined leached A_2 horizons predominate. Where the parent material is high in carbonate, the de-

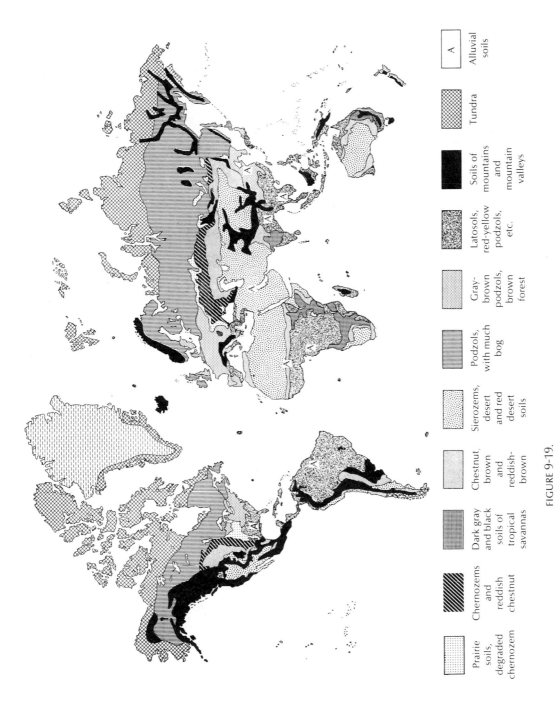

FIGURE 9-19.
Generalized world soils map. (From "Soil" by Charles E. Kellogg.
Copyright © 1950 by Scientific American, Inc. All rights reserved.)

Prairie
soils,
degraded
chernozem

Chernozems
and
reddish
chestnut

Dark gray
and black
soils of
tropical
savannas

Chestnut,
brown
and reddish-
brown

Sierozems,
desert
and red
desert
soils

Podzols,
with much
bog

Gray-
brown
podzols,
brown
forest

Latosols,
red-yellow
podzols,
etc.

Soils of
mountains
and
mountain
valleys

Tundra

A

Alluvial
soils

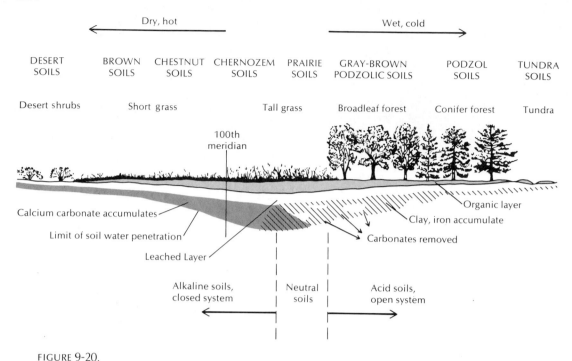

FIGURE 9-20.
Schematic representation of major changes in soil structure and characteristics along environmental gradients in arctic and temperate regions. (From *Geology of Soils* by Charles B. Hunt. W. H. Freeman and Company. Copyright © 1972.)

velopment of a leached zone is inhibited, and the resulting soil is commonly termed a **brown forest soil.** As one proceeds southward, still in more humid regions, these podzolic soils give way in stages to lateritic soils.

In drier regions, where calcification is the dominant soil formation process, we can distinguish a number of more specific soil types. In areas of tall-grass prairie, close to the forest boundary, where precipitation is still great enough to leach away carbonates and to concentrate clays in a well-defined B horizon deposit, we find the so-called **prairie soils.** To their west lie the **chernozem soils,** which have a deep, dark A horizon, and in which a smaller amount of precipitation allows carbonate deposition in the deeper soil layers. In areas of still less precipitation, and with vegetation types that vary from short grasses

to semiarid woodlands and chaparral, a variety of **chestnut, brown,** and **reddish-brown soils** are developed. Finally, in desert regions, because of sparse vegetation, the organic layer can be very weakly developed or discontinuous, and the carbonate deposition layer can lie within one foot of the ground surface. Soils of this sort are termed **desert** or **red desert soils.** In a very general way, some of the basic changes that occur along a transect from arctic tundra through temperate forests to grassland and desert can be summarized in a single diagram (Figure 9-20).

SUMMARY

Soils are complex ecosystems in which physical and chemical components and various groups of living organisms interact through processes that involve interchanges of nutrients and energy. Soils exist in nature as three-dimensional, open systems occupying erosional drainage basins. These units, which may be termed soil landscape systems, develop in response to the influences of parent material, topography, climate, and biota.

The parent material of a soil might be the underlying bedrock, in the case of residual soils, or it might consist of material transported to the location by wind, water, ice, or gravity. This parent material is subject to various processes of mechanical and chemical weathering. Mechanical weathering includes various processes that physically reduce consolidated materials to smaller fragments; chemical weathering includes processes in which components of the parent material are dissolved, modified to different chemical form, and translocated to particular places within the soil. A variety of clay minerals, including the silicate and hydroxide clays, are produced secondarily by chemical weathering. These processes create soil texture, which is described according to the distribution of sand, silt, and clay particle sizes.

Biotic processes add organic matter to the soil, and organic matter has considerable effect on soil structure and fertility. Humus, a complex of long-chain organic polymers, humic acids, lipids, and other decomposition residues, creates stable crumb structure by cementing individual mineral particles into aggregates a few millimeters in diameter. Well-developed crumb structure creates conditions favorable to aeration and water infiltration, eases cultivation and root penetration, and increases the soil's resistance to wind and water erosion. Granular structure, in clay soils, can occur as the result of ionic relationships in which calcium ions favor the aggregation of clay particles.

Under particular climatic conditions, pedogenesis creates characteristic soil profiles. In the arctic, for example, pedogenesis produces a soil with a thick organic surface layer and a dense lower gley layer consisting of waterlogged mineral soil colored blue-gray by reduced iron compounds. In regions of the northern coniferous forests and temperate deciduous forests, abundant precipitation tends to leach clay minerals and a portion of the organic matter from a surface horizon (A) and deposit them in a lower horizon (B). Carbonates may be leached completely out of the soil system. In semiarid regions, where evapotranspiration exceeds precipitation, soils are closed systems; leaching is weaker, pH is alkaline, and carbonates and other soluble minerals are deposited in lower parts of the profile. The soil formation process of the humid tropics is termed laterization. Lateritic soils, formed under conditions of high precipitation and year-round high temperatures in tropical lowlands, tend to lack silicates, which are removed by leaching, and to possess high concentrations of hydroxide clays. The fertility of these soils tends to be low, because hydroxide clays have a low cation retention capacity and because rapid decomposition allows nutrients in organic matter to be released and leached away rapidly. Only a fraction of the lateritic soils of the lowland tropics solidify into bricklike laterite upon exposure, however.

Literature Cited

Abaturov, B. D. 1972. The role of burrowing animals in the transport of mineral subtances in the soil. *Pedobiologia* 12:261–266.

Barley, K. P. 1961. The abundance of earthworms in agricultural land and their possible significance in agriculture. *Adv. in Agron.* 13:249–268.

Baver, L. D. 1968. The effect of organic matter on soil structure. In *Organic Matter and Soil Fertility*. pp. 383–403. New York: John Wiley and Sons.

Borner, H. 1960. Liberation of organic substances from higher plants and their role in the soil sickness problem. *Bot. Rev.* 26:393–424.

Buol, S. W., F. D. Hole, and R. J. McCracken. 1973. *Soil Genesis and Classification.* Ames: Iowa State University Press.

Buckman, H. O., and N. C. Brady. 1974. *The Nature and Properties of Soils,* 8th ed. New York: Macmillan Publishing Company, Inc.

FitzPatrick, E. A. 1972. *Pedology: A Systematic Approach to Soil Science.* New York: Hafner Pub. Co.

Gillott, J. E. 1968. *Clay in Engineering Geology.* New York: Elsevier Pub. Co.

Gradusov, B. P. 1974. A tentative study of clay mineral distribution in soils of the world. *Geoderma* 12:49–55.

Harris, R. F., G. Chesters, and O. N. Allen. 1966. Dynamics of soil aggregation. *Adv. in Agron.* 18:107–169.

Huggett, R. J. 1975. Soil landscape systems: a model of soil genesis. *Geoderma* 13:1–22.

Hunt, C. F. 1972. *Geology of Soils.* San Francisco: W. H. Freeman and Company.

Karavayeva, N. A. 1974. Major kinds of gley soils of the tundra and northern taiga regions of the Soviet Union. *Geoderma* 12:91–99.

Kellogg, C. E. 1950. Soil. *Sci. Amer.* 183(1):30–39.

Kendeigh, S. C. 1974. *Ecology.* Englewood Cliffs, N.J.: Prentice-Hall, Inc.

Kononova, M. M. 1966. *Soil Organic Matter.* Oxford: Pergamon Press.

McNeil, M. 1964. Lateritic soils. *Sci. Amer.* 211(5):97–102.

Russell, E. W. 1971. Soil structure: its maintenance and improvement. *J. Soil Sci.* 22:137–151.

Sanchez, P. A., and S. W. Buol. 1975. Soils of the tropics and the world food crisis. *Science* 188:598–603.

Stefanson, R. C. 1973. Polyvinyl alcohol as a stabilizer of surface soils. *Soil Sci.* 115:420–428.

Taylor, S. A. 1972. *Physical Edaphology.* San Francisco: W. H. Freeman and Company.

Thomas, W. A. 1969. Accumulation and cycling of calcium by dogwood trees. *Ecol. Monog.* 39:101–120.

Thompson, L. M. 1957. *Soils and Soil Fertility.* New York: McGraw-Hill Book Co.

10

SOIL WATER AND NUTRIENTS

The fertility of the soil ecosystem is determined by the extent to which it can retain water and nutrients in forms readily available to plants. Water itself is a plant nutrient, since it is one of the basic raw materials for photosynthesis, but it is also the active medium in the soil from which various mineral nutrients are absorbed into root tissues. Within the plant, the water stream, which enters the roots, passes upward through xylem tissues, and exits via the stomata, is also the vehicle by which dissolved materials are transported from below-ground to above-ground portions of the plant. Finally, of course, the flux of water through the soil ecosystem carries with it certain quantities of mineral nutrients and other dissolved materials. Thus soil water and nutrient relationships are closely interrelated and should be examined together.

With these general observations in mind, let us first examine in detail the relationships of water to soil structure and plant uptake during its relatively brief period of transit through the soil ecosystem.

SOIL WATER

General Concepts

The fraction of the earth's water present in the soil and subsoil is exceedingly small (Table 10-1); it corresponds to a depth of about 44 centimeters, however, and is replaced about once every 280 days. By contrast, the amount of water in the atmosphere at any one time corresponds to a depth of only about 3 centimeters. From these and other data (Kalinin and Bykov 1969; Penman 1970) we can construct a generalized budget of the flux of water through a "typical" soil ecosystem unit (Figure 10-1). This budget shows, first, that the amount of water lost from the soil system by evaporation is usually greater than that lost by surface and subsurface drainage, both of which eventually show up as stream flow. Surface runoff appears to exceed subsurface drainage somewhat. However, if the value for infiltration were 10 percent higher, the reverse would be true. Since these estimates might easily contain a

TABLE 10-1
Major quantitative features of the global hydrologic cycle.

Quantity or flux rate	Volume (1000 km³)	%	Annual flux rate (1000 km³)	Renewal period
Total water on earth	1,460,000	100.00		2,800 years
Evaporation or precipitation			520.0	
Total water in oceans	1,370,000	93.84		3,100 years
Oceanic evaporation			448.9	
Oceanic precipitation			411.6	
Net river discharge			37.3	
Water in soil and subsoil	65	0.004		280 days
Continental precipitation			108.4	
Continental evaporation			71.1	
Surface flow to oceans			37.3	
Evaporation and subsurface drainage			85.0	
Freshwater lakes	750	0.05		?
Active groundwater	4,000	0.27		300 years
Exchange with surface	13	0.001		
Total water in continental crust	60,000	4.11		4,600 years
Glaciers and permanent snow	29,000	1.99		16,000 years
Melting			1.8	

Source: Based on Kalinin and Bykov 1969.

10-percent error, we should probably suggest only that, on a land area of average climate and soil conditions, both surface and subsurface runoff are of major importance.

This general water budget serves only as a point of departure for examining conditions that prevail in more specific situations. Our objective, in fact, is to understand the factors that control various components of the soil water budget, and we shall begin by examining the factors that influence infiltration of water into the soil from the surface.

Infiltration rate is affected by such factors as the initial water content, surface permeability, internal structural characteristics, degree to which soil colloids swell, temperature of the soil and water, and duration of rainfall (Kramer 1969). The drier the soil, the more rapid the rate of infiltration, other factors being equal; so when water is applied to a soil, infiltration is very rapid at first, as the water percolates into a dry surface. Thereafter, infiltration declines appreciably. Water moves through the soil by the advance of a **wetting front** at which a very sharp, visible change in moisture content occurs. The drier the soil, the more slowly this front advances, because more water is required to saturate the soil behind this front.

Surface permeability is determined by the existence of structural pores that allow water to enter and by their persistence under conditions of rainfall or surface wetting. Structural pores that are created by granular or crumb aggregation of the soil favor infiltration. The beating action of rain can destroy such aggregates, however, and can block pore spaces at the surface with the finer soil particles, thus reducing infiltration (Figure 10-2). Muddy irrigation water applied to the surface can cause the same problem by depositing fine silt and

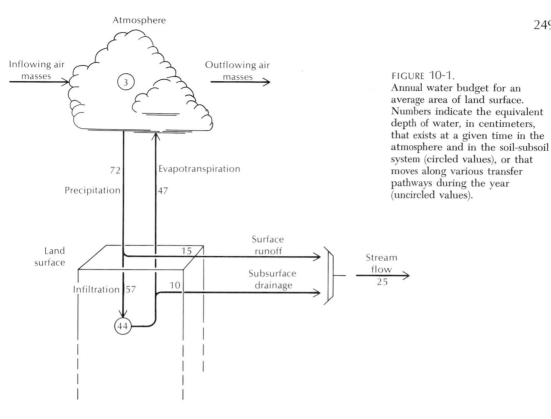

Atmosphere

Inflowing air masses →

Outflowing air masses →

(3)

72 Precipitation

Evapotranspiration 47

Land surface

15 Surface runoff →

Stream flow 25 →

Subsurface drainage →

Infiltration 57

10

(44)

FIGURE 10-1.
Annual water budget for an average area of land surface. Numbers indicate the equivalent depth of water, in centimeters, that exists at a given time in the atmosphere and in the soil-subsoil system (circled values), or that moves along various transfer pathways during the year (uncircled values).

FIGURE 10-2.
Heavy rainfall on an exposed soil surface tends to break down soil aggregates and seal the surface to infiltration. As a result, runoff can be great even on gently sloping surfaces, as in this cotton field in Oklahoma. (Photograph courtesy of the USDA Soil Conservation Service.)

clay particles in infiltration channels. Protection of the soil surface by vegetation, as in a forest, or by an organic mulch, in agroecosystems, is an effective way to preserve the structurally related infiltration capability of the surface.

Below the surface, rapid infiltration is encouraged by structural conditions that create a high proportion of pore space—abundant organic matter, stable crumb structure, coarse texture. The presence of swelling clays can interfere with infiltration as wetting proceeds, however, and the expansion of clay minerals, such as montmorillonite, can block infiltration channels and reduce pore space. Infiltration also tends to proceed more rapidly in warm temperatures, apparently because the viscosity and surface tension of water are lower under these conditions (Taylor and Ashcroft 1972). Furthermore, the longer the rain, the slower the rate of infiltration becomes. Conditions that favor rapid infiltration tend usually to diminish with depth in the soil; when the wetting front reaches more impervious, deeper layers, infiltration slows markedly.

Once in the soil, water is held in many ways and with greatly varying force. The specific mechanisms of water retention influence both the quantity of water present and its availability to plants.

Water exists in the soil in three major forms. The water that occupies the larger pore spaces, which tends to drain away under the influence of gravity, is termed **free water.** While it is present, during the periods immediately after rain or irrigation, it is entirely available to uptake by plants. In the smaller pore spaces, however, water can be retained by surface tension that prevents its being drained away by gravity. This fraction is termed **capillary water.** Capillary water is in equilibrium with water vapor in the air-filled portion of the pore space of the soil, and it is capable of moving from place to place; that is, as we shall see, from locations where it is held less strongly

to places where it is held more strongly. Some of this water is available to plants, but the water retention forces in the smallest of these capillary spaces are great enough that plants are unable to extract the water involved. The third water fraction, **hygroscopic water,** is that which is held adsorbed on surfaces of mineral and organic particles in the soil. Nonfluid in nature, hygroscopic water does not move in response to the wetting and drying of the soil, and because of the strength of the forces by which it is held, most of it is unavailable to plants. In addition, of course, **water of hydration** exists in the chemical structure of clay minerals and other compounds, another form essentially unavailable to plants.

The water-holding capacity of a soil is thus related primarily to the quantity and the characteristics of the pore space of the soil and to the hygroscopic characteristics of its constituents. The total amount of pore space in a soil system varies from 25 or 30 percent to about 50 or 60 percent, depending primarily upon textural composition and degree of crumb or granular macrostructure. Sandy soils, although they permit relatively free infiltration of water, generally have smaller percentages of pore space, which limits their water-holding capacity; clayey soils usually have the highest percentage of pore space.

As implied earlier, some of the internal space of soils consists of larger openings that cannot retain water in the face of gravity. This is termed **noncapillary pore space** and is essential to the aeration of soils and the infiltration of water to deeper soil layers. In sandy soils, much of the pore space falls in this category, and, although they allow easy infiltration, the relatively small, noncapillary pore space results in poor water retention, a condition referred to by farmers as "droughty" soils. In clayey soils, on the other hand, most pore spaces are of capillary size, and so, once the soil is wetted, water retention is high, so high, in fact, that inadequate aeration can often become a problem. Thus, the intermediate characteristics of a loamy soil with good macrostructure—that is, a relatively large pore space with a reasonable balance of capillary and noncapillary spaces—provide the best combination of infiltration, water retention, and soil aeration. Chapters 11 and 12 examine the impacts of various agricultural practices upon these soil characteristics. Here we will simply note that the intensive cultivation of, for example, a forest soil tends to reduce the total pore space in the surface soil layers, principally by reducing the noncapillary space fraction (Figure 10-3).

As we noted earlier, all the water retained in a soil is not equally available to plants; some is held with such force that it is virtually inaccessible to plants. Field capacity and permanent wilting point are the concepts that have been used most often to describe a soil's capacity to hold water in a form available to plants. **Field capacity** is the quantity of water that the soil still holds after it has been wetted thoroughly and after gravitational water has been allowed to drain away. Soils with good drainage generally hold this amount of water at about 1 to 3 days after rain or irrigation. Field capacity is measured by ponding water on a soil and permitting it to drain for 1 (for coarse soils) to 3 (for fine-textured clayey soils) days. Soil cores are then taken and dried in an oven; the difference in weight before and after drying represents the quantity of water present at field capacity. Not all of this water is available to the plant, however.

To determine the **permanent wilting point** it is necessary to grow an index plant, usually the common sunflower, *Helianthus annuus*, in a sample of the soil being studied. Once the plant is well established, the soil is allowed to dry until permanent wilting occurs, which is wilting from which the plant does not recover when it is placed overnight in a chamber containing saturated air. After this point has been reached, a soil sample is taken and oven dried,

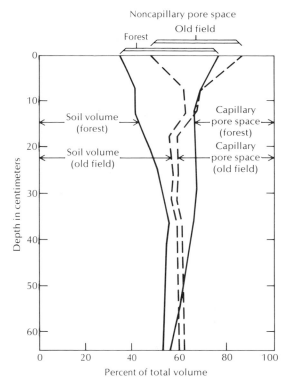

FIGURE 10-3.
Intensive cultivation, by gradually degrading the crumb structure of the surface soil, tends to severely reduce the noncapillary pore space of the soil. These data pertain to a North Carolina soil. (Reproduced with modification from M. Hoover, *Soil Science Society of America Proceedings*, Vol. 14, p. 354, 1949, by permission of the Soil Science Society of America.)

and the difference in its weight before and after drying is recorded. This figure, the permanent wilting point, is subtracted from the field capacity, and the result is an estimate of the soil's capacity to hold water in forms available to the plant.

Available soil water, relative to total field capacity, is limited by structural features of the soil, such as pore size, and by the hygroscopic characteristics of various soil materials, as we have already mentioned. The presence of solutes can also limit the amount of the soil water that can be taken up by a plant. In many coastal and semiarid regions, for example, soil salinity prevents the use of lands that are otherwise favorable for agriculture. Improperly designed irrigation systems can sometimes salinate soils to the point where crop agriculture becomes impossible. These problems are examined in detail in Chapter 12.

Soil Water Potential

In recent years, a new system has been developed for describing and predicting soil water behavior. This system is based on the analysis and description of potential energy relationships in soil-water or soil-water-plant systems. Any quantity of water possesses a capacity to do work in moving from one point to another. Man has learned to harness flowing waters, and it is obvious, for example, that the water behind a hydroelectric dam possesses, because of its location, a large amount of potential energy. This potential exists because other forms of energy throughout the hydrologic cycle have been invested in evaporation, transport, and condensation, eventually trapping the water behind the dam.

In thinking about soil water and potential energy, the first point to keep in mind is simply that water tends to move from a point of high potential energy to a point of low potential energy. The only way for it to do otherwise is for work to be applied from outside the system. Second, we should note that potential energy associated with soil water is usually expressed in comparison to the potential of a unit of pure water in the same location. Because soil water is constrained in its behavior by a great many factors, its ability to do work—that is, its potential energy—is less than that of pure water in almost all cases. The values in

which the potential energy of soil water is expressed are therefore almost always negative. Finally, we should note that we will often speak of water potential, and of other potentials, as a convenient shorthand for potential energy.

Potential energy levels can be expressed in units of pressure, or force per unit area. The unit used most often is the **bar,** equal to 10^6 dynes per cm^2. The water potential of a particular system is usually symbolized by the Greek letter *psi,* ψ. If we accurately determine the water potential at two points within a soil system, or for the soil and the plant tissues in contact with it, we can predict the direction in which water should move, which will be from the point of higher potential to that of lower potential.

Soil water potential, ψ_{soil}, is the sum of a number of component potentials: matric, solute, pressure, and gravitational. **Matric potential,** ψ_m, is the force with which water is held in the soil by forces of adsorption and capillarity. It is measured by the suction that must be applied in order to extract water from a soil with particular structural properties, and thus has a negative value. The **solute potential,** ψ_s, describes the potential energy of soil water as it is influenced by the concentration of solute particles. This potential is lower than that of pure water, and so the solute potential is also negative in value. This simply means that when a solute concentration difference exists, water will tend to move from the point of lower solute concentration to higher solute concentration.

Matric and solute potentials are the two most important components of soil water potential. Two other components, however, can play appreciable roles under certain circumstances. **Pressure potential,** ψ_p, summarizes the forces created by atmospheric and hydraulic relationships—that is, the pressure of air and water in the soil system. Normally, in an open soil system, no appreciable differences in air pressure will exist. Under certain circumstances, associated with springs, aquifers, and some irrigation systems, differences in hydraulic pressure can exist between two or more points in a soil system. Pressure potential is expressed relative to the normal atmospheric pressure at the location involved, and because the value reflects the atmospheric pressure plus any hydraulic pressure, this potential will range from zero to slightly positive in value. Hydraulic pressure can exist only under saturated soil conditions. **Gravitational potential,** ψ_z, describes, relative to a chosen reference level, any energy potential due to height differences of soil water. Gravitational potential can therefore be positive or negative, depending on the location of the point in question relative to the reference point. Still other potentials, of less importance, can occur under special circumstances.

Soil water potential is thus the sum of matric, solute, pressure, and gravitational potentials:

$$\psi_{soil} = \psi_m + \psi_s + \psi_p + \psi_z$$

Determinations of soil water potential for particular soils give curves that relate total water content per unit weight of dry soil to soil water potential measured in bars (Figure 10-4). When the principal difference between soils is in matric potential, we can make some important generalizations about the availability of soil water. First, soil water potential can be related to the concepts of field capacity and permanent wilting point discussed earlier. Field capacity equals roughly a soil water potential of –0.3 bars, the value at the lower endpoints of the curves in Figure 10-4. Permanent wilting point varies widely from plant to plant, but lies generally between –10 and –20 bars; arbitrarily, it is considered to be –15 bars, the upper endpoint for the curves in Figure 10-4. The three soils shown in this figure have very different water-holding and availability features. At any given value of soil water potential, clay soil holds more water than sandy loam

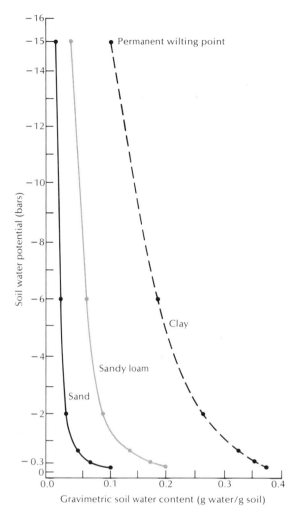

FIGURE 10-4.
Soil water content in relation to water potential for various soil types. The difference, on the abscissa, between the soil water content at −0.3 bars and that at −15 bars represents the approximate capacity of the soil to hold water in a form available to plants. (From Slatyer 1967.)

or sand. At the same time, however, the percentage of total water that is unavailable, or beyond the permanent wilting point, is also greater for clay than for sandy loam and sand. Nevertheless, the total amount of water that the plant can extract—that is, the difference between field capacity and permanent wilting point—is greatest for the clay soil. Clearly, the clay component of soils plays an important role in the retention of available water. As we have noted, however, this must be balanced by characteristics related to aeration, infiltration, ease of cultivation, and other qualities.

Water Uptake by Plants

These concepts of water potential can also be applied to plant tissues, particularly those of root systems (Newman 1974). In this application of the concept, matric and solute potentials are closely analogous to those defined for the soil. Matric potential reflects the forces in plant tissues that hold water molecules adsorbed to colloidal components of the cell and to other structural parts. Solute potential refers to the osmotic forces that hold water in plant tissues. Pressure potential, however, can be much more important in plants than in the soil, because plant cells are enclosed by somewhat rigid cell walls. When water enters a plant cell it tends, of course, to enlarge the cell volume and to create pressure against the inside of the cell wall. This pressure, commonly termed **turgor pressure,** can become great enough to counteract movements of water into the cell that would otherwise occur. Gravitational potential comes into play in the analysis of water movements in soil-plant systems when the points under consideration lie at different heights; for example, when aboveground tissues are considered relative to belowground tissues or the soil. As in the case of the soil system, **plant water potential** is the sum of these four component potentials, with due regard being taken of negative and positive signs of the values involved (negative for matric and solute, positive for pressure).

With this in mind, we can characterize the pattern of water movement from the soil into a root system (or in the opposite direction!)

according to the difference between the soil and root water potentials, the surface area of roots per unit volume of soil, and the resistance of the soil-root system to water movement (Gardner 1968). As plants take up water, of course, the water potentials of both soil and plant change. Assuming that the flow is into the root system from the soil, the soil water will increasingly be depleted, and the soil water potential will become more and more negative. The water potential of plant tissues, on the other hand, will change in a manner that reflects both the uptake of water from the soil and the loss of water by transpiration. The soil and plant water potentials, for a plant initially existing in a saturated soil, can be expected to follow, over 5 days' time, the pattern shown in Figure 10-5. At the beginning of the day, the water potentials of both soil and plant tend to be equal or nearly so. During the day, water lost from the plant by transpiration tends to reduce the water potential inside the plant, thus encouraging inflow from the soil system. At night, for at least the first few days, this continued inflow restores the water content of the plant until the water potentials of plant and soil are again nearly equal. Eventually, however, daytime water loss carries the water potential of the plant below the permanent wilting point, beyond which the plant is unable to recover its water loss by nighttime absorption.

It should also be noted (Figure 10-5) that change in soil water potential does not show a straight-line relationship to actual water content. The change in soil water potential caused by a single day's uptake of water by a plant is much greater when soil moisture levels are near the permanent wilting point than when the soil is nearly saturated. Furthermore, we should note that the kind of macrostructure of a soil significantly influences its field capacity and the form that the water potential curve takes as the soil dries (Figure 10-6). A com-

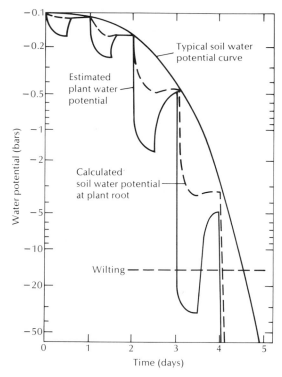

FIGURE 10-5.
Schematic representation of soil and plant water potentials for a 5-day period of drying conditions. During the first half of each day period (daylight hours), plant water potential declines as water is lost through transpiration; this creates a water potential gradient from the soil into the root system. During the second portion of the day period (night hours), transpiration ceases and the water potential of the plant rises, except on the last day, when the daytime loss has carried the plant water potential beyond the permanent wilting point. (From *Physical Edaphology* by Sterling A. Taylor, revised and edited by Gaylen L. Ashcroft. W. H. Freeman and Company. Copyright © 1972.)

relatively little during most of the early stages of drying but suddenly increase very rapidly as the soil approaches the permanent wilting point (Hadas 1973).

Black (1968) has offered a number of criticisms of the potential energy system of describing and analyzing soil water relationships. The two most important criticisms deal with whether matric and solute potentials are truly additive, and with the weakness of the poten-

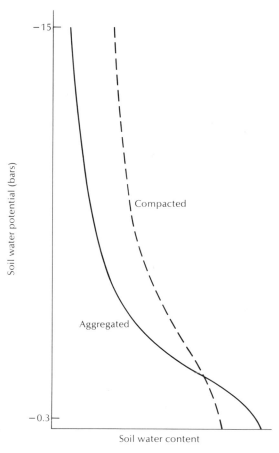

FIGURE 10-6.
Schematic representation of the relationship between soil water content and water potential for aggregated soils with good crumb structure and for compacted soils with poor crumb structure.

pacted, fine-textured soil might have a lower field capacity than the same soil with a well-developed crumb structure. Furthermore, a greater percentage of the available water might be taken up by the plant with a change in the soil's water potential of only a few bars. In other words, with well-developed crumb structure, the water potential of a soil might change tial energy relationship in providing information on quantitative water relationships. Some experiments have compared water uptake in plant seedlings under situations of varying water potentials that represent matric potential on the one hand and solute potential on the other; these experiments suggest that significant differences in uptake exist even when these different potentials are equal (Figure 10-7).

Second, it is clear that both the quantity of water (total or available) per unit soil varies with factors such as soil texture, even when matric potentials are adjusted to be equal (Figure 10-8). Experiments have been made on root elongation of seedlings in soils with equal matric potentials but different actual percentage contents of water; these experiments demonstrate that growth is more rapid in the soil of higher percentage water content. In all probability, this means that the soil with

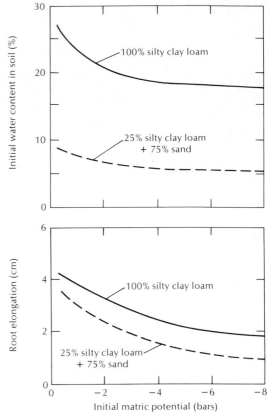

FIGURE 10-8.
Soil water content and corn root elongation in relation to soil matric potential for soils with different textures. At any given matric potential the silty clay loam has a higher water content and permits more rapid root elongation than does the sandier soil. (From Black 1968.)

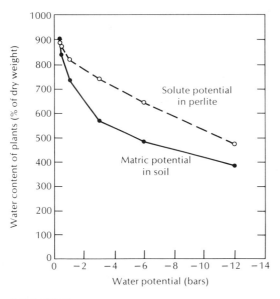

FIGURE 10-7.
Water content of young corn plants grown under a range of soil water potentials, but for which potentials were entirely matric (*solid line*) on the one hand, or entirely solute (*dotted line*) on the other. (Modified from Black 1968.)

finer texture and higher absolute water content was able to supply water to the seedling roots more rapidly than the coarser soil could. This is equivalent to saying that significant differences exist among soils in their resistance to water movement along gradients of water potential.

The water potential of soils is, of course, greatly important to the productivity of crop species, as well as to their survival. Photosynthetic and respiratory responses differ markedly (from each other) as stress from drying soil conditions increases. In most crop species photosynthesis begins to decline with water potential stresses of as little as –1 to –3 bars. As stress increases, even though the permanent wilting point may still be far away, photosynthesis declines relative to respiration until respiration exceeds photosynthesis. At this point, of course, growth stops, and the plant begins to use stored food materials to maintain itself. Thus soil water potential is obviously more critical for maximum productivity than for simple survival. It is also a significant factor in root disease, because of influences on the physiology of disease agents and on the ability of roots to resist infection (Cook and Papendick 1972).

SOIL NUTRIENTS

General Concepts

Soil nutrients include mineral and nonmineral elements that higher plants must obtain by uptake from the soil. The major nutrients, or macronutrients, include nitrogen, phosphorus, potassium, sulfur, calcium, magnesium, and, for a few plants, sodium. Several of these, such as potassium, calcium, and magnesium, are taken up by the plant as cations. One, nitrogen, can be taken up either in the form of positively charged ammonium ions, NO_4^+, or as negatively charged nitrate ions, NO_3^-. Both phosphorus and sulfur are absorbed as anions. In addition, a number of trace elements, or micronutrients, are also required by plants. These include iron, manganese, copper, zinc, molybdenum, boron, chlorine, and cobalt. Domestic animals, in addition, require iodine and selenium. Several other elements may or may not prove to be essential micronutrients, but if so, they are of very minor importance and only rarely affect the success of agriculture. Most of these micronutrients are absorbed as cations, but boron, chlorine, and iodine are taken up as anions.

Thus in dealing with soil nutrients we are concerned primarily with how these various cations and anions are held in the soil and with factors that determine what portions of the quantities present are available to plant uptake processes. The supply of available nutrients is a major determinant of ecosystem productivity, and the particular nutrient available in the least quantity, relative to need, is often called the **limiting nutrient.**

The soil system has components that represent all three phases of matter: solid, liquid, and gas. Nutrient ions are taken up directly from the liquid phase; they pass from the soil solution through the walls and membranes of the root hairs—the fine, tubelike extensions of epidermal cells of the smallest roots of the plant. However, if we determine the concentration of various ions in the soil solution, and calculate the quantities of nutrients per hectare that these concentrations represent, we quickly see that the liquid phase cannot be where most soil nutrients are stored. Concentrations of most nutrients in the soil solution lie in the range of 10^{-3} to 10^{-4} molar, and for phosphorus (one of the most important) this concentration is even lower—10^{-5} or 10^{-6} molar. Because the gas phase of the soil cannot possibly be a major storehouse, we are forced to conclude that most of the nutrients usable by plants are retained in the solid phase of the system.

As we have seen, the solid phase of the soil system is extremely complex, consisting of a great many primary minerals, secondary minerals, and organic substances. If we examine the gross abundance of various nutrient elements contained in the solids that make up the soil matrix, we find large amounts of nearly all nutrients (Table 10-2). Greater amounts, in terms of percentage of the total weight of a component, are usually present in the finer textural components. In spite of the absolute abundance of such nutrients, we commonly find that, when crops are grown in it con-

TABLE 10-2
Percentage abundance of phosphorus, potassium, and calcium in various textural components of a glacial loess soil.

Texture group	%P	%K	%Ca
Sand	0.07	1.43	0.91
Silt	0.10	2.00	0.93
Clay	0.38	2.55	1.92

tinuously, a soil shows deficiencies in one or more major nutrients. This indicates simply that only part of the gross quantity of a particular nutrient is actually available to the plant. Another part might still be tied up in unweathered primary minerals and thus be unavailable to a particular plant during its growth period. This portion is called the **structural component.** A second component might be located in secondary minerals where it is "fixed" or held very strongly, as potassium ions are sometimes locked up in the internal structure of clay minerals. Slow exchanges do take place between this component and those that are more readily available, but the rates involved are significant only over years or decades; we can therefore generally regard this component as unavailable. A third component, associated with the solid phase, is in short-term equilibrium with the concentration of its ions in the soil solution. Exchanges between this component and the soil solution occur within seconds or minutes or, at most, over weeks or months. This readily exchangeable component, together with the amount in the soil solution, constitute the available quantity of a particular nutrient, although there is no sharp distinction between available and unavailable quantities. A continuous spectrum, which can be divided arbitrarily into two portions, actually exists for the rate at which different components of the solid phase can replenish the nutrient supply of the soil solution.

Cation Exchange Capacity

The mechanisms by which nutrients are held in available form are somewhat different for cations and anions. Most available nutrient cations—over 99 percent—are held adsorbed on the surfaces of organic matter and clay minerals, as we noted in Chapter 9. The capacity of the soil to hold cations in a form available to higher plants is termed the **cation exchange capacity,** and is generally measured in milligram equivalents per unit weight of soil.

Milligram equivalents, or milli-equivalents (usually abbreviated meq) state the number of milligrams of hydrogen ions that would be present if all the adsorption sites concerned were occupied by this cation. For calcium, a divalent cation (Ca^{++}), one meq would thus have half the number of ions that one meq of hydrogen would have. Because the atomic weight of calcium equals 40, this also means that the actual weight of one meq of calcium would be 20 milligrams.

Cation exchange capacity is determined by flooding a soil sample with a salt solution, usually 1 N ammonium acetate, in such a way that all the adsorption sites are taken over by the cation of the salt involved (ammonium ions in this case). In a second step, these ions are then displaced and leached from the sample using a solution of salt of another cation; their quantity is then determined by titration.

Cation exchange capacity differs greatly for organic matter and various clay minerals (Table 10-3). It is greater for organic matter than for any of the clay minerals, equaling 200 to 400 meq per 100 grams. In general, cation exchange capacity is greater for the clay minerals that are capable of expansion, since this expansion permits the adsorption and exchange of ions between the soil solution and both the internal and external surfaces of clay crystals. It is also high for amorphous silicate clays, such as allophane, because of the high surface-to-

TABLE 10-3
Cation exchange capacity of various
colloidal components of the soil system.

Material	Cation exchange capacity (meq/100g)
Organic matter	200–400
Vermiculite	100–150
Montmorillonite	60–100
Illite	20–40
Kaolinite	2–16
Hydroxide clays	<2

volume ratio of the particles of these clays. The cation exchange capacity of hydroxide clays typical of the tropics and subtropics is very low, generally less than 2 meq per 100 grams.

Because the distribution patterns of organic matter and clay minerals are differentiated in the soil profile as a result of pedogenic processes, different soil types are characterized by different profiles of cation exchange capacity (Figure 10-9). For example, in podzolic soils with highly differentiated profiles, which are typical of regions with a climax vegetation of temperate forest, cation exchange capacity in the surface layers may be primarily a function of organic matter content. A foot or so below the surface, however, it may become largely a function of the clay content and type. In grassland soils, on the other hand, clay content and organic matter content might both be of major importance near the surface, with organic matter becoming gradually less important with increasing depth (Thompson 1957).

Cation exchange capacity is also affected by soil pH, but the effects are different in clay minerals and organic matter (Figure 10-10). Below pH 6.0, the exchange capacity for most clay minerals is essentially constant. Above this level, exchange capacity increases slightly—apparently because the ionization of exposed OH groups along crystal edges releases an H^+ ion and produces an additional negatively charged site on the crystal surface. This increase is termed the **pH dependent** cation exchange capacity. For organic matter, the effect of pH is much stronger: Below pH 4.0, in very acid soils, the cation exchange capacity of or-

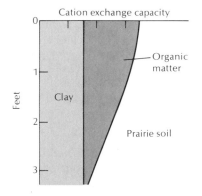
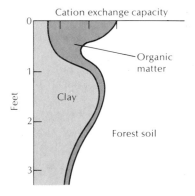

FIGURE 10-9.
Profiles of cation exchange capacity attributable to clays and to organic matter for typical grassland and forest soils. (Modified from *Soils and Soil Fertility* by L. M. Thompson. Copyright © 1957 by the McGraw-Hill Book Company. Used with permission of McGraw-Hill Book Company.)

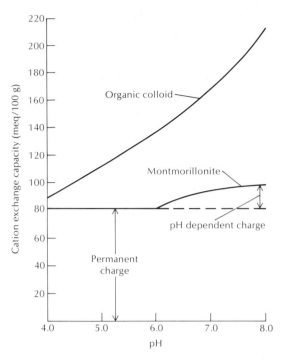

FIGURE 10-10.
Influence of pH on cation exchange capacity
of organic matter and of montmorillonite clay.
(From Buckman and Brady 1974.)

ganic matter is extremely small; from pH 4.0 to 8.0 or greater the capacity increases greatly. In a study of pH influences upon cation holding capacity of a number of California soils, Pratt and Bair (1962) found that between pH 3.0 and 8.0 cation exchange capacity of clay changed by 15.8 meq per 100 grams, while that for organic matter changed by 370 meq per 100 grams. Allophane, a noncrystalline silicate clay, differs from other clays by having a relatively large pH dependent cation exchange capacity.

A variety of different cations can be held by this cation exchange system, including hydrogen ions, H^+. From the standpoint of plant nutrition, however, the abundance of four mineral cations—calcium, magnesium, potassium, and sodium—is particularly important. These elements are termed the exchangeable bases. Total **exchangeable bases** refers to the sum, in milligram equivalents, of these four cations; **percentage base saturation** refers to the percen-

tage that these four make up relative to the total cation exchange capacity. Of these bases, calcium is typically the most abundant, followed by magnesium; potassium and sodium are present in smaller amounts, except in saline soils, where sodium may be more abundant than all of the others. For a typical humid-region soil, 80 to 85 percent of the exchangeable bases tend to be calcium; 15 to 20 percent, magnesium; and less than 1 percent each potassium and sodium.

Soil pH is a good indication of the percentage base saturation. This is, of course, intuitively reasonable, because pH is a measure of the concentration of hydrogen ions in the soil solution. As we noted in Chapter 9, various cations are held with different forces by clays and organic matter, the relation being roughly:

$$Al^{3+} > Ca^{2+} = Mg^{2+} > K^+$$

$$= NH_4^+ > Na^+$$

TABLE 10-4

Cation exchange capacity of various soil types, together with influence of pH on percentage base saturation.

Major soil group	Textural class	Location	Cation exchange capacity (meq/100g)	Percentage base saturation at pH values of:				
				4.8	5.0	5.5	6.0	6.5
Prairie soil	Silt loam	Illinois	26.33	57	60	69	80	91
Podxolic	Silt loam	Wisconsin	9.79	43	50	63	72	82
Lateritic	Clay loam	Alabama	4.85	6	23	41	58	74
Lateritic	Sandy loam	Alabama	1.83	9	16	32	44	60

Source: Modified from Thompson 1957.

However, where leaching rates are high and the pH of the through-flowing water is acid, hydrogen ions tend to displace other cations and to reduce the number of adsorption sites by forming covalent bonds with exposed OH groups. This leaves other cations free to be carried away in the water passing through the soil. The decline in base saturation percentage with increasing acidity is illustrated in Table 10-4. When base saturation drops below about 80 percent, too few nutrients are available and most field crops suffer.

In the acid mineral soils of the tropics and subtropics, pH values of 5.0 or less generally indicate large amounts of exchangeable aluminum rather than hydrogen ions (Kamprath 1972). Aluminum cations can occupy 60 percent or more of the total base exchange capacity. Acidity results from the fact that, under these conditions, aluminum in the soil solution is hydrolyzed, and the hydrolysis is accompanied by a release of hydrogen ions. Aluminum is toxic to many crop plants, so that the apparent infertility of many of these soils is in part caused by toxicity.

Cations are taken up by plants from the soil solution. The mechanism involved is basically one of cation exchange: The uptake of a nu-

trient cation is accompanied by the release of a hydrogen ion into the soil solution; these changes lead ultimately to readjustments in the equilibrium between the soil solution and the cation holding system of the solid phase of the soil.

The use of ammonia fertilizers and lime is designed to influence the quantities of available nutrient cations and the base saturation capacity of the soil (Figure 10-11). We shall discuss these processes in more detail in Chapter 12.

Anion Retention In Soils

The retention of anions, including those of nitrite and nitrate, phosphate, and sulfate, involves complex mechanisms that are not yet well understood. The concept of available and unavailable quantities still applies, but it is difficult to obtain reproducible measurements of available quantities of these elements, especially phosphorus. We shall begin by examining nitrogen, since we have already seen that it is retained in part as a cation.

As we noted in Chapter 2, the nitrogen cycle is perhaps the most complex of the nutrient cycles (see Figure 2-9), and much of this com-

262

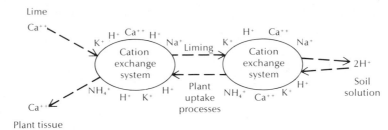

A LIMING

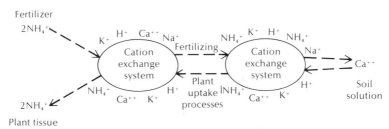

B FERTILIZATION

FIGURE 10-11.
Influence of liming, of the addition of ammonium sulfate fertilizer, and of plant
growth processes on the cation exchange system. Liming tends to displace H⁺ ions
from adsorbtion sites, making them subject to removal by leaching. Adding ammonium ions results in the release of other cations, such as Ca⁺⁺, into the soil solution,
from which they can be taken up by plants. The release of H⁺ ions by roots can also
lead to the displacement of Ca⁺⁺ and other cations from adsorption sites, which permits their uptake from the soil solution. (Modified from Hunt 1972.)

plexity involves relationships among different forms of nitrogen in the soil. Most nitrogen in the soil is in the form of organic matter; inorganic nitrogen (NH_4^+, NO_2^-, and NO_3^-) usually amounts to less than 2 percent of the total present (Black 1968). The quantity of organic matter, and thus of organic nitrogen, varies greatly, however. Kononova (1966) gives, for the humus content of the surface 20 to 25 centimeters of a wide series of soils, a range from about 15 metric tons per hectare (1 metric ton = 1000 kilograms) for a desert soil, to almost 300 metric tons per hectare for a rich chernozem soil. Roughly speaking, nitrogen constitutes 4 to 5 percent of humus in soils of the more humid regions, and somewhat more for desert soils. For nitrogen content, Kononova

(1966) gives values ranging from 2.3 to 11.3 tons per hectare, lowest for desert soils and highest for chernozems.

Nitrogen is regarded as unavailable in its organic forms but available in the form of ammonium and nitrate ions, both of which can be taken up by higher plant roots. The release of these forms of nitrogen from organic matter is termed **mineralization.** This release, accomplished by the decomposition of soil organic matter and the subsequent conversion of ammonium ions to nitrate, generally amounts to only 1 to 4 percent of the total organic nitrogen present annually. Thus for a soil having 5 tons per hectare of organic nitrogen we might expect an annual mineralization rate of 50 to 100 kilograms per hectare. Ammonium ions are released by the action of a variety of decomposer organisms, which are essentially engaged in the respiratory consumption of dead organic matter as their food source. The ammonium ions released may be retained in exchangeable form by the cation exchange system that we discussed earlier, and in this form they are readily available to plants.

Ammonium ions, however, can be oxidized in a two-step sequence to nitrate ions: Nitrite bacteria, *Nitrosomonas*, oxidize ammonium to nitrite; nitrate bacteria, *Nitrobacter*, complete the conversion, oxidizing nitrite to nitrate. In this process, little nitrite accumulates in the soil. The nitrate anion that is formed is highly soluble and is held very weakly in the soil. There appears to be an anion exchange system analogous to the cation exchange system in most soils, but this system has nowhere near the capacity of the cation exchange system, and its exact nature remains in doubt. Consequently, when nitrate is produced in large quantities, it is very susceptible to being leached out of the system.

Nitrogen also enters the soil system through fixation by a number of groups of bacteria, actinomycetes, and blue-green algae. These fixation activities involve the reduction of molecular nitrogen to ammonium, and they serve simply as the nitrogen source for the organisms involved. Bacteria of several different heterotrophic and autotrophic groups, free-living in the soil, have a nitrogen-fixing ability. Several groups of these organisms seem to occur only in tropical and subtropical regions, however. Blue-green algae, with nitrogen-fixing ability, are also important in aquatic systems, paddy agroecosystems, and damp soils. More important, at least at present, are the symbiotic bacteria of the genus *Rhizobium* that invade root, and occasionally leaf, tissues of a number of plants and form nodular growths in which the bacteria exist and carry on nitrogen fixation (Figure 10-12). Much of the nitrogen fixed becomes available to the host plant.

FIGURE 10-12.
Root nodules on Austrian winter peas, a leguminous plant used as a green manure crop. Nitrogen-fixing bacteria of the genus *Rhizobium* invade the root tissues of most legumes, stimulating nodule formation and creating a symbiotic relationship with the host plant. (Photograph by John McConnell, courtesy of the USDA Soil Conservation Service.)

This symbiotic association exists between *Rhizobium* and virtually all members of the legume family, but similar symbioses occur for other higher plants, including certain cycads, gymnosperms, monocots, and dicots. The contribution of fixed nitrogen to the soil system by such species varies, of course, with the abundance of plants capable of hosting the bacterial fixers. Moreover, fixation tends to be inhibited by high existing levels of mineral nitrogen in the soil. In the case of leguminous crops, fixation rates can be as great as 300 kg/ha/yr (Bell and Nutman 1971). The role of these microbial agents in soil nutrient dynamics is discussed more completely in Chapter 13.

We should also note that, under anaerobic conditions, nitrogen is lost from the soil by denitrification, a process carried out by still another specific bacterial group. In dense, damp soils, denitrification can be a major route by which nitrogen, including that added in the form of fertilizers, is lost.

The second anion of major importance is phosphate. Phosphorus frequently becomes a limiting nutrient in agricultural systems; it exists in the soil in the form of organic matter, in combination with calcium, iron, aluminum, and other metalic ions, and in association with the anion exchange system of the soil. Phosphate is released from organic matter as it decomposes. As we have seen, the fraction of the organic matter present that undergoes decomposition is 1 to 4 percent annually, and so this source has a relatively limited capacity to provide large quantities of phosphorous required for crop growth. Phosphates of calcium, iron, aluminum, and other materials tend to be of low solubility, and the phosphorous in these forms is regarded as being fixed; nevertheless, phosphorus in these minerals is in equilibrium with that in the soil solution, and the depletion of available phosphorus is accompanied by slow release from these portions of the system. These minerals also provide a powerful mechanism for retaining phosphorus in the system; phosphate fertilizers added to the soil are generally retained almost completely. Various microorganisms, for example, tend to take up phosphorus rapidly, "storing" it, in a sense, in quantities greater than the minimum necessary for their immediate needs. This organic phosphorus can eventually be released by the death and decomposition of the organisms involved. Other quantities, not taken up by higher plants or by the soil microflora, are fixed as phosphates of the elements noted above. Leakage

of phosphates from agroecosystems is obviously very small.

Certain soil microorganisms seem to be able to extract phosphorus from fixed inorganic phosphate compounds. Nevertheless, this component of the soil pool must be regarded as having only a limited ability to replenish available phosphorus supplies. With respect to the anion exchange system in the soil, it appears that the hydroxide clays might play a significant role; that is, that phosphate ions may displace hydroxide ions from these minerals and be held in a manner that makes them relatively available.

It is very difficult to determine the quantities of available phosphorus in a given soil. Generally, values for available phosphorus are based on the results of extraction of the soil with weak acids, the assumption being that the mechanisms of nutrient uptake by roots tend in general to acidify the immediate environment of the root. More satisfactory, however, are empirical tests for a particular soil type that simply use a standard extraction method and correlate the results obtained for the soil with the quantities of phosphorus present in the tissues of plants grown on that soil. A great deal obviously remains to be learned about phosphorus relationships in soil and its management for agricultural productivity.

Sulfur is another necessary element that is present in soil systems as an anion. It is essential to the structure of proteins, providing the atoms between which the cross-linkages occur that give protein molecules their characteristic and essential three-dimensional shape. Sulphates are retained in the soil primarily in the form of organic matter and ions in the soil solution. In acid soils, a certain quantity can be retained by clay minerals in exchangeable fashion; in calcareous soils, largely insoluble calcium sulfates may retain sulfur in fixed form.

Man's effect on the sulfur cycle has been considerable (Kellogg et al. 1972). He adds sulfur compounds to the atmosphere at about half the natural rate for the earth as a whole, and in industrialized regions his impact greatly exceeds the rate of natural additions. Downwind from major industrial centers, the increased fallout of sulfur compounds is augmented for several thousand kilometers to the extent that sulfur may only rarely become deficient in the agricultural soils affected. In England, for example, sulfur deficiency is unknown, even in intensively farmed areas (Cooke 1967). In other areas, however, deficiencies do appear, and as progress is made in air pollution control, the need for sulfur-containing fertilizers could increase.

Trace Elements

For the various trace elements, we will simply note that the major retention mechanisms are the cation exchange system, the soil organic component, and various other specific chemical combinations. The most important of these other mechanisms is chelation. **Chelates** are organic substances, natural or synthetic, that are capable of holding metalic atoms in loose association, and chelation may be an important mechanism for the retention of nutrients such as iron, copper, manganese, and zinc in forms more or less available to higher plants.

GENERAL PATTERN OF NUTRIENT RETENTION

We can summarize the mechanisms by which nutrients are retained and exchanged in the soil in terms of a relationship such as that shown in Figure 10-13. Growing plants or other soil organisms extract nutrients from the soil solution. The soil solution in turn is in equilibrium with various nutrient holding systems that are part of the solid phase of the soil. For the cation and anion exchange systems that involve secondary minerals—mostly clays and soil organic matter—these equilibrium responses occur on a time scale of perhaps hours to years. In addition, over years, decades, or centuries, new nutrients are released into the

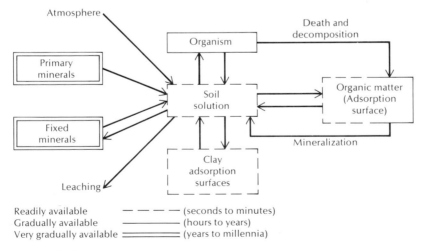

FIGURE 10-13.
Schematic relationship of exchange routes and availability status of soil nutrients.

soil solution by the decomposition of organic matter, by the weathering of primary minerals. and by equilibrium responses that involve fixed forms of the nutrient in secondary minerals. For specific nutrients such as nitrogen and sulfur, appreciable inputs enter from the atmosphere, by means of either the active processes of soil organisms or passive importation via precipitation. Likewise, losses occur by leaching to locations from which return is impossible. For nutrients that possess volatile compounds, losses to the atmosphere are also possible.

SUMMARY

The abundance and availability of water and nutrients in a soil ecosystem determine its fertility. Several of the basic elements of living systems, including carbon, hydrogen, oxygen, nitrogen, and sulfur, have truly cyclical movements through the biosphere and can be considered to constitute a single volatile element cycle. In particular, all these elements, because they have gaseous compounds, can return from the oceans to the continents over short periods of geological time. The mineral nutrients, however, generally flow in a one-way manner from the continents to the ocean basins, carried by the liquid-phase movements of the hydrologic cycle.

Water is held in soils in different forms and by different forces. Gravitational water, which eventually drains away under the force of gravity, is available to higher plants as long as it is present. Some of the remaining water, occupying capillary spaces and existing as a film or molecular layer on soil solids, is available to plants. Its availability is influenced by its content of solutes and by the capillary and adsorptive forces associated with the soil matrix. Generally speaking, the water that is available by a suction force of about -15 bars can be taken up by plants. Beyond this point, most plants undergo permanent wilting; but before they reach this level their net photosynthetic activity becomes seriously reduced.

Soil nutrients are held in the soil in a variety of ways. Major differences in the retention mechanisms exist for cations and anions, however. For cations, a well-defined cation exchange system exists. This system involves the clay minerals and organic matter of the soil, both of which bear net negative electrical charges and thus attract and adsorb a variety of cations, including hydrogen ions, various mineral cations, and the ammonium ion.

The negative nutrient ions, principally nitrate, sulfate, and phosphate, are held by an anion exchange system, but this system is only of minor importance to total nutrient availability. The most important reservoir for anions is the organic matter of the soil, from which they are released by decomposition. For all nutrients, both cations and anions, exchanges with plant roots take place between the root and the soil solution.

Literature Cited

Bell, F., and P. S. Nutman. 1971. Experiments on nitrogen fixation by nodulated lucerne. In Biological Nitrogen Fixation in Natural and Agricultural Habitats. T. A. Lie and E. G. Mulder (eds.). *Plant and Soil* (Special vol.):231–264.

Black, C. A. 1968. *Soil-Plant Relationships.* New York: John Wiley and Sons.

Buckman, H. O., and N. C. Brady. 1974. *The Nature and Properties of Soils,* 8th ed. New York: Macmillan Publishing Company, Inc.

Cook, R. J., and R. J. Papendick. 1972. Influence of water potential of soils and plants on root disease. *Ann. Rev. Phytopathol.* 10:349–374.

Cooke, G. W. 1967. *The Control of Soil Fertility.* New York: Hafner Publishing Co.

Deevey, E. S., Jr. 1970. Mineral cycles. *Sci. Amer.* 223(3):149–158.

Gardner, W. R. 1968. Availability and measurement of soil water. In *Water Deficits and Plant Growth,* vol. 1, *Development, Control, and Measurement.* T. T. Kozlowski (ed.). p. 107–135. New York: Academic Press.

Hadas, A. 1973. Water retention and flow in soils. In *Arid Zone Irrigation,* B. Yaron, E. Danfors, and Y. Vaadia (eds.). *Ecological Studies,* vol. 5. pp. 89–109. New York: Springer-Verlag.

Hunt, C. B. 1972. *Geology of Soils: Their Evolution, Classification, and Uses.* San Francisco: W. H. Freeman and Company.

Kalinin, G. P., and V. D. Bykov. 1969. The world's water resources, present and future. *Impact of Sci. on Soc.* 19:135–150.

Kamprath, E. J. 1972. Soil acidity and liming. In *Soils of the Humid Tropics,* National Research Council. pp. 136–149. Washington, D.C.: National Academy of Sciences.

Kellogg, W. W., R. D. Cadle, E. R. Allen, A. L. Lazrus, and E. A. Martell. 1972. The sulfur cycle. *Science* 175:587–596.

Kononova, M. M. 1966. *Soil Organic Matter.* Oxford: Pergamon Press.

Kramer, P. J. 1969. *Plant and Soil Water Relationships: A Modern Synthesis.* New York: McGraw-Hill Book Co.

Newman, E. I. 1974. Root and soil water relations. In *The Plant Root and Its Environment.* E. W. Carson (ed.). pp. 363–440. Charlottesville: University of Virginia Press.

Penman, H. L. 1970. The water cycle. *Sci. Amer.* 223(3):98–108.

Pratt, P. F., and F. L. Bair. 1962. Cation-exchange properties of some acid soils of California. *Hilgardia* 33:689–706.

Slatyer, R. O. 1967. *Plant-Water Relationships.* New York: Academic Press.

Taylor, S. A., and G. L. Ashcroft. 1972. *Physical Edaphology.* San Francisco: W. H. Freeman and Company.

Thompson, L. M. 1957. *Soils and Soil Fertility.* New York: McGraw-Hill Book Co.

Townsend, W. N. 1973. *An Introduction to the Scientific Study of the Soil.* New York: St. Martin's Press.

11

IMPACTS OF CULTIVATION AND GRAZING ON SOIL

When man converts natural ecosystems into agroecosystems, he modifies many specific features of their structure and dynamics. In the soil subsystem, as we noted in Chapters 2 and 7, these modifications affect two basic ecosystem functions: They reduce the importance of detritus food chains and increase the importance of nutrient and energy exports from the system.

In almost all natural ecosystems, most of the net primary production of the ecosystem passes along detritus food chains (Figure 11-1). In forest ecosystems, where most herbivores are leaf and fruit eaters, over 99 percent of the energy flow can involve detritus pathways. In a grassland ecosystem, which has a variety of rodent and ungulate species, this percentage can fall to around 75 percent. Man's farming and herding activities, in almost all cases, have reduced the proportion of energy that flows along detritus pathways; in extreme cases he has converted forest areas to monocultures of annual plants, such as wheat, from which he takes both grain and straw and leaves the roots and stem bases as the only portion of net primary production that enters the detritus food chain. In managing pastures and rangelands, his objective has been to maximize the amount of plant growth harvested by his animals and to minimize the production of inedible species that would otherwise contribute to detritus food chains.

In addition, except in a few patterns of subsistence agriculture, human activities have increased the amount of energy and nutrients that flow from food-producing systems to food-consuming systems. In other cases, the disruption of natural processes has allowed nutrients to leak out of agricultural systems along natural pathways, but at increased rates. To balance this, of course, man has been forced to use various systems of artificial fertilization, and the net result is a shift from nutrient cycling patterns that tend to be closed and local to patterns that are open and biospheric.

In this chapter, we shall examine the specific impacts of these two modifications on specific aspects of soil structure and fertility. In Chapter 12, we shall consider in detail the effects that particular techniques of irrigation and fertilization have had on the dynamics of the soil ecosystem.

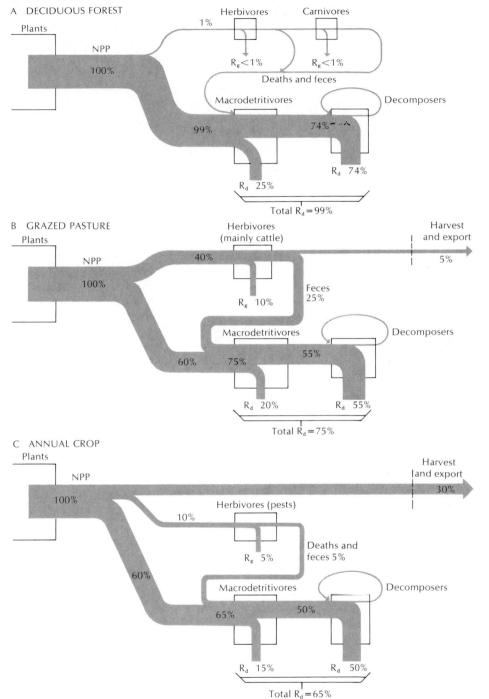

A DECIDUOUS FOREST

Plants

NPP
100%

Herbivores   Carnivores

1%

$R_g < 1\%$ $R_g < 1\%$

Deaths and feces

Macrodetritivores   Decomposers

99% 74%

R_d 74%

R_d 25%

Total $R_d = 99\%$

B GRAZED PASTURE

Plants

NPP
100%

Herbivores
(mainly cattle)

Harvest
and export

40%   5%

R_g 10%

Feces
25%

Macrodetritivores   Decomposers

60%   75%   55%

R_d 20% R_d 55%

Total $R_d = 75\%$

C ANNUAL CROP

Plants

NPP
100%

Harvest
land export

30%

Herbivores (pests)

10%

R_g 5%

Deaths and
feces 5%

Macrodetritivores   Decomposers

60%   65%   50%

R_d 15% R_d 50%

Total $R_d = 65\%$

FIGURE 11-1.
Schematic representation of energy flow through grazing and detritus food chains in deciduous forest, grazed meadow, and annual crop ecosystems. Diagrams show the portions of net primary production (NPP) that are consumed through the respiration of consumers in the grazing food chain (R_g) or by those in the detritus food chain (R_d) or that are exported from the system.

AGRICULTURAL ACTIVITY AND SOIL STRUCTURE

In a natural ecosystem, the organic content of the surface mineral soil might range from a trace amount to as much as 15 to 20 percent, varying with climate and natural vegetation type. In general, soil organic matter levels are higher in areas of cooler climatic conditions or greater moisture (Figure 11-2). In most temperate zone and tropical soils, however, the

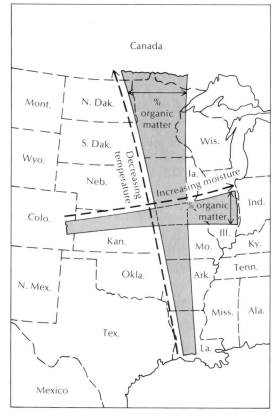

FIGURE 11-2.
General pattern of variation of soil organic matter along gradients of temperature and moisture availability in the central United States. (Modified with permission of Macmillan Publishing Co., Inc., from *The Nature and Properties of Soils*, 7th ed., by H. O. Buckman and N. C. Brady. Copyright © Macmillan Publishing Co., Inc., 1969.)

organic matter content ranges from about 1 to 5 percent (Buckman and Brady 1974; Sanchez and Buol 1975). Organic matter content is closely related to both carbon and nitrogen content of the soil, the general relationship being that organic content is about $2\times$ carbon content and $20\times$ nitrogen content.

Once a soil with a relatively high organic content is put under intensive cultivation, the amounts of organic matter, carbon, and nitrogen begin to decline unless specific techniques are used to add organic matter. The initial decline in organic content is rapid; after the first few years of cultivation the rate of decline slows (Figure 11-3). The extent of decline in organic content depends a great deal on the crop or crop rotational sequence employed, but for continuous grain cultivation it commonly leads to a depletion of 30 to 60 percent (Table 11-1). In chernozem soils in Saskatchewan, for example, 60 years of continuous cultivation (primarily of small grains) depleted soil carbon by 33 to 52 percent (Martel and Paul 1974). With corn, a crop that provides very poor soil protection and organic matter return, losses tend to be greater. The percentage loss tends to be greatest in soil layers near the surface (Figure 11-4).

When soils in arid lands are brought into cultivation under irrigation, the organic content usually increases. The amount of this increase varies and is governed by the principles that apply in more humid regions. Our concern, therefore, is with the level of organic matter that develops in irrigated, arid-region soils, compared to that which develops in virgin or well-managed soils with comparable moisture regimes, in more humid areas.

The fact that the rate of decline in the organic content of intensively cultivated, rain-fed soils eventually slows indicates that not all components of the organic matter originally present are equally susceptible to decomposition. The organic matter present after many years of cultivation, termed **old organic matter,** is highly

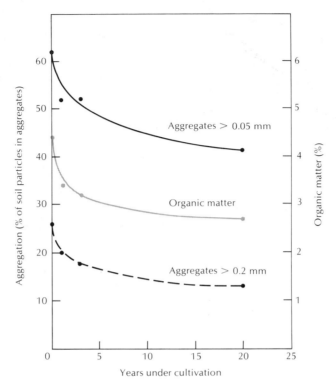

FIGURE 11-3.
Patterns of decline in soil organic content and in the percentage of soil particles in crumb aggregates of various sizes as a function of the length of continuous cultivation for a rain-fed soil. (Modified from Baver 1968.)

TABLE 11-1
Quantities of organic matter and nitrogen in Ohio soils after 32 years of cultivation according to various systems.

Cropping system	Organic matter (tons/ha)	Nitrogen (kg/ha)
Original crop land	39.2	2439
Continuous corn	14.3	941
Continuous wheat	24.6	1474
Continuous oats	25.5	1597
Corn, oats, wheat, clover, timothy	30.0	1733
Corn, wheat, clover[a]	33.2	1995

[a]29 years of cultivation only.
Source: From R. M. Salter, R. D. Lewis, and J. A. Slipher. 1941. *Our Heritage—The Soil.* Ohio Agr. Expt. Sta. Bull. 175.

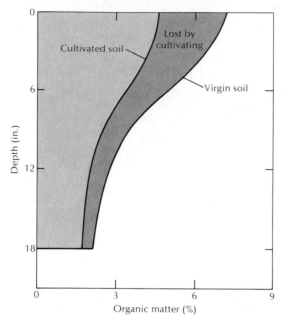

FIGURE 11-4.
Organic matter profiles in virgin grassland soil and in soil cultivated for 43 years in North Dakota. (Modified from Buckman and Brady 1974. USDA drawing.)

into grass, however, the redevelopment of crumb structure was rapid, particularly in the surface 5 centimeters where most of the grass roots are concentrated. At Jealott's Hill Research Station in England, similar results were obtained for clayey soils (Low 1972). In these studies an old grassland (put into grass in the 1840s) was compared to a soil of the same type that had been cultivated for over one hundred years. A portion of the old grassland itself was also plowed and its characteristics followed over several years. Under the old grassland, 70 to 80 percent of the soil particles were in water stable aggregates greater than 2 milli-

resistant to microbial attack and to decomposition. This organic matter is truly old: Radiocarbon dating of the organic matter present in cultivated soils commonly gives average ages of several hundred years, and occasionally ages up to 3000 years (Russell 1971).

The decline in organic content is paralleled by a deterioration of crumb structure in soils with an appreciable content of clay and silt (Figure 11-3). Studies by Greacen (1958) of aggregation in virgin grassland, pasture, and cultivated soil in Australia show dramatically the extent to which this structure can vary (Figure 11-5). In the upper 6 inches of the virgin soil, 80 to 100 percent of the mineral particles were tied up in crumb aggregates. Under continuous cultivation, this was reduced to less than 5 percent. When cultivated soils were put

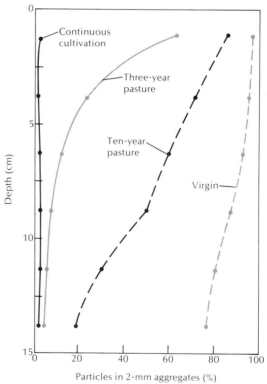

FIGURE 11-5.
Profiles of the percentage of soil particles in crumbs of 2 mm diameter for Australian soils with various cropping histories. (Modified from Baver 1968.)

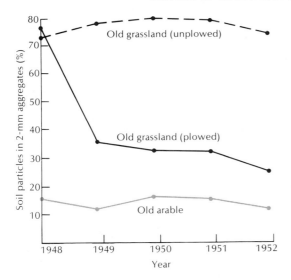

FIGURE 11-6.
Percentages of soil particles in 2-millimeter, water stable aggregates in old grassland and old arable soils in England, showing the pattern of change after the plowing of an area of old grassland. (Modified from Low 1972, *Journal of Soil Science*, published by Oxford University Press, Oxford, England.)

meters in diameter, whereas in the old cultivated field this percentage was 10 to 20 percent. In the newly plowed areas of old grassland, the aggregation percentage dropped about 40 percent in the first year and declined slowly thereafter (Figure 11-6).

In moist temperate-zone soils, decline in organic content and deterioration of crumb structure are frequently correlated with major changes in earthworm populations; these changes are partly cause and partly effect. In the Jealott's Hill studies mentioned above, the old grassland had ten times as many earthworms as the old cultivated soil (Table 11-2). The annual deposition of castings on the soil surface in English pastureland has been estimated at 5 to 9 kilograms dry weight per square meter (Barley 1961). That earthworm activity, other factors being equal, is important to soil macrostructure has been demonstrated by a number of studies. For example, van de Westeringh (1972) examined soil structure in two orchards in the Netherlands, one of which had abundant worm populations, the other of which lacked earthworms as a result of the intensive use of certain pesticides. In the latter orchard the fact that the soil structure was beginning to deteriorate was apparent in the more cloddy, less porous condition of the surface 20 centimeters of the soil.

With deterioration of crumb structure, changes in bulk density, pore space, and degree of soil aeration also occur. **Bulk density** is simply the dry weight, in grams per cubic centimeter, of an undisturbed field sample of soil. Because the specific gravity of the various soil particles themselves varies little, bulk density is a good index of soil porosity: The higher the bulk density, the lower the porosity. Data on bulk density, pore space, and the portions of the pore space occupied by water and air were

TABLE 11-2
Earthworm populations in old grassland and old cultivated soils at Jealott's Hill Research Station, England.

Cultivation history	Earthworm populations (numbers/ha)
Old grassland (grass since 1840s)	1.25 to 1.80 $\times$ 10^6
Cultivated old grassland (plowed 3 years earlier)	4.0 to 7.5 $\times$ 10^5
Old cultivated (continuous since 1840s)	1.0 to 2.0 $\times$ 10^5

Source: Low 1972.

TABLE 11-3

Bulk density, total pore space, and pore space occupied by air and by water for old grassland and old cultivated soils at Jealott's Hill Research Station, England.

Cultivation history	Bulk density	Total pore space (%)	Pore space occupied by	
			water (%)	air (%)
Old grassland (grass since 1840s)	1.08	55.6	35.3	20.2
Cultivated old grassland (plowed 4 years earlier)	1.09	55.3	47.5	7.8
Old cultivated (continuous since 1840s)	1.47	41.8	36.8	5.0

Source: Low 1972.

determined by Low (1972) for the three soils at Jealott's Hill Research Station that were described earlier (Table 11-3). These samples were taken during the winter, when the soils were almost certainly at field capacity. Bulk density determinations gave values 1.3 to 1.4 times higher for the old cultivated soil than for the old grassland or newly-plowed old grassland soils. As expected, total pore space was greater by 13 to 14 percent for the grassland and the newly-plowed grassland soils than for the old cultivated soil. Striking differences were also shown in the measurements of pore space occupied by air; this was equal to 20.2 percent of the soil volume for the old grassland and only 5.0 percent for the old cultivated soil. Other data showed that the water present in the old cultivated soil was retained with considerable force. Not only did the old grassland soil have nearly the same volume percentage occupied by water as the old cultivated soil, it also had an additional major fraction of pore space occupied by air. Furthermore, the water present was more readily available to growing plants because it was held in the pore space with less force. Overall, these data indicated that under wet weather conditions aeration in the old cultivated soil was likely to be inadequate.

Crumb or granular structure also has a major influence on the degree to which water can infiltrate at the soil surface. Fine-textured soils that lack effective water stable aggregate structure tend to form crusts and to become sealed to water infiltration under the beating action of rain. Soils with greater bulk densities also show slower rates of water movement within the soil. Slower infiltration, of course, means increased surface runoff and thus a greater potential for erosion.

Increased bulk density also results in greater problems of cultivation. In the comparison of recently-plowed old grassland soils and old cultivated soils at Jealott's Hill, for example, the mean force, or draw-bar pull, required to operate a four-furrow plow was 1.77 times greater for the old cultivated soil. Essentially, this means that as the soil structure deteriorates, more powerful equipment and greater fuel inputs are required for tilling. Furthermore, as aggregate structure deteriorates, the moisture level of the soil becomes more critical. The critical moisture level for tillage is frequently measured by determining the **sticky point** of the soil. This is simply the moisture level at which a thoroughly kneaded mass of soil first begins to feel sticky to the fingers, or first tends to stick to a knife blade. Moisture levels must be below this point for plowing and mechanical cultivation. In many cases, plowing in wetter conditions causes the soil to puddle, that is, to form massive, resistant clods. In the studies at Jealott's Hill (Low 1972), the sticky point of the old grassland soil (unplowed) was at 63.3 percent moisture, while that of the old cultivated soil was at 45.1 percent. In other words, the old grassland soil could be cultivated at a moisture level up to 18 percent higher than that of the soil lacking good crumb structure.

These factors all interact, of course, with the use of mechanical equipment or with any trampling by livestock. As crumb structure deteriorates, in fact, it is primarily the action of heavy machinery and trampling that leads to the compaction and increase in bulk density (Figure 11-7).

Plowing to the same depth year after year can also create an artificial, consolidated layer termed a **plow pan** at the average maximum depth reached by the plow (Figure 11-8). Plow pans are most common in heavy clay soils, and when present, they may interfere with drainage, soil aeration, and root growth.

Not all aspects of plowing and other forms of mechanical cultivation are detrimental, of course. Preparation of a seedbed by plowing creates a loose, absorbant, well-aerated soil condition favorable to the germination and early growth of most crop plants. In many cases it also mixes fertilizers or manures into the soil where they become available to the root systems of the crop. Mechanical cultivation also reduces weeds. Under special circumstances, deep plowing can beneficially influence soil conditions by breaking up hardpan layers, bringing clay material into surface soil layers that may be deficient in it, and mixing lime or fertilizer into deeper soil layers with good effects on overall soil fertility.

All these benefits of mechanical tillage are increased by a good aggregate structure, however; and they are all diminished by intensive mechanical cultivation with heavy equipment

FIGURE 11-7.
The increasing use of heavy equipment, such as these mechanical pea harvesters in the San Joaquin Valley of California, is a major cause of compaction problems in soils with weak crumb structure. (Photograph by J. C. Dahilig, courtesy of the U.S. Bureau of Reclamation.)

FIGURE 11-8.
Effect of a plow pan on the root development of cotton plants in Louisiana. Cotton normally forms a tap root that penetrates to a depth of over two feet. Compaction of this heavy clay soil at a depth of about four inches has effectively prevented root penetration beyond this level. (Photograph by L. L. Loftin, courtesy of the USDA Soil Conservation Service.)

and by practices that severely deplete the organic content of soils. The availability of powerful cultivation equipment, and the low fuel costs that have prevailed until recent years, have encouraged excessive tillage in many situations (Gill and Vanden Berg 1967). The compaction caused by the extensive use of heavy machinery can extend to a depth of a meter or more and can be difficult and slow to correct. This fact, together with sharp increase in fuel costs, has encouraged experimentation with new practices of minimum tillage and zero-tillage. We shall examine several of these approaches in Chapters 14 and 24.

Grazing, like cultivation, also has significant effects on soil structure, and damaging ones are particularly likely in dry rangelands subject to considerable fluctuations in annual rainfall. Avoiding such impacts requires careful and continuous adjustment of grazing intensity (Albertson, Tomanek, and Riegel 1957). In the central Great Plains, for example, the amount of precipitation varies greatly; but the resulting annual range-grass production varies much more. Furthermore, under the influence of a series of dry seasons, the composition of the grassland community tends to shift from a predominance of taller bunchgrasses to a predominance of short bunchgrasses and eventually of annual grasses and forbs. Higher intensities of grazing reinforce this tendency, of course, and the result is often the degradation of grassland cover, which leaves the soil surface exposed to sun, rain, and wind.

In mountain rangelands, the condition of the soil ecosystem is perhaps the most practical index of the degree to which grazing has been successfully integrated into the ecosystem complex dominated by topography, climate, and vegetation (Ellison 1949). Satisfactory range condition assumes a plant cover with a good proportion of choice forage species; but it is also necessary for this plant cover to be dense enough to maintain soil stability. The slowness of the processes that developed a soil profile capable of supporting vegetation that is good for grazing implies that, under natural conditions, annual losses from the system must be very low. This means that any visible evidence of regular, general erosion, even though it may not seem to be severe, is evidence of deterioration for most mountain rangelands.

The feeding and trampling by grazing animals on rangelands are analogous to the mechanical disruption and compaction of the soil in cultivated lands. Although fewer nutrients and less organic matter are generally exported from rangelands than from croplands, these exports can also be significant, because most rangelands possess smaller ecosystem nutrient and energy budgets than do farmed agroecosystems. The results of all these factors on soil structure, in both cultivated and grazed systems, are accelerated erosion and reduced productivity.

SOIL EROSION

Erosion of agricultural soils by water and wind continues to be one of the most serious problems of modern agriculture. The population increases among peoples who practice subsistence forms of agriculture, such as pastoralism and shifting cultivation, make erosion an increasingly serious threat in developing nations, but in the developed countries current trends toward the intensification of crop production through mechanized practices threaten to create particularly serious problems. In the 1973/1974 crop season in the United States, for example, a total of 3.6 million hectares of land were shifted from idle cropland, grassland, or woodland to active cultivation (Grant 1975). Of this total, 1.6 million hectares had inadequate systems of erosion control. As a result, on this land area alone, loss of soil by wind and water erosion, above that which would otherwise have occurred, equalled 55 million tons, 12 million by wind and 43 million by water. The areas affected were spread widely over the country, from the Atlantic Coastal Plain to the Palouse Grasslands of the Northwest. In the parts of this converted acreage that lacked adequate measures for soil conservation, extra losses averaged 26.9 tons per hectare—more than double the 11.2 tons per hectare allowable loss limit specified by the Soil Conservation Service. Considering that land converted from grass to cultivation generally resists erosion because it has good structure, it is clear that erosion represents a very serious concern for U.S. agriculture in the immediate future.

Water erosion represents the effect of the energy developed by water as it falls and as it strikes and flows over inadequately protected soil (Beasley 1972). A 5-cm thunderstorm deposits slightly over 506 tons of water on each hectare of land in the form of raindrops averaging 3 millimeters in diameter and falling at a speed of about 7.62 meters per second. In heavy storms, it is not unusual for this amount of water to fall in one-half hour.

Erosion by water is generally classified as sheet, rill, or gully erosion, although all of these grade gradually into one another. **Sheet erosion** is, in a sense, the earliest stage in the erosion process. It is caused by the beating action of raindrops on the soil surface and by the transport of soil particles by water flowing in a thin sheet over this surface. When a drop strikes the surface, it tends to shatter small lumps or aggregates of soil into smaller units or individual soil particles. The weaker the crumb structure of

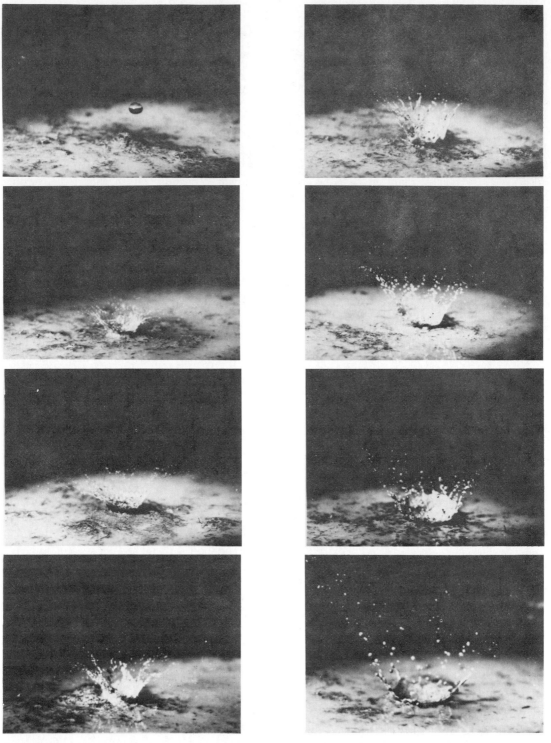

FIGURE 11-9.
Photo sequence illustrating the impact of a water droplet on the soil surface. Note the wide scattering of particles after impact. (Photograph courtesy of the USDA Soil Conservation Service.)

FIGURE 11-10.
The erosive action of raindrop impact is frequently shown in pebble-capped soil pedestals. The exposed soil has been loosened by raindrop impact and eroded by surface runoff. The soil beneath small pebbles, protected from impact, has remained in place. (Photograph by E. A. Hodson, courtesy of the USDA Soil Conservation Service.)

the soil, the greater is this effect, although even the most water stable aggregate structure lacks the ability to withstand the intense beating of a heavy convectional storm. The impact of the drop on the surface can splash these loosened particles to a distance of several feet (Figure 11-9, and if the surface has an appreciable slope, this action alone will cause a net downslope movement of soil particles. The impact of the raindrops also compacts the surface layer and fills infiltration pores with loosened particles, thus sealing the soil surface to some degree. Although much of the water deposited initially can be absorbed if the soil is dry to begin with, reduced infiltration can eventually lead to the accumulation of water on surface and to surface runoff. This runoff carries with it many of the soil particles placed in suspension by the splashing of raindrops (Figure 11-10). The finest of these, the clays, are very slow to settle out, even in still water, and larger particles are also carried, even by relatively slow surface flow, because splashing tends to resuspend particles as long as the rain continues (Figure 11-11). Some particles are large enough to be carried by a rolling or sliding action sometimes called **surface creep;** smaller particles can move by **saltation,** effectively a bouncing, downflowing movement; and still others, the smallest, are carried in suspension. The overall erosiveness of flowing water depends on its velocity, of course, but is also influenced by the degree of turbulence and the abrasiveness of the materials already in suspension.

Sooner or later, water flowing over a soil surface finds small channels in which its flow be-

FIGURE 11-11.
Severe sheet erosion on a Southern California field with a slope of less than two percent.
(Photograph courtesy of the USDA Soil Conservation Service.)

comes concentrated. The erosive action of the water then becomes more concentrated, eroding downward to increase the depth of these channels. This early stage of channel erosion is called **rill erosion** as long as the channels remain small enough not to interfere with the use of mechanized cultivation equipment (Figure 11-12). Damage by sheet and rill erosion tends to be inconspicuous, even when actual losses are considerable, because their visible effects are obscured by subsequent cultivation activity. Nevertheless, these two forms of erosion are probably more significant in relation to agricultural productivity than more conspicuous ones.

When channel erosion becomes extensive enough to interfere with equipment use, it is termed **gully erosion** (Figure 11-13). At this point, the channels have been cut so deep that

erosion of the subsoil as well as of the topsoil takes place.

A great many factors affect the amount of soil lost by water erosion. Many of these are related to the crop and the agricultural techniques used on cultivated land and to the grazing practices followed on rangeland. In Missouri, comparisons of average soil losses in five different tillage and cropping systems showed more than a hundredfold variation in average annual losses. The data, covering a 14-year period, came from fields with slopes averaging between 3 and 4 percent (Table 11-4). As might be expected, losses were greatest from plowed but unplanted soil, over 92 tons per hectare annually. However, erosion losses under corn cultivation were almost half this maximum. Under perennial wheat cultivation, losses were reduced to about 23 tons

FIGURE 11-12.
Severe sheet and rill erosion on winter wheat land in eastern Washington. The soil loss from this area ranged from 50 to 200 tons per hectare from a single storm. (Photograph by G. R. McDole, courtesy of the USDA Soil Conservation Service.)

FIGURE 11-13.
Severe gully erosion in Wisconsin. (Photograph courtesy of the USDA Soil Conservation Service.)

TABLE 11-4

Annual soil losses by water erosion for agricultural land subject to various uses in Missouri (average for 14 years for slope of 3.68% and average precipitation of 102.5 cm).

Statistic	Bluegrass sod	Corn-wheat-clover rotation	Continuous wheat	Continuous corn	Plowed + uncultivated
Annual soil loss (tons/ha)	0.76	6.23	22.63	44.19	92.06
Annual runoff (% of precipitation)	12.0	13.8	23.3	29.4	30.3
Loss per unit runoff (kg/runoff cm)	164	1144	2476	3818	7758
Time to erode 10 cm (years)	1711	207	56	28	13

Source: Data from M. F. Miller and H. H. Krusekopf. 1932. *The Influence of Systems of Cultivation and Methods of Culture on Surface Runoff and Soil Erosion.* Missouri Expt. Sta. Res. Bull. 177.

per hectare, and under bluegrass sod, erosion removed less than 1 ton per hectare. Most interesting, however, was the result obtained in a rotation of corn, wheat, and clover: Here the annual loss rate was only 6.23 tons per hectare, well below not only the rate for continuous corn cultivation but also that for continuous wheat cultivation. As we shall see, rotations such as this have a strong potential for preserving the organic matter level, and presumably the aggregate structure, of cultivated soils.

Heavy grazing levels are also associated with reduced rates of water infiltration, increased runoff, and accelerated erosion (Rauzi and Smith 1973). A heavier plant cover, of course, protects the soil surface against the sealing and eroding effects of beating raindrops, and compaction by the trampling of grazing animals can reduce the permeability of the surface layer of the soil.

In windy, semiarid regions, wind erosion takes a great toll on agricultural land; and it is a problem of increasing importance in more humid regions with fine-textured soils whose aggregate structure is deteriorating. In the United States, wind erosion is the predominant erosion problem on about 22.2 million hectares of cropland and 3.6 million hectares of rangeland (Woodruff et al. 1972) (Figure 11-14). In a number of other world regions, including parts of Africa and central and eastern Asia, dust in the atmosphere, due in part to agricultural activity, is a chronic problem that exists on a scale unknown in Europe and North America.

The dynamics of wind erosion are similar to those of heavy sheet erosion by water. Wind erosion is most intense when strong, turbulent winds blow across an unprotected soil that is dry, loose, and fine textured. Soil particles are transported in much the same manner as they are by water; that is, by **surface creep, saltation, and suspension.** Under suitable conditions, wind speeds as low as 5.8 meters per second at a height of 30 centimeters can start the soil blowing. Soil particles less than 0.1 millimeters in diameter are usually carried in suspension; those that range from 0.1 to 0.5 millimeter in diameter move largely by saltation; and larger particles, up to 1 millimeter or more in diameter move mainly by surface creep. Intense winds, of course, can move even larger particles.

FIGURE 11-14.
In the United States, wind erosion accounts for about one-fourth of all soil erosion. This photo shows a newly planted sugar beet field near Boise, Idaho. (Photograph by J. D. Roderick, Courtesy of the U.S. Bureau of Reclamation.)

The vulnerability of agricultural land to wind erosion varies with a number of factors. Soils with good crumb structure, for example, are highly resistant to erosion. One measure of erodibility is the percentage of the soil that consists of particles greater than 0.84 millimeter in diameter; a soil that is two-thirds such particles, including those in crumb aggregates, is essentially nonerodible. Surface roughness, including ridges and troughs created by plowing, also reduces vulnerability to wind erosion; so do a high moisture level and a vegetative cover or an organic mulch. The shape and orientation of a field also affect its erodibility: A long field oriented parallel to the direction of prevailing winds is most vulnerable.

The total amount of soil lost annually through erosion in the United States has been estimated at about 4 billion tons by water (three-fourths from agricultural land) and 1 billion tons by wind (Pimentel et al. 1976). Of the 4 billion tons carried by water, about 1 billion tons enter the ocean, and the rest is deposited in reservoirs, lakes, streams, and other locations. Combined annual losses through wind and water erosion are approximately 25 tons per hectare of cropland.

Wind and water erosion together have ruined or seriously damaged an estimated 114 million hectares of land in the United States. About 20 million hectares of land once used for crop cultivation have been rendered useless and an additional 20 million are nearly so. An estimated 40 million hectares of cropland have lost at least one-half the topsoil they originally had. Considering that there is a total in the United

States of about 225 million hectares of land suitable for either regular or limited cultivation, the extent of erosion damage is considerable (Millar et al. 1965).

The net effect of sheet erosion is essentially the truncation of the soil profile; that is, the removal of surface horizons that possess the highest level of organic matter and the greatest degree of aggregate structure. This effect, together with the deterioration of soil structure, leads to structural changes that reduce productivity. These changes include increased bulk density, reduced aeration, and, at least in fine-textured soils, reduced water availability. When soil is lost by erosion considerable quantities of nutrients are also lost.

SOIL STRUCTURE AND CROP YIELD

In the midwestern United States, topsoil losses by erosion are directly translatable into declines in productivity. One study, for example, has shown that loss of the surface 5.1 centimeters leads to a 15-percent reduction in corn yield; losses of 15.2 centimeters reduced yield by 30 percent; and losses of 30.5 centimeters caused a 75-percent decline. In a classic study conducted at the University of Missouri, corn yields were compared on two areas of silt loam soil, one of which had the entire topsoil layer removed. Both areas were subsequently fertilized so that fertility limitations did not influence yields. Over 18 years, from 1950 to 1968, yields for the plot with topsoil present (only slightly eroded) averaged 227 bushels per hectare, but yields for the plot from which the topsoil had been stripped averaged only 119 bushels per hectare. In several years, the crop on the stripped plot failed completely or nearly so.

The effects of soil compaction, alone, seem to vary with soil type and crop (Rosenberg 1964). In general, however, it seems that for a particu-

lar soil-crop combination an optimal degree of compaction exists. In sandy soils, in particular, increased compaction and bulk density can improve crop yields, probably because the capillary pore space in which water can be retained at field capacity is simultaneously increased without reducing air-filled noncapillary space to critically low levels. In other words, increased compaction can improve moisture relationships in the soil.

In finer-textured soils, however, compaction usually reduces yields. In California alfalfa fields, for example, vehicle traffic is responsible for severe reductions in yield and in stand life. Compaction in areas regularly affected by wheel traffic led to a 57 percent reduction in root density (Sheesley et al. 1974). In extremely compact soils, it becomes impossible for the roots of growing plants to penetrate the soil. Soil compaction may also lead to poor aeration, which creates a variety of problems that involve a lack of oxygen, reduced availability of nutrients in the soil solution, accumulation of toxic chemicals in the root environment, and increased susceptibility of the root system to invasion by parasitic microorganisms (Grable 1966).

Finally, of course, the reduction in organic matter that is either a cause or effect of structural deterioration represents a weakening of the cation exchange system by which certain nutrients are held in available form. Structure and fertility are thus intimately related, and it is in this way instead of as a source of specific nutrients that soil organic matter probably plays the most important role in the soil ecosystem.

The recent studies by Low (1973) at Jealott's Hill Research Station and at a number of other locations in England have shown the nature of some of the interactions that take place among crop species, soil type, and aggregate structure. First, let us consider the different patterns in yields of winter wheat on soils with differ-

TABLE 11-5
Yields of winter wheat in the Vale of Pewsey, England,
in relation to soil texture and aggregate structure.

		Textural composition (%)		Relative aggregate stability	Wheat yield (kg/ha)
Soil series and cultivation history		Coarse sand	Clay		
Wantage series	Third year of cultivation after six years grass	7–11	33–35	24.4	1970
	Third year of cultivation after 80 years grass	7–11	33–35	45.0	3810
Upper Greensand series	Cultivated for many years	36.6	14.2	1.0	3810

Source: Low 1972.

ent structures (Table 11-5). This table shows that the yield of wheat on the clayey soils of the Wantage Series was strongly influenced by aggregate structure: The soil with stronger aggregate structure showed about twice the yield of that with poorer structure. Yield on the sandy soils of the Greensand Series, however, was equal to the yield on the Wantage Series soil with good aggregate structure, despite the fact that little aggregation existed in the former. This suggests that favorable structure might be provided by textural composition, given either a desirable level of sand or, in finer-textured soils, a well-developed crumb structure.

A second series of studies, carried out in 4-square-meter plots lined with concrete and filled with experimental soils, revealed an important difference between two crops in their sensitivity to soil structure. The soils used here were of the Berkhamsted Series and had about 30.6 percent coarse sand (0.02 to 2.0 millimeters) and 18.3 percent clay. Soil samples were taken from two locations, one of which had been under continuous cultivation for many years, the other under grass for eight years. Oats and beets were the experimental

crop species, and different levels of nitrogen fertilizer were applied so that the performance of each species could be observed under the best conditions of nitrogen availability (Table 11-6). For oats, the difference between best performance on the two soils was small, about 12 percent higher on the soil sample from under grass. For beets, however, yields were over $2.6\times$ greater on the soil from under grass, which had better aggregate structure. One of the major differences in these two species is in diameter of the root radicle, which is much smaller for wheat than for beets. Peas and onions have even coarser roots than beets do, and their yields are also severely reduced on soils that lack good aggregate structure. Grains such as wheat and oats, on the other hand, have very fine rootlets that are able to penetrate the soil more easily; these crops are inhibited only in soils with both a very low sand content and impaired aggregate structure.

One of the experimental techniques that has become available for the study of soil structure in relation to plant yield is the use of synthetic **soil conditioners.** One of these, Krillium, is a synthetic polymer, a derivative of polyacrylic acid. Although it is too expensive for

TABLE 11-6
Yields of oats and beets (g/plot) on 2 × 2-m experimental areas at Jealott's Hill Research Station, England, in relation to soil cultivation history and level of nitrogen fertilization.

Crop	Soil cultivation history[a]	kg N/ha			
		0	46	92	136
Oats	Old cultivated	210	277	286	261
	After 8 years grass	264	320	316	284
Beets	Old cultivated	58	172	144	148
	After 8 years grass	457	420	452	393

[a]Soil texture = 30.6% coarse sand, 18.3% clay.
Source: Low 1972.

TABLE 11-7
Yields and relative head size for cauliflower on experimental plots receiving different applications of a soil conditioner (Krillium).

Soil treatment	Yield (kg/plot)	Predominant head size
Control	35.7	Small to medium
678 kg Krillium/ha	43.9	Medium
2550 kg Krillium/ha	80.5	Medium to large

Source: Low 1973.

general use, it can be used experimentally to produce a soil aggregate structure comparable to that which is created by natural organic polymers in the soil. Since Krillium contains no phosphorus, nitrogen, or mineral nutrients, it allows the effects of organic matter on soil structure to be studied with fertility held constant. Experiments carried out by Low (1973) using cauliflower show how important soil structure is to the yield of this species (Table 11-7). The soil used in this set of experiments contained only 6.3 percent sand and had 25.5 percent clay; it was thus a fine-textured soil for which aggregate structure would be expected to be of considerable importance in aeration, water holding capacity, and root penetration. In soil that received slightly over 2½ tons of Krillium per hectare, the yield of cauliflower more than doubled, and the average head size increased considerably.

Aggregate structure obviously has a major effect on the productivity of many crops, and the degree of aggregate structure is most important in fine-textured soils and for crop species with coarse roots. It appears, moreover, that good aggregate structure can be obtained through either management strategies that depend upon naturally produced organic

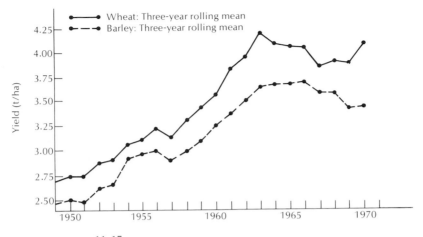

FIGURE 11-15.
Trends of barley and wheat yields in England and Wales from 1950 through 1971. (Modified from Fiddian 1973.)

matter or the application of more expensive, synthetic soil conditioners.

There is serious concern in a number of grain-producing regions that the deterioration of soil structure might be imposing considerable limitations on increased production (Fiddian 1973). In England, for example, average yields of wheat and barley plateaued, or perhaps even dropped somewhat, during the 1960s (Figure 11-15). This leveling of the yield-per-hectare curve occurred despite the continued development of improved crop varieties and cultivation practices. During the 1960s the average annual acreage of grain crops also increased, meaning, in practice, that the tendency to plant grains year after year and to eliminate rotational sequences with grass and other crops was greater. Deterioration of soil structure, evidenced also by an increase in wind erosion in eastern England, was identified as one of the factors involved. A second was probably the accumulation in the soil of crop pathogens that were specific for the grains involved. We shall consider this relationship in more detail in Chapter 13.

IMPACTS ON SOIL NUTRIENT POOLS

Cultivation and grazing, together with the export of the food materials produced, deplete the pools of available nutrients within agro-ecosystems, and many of these agricultural activities reduce the capacity of the soil to hold nutrients in available form. These factors all combine to make fertilization essential to permanent cultivation and intensive grazing.

One of the processes basic to the fertility of ecosystems, yet still poorly understood, is that by which nutrients are released through the weathering of primary materials. The rate of weathering no doubt varies with the nature of the parent material, the climate, and many other factors, but few careful estimates are actually available. Johnson et al. (1968) utilized data from a comprehensive analysis of the nutrient dynamics of forested watersheds in New Hampshire to obtain such an estimate. Their estimate, for parent material consisting of bouldery glacial till, was that weathering released nutrients from about 800 kilograms/ha/yr. For the major nutrient elements present, this corresponds to 24.8 kilograms of iron, 11.2 of calcium, 12.8 of sodium, 23.2 of potassium, and 8.8 of magnesium. In a climax system, these releases would essentially balance losses of nutrients via water outflow. In addition, it should be noted that, in this New Hampshire example, neither nitrogen or phosphorus entered the system in appreciable quantity by weathering.

The quantities of nutrients exported from an agroecosystem in the materials harvested by man also vary, depending on whether the food material is plant or animal, seed or leafy forage, and so on. For a corn crop of 10,000 kilograms per hectare (a good yield) the nutrient quantities removed in grain alone equal 128 kilograms of nitrogen, 23 of phosphorus, 13 of sulfur, 20 of magnesium, 31 of potassium, and 14 of calcium (Delwiche 1971). The removal of nutrients in the harvest, of course, does not eliminate nutrient outflows from the system in other ways, though it is obvious that, compared to the rates at which ecosystem nutrient pools are replenished by weathering, agricultural harvest is a major route of nutrient depletion. We have also observed that atmospheric inputs, though appreciable, are limited and (at least in the case of nitrogen) dependent on specific organisms that may or may not be present. The replacement of nutrients from sources outside the agricultural system is therefore absolutely essential to the maintenance of original productivity levels.

In 1938 an experimental comparison of organic and other types of farming strategies was begun at Haughley Research Farms in England (Alther 1972). On one section of this farm, 27.5 hectares in size, only organic farming techniques were used. A carefully designed system of crop and pasture rotation was followed, and all animal wastes and crop residues were composted and returned to the soil. Nevertheless, certain products such as eggs, milk, meat, and vegetables were removed without their nutrient content being replaced by

inorganic or other imported fertilizers. Although this experiment showed that many of the techniques followed were sound, practical, and efficient, it also showed eventually that they were inadequate. By 1965 crop yields on the organic section of the farm began to drop and mineral deficiencies appeared in the livestock. Except under very unusual circumstances, even moderate exports of nutrients in agricultural products require replacement of nutrients from outside.

Both erosion and the destruction of nutrient retention systems in the soil also reduce the size of soil nutrient pools. The soil removed by erosion in the United States has been estimated to contain an average of 0.10 percent nitrogen, 0.15 percent phosphorus, and 1.5 percent potassium (Beasley 1972). For a severe erosion loss of 224 tons per hectare, this would amount to 224 kilograms of nitrogen, 336 of phosphorous, and 3362 of potassium. A significant portion of this is likely to have been present in unavailable form, of course; nevertheless, these quantities multiplied by the current cost of fertilizers that contain them give an idea of the monetary losses involved. In this case, replacing the losses by fertilizer additions would, at current prices, cost well over $494.00 per hectare.

If we examine fertility relationships over large geographical areas in the light of agricultural impacts, we are forced to conclude that man has, in general, considerably depleted the capital of available nutrients. Unfortunately, there have been few serious attempts to work out detailed nutrient budgets for agricultural regions or major continental areas. Lipman and Conybeare (1936) attempted one of the few analyses of nutrient balance for agricultural systems in the United States (Table 11-8), taking into account the input of nutrients by fertilizer and liming practices, nitrogen fixation by natural mechanisms, inputs from rainfall and irrigation water, and the nutrients in seeds. The outputs they considered included

TABLE 11-8
Nutrient balance for United States agricultural land for 1930.

	Quantities (millions of metric tons)					
	N	P	K	Ca	Mg	S
Losses due to harvesting, grazing, erosion, and leaching	20.8	3.8	45.4	61.8	22.3	10.9
Gains due to liming, organic, and inorganic fertilizers, rainfall, irrigation, natural fixation, and seeds	14.8	1.3	4.6	11.4	3.6	8.2
Net annual deficit	6.0	2.5	40.8	50.4	18.7	2.7

Source: Lipman and Conybeare 1936.

the crop and animal tissues harvested and losses from erosion and leaching. Net losses were calculated for all six of the major nutrient elements considered: nitrogen, phosphorus, potassium, calcium, magnesium, and sulfur. Since 1930, when these calculations were made, farming practices and fertilizer use have changed greatly. Total use of fertilizers has more than tripled; but this does not necessarily mean that inputs now balance outputs. The increased use of fertilizers has also increased the export of nutrients in harvested materials, as well as their loss by leaching, erosion, and volatilization.

It is unquestionably the case that soil organic matter and its contained nutrients have been greatly depleted in virtually all agricultural soils of the midwestern United States since they were put into cultivation. Viets (1971) estimates the loss of nitrogen from the surface meter of U.S. soils at 1.59 billion tons for the 100 years of their cultivation. Thus the major increases in use of nitrogen fertilizers have, in essence, been made necessary by the exhaustion of natural soil reserves; and although these increases may have come closer to balancing the input-output equation, it must be recognized that this balance exists at a new level of nutrient capital, if it exists at all. Chapter 12 examines the implications of fertilizer use in greater detail.

The depletion of nutrient capital that accompanies cultivation is nowhere more important than in the humid tropics, and as we saw in Chapter 5, the diverse systems of shifting cultivation are to a great extent adjusted to this fact. Their strategy is to "mine" the accumulated nutrient capital from natural ecosystems—a nutrient capital that has accumulated over a period of ecological succession.

Tropical forest ecosystems of the wet lowlands appear to function according to a **direct nutrient cycling** process (Stark 1971), which involves the recycling of nutrients from dead organic matter to living plants without the release of major quantities into the mineral soil. The climax forest trees have a few deep roots that provide support and acquire new nutrients from subsoil layers in which the weathering of primary minerals may be taking place (Figure 11-16). Most of the root system, however, consists of a dense feeder-root system that extends through the surface few centimeters of soil and even penetrates the layer of organic matter on the surface of the mineral soil. The roots of many of these species exist in symbiotic association with fungi, forming a combination termed **mycorrhizae**, or "fungal roots." Along with other soil organisms, the fungal partner of this symbiosis participates in the decomposition of dead matter. Some of

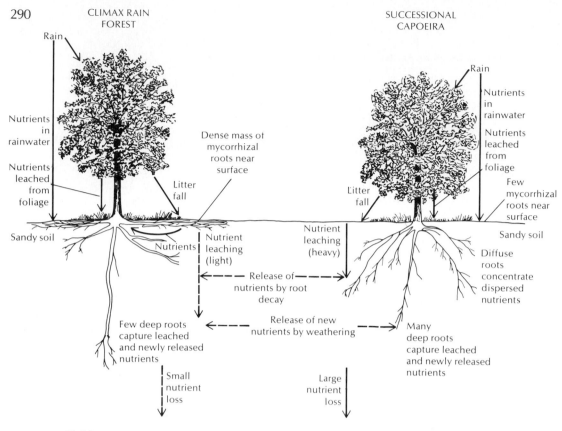

CLIMAX RAIN FOREST

Rain

Nutrients in rainwater

Nutrients leached from foliage

Sandy soil

Dense mass of mycorrhizal roots near surface

Litter fall

Nutrients | Nutrient leaching (light)

Release of nutrients by root decay

Few deep roots capture leached and newly released nutrients

Small nutrient loss

SUCCESSIONAL CAPOEIRA

Rain

Nutrients in rainwater

Nutrients leached from foliage

Few mycorrhizal roots near surface

Sandy soil

Litter fall

Nutrient leaching (heavy)

Diffuse roots concentrate dispersed nutrients

Release of new nutrients by weathering

Many deep roots capture leached and newly released nutrients

Large nutrient loss

FIGURE 11-16.
Patterns of nutrient cycling in the climax tropical rain forest of the Amazonian lowlands and in the capoeira community of ecological succession after clearing of the forest. (Modified from Stark 1971.)

these fungi are able to fix nitrogen, as well. A large percentage of the nutrients obtained by decomposition or fixation passes into the cells of the tree root. The leakage of nutrients into the mineral soil in the process is much less than that which occurs in other types of ecosystems, such as temperate forests or grasslands, although mycorrhizal associations are known for various temperate zone trees, particularly conifers.

The mineral soil of humid tropical regions has a very low cation exchange capacity, and nutrients that do enter the mineral soil are highly susceptible to leaching. In the Amazon Basin, where the surface soils are extremely sandy, the cation exchange capacity of the surface soil ranges from 7.0 to about 32 meq per 100

grams, most of which is due to the organic matter present and not to clay minerals (Stark 1971). Moreover, these soils are strongly acid, and the base saturation is less than 25 percent in almost all cases. These Amazonian soils are thus podzolic in their characteristics. Elsewhere in the tropics, upland soils in many places are lateritic in nature, strongly leached, acid, and dominated by hydroxide clays with low cation exchange capacity and very little organic matter.

The strategy of shifting agriculture is to use the nutrients accumulated in organic matter by cutting the forest vegetation, allowing it to dry, and burning it. The resulting ash provides abundant mineral bases for a short period of time and permits high levels of productivity.

Depending on the ability of the soil to retain these nutrients, they become depleted in two to several years, and the land must be allowed to undergo secondary succession to rebuild a high nutrient capital.

It is interesting that the successional plant species that invade such land have specific adaptations to the nutrient conditions that prevail. This successional community, known in the Amazon Basin of Brazil as the **capoeira community,** is characterized by tree species that have extensive, deeply and widely penetrating root systems quite different from those of the climax forest species (Figure 11-16). Mycorrhizal symbioses are less frequent, and root systems seem to be adapted to recovering nutrients that have been dispersed widely through deeper layers of mineral soil. The plant species of the capoeira are relatively short-lived, and by the time they have achieved their maximum growth, species with nutrient retention mechanisms of the mature forest begin to invade. Nevertheless, the recovery of the nutrient capital of the original climax forest may be a very slow process.

The classic studies of the change in soil fertility during shifting cultivation in the humid tropics are those of Nye and Greenland (1964). These studies, carried out in Ghana, dealt with a system in which the forest was cut and burned but the surface soil was not plowed. Crops, including corn, cassava, yams, and plantains, were planted in small holes made with a stick in the loose surface soil. Crop cultivation was carried on for two to four years. Nye and Greenland compared soil nutrient relationships under a number of crop-growing systems, but the basic trends in fertility were much the same in all cases. Comparing the quantities of calcium, potassium, and magnesium in the soil before and after burning showed increases of $2\times$ to $3\times$ for calcium and potassium and about $1.5\times$ for magnesium (Table 11-9). Thereafter, regardless of whether the soil was actually cultivated or left fallow, the concentrations of these bases declined. After two years, levels of potassium and magnesium had returned almost to preburning values.

Nutrient losses tended to be somewhat higher in the cultivated than in the uncultivated soils, indicating that export of nutrients in harvested materials was of some importance. This was

TABLE 11-9
Changes in nutrient levels in tropical soils in Ghana during a cycle of shifting cultivation.

Nutrient element	Soil use	Before clearing	After burning	After 1 year	After 2 years
Calcium	Fallow	1.4	4.4	3.1	3.0
(meq/100g)	Planted[a]	1.5	4.5	2.8	2.8
Potassium	Fallow	0.25	0.49	0.38	0.29
(meq/100g)	Planted[a]	0.23	0.70	0.30	0.20
Magnesium	Fallow	0.95	1.50	1.12	0.93
(meq/100g)	Planted[a]	1.05	1.43	1.05	0.93
Carbon	Fallow	2.48	2.19	1.94	1.71
(%)	Planted[a]	2.26	2.06	1.64	1.60
Nitrogen	Fallow	0.228	0.205	0.170	0.150
(%)	Planted[a]	0.241	0.187	0.148	0.145

[a]Minimum cultivation only
Source: Nye and Greenland 1964.

TABLE 11-10

Total nutrient losses and losses in harvested crops for calcium, potassium, and magnesium after two years of shifting cultivation in tropical soils in Ghana.

		Losses (kg/ha)	
	Ca	K	Mg
Follow after clearing and burning	1199	504	291
Planted with minimum cultivation	1457	829	258
Estimated removal in harvested crops	101	392	45

Source: Nye and Greenland 1964.

particularly true for potassium, the estimated removal of which by the the crop was nearly half the total loss during the two-year period of experimental cultivation (Table 11-10). The pattern of change in carbon and nitrogen in the soil was, of course, somewhat different. The percentage of both of these elements was reduced by burning, after which this percentage continued to decline, especially in the bare fallow soils. This observation, of course, demonstrates that crop cultivation itself serves to return organic matter to the soil.

It is thus apparent that the mechanisms of nutrient retention within lowland ecosystems of the humid tropics are specialized and fragile. The complex, highly productive rain forest vegetation exists because it has enjoyed a long, evolutionary history less disrupted than many of the world's other ecosystems in other regions of extreme environmental conditons. This evolution has shaped a system of efficient nutrient conservation and utilization, and this system depends on the integrity of the native tropical forest. Forests such as those in the Amazon Basin clearly do not indicate rich lands that can be converted to traditional agriculture (Figure 11-17).

Not all tropical soils are as fragile and as limited in agricultural potential as this, however. Some soils, especially those of volcanic origin, are suitable for permanent agriculture

FIGURE 11-17.

Tropical forest being cleared for permanent cultivation in the Amazon Basin near Leticia, Colombia. The sandy soils in this locality are unable to retain adequate quantities of nutrients in the absence of the biotic retention mechanisms that operate in the mature rain forest. (Photograph by G. W. Cox.)

even in climates with high rainfall. In Costa Rica, for example, Krebs (1975) found that conversion of forest to permanent cultivation led to reduction in organic matter, nitrogen, and certain cations to lower levels, but not to a catastrophic loss of fertility. The most serious long-term effects were a gradual increase in aluminum, a decline in pH, and the depletion of soil calcium, all of which can be remedied by careful soil management.

Elsewhere, in quite different environments, cultivation and the grazing of domesticated herds have influenced fertility and productivity in other ways: The Rajputana Desert of northwestern India and Pakistan is one of the dustiest deserts in the world. It is heavily populated for a desert, and evidence exists that the region was once agriculturally productive (Bryson 1972). Moisture levels in the air, furthermore, compare to those of regions with much greater rainfall.

It is very possible that the present condition of the climate and desert is a product mainly of agricultural activity, particularly of overgrazing by sheep, goats, and other animals (Figure 11–18). Overgrazing and the resulting wind erosion of the fine clay soils (largely montmorillonite), which have a relatively high intrinsic fertility, are responsible for a quantity of dust in the atmosphere over much of the region that ranges from 300 to 800 micrograms per cubic meter of air. This is several times greater than the particulate loads in the lower atmosphere of industrial cities like Chicago.

This great quantity of atmospheric dust influences the energy exchange pattern of the Rajputana region. The albedo of these particulates is higher than that of the vegetated ground surface, and so it reflects a greater fraction of incoming radiation back to space. Another fraction is absorbed by the particles themselves, warming the layer in which they are suspended instead of the ground surface. At night these particulates act as a dusty blanket and reduce radiative cooling of the ground surface. The combined effects of these changes in the radiation balance seem to reduce the development of atmospheric instability that is essential to induce convectional rainfall. Heating of the atmospheric system as a whole is reduced, and the ground-level heating that can form bubbles and columns of intensely heated air is weakened. Hence, although there is enough moisture present to produce appreciable rainfall, the conditions that induce it are absent. Similar factors might also contribute to drought conditions in the Sahel Zone of sub-Saharan Africa.

Agricultural activities have profoundly affected the physical structure and fertility of cultivated soils and grazing lands. Although man has improved soil conditions in many areas, the intensification of mechanized cultivation in parts of the developed world and the increased pressures of farming and grazing on fragile ecosystems of the tropics and arid zones are increasingly damaging production.

SUMMARY

Human interventions into natural systems through the cultivation of crops and the pasturing of domestic animals have (1) reduced the importance of detritus food chains, and (2) increased the net output of nutrient elements from agroecosystems.

Reducing the importance of detritus food chains has lessened the rate at which organic matter is reincorporated into agricultural soils. This change, which appears in the form of a decrease in organic matter, carbon, or nitrogen in the soil, is detrimental to several important features of ecosystem productivity. Crumb structure deteriorates, and for fine-textured soils this leads to greater problems of compaction, aeration, water availability, root penetration, and susceptibility to wind and water erosion. Conditions required for mechanical cultivation become more critical, and the energy required for cultivation increases.

FIGURE 11-18.
Areas exposed to and protected from grazing by livestock in the Rajputana Desert of western India. The desertlike aspect of much of this region, together with the severe wind erosion, is partly attributable to this overgrazing. (Photographs courtesy of Reid Bryson, Institute for Environmental Studies, University of Wisconsin.)

With deterioration of structure, and exposure of the bare soil to wind and water through plowing or heavy grazing, erosion can increase to a serious degree. Sheet and rill erosion by water can remove large quantities of the most valuable surface soil layers and still be inconspicuous. Wind erosion, a perennial problem in dry, windy regions, is an increasingly serious problem affecting fine-textured soils in more humid regions, partly because of the deterioration of crumb structure.

These changes lead both to the loss of nutrients from agricultural systems and to a reduced ability of the remaining portion of the soil to hold nutrients. The need for fertilization becomes greater, and soil nutrient cycles change from being local and internal to being open and biospheric.

In no large area has a perfect balance been achieved between nutrient input and removal. In the humid tropics, the most successful subsistence systems are those that "mine" accumulated nutrient capital until it is exhausted by harvest and by leaching. In some areas, such as semiarid regions of the Middle East, even the input of moisture to agricultural regions may have been reduced by the side effects on regional climate of past agricultural activity.

Literature Cited

Albertson, F. W., G. W. Tomanek, and A. Riegel. 1957. Ecology of drought cycles and grazing intensity of grasslands of the Central Plains. *Ecol. Monog.* 27:27–44.

Alther, L. 1972. Organic farming on trial. *Nat. Hist.* 81(9):16–24.

Barley, K. P. 1961. The abundance of earthworms in agricultural land and their possible significance in agriculture. *Adv. in Agron.* 13:249–268.

Baver, L. D. 1968. The effect of organic matter on soil structure. In *Organic matter and Soil Fertility.* pp. 383–403. New York: John Wiley and Sons.

Beasley, R. P. 1972. *Erosion and Sediment Pollution Control.* Ames: Iowa State University Press.

Bryson, R. A. 1972. *Climatic Modification by Air Pollution.* Report No. 1, Institute of Environmental Studies. Madison: University of Wisconsin.

Buckman, H. O., and N. C. Brady. 1969. *The Nature and Properties of Soils.* New York:Macmillan Publishing Company, Inc.

Delwiche, C. C. 1971. Man and mineral cycles. *Yale Sci.* 46(2):12–20.

Ellison, L. 1949. The ecological basis for judging condition and trend on mountain range land. *J. Forestry* 47:786–795.

Fiddian, W. E. H. 1973. The changing pattern of cereal growing. *Ann. Appl. Biol.* 75:123–149.

Gill, W. R., and G. E. Vanden Berg. 1967. *Soil Dynamics in Tillage and Traction.* Agr. Handbook No. 316. Washington, D.C.: USDA.

Grable, A. R. 1966. Soil aeration and plant growth. *Adv. in Agro.* 18:58–106.

Grant, K. E. 1975. Erosion in 1973–74: The record and the challenge. *J. Soil and Water Cons.* 30(1):29–32.

Greacen, E. L. 1958. The soil structure profile under pastures. *Austral. J. Agr. Res.* 9:129–137.

Johnson, N. M., G. E. Likens, F. H. Bormann, and R. S. Pierce. 1968. Rate of chemical weathering of silicate minerals in New Hampshire. *Geochim. Cosmochim. Acta* 32:531–545.

Krebs, J. E. 1975. A comparison of soils under agriculture and forests in San Carlos, Costa Rica. In *Tropical Ecological Systems.* F. B. Golley and E. Medina (eds.). p. 381–390. New York: Springer-Verlag.

Lipman, J. G., and A. B. Conybeare. 1936. *Preliminary Note on the Inventory and Balance Sheet of Plant Nutrients in the United States.* Bull. N. J. Agr. Expt. Sta. 607.

Low, A. J. 1972. The effect of cultivation on the structure and other physical characteristics of grassland and arable soils (1945–1970). *J. Soil Sci.* 23:363–380.

———. 1973. Soil Structure and crop yield. *J. Soil Sci.* 24:249–259.

Martel, Y. A., and E. A. Paul. 1974. Effects of cultivation on the organic matter of grassland soils as determined by fractionation and radiocarbon dating. *Can. J. Soil Sci.* 54:419–426.

Millar, C. E., L. M. Turk, and H. D. Foth. 1965. *Fundamentals of Soil Science.* New York: John Wiley and Sons.

Nye, P. H., and D. J. Greenland. 1964. Changes in the soil after clearing tropical forest. *Plant and Soil* 21(1): 101–112.

Pimentel, D., E. C. Terhune, R. Dyson-Hudson, S. Rochereau, R. Samis, E. A. Smith, D. Denman, D. Reifschneider, and M. Shepard. 1976. Land degradation: Effects on food and energy resources. *Science* 194:149–155.

Rosenberg, N. J. 1964. Response of plants to the physical effects of soil compaction. *Adv. in Agron.* 16:181–196.

Russell, E. W. 1971. Soil structure: its maintenance and improvement. *J. Soil Sci.* 22:137–151.

Rauzi, F., and F. M. Smith. 1973. Infiltration rates: three soils with three grazing levels in northeastern Colorado. *J. Range Manag.* 26:126–129.

Sanchez, P. A., and S. W. Buol. 1975. Soils of the tropics and the world food crisis. *Science* 188:598–603.

Sheesley, R., D. W. Grimes, W. D. McClellan, C. G. Summers, and V. Marble. 1974. Influence of wheel traffic on yield and stand longevity of alfalfa. *Calif. Agr.* 28(10):6–8.

Stark, N. 1971. Nutrient cycling: I. Nutrient distrubition in some Amazonian soils. *Trop. Ecol.* 12(1):24–50.

van de Westeringh, W. 1972. Deterioration of soil structure in worm free orchards. *Pedobiologia* 12:6–15.

Viets, F. G., Jr. 1971. Water quality in relation to farm use of fertilizer. *BioSci.* 21(10):460–467.

Woodruff, N. P., L. Lyles, F. H. Siddoway, and D. W. Fryrear. 1972. *How to Control Wind Erosion.* Agro. Inf. Bull. 354. Washington, D.C.:USDA.

12

IMPACTS OF IRRIGATION AND FERTILIZATION

As we have seen, the harvest of plant and animal products, combined with the side effects of cultivation and grazing, increase the outflow of both nutrients and water from agroecosystems. Nutrient losses almost always deplete the available pools of several nutrients until, in order to maintain high productivity, it becomes necessary to add fertilizers that contain major quantities of the lost elements in quickly available form. Water losses, caused by the increased surface runoff that is promoted by intensive cultivation and grazing, often increase erosion and compound the problems of nutrient loss. In some areas, where precipitation is high and drainage poor, ditch and tile systems are constructed to speed subsurface drainage.

Furthermore, agriculture has been extended to regions where temperature and soil structure are favorable but precipitation is deficient. Here, even to begin agricultural activity, large quantities of both water and nutrients must be supplied from outside the system, and in order to remove or avoid the accumulation of harmful substances in the soil, allowances must be made for the outflow of water and its contained materials.

Irrigation and fertilization thus represent major changes in ecosystem function, and generate their own particular ecological and economic problems. These problems relate to several factors: the availability and cost, in both dollars and energy, of the water and fertilizer materials; the quality and effect of these materials on processes within the soil ecosystem; and the secondary effects of outflows of water and nutrients from the agricultural areas to which they are applied. The purpose of this chapter is to examine these sorts of problems. In later chapters we shall consider specific ways of dealing with them.

IRRIGATION

Use of Freshwater Resources

Of the total 157 million hectares of cultivated cropland in the United States, approximately 16 million, or roughly 10 percent, are irrigated (EPA 1973). Most of this area, 14 million hectares, lies in the western states. Crop production on much of this area, including central and southern Arizona and southeastern California,

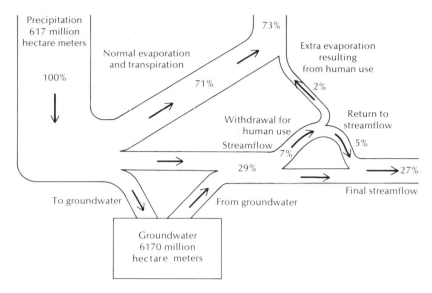

FIGURE 12-1.
Annual human water use in the United States (excluding Alaska and Hawaii) in relation to the hydrologic cycle. (Based on data from Revelle 1963.)

depends entirely on irrigation. Total use of water in the United States now equals about 46.9 million hectare-meters per year (one hectare meter = the amount of water that covers one hectare to the depth of one meter). About 35 percent of this, or 16.4 million hectare meters, is used for irrigation.

In a number of other countries, irrigation is even more important to crop production. For example, according to Wortman (1975) about one-third of the cultivated land of mainland China, 33.5 million hectares, is irrigated. In China, of course, paddy rice is a major grain crop, but much of the wheat acreage is also irrigated, unlike the situation in the United States. Worldwide, about 160 million hectares, which is about 1 percent of the earth's total land surface, are now under irrigation (Revelle 1974).

The demand for freshwater resources is continually growing. Revelle (1963) estimated that by 2000 A.D. the total withdrawal from stream flow in the contiguous United States for various human uses could triple the early 1960s level (Figure 12-1). At the time of his estimate, withdrawal from stream flow equalled about 7 percent of total precipitation and nearly one-fourth of the stream flow volume. Of this 7 percent, about 2 percent was consumed during use, that is, its use led to evaporation that would not otherwise have occurred, and thus reduced the fraction returning to stream flow. Most of this was used for irrigation. The remaining 5 percent was returned to stream flow, but of course with a heavier load of solutes from various agricultural and nonagricultural uses. By 2000 A.D., total withdrawals are projected to be about 21 percent of total precipitation, or nearly three-fourths of stream flow. In all probability, the fraction of this withdrawal that is consumed during use will be greater, because more of this amount will go for irrigation, power plant cooling, and other uses that involve considerable evaporation.

TABLE 12-1
Estimated total world water requirements for human use by 2000 A.D.

Quantity or use	Quantity in cubic kilometers			
	Total	%	Lost by evaporation	% of human use quantity
World stream and river flow	37,300			
Human use	18,700	50.1	5,570	29.8
Irrigation	7,000	18.8	4,800	68.6
Domestic	600	1.6	100	16.7
Industrial	1,700	4.6	170	10.0
Dilution of wastes	9,000	24.1		0
Other	400	1.1	400	100.0

Source: From Kalinin and Bykov 1969.

FIGURE 12-2.
Major drainage basins and water supply regions of the United States with an indication of those that are expected to show water supply deficiencies, relative to demand, by 2000 A.D. (Data from Revelle 1963.)

Kalinin and Bykov (1969) have projected worldwide freshwater use for the year 2000 A.D. Their estimates suggest that withdrawal will equal about one-half of total stream flow, and that over one-fourth of this withdrawal will be lost as extra evaporation (Table 12-1).

For the United States, Revelle's (1963) estimates indicate that by 2000 A.D. demand will exceed both total surface flows and sustainable groundwater withdrawals in a number of water basins and regions (Figure 12-2). This is true in Southern California now, and in the near future deficiency problems are likely to develop in areas such as the Great Basin, the entire Colorado Basin, the Texas Gulf region, the Rio Grande and Pecos basins, the Arkansas-

White-Red River Basin, the Upper Missouri Basin, the Great Lakes region, and the Delaware and Hudson river basins. For Southern California and the two regions of the Great Lakes and Northeast, this excess demand is likely to be due primarily to urban and industrial needs. For the other regions, it will come primarily from increased demands for irrigation.

Heady et al. (1973) have recently analyzed relationships among water demand, land use, and agricultural policy through 2000 A.D. for the United States. Their conclusion is that national water needs, including those of agriculture, can be met through this period, and that more than adequate levels of farm productions can be maintained. However, this will require major changes in patterns of crop and livestock production: Larger acreages of dryland production of certain crops will have to be substituted for smaller acreages now under irrigation; and regional shifts of certain crops from drier areas of the West to wetter areas of the East are likely. The uncertainties of population growth, industrial expansion, and energy cost make the details of any such analysis unreliable, but it is clear that limitations on regional water supplies, as reflected in irrigation water cost, will become increasingly important.

Irrigation Water Quality

It is also obvious that the importance of water quality will grow. With population and industrial growth, and with an increasing fraction of total stream flow being diverted for human use, the solute loads of streams and rivers will almost inevitably increase because of urban and industrial waste discharges, drainage from irrigated cropland, runoff from urban and rural land surfaces, and increases in solutes in rainfall caused by air pollution. Increased water losses by evaporation from reservoirs and canals also create higher solute concentrations.

Higher solute concentrations in irrigation water create more problems of soil salinity within agroecosystems. Total solute concentrations can reach levels that interfere with crop growth through their influence on soil water potential, and particular ions can have toxic effects on crop species and can damage soil structure.

Disregarding toxicity, solute problems exist in three major kinds of soils: saline soils, saline-sodic soils, and sodic soils. These three groups are characterized by their total solute content and by the relative importance of sodium compared to other cations. Solute content is usually determined by measuring the electrical conductivity of a saturated soil sample, which is expressed in units of millimhos per centimeter. **Conductivity** is actually determined as the reciprocal of resistance, measured in ohms (mho is ohm spelled backwards) for a cubic centimeter of soil (or any other material). The relative importance of sodium is expressed as the percentage of **exchangeable sodium,** that is, the percentage of the total cation exchange capacity that is occupied by sodium:

$$\text{exchangeable sodium \%}$$
$$= \frac{\text{exchangeable sodium in meq}}{\text{cation exchange capacity in meq}} \times 100$$

Saline soils have a conductivity of greater than 4 millimhos per centimeter and less than 15 percent exchangeable sodium; their pH is usually less than 8.5. The principal cations present are generally calcium, magnesium, potassium, and up to 15 percent sodium; anions may include chloride, sulfate, and occasionally bicarbonate or nitrite.

Saline-sodic soils also have conductivity greater than 4 millimhos, but they have more than 15 percent exchangeable sodium. Both saline and saline-sodic soils interfere with plant growth primarily by affecting water uptake and nutrient exchange processes, but only saline-sodic soils are liable to be converted into **sodic** (nonsaline) **soils** by improper management. If water low in calcium salts is used to leach a saline-sodic soil, it removes the solutes in the soil solution. When this occurs, sodium leaves the cation exchange sites on clay minerals and is replaced by hydrogen ions; this forms sodium hydroxide, NaOH, in the soil solution. Thus through leaching the conductivity of the soil is reduced below 4 millimhos per centimeter, the exchangeable sodium stays above 15 percent, and the pH rises to between 8.5 and 10.0. This produces a sodic soil. Under these conditions, the clays tend to defloculate and form massive, hard clods, and the soil surface develops a heavy crust. Under the strongly alkaline conditions, organic materials dissolve and are transported by capillary movement to the surface, where they are deposited as a dark surface layer known commonly as **black alkali.** High pH is bad for most plants because their organic tissues are affected much the way dead organic material is affected in the soil, and the structure of the alkaline soil inhibits the percolation of water and the growth of roots. Thus despite a lower solute content the soil may be less favorable than before.

Water used to leach saline-sodic soils must be naturally high in calcium salts or used along with applications of calcium sulfate (gypsum). The abundant supply of calcium ions under these circumstances causes the sodium to be almost completely displaced from cation holding sites and then leached away with the bulk of the solutes. Substances such as ferrous sulfate that acidify the soil and dissolve calcium carbonate (which is often present in insoluble form) can also be used to reclaim sodic soils.

The direct effects of soil solutes on crop species are of two types: **osmotic effects,** caused by the overall concentration of solute materials; and **specific ion effects,** caused by the toxicity of particular ions. Osmotic effects are generally

the result of several specific influences (Meiri and Shalhevet 1973): First, the gradient of water potential from soil to plant is reduced and the rate of water uptake is therefore slowed. In addition, relative production rates of root and shoot hormones are modified. Cytokinin, a hormone that is produced in the root and transported to the leaf, where it stimulates transpiration and promotes protein synthesis, is inhibited by high soil salinities. Absissic acid, which inhibits transpiration and is produced in the leaves under conditions of water stress, tends to accumulate when salinity is high. High solute concentrations in plant tissues, caused by high salinity of the soil, can also interfere with normal chloroplast and mitochondrial function, in general producing effects that reduce photosynthesis while increasing respiration.

Plant species vary widely in their sensitivity to soil salinity (Figure 12-3). No specific threshold for effect appears to exist, however, and decline over much of the yield range follows a fairly straight line. Various grasses, grains, and a few other species such as sugar beets, cotton, safflower, and red beets have a relatively high tolerance to salinity (Figure 12-4), whereas

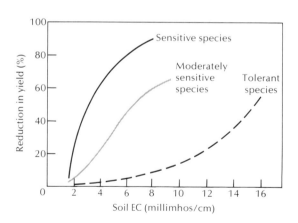

FIGURE 12-3.
Relationship of crop yield to soil salinity for sensitive, moderately tolerant, and tolerant species (schematic). (Based on various sources.)

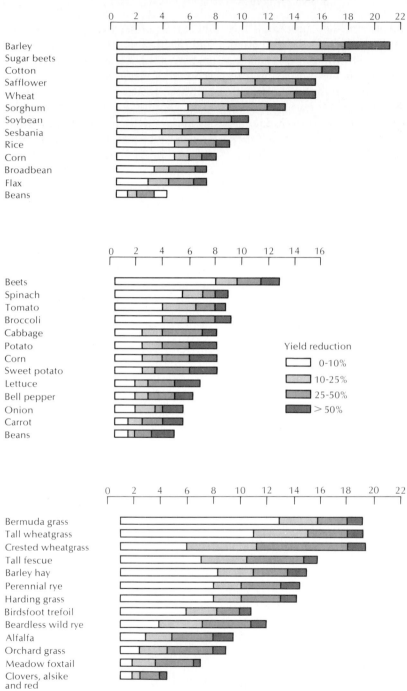

Soil EC in millimhos/cm at 25°C

FIGURE 12-4.
Yield reductions caused by salinity in various major crop species. (From L. Bernstein 1964. *Salt Tolerance of Plants*. USDA Information Bull. No. 238.)

most traditional temperate zone vegetables and certain grains such as corn and rice have a much lower tolerance.

The harmful effects of specific ions present in irrigation water can also be greater than one would expect, given their total concentrations. These effects are especially damaging to many orchard species. Boron, a required micronutrient for plants, is one of the ions for which this is true; toxicity effects show up at concentrations of only a few parts per million in the soil solution. Again, species vary considerably in their sensitivity to this substance (Table 12-2), and many orchard species are sensitive to concentrations even below one part per million. Boron problems usually arise in irrigation operations that use water from deep wells, since surface waters generally have very low levels (Allison 1964).

TABLE 12-2
Relative sensitivity of major crop species to boron.

Tolerant (2–4 ppm)[a]	Semitolerant (1–2 ppm)[a]	Sensitive (0.3–1 ppm)[a]
Asparagus	Potato	Pecan
Date palm	Cotton	Walnut
Sugar beet	Tomato	Navy bean
Garden beet	Radish	Plum
Alfalfa	Pea	Pear
Onion	Olive	Apple
Turnip	Barley	Grape
Cabbage	Wheat	Fig
Lettuce	Corn	Cherry
Carrot	Oats	Peach
	Pumpkin	Apricot
	Bell pepper	Blackberry
	Sweet potato	Orange
	Lima bean	Avocado
		Grapefruit
		Lemon

[a]Limiting concentration in irrigation water.
Source: From Allison 1964.

Other ions, including sodium, chloride, sulfate, bicarbonate, and magnesium can also have toxic effects, again most frequently on fruit trees.

The quality of irrigation water thus depends on three factors: salinity, sodicity, and toxicity (Rhoades 1972). In evaluating the salinity hazard of irrigation water, one must first take into account the tolerance of the crop species involved and the degree of yield decrease that becomes significant for economic reasons. For most practical purposes, salinity levels that produce a 50-percent decrease in the yields of field crops, or a 10-percent decrease for fruit, are limiting. Because of the leaching action of water applied to the surface, salinity is usually greater at greater depths in the soil, and the critical salinity level usually occurs at the bottom of the root zone. Given the critical salinity tolerance of a crop species, one must then consider the characteristics of the soil and of the water available. These factors are interrelated because the salinity of the water and the characteristics of the soil determine the quantity of water that must be applied in order to achieve enough leaching to prevent salt accumulation. Degree of leaching is expressed as the **leaching fraction,** or decimal fraction of the applied water that passes through the upper soil to a point below the bottom of the root zone. If we characterize the solute content of the water and the soil solution in terms of electrical conductivity (*EC*), and represent the required leaching fraction by *LF*, this relation can be expressed:

$$EC_{\text{irrigation water}} = EC_{\text{critical at root zone bottom}} \times LF$$

Thus if the critical salinity at the bottom of the root zone is 10 millimhos per centimeter, and the irrigation water has a conductivity of 5 millimhos per centimeter, enough water must be applied so that a fraction of 0.5 (one-half) passes through the root zone. If the water available has a conductivity of 1 millimho per centimeter, less water is necessary and only 10 percent must pass through the root zone. This relationship is based upon the assumption that the salinity of water that passes beyond the root zone bottom will be equal to that of the soil solution at that level, rather than to the salinity of the irrigation water input.

The sodicity hazard, that is, the danger of converting the soil into a saline-sodic or sodic system, is reflected primarily in the ratio of sodium to calcium and magnesium in the water. The **exchangeable sodium percentage** (*ESP*) in the soil is estimated by a ratio called the **sodium adsorption ratio** (*SAR*), which is calculated:

$$SAR = \frac{Na}{\left(\dfrac{Ca + Mg}{2}\right)^{1/2}}$$

In this calculation, values for the different ions are expressed as milli-equivalents per liter. Values of *SAR* can be calculated, of course, for the soil solution or for irrigation water. In practice, the *ESP* must be calculated by equations that take into account not only the determined *SAR* of irrigation water, but also various other values that reflect molar concentrations of base ions and dissociation constants for carbonate minerals (Rhoades 1972).

The toxicity hazard is evaluated empirically on the basis of crop tolerances and concentrations of toxic ions present in the available water. Recommended maximum concentrations of a number of trace elements are given in Table 12-3.

It should be clear from this discussion that irrigation cannot be carried out except under conditions in which adequate natural or artificial drainage permits excess salts to be drained out of the productive portion of the soil system (Figure 12-5). The natural consequence of water use involving appreciable quantities of solute is that the water leaving the soil system has a higher concentration of salts than the water applied, and so the systems that receive the outflows, either groundwater or surface water systems, are salinated.

TABLE 12-3
Recommended maximum concentrations of trace elements in irrigation waters.[a]

Element	For waters used continuously on all soil	For use up to 20 years on fine-textured soils of pH 6.0 to 8.5
	mg/l	
Aluminum	5.0	20.0
Arsenic	0.10	2.0
Beryllium	0.10	0.50
Boron	0.75	2.0–10.0
Cadmium	0.010	0.050
Chromium	0.10	1.0
Cobalt	0.050	5.0
Copper	0.20	5.0
Fluoride	1.0	15.0
Iron	5.0	20.0
Lead	5.0	10.0
Lithium	2.5[b]	2.5[b]
Manganese	0.20	10.0
Molybdenum	0.010	0.050[c]
Nickel	0.20	10.0
Selenium	0.020	0.020
Vanadium	0.10	1.0
Zinc	2.0	10.0

[a]These levels will normally not affect plants or soils. No data available for mercury, silver, tin, titanium, or tungsten.
[b]For citrus, 0.075 mg/l.
[c]Only for fine-textured acid soils or acid soils with relatively high iron oxide content.
Source: Reproduced from Branson, *Journal of Environmental Quality,* vol. 4, 1975, by permission of the American Society of Agronomy, Crop Science Society of America, and Soil Science Society of America.

FIGURE 12-5.
Extensive crop damage caused by saline soils and inadequate drainage in the lower Gila River Valley of Arizona in 1959. (Photograph by R. C. Middleton, courtesy of the U.S. Bureau of Reclamation.)

External Impacts of Irrigation

Intensive irrigation has the potential to create secondary effects outside the agricultural lands to which water is applied; it can, for example, salinate ground and surface waters, cause land from which water is pumped to subside, and generally modify regional climate.

Nightingale (1974), for example, has examined soil and groundwater salination in relation to agricultural practice in the San Joaquin Valley of California. Examination of salinity profiles in 70 sites showed increases in salinity to a depth of about 3 meters beneath row and truck crops, and to over 5 meters beneath orchards (Figure 12-6). Salts leached to these depths appear in the groundwater and thus affect the salinity of water obtained from wells.

Nightingale also found that salinities of well water showed a strong correlation with the salinities observed between about two and six meters deep in adjacent agricultural land.

Domestic animal wastes are also high in salts, and methods used to dispose of these wastes can also introduce large quantities of salts locally into groundwater systems (Branson et al. 1975). Table 12-4 shows the salt output by a number of animals and estimates the portion of this output that enters groundwater systems in a California locality.

Surface waters also receive waste irrigation water carrying salt loads. The concentration of dissolved materials is inevitably higher in waste water than in the irrigation water, but the volume of waste water is lower; thus the total volume of salts leaving the irrigated land area

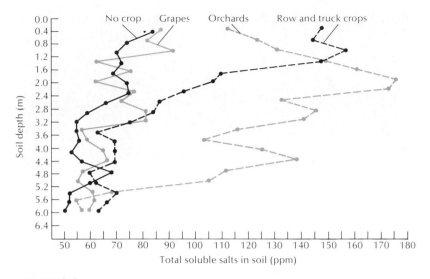

FIGURE 12-6.
Profiles of soil salinity beneath various crop systems in the San Joaquin Valley of California. (Modified from Nightingale 1974. Copyright © 1974 by the Williams & Wilkins Co., Baltimore.)

TABLE 12-4
Estimates of the quantities of salts excreted by farm animals, and the approximate portion of these quantities currently entering groundwater systems in California.

Animal	Unit number	Total salt production (kg/yr)	Contribution to groundwater (kg/yr)	%
Dairy cows	1	599	212	35.4
Beef cattle				
Steers	1	281	82	29.2
Calves	1	166	56	33.7
Poultry				
Broilers	1000	2611	1356	51.9
Layers	1000	6002	2834	47.2
Turkeys	1000	8964	4470	49.9

Source: Reproduced from Branson. *Journal of Environmental Quality*, vol. 4, 1975, by permission of the American Society of Agronomy, Crop Science Society of America, and Soil Science Society of America.

TABLE 12-5
Net quantities of extra salts leaving agricultural lands in various river basins, and entering surface stream flow, expressed as a percentage of the quantity that enters in irrigation water for various leaching fractions.

River basin	Leaching fraction 0.1	0.2	0.3
	% extra salts		
Feather	140	232	313
Missouri	−11	6	10
Salt	−10	0	11
Colorado	−36	−2	2
Gila	−4	2	6
Pecos	−38	−23	−5

Source: Reproduced from Branson, *Journal of Environmental Quality*, vol. 4, 1975, by permission of the American Society of Agronomy, Crop Science Society of America, and Soil Science Society of America.

can be either less or greater than that entering in irrigation water. This relationship depends on salinity characteristics of the irrigated land, but also varies with the leaching fraction (Table 12-5). With higher leaching fractions, more salts tend to be removed in wastewater than introduced in irrigation water. This is obviously the basic technique for preparing saline soils for cultivation. In recent years, the expansion of irrigated agriculture in areas with relatively saline soils in the Upper Colorado River Basin and the Gila River Valley of Arizona (Figure 12-7) has led to large salt load increases downstream from the areas involved, and has created especially severe problems in the Mexicali Valley of Baja California, Mexico, which is last to receive water from the Colorado River (Figure 12-8). The limited allotment of water,

FIGURE 12-7.
Irrigated farmland in the Wellton-Mohawk Irrigation and Drainage District, Gila River Valley, Arizona, in 1975. Portions of the irrigated land in this district suffered salinity problems such as those shown in Figure 12-5 until the construction of modern irrigation and drainage systems. (Photograph by E. E. Hertzog, courtesy of the U.S. Bureau of Reclamation.)

FIGURE 12-8.
Salt accumulation in a field in the Mexicali Valley of Baja
California, Mexico. Increasing salinity in irrigation water
from the Colorado River, combined with inadequate field
drainage systems, have created severe soil salinity problems in parts of this area. (Photograph by Charles O'Rear,
courtesy of EPA-DOCUMERICA.)

rising salt load, and inadequate drainage systems
have combined to create severe soil salination
problems in this area.

Irrigation has other problems of quite a different nature—for example, land subsidence in
areas from which large groundwater withdrawals have occurred (Figure 12-9). In the
area near Eloy, Arizona, between Phoenix and
Tucson, lands have subsided more than 2 meters
over a 20-year period (McCauley and Gum
1975). This is caused by the withdrawal of irrigation water from wells at rates far greater than
replacement rates and is associated with a rapid

increase in pumping depth. As the pore space
of the sediments is emptied of water, compaction takes place under pressure of overlying
sediments, and the settling that results damages
roads, buildings, and irrigation structures themselves. Wells, for example, seem to "grow"
up from the surface as the soil surface sinks
and the pump and concrete pump pad are
left elevated above the surface, held up by the
well casing.

Another potential effect, whose extent and
importance are not certain, is that of widespread irrigation on regional climate (Fowler
and Helvey 1974; Stidd 1975). Central Washington State has been the site of recent attempts
to evaluate this influence. The development of
irrigation in this area began in about 1950 as
part of the Columbia Basin Project. By 1970
over 180,000 hectares were irrigated, and most
of this development had already been accomplished by 1965. It seems that if major changes
in precipitation patterns can be induced by
such projects, either directly or indirectly
through extra evaporation from irrigated land,
they should be apparent here. Stidd (1975) has
analyzed precipitation for a large number of
weather stations in a 240-kilometer radius of
the project center, calling the stations in the
central part of this area a target zone and those
in the outer part a control zone. His conclusion
is that from 1955 to 1973 precipitation was 50
percent higher in the target area than for the
period 1931 to 1950; it was only 23 percent
higher in the control area. Fowler and Helvey
(1974), however, attribute this difference to
another factor—the generally drier climatic
conditions of the 1930s and 1940s. Their analysis, designed to compare stations by a method
independent of influences of weather cycles,
suggests that there is no good evidence for increased precipitation due to irrigation effects.
This conclusion is a compelling one, and it remains to be demonstrated that irrigation leads
to significant increases in precipitation in downwind regions.

FIGURE 12-9
Groundwater withdrawals near Eloy, Arizona, have led to a variety of land subsidence problems. Here poorly drained depressions and fissures have formed in the land surface. Some of the fissures, 2 meters or more in depth, are associated with damage to highways and irrigation systems. (Photograph by E. E. Hertzog, courtesy of the U.S. Bureau of Reclamation.)

Irrigation Systems

Many kinds of irrigation systems have evolved throughout the world in many kinds of environments. In Iran, for example, a distinctive system of horizontal wells, known as **qanats,** supplies a major portion of irrigation water (Wulff 1968). These wells, actually sloping tunnels, tap deep water tables in the areas where alluvial fans abut mountain ranges. These tunnels sometimes extend for many miles before bringing the water to the surface in an irrigated area (Figure 12-10). Some 22,000 qanats are in use in Iran, and others exist in nearby areas of the Middle East, as well as in parts of North Africa.

More extensive irrigation systems were de-veloped in prehistoric times, however. In the Salt River Valley of Arizona, early Indian residents developed a system of 1609 kilometers of canals that irrigated about 100,000 hectares of land (Cantor 1970). In northern Luzon, Philippines, systems of irrigated paddy rice terraces, apparently several thousand years old, cover more than 648 square kilometers of mountainous land.

Irrigation systems in modern agriculture, however, depend on damming major streams to store and control the flow of water and to allow delivery in the desired amount whenever it is needed. These systems offer great flexibility in this regard, but of course they achieve this at the cost of evaporation and seepage losses

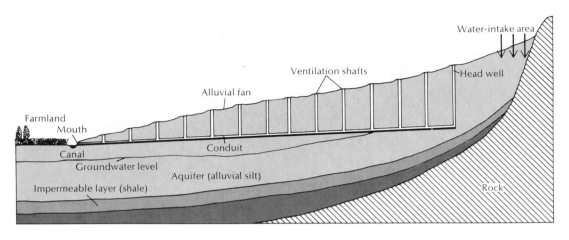

FIGURE 12-10.
Design of a qanat, or horizontal well, in Iran. These wells tap groundwater deep in alluvial fans along mountain fronts and transport this water with the aid of gravity to distant areas of low-lying irrigable land. (From "The qanats of Iran" by H. E. Wulff. Copyright © 1968 by Scientific American, Inc. All rights reserved.)

that at times can be great. Moreover, storage reservoirs act as traps for sediment, which might otherwise be valuable as fertilizer.

Various techniques have been devised for applying water to the land surface (Figure 12-11). Those in most general use are systems of surface irrigation including **flood, border** or **strip, corrugation,** and **sprinkler irrigation** (Figures 12-12 through 12-15). These systems, of course, are all suitable for relatively level land areas with low or moderate porosity. Sprinkler irrigation, on the other hand, is the only one that can be used on very sandy land, where application by surface flow would lead to heavy infiltration near the input point and poor transport to distant parts of a field. Sprinklers can also be used on more uneven land surfaces. Sprinkler application is more subject to disturbance by wind, however, and the water tends to be distributed unevenly over the surface. A number of other newer systems have also been devised; we shall examine some of these in Chapter 14.

The principle that the development of efficient drainage systems must go hand in hand with irrigation has been violated in both ancient and historical times. In the Indus Valley of Pakistan, extensive areas of deep alluvial soils were introduced to irrigation beginning in the mid-1800s. By 1959, some 9.2 million hectares had been irrigated. Unfortunately, however, the natural water table in this region lies relatively close to the surface, and natural drainage is weak. By 1959, about 2 million hectares of irrigated land had become salinated and waterlogged to the point where cultivation was impossible. The water table over the irrigated region was rising at the rate of 15 to 25 centimeters per year, and from 20,000 to 40,000 additional hectares per year were being affected by salt accumulation and waterlogging. The cost of reclamation activities from 1965 to 1975 was $1.1 billion, yet the problem is far from a final resolution.

Much of the land in the Helmand Valley of Afghanistan, brought under irrigation in the 1950s after the construction of the Kajaki Dam on the Helmand River, also suffered waterlogging because of inadequate natural drainage. Here, some 16,000 hectares became water-

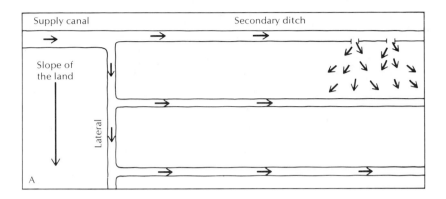

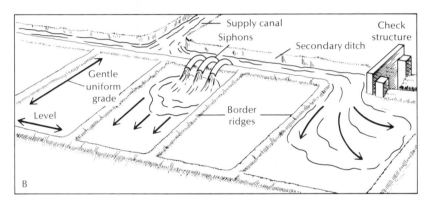

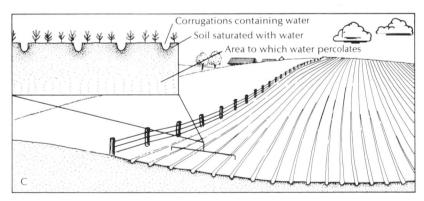

FIGURE 12-11.
Various systems of irrigation with flowing water: (A) flood; (B) border; (C) corrugation. (From Cantor 1970.)

FIGURE 12-12.
Flood irrigation of oats in Montana. (Photograph by B. C. McLean, courtesy of the USDA Soil Conservation Service.)

FIGURE 12-13.
Border or strip irrigation of cotton in Oklahoma. (Photograph courtesy of the USDA.)

FIGURE 12-14.
A field in Idaho prepared for corrugation irrigation. (Photograph courtesy of the USDA Soil Conservation Service.)

FIGURE 12-15.
A mobile sprinkler system for irrigating cotton in Texas. The sandy soil here makes surface application of water impractical because of infiltration problems. (Photograph by Fred Witte, courtesy of the USDA.)

logged within three to four years because of the unrecognized existence of an impervious conglomerate layer at a shallow depth in the soil. This land is now being reclaimed, but the costs of reclamation greatly exceed those of developing satisfactory drainage systems to begin with. It should be noted that this problem has not characterized the entire irrigated area developed as part of the Helmand Project, only the portion with particular drainage limitations that should have been identified and dealt with by the American contractors who worked on the project. Cases of this sort clearly demonstrate the need to examine the soil system thoroughly and to identify all the problems of salinity related to the increased flux of water through the agroecosystem.

FERTILIZATION

Need for Fertilization

The intensive use of inorganic fertilizers, particularly those that contain nitrogen and phosphorus, is a key factor in the high yields obtained per hectare in modern intensive agriculture. In the United States, increases in yield per hectare, due in part to the use of such materials, led to a decline in the area of cultivated land by 27 million hectares between 1944 and 1969 (Viets 1971). During this period, fertilizer use climbed from about 10.9 million metric tons to nearly 36.3 million tons per year. This increase was necessary partly because of the depletion of natural fertility that had occurred since American farmlands were first put into cultivation. However, it also permitted some agricultural land to return temporarily to grass or woodland, which protected it from erosion and allowed fertility to build up again.

It is easy to demonstrate that the addition of inorganic fertilizers makes rapid and significant improvements in crop yields. Ibach and Adams (1968) analyzed the effect of fertilizers on crop yields for various crops and forages in different parts of the country for 1964. They concluded that, in general, about 20 percent of total crop and forage production is attributable to fertilizer use, although the specific effect varies with crop and location (Table 12-6). For corn grown in rich prairie soils, developed on Wisconsin glacial till in Iowa, the fertilizer effect appears to be about 20.3 percent. In other words, if present use of fertilizer were discontinued, the next season's yield would be expected to drop by that amount (all other factors being equal). In east-central Illinois, where more podzolized soils, developed on older Illinoian till, are farmed, the fertilizer effect is about 36.8 percent; and in the southeastern states, including Florida, it is often more than 50 percent. On the other hand, a number of basic crop species, such as soybeans and wheat, neither receive much fertilizer nor show strong responses to it.

It is clear that fertilization is essential to permanent agriculture and just as clear that the fertilizing techniques in mechanized agriculture are highly successful. Nevertheless, there are detrimental side effects; moreover, the increasing energy costs involved and the important considerations of raw materials availability make it essential to examine present use patterns critically.

Nitrogen

More nitrogen fertilizers are used than any other. World nitrogen fertilizer consumption, in metric tons of elemental nitrogen, rose from 6.5 million in 1955 to 28.5 in 1970 (Harre et al. 1971). Of this total, 6.9 million tons, or nearly one-fourth, were consumed in North America. The annual increase in rate of production has been about 5 percent (Institute of Ecology 1972). By 2000 A.D., according to estimates made before major price increases were im-

TABLE 12–6
Examples of current fertilizer use rates and probable declines in production in a growing season immediately following discontinuation of fertilizer use.

Crop	Yield unit	Region	Fertilizer used		Yield and fertilizer effects		
			N (lb/acre)	P (lb/acre)	Present average	Without fertilizer	% decline
Corn	bu	Iowa (SE)	67.7	17.7	88.8	70.8	20.3
Soybeans	bu	Iowa (SE)	3.9	9.0	31.1	31.0	0.3
Vegetables	tn	Ala. (coastal)	156	68	3.3	1.5	55
Cotton	lb	Ga. (coastal plain)	88	29	346	212	38.7
Tobacco	lb	Ga. (coastal plain)	85	62	1926	500	70
Sugarcane	tn	Fla. (central)	0	8.7	33.0	10.0	69.7
Grapefruit	tn	Fla. (central)	200	0	15.7	1.0	93.6
Celery	cr	Fla. (central)	100	87	800	250	68.8
Vegetables	tn	Delaware	35	31	2.1	1.2	42.9
Corn	bu	Ill. (east central)	73	19	93.4	59.0	36.8
Improved pasture	tn	Ohio (east central)	35	20	1.3	1.1	15.4
Wheat	bu	Kan. (west and central)	30	8	20.2	18.0	10.9
Rice	lb	Calif. (cent. vall.)	91	14	4910	2500	49.1
Sugar beets	cwt	Calif. (cent. vall.)	132	20	68.3	50.3	26.4
Corn	bu	Calif. (cent. vall.)	128	17	82.9	65.0	21.6
Vegetables	tn	Calif. (coastal)	160	37	11.2	5.8	48.2
Vegetables	tn	Wash. (Paloose)	25	12	2.0	1.5	25.0
Wheat	bu	Wash. (Paloose)	20	12	48.8	40.0	18.0
Alfalfa	tn	Ariz. (S)	0	32	5.0	3.3	34.0
Fruits and nuts	tn	Ariz. (S)	150	27	7.0	4.0	42.9
Potatoes	bu	Idaho (S)	125	40	330	212	35.8

Source: Ibach and Adams 1968.

posed by petroleum exporting countries, nitrogen use in fertilizers will increase to 100 million tons annually.

Now, however, a serious problem has developed in the form of increasing cost of nitrogen fertilizer. Virtually all synthetic nitrogen fertilizer is manufactured by some variation of the **Haber-Bosch process** in which hydrogen is combined with nitrogen at a temperature of about 500°C and a pressure of 1000 atmospheres over a bed of iron oxide catalyst. Furthermore, in most fertilizer plants, hydrogen is obtained by a steam hydrocarbon-reformation process that uses natural gas as the raw feedstock. Hydrocarbon fuels are required in quantity both as raw material and as fuel to create the high temperature and pressure conditions needed for ammonia synthesis. Ammonia production is thus expensive in terms of energy-rich fuel, costing about 17,600 kcal/kg (Pimentel 1974). For the 28.5 million tons of fertilizer nitrogen produced in 1970, this energetic cost equals the energy content of over 10 billion gallons of gasoline. As a result, nitrogen fertil-

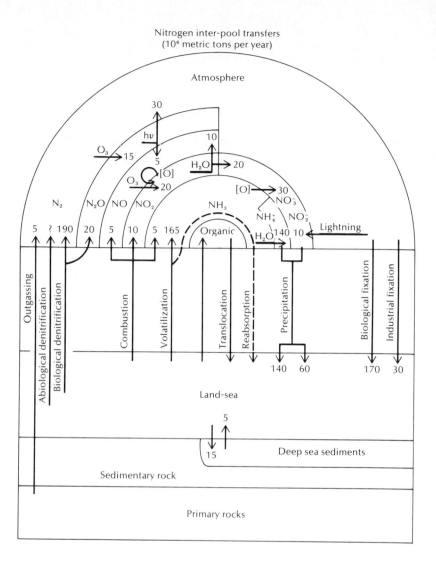

Nitrogen inter-pool transfers
(10^6 metric tons per year)

FIGURE 12-16.
Quantitative features of
the global nitrogen cycle.
(Modified from Hardy
and Holsten 1974.)

izer costs sharply reflect changes in petroleum prices and availability. Alternative techniques of obtaining hydrogen, such as the electrolytic breakdown of water molecules, at present used only in a few plants, also have high energy costs.

Much of the nitrogen fertilizer produced is applied as anhydrous ammonia, a liquefied gas. It is also converted to a number of other ammonium salts, nitrates, and other compounds, such as urea, for use as fertilizer. These materials, applied to the soil ecosystem, create major changes in ecosystem dynamics.

Quantitatively describing man's effect upon the nitrogen cycle, either at the local or the global level, is difficult. Many of the important processes are mediated by groups of microorganisms whose activities are still not well understood and whose operation in the field is very difficult to measure. The most recent comprehensive estimates of the transfers occurring in the nitrogen cycle at the global level are those of Burns and Hardy (1974), outlined in Figure 12-16. These estimates show that industrial fixation of nitrogen, now about 30 million

TABLE 12-7
Quantitative estimates of biological nitrogen fixation in major agricultural and nonagricultural environments.

Environmental type	Area (10^6 ha)	Fixation per ha (kg N)	Global fixation (metric tons $\times 10^6$ per yr)
Agricultural	4,400		
Cultivated cropland	1,400		
Legumes	250	140	35
Rice	135	30	4
Other	1,015	5	5
Pasture and range	3,000	15	45
Nonagricultural			
Forest and woodland	4,100	10	40
Unused	4,900	2	10
Ice covered	1,500	0	0
Total land	14,900		139
Oceans	36,100	1	36
Total earth	51,000		175

Source: From Burns and Hardy 1974.

tons per year (for all purposes, but primarily for nitrogen fertilizers) is still well below the rate of biological fixation, for which the current estimate is 175 million tons. Incidentally, these estimates take into account recent recognition that a significant amount of fixation occurs in the sea as well as on land and that rates of fixation are higher than formerly thought for various terrestrial environments (Table 12-7). In 1970, for example, Delwiche (1970) estimated biological fixation at only 44 million tons for the whole biosphere.

A second intervention by man (Figure 12-16) is the increased rate at which nitrogen oxides are introduced into the atmosphere by combustion, a current rate of about 20 million tons per year. Most of these oxides are returned to the land surface by precipitation at about 5 to 15 kilograms per hectare in areas that are downwind from major industrial regions.

Human impacts upon the nitrogen cycle are also difficult to quantify at the local level, too; but we do know that intensive cultivation and

livestock production have significantly increased the outflow of nitrogen from agroecosystems. We have discussed the ways in which food harvests and accelerated erosion deplete the original soil nitrogen pools (Chapter 11); two other factors also contribute to this depletion: the application of large quantities of inorganic fertilizers, and the development of feedlots, poultry factories, and other sources of concentrate animal wastes. It has been estimated that cropland systems contribute a total of about 5.5 million tons of nitrogen to surface waters by erosion (2.7 million tons), by leaching of native nitrogen (1.8 million tons), and by surface and subsurface losses of nitrogen fertilizers (1.0 million tons). Fertilizer losses thus amount to 10 to 15 percent of the amounts applied (National Academy of Sciences 1972).

It is difficult to describe the nitrogen balance of particular ecosystems, expecially exploited and manipulated systems. Nevertheless, we have prepared a set of simplified nitrogen cycle diagrams (Figure 12-17) that show hypothetical

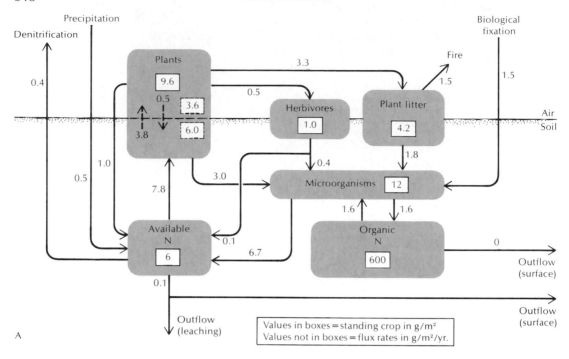

Values in boxes = standing crop in g/m²
Values not in boxes = flux rates in g/m²/yr.

A

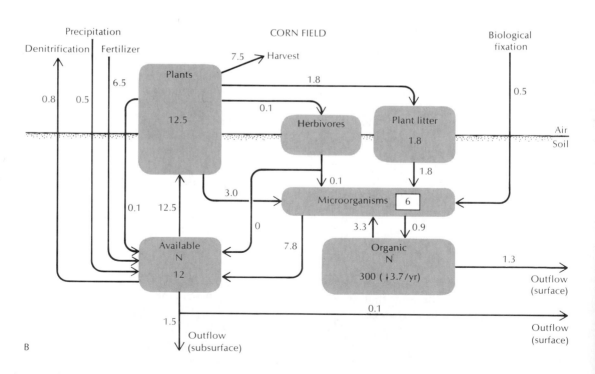

B

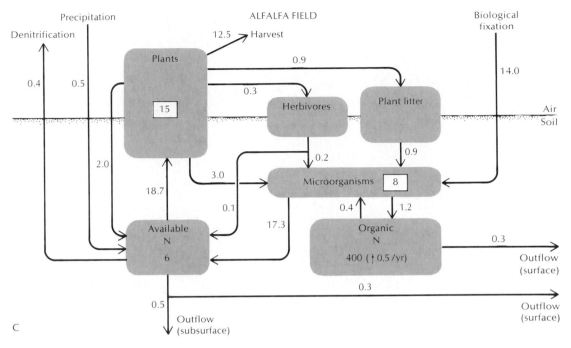

FIGURE 12-17.
Hypothetical annual nitrogen budgets for a midwestern United States native prairie (A), a corn field (B), and an alfalfa field (C). Values are grams per square meter (tens of kilograms per hectare).

and approximate data derived from various sources, including Dahlman et al. (1969), Hardy and Holsten (1974), and Viets (1971). These diagrams begin with a budget describing nitrogen flow in a virgin grassland area in the midwestern United States. In this system we assume only a small loss of nitrogen through surface and subsurface water movement; moderate inputs by biological fixation and losses by fire equal to these inputs; and a high, stable level of soil organic nitrogen. The mineralization of old organic nitrogen is assumed to be balanced by organic matter added to the soil pool by microbial conversion of dead root material; and leaching seems to return a significant amount of nutrients from foliage to available soil pools, especially during the autumn.

In the corn field system, on the other hand, we assume lower rates of biological fixation; major losses by erosion, runoff, and leaching; and the removal of the nitrogen in an average yield of 222 bushels of corn per hectare. Fertilization is assumed to take place at the rate of about 78 kilograms per hectare. Higher denitrification rates are likely, due to poorer aeration, and despite the level of fertilizer application, a slight decline in organic nitrogen is also likely.

In the alfalfa field, assumed to be an area of previously mixed crop cultivation that is less depleted in organic nitrogen than the corn field, no use of nitrogen fertilizers is assumed. The high nitrogen content of harvested alfalfa, which, it seems, can be achieved without depleting soil organic nitrogen, requires a high

FIGURE 12-18.
Part of a large cattle feedlot near Hereford, Texas. With a capacity of 30,000 animals, this operation produces nitrogen wastes equal to those from a city of about 250,000 people. (Photograph by L. C. Geiger, courtesy of the USDA Soil Conservation Service.)

level of biological fixation, largely by symbiotic bacteria in root nodules. Moderate amounts of nitrogen are also lost through surface and subsurface flows; for example, the spring runoff can contain high levels of nitrogen leached from the metabolically inactive alfalfa shoots.

Intensive livestock operations are also major agricultural sources of nitrogen-containing pollutants in natural waters. Between 1962 and 1970, the number of cattle feedlots in the United States increased from 1674 to 2403 (National Academy of Sciences 1972); and the number of extremely large feedlots, those with over 32,000 animals, increased from 5 to 41. Each animal in such a large feedlot produces an average of 43 kilograms of nitrogen per year in the form of excretory and digestive wastes. For a feedlot with 32,000 animals, this adds up to 1400 metric tons of nitrogen per year, which is equal to the output of a city of 260,000 people.

In the United States, feedlot operations produce a total of 6.0 million metric tons of nitrogen per year in the form of wastes. The soil beneath the areas used for such operations naturally contains a greatly elevated level of nitrogen compounds, and in many instances these materials find their way into ground- and surface-water systems (Figure 12-18).

Nitrogen compounds derived from agricultural sources are combined with those from municipal, industrial, and household waste disposal and with those that entered the air from combustion processes of all sorts. Nitrogen, being an important nutrient in aquatic systems, contributes to problems of eutrophication, which we shall discuss later in connection with phosphorus fertilization.

Because of their toxicity, however, high concentrations of particular nitrogen compounds in crops and in water supplies can be a direct concern to health. The health hazard to both humans and livestock is **methemoglobinemia,** a condition in which nitrite in the blood combines with hemoglobin and thus reduces its oxygen-carrying capacity. The nitrite involved is generally derived from nitrate in water or food that becomes reduced to nitrite. For livestock, this usually occurs in the digestive tract of the animal itself.

Water from wells, particularly shallow wells, that draw from groundwater contaminated by heavy fertilizer applications, feedlot wastes, septic tank discharge, or any other source of nitrogen, is one cause of methemoglobinemia. The allowable limit imposed by the U. S. Public Health Service for nitrate nitrogen in drinking water is 10 ppm (45 ppm nitrate), although this limit is currently exceeded in many rural water sources. A number of human deaths have been recorded from methemoglobinemia, generally among infants.

Nitrate poisoning is much more common in livestock, however. Both feed and water are major potential sources of nitrate. Plants generally contain only small quantities of nitrate, and most of what is taken up is reduced to ammonium and incorporated at this reduced state into protein and other organic compounds. Under a variety of conditions, however, certain forage plants and weeds can accumulate high levels of nitrate. Overfertilization is one cause, but stress from weather conditions and a number of other factors can also trigger nitrate accumulation. Ruminants and young swine seem particularly sensitive to nitrates in food, and the most serious problems usually involve cattle. Information on what constitutes a dangerous dose is still somewhat confusing, but the equivalent of 50 milligrams per kilogram of body weight, in food or water, seems to be a dangerous intake level for most livestock, and doses of half that amount can significantly affect some animals.

The growing use of nitrogen fertilizers has recently become a matter of concern in relation to the ozone layer of the stratosphere (McElroy et al. 1976). Some of the nitrogen in these fertilizers is returned to the atmosphere as N_2 or N_2O (nitrous oxide) by denitrification. Nitrous oxide is the less abundant product, but it can equal as much as 30 percent of the quantity of molecular nitrogen produced. Once it has been carried to the upper atmosphere, N_2O participates in a series of photochemical reactions that ultimately cause the breakdown of ozone (O_3). The extent to which increasing fertilizer use will promote this process is highly uncertain, but some scientists believe that the ozone layer might be reduced by roughly 20 percent during the first quarter of the twenty-first century as a result of current and future use of nitrogen fertilizers. The ozone layer of the stratosphere acts as a shield to absorb incoming ultraviolet radiation; and it is now thought that a 5-percent reduction in the ozone layer will translate into about 10-percent increase in the ultraviolet radiation that reaches the earth's surface. Among other effects, increased ultraviolet radiation raises the incidence of skin cancer. A significant health hazard could thus be created, and close attention must be given to the movement of nitrogen compounds through the atmosphere as well as other parts of the biosphere.

Phosphorus

The second plant nutrient of critical importance to modern agriculture is phosphorus. World consumption of phosphorus fertilizers has been increasing rapidly (Harre et al. 1971); from 1955, when consumption was only 3.3 million metric tons, it had jumped to 8.1 million tons per year by 1970. About 2.0 million tons of this are consumed in North America; and North America and Europe together consume three-fourths of all phosphorus fertilizers.

Growth in the rate of phosphorus fertilizer consumption, and its essential nature in modern, intensive cultivation, raises serious questions about future supplies. Currently, these fertilizers come almost entirely from the mining and processing of sedimentary phosphate rock, of which there is a limited supply. Mined phosphate rock deposits now contain 12 percent or more phosphorus, but if we consider deposits as low as 8 percent phosphorus to be successfully mineable, total world reserves are currently estimated at 19.8 billion metric tons of phosphorus (Institute of Ecology 1972). At present rates of use, these reserves would last 1750 years; but present rates of use reflect current population and current technology, both of which are changing. World population is growing at a rate of 1.9 percent per year, and the use of phosphorus fertilizer is growing even faster, at a rate of 5.25 percent per year, which is 2.76 times the rate of population growth. Projecting these two trends into the future—always a dubious undertaking—suggests that the reserves mentioned above will run out in 90 years, at a time when world population will be 20 billion. The static and simplistic nature of this analysis has recently been criticized by Wells (1976), whose analysis of the existing quantities of high-grade ore, together with the potential for recycling wastes, using lower-grade ores, and reducing per capita consumption of phosphorus, suggests that reserves will be adequate for a very long time. Nevertheless, it is clear that high-quality reserves are limited, and that costs will increase significantly as lower grades of ore are exploited. Unlike petroleum, phosphorus, is not a resource for which man can substitute as he does for many other industrial raw materials. It is an essential nutrient in the biochemistry of living systems.

As we saw in Chapter 10, our use of phosphorus, mostly for fertilizers, acts basically to speed the flow of phosphorus from the continents to the ocean basins. The amount of phosphorus involved is about 14 million tons per year; but the amount of phosphorus that returns from the seas to the continents, in guano

and marine fisheries harvests, is only 100,000 metric tons per year. In the ocean, moreover, phosphorus becomes so widely dispersed that no reasonable probability exists for extracting it for fertilizer.

We must therefore carefully husband the reserves of phosphate rock that are available for use as fertilizers. We must curtail the use of phosphorus for nonessential purposes such as detergent use, which in 1970 was about 15 percent of total use. This type of use only speeds the flow of valuable agricultural resources to a nonrecoverable sink, the ocean. Improved recovery of phosphorus in mining and processing; the recovery of phosphorus-rich materials from waste water, in which they are still concentrated enough that recovery is economically possible; and improved efficiency in the agricultural use of phosphorus fertilizers are all called for in this case.

Figure 12-19, based on data from Stumm (1975), shows the pattern of phosphorus movement on the global scale. We are more interested, however, in examining the behavior of phosphorus in agricultural ecosystems, and especially in determining the extent and cause

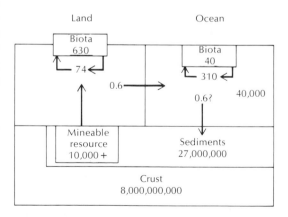

FIGURE 12-19.
Global phosphorus cycle, with estimates of quantities in major pools and of annual transfer rates. Units are 10^{11} moles (quantity in pool) or 10^{11} moles per year (transfer rate). (Modified from Stumm 1975.)

TABLE 12-8

Measured sediment loss and calculated phosphorus (available) loss under four crops and two farming systems on small Indiana watersheds from 1947 to 1950.

| | | Annual losses | | | |
| | | Sediment | | Phosphorus | |
Crop	Farming system	(kg/ha)	% of prevailing	(kg/ha)	% of prevailing
Corn	prevailing[a]	7060		2.86	
	conservation[b]	1670	(23.6)	0.86	(30.1)
Soybeans	prevailing	9440		3.82	
	conservation	3740	(39.6)	1.93	(50.5)
Wheat	prevailing	2020		0.82	
	conservation	930	(46.0)	0.48	(58.5)
Pasture	prevailing	210		0.09	
	conservation	130	(61.9)	0.07	(77.8)

[a]Moderate fertilization, liming to pH 6.0, straight row planting and cultivation.
[b]Higher fertilization, liming to pH 6.5, manure before corn, contour planting and cultivation.
Source: From Stoltenberg and White 1953.

of phosphorus losses from such systems. Most field crops take up between 10 and 50 kilograms per hectare of phosphorus in the harvested plant material. Rates of phosphorus application vary greatly, of course, but they range upward to nearly 100 kilograms per hectare. In the United States, the average application rate on cultivated cropland in 1970 was 12 to 13 kilograms per hectare.

As we have seen, phosphorus is retained in the soil largely as organic phosphorus, in more or less available form, and as fixed iron, aluminum, and calcium phosphates that have low availability. It is therefore not surprising that much less phosphorus than nitrogen is lost from agricultural systems.

Ryden et al. (1973) have recently summarized data on phosphorus inputs to streams from runoff and subsurface drainage. To place losses from agroecosystems in perspective, we will first note that the losses from natural ecosystems tend to be very low. For a series of forest watersheds, total loss (both in dissolved and particulate form) tends to be less than 1 kg/ha/yr. Where appreciable erosion is absent, more-

over, these losses are generally less than 0.1 kg/ha/yr. Natural ecosystems thus retain phosphorus very well.

In agricultural systems, losses from rangelands that receive no phosphorus fertilizers and are not subject to serious erosion appear to correspond closely to those from natural ecosystems; they are generally less than 0.1 kg/ha/yr. In cultivated cropland, the phosphorus losses are higher than in natural systems and are caused by erosion, runoff, and subsurface leaching. Unfortunately, many studies have failed to evaluate all these routes, and many have failed to separate particulate losses (through erosion) from the losses of dissolved phosphorus in runoff. Stoltenberg and White (1953) determined, by soil analyses and measurements of erosion (Table 12-8), the losses of available phosphorus from a series of cropland plots in Indiana. Total losses of phosphorus would, of course, be somewhat greater. These values show that erosion can be a major route by which this nutrient is lost. Losses of phosphorus dissolved in surface runoff appear, from the limited studies available, to be about 5 to

40 percent of those of particulate matter (Ryden et al. 1973), depending on how recently fertilizer was applied and on other factors. Subsurface drainage has generally been considered a minor route of phosphorus loss because phosphorus has a strong tendency to be chemically fixed in most soils. When an appreciable amount of water does flow through the soil, however, this can become a significant route of loss for applied phosphate fertilizers. In nonirrigated areas that receive moderate amounts of phosphorus, such losses appear to be 0.1 to 0.3 kg/ha/yr, but in irrigated systems they can rise as high as 4 to 14 kg/ha/yr, thus becoming greater than any other loss route (Ryden et al. 1973).

All in all, except where there is major erosion, moderate application rates of phosphorus fertilizers seem to more than balance losses from crop ecosystems, although they can accelerate somewhat the actual rates of loss. Thus in many cases the total phosphorus content of agricultural land is actually increasing; but this increase, stored in fixed form, is only slowly made available to crop plants. A major objective in studying phosphorus, therefore, is how to use phosphorus fertilizers so that they are retained in the soil in available form without being tied up in fixed minerals.

Eutrophication

Phosphorus and nitrogen fertilizers, upon entering surface waters, act as fertilizers in these aquatic systems. Higher productivity in fresh and coastal marine waters, created by increased nutrient inputs, is termed **eutrophication.** Although this is desirable in some situations, in others it leads to detrimental changes in the ecosystem. The increased production of organic matter in the surface water of lakes, for example, can increase the amount of dead matter transported to the bottom, thus increasing decomposition and oxygen exhaustion in

bottom waters. The chemistry of eutrophic lakes also favors the growth of blue-green algae of various types, many of which produce blooms that create nuisances such as bad water taste and foul-smelling windrows of decaying algae that accumulate on shore. The fauna of lakes, as well as the flora, are also modified by eutrophication. In a number of lakes, including Lake Erie, for example, clear-water faunas dominated by such species as trout and whitefish have been replaced by less desirable "trash" fish such as perch and carp.

In freshwater systems, it appears that phosphorus is generally the limiting nutrient. Additions of nitrogen alone rarely induce symptoms of eutrophication, but adding both nitrogen and phosphorus increases productivity levels to a highly eutrophic condition. In coastal marine areas, on the other hand, nitrogen appears to be the limiting nutrient in most situations.

Eutrophication is of some concern to agriculture, and particularly to agricultural ecology, because it affects other ecosystems from which human beings obtain food, such as freshwater lakes and coastal marine waters. It is possible that agriculture is a major source of nitrogen, and a somewhat less important source of phosphorus, contributing to the total nutrient load of surface waters. Policies of fertilizer use in the agricultural sector must therefore take such effects into account.

Other Fertilizer Nutrients

Of other fertilizer nutrients, potassium is quantitatively the most important. In 1970, the world use of potassium fertilizer was 13.1 million metric tons per year (Harre et al. 1971), of which 84 percent was used in North America and Europe. Supplies appear not to be limited for this element, and its use seems to create few side effects within or outside agroecosystems. Sulfur is also gaining importance as a plant fertilizer (McLachlan 1975). Sulfur deficiencies are being recognized as more frequent and widespread and are especially prevalent in tropical and subtropical regions. In part, the growing need for sulfur as a fertilizer parallels the reduction in the organic matter levels in soils, since much of the sulfate required was once supplied by the mineralization of organic sulfur. With higher crop yields per acre, greater amounts of sulfur are also being withdrawn in crop harvests. A harvest of 200 bushels of corn per hectare, for example, removes about 49 kilograms of sulfur.

There is also a growing need for various micronutrients to be added in particular situations. A great deal of care must be exercised, however, since many of the elements that are necessary in trace quantities become toxic in large amounts. This true for boron, chlorine, copper, manganese, molybdenum, and occasionally zinc.

For most nutrient elements other than nitrogen and phosphorus, little is known about the impact of agricultural fertilization on outflow rates. However, studies of both natural and agricultural ecosystems demonstrate that the biotic structure of the system has a profound influence on the chemical composition of flowing waters associated with its terrestrial part (Reichle 1975). These influences are important with respect to both environmental quality and the efficiency with which valuable resources are used. Our objective in managing the natural and exploited landscape must be to make use of the parts that can yield food and fiber, but to avoid the deleterious side effects that this use can have on food and fiber, but to avoid the deleterious side effects that this use can have on the other parts of the system.

SUMMARY

Irrigation and fertilization are basic to intensive agriculture. Both, however, have associated problems of supply and cost, impact upon the agroecosystem, and side effects outside agricultural areas. Man's total use of water, most of which is for irrigation, is beginning to approach the limits of supply. In the near future, this will create regional water deficiencies along with greater problems of water quality, caused by the use and reuse that add solutes to exploitable surface waters.

The quality of water for irrigation is determined by its salinity (total solute content), its sodicity (relative sodium content), and its content of toxic materials. The lower the quality of the available water, the greater the amount that must be applied to a soil to achieve adequate leaching of salts from the soil system. Irrigation must therefore be carried out only when adequate provision has been made for drainage. Even then, the problem of excessive solutes might only be shifted downslope and thus create higher salinities of ground or surface water for other users.

The use of nitrogen and phosphorus fertilizers has been increasing at a rate of about 5 percent or more per year. Profitable yields on many soils are achieved only through the use of such materials. For nitrogen fertilizers, most of which are industrial products made from natural gas, rapidly escalating costs demand that major attention be given to efficiency of use. Current cultivation and fertilization practices permit major losses of nitrogen from agroecosystems, primarily as nitrate in runoff and seepage, but occasionally through denitrification. Feedlots can also be a major source of nitrogen pollution.

Phosphorus fertilizers are derived by mining and processing phosphate rock, the total supply of which is severely limited. Although losses of phosphorus from agroecosystems are lower than those of nitrogen, they are significant. Combined with nonagricultural losses, and with nitrogen pollution, they create problems of eutrophication in many freshwater and coastal marine areas.

Literature Cited

Allison, L. E. 1964. Salinity in relation to irrigation. *Adv. in Agron.* 16:139–180.

Branson, R. L., P. F. Pratt, J. D. Rhoades, and J. D. Oster. 1975. Water quality in irrigated watersheds. *J. Env. Qual.* 4:33–40.

Burns, R. C., and R. W. F. Hardy. 1974. *Nitrogen Fixation in Bacteria and Higher Plants.* New York: Springer-Verlag.

Cantor, L. M. 1970. *A World Geography of Irrigation.* New York: Praeger Publishers.

Dahlman, R. C., J. S. Olson, and K. Doxtader. 1969. The nitrogen economy of grassland and dune soils. In *Biology and Ecology of Nitrogen.* pp. 54–82. Washington, D. C.: National Academy of Sciences.

Delwiche, C. C. 1970. The nitrogen cycle. *Sci. Amer.* 223(3):136–146.

EPA. 1973. *Methods for Identifying and Evaluating the Nature and Extent of Non-point Sources of Pollutants.* Report 430/9–73–014. Washington, D. C.: U.S. Environmental Protection Agency.

Fowler, W. B., and J. D. Helvey. 1974. Effect of large-scale irrigation on climate in the Columbia Basin. *Science* 184:121–127.

Hardy, R. W. F., and R. D. Holsten. 1974. Global nitrogen cycling: pools, evolution, transformation, transfers, quantitation, and research needs. In *The Aquatic Environment: Microbial Transformations and Water Management Implications.* Report 430/G–73–008. Washington, D. C.: U.S. Environmental Protection Agency.

Harre, E. A., W. H. Garman, and W. C. White. 1971. The world fertilizer market. In *Fertilizer Technology and Use.* pp. 27–55. Madison, Wisc.: Soil Science Society of America.

Heady, E. O., H. C. Madsen, K. J. Nicol, and S. H. Hargrove. 1973. National and inter-regional models of water demand, land use, and agricultural policies. *Water Resources Res.* 9:777–791.

Ibach, D. B., and J. R. Adams. 1968. *Crop Yield Response to Fertilizer in the United States.* U.S. Dept. USDA Agr. Sta. Bull. 431:295 pp.

Institute of Ecology 1972. Cycles of elements. In *Man and the Living Environment.* pp. 41–89. Madison: University of Wisconsin Press.

Kalinin, G. P., and V. D. Bykov. 1969. The world's water resources, present and future. *Impact of Sci. on Soc.* 19:135–150.

McCauley, C., and R. Gum. 1975. Land subsidence: An economic analysis. *Water Res. Bull.* 11:148–154.

McElroy, M. B., J. W. Elkins, S. C. Wofsy, and Y. L. Yung. 1976. Sources and sinks for atmospheric N_2O. *Rev. Geophys. Space Sci.* 14:143–150.

McLachlan, K. D. (ed.) 1975. *Sulphur in Australian Agriculture.* Sydney, Australia: Sydney University Press.

Mieri, A., and J. Shalhevet. 1973. Crop growth under saline conditions. In *Arid Zone Irrigation.* B. Yaron, E. Danfors, and Y. Vaadia (eds.). *Ecological Studies,* vol. 5. pp. 227–290. New York: Springer-Verlag.

Michel, A. A. 1972. The impact of modern irrigation technology in the Indus and Helmand Basins of southwest Asia. In *The Careless Technology.* M. T. Farvar and J. P. Milton (eds.). pp. 257–275. Garden City, N. Y.: Natural History Press.

National Academy of Sciences. 1972. *Accumulation of Nitrate.* Washington, D. C.: National Research Council.

Nightingale, H. I. 1974. Soil and ground-water salinization beneath diversified irrigated agriculture. *Soil. Sci.* 118:365–373.

Phillips, A. B., and J. R. Webb. 1971. Production, marketing, and use of phosphorus fertilizers. In *Fertilizer Technology and Use.* pp. 271–301. Madison, Wisc.: Soil Science Society of America.

Pimentel, D. 1974. *Energy Use in World Food Production.* Report 74–1, Department of Entomology and Section of Ecology and Systematics. Ithaca, N. Y.: Cornell University.

Reichle, D. E. 1975. Advances in ecosystem analysis. *BioSci.* 25:257–264.

Revelle, R. 1963. Water-resources research in the federal government. *Science* 142: 1027–1033.

——. 1974. Food and population. *Sci. Amer.* 231(3):160–170.

Rhoades, J. D. 1972. Quality of water for irrigation. *Soil Sci.* 113:277–284.

Ryden, J. C., J. K. Syers, and R. F. Harris. 1973. Phosphorus in runoff and streams. *Adv. in Agron.* 25:1–45.

Stidd, C. K. 1975. Irrigation increases rainfall? *Science* 188:279–280.

Stoltenberg, N. L., and J. L. White. 1953. Selective loss of plant nutrients by erosion. *Soil Sci. Soc. Amer. Proc.* 17:406–410.

Stumm, W. 1975. Man's acceleration of hydrogeochemical cycling of phosphorus: eutrophication of inland and coastal waters. *Water Poll. Cont.* 74:124–133.

Viets, F. G., Jr. 1971. Fertilizer use in relation to surface and ground water pollution. In *Fertilizer Technology and Use.* pp. 517–532. Madison, Wisc.: Soil Science Society of America.

Wells, F. J. 1976. *The Long-run Availability of Phosphorus: A Case Study in Mineral Resource Analysis.* Baltimore: Johns Hopkins University Press.

Wortman, S. 1975. Agriculture in China. *Sci. Amer.* 232(6):13–21.

Wulff, H. E. 1968. The qanats of Iran. *Sci. Amer.* 218(4):94–105.

13

SOIL AND PLANT MICROBIOLOGY

The microflora and microfauna of the soil, which range from the size of an earthworm to the size of a single-celled organism, together with the microbial flora of aboveground plant surfaces, constitute one of the most complex and poorly understood components of natural and agricultural ecosystems. They also carry out some of the most important functions related to the basic aspects of ecosystem fertility and productivity.

The adjustments of the microbiota in the soil clearly reflect two major environmental characteristics: density and darkness. The density of the soil means, for obvious reasons, that its inhabitants must be small or even microscopic. Such small organisms have certain metabolic advantages because of the rapid and efficient exchange of materials (water, gases, nutrient ions, and food molecules) with their surroundings. Thus they have a surprisingly high metabolic rate per unit biomass and a very high reproductive potential. The density of the soil also means that atmospheric gases can reach limiting concentrations, something that rarely occurs aboveground; and, along with the chemical diversity of the matrix materials, this also means that a great variety of substances can

exist and be retained within the soil. Chemical interactions among different species might therefore be more important to organisms in the soil than to those that live in the open air.

The soil environment is also dark, except at its very surface, and so light energy is unavailable within the soil. Energy for the functions of organisms must therefore come from chemical sources, such as the oxidation of inorganic compounds in which a net energy release occurs, or the catabolism of organic molecules transported into the soil from its surface.

The density of the soil shields and buffers the internal soil environment from the extremes of aboveground conditions, especially of temperature and humidity. All these factors combine to allow the existence of an incredible variety of distinctive, chemically characterized microenvironments, the conditions of which are not static, but change slowly in response to seasonal changes, successional trends, and human manipulations.

We shall attempt to describe the most important processes mediated by the soil inhabitants, considering the soil biota in its entirety, but concentrating on the microscopic and near-microscopic members of the system. We shall find it useful to distinguish the organisms that live very near the surfaces of roots of higher plants from those that live elsewhere in the soil. The root surface zone, termed the **rhizosphere,** is particularly important because many of the activities that occur there strongly influence the health of these higher plants, including cultivated crop species.

The microbiota of ecosystems is not restricted to the soil, however. Increasingly, scientists have become aware that a significant microflora occupies the surfaces of plant leaves and stems. The activities of these microorganisms, in the zone termed the **phyllosphere,** are apparently less important than those living in the soil but still have some significance.

We shall begin our discussion by identifying and characterizing the major groups of organisms that make up the microbiota of ecosystems.

THE MICROBIOTA

The true microbial component of the soil biota includes the bacteria, actinomycetes (considered by most microbiologists as an order of bacteria), fungi, algae, and protozoa. Among the smaller multicellular animals, the nematodes, earthworms, and various groups of arthropods are also important members of the soil ecosystem. The phyllosphere is, of course, occupied by a variety of organisms; but here we shall consider only the microbial inhabitants of this zone, the bacteria and fungi, and discuss the arthropods in later chapters that deal with crop pests. We shall also defer discussion of viruses and specific plant and animal diseases to later chapters. In describing the taxonomic relationships of various groups of organisms, we shall use the five-kingdom system proposed by Whittaker (1969) and outlined in Figure 13-1.

The bacteria comprise various groups belonging to the kingdom Monera (Table 13-1). These organisms are single-celled or colonial forms that lack organized nuclei, plastids, or mitochondria. Two phyla, the Eubacteriae or true bacteria, and the Actinomycota or actinomycetes, are of major importance in soil ecosystems. The true bacteria are single-celled organisms that vary in size from about 0.1 to 10 (rarely 60) microns (1 micron = 10^{-3} millimeter), and in shape from spherical **cocci** to cylindrical **rods** and helical **spirilla** (Figure 13-2). Some of these individual cells are nonmotile; some possess one or more flagella of primitive structure that enable them to swim actively. Nonmotile forms are occasionally grouped in chains or filaments. The organisms of the second phylum, the actinomycetes, develop first as a branching, filamentous **mycelium** that only later fragments into individual rods or cocci.

True bacteria and actinomycetes have very diverse nutritional needs: Some are **aerobic,** requiring oxygen for their life activities; others are **anaerobic** and flourish only in the absence of oxygen. Some are **facultatively anaerobic;** that is, they can adjust their metabolic systems to operate in either the presence or absence of oxygen. Many bacterial groups are **heterotrophic,** which means that they require a supply of preformed organic matter that they decompose and catabolize to obtain energy-rich molecules and the necessary organic building blocks for new protoplasm. Other bacteria, which are **autotrophic,** use some source of energy to synthesize complex organic compounds from simple inorganic ones; in this way they obtain energy-rich molecules for later use in respiration as well as the organic constituents needed for the growth of new protoplasm. Some of these autotrophic bacteria are **photosynthetic** and use green or purple pigments to capture light energy; others are **chemosynthetic,** deriving energy from the oxidation of simple inorganic compounds and using this energy in much the same way as photosynthetic forms. Other specialized metabolic patterns exist among both autotrophs and heterotrophs; many of these lead to important chemical conversions of nitrogen, sulfur, iron, and organic constit-

330

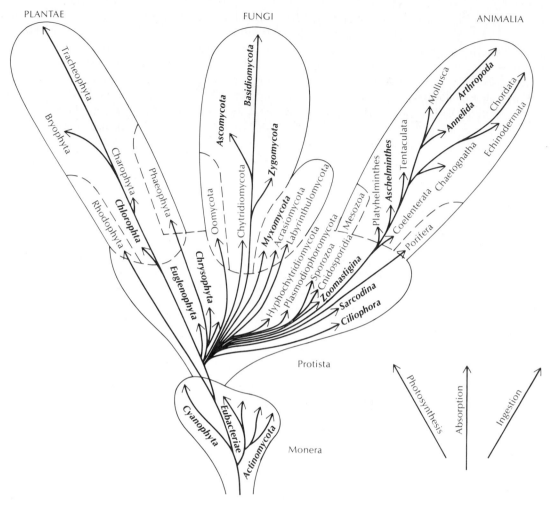

FIGURE 13-1.
Kingdoms and phyla of organisms that are important members of the soil ecosystem. Groups important in the soil are in bold type. (Modified from Whittaker 1969. Copyright © 1969 by the American Association for the Advancement of Science.)

uents of the soil system. All these patterns of nutrition occur among members of the phylum Eubacteriae; but members of the actinomycota, some of which are aerobic and others anaerobic, are all heterotrophic. Both phyla include disease agents of both plants and animals. Some of the specific activities of certain groups of these bacteria are indicated in Table 13-2.

Soil algae belong to several phyla that are apparently not closely related and that are grouped in three different kingdoms (Table 13-1). The phylum Cyanophyta, or blue-green algae, are grouped in the Kingdom Monera with the bacteria, which they resemble in subcellular structure. The Euglenophyta, or green flagellates, and the Chrysophyta, or diatoms, are placed in the kingdom Protista with proto-

TABLE 13-1
Taxonomic relationships of the major groups of soil organisms.

Kingdom	Phylum	Class	Common Name
Monera	Cyanophyta		Blue-green algae
	Eubacteriae		True bacteria
	Actinomycota		Actinomycetes
Protista	Euglenophyta		Euglena-like organisms
	Chrysophyta		Diatoms
	Zoomastigina		Animal flagellates
	Sarcodina		Amoeba-like protozoans
	Ciliophora		Ciliate protozoans and suctorians
Plantae	Chlorophyta		Green algae
Fungi	Myxomycota		Slime molds
	Zygomycota		Conjugation fungi
	Ascomycota		Sac fungi
	Basidiomycota		Club fungi
Animalia	Aschelminthes		Nematodes and relatives
	Annelida		Segmented worms
		Oligochaeta	Earthworms and enchytraeids
	Arthropoda		Arthropods
		Arachnida	Mites, spiders, and relatives
		Chilopoda	Centipedes
		Diplopoda	Millipedes
		Crustacea	Crustaceans
		Insecta	Insects

Source: Whittaker 1969.

zoans, which they resemble in details of structure and reproduction. The Chlorophyta, or green algae, appear most closely related to other groups in the kingdom Plantae. Many approach multicellularity, advanced cell organization, and alternation of generations in reproduction. All, of course, possess chlorophyll, but the fact that they appear at surprising depths in the soil, together with the demonstrated ability of some to multiply under heterotrophic conditions, suggests that they may not be strict autotrophs under all conditions (Lund 1967). Certain blue-green algae are important in nitrogen fixation, especially in rice paddy systems, and the oxygen that they liberate in flooded or saturated soils may be valuable to the roots of higher plants (Alexander 1961). Otherwise, the role of algae in the soil system is still little understood.

The soil protozoa belong to three phyla in the kingdom Protista: the Zoomastigina, or animal flagellates; the Sarcodina, or rhizopods (amoebalike forms); and the Ciliophora, or ciliates*. A few of the flagellates possess chloro-

---

*Many classifications consider these to be classes within the phylum Protozoa.

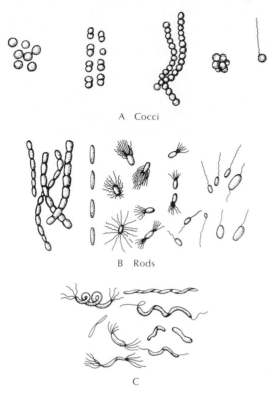

A   Cocci

B   Rods

C

FIGURE 13-2.
Common patterns of shape, flagellation, and arrangement of cells among the true bacteria. (A) Cocci are spherical cells that occur singly, in small groups, or in chains; some bear flagella. (B) Rods also occur singly or in colonies; many are flagellated. (C) Spirillae are generally solitary; some have flagellae, some do not.

phyll, but by and large the members of these phyla are heterotrophic. Some are saprophages, feeding on particular organic materials, but most are bacteria-feeders (Stout and Heal 1967). Although they seem to feed only on certain species of bacteria, the impact on bacterial populations remains uncertain. Some protozoa are predaceous on other protists, and, of course, some are agents of plant and animal diseases.

The Fungi are best thought of as a Kingdom of their own (Table 13-1). They have well-organized nuclei that occur in separate cells, in multinucleate filaments lacking cellular parti-

tioning, or in naked protoplasmic sheets (**plasmodia**) that lack any wall structure and that move about in amoeboid fashion. All are heterotrophic. The Myxomycota, or slime molds, are a phylum whose members all exist in the form of a plasmodium at some point in their life cycles. This plasmodium is capable of moving slowly over surfaces, and is often found in damp, highly organic situations. They are believed to feed to some extent upon bacteria that live on the substrates over which they pass (Alexander 1961). Eventually, they produce a multicellular, macroscopic fruiting structure, or spore case, that is made up of distinct, multinucleate filaments that have no septa between adjacent nuclei. Three other phyla of fungi, the Zygomycota, Ascomycota, and Basidiomycota, exist abundantly in the soil as mycelia—thin, tubelike filaments that average perhaps 5 microns in diameter (Figure 13-3). Some of these mycelial tubes are septate—that is, they have cell walls separating individual nuclei; others have no septa. Most of the members of these phyla produce a specialized fruiting body at some stage in their life cycle, the differences in these reproductive structures being the primary basis for distinguishing phyla. Several members of one phyla, the Basidiomycota, form intimate associations with roots of higher plants. We shall consider this relationship, called **mycorrhiza** (fungus-root), in detail. Other fungi participate in the complex processes of decomposition of organic matter in the soil; many are agents for plant disease, and a few are agents of diseases in animals.

Three phyla belonging to the kingdom Animalia are also represented abundantly in the soil: the Aschelminthes, or nematode worms; the Annelida, or segmented worms; and the Anthropoda, or arthropods. Soil nematodes are multicellular, wormlike animals, most of which are less than 1 millimeter in size. A great many species, difficult both to identify and classify, exist in the soil and in association with plant

TABLE 13-2
Major groups of soil bacteria.

Order and family	Nutrition	Shape	Other	Important genera
Pseudomoniales				
Thiorhodiaceae (Purple sulfur bacteria)	Photosynthetic, anaerobic	Flagellated rods, spirals	Reduce S	*Chromatium, Thiospirillum*
Athiorhodaceae (Nonsulfur purple bact.)	Photosynthetic, aerobic	Rods, spirals		*Rhodopseudomonas, Rhodospirillum, Rhodomicrobium*
Chlorobacteriaceae (Green bacteria)	Photosynthetic, anaerobic	Immobile rods	Oxidize H_2S	*Chlorobium*
Nitrobacteriaceae (Nitrifying bacteria)	Chemosynthetic, aerobic	Flagellated rods	Oxidize N compounds	*Nitrosomonas, Nitrobacter*
Methanomonadaceae (Methane-oxidizing bacteria)	Heterotrophic, aerobic	Flagellated rods	Oxidize CH_4	*Methanomonas*
Pseudomonadaceae	Heterotrophic, aerobic or facultatively anaerobic	Flagellated rods		*Pseudomonas, Acetobacter, Aeromonas, Photobacterium, Zymomonas*
Spirillaceae	Heterotrophic, aerobic	Comma-shaped, flagellated		*Vibrio*
	Heterotrophic, anaerobic	Flagellated spirals	Reduce S (° only)	*Desulfovibrio°, Spirillum*
Eubacteriales				
Azotobacteriaceae	Heterotrophic, aerobic	Flagellated rods	Nonsymbiotic N fixers	*Azotobacter*
Rhizobiaceae	Heterotrophic, aerobic	Flagellated rods	Symbiotic N fixers	*Rhizobium*
Enterobacteriaceae	Heterotrophic, facultatively anaerobic	Flagellated rods		*Escherichia, Erwinia, Shigella, Salmonella, Agrobacterium*
Corynebacteriaceae	Heterotrophic, aerobic	Immobile rods		*Corynebacterium, Arthrobacter*
Bacillaceae	Heterotrophic, aerobic	Flagellated rods		*Bacillus*
	Heterotrophic, anaerobic	Flagellated rods		*Clostridium*

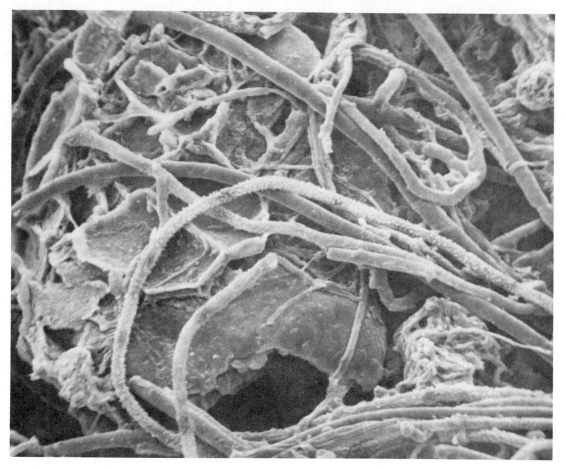

FIGURE 13-3.
Scanning electron micrograph of fungal mycelia in association with dead plant material of the forest floor. Magnification is about 1000×. (Photograph by R. Todd, K. Cromack, and R. Knutson, University of Georgia. *Bull. Ecol. Res. Comm.* 17:109–118.)

roots. They employ three major modes of feeding (Nielsen 1967). Many are root parasites and feed on plant juices obtained by piercing the roots and the walls of individual cells. Others feed on bacteria and algae, and still others are predators or parasites of soil animals such as protozoa, other nematodes, and, in the case of parasites, earthworms and arthropods. The segmented worms are represented in many soils by the lumbricid earthworms, which we have dis-

cussed in connection with the formation of soil aggregate structure. A second family of this phylum, the Enchytraeidae, or pot worms, are also important in some soils. These worms range in size from less than 1 millimeter to about 5 centimeters and are especially abundant in cool, acid soils. Like earthworms, they feed on organic matter in soils. Some forms appear to feed selectively on fungal material, and it is thought that others kill masses of root-feeding nematodes and feed upon them (O'Connor 1967). The arthropods of the soil are an extremely diverse group including crustaceans, arachnids, centipedes, millipedes, and insects. Many of these organisms are **saprotrophic;** that is, they feed on dead organic matter. Others feed upon bacteria and fungi, and still others are predators or forms that are parasitic upon plants and animals. Soil animals include many forms that are crop pests; we shall consider these in detail beginning in Chapter 15.

The microbiota of the soil thus forms one of the most diverse assemblages of organisms in nature, including members of all Kingdoms and quite a number of phyla. It is an ancient assemblage, containing the descendants of some of the first forms of life ever to evolve; it is likely that the survival of these organisms reflects their basic importance in essential processes of the soil ecosystem.

NUMBERS AND BIOMASS

Determining the abundance of soil microorganisms, especially the bacteria, fungi, and algae, continues to be one of the most difficult tasks in soil microbiology. Even under the best conditions, estimates of numbers and biomass made by different techniques vary greatly. Witkamp (1973), for example, found that estimates of fungal biomass in cultures grown in the laboratory in sand varied by a factor of four, even when two species of innoculated fungi were the only organisms present and all organic matter consisted of fungal spores and mycelia. This is complicated by the fact that in nature it is very hard to determine what fraction of directly observed cells or tissue is living or actively metabolizing at a particular time. This factor is apparently responsible for major differences in biomass estimates for bacterial and fungal populations.

The predominant organisms in the leaf phyllosphere are the bacteria, although in particular situations actinomycetes, blue-green algae, fungi, and protozoa have been recorded (Clark and Paul 1970; Weaver et al. 1974). The bacterial forms present include some that are capable of nitrogen fixation along with many others, some of which may be present only accidentally. Certain forms, however, appear to be phyllosphere residents, dependent on organic leaf exudates; and there seems to be some degree of specificity with respect to higher plant species. The abundance of the organisms present varies greatly, but appears to be of the order of magnitude of 10^7 cells/cm^2. For a vegetation with a leaf area index of 2.0, this would translate into a bacterial biomass of perhaps 0.04 g/m^2, dry weight.

It is not surprising that the abundance of microorganisms in the soil usually decreases with depth (Table 13-3). In bare cultivated land, the pattern might be somewhat different. Here, the population densities close to the surface might be reduced because of the heating and drying of the surface and perhaps because of the bactericidal action of sunlight (Alexander 1961). Roughly speaking, bacterial numbers can be converted to biomass based on the relationship that 10^{12} bacteria equal 1.0 gram wet weight or 0.2 gram dry weight. Applying this relationship to the data from Table 13-3, for both true bacteria and actinomycetes, and assuming that the data for the A_1 horizon actually apply over the first 20 centimeters of depth (and that soil density = about 1 g/cm^3), total bacterial dry weight is between 0.5 and 1.0 g/m^2. Witkamp (1971) suggests that bacterial biomass is generally less than 1 g/m^2, although he indicates that, in calcareous forest soils that are especially favorable to bacterial growth, values can reach 30 g/m^2. The estimates by Clark and Paul (1970), based on the assumption that average numbers of bacteria per gram of agricultural soils were of the order of 10^9 cells, are certainly much too high and probably reflect the inclusion of many dead cells in direct counts.

TABLE 13-3
Abundance of various groups of microorganisms in a Temperate Zone soil profile.

Horizon	Depth (cm)	10^3 organisms per gram of soil				
		Aerobic bacteria	Anaerobic bacteria	Actinomycetes	Fungi	Algae
A_1	3–8	7800	1950	2080	119	25
A_2	20–25	1800	379	245	50	5
A_2–B_1	35–40	472	98	49	14	0.5
B_1	65–75	10	1	5	6	0.1
B_2	135–145	1	0.4		3	

Source: From Alexander 1961.

The fungal biomass is usually estimated by measuring the length of fungal mycelium per unit of soil. Once again the problem is to determine what fraction of the quantity observed is actually living. The percentage of dead mycelium tends to increase with depth, as indicated by the use of vital staining techniques (techniques using stains that affect only living tissues). The biomass of fungi in almost all cases exceeds that of bacteria and is of the order of 10 g/m^2, dry weight (Witkamp 1971). Again, much higher values have been reported that in some instances have exceeded 150 g/m^2, dry weight, which probably reflects a failure to distinguish living from dead mycelia. Tentatively, we can suggest that soils with good organic content possess on the order of 1 gram of bacteria and 10 grams of fungi per square meter, dry weight.

Greater quantities of algae exist on and in the soil than have been generally recognized. Shtina (1974), for example, indicates that over one thousand forms have been identified in soils of the USSR. These forms range from unicellular and spherical in shape to colonial and filamentous. Their abundance varies greatly with seasonal conditions, more so than that of bacteria and fungi, and although they are most abundant near the surface, appreciable quantities are carried to greater depths, apparently by the percolation of water and activities of organisms such as earthworms. Under favorable conditions—that is, in moist, nutrient-rich, unshaded grasslands and cultivated areas—the biomass of soil algae can reach 30 to 50 g/m^2, wet weight, or about 6 to 10 g/m^2, dry weight (Shtina 1974). In desert regions, moreover, surface films of green and blue-green algae and lichens (symbiotic assemblages of algae and fungi) appear to be well developed, although they are obviously inactive much of the time (Allison 1973). Soil algae are important both because they are primary producers and because certain of the blue-green algae, and possibly other forms, are capable of fixing major quantities of nitrogen.

The numbers and biomass of soil protozoans are much smaller, however (Table 13-4). Estimates of total biomass for various grassland and forest ecosystems range up to 20 g/m^2, wet weight; but according to Stout and Heal (1967) the quantity present in most temperate zone grasslands and cultivated soils is probably less than 5 g/m^2, wet weight, or about 1 g/m^2, dry weight.

The numbers and biomasses of metazoan groups present in the soil fauna depend to a great extent on local environmental characteristics; but nematodes usually make up 80 to 90 percent of the individuals present (Oosten-

TABLE 13-4

Estimated numbers and biomass of various groups of protozoa in an agricultural soil in England.

	Zoomastigina	Sarcodina	Ciliophora	Total
Numbers/g wet soil	70,500	41,400	377	112,300
Biomass (g/m^2)	0.35	1.60	0.12	2.07

Source: From B. N. Singh 1946. *Ann. Appl. Biol.* 33:112–119.

TABLE 13-5

Representative biomass and metabolism estimated for soil faunas in three European ecosystems.

Faunal group	Grassland g/m^2	Grassland kcal/m^2/yr	Beech forest g/m^2	Beech forest kcal/m^2/yr	Spruce forest g/m^2	Spruce forest kcal/m^2/yr
Predators						
Arthropods	9.6	124.7	1.1	93.5	1.1	99.3
Herbivores						
Nematodes	6.0	183.2	2.0	61.1	2.0	61.1
Molluscs	10.0	61.5	3.2	19.7	0.2	1.0
Arthropods	1.4	9.2	11.0	76.7	3.7	26.8
Large decomposers						
Lumbricids	120.0	178.1	5.4	8.0	5.1	7.5
Isopods	5.0	38.2	0.1	0.4	—	—
Diplopods	12.5	95.4	1.1	8.6	0.4	2.8
Small decomposers						
Nematodes	6.0	183.2	1.0	30.5	1.0	30.5
Enchytraeids	12.0	102.0	1.6	13.2	0.1	0.2
Oribatid mites	2.0	10.2	0.2	11.2	0.3	14.2
Collembola	5.0	152.6	0.2	168.0	0.1	68.8
Total	189.5	1139.3	26.9	490.0	13.9	311.7

Source: From Macfadyen 1963a.

brink 1971). Some idea of the biomass relationships of the organisms present is given by data summarized by Macfadyen (1963a) regarding wet weight biomasses of smaller invertebrates in grassland and forest areas in Europe (Table 13-5). The biomass of the larger members of the soil biota is considerably greater than that of the smaller organisms; but this relationship does not extend to their metabolic importance.

The abundance of soil microorganisms, especially of bacteria, fungi, and nematodes, is different in the rhizosphere than in the soil away from the roots of higher plants (Rovira and McDougall 1967). Although the area ascribed to the rhizosphere is somewhat arbitrary, there are usually 5 to 20 (or more) times as many of these organisms within a few millimeters of root surfaces than at greater distances.

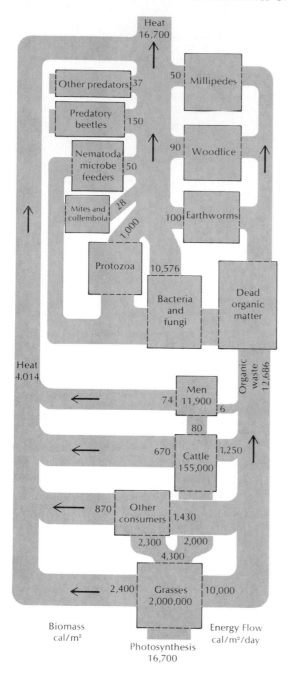

FIGURE 13-4.
Energy flow through a grazed pasture ecosystem (values in gram calories/m²/day). (Modified from Macfadyen 1963b.)

METABOLISM OF THE SOIL BIOTA

As we have noted in other chapters, most of the energy flow in natural and seminatural ecosystems actually passes through the soil and detritus food web. Macfadycn (1963b) illustrates this fact with data on energy use by various groups of soil and aboveground organisms (Figure 13-4). These data, synthesized to give a picture of the energy flow through a pasture ecosystem, show that 76 percent of the gross primary production of this system is consumed in the soil and detritus food web, and 63 percent by the bacteria and fungi in this portion of the system.

This relates to the metabolism associated with small body size (Figure 13-5). Bacteria, with an average body weight of 10^{-12} gram, metabolize at the rate of about 1 kcal/g/hr; protozoa, with a body weight of 10^{-6} gram, metabolize at 0.001 kcal/g/hr; and earthworms, with a weight of 1 gram, have a metabolic rate of 0.0001 kcal/g/hr. With a few calculations, one can see that the estimates of energy flow through the soil components of a pasture ecosystem such as that shown in Figure 13-5 are quite reasonable. One can also see that the metabolic rate of an active 100-gram biomass of bacteria would exceed the primary productive capacity of a square meter of even the most productive world environment.

Several other points should be noted about metabolic relationships with regard to total production and respiration in the soil. First, the productive contribution of soil algae populations is inadequately understood. Because of their small size, the productive potential of these organisms must be high, just as the respiratory potential of small soil heterotrophs is high. Shtina (1974) suggests that, under favorable conditions, the standing crop of soil algae might renew itself at least three times per month, meaning that in soils with a relatively high algal biomass, algal productivity might be a significant part of total primary production.

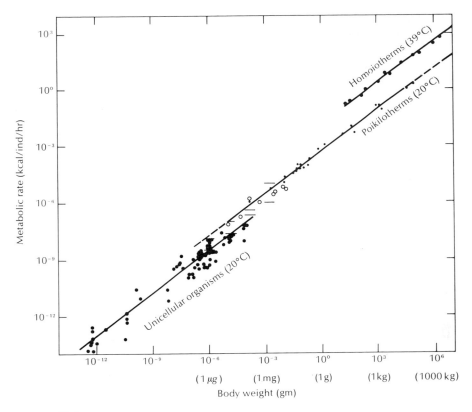

FIGURE 13-5.
Relation between body size and metabolic rate for organisms of a wide range of sizes. Metabolic rate equals kcal/hr/individual. The slope of the lines, corresponding to the exponent relating weight to metabolism, equals 0.75.

Second, it is clear that, in addition to the organic matter introduced into the soil and detritus food web by the death of shoot and root tissues, a significant amount of organic material is introduced into the soil during the growth of live roots. This consists of surface tissues sloughed from the root and organic solutes exuded by the root. It is difficult to estimate the amount of material involved, but both tissues and exudates seem to be important, apparently equalling 1 to 2 percent of the total biomass of live tissues that develops (Clark and Paul 1970). These materials of course are concentrated in the immediate vicinity of the roots—the rhizo-

sphere—and they presumably play a role in the nutrition of bacteria, fungi, and other organisms that live selectively in this zone.

DECOMPOSITION PROCESSES

We can begin our discussion of decomposition processes by distinguishing between litter breakdown and litter decomposition. **Litter breakdown** is the primarily mechanical process by which organic materials are reduced to smaller particles; it is largely the function of the invertebrate animals that populate soil

TABLE 13-6
Annual litter fall (aboveground material only) and fall:accumulation ratio for various ecosystems.

Plant community	Rainfall (cm/yr)	Mean temperature (°C)	Litter fall (kg/ha/yr)	Annual fall: accumulation ratio
Tropical				
Mixed rain forest	170	24.5°C	12,300	2.9:1
Mixed dry lowland forest	120	27.0°C	5,600	2.2:1
Temperature				
Mixed woodland	85	10.0°C	3,100	1:1.2
Oak woodland	75	10.0°C	1,380	1:29
Pine stand	75	10.0°C	2,800	1:60

Sources: Data from various studies.

and litter layers. Nevertheless, some initial action of bacteria and fungi on newly dead material seems necessary before it is palatable to members of the invertebrate fauna (Clark and Paul 1970). More and more of these organisms appear on the surfaces of senescent material, and they increase even more rapidly after death. As small soil animals feed they create more surface area for attack by bacteria and fungi and modify it chemically to favor more rapid decomposition.

Decomposition, of course, refers to the chemical degradation and reorganization of the dead organic matter. Here too we can distinguish two sets of processes: those that reduce the original matter to a set of relatively resistant residues, and those that involve the synthesis of organic components derived from dead matter into the tissues of microorganisms themselves. The latter can lead ultimately to the production of a distinctively different set of organic residues. There is good evidence that both these processes are major contributors to the resistant organic substances that constitute soil humus (Allison 1973).

The quantity of litter deposited upon the soil surface in natural ecosystems usually varies in relation to the level of primary production (Table 13-6). In wet tropical forests, values over 1200 g dry weight/m^2/yr can occur. The rates

at which this material is broken down and decomposed also vary greatly, as reflected by the ratio of annual fall to amount present at a given time. In the wet tropical forests, the amount of litter present is typically less than the annual litter fall, whereas in temperate regions it is more—often much more. The litter deposited in the soil by death of roots is less easy to determine. It has been estimated that, in grasslands, about 25 percent of the root biomass is replaced annually. For a series of grassland systems in North America, this relationship yields estimates of 137 to 475 grams dry weight per year (Clark and Paul 1970). In annual plant communities, such as certain crop ecosystems, the annual contributions both above ground and within the soil would equal maximum growth minus herbivore consumption.

Litter breakdown and decomposition are influenced by the composition of the materials and by environmental conditions. The relationships of a number of the more important processes are diagramed in Figure 13-6. Certain litter components, particularly carbohydrates and proteins, decompose rapidly, while others, such as lignin and various waxes and resins, are broken down only very slowly. In general the carbohydrate and protein components that decompose rapidly are utilized by microorganisms for their own respiration or for produc-

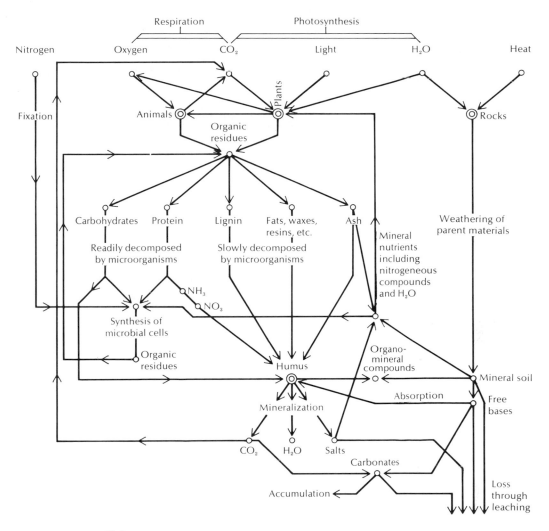

FIGURE 13-6.
The relationship of various processes concerned with litter breakdown, humus formation, and decomposition of organic matter in the soil. (Modified from Allison 1973.)

tion of new protoplasm. The more resistant components appear as residues that make up part of the humus component of the soil. The decomposition rates of some of these major components appear in Figure 13-7.

The relative importance of different groups of microorganisms is influenced by environmental conditions, especially by pH and by oxygen availability. Under conditions of near-neutral pH, bacteria tend to predominate over fungi, which become more important as acidity increases. Where oxygen is reduced to very low levels, aerobic bacteria are largely replaced by anaerobic bacteria, unless conditions become strongly acidic at the same time, as in a peat bog. These anaerobic, acidic conditions result in the continuous accumulation of only slightly decomposed material.

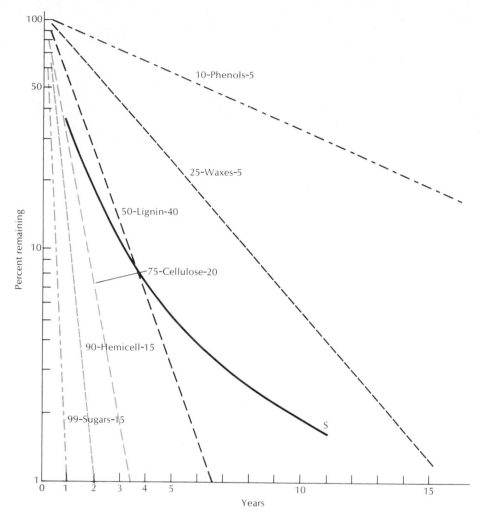

FIGURE 13-7.
Decomposition curves for various constituents of plant litter. The number before the name of the constituent indicates the percent loss in weight after one year; the number following the name is the approximate percentage of the constituent in the original litter. The line S is a summation curve calculated from the residual values for the several components. (Modified from Clark and Paul 1970.)

Rapid decomposition is also favored by moist conditions. As the soil or litter dries, bacteria and fungi tend to be replaced by actinomycetes, which are more tolerant of dry conditions. The pleasant mustiness associated with dried hay is produced by the action of actinomycetes.

The rate of decomposition, and the qualitative and quantitative impact of decomposition processes on other components of the system, such as higher plants, also depends on the carbon:nitrogen ratio of the litter material. When large quantities of organic matter, such as animal or green manures, are added to a soil, they can increase the level of microorganismal activity 10 to 50 ×. Under these conditions, populations of decomposer organisms, especially of bacteria, can grow rapidly. Because bacteria possess between 6 and 13 percent nitrogen by dry weight, this growth requires large quantities of nitrogen, and as a result the nitrogen present or available in the soil can become tied up, or **immobilized,** by these organisms. If the carbon:nitrogen ratio exceeds about 25 to 30 (i.e., 25 to 30 times as much carbon as nitrogen), but if considerable decomposable material is present, the nitrogen, including any that might have been available in the soil, will be tied up almost completely. This can both limit the decomposition rate and restrict the availability of nitrogen to higher plants such as crop species. Adding nitrogen-rich manures, or manures augmented with inorganic nitrogen fertilizer, in which the C:N ratio is lower than 25:1, may allow maximum decomposition and the release of nitrogen to higher plants. Most crop residues, such as straw, cornstalks, and weeds, have C:N ratios that are greater than this critical value; so do certain other materials, such as sawdust, that are sometimes recommended for addition to soil.

Humus formation can be considered as the final stage of decomposition. We shall not attempt to examine the several theories of humus formation (Felbeck 1971), except to indicate that they all assume the active participation of soil microorganisms in complex processes of biochemical synthesis. The materials involved represent a soil organic fraction that is highly resistant to further decomposition. The molecules are large, and most of them are complex polymers. We have already discussed the chemical nature and properties of these materials in connection with their role in soil aggregate structure (Chapter 9).

Microorganisms have an unusual ability to degrade diverse and complex organic compounds, both natural and synthetic. With respect to molecules of natural origin, we can assume that the so-called "principle of microbial infallibility" applies (Alexander 1964). That is, we can assume that degradation of all such materials is possible, although, as we have seen in the case of humic materials, it may be very slow. This principle reflects a diverse and plastic biochemical capacity among microorganisms. In many cases, their metabolic mechanisms are adapted to operate in alternative pathways under different environmental conditions; in other cases, they adapt to specific substrates by **enzyme induction,** the process by which the production of enzymes appropriate to the breakdown of particular substrates is stimulated by the presence of the substrate. The capacity to produce such enzymes, of course, must exist in the genetic material of the organism, but it need not produce quantities of the enzyme unless the situation is appropriate.

The ability of soil microorganisms to degrade complex organic molecules extends to most synthetics. In the laboratory, at least some microorganisms have been found that can degrade even the most stable and persistent pesticides (Matsumura and Boush 1971). Nevertheless, some groups of pesticides are highly resistant to microbial degradation, and the extent to which such degradation takes place in nature is very uncertain. Among the insecticides, many of the organochlorines, dieldrin in particular, are strongly resistant to microbial attack, and the fungicide captan seems to be, too (Woodcock 1971). The particular features that confer such resistance are still poorly understood, but the presence of chlorine (or other halogens) in certain phenols or phenoxy groups, together with certain specific molecular linkages in complex compounds, seem to be consistently effective.

Pesticide chemicals are degraded in a very complicated fashion by processes that involve not only microbial processes but also physical and chemical reactions in the soil as well as photochemical reactions. Microbial degradation occurs largely by means of reactions that involve oxidation and hydrolysis, although there are others (Crosby 1973). It should be noted that some compounds are initially converted by microbial action to highly toxic and even more stable forms, for example, aldrin to dieldrin and heptachlor to heptachlor epoxide. Because of the complexity of the conversions and the complicated movements of pesticide chemicals through soil, air, and water, it is understandable that our knowledge of pesticide degradation in nature is very incomplete.

Decomposition processes are important in the nutrient dynamics of the soil ecosystem, because it is through these mechanisms that nutrients are mineralized, or freed from organic compounds, and made available for uptake by higher plants. A variety of other processes, which we can call **nutrient conversion processes,** and which are not directly involved with decomposition, are also important with respect to nutrients. Most of these are also caused by metabolic activities of the soil microbiota.

NUTRIENT CONVERSION PROCESSES

The nitrogen cycle (see Figure 2-9) is extraordinary in the degree to which chemical conversions are mediated by particular groups of microorganisms. Each of the groups involved derives some benefit from the particular conversion, although the particular benefit varies greatly (Table 13-7). We shall describe the more important of these conversions and consider their implications for agroecosystems.

Nitrogen fixation involves the reduction of nitrogen from molecular form N_2 to NH_3, in which form it can be incorporated into organic molecular structure. The capacity for biological nitrogen fixation is limited to certain members of the kingdom Monera, including various bacteria, actinomycetes, and blue-green algae. These organisms possess a specific enzyme,

nitrogenase, which mediates this conversion. This enzyme is inactivated by high oxygen concentrations, so the nitrogen fixation process must occur under anaerobic or low oxygen conditions, or under the influence of biochemical mechanisms that shield the enzyme from oxygen. Many nitrogen-fixing blue-green algae, for example, possess special cells, called **heterocysts,** which lack chlorophyll and in which nitrogen fixation occurs. Oxygen produced in high concentration by photosynthesis thus does not inhibit nitrogen fixation. While the reaction that combines hydrogen and nitrogen yields energy, the biological conversion process requires a large energy input and in this sense is highly inefficient. The reduction of one molecule of nitrogen (N_2) to two molecules of NH_3 requires 24 molecules of ATP (adenosine triphosphate, the cellular energy transfer compound). This requires the consumption of carbohydrate, with measured values indicating that about 6 grams of carbon are respired per gram of nitrogen fixed (Hardy and Havelka 1975).

The organisms capable of nitrogen fixation vary from free-living forms to those that form associative or obligatory symbioses with higher plants (Figure 13-8). Free-living nitrogen fixers include a variety of blue-green algae, photosynthetic bacteria, and heterotrophic bacteria. Among the blue-green algae, fixation has been demonstrated for quite a number of filamentous forms that bear heterocysts, along with a few other nonheterocystous and unicellular forms (Stewart 1973). The relative abundance of these nitrogen-fixing forms appears to be greater in lower latitudes, where, in samples taken from rice paddy environments, they can constitute almost 12 percent of all algae present (Watanabe and Yamamoto 1971). The contribution of fixed nitrogen to the soil or rice paddy environment by these organisms may be significant. In open field and meadow environments in Sweden, the annual contribution by these organisms was estimated to range from 4 to 51 kg N/ha

TABLE 13-7

Conversion processes, reaction patterns, energy input or output, and value to organism for various nitrogen conversions media organisms involved in the nitrogen cycle.

Group of organisms	Reaction type	Typical equation	Energy gain or loss (kcal/mole N)	Value to organism
Nitrogen-fixing monerans	Reduction of N_2	$2N + 3H_2 \rightarrow 2NH_3$	-147	NH_3 for synthesis of organic molecules
Nitrite bacteria (Chemosynthetic)	Oxidation of NH_3	$NH_3 + 1\frac{1}{2} O_2 \rightarrow HNO_2 + H_2O$	66	Energy yield for chemosynthesis
Nitrate bacteria (Chemosynthetic)	Oxidation of nitrite	$KNO_2 + \frac{1}{2} O_2 \rightarrow KNO_3$	17.5	Energy yield for chemosynthesis
Denitrifying bacteria (Anaerobic)	Reduction of nitrate	$5C_6H_{12}O_6 + 24KNO_3 \rightarrow$ $30CO_2 + 18H_2O + 24KOH + 12N_2$	570[a]	Nitrates act as hydrogen acceptor, permitting carbohydrate respiration[b]
Denitrifying bacteria (Chemosynthetic)	Reduction of nitrate	$5S + 6KNO_3 + 2CaCO_3 \rightarrow$ $3K_2SO_4 + 2CaSO_4 + 2CO_2 + 3N_2$	132[c]	Nitrate reduction permits sulfur oxidation with net energy release for chemosynthesis
Decomposer organisms	Ammonification of organic N	$CH_2NH_2COOH + 1\frac{1}{2} O_2 \rightarrow$ $2CO_2 + H_2O + NH_3$	176	Oxidative metabolism of organic molecules
All nonfixers of N	Nitrate reduction and amination	complex	variable cost	Production of NH_3 and its incorporation in organic molecules

[a]Per mole of glucose respired.
[b]In some organisms, N_2O is produced instead of N_2, with lower energy yield.
[c]Per mole of S.

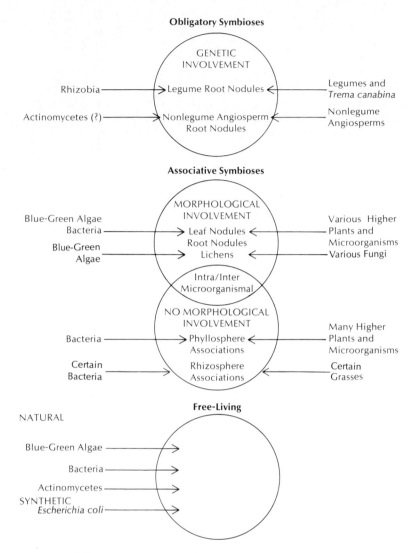

FIGURE 13-8.
Biological nitrogen-fixing relationships of free-living and symbiotic monerans.
(Modified from Hardy and Havelka. Copyright © 1974 by American Association
for the Advancement of Science.)

(Hendricksson 1971). In paddy soils of western Africa, algal fixation can contribute 15 to 20 kilograms of nitrogen per hectare per month to the soil (Rinaulde et al. 1971). Even in desert environments the algae present in surface soil crusts can fix a significant amount of nitrogen during periods when moisture is available. In the Great Basin deserts of the United States, this can equal as much as 10 to 100 kg N/ha/yr (Rychert and Skujins 1974).

Certain species of photosynthetic bacteria, but perhaps not all, also fix nitrogen, always under anaerobic conditions. The combination of anaerobic conditions and light creates a specialized situation that usually occurs in highly organic muds covered by shallow water; hence it is of little interest to agriculture.

Heterotrophic bacteria capable of nitrogen fixation include both aerobic and anaerobic forms, all of which must of course use energy from organic matter of some sort to carry out fixation. Although these forms may be free-living in the soil at large, their dependence on a source of organic matter frequently leads to a facultative association with the root or leaf surfaces of higher plants, where tissues or exudates appear to be a food source. Some of the aerobic forms appear only in tropical regions; others are widely distributed. The actual and potential contribution of fixed nitrogen by such forms is still incompletely evaluated, but it has been estimated at as much as 10 to 15 kg/ha/yr for soil forms (Dobereiner 1969) and 10 kg/ha/yr for phyllosphere species (Weaver et al. 1974).

A more complex form of symbiosis involves various blue-green algae, which themselves are capable of free life, with other plants in symbiotic relationships that involve some degree of morphological specialization. Certain lichens that consist of a species of fungus living in association with one or more species of algae, one of which is a blue-green alga, consequently show nitrogen-fixation ability. In addition, symbiotic associations exist between blue-green algae and certain liverworts, one genus of aquatic ferns, many cycads (where root nodules containing the algae are formed), and one large-leaved flowering plant of tropical regions, in which the algae live in swollen leaf bases.

The symbiotic relationships of greatest importance, however, are those in which bacteria or actinomycetes live in obligatory symbiosis with the roots of higher plants. Certain actinomycetes form root nodules with members of several genera of angiosperms, most of which are woody shrubs or trees. Many of the species involved are characteristic of early stages of ecological succession or else occupy sterile habitats such as sand dunes, where nutrient supplies would be expected to be low. In the natural ecosystems that contain these species, the inputs of fixed nitrogen are sometimes quite large. In the case of various species of alders, *Alnus* spp., the quantities fixed might exceed 100 kg/ha/yr and be very important to the growth of associated tree species (Tarrant and Trappe 1971). Certain other angiosperms bear leaf nodules that also seem to harbor nitrogen fixing organisms, possibly actinomycetes.

Bacteria of the genus *Rhizobium* are the most important to agriculture, however. These organisms induce the formation of root nodules on most species of legumes (Figure 13-9). Colonies of the bacteria, whose cells show a distinctively modified form, exist and carry out nitrogen fixation within the cells of these nodules (Figure 13-10). A legume crop commonly fixes 100 to 200 kg N/ha/yr, and values even higher have been recorded in several cases. Considerable interest surrounds the mechanism that generates the formation of nodules in members of this one plant family (although *Rhizobium* nodules have been found on at least one nonlegume species.)

Nitrogen fixation by *Rhizobium* bacteria living in contact with or close to cells of a number of nonlegumes in tissue culture has recently been demonstrated (Child 1975). Several workers have also succeeded in culturing certain *Rhizobium* strains on a defined medium (one containing known organic and inorganic compounds) and have observed fixation under these conditions (e.g., Pagan et al. 1975). This indicates that the genetic instructions for synthesizing the enzyme nitrogenase reside entirely in the bacterium, and that production and action of the enzyme might be triggered by one or more common, diffusable plant substances. Likewise, the specificity of *Rhizobium* for legumes may be related to secondary factors that influence the infection or nodulation of roots, and not to the fixation process itself.

FIGURE 13-9.
Soybean root system, showing nodules occupied by nitrogen-fixing bacteria of the genus *Rhizobium*. (Photograph by Murray Lemmon, courtesy of the USDA.)

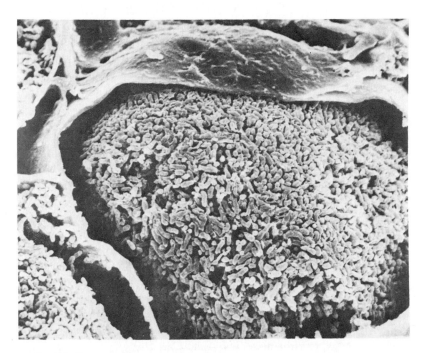

FIGURE 13-10.
Scanning electron micrograph of cells from the root nodule of a soybean plant, showing the mass of *Rhizobium* bacteria occupying these cells. Magnification about $5000\times$. (Photograph by P. Sihanonth and R. Todd, University of Georgia.)

Of the other nitrogen conversions mediated by microorganisms, nitrification and denitrification are the two most important groups. Nitrification results in the conversion of ammonia to nitrate and is mediated by two groups of autotrophic bacteria (Table 13-7). This conversion is considerably important, because nitrate is highly soluble and susceptible to being leached out of the soil system. It is interesting to note that, during secondary succession leading to a pine-oak climax in eastern Oklahoma, both these groups of bacteria were inhibited by chemicals that were released by higher plant members typical of later successional stages (Rice 1964; Rice and Pancholy 1974). A variety of tannins, tannin derivatives, and other organic compounds seem to be involved. The advantage of such mechanisms is obviously the retention of mineral nitrogen as ammonium ion, so that it can either be taken up by organisms or held by the cation exchange system in the soil.

When nitrate is present in any significant amount, and when conditions of low oxygen tension prevail, denitrification can occur, releasing nitrogen from the soil as N_2 or as nitrous oxide (N_2O). In certain cases, denitrification can be a major route by which inorganic nitrogen fertilizers are lost. In some of the more poorly drained clay soils of the Imperial Valley in California more than half the applied fertilizer may be lost in this way.

Soil microorganisms also seem to play an important role in the uptake of phosphorus, another important nutrient. Some workers have suggested that bacteria might promote phosphorus uptake by solubilizing phosphorus from fixed forms in the soil; and certain experiments in artificial culture also suggest that this is possible. In the soil, however, there is no good evidence for this action, and, if anything, rhizosphere bacteria seem to compete with roots for the phosphorus present in available form (Tinker and Sanders 1975).

Various fungi of the phylum Basidiomycota grow in close association, or symbiosis, with the roots of many plants; here the evidence is strong for a positive role in phosphorus nutrition of the higher plant. Two relatively distinct forms of this association, known as mycorrhiza, exist. **Ectotrophic mycorrhizae** are those in which the fungus forms a dense sheath over the surface of the smaller roots and sends some mycelia into the tissues of the root itself without penetrating the living cells. These mycorrhizae are typical in a number of groups of trees, including pines, oaks, and beeches. The relationship is especially prevalent in nutrient-poor soils and is often essential to the survival and growth of the tree species.

More widespread, however, are **endotrophic mycorrhizae,** in which the mycelia of the fungal member actually penetrate the living cells of the plant root. Many plants, including legumes, grasses, and other crop species, form mycorrhizae with fungi of the genus *Endogone.* These mycorrhizae, known because of their appearance as vesicular-arbuscular (VA), have existed since early in the evolution of land plants and are of major importance to the nutrition of the plants that possess them (Baylis 1972). In general, they appear to fulfill the function of root hairs in the acquisition of phosphorus, zinc, sulfur, and perhaps other mineral nutrients. LaRue et al. (1975) found that innoculating the soil in a peach seedling nursery with VA mycorrhizae led to faster growth and a higher zinc content of leaf tissues.

Endotrophic mycorrhizae can interact strongly with other microbial nutrient processes, as they do in legumes. Crush (1974), for example, has shown that innoculating legumes with VA mycorrhizal fungi, especially in phosphorus deficient soils, promotes the development of *Rhizobium* nodules in which nitrogen fixation occurs. Uninnoculated plants showed weak nodule development and slower growth. Yields for various plants, under laboratory conditions, have been increased as much as several hundred percent (Tinker and Sanders 1975); thus endotrophic mycorrhizae might prove to be of considerable importance to agriculture.

Stark (1971) has found that endotrophic mycorrhizae are abundant in association with the roots of most climax tree species in the Brazilian rain forest. Here, their importance might be in acquiring nutrient ions directly from decomposing litter and translocating them through the mycelial tube directly into the living cells of tree roots. This seems to be one of the important mechanisms that permit the development of a diverse and highly productive ecosystem in an environment that is very susceptible to nutrient outflow by leaching.

Microorganisms participate in the conversion of a number of other elements in the soil, most notably sulfur. Sulfur is taken up by higher plants mainly in the form of the sulfate ion SO_4^{2-}. It is reduced in the process of incorporating sulfur into organic compounds, however, and when decomposition occurs, this sulfur is released as hydrogen sulfide, H_2S. In soils, one genus of rodlike bacteria, *Thiobacillus,* oxidizes such reduced forms of sulfur to sulfate, in the process obtaining energy for chemosynthesis. Thus in most soils the action of these sulfur bacteria increases sulfur availability to higher plants. In highly organic soils, such as those encountered in newly drained swampland areas, this conversion, which also increases the acidity of the soil, can cause overacidity.

In poorly oxygenated soils a variety of other sulfur conversions can occur. At least one member of the genus *Thiobacillus* is capable of oxidizing sulfur under such conditions as long as nitrates are available. This reaction is actually one of denitrification, leading to the release of molecular nitrogen. Other bacteria can reduce sulfur under such conditions, using sulfate as a hydrogen acceptor for the oxidation of organic materials. The action of microorganisms in sulfur conversion is obviously important to

agriculture. Sulfur deficiencies are becoming an increasing concern in many areas of the world, and the beneficial action of aerobic sulfur-oxidizing bacteria should be recognized as one of the valuable contributions of the soil microflora.

CHEMICAL INTERACTIONS AMONG SOIL ORGANISMS

At the beginning of this chapter we noted that the soil environment favored the retention of many chemical substances. This allows biochemical interactions to take place among various macroscopic and microscopic members of the soil ecosystem. Scientists are only beginning to appreciate the diveristy and importance of such interactions.

One form of chemical interaction that involves higher plants, is **allelopathy,** the production of chemical substances by one species that inhibit the germination or growth of other species nearby (Whittaker 1970). Inhibiting substances can enter the soil as root exudates, as leached material from shoot tissues, or by movement through the air in volatile form. Most attention so far has focused on the influence these substances have on other higher plants, but as we noted earlier some of these materials have been shown to inhibit nitrification by soil bacteria.

Antibiosis refers to the inhibition of one microorganism by chemical substances released by another microorganism, and so in principle it is not really different from allelopathy among higher plants. Antibiosis appears to be very common among soil bacteria and fungi, although the number of antibiotic substances that have been isolated and identified is still rather small. The most powerful antibiotics, those derived for medical use, appear to be produced by soil fungi; but we should note that most of our knowledge about these materials comes from laboratory studies, and we are still very uncertain about the extent to which antibiotic relationships function in the soil environment.

Another phenomenon that is still not well understood is **fungistasis,** the maintenance of fungi in inactive state in the soil under conditions that generally favor the activity of other soil organisms. Examination of the soil shows that most of the spores, as well as a good many of the living mycelia, are metabolically inactive at any given moment. The mechanism of fungistasis seems to be general; that is, it affects many species at the same time. It also appears to be destroyed by the addition of nutrients to the soil and to require the presence of other soil microorganisms. Beyond this, however, we know very little about the precise mechanisms involved. Fungal dormancy may be due in part to competition from organisms such as bacteria for inorganic nutrients or for other, possibly organic, growth stimulators. Chemical inhibitors may also be involved. Watson and Ford (1972) have recently concluded that fungistasis is most likely a complex phenomenon that reflects the balance of stimulatory and inhibiting substances of varied origin, abiotic as well as biotic.

Antibiotic action by root-symbiotic soil organisms might also protect plants against root pathogens. Several species of fungi that form mycorrhizae also produce antibiotics, and in a number of cases this appears to permit higher plants to survive in soils infected by serious pathogens.

BIOTIC INTERACTIONS
AND DISEASE RELATIONSHIPS

Soil organisms exhibit food chain interactions that are at least as diverse as those of the aboveground biota. Most of these contribute to basic patterns of soil ecosystem function and are beneficial to agricultural interests, but some generate diseases in certain crop species.

The major groups of soil organisms that include disease agents are the bacteria, actinomycetes, fungi, and nematodes. Crown gall in apples is caused by the bacterium *Agrobacterium tumefaciens;* potato scab, by an actinomycete, *Streptomyces scabies*. By far the greatest number of plant diseases are caused by fungi, however. Fungal diseases vary from nonspecific root-system infections caused by species of fungi capable of attacking many different crop plants to highly specific relationships between particular fungal species or strains and particular crops. The unspecialized diseases frequently attack germinating seeds and seedlings. Dampness for example, can favor the spread, at or above the ground surface, of fungi that attack the hypocotyls of germinating seeds; this condition is called **damping off.** Other diseases, frequently termed **seedling blights** are caused by nonspecific fungi that infect developing root systems. In older plants, various **decline diseases** are caused by nonspecific fungal agents.

The susceptibility of plants to such infections is the result of many factors, including the conditions of the physical environment, the quantity of fungal innoculum present, the age and health of the plant, and the influence of the growing root upon spores and inactive mycelia in the soil. Exudates from plant roots appear to be important in stimulating fungal activity. The state of the fungus in the soil, specifically its abundance and metabolic condition, can be considered to define an **innoculum potential** that is balanced against the resistance of the host plant. This resistance in turn depends on the nutrition of the plant: Important deficiencies make plants more susceptible. In addition, young plants that are not yet capable of producing antifungal agents are likely to be more susceptible to infection.

The more specialized fungal agents include forms that enter and then spread throughout the plant, as in the case of vascular wilts that spread through the xylem tissues, blocking their normal function in upward transport of water and nutrients. Other specialized forms produce dense "ectotrophic" mycelial coverings over root surfaces; these resemble ectotrophic mycorrhizae but lack their beneficial influences. In fact, mycorrhizal and other symbiotic relationships have probably evolved out of relationships of this sort. As a general rule, evolution tends to proceed toward the conversion of detrimental interactions into relationships that are more mutually beneficial.

Many species of nematodes are also root parasites; these organisms feed primarily on the cytoplasm of root cells. One important group, the **root knot nematodes,** stimulate the formation of gall-like structures on the roots (Figure 13-11), which makes their presence relatively easy to diagnose. Other groups of nematodes cause generalized damage and deterioration of

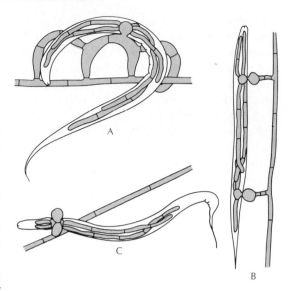

FIGURE 13-12.
Nematode trapping systems of predatory *Hyphomycetes* fungi: (A) adhesive knobs and networks of *Arthrobotrys oligospora*; (B) adhesive knobs of *Dactylella ellipsospora*; (C) constricting ring of *Dactyllela bembicoides*. (From Duddington 1963.)

FIGURE 13-11.
Root system of a tomato plant heavily infected by root knot nematodes. (Photograph courtesy of the USDA.)

root systems resembling that caused by bacterial or fungal agents.

Not all fungi, nor all nematodes, are detrimental to agriculture, however. As we noted earlier, many nematodes feed upon fungi and bacteria. Likewise, in what constitutes one of the more remarkable food chain relationships in the soil ecosystem, some fungi are predators of nematodes. The mechanism of nematode capture varies with the fungal species, but in some cases consists of a knoblike cell or a series of cellular loops that have an adhesive surface substance (Figure 13-12). Other species have cells that form a ring, which in some cases constricts when a nematode enters, the constriction being caused by a rapid increase in cell size that simultaneously reduces the loop opening. Once a nematode is caught, hyphae of the fungus grow into the body of the animal and digest it (Duddington 1963).

SUMMARY

The soil, as well as the aboveground surface of plants, bears a complex and important microbiota consisting of bacteria, actinomycetes, several groups of algae, fungi, protozoa, and members of several phyla of metazoan animals. Bacteria predominate in the aboveground plant surface zone, or phyllosphere. In the soil, especially in the root surface zone, or rhizosphere, many of these organisms are abundant and interact in complex and influential ways.

Although the biomass of many of these groups is quite small, especially the biomass of metabolically active tissue, the small size of the individual organisms is reflected in a very high metabolism per unit biomass. This is particularly true for bacteria, for which one gram can equal in metabolic effect more than a kilogram of tissue of higher plants. This means that, in almost all terrestrial ecosystems, the respiratory metabolism of the soil microbiota exceeds that of the aboveground organisms and higher plant root systems.

Litter breakdown and decomposition are two of the major processes mediated by the soil biota. The larger invertebrates participate primarily in the mechanical breakdown of the materials involved; the microflora carries out the ultimate decomposition, breaking organic materials down and leaving certain highly resistant residues, simultaneously resynthesizing materials into microbial tissues. These eventually die and are converted into secondary residues. Both primary and secondary residues constitute humus.

Microorganisms are also important in processes of nutrient conversion, especially of nitrogen. Fixation of molecular nitrogen is carried out by various members of the kingdom Monera—bacteria, actinomycetes, and blue-green algae. This fixation involves the activities of free-living as well as symbiotic organisms and is of the greatest importance to agriculture.

The soil biota is involved in complex patterns of chemical and biotic interaction, most of which are only beginning to be identified, but which appear in many cases to influence the vulnerability of higher plants to disease.

Literature Cited

Alexander, M. 1961. *Introduction to Soil Microbiology.* John Wiley and Sons, New York.
——. 1964. Biochemical ecology of soil microorganisms. *Ann. Rev. Microbiol.* 18:217–252.
Allison, F. E. 1973. *Soil Organic Matter and Its Role in Crop Production.* New York: Elsevier Scientific Publishing Co.
Baylis, G. T. S. 1972. Fungi, phosphorus, and the evolution of root systems. *Search* 3:257–258.
Burns, R. C., and R. W. F. Hardy. 1974. *Nitrogen Fixation in Bacteria and Higher Plants.* New York: Springer-Verlag.
Child, J. J. 1975. Nitrogen fixation by a *Rhizobium* sp. in association with non-leguminous plant cell cultures. *Nature* 253:350–351.
Clark, F. E., and E. A. Paul. 1970. The microflora of grassland. *Adv. in Agron.* 22:375–435.
Crosby, D. G. 1973. The fate of pesticides in the environment. *Ann. Rev. Plant Physiol.* 24: 467–492.
Crush, J. R. 1974. Plant growth responses to vesicular-arbuscular mycorrhiza. VII. Growth and nodulation of some herbage legumes. *New Phytol.* 73:743–749.

Dobereiner, J. 1969. Nonsymbiotic nitrogen fixation in tropical soils. In *Biology and Ecology of Nitrogen*. pp. 114–128. Washington, D.C.: National Academy of Sciences.

Duddington, C. L. 1963. *Predacious Fungi and Soil Nematodes*. In *Soil Organisms*. J. Doeksen and J. van der Drift. pp. 298–303. Amsterdam: North Holland Pub. Co.

Felbeck, F. T., Jr. 1971. Chemical and biological characterization of humic matter. In *Soil Biochemistry*, vol. 2. A. D. McLaren and J. Skujins (eds.). pp. 36–59. New York: Marcel Dekker.

Garrett, S. D. 1970. *Pathogenic root-infecting fungi*. Cambridge: Cambridge University Press.

Hardy, R. W. F., and U. D. Havelka. 1975. Nitrogen fixation research: a key to world food? *Science* 188:633–643.

Hendriksson, E. 1971. Algal nitrogen fixation in temperate regions. *Plant and Soil* (special vol.):415–419.

LaRue, J. H., W. D. McClellan, and W. N. L. Peacock. 1975. Mycorrhizal fungi and peach nursery nutrition. *Calif. Agr.* 29(5):6–7.

Lund, J. W. G. 1967. Soil algae. In *Soil Biology*. A. Burges and F. Raw. pp. 129–147. New York: Academic Press.

Macfadyen, A. 1963a. The contribution of the fauna to the total soil metabolism. In *Predaceous Fungi and Soil Nematodes*. J. Doeksen and J. van der Drift. pp. 3–16. Amsterdam: North Holland Pub. Co.

———. 1963b. *Animal Ecology: Aims and Methods*. New York: Pitman Pub. Co.

Matsumura, F., and G. M. Boush. 1971. Metabolism of insecticides by microorganisms. In *Soil Biochemistry*, vol. 2. A. D. McLaren and J. Skujins (eds.). pp. 320–336. New York: Marcel Dekker.

Nielsen, C. O. 1967. Nematoda. In *Soil Biology*. A. Burges and F. Raw. pp. 197–211. New York: Academic Press.

O'Connor, F. B. 1967. The enchytraeidae. In *Soil Biology*. A. Burges and F. Raw. pp. 213–257. New York: Academic Press.

Oostenbrink, M. 1971. Nematodes. In *Methods of Study in Quantitative Soil Ecology: Population, Production, and Energy Flow*. J. Phillipson. pp. 72–82. Oxford: Blackwell Scientific Pub.

Pagan, J. D., J. J. Child, W. R. Showcroft, and A. H. Gibson. 1975. Nitrogen fixation by *Rhizobium* on a defined medium. *Nature* 256:406–407.

Rice, E. R. 1964. Inhibition of nitrogen fixing and nitrifying bacteria by seed plants. *Ecology* 45:824–837.

Rice, E. R., and S. K. Pancholy. 1974. Inhibition of nitrification by climax ecosystems. III. Inhibitors other than tannins. *Amer. J. Bot.* 61:1095–1103.

Rinaudo, G., J. Balandreau, and Y. Dommergues. 1971. Algal and bacterial non-symbiotic nitrogen fixation in paddy soils. *Plant and Soil* (special vol.):471–479.

Rovina, A. D., and B. M. McDougall. 1967. Microbiological and biochemical aspects of the rhizosphere. In *Soil Biochemistry*. A. D. McLaren and G. H. Peterson (eds.). pp. 417–463. New York: Marcel Dekker.

Rychert, R. C., and J. Skujins. 1974. Nitrogen fixation by blue-green algae—lichen crusts in the Great Basin Desert. *Soil Sci. Soc. Amer. Proc.* 38:768–771.

Shtina, E. A. 1974. The principal directions of experimental investigations in soil algology with emphasis on the USSR. *Geoderma* 12:151–156.

Stark, N. M. 1971. *Mycorrhizae and Nutrient Cycling in the Tropics*. pp. 228–229. Misc. Publ. 1189, USDA Forest Service.

Stewart, W. D. P. 1973. Nitrogen fixation by photosynthetic microorganisms. *Ann. Rev. Microbiol.* 27:283–316.

Stout, J. D., and O. W. Heal. 1967. Protozoa. In *Soil Biology*. A. Burges and F. Raw. pp. 149–195. New York: Academic Press.

Tarrant, R. F., and J. M. Trappe. 1971. The role of *Alnus* in improving the forest environment. *Plant and Soil* (special vol.):335–348.

Tinker, P. B. H., and F. E. Sanders. 1975. Rhizosphere microorganisms and plant nutrition. *Soil Sci.* 119:363–368.

Watanabe, A., and Y. Yamamoto. 1971. Algal nitrogen fixation in the tropics. *Plant and Soil* (special vol.):403–413.

Watson, A. G., and E. J. Ford. 1972. Soil fungistasis—a reappraisal. *Ann. Rev. Phytopath.* 10:327–348.

Weaver, T., F. Forcella, and G. Strobel. 1974. *Phyllosphere nitrogen fixation in vegetation of the northern Rocky Mountains: a preliminary study*. Tech. Rept. No. 253, Grassland Biome, U.S. Int. Biol. Prog.

Whittaker, R. H. 1969. New concepts of kingdoms of organisms. *Science* 163:150–160.

——. 1970. The biochemical ecology of higher plants. In *Chemical Ecology*. E. Sondheimer and J. B. Simeone. pp. 43–70. New York: Academic Press.

Witkamp, M. 1971. Soils as components of ecosystems. *Ann. Rev. Ecol. Syst.* 2:85–110.

Woodcock, D. 1971. Metabolism of fungicides and nematocides in soils. In *Soil Biochemistry*, vol. 2. A. D. McLaren and J. Skujins (eds.). pp. 337–360. New York: Marcel Dekker.

14

MAINTENANCE OF THE SOIL ECOSYSTEM

The origin of agriculture set the stage for the development of urban society and for the eventual rise of all the world's great civilizations. Throughout most of civilized history, cities and major regional populations have grown up in direct response to the patterns by which locally available natural resources have been exploited. Only in the past few hundred years, particularly since the industrial revolution spurred advances in production and transportation, has it been possible to build cities whose economies rest on government, the military, or tourism. Regardless of the forces responsible for a city's origin, however, food must somehow be supplied, and through most of human history, urban areas have depended on local sources of food.

Many major cities and some entire civilizations have declined and disappeared. Wars and conquests have often destroyed cities and disrupted society. But where a productive natural resource base has continued to exist, the reconstruction of both cities and societies has been almost inevitable. As Dale and Carter (1955) have argued, the decline and disappearance of urban civilizations is due mainly to deterioration of the natural resource base

on which they depended (Figure 14-1). Most historians attribute such declines to wars and conquests, but these are simply dramatic events in the longer process of decline. Historical analyses have been inadequate through their failure to search for and consider the relationship between human societies and their natural resource environment.

No natural resource has been more important in determining the rise, duration, and fall of human civilizations than the quantity and quality of agricultural land. No relationship has been more critical to land quality than the impact of cultivation techniques upon arable and potentially arable land, whether the productivity has been enhanced by irrigation and protective management or depleted by nutrient removal, salination, or erosion (Hyams 1952; Dale and Carter 1955). Deterioration of agricultural soils has been the principal factor in the decline and disappearance of major civilizations in a number of world regions, especially in the dry lands that straddle the subtropical deserts. In some cases, such as in the Euphrates, Tigris, and Indus valleys of the Middle East, much of the damage involved the salination of poorly drained soils and the siltation of irriga-

FIGURE 14-1.
Roman ruins backed by barren, eroded hillsides at Djemela, Algeria. The decline of many
of the Roman cities of North Africa has been attributed to destructive land use.
(Photograph by P. Pittet, courtesy of the FAO, Rome.)

tion systems. It may be possible to reclaim such lands by modern technology, but the costs will certainly be great. In upland areas, where most of the land was originally covered by woodland, forest, or other natural vegetation that provided erosion protection, the clearing and cultivation of the land has been followed by massive erosion that has made it impossible to reclaim many areas. Parts of North Africa, Spain, southern France, Italy, Greece, Jordan, Israel, Turkey, Syria, Iraq, and Pakistan have been rendered virtually useless by erosion through improper cultivation and over-grazing. This process is taking place now in many parts of Central and South America, sub-Saharan Africa, and eastern Asia (Figure 14-2).

Still other fragile lands have been severely damaged by cultivation; for example, the dry loess plains of China have been so eroded in many areas that it is doubtful whether their reclamation would be practical. The history of cultivation in mountainous regions, with few exceptions, is full of examples of serious erosion.

We know little about the history of the humid tropics; civilizations such as the Maya of southern Mexico and Central America have disappeared for reasons we do not know. It would not be surprising, however, if the decline of these civilizations had been triggered by the deterioration of soils in the humid tropical environments that surrounded them.

Now, perhaps more than ever in history, maintaining the fertility and condition of cultivated and grazed soils is essential to human food supply. Unfortunately, this need is poorly sensed. Since World War II, we have by and

FIGURE 14-2.
The barrenness of these hills in the southern part of the Atacama Desert near Copiapo, Chile, is at least in part the result of intensive grazing by goats. The hillsides are criss-crossed by innumerable animal trails. (Photography by G. W. Cox.)

large escaped having to come to grips with soil protection. We have met our expanding food needs through an increasingly energy-intensive approach to production on limited areas of prime agricultural land, much of which has been made arable through irrigation. The increasing costs of these methods and the exploding demand for food mean that intensive production will tend to spread into less suitable land areas, and that costs are likely to be cut by reducing protective activities on much of the area now farmed intensively. The maintenance of soil ecosystems is therefore essential to the sustained exploitation of both cultivated cropland and grazing land.

CULTIVATED CROPLAND

Treating Symptoms of Water Erosion

The potential soil loss from an area by water erosion is predictable by relatively straight-forward means. The **water erosion equation** (Beasley 1972) is:

$$A = (R)(K)(L)(S)(C)(P)$$

where

A = soil loss in tons per acre
R = rainfall influence factor
K = soil erodibility factor
L = slope length factor
S = slope gradient factor
C = cropping practice factor
P = erosion control factor

Actual values can be determined for each of these variables, and the potential loss of soil can be calculated for a single storm, with particular rainfall characteristics, or an entire season, with particular overall rainfall pattern. Here, however, we are interested only in understanding the importance of these variables and in considering how certain ones can be influenced.

Rainfall and slope gradient are beyond the control of the farmer, although he can choose whether or not to cultivate a slope of given steepness. The rest of the variables in the righthand portion of the equation can be influenced. Soil erodibility, K, varies considerably for soils of different texture, tending to be quite low for porous, sandy soils and high for certain silt loams and clay loams. Well developed crumb structure, however, which can be achieved through careful soil management, promotes infiltration, thus minimizing surface runoff and protecting against erosion under conditions of light or moderate rainfall. When such soils are exposed to heavy storms, however, the loose, granular structure can make them more vulnerable to erosion than a soil of poorer structure.

Slope length, L, affects erodibility because of the greater volume of runoff that reaches the lower portions of longer slopes. This of course can be modified by the design of cultivated fields and by the construction of artificial drainage channels that collect accumulated runoff from various sections of a slope, instead of allowing it to pass undiverted to the bottom of the slope over field surfaces.

Erosion control practices, P, obviously represent one of the major ways in which the equation can be modified. **Contour plowing and planting** is perhaps the most widely used practice of this type, and it is nothing more than plowing, cultivating, and planting crop rows on the contour, so that the overall ridging pattern of the surface acts as a set of barriers to the downslope surface movement of water and thus promotes infiltration. **Contour listing** can increase this effect. Listing—that is, plowing so that adjacent plow cuts are turned toward each other—creates a strong ridge and furrow system. However, contour cultivation reaches its maximum effectiveness on slopes of only about 3 to 7 percent. Furthermore, protection is given only when rainfall is light or moderate. On steeper slopes, and with heavier rains, surface runoff breaks through contour ridges and tends to lead quickly to gully erosion. In such cases, contoured fields might erode even more than those that are not contoured. This breakthrough tendency can be mitigated somewhat by the use of a **graded furrow** system of cultivation and planting (Richardson 1973), in which furrows are created nearly on contour, but with a slight gradient that tends to divert surface runoff toward the field edge where it can be discharged into protected channels.

Terraces tend to be more effective in preventing erosion and are adaptable to steeper slopes. In fact, in some parts of the world terraced fields, usually small and supported by handmade rock walls (Figure 14-3), have permitted both dry-land and irrigated agriculture to be practiced successfully on very steep, mountainous terrain (Figure 14-4). Terraces used in mechanized agriculture for cultivated land or pastures are of a somewhat different type (Figure 14-5). In all cases, terraces involve the construction of widely separated ridges that act as barriers to the downslope movement of water and soil. These barriers are either level, where rainfall is small enough to allow complete infiltration into the soil, or graded slightly to permit gradual surface flow along the terrace channel to one end, where it can be discharged into a protected channel. Some terrace systems, such as the **broad-base terrace,** allow cultivation over the entire land surface (Figure 14-6); others, such as **steep backslope terraces,** require that a sod-covered, relatively steep bank be left on the steep downslope side of the terrace ridge (Figure 14-7).

Maintaining terraces can be a problem. Cultivation often tends to create a gradual flattening of the ridges and filling of the channels, as does erosion in the area between terraces. In some cases, however, plowing to turn each furrow uphill is adequate to maintain terraces and offset erosive movements between terrace ridges. Unless terraces are the same distance apart, an arrangement that is usually difficult to engineer, problems in cultivation can arise. If the distances between terraces vary, cultivating the widest section requires turning machinery between terrace ridges, and the plowing of short rows in the wider-than-average area.

360

FIGURE 14-3.
Terracing has proved an effective means of soil protection throughout the world. In Lebanon, some of these steep terraced slopes, here with vineyards, have probably been farmed for 3000 years or more. (Photograph by W. C. Lowdermilk, courtesy of the USDA Soil Conservation Service.)

FIGURE 14-4.
Terracing, combined with skillful irrigation, has permitted intensive farming of the steep volcanic hillsides of the Canary Islands. Bananas are being grown here, and irrigation is from small reservoirs like the one in the foreground. (Photograph by G. W. Cox.)

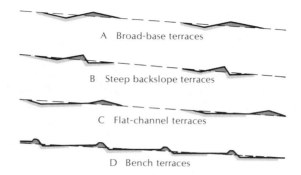

FIGURE 14-5.
A variety of terrace designs are compatible with the cultivation requirements of mechanized agriculture.
(Reprinted by permission from *Erosion and Sediment Pollution Control* by R. P. Beasley. Copyright © 1972 by Iowa State University Press, Ames, Iowa 50010.)

FIGURE 14-6.
Broad-base terraces like those on this Iowa farm permit cultivation of the entire land surface area. (Photograph courtesy of the USDA.)

FIGURE 14-7.
Steep backslope terraces require that a sodded bank be left on the downslope side of the terrace ridge, as on this farm near Silver City, Iowa. (Photograph by Lynn Betts, courtesy of the USDA Soil Conservation Service.)

The effectiveness of contouring and terracing can be illustrated by the results of an eight-year study in southeastern Iowa (Figure 14-8). In this study water yields and erosion losses due to gully and sheet-rill erosion were measured in a series of watersheds that ranged in area from 30.4 to 157.4 hectares (Spomer et al. 1973). The areas involved received about 81 centimeters of rainfall annually and varied in slope up to 18 percent. The contour cultivated corn fields were clearly suffering serious erosion, a significant fraction of which involved gullying. Soil losses from the terraced corn areas, on the other hand, were quite small, about the same as losses from grassed watersheds.

This study also demonstrates that cropping practice, C, influences erosion. There are many specific types of cropping practices. One of the most common is **contour strip cropping**, in

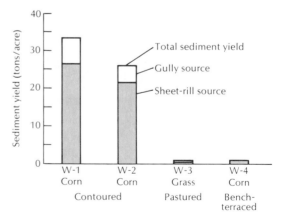

FIGURE 14-8.
Average annual sediment yields for contour-plowed corn, bench-terraced corn, and grass pasture in Iowa. (Modified from Spomer et al. 1973.)

FIGURE 14-9.
Contour strip-cropping of corn, a crop with a high erosion risk, and alfalfa, a crop with
low erosion risk, near West Concord, Minnesota. (Photograph by E. Cole, courtesy of the
USDA Soil Conservation Service.)

which a crop susceptible to relatively high erosion is alternated, in strips that follow the land contour, with one for which the erosion risk is low (Figure 14-9). The width of the strips varies with land slope, being narrower for steeper slopes. The reduction in observed erosion in this case is due largely to the fact that the effective slope length (L in the water erosion equation) becomes the width of the strip occupied by the more erodible crop, and to the fact that much of the material eroded from the vulnerable crop is deposited in the resistant strip.

Rotation of crops, another means of reducing the danger of soil erosion, usually integrates a high erosion risk species, such as corn or sorghum, with small grains and legumes, or it alternates row crops and grassland. Much of its

success is due to the beneficial influences of the small grains, grass, and legumes on soil aggregate structure and porosity. The benefits of such rotations for structure and resistance to erosion can last several years after rotations are abandoned (Adams 1974), but a decline in organic content and soil aggregate stability gradually occurs during such periods.

The planting of cover crops and the use of mulches that consist of crop residues, manures, or other organic wastes, help to prevent erosion by protecting the soil surface against the beating action of raindrops, thus maintaining a porous surface condition and retarding the surface flow of water. Crop residue mulches consist either of the rooted basal portions of a previous crop, or of loose fragments lying on the surface. The protective effect is generally greatest when the organic matter remains on the surface instead of being worked into the soil. Cover crops, such as annual rye or winter vetch, also provide effective protection against erosion, generally within two weeks after planting (Pimentel et al. 1976). These plants also trap nutrients against loss by runoff or leaching, and they return organic matter to the soil when they are plowed under. A cover crop planted in the fall means, however, that the field cannot be cultivated in the fall in preparation for planting in the early spring.

Treating Wind Erosion Symptoms

Wind erosion can be described by an equation similar to that for water erosion, using somewhat different variables. Five major fac-

tors determine the average annual loss of soil (Woodruff et al. 1972): the relationship between kind of soil and soil moisture, soil aggregate structure, soil surface roughness, field length, and vegetative cover. Soils with good aggregate structure are quite resistant to wind erosion, and as we have seen this can be improved by deliberate practice. Wind conditions are beyond the farmer's control, and of course the erosive action of wind is greater when soils are dry. Thus, where irrigation is possible it can be used to prevent wind erosion at critical times. Moreover, the rougher the field surface, the less danger there is of wind erosion. A cloddy surface can inhibit erosion in otherwise favorable conditions, and plowing a soil when a blowing threat is approaching can prevent serious erosion (Figure 14-10). Such a procedure is of only temporary value in most cases

FIGURE 14-10.
Plowing to increase the surface roughness of a field can serve as an emergency treatment Severe gully erosion in California due to heavy grazing and the formation of trails by cattle on highly erodable soils. (Photograph courtesy of the USDA Soil Conservation Service.)

FIGURE 14-11.
Windbreaks in the rich Lake Agassiz bottomlands of the Red River near Grand Forks, North Dakota. Note that the fields are oriented crosswise to the direction of prevailing winds, as indicated by windbreak locations. (Photograph by B. C. McLean, courtesy of the USDA Soil Conservation Service.)

because the exposed clods eventually dry and disintegrate.

The length of a field also influences wind erosion risk. What is important is the length of the unsheltered field in a direction parallel to the wind direction. This can be dealt with initially by field layout so that the long axis is crosswise to prevailing winds, and windbreaks can be planted at the edges of, or within, exposed fields (Figure 14-11). The effectiveness of windbreak vegetation depends on a number of factors, including the height, thickness, and porosity of the barrier planting. Various types of windbreaks can be characterized in terms of the relationship between the height of the windbreak and the downwind distance to which it provides shelter. For example, a windbreak that consists of a single row of Osage orange, a tree widely used for this purpose in the southern Great Plains, protects a distance about 12 times its height when a 65 kph wind is blowing above it. Thus a 10-meter high windbreak would protect a field 120 meters wide. Within fields, narrow strip plantings of annuals can also give important protection.

Finally, covering the soil with living plants, crop stubble, plant debris, or other organic material can give considerable protection against wind erosion as it can against water erosion (Figure 14-12). The amounts of such materials required to protect the soil under various conditions can be determined quite accurately from tables that take into account soil texture, climate, and type of mulch or cover. For organic

FIGURE 14-12.
Cultivation with a spiked-tooth cultivator leaves most of the crop stubble at the surface, which reduces the danger of wind erosion and promotes infiltration of rain and melting snow. This location is near Hillsboro, eastern North Dakota. (Photograph by E. W. Cole, courtesy of the USDA Soil Conservation Service.)

mulches, of course, the fact that decomposition constantly reduces the amount present must also be recognized.

Tillage Systems

Although there are a great many different tillage sequences, the basic pattern for conventional practice is (1) an initial deep plowing that loosens and turns over the soil, burying old crop residues and other materials; (2) a secondary tilling for the preparation of a fine seed bed; and (3) one or more pre-emergence or post-emergence cultivations or herbicide treatments to eliminate weeds. In Chapter 11 we examined in detail the impact of conventional mechanized practices on the soil ecosystem. These practices have both problems and limitations, and erosion and compaction are among the major ones in conventional tillage systems. In addition, weather conditions sometimes prevent tilling from being accomplished at the correct time, and for various reasons heavy, shallow, stony, or peaty soils cannot be cultivated successfully by these means. Finally, of course, conventional practices require several machine operations, each with its fuel costs.

Interest has therefore arisen in a variety of reduced tillage systems. **Zero tillage** (sometimes termed "direct drilling") limits the mechanical disturbance of the soil to the preparation of a very restricted seedbed, which is done at the same time that seeds are planted (see Figure 24-18). Fertilizers and herbicides can also be applied at this time, which greatly reduces soil disturbance and the number of vehicle passages over the soil surface. This practice, however, relies more heavily than others on chemical herbicides and on consistently satisfactory performance of weed control techniques that replace the control formerly achieved by plowing and mechanical cultivation (Figure 14-13). Before planting, any vegetation present must be killed with a broad-spectrum herbicide, the effects of which are nonpersistent; after planting, more specific and more persistent herbicides are usually required to control specific weeds peculiar to the crop situation. Savings in fuel costs and erosion control practices are therefore partially offset by increased weed control costs.

Studies of zero tillage have shown that it leads to significant changes in the physical and biotic characteristics of the soil environment. Most studies have shown that the soil becomes somewhat more compacted, primarily because the number of larger pore spaces in the soil is reduced and the number of smaller spaces is simultaneously increased. This reduces aeration somewhat, but tends to increase the water-holding capacity of the soil. Zero-tilled soils tend to be cooler than others, partly because a surface layer of plant residues is present. Several studies suggest that the rates at which organic compounds in the soil are mineralized may be decreased and the level of soil organic matter raised. Where earthworms exist, they usually become more abundant. Earthworm burrows, as well as root channels, remain undisturbed and serve as effective channels of water infiltration in spite of the increased density of the soil. Because nutrients are released mostly at the soil surface by the decomposition of organic matter or by the application of fertilizers, they tend to

FIGURE 14-13.
In this Illinois field, corn was planted directly in wheat stubble without tillage. Chemical herbicide was applied at the same time to kill weeds. (Photograph by E. B. Trovillion, courtesy of the USDA Soil Conservation Service.)

develop higher concentrations near the surface and lower concentrations in deeper layers. It is possible that some nutrients, especially phosphorus, are retained effectively in more readily available form under zero tillage than under other systems of cultivation (Moschler et al. 1975).

Zero tillage has been used most successfully for corn cultivation, and it appears to be especially useful where erosion is a serious threat or where increased water infiltration is of particular benefit to the crop. It has proven very effective in reducing erosion, and in this regard its use in corn cultivation seems to be successful in humid tropical areas as well as in temperate regions (Lal 1974). It is potentially valuable to a few other crops, including certain cereals and fodder crops and perhaps cotton. Nevertheless, zero tillage has not proven as successful as first imagined. Little is actually known about a number of important impacts it might have on soil ecology, particularly the effects of intensive herbicide usage on components of the soil biota. In addition, weed control is often inadequate because of herbicide application problems and the existence of weed species that are very difficult to control with currently available herbicides. Like other aspects of energy-intensive technology, chemical herbicides are subject to sudden price changes, and there is also the risk that the detection of harmful health or environmental effects will lead to their withdrawal from use. Thus cultivation systems that rely on herbicides might quickly change from being practical and profitable to just the opposite. Furthermore, like other agricultural pests, weeds could develop resistances to various herbicides, a phenomenon that has not yet become serious. Nevertheless, the incidence of persistent weeds is likely to increase rather than decrease.

Zero tillage and other modifications of conventional practice deserve careful study and are likely to prove valuable in certain situations.

However, zero tillage does not seem likely to be well suited to all situations, and its great dependence on herbicide use raises serious ecological questions.

Fertilizer Efficiency

The traditional measure of a fertilizer's efficiency is the percentage of applied chemical fertilizer that is recovered in the crop harvested at the end of the season in which it was applied. Under good practice, this percentage is about 50 to 60 percent for nitrogen, 5 to 25 percent for phosphorus, and 40 to 70 percent for potassium. We have discussed the fate of various portions of applied fertilizers in Chapter 12. The main problems associated with nitrogen are losses by leaching of nitrates and the microbial conversion of nitrate to gaseous molecular nitrogen, sometimes combined with the tie-up of nitrogen in microbial tissues. For phosphorus, the principal problem is one of chemical fixation in compounds not readily available to plants.

The basic problem in the use of fertilizers is that the chemical form of the nutrient that is available to the plant is also one that is subject to other conversions or transport processes, while logistical considerations demand that enough nutrient for a major portion of the growing season must be applied in a single operation. Fertilizer application procedures must be devised that make nutrients available over a prolonged period without simultaneously triggering major losses such as those noted above. Two approaches have been taken to do this: the development of **slow-release fertilizers** and the inhibition of mechanisms that render applied fertilizers unavailable to plants.

Several means are used to achieve the slow release of fertilizer substances: fertilizer pellets are coated with materials that delay the release of the internal chemical; nutrient elements are

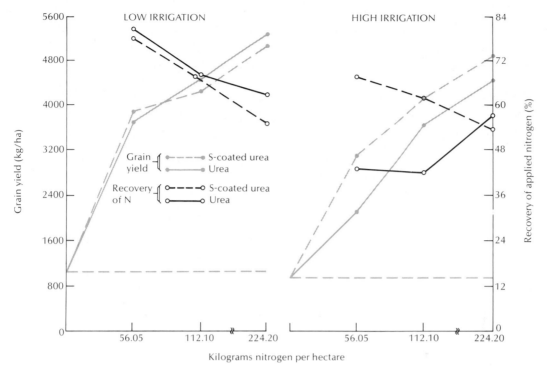

FIGURE 14-14.
Grain yield and nitrogen fertilizer recovery (i.e., the percentage of the applied amount recovered in grain) for plain and S-coated urea applications under high and low irrigation. (From Prasad et al. 1971.)

incorporated in inorganic compounds of low solubility; and nutrients are combined with organic materials so that they are released gradually by microbial decomposition. Coated fertilizer pellets have been used most successfully for nitrogen fertilizers, although tests of similar preparations for phosphorus and potassium suggest that similar preparations might be developed for them (Hauck and Koshino 1971). The use of **sulfur-coated urea** pellets has been particularly successful. In one set of experiments, single applications of S-coated urea proved equal in effect and percent recovery to three split applications of uncoated urea, when irrigation and leaching effects were low (Prasad et al. 1971). When irrigation was heavy, how-

ever, both effect and percent recovery were greater for the coated urea fertilizer (Figure 14-14). Sulfur coating might be even more valuable in rice paddies, where chemical conditions can inhibit virtually all urea release except in the vicinity of rice plant roots. Sulfur itself is in many cases a valuable secondary fertilizer, and so this approach seems both sound and practical.

Compounds such as magnesium ammonium phosphate ($MgNH_4PO_4$) or magnesium potassium phosphate ($MgKPO_4$) are examples of nutrients in relatively insoluble inorganic form. Their low solubility leads to their very gradual depletion as ammonium and phosphate ions are withdrawn from the soil solution, and withdrawals are replaced by dissolution of the fertilizer substances. Ureaforms, on the other hand, are organic compounds formed by reactions between urea and formaldehyde. Actually, they consist of polymers of various size that decompose at various rates, releasing urea as they do so.

The second approach, inhibiting processes of nutrient loss of fixation, has been attempted mainly for nitrogen. The objective here has been to inhibit the nitrification process, thus reducing conversion of ammonia to nitrate that can be lost by leaching or denitrification. Several potential inhibitors, which act primarily on nitrifying bacteria of the genus *Nitrosmonas*, have been discovered and shown to be effective in the field. The effects of these substances on other components of the soil biota are still poorly known, however, and the ecological soundness of an approach that tries by artificial means to maintain large quantities of a major nutrient in active state is questionable. For example, such conditions might influence cation exchange relations in unsuspected ways and evolutionary processes are likely to allow *Nitrosmonas* eventually to escape inhibition by such substances.

In general, improvements in fertilizer composition seem to be possible, with the most sound approach being the formulation of slow-release materials that ultimately decompose into natural constituents of the soil ecosystem.

Irrigation Efficiency

In individual fields, the amount of water used in agriculture must be regulated to maintain favorable conditions for plant growth while minimizing salinity problems, soil erosion, and the leaching of nutrients. These local considerations, however, grade into regional problems that involve the cost and efficiency of water transport, the climatically related level of evapotranspiration in various regions, and the impacts of irrigation waste water on downstream users. We shall consider the latter problems in Chapter 25, but we may note that tendencies in the past have been to regard land that can be irrigated as land that should be irrigated, with little consideration of other factors, and to design irrigation systems for traditional salt intolerant crops instead of trying to develop crops suited to more challenging climatic and water conditions.

Maintaining favorable conditions in the soil environment relates directly to the quality of the water used for irrigation. The higher the salinity of such water, the greater the leaching fraction that must be employed. The use of irrigation water in amounts that exceed the minimum possible leaching fraction tends to reduce the salt load of the land irrigated, but it does so at the expense of the areas to which drainage water passes (Rhoades et al. 1974) and it contributes to the loss of soluble nutrients. Approaches to irrigation that maximize the quality of irrigation water, where it is used, must therefore be emphasized whenever possible.

Storage and transport systems for irrigation water must be designed for minimum losses of water by evaporation, where deliberate modifications to this end do not damage other values; and the irrigation of lands that contribute excessive salt loads into waters used for irrigation downstream can properly be discontinued. The benefits of such modifications are multiplicative, since higher water quality permits the use of more efficient irrigation systems.

Trickle or drip irrigation is one of these efficient systems (Heller and Bresler 1973); it involves the discharge, at short intervals, of small amounts of water from a small opening in a pipe at a point close to the growing plant (Figure 14–15). This discharge point can be on the surface or slightly below it, and the objec-

FIGURE 14-15.
Drip irrigation system in a young citrus orchard in Monterey County, California. The white emitter releases water mixed with liquid fertilizer at the base of the young tree. When trees become larger, a lateral system must be constructed with additional emitters for each tree to encourage more extensive root growth. (Photograph by Linden Brooks, courtesy of the USDA Soil Conservation Service.)

tive is to maintain optimal moisture in the immediate root zone of the plant. Obviously, soluble nutrients can be distributed the same way and their concentrations maintained in the soil at optimal levels. Although such discharges tend to leach salts away from the water discharge point to the outer edge of the root zone, leaching effectiveness diminishes with increased soil and water salinity. Moreover, the presence of large quantities of solutes in the irrigation water can clog the pipe openings, which must function precisely if the system is to work satisfactorily.

The failures that have characterized the long history of irrigation demonstrate that we still have a great deal to learn about irrigation practice, both in improving the efficiency of water use and in avoiding salinization problems (Marshall 1972).

Comprehensive Soil Management

Several long-term experiments designed to compare the effects of various cultivation and fertilization practices have been carried out in Europe and North America (Cooke 1967). In North America, the oldest and most interesting are the Morrow Plots (University of Illinois 1957), which were established in 1876; in 1904 cultivation systems comparing continuous corn culture with a two-year rotation of corn

TABLE 14-1

Levels of soil organic matter and yields of corn under various rotation and fertilization schemes on the Morrow Plots, University of Illinois, between 1904 and 1953.

Organic matter (t/ha) 1904	Rotational sequence	No fertilization		Farmyard manure[a] + lime + P	
		Maize yield (kg/ha)	1953 organic matter (t/ha)	Maize yield (kg/ha)	1953 organic matter (t/ha)
90	Continuous corn	1640	54	3380	70
96	Corn-oats rotation	2200	74	4460	99
112	Corn-oats-red clover rotation	3580	90	5020	110

[a]Wet weight of manure equal to dry weight of crops removed (1909–1953); 1.6 t/ha (1904–1909).

Source: University of Illinois 1957.

and oats and a three-year rotation of corn, oats, and red clover were initiated. These systems were followed until 1955. Each of the cultivation systems was subjected to two fertilizer treatments: (1) no fertilizers, and (2) manure equal in wet weight to the dry weight of the crop removed, plus lime and phosphorus.

The results of these studies through 1955 (Table 14-1) indicate the influence of each practice on yields of corn and on the levels of organic matter and nitrogen in the soil. Since the plots initiated in 1904 had been under similar treatment since about 1876, initial differences existed in both soil conditions and productivity. Soil organic matter equalled 114 tons per hectare and soil nitrogen to 5500 kilograms per hectare in the fertilized, corn-oats-clover plot in 1904; at the same time, organic matter equalled only 90 tons per hectare and nitrogen 4700 kilograms per hectare on the continuous corn plot. In 1955, analyses of soil organic matter and nitrogen showed that the only treatment that had come close to retaining orginal conditions was that which involved crop rotation plus the addition of both

manure and chemical fertilizers. This treatment also gave by far the highest average corn yield. Without rotation or fertilization, the levels of organic matter and nitrogen dropped to less than 50 percent of what they had probably been in 1876. Corn yield on what is some of the best natural farming land in the United States was less than one-third that achieved with rotation and fertilization.

Even more interesting is the fact that the three-crop rotation, without fertilization, was more effective in maintaining fertility than the use of both organic and inorganic fertilizers on continuously cultivated corn. Altogether, these results suggest that sound management of cultivated soils must take into account, in an integrated fashion, several major aspects of the soil system. At least four critical features can be identified: (1) organic content, which influences physical structure and cation exchange processes, as well as being a nutrient source; (2) the supply of available nutrients within the soil; (3) soil moisture availability; and (4) soil salinity.

GRAZING LAND

The management of the soils of grazing lands, like the management of cultivated soils, must consider many factors besides those that exist within the soil itself. The extent and composition of the plant cover influences the dynamics of the soil ecosystem, and this vegetation cover is subject to extensive modification by the action of grazing animals, which also directly affect the surface soil by trampling it.

Erosion, both by wind and water, can be a major problem on rangelands. Excessive grazing depletes the vegetational cover, reduces the reincorporation of organic matter into the soil, compacts the soil, and initiates soil displacement on sloping areas. Studies of the rate at which water infiltrates range soils show generally that this rate is closely related to the amount of living and dead plant material present (Meeuwig 1970). In northeastern Colorado, studies on infiltration rates for three soils that had been grazed lightly, moderately, and heavily for 30 years showed that infiltration rates were significantly reduced by heavy grazing on all but the sandiest of the soil types.

Trampling displacement, the direct downslope movement of soil caused by livestock activity, shows up in various ways, one of which is the extensive development of terraced trails on slopes (Figure 14-16). Such trails are evidence of the downslope displacement of soil, and they leave the slope soils more vulnerable to water erosion—the trail itself is likely to

FIGURE 14-16.
Severe gully erosion in California due to heavy grazing and the formation of trails by cattle on highly erodable soils. (Photograph courtesy of the USDA Soil Conservation Service.)

become a watercourse during rainstorms. Generalized trampling effects can be even more serious, however, because they may involve the entire land surface (Anderson 1974). General land displacement by trampling is indicated by raised areas downslope from individual hoofprints and mounded soil accumulations on upslope sides of boulders and plants.

Active erosion on rangelands shows up as rills and gullies the way it does on cultivated land, but it can take other distinctive forms that are peculiar to rangeland. These include **soil pedestals, lichen lines, erosion pavements,** wind-scoured depressions, and deposits of wind and water carried soils. Soil pedestals are simply elevated masses of soil, protected on top by rocks, plants, or other objects; their height indicates the former soil surface level (see Figure 11-10). Similarly, lichen-covered rocks might show sharp transition lines to lichen-free lower sides that have recently been exposed by erosion. Wind erosion can also cause a surface accumulation of pebbles that were originally scattered through a greater depth of soil, and can scour shallow depressions that contain accumulated gravel and rocks too heavy to be removed by wind. Deposits of fine soil material downwind from rocks and plants indicate wind erosion, whereas mud "deltas" at the ends of rills and gullies give evidence of water erosion.

Erosion on rangelands is generally remedied by adjusting the livestock load to range carrying capacity. Here, the objective is to retain an adequate vegetation cover and to keep direct trampling effects within the natural recovery capacity of the soil system. Studies carried out at the Central Plains Experimental Range in Colorado indicate the livestock intensity that allows good range condition to be maintained (Hyder 1969). These studies suggest that individual animals achieve optimum growth at stocking rates that allow 390 kilograms per hectare of plant material to remain ungrazed at the end of the summer growing season; maximum total growth of a herd of animals

occurs when 280 kilograms per hectare remain. In an average year, the former value corresponds to 65 percent of the total forage production during the season, the latter to 48 percent of total forage growth. We can therefore conclude that, other factors being equal, harvesting much more than half the forage produced during a season does not enable the vegetation to reach the same level of production in the next season. Data like these can be translated into stocking rates, based on the forage consumption of the animals in question and of course on the weather conditions that affect forage plant growth. We should note that this can sometimes give a stocking rate of zero, if weather conditions, particularly drought, do not allow production, without grazing, to reach the value of 280 kilograms per hectare.

Optimizing productivity under given conditions guarantees ecosystem health only if it is not accomplished at the expense of future productivity. In addition to selecting appropriate stocking rates, it is essential to long-term management to assess range condition and trend by other means. There are a number of standardized methods for evaluating range condition (Humphrey 1962). These methods are quantified and usually take into account the amount of vegetation present, its composition in terms of desirable and undesirable species, and the vigor or health of the desirable species present (Table 14-2). **Range condition,** of course, is the state that exists at the time these factors are assessed; **trend** is the pattern of change in range condition over time. The object of ecologically sound management is to improve range condition to the point where maximum sustainable forage can be harvested by grazing animals.

Because livestock graze selectively on some species of forage plants and not others, it is difficult to prevent the less desirable species from proliferating at the expense of desirable ones. When the desirable forage species are native to an area, certain systems of rotational

TABLE 14-2
Scoring system for representative range conditions in mountain grassland areas in Montana.

A. Forage density index[a]

Condition	Score
55–65 hits	9–10
45–54 hits	7–8
35–44 hits	5–6
20–34 hits	3–4
0–19 hits	0–2

B. Stand composition index

Condition	Score
Desirable perennial grasses, weeds, and shrubs make up 70% of plant cover. *Festuca scabrella* forms 10% or more of forage.	13–15
Desirable perennials dominant (50–69%). *Festuca scabrella* less than 10% of vegetation (may be absent).	9–12
Desirable perennials make up 30–40% of cover, with most of the remainder consisting of species of intermediate value.	6–8
Desirable perennials make up 5–29% of cover; species of intermediate value make up 35–60%.	3–5
Desirable perennials make up less than 5% of cover; undesirable species constitute at least 65% of cover.	0–5

C. Desirable species vigor index

Condition	F. scabrella	F. idahoensis	Score
Leaves exceed length of	16 in.	6 in.	9–10
Leaves show length range of	12.0–15.9 in.	5.0–5.9 in.	7–8
Leaves show length range of	10.0–11.9 in.	4.0–4.9 in.	5–6
Leaves show length range of	8.0–9.9 in.	3.0–3.9 in.	3–4
Leaves shorter than	8 in.	3 in.	0–2

Range condition index (A + B + C)

Excellent	30–35
Good	25–29
Fair	20–24
Poor	15–19
Very Poor	0–14

[a]Number of points (out of 65 possible) at which a needle lowered through frame above vegetation toward ground surface strikes plant tissues.
Source: From Robert R. Humphrey. *Range Ecology*. Copyright © 1962, the Ronald Press Company, New York.

grazing can help to maintain good condition over the long run. Various "formulas" have been devised for the rotation of grazing activities among different range or pasture areas. These systems generally combine intensive grazing, where livestock crop all the species present instead of just the ones they prefer most strongly, with periods of range resting, during which natural processes of ecological succession reestablish a healthy sward with a desirable composition.

Systems of **rotational grazing** can be adjusted to the life cycles of the desirable dominant species (Figure 14-17). An example of such a

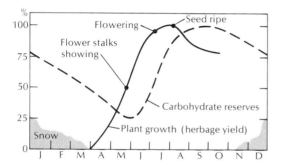

FIGURE 14-17.
Seasonal pattern of life cycle activities of Idaho fescue grass in the northern Great Basin region. Systems of rotational grazing must be designed that will allow the periodic reproduction of range species as well as the adequate accumulation of carbohydrate reserves to enable plants to overwinter. (From Hormay 1970.)

system is the plan of rest-rotation grazing outlined by Hormay (1970), in which a five-year grazing sequence can be carried out in a range area partitioned into five sections (Figure 14-18). Under this system, the range is free of animals roughly 40 percent of the time, during which the plant community regains its normal structure, individual species are able to store carbohydrates, litter accumulates to protect the soil, and seed production is unhindered. Livestock can be introduced after seeds are

ripe and can serve a useful purpose by trampling the seeds into the soil.

Not all rangelands can be managed successfully by such rigid systems, however (Hyder 1969). In more arid regions, where rainfall is greatly variable and where most range plants grow in short spurts correlated with short-term moisture conditions, more flexible patterns are probably more suitable. In any case, a rest-rotation grazing system does not eliminate the need to adjust average stocking levels to the fluctuating of range conditions created by favorable and unfavorable years.

The use of synthetic fertilizers is an increasingly important practice in pasture and rangeland management, but in general it is economically practical only on well-watered pasture lands with high potential productivity, or on rangelands with trace element deficiencies that can be corrected with inexpensive applications. The other considerations related to the use of fertilizer on rangeland are generally the same as those that apply to cultivated soils. However, the choice is complicated by the fact that the species composition of the range might be greatly affected by the use of fertilizer.

The application of nitrogen fertilizers to rangeland almost invariably causes a decline in the abundance of legumes, for example, although total forage and plant productivity might increase (Rossiter 1966). Phosphate fertilizers, on the other hand, often favor legumes. Considerable care must therefore be used in planning fertilizer applications, since, in effect, certain patterns of applications can reduce the contribution of the same nutrient by natural processes.

In many, but not all, grassland areas, fire is an important natural factor in the environment, and although this has been known for a long time, the use of fire in management is still not common. Fire can be a valuable tool in maintaining a desirable species composition of grassland areas and in stimulating the productivity of major species.

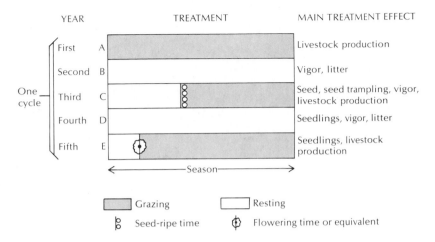

FIGURE 14-18.
An example of a rotational grazing system covering a five-year cycle and geared to the flowering and seed-ripening periods of the dominant grasses. (From Hormay 1970.)

Humphrey and Mehrhoff (1958), for example, have described the extensive change that has occurred on many southern Arizona rangelands since ranchers first moved into the area in numbers in the late 1800s. Before southern Arizona was settled, fire, both set by Indians and sparked by lightning in mountain areas, was a regular phenomenon in desert grasslands. One of the less obvious effects of man on this environment has been a reduction in the number of range fires, partly because of deliberate control, but largely because of grazing, which reduces the amount of grass fuel to the point where fires do not spread well. Many of the desert scrub species that have invaded these ranges are fire intolerant.

On the Santa Rita Experimental Range, south of Tucson, maps of the range vegetation in 1904 and 1954 showed that the area dominated by bunchgrasses declined from more than one-half the total area to less than one-fifth. The lower, drier portions of the range were progressively invaded by mesquite, creosote bush, burroweed, cholla cactus, and a variety of other

shrubs and cacti typical of desert scrub communities (Figure 14-19). With the decline in cover of perennial grasses, more of the soil surface has become exposed, and with greater exposure come increased compaction and crusting from livestock trampling and the beating action of raindrops, as well as greater surface runoff. Increased runoff, with attendant erosion, is evidenced over much of the Southwest by the deeper downcutting of drainage channels.

Humphrey and Mehrhoff (1958) were forced to conclude that, although the invasion of various species was partly a result of seed distribution by cattle and rodents, and by selective grazing on grasses, the main factor that maintained these areas as grassland was fire, and the principal cause of shrub invasion has been the reduction in range fires.

This general pattern has been recognized in many rangeland areas of the Southwest, where grasslands exist in close proximity to other plant communities that have aggressive shrub and cactus species. Even in the Midwest

FIGURE 14-19.
A portion of the Santa Rita Experimental Range, south of Tucson, Arizona. The large
shrub is mesquite; scattered cholla cacti can also be seen. These species tend to increase
rapidly with overgrazing and reduction in fire frequency. (Photograph by D. G. Cox.)

border region between the tall-grass prairie
and eastern deciduous forest, fire seems to have
been the critical factor in preventing invasion
of the grassland by woody plants.

Prescribed burning, or **controlled burning,**
is being used more and more to maintain the
natural composition of grassland ranges (Wright
1974). In western Texas, for example, tobosa
(*Hilaria mutica*) grasslands tend to be invaded
by mesquite and a number of other shrubs and
cacti. Prescribed burning is more effective than
herbicide treatments in ridding range areas
of these invaders. Fire damages prickly pear
and cholla cacti so that they are invaded by
insects and most of them are killed, whereas
herbicides such as 2,4,5–T will not kill these
species. Fire also kills the crowns of mesquite
and leaves the dead tissues susceptible to invasion by wood-boring insects. Mesquite crowns
killed by herbicides tend to resist such invasion.
In other areas, fire can also be an effective
weeding agent against sagebrush, juniper, and
other rangeland invaders. Of course, not all
shrubby plants vulnerable to fire are undesir-

able; for example, bitterbrush, *Purshia triden-
tata*, which is a prime browse species used
heavily by deer in the Great Basin region, is
sensitive to fire.

Fire frequently stimulates the productivity
of native prairie or grassland communities,
a fact that was demonstrated dramatically in
studies of burning in tall-grass prairies of Illinois
(Hadley and Kieckhefer 1963). These studies,
concentrating on the productivity of two of
the dominant grasses—big bluestem, *Andropo-
gon gerardi*, and Indian grass, *Sorghastrum
nutans*—compared several areas that had been
burned from 1 to 19 years earlier (Table 14-3).
The net production of shoots and the abundance
of flowering stalks showed dramatic increases
in the growing season immediately following
burning, which was carried out in the spring.
This increase was not at the expense of root
material, because the biomass of roots increased
also. The residual effect of burning was short,
lasting only about two additional seasons, and
in the absence of burning productivity quickly
dropped to levels comparable to those in areas

TABLE 14-3
Shoot production, flower stalk density, and root biomass of two Illinois perennial grasses in relation to fire frequency.

Species	Burn[a]	Shoot production (g/m²/day)		Flowering stalks (number/m²)		Root biomass (g/m²)	
		1961	1962	1961	1962	1961	1962
Big bluestem,	0	0.83	0.99	10.4	27.8	886.6	900.4
Andropogon	2	1.00	0.98	15.5	29.5	1099.5	1123.1
gerardi	3	3.62	1.62	117.7	52.9	1211.9	1261.9
	4		3.73		129.5		1309.8
Indian grass,	0	1.34	1.30	17.3	32.1	781.9	786.5
Sorghastrum	2	1.38	1.46	28.4	38.4	923.5	965.4
nutans	3	4.04	1.74	132.0	80.8	949.6	980.5
	4		4.21		162.2		1029.1

[a]Burn 0, unburned since 1943; burn 2, burned in February 1952 and again in April 1959; burn 3, burned in 1961; burn 4, burned in May 1962.
Source: Hadley and Kieckhefer 1963.

not burned for many years. A significant weeding action was performed by fire in this study, as well.

In drier regions, burning might not increase productivity in all years (Wright 1974). During dry years, burning can actually intensify drought stress and reduce productivity to a lower level than that for unburned areas.

Thus we can see that in managing rangelands to maintain an optimal soil environment it is important to consider the intensity and timing of grazing, the quantities and types of fertilizers used, and, in some situations, the manner in which prescribed burning is incorporated into a comprehensive management plan.

INTEGRATED SOIL MANAGEMENT

Maintaining the soil ecosystem clearly requires that a number of chemical, physical, and biotic relationships be taken into account by a system that relates agricultural practice to the entire environmental complex. A regional or national system for achieving what we may call **integrated soil management** has two essential requisites: (1) a means of classifying land on the basis of its intrinsic ecological nature and its response to agricultural use, and (2) a means of enforcing the maintenance of basic standards of soil ecosystem health on land under agricultural (or other) use.

A number of land classification systems have been developed and put into use (Olson 1974), of which the most widely used and adaptable is probably the system of **Land Capability Classification,** employed by the Soil Conservation Service of the U.S. Department of Agriculture. This system is based on soil surveys that take into account physical properties, degree of development, and vulnerability to erosion; it employs a series of eight land capability classes (Figure 14-20), which represent, in theory, soil situations that have a particular potential for the production of cultivated crops and pasturage over an indefinite time and that

FIGURE 14-20.
Landscape illustrating the various classes in the Soil Conservation Service's Land Capability Classification system. (Photograph courtesy of the USDA Soil Conservation Service.)

are characterized by particular limitations and hazards (Table 14-4). Class I soils, for example, have few or no limitations and can safely be used for cultivated crops, pastures, or other nondestructive purposes. Class VIII lands, at the other extreme, possess severe limitations, such as erosion hazards, wetness or salinity, poor soil development, and other characteristics that preclude their use for agricultural purposes.

The Bureau of Reclamation of the U.S. Department of the Interior has also developed a system of **Irrigation Suitability Classification**

based on the requirement that federal funds cannot be used for irrigation development unless it has been demonstrated by land survey and classification that the areas to be irrigated are suitable to the production of agricultural crops. This classification consists of a series of six classes ranging from highly suitable for a wide range of crops to unsuitable because the costs of their reclamation cannot be expected to be offset by their productivity after reclamation. The criteria used in this analysis and classification are strongly economic; the application of this classification to a proposed project area

TABLE 14-4
Characteristics of major categories in the U.S. Department
of Agriculture Land Capability Classification.

Class I.	Soils with few limitations.
	Deep, well-drained, and easily cultivated soils on level or gently sloping land. Possess high levels of natural fertility or fertilizer responsiveness and an assured moisture supply from natural sources or irrigation. Suitable for a wide range of agricultural activities.
Class II.	Soils with some limitations that reduce the choice of plants or require moderate conservation practices.
	Slight, easily correctable limitations involving erosion hazard, salinity or waterlogging, or depth and ease of cultivation.
Class III.	Soils with severe limitations that reduce the choice of plants and/or require special conservation practices.
	Major limitations that do not preclude cultivation but require special practices to prevent erosion, waterlogging, salination, or other damage, or to maintain fertility and cultivability.
Class IV.	Soils with very severe limitations that restrict the choice of plants and/or require very careful management.
	Soils in this group are generally suited to only a few crops and usually exhibit low yields due to limitations of the sort noted above.
Class V.	Soils with little or no erosion hazard but other limitations that are impractical to remove and that limit use largely to pasture, range, or other noncultivation purposes.
	Soils in this group tend to be rocky, permanently waterlogged, subject to frequent flooding, or limited for crop production because of short growing season.
Class VI.	Soils with severe limitations that make them generally unsuited to cultivation.
	Soils of this group possess limitations, including erosion danger, that make them unsuited to cultivation but, with the application of special practices, allow their use for grazing and other noncultivation purposes.
Class VII.	Soils with very severe limitations that make them unsuited to cultivation and restrict their use to grazing, woodland, or wildlife.
	Limitations on soils of this group are severe enough that few improvements are possible even for grazing use.
Class VIII.	Soils of this group have limitations that preclude their use for agriculture and restrict it to recreation, wildlife, watershed, and esthetic purposes.

thus contributes to an overall cost-benefit analysis of the project.

These land classification systems, however, serve only as guidelines for the agricultural use of land areas, or as tools for decision making in the development of irrigation projects. The system of Land Capability Classification, an analysis provided free to the farmer by the Soil Conservation Service, does not specify conditions that must be maintained when the classified areas are subject to use. We suggest that a series of ecological indicators, such as annual erosion loss, soil organic content, and range condition, must eventually be developed and applied to lands subject to exploitation. These indicators should specify minimum conditions that must be maintained as long as the lands are exploited. The activities carried out by an individual farmer increasingly affect the well-being of people in other areas of society. Using land in a way that degrades its quality or damages surrounding areas (e.g., by sediment deposition) is under growing attack. Just as urban zoning systems restrict the uses of the land areas involved, ecologically-based systems of agricultural land use control are likely to develop in rural areas.

Erosion and sedimentation have increasingly been recognized as problems that require legislation. In general, it has been the deposition of eroded materials that has triggered the enactment of erosion control legislation. This has been accomplished in ten states and is under study in many others (Gray 1974). Typical legislation defines situations or levels of erosion that constitute nuisance conditions and provides some mechanism of nuisance abatement.

It is difficult to determine the impact of this sort of national soil conservancy law. Nicol et al. (1974) have attempted to do this by means of a large-scale mathematical model of agricultural relationships for the United States. This model considers land, water, crop, and cultivation technique in relation to erosion and agricultural production, and it predicts conditions that could prevail in 2000 A.D. with a U.S. population of 284 million. Current trends in technological inputs to production, as well as the productivity per hectare of various crops, are assumed to continue through this projection date. A soil conservancy law was interpreted as limiting the annual soil loss per hectare. Within the model, responses could occur by regional crop shifts, shifts of high-erosion crops from areas of higher to areas of lower erosion risk, and the implementation of erosion-reducing practices such as terracing, strip cropping, and

so forth. The results of this simulation (Table 14-5) suggest that without erosion restrictions, the average loss of soil per hectare in 2000 A.D. will be about 22.2 tons. However, by instituting the sorts of changes noted above, this loss could be reduced to about 4.2 tons per hectare, even though the number of hectares in cultivation would be reduced. The actual loss limitation would be 6.7 tons per hectare. This assumes that food commodity exports would remain at 1969-to-1971 levels. If these levels double, however, average losses can be kept at 6.5 tons per hectare with only a small acreage increase. Furthurmore, except in the case of doubled exports, the adjustments are predicted to be possible without major farm-level increases in food costs.

Unfortunately, these projections assume continued increases in technological inputs, which may or may not happen, and they ignore the possibility that intensified use of certain technologies, such as herbicides, in programs to reduce erosion can create new serious problems or can at least become less effective than they appear to be at present. On the other hand, the study assumes no benefits of increased production, over present levels, due to reduction of erosion effects. Clearly, however, major changes in soil ecosystem management are possible without sacrificing high levels of productivity.

SUMMARY

Deterioration of agricultural soils has been a primary or contributing factor in the decline of civilizations in various world regions. In some upland areas the soils have eroded to the point where their reclamation is probably impossible. As the intensity of modern agriculture increases concern over the problems of erosion increases correspondingly.

In cultivated cropland, erosion can be reduced by practices such as contour plowing and strip cropping, terracing, mulching, and the use of rotations that maintain soil organic content

TABLE 14–5
Summary of cultivated acreage, erosion, and tillage and conservation practices projected for the U.S. under soil conservancy laws of varying intensity and food commodity export levels equal to and above those of 1969–1971.

Item	Soil loss restriction and food export level				
	A Unrestricted ave 1969–1971	B 22.4 tons/ha ave 1969–1971	C 11.2 tons/ha ave 1969–1971	D 6.7 tons/ha ave 1969–1971	E 11.2 tons/ha 2 × 1969–1971
Erosion per hectare (tons)	22.2	9.6	6.3	4.2	6.5
Total erosion (10^9 tons)	2.4	1.0	0.6	0.4	0.7
Cultivated land (10^6 hectares)	109	106	105	104	111
Percent of land by tillage method					
Conventional	92.1	83.1	77.7	71.7	75.2
Reduced	7.9	16.9	22.3	28.3	24.8
Percent of land by conservation practice					
Contouring	4.6	17.3	21.6	25.2	23.3
Strip crop and terracing	1.2	8.7	18.9	26.0	21.9
Straight row	94.2	74.0	59.5	48.8	54.8

Source: Nicol et al. 1974.

and aggregate structure. Wind erosion can likewise be reduced by the use of mulches and by maintaining the aggregate structure, as well as by laying fields out properly and by using windbreaks. Reduced tillage systems reduce erosion considerably, although their practicality for most crops remains to be demonstrated; their heavy reliance on chemical herbicides, due to their cost and the potential counter-evolution by weeds, constitutes a major weakness.

Maintaining soil fertility is also essential to soil protection. Slow-release fertilizers seem to offer the possibility of increased fertilizer efficiency. Long-term studies of cultivation practices, however, suggest that comprehensive schemes that involve crop rotation and the use both of organic and inorganic fertilizers are necessary to maintain cultivated soils in the most favorable state.

Grazing land soils have similar problems. Trampling and vegetation depletion can induce erosion. Studies of maximum sustained harvest from rangeland suggest that at least half the aboveground net production must remain unharvested to achieve maximum growth of livestock. Deferred rotational grazing systems allow rangeland recovery, and fire can also be an important management tool.

Systems of integrated soil management are now beginning to evolve; they incorporate an ecologically-based scheme of land classification, incorporating ecological indicators of soil ecosystem health, and a legislative means for enforcing maintenance of soil standards above critical levels.

Literature Cited

Adams, J. E. 1974. Residual effects of crop rotations on water uptake, soil loss, and sorghum yield. *Agron. J.* 66:299–304.

Anderson, E. W. 1974. Indicators of soil movement on range watersheds. *J. Range Manag.* 27(3):244–247.

Baeumer, K., and W. A. P. Bakerman. 1973. Zero-tillage. *Adv. in Agron.* 25:27–123.

Beasley, R. P. 1972. *Erosion and Sediment Pollution Control.* Ames, Iowa: Iowa State University Press.

Cooke, G. W. 1967. *The Control of Soil Fertility.* New York: Hafner Pub. Co.

Dale, T., and V. G. Carter. 1955. *Topsoil and Civilization.* Norman, OK.: University of Oklahoma Press.

Gray, R. M. 1974. A national soil conservancy law: implications for public soil conservation programs. *J. Soil and Water Cons.* 29:210–212.

Hadley, E. B., and B. J. Kieckhefer. 1963. Productivity of two prairie grasses in relation to fire frequency. *Ecology* 44:389–395.

Hauck, R. D. and M. Koshino. 1971. Slow-release and amended fertilizers. In *Fertilizer Use and Technology.* pp. 455–494. Madison, Wisc.: Soil Science Society of America.

Heller, J., and E. Bresler. 1973. Trickle irrigation. In *Arid Zone Irrigation.* B. Yaron, E. Danfors, and Y. Vaadia (eds.). pp. 339–351. New York: Springer-Verlag.

Hormay, A. L. 1970. *Principles of Rest-Rotation Grazing and Multiple-use Land Management.* Training Text 4 (2000). Washington, D.C.: USDA Forest Service.

Humphrey, R. R. 1962. *Range Ecology.* New York: Ronald Press.

Humphrey, R. R., and L. A. Mehrhoff. 1958. Vegetation changes on a southern Arizona grass-land range. *Ecology* 39:720–726.

Hyams, E. 1952. *Soil and Civilization.* London: Thames and Hudson.

Hyder, D. N. 1969. The impact of domestic animals on the function and structure of grassland ecosystems. In *The Grassland Ecosystem.* R. L. Dix and R. G. Beidlemen (eds). pp. 243–260. Range Science Department Science Series, No. 2. Fort Collins, Colo.: Colorado State University.

Lal, R. 1974. No-tillage effects on soil properties and maize (*Zea mays* L.) production in western Nigeria. *Plant and Soil* 40:321–331.

Marshall, T. J. 1972. Efficient management of water in agriculture. In *Optimizing the Soil Physical Environment Toward Greater Crop Yields.* D. Hillel (ed.). pp. 11–22. New York: Academic Press.

McCalla, T. M., and T. J. Army. 1961. Stubble mulch farming. *Adv. in Agron.* 13:126–196.

Meeuwig, R. O. 1970. Infiltration and soil erosion as influenced by vegetation and soil in north-ern Utah. *J. Range Manag.* 23:185–188.

Moschler, W. W., D. C. Martens, and G. M. Shear. 1975. Residual fertility in soil continuously field cropped to corn by conventional tillage and no-tillage methods. *Agron. J.* 67:45–48.

Nicol, K. J., H. C. Madsen, and E. O. Heady. 1974. The impact of a national soil conservancy law. *J. Soil and Water Cons.* 29:204–210.

Olson, G. W. 1974. Land classifications. *Search—Agri.* 4(7):3–33.

Rhoades, J. D., J. D. Oster, R. D. Ingvalson, J. M. Tucker, and M. Clark. 1974. Minimizing the salt burdens of irrigation drainage water. *J. Env. Qual.* 3:311–316.

Richardson, C. W. 1973. Runoff, erosion, and tillage efficiency on graded-furrow and terraced watersheds. *J. Soil and Water Cons.* 28:162–164.

Rossiter, R. C. 1966. Ecology of the Mediterranean annual-type pasture. *Adv. in Agron.* 18:1–56.

Pimentel, D., E. C. Terhune, R. Dyson-Hudson, S. Rochereau, R. Samis, E. A. Smith, D. Den-man, D. Reifschneider, and M. Shepard. 1976. Land degradation: Effects on food and energy resources. *Science* 194:149–155.

Prasad, R., G. B. Rajale, and B. A. Lakhdive. 1971. Nitrification retarders and slow-release nitrogen fertilizers. *Adv. in Agron.* 23:337–383.

Spomer, R. G., K. E. Saxton, and H. G. Heinemann. 1973. Water yield and erosion response to land management. *J. Soil and Water Cons.* 28:168–171.

University of Illinois. 1957. *The Morrow plots.* Circ. 777. Updated by annual mimeographed reports.

Woodruff, N. P., L. Lyles, F. H. Siddoway, and D. W. Fryrear. 1972. *How to Control Wind Erosion.* Agriculture Information Bulletin No. 354. Washington, D.C.: USDA.

Wright, H. A. 1974. Range burning. *J. Range Manag.* 27(1):5–11.

15

THE NATURE OF
AGRICULTURAL PEST PROBLEMS

Most organisms play very beneficial roles in the natural ecosystems in which human beings themselves are simply a part of natural history; but in more and more cases human beings interact with plants, animals, and pathogens as managers instead of as another well-integrated biological component of a larger system. Furthermore, the number of species under human management is bound to increase as man strives to meet his increasing needs by further exploiting natural ecosystems or by replacing them with agroecosystems. When man becomes a manager, he frequently adopts strategies and objectives that fall into conflict with natural processes. Attempts to maximize the productivity of managed systems force some species into more competitive roles than they would play under natural systems. This brings them into disfavor and causes them to be designated as **pests**—a designation based purely on human judgment.

In recent times we have become so conscious of the conflicts between certain organisms and ourselves that we tend to think of some groups only as pests and to assume that they could never be anything else. This is unfortunate from an ecological point of view because it

leads far too frequently to premature or unnecessary pest control programs that are more disruptive than beneficial. We should realize that situations change; and organisms should be considered pests only when and where they cause significant damage or, in our best judgment, will do so if we do not intervene. This, of course, requires some understanding of the relationship between the population level of an organism and the damage it causes, as well as an awareness of factors that cause populations to increase to the point where significant harm will result.

Many pest situations are caused by natural phenomena. For example, large numbers of mosquitoes and biting flies emerge more or less simultaneously in areas where the climate synchronizes their development. When most of the adult population of these insects is present at one time and many individuals are competing for food, they can cause quite severe discomfort. Similarly, in the ecologically simple and relatively unstable northern forests there are large-scale, climatically released outbreaks of defoliating insects. The resulting tree mortality is one form of natural management in such systems, but it is not always in our best interest if we planned to harvest the fiber at some future date. We could list many other similar examples, but the point we wish to make is that pest problems arise out of natural conditions and relationships and become problems only because of the emphasis we place on human needs and comfort.

ORIGINS OF PEST SPECIES

Some species are potential pests because of their coevolution with the ancestors of modern crops and domesticated animals. The crop plants and animals that form the main components of modern agroecosystems had their origin in the natural ecosystems from which man originally gathered his food and fiber. In their ancestral form, these species interacted with other components of their environment according to the adaptations and relationships that developed in the course of their evolution. When the species that man selected to fill his requirements were first cultivated or domesticated, they nevertheless remained in close contact with their natural environs, and many of their normal interactions with associated wild species were perpetuated. In the 10,000-year history of agriculture, many domesticated species have been improved through selection and natural hybridization, but for the most part, until quite recently, the changes were gradual and many symbionts were thus able to adapt quite favorably to new situations.

Harmless associates of crops or domesticated animals often become pests when they are transported to new regions. Once organized agriculture permitted man to establish centers of civilization, he began to explore new areas into which he could expand. As he moved about, he relied heavily on the crop plants and domesticated animals with which he was most familiar, and carried them with him to new areas. Once he was established in new regions he cultivated desirable local species

as well. By the Renaissance, the transport of domesticated species from region to region was commonplace and continued at a high rate through the colonial period. A number of insects, weeds, and diseases associated with crop plants and domestic animals were accidentally transported along with their hosts. A good example is the common white cabbage butterfly, or imported cabbageworm, *Pieris rapae*, which was first introduced to North America from Europe about 1860 and is now a cosmopolitan pest of crucifers (Figure 15-1).

In other cases crop plants were often introduced to new areas without their more important natural associates. Some of the most serious pest problems have arisen when these associates have subsequently been introduced and reunited with their host. The Hessian fly, *Mayetiola destructor*, frequently a serious pest of wheat in the United States, was not introduced to America when wheat was, but arrived later in the straw used for animal bedding by the Hessian troops who came to fight in the Revolutionary War.

The cereal leaf beetle, *Oulema melanopus*, is another pest of wheat that was introduced around 1958. The cereal leaf beetle is endemic to most Old World areas where small grains are grown. In areas where the climate is moderated by the oceans, it is rarely or only sporadically a pest, but in areas of continental climate, such as Hungary and the Ukraine, it causes measurable economic damage. Since its introduction into the United States, the cereal leaf beetle has spread throughout Michigan, Illinois, Indiana, Ohio, Pennsylvania, New York, West Virginia, and Kentucky. In its new habitat the beetle emerges in April to feed on native grasses, winter grains, and later spring grains, and it clearly poses a considerable threat to small grain production. At present, the only effective protection against cereal leaf beetle damage is the costly application of insecticides. Many other pest problems of similar origin could be cited.

FIGURE 15-1.
Larvae of the cabbage butterfly, *Pieris rapae*, causing damage to cabbage leaves.
(Photograph courtesy Plant Protection Ltd., ICI issued by FAO.)

Pest problems, particularly in forestry and agriculture, are mainly population problems. There is often something about a new environment that allows populations of introduced species to increase much more rapidly than in their places of origin (Elton 1958). This has happened on numerous occasions after the introduction of species that encounter hosts related to those in their native regions. These introduced species are often not serious pests in their places of origin because of environmental constraints such as climate or natural enemies; but in a new environment they are sometimes freed from such constraints and thus become pests. For example, the European corn borer, *Ostrinia nubilalis*, was imported into the United States in broom corn, probably from Hungary. The insect is not a serious pest in Europe, partially because its populations are held down by a complex of native parasites and predators, and partly because it produces only one generation per year under the European climate. After its introduction into the northeastern United States, the corn borer spread quite rapidly in the absence of natural enemies and became established throughout the corn-growing areas of this country. After about twenty years, a second generation of corn borers began to appear annually in warmer areas. The two-generation biotype soon became widespread, and now the characteristic of two generations per year is the basis of the European corn borer problem in the Western Hemisphere.

Far too often species have been deliberately introduced to new areas in the belief that they will provide either tangible or intangible benefits. This has frequently been the case with plants and birds imported for aesthetic reasons. *Lantana camara*, an attractive perennial shrub native to Central America, has been transported to many parts of the world as an ornamental; but when *Lantana* escapes cultivation it can become a serious weed pest in rangelands and plantations. Similarly, the prickly pears, *Opuntia* spp., have been imported to a variety of places either as hedge plants, for their fruit, or for use as cattle feed; but in many areas, particularly Australia, prickly pears are serious rangeland pests (Figure 15-2).

The gypsy moth, *Porthetria dispar*, was imported to Massachusetts from Europe in 1869 by a misguided naturalist, Leopold Trouvelot, who hoped to crossbreed the gypsy with the silkworm moth as the basis for a new textile industry. The gypsy moth escaped and became one of the most dreaded pests in the eastern United States (see Graham 1972). Because of its hardiness and its wide variety of host plants, ranging from pine to holly, the gypsy moth is

a threat to forest and ornamental trees throughout the country. Vast sums of money are being spent on efforts to contain the pest within presently infested areas.

The African bee, *Apis mellifera adansonii*, is more aggressive, produces stronger colonies, and gathers more honey than its Italian and Caucasian relatives *A. mellifera ligustica* and *A. mellifera caucasica*, and so it was imported to South America by a group of Brazilian scientists interested in bee genetics and in the development of a superior hybrid honey producer. Unfortunately, the African bee escaped. It has since become established over a wide area and because of its aggressiveness has replaced the Italian and Caucasian subspecies in some areas. This is a matter of some concern because the African bee's vicious nature and highly toxic venom make it less suited to domestication. As is so often the case with introduced

FIGURE 15-2.
The invasion of severely grazed range by brush and prickly pear cactus near Cotulla, Texas. In some parts of the world the prickly pear cactus has adversely affected the ecology of quality grazing land. (Photograph courtesy of the USDA.)

390

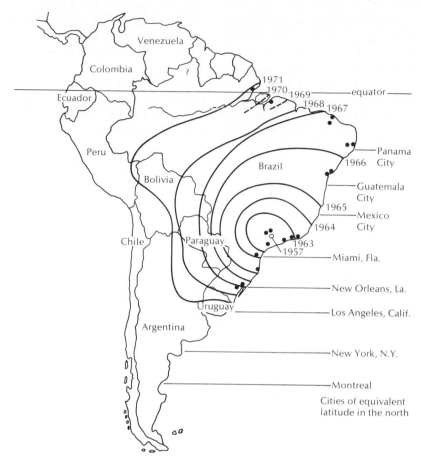

FIGURE 15-3.
The spread of the African honeybee and hybrid descendants (Brazilian bee) from the
point of accidental introduction near São Paulo, Brazil in 1957. (Reproduced, with
permission, from *Annual Review of Entomology*, Volume 20. Copyright © 1975 by
Annual Reviews, Inc. All rights reserved.)

species, the African bee is spreading rapidly
(Figure 15-3) and poses problems for the apicul-
tural industry throughout the Western Hemisphere (Michener 1975).

Many pests originate as components of
natural communities in areas to which crops or
domestic animals are introduced. Weeds, for
example, are simply plants that grow where
they are not wanted, so almost any native plant
with the capacity to invade a disturbed site, or

to leave seeds that germinate in the soil of cropland, can become a weed pest. Annual plants,
such as cheat, chickweed, or wild oats, plus
a number of perennials, such as Canadian
thistle, readily invade small-grain fields. Their
seeds germinate along with the crop seeds, and
new seeds are shed before the grain is harvested.

Insects that feed on the foliage of native
plants switch over to related introduced crop
plants, and they respond to the increased food

supply with a marked increase in population. The alfalfa caterpillar, *Colias eurytheme,* a California native that fed naturally on scattered wild legumes, was once a rare species; but the widespread cultivation of alfalfa throughout the state has made the butterfly a common sight in agricultural areas where it often attains pest status. The Colorado potato beetle, *Leptinotarsa decemlineata,* was also a little known native of the eastern slopes of the Rocky Mountains where it fed on buffalo bur, a member of the family Solanaceae. The pioneer settlers introduced the potato, another solanaceous plant, along the trail westward, however, and the beetle found the potato to be a suitable host. It soon spread eastward by dispersing from farm to farm, often totally destroying the potato crop. The Colorado potato beetle remains a major pest of potatoes in the United States and also attacks a variety of other solanaceous crops, including tobacco, peppers, tomatoes, and eggplant.

Although numerous native species that become pests of introduced crops have a limited range of hosts, as the alfalfa caterpillar and Colorado potato beetle do, others are rather general feeders that increase in numbers because of the favorable conditions provided by crop ecosystems. Once the crop is harvested, such insects go back to feeding on native plants that provide a hold-over site (**reservoir**) until the crop is replanted. Many aphids have evolved this lifestyle; their movement back and forth between native vegetation and cultivars not only causes direct damage to the crop, but also facilitates the transmission of plant pathogens. The local plants are not only an alternate food supply for the aphids but are also reservoirs for pathogens when the cultivated fields are bare.

Storage pests also seem to have developed from organisms that have made rather predictable transitions from natural situations to ones in which they come into conflict with purely human values. Many common storage pests are insects that normally feed on fruits or seeds. For example, when apples or walnuts infested with larvae of the codling moth, *Laspeyresia pomonella,* are placed in storage, the insects continue to feed until they complete their development. There are some circumstances under which storing produce allows an adaptive change to take place in an insect's life cycle. Pests of seeds generally have life cycles that are synchronized with the seasonal availability of their food supply. If infested seeds are placed in storage facilities with a more even and constantly favorable environment, the pests lose their dependency on the synchronization and multiply continuously within the stored product.

Some storage pests have had unusual origins because of fortuitous circumstances. The earliest method of storing produce was to place filled crockery vessels or baskets in trees or caves. Although this protected the food from some pests, it subjected it to infestation by the insects that inhabited such places. For example, insects such as booklice and a variety of small beetles that normally fed on detritus associated with bird nests or animal lairs probably gained access to man's caches of food and were subsequently carried from place to place. Many of these insects have been spread by world trade and are recognized as cosmopolitan species of general concern.

There are a few insects whose mere presence can generate concern; most of these have histories of causing economic damage throughout their known range, and regulations are therefore enforced to keep them from spreading to new areas. If a single specimen is encountered within a restricted zone, an eradication program is usually proposed. This was the case after the recent discovery in southern California of both the Japanese beetle, *Popillia japonica,* and oriental fruit fly, *Dacus dorsalis.* Most pest problems, however, develop only when the population of an injurious species reaches or exceeds the level at which human beings cannot tolerate the resulting damage, discomfort, or health hazard.

SOURCES OF PEST PROBLEMS

The populations of all animals, with the apparent exception of man, tend to fluctuate around some level that we refer to as the animal's **population equilibrium.** Some species display population oscillations of greater amplitude than do others; such populations are usually described as unstable, and communities in which the populations of some species oscillate widely are often called unstable systems. Many ecologists have drawn attention to the fact that stability seems to be related to diversity, and instability to a lack of diversity, while others see no causal relationship between them. The question of stability and diversity in agroecosystems has been reviewed recently by van Emden and Williams (1974). We do not wish to discuss the controversy in detail here, but there seems to be ample evidence that the management practices employed in agriculture and forestry, which are basically designed to increase productivity by stalling succession, often reduce diversity and at the same time stimulate the development of pest problems.

Periodic outbreaks of pest populations, such as those of forest defoliators, seem to occur more often in relatively simple ecosystems, such as boreal forests, than they do in complex tropical rain forests. The ecological effect of the defoliators of northern forests is to disturb the uniform climax community and to encourage diversity by opening up parts of the stand to succession. This of course can have a number of very valuable effects on wildlife populations, as well as maintaining populations of species that permit the ecosystem to respond to disasters of various kinds.

We noted in Chapter 6 that most agroecosystems, particularly mechanized ones, are characterized by the simplicity of their structure; that is, they have fewer trophic levels and lower species diversity than most natural ecosystems do. The cultivation of a single crop over an extensive area creates conditions that favor population increases among potential pests. As Marchal (1908) stated, "Man, in planting over a vast extent of country certain plants to the exclusion of others, offers to the insects which live at the expense of these plants conditions eminently favorable to their excessive multiplication." Pimentel (1961) demonstrated the wisdom of Marchal's statement by showing experimentally that several insect species attained pest status in single-species plantings of crucifers, but did not do so in mixed-species plantings.

Although Pimentel's study indicated that increased species diversity was correlated with reduced herbivore outbreaks, it did not clearly reveal the nature of the regulatory process. Complexity can contribute to stability only if there are more or less direct interrelations among the species in a system. Certainly the cultivation of a single plant species increases the ease with which herbivores that feed upon it can locate food; and large quantities of available food decrease both competition and density related mortality. The addition of different plant species, therefore, could, by disrupting the continuity and abundance of its food supply, reduce the population of an herbivore that feeds only on the original plant; but this increased diversity need not affect the herbivore in any other way. If new herbivores that enter the system to feed on the added plant species could serve as alternate hosts for enemies of the pest herbivore, these natural enemies might maintain higher populations and thus have a stabilizing effect on the pest. In contrast, Watt (1965) contends that, although an increase in the diversity of parasites and predators of a pest might lead to stability on their trophic level, it might also impede their ability to respond to a rapid increase in the pest population, thus contributing to instability.

As van Emden and Williams (1974) pointed out, adding species of herbivores might damp the oscillations of the populations that operate on a particular trophic level, but in an agricultural situation just as much or more of the crop biomass would probably still be lost. Such a procedure could only be justified, therefore, if adding an insect that does not vector a crop disease reduced the population of a vector by competing with it. Obviously this is a complex subject for which generalizations are inappropriate, as each situation needs to be evaluated separately.

Most of the highly productive agricultural acreage in the world is under monoculture, and this practice is often cited as a major cause of pest problems. Certainly, the cultivation of a single crop with a fairly long maturation period, or the intensification of cropping practices, provides an ideal environment for certain kinds of pest problems to develop. The dangers are greater in areas where the climate favors the continuous cultivation of the same crop, but these have not been numerous until recently. In many regions with favorable year-round temperatures, a dry season prevented the planting of multiple crops. But now many of these areas use irrigation to permit continuous crop growth and some pest problems have resulted. In India, for example, recent leaf hopper outbreaks in rice have been attributed to irrigation (Pradhan 1971), and in Ghana a second annual crop of yams is often seriously damaged by increased populations of yam beetles (Nye and Greeland 1960).

Monoculture and continuous cropping also intensify some crop disease problems. The continuous cultivation of cotton, for example, can result in reduced yields due to *Verticillum* and *Fusarium* vascular wilt diseases. Continuous culture can also produce more severe virus problems because virus reservoirs build up and the populations of vector species increase. Although breeding corn for disease resistance has kept pace with some disease problems, the dwarf mosaic of corn spreads rapidly after it is introduced, primarily because of monocultures of corn. Likewise, in vast contiguous plantings of wheat, soybeans, and other crops fungal and

viral diseases often spread rapidly (National Research Council 1968).

Nevertheless, we cannot label monoculture as the sole cause of pest problems; in fact, monocultures sometimes seem to have advantages over mixed culture, especially when other crops provide alternate hosts for potentially damaging species. The southern masked chafer, for example, is a more serious pest of cotton and soybeans when these two crops are planted next to wheat because the larvae feed on the roots of the wheat and the adult beetles feed on cotton and soybean leaves.

The life cycle of a crop under cultivation is an important consideration. For example, a mature stand of alfalfa grown as a perennial might be a highly diverse ecosystem, but crops such as radishes, on the other hand, might progress from a barren seedbed to harvest in a matter of weeks, which is scarcely sufficient time for pest populations to build up except by immigration. Many cereals can be grown continuously as monocultures without suffering economic disease damage; and some pathogens that cause severe damage for several years seem eventually to achieve an equilibrium with the host crop. The *Cercosporella* disease of wheat has apparently adapted to monoculture with a decline in virulence that now permits wheat to be grown as a monoculture without significant losses.

Shifting cultivation and mixed cropping as practiced throughout the tropics seems to reduce the degree to which pest populations oscillate. However, low productivity is a fairly common disadvantage to this system of farming. The human population increase that is taking place in developing tropical nations and the desire of the people for a better standard of living are stimulating changes in their agricultural systems that they hope will increase productivity. These changes, which could be considered together as means of intensifying cultural practices, often take the form of the mechanized agriculture of developed countries in the Temperate Zone. However, the fact that mechanized agriculture, although it is highly productive, often generates pest problems should be kept in mind whenever changes are contemplated in the cultural practices of other regions.

Many aspects of the way in which crops and forests are managed enhance the buildup of plant-feeding insects. In harvesting forests, for example, both the damage done to residual trees and the slash left behind can provide abundant suitable breeding sites for bark beetles and thus can cause a marked population increase during the following year. In agriculture, the practice of fertilizing crops, which is designed to increase productivity, can also increase pest populations because fertilized crop plants are often a more nutritious and attractive source of food for insects. Increases in insect herbivore populations have been recorded or observed, for example, for leafhoppers on rice fertilized heavily with nitrogen (Nene 1971), for mites on a variety of fertilized crops (LeRoux 1954; Cannon and Connel 1965), and for aphids and a variety of other insects. On the other hand, there are reports that fertilizer applications create less favorable conditions for some phytophagous pests, and the more prolific growth of fertilized crops can raise the level of foliage damage by insects or pathogens that can be tolerated.

Many plants have evolved mechanisms that prevent or discourage the feeding of herbivores or provide resistance to disease. These mechanisms include stout spines that deter large browsers from feeding, leaf pubescence or sticky exudates that inhibit insect attack, and the complex of chemical substances that act as poisons, repellents, or hormone mimics. Sometimes, however, the characteristics that make a plant undesirable to its natural enemies also make it undesirable to us, and so these characteristics have often been selected out in the process of developing crop plant varieties. This is particularly true of certain groups of

chemicals such as the terpenes and alkaloids that affect the flavor of plants. In other cases, resistance factors that do not affect flavor are accidentally lost in breeding programs designed to improve other characteristics (van Emden 1966).

One of the more exciting developments in modern agriculture has been the breeding of insect- and disease-resistant varieties with high yields. Developing resistance can mean breeding back into a crop plant some of the characteristics that were previously bred out intentionally or inadvertently. For example, wheat varieties with pubescent leaves seem to be less desirable to the cereal leaf beetle, but pubescence is no longer a characteristic of commercial wheat varieties (Hayes 1972). There are many examples of pest problems that have been diminished by the introduction of resistant varieties, but it is important to realize that introducing new high-yielding varieties in place of traditional varieties carries the danger that pests will develop from species that were of minor importance in more standard crops. This proved to be the case for IR-8 rice, which was high yielding but more susceptible than other varieties to tungro virus.

The high degree of genetic diversity that characterizes certain crop plants provides some degree of protection against potential local pests. For example, Smith (1971) reported that more than 600 varieties of Bulu rice were grown throughout Indonesia, whereas the new high yielding varieties that now dominate rice culture in the Far East are much less diverse in their genetic makeup. New pest problems have developed as a result. Although it is necessary to be aware of such problems, it would be unfair not to mention that the plant breeders who develop new crop plants are well aware of the potential dangers and are usually ready to respond with new crosses.

Irrigation permits sequential cultivation in areas with suitable temperatures for most of the year, and, as we have mentioned, this can create pest problems. Moreover, irrigation produces more succulent plants that are more attractive to both feeding and reproducing insects, and it can encourage the spread of plant pathogens. Rivnay (1964) lists a number of instances in which increased insect attack is associated with irrigation, particularly in cotton and peanuts. In other ways, however, insect populations and damage can actually be reduced by irrigation; for example, flooding is sometimes a practical way to reduce populations of certain soil insects and to combat pathogens such as bacteria, fungi, and nematodes in the soil.

Many important crops are annuals planted in well prepared seedbeds; in these situations many pest problems must begin with the invasion of the crop from other less disturbed areas. Interestingly, the tendency for insects to disperse, and the potential population growth rate after dispersal, are related to the stability of their habitat. Insect populations associated with unstable vegetation, such as annual crops, show a greater tendency to disperse and have a greater potential to increase than those associated with stable vegetation (Southwood 1962). Many species associated with temporary habitats disperse between generations as a characteristic of their life cycle, and they might also disperse readily when their habitat is disturbed by cultivation. This can affect crops grown along a transect through areas ranging from early to late planting dates, because cultivation of an early crop can generate potential invaders of the late crop. If the prevailing wind tends to blow along the transect in the direction of the later plantings, serious pest problems can result.

Population increases within an agroecosystem that are caused by invasion are also influenced by the distance between the crop and the source of the invaders, as well as by the overall attractiveness of the crop to the invading species. Wind can play a significant role in determining the movement of potential pests between crop areas, but it might not always be the prevailing winds that cause problems. Cook (1967) showed, for example, that the beet leafhopper in the San Joaquin Valley of California often spreads against the prevailing wind because the leafhopper's behavior patterns make it more likely to be dispersed on local winds that are related to the valley's topography.

Actively dispersing insects are able to respond to a variety of stimuli, including the colors, forms, and odors of crops or the microclimatic gradients associated with cultivated land. This ability can lead to preferential settling in desirable vegetation. Furthermore, many agroecosystems contrast with the surrounding vegetation in ways that can increase a crop's attractiveness to dispersing insects, and passively dispersed organisms can swell resident populations, thus contributing to the creation of pest problems. The number of wind-blown individuals that reach a crop planting will also depend, of course, on its location relative to the source of invaders, and on the size of the target formed by the crop. The vast wheat fields throughout the northern United States and southern Canada form an enormous drop-zone for wheat rust spores that are generated in Mexico during the winter and carried north by the prevailing atmospheric circulation in the spring.

In the preceding paragraphs we have enumerated ways in which agricultural and forestry practices create environments that favor the increase of potential pests. Once it has been determined that a pest problem exists the usual response is an effort to make the environment less favorable to the pest and thereby to reduce its numbers. But although direct pest control (e.g., through the use of pesticides) makes the environment very unfavorable for a short time, it can actually make it more favorable in the long run and can set up a reaction elsewhere in the system or in a neighboring system. Blair (1964) suggested, for example, that eliminating the screwworm, *Cochliomyia hominivorax*, in the southeastern United States might allow deer and jackrabbit populations to grow, thereby increasing the competition for range vegetation between these and domestic species. In other cases, the use of systemic insecticides against livestock pests could contaminate their dung and destroy the insects that aid in its decomposition (Anderson 1966). These examples are rather subtle, however, compared to the impact of nonspecific pesticides on crop systems.

Although pesticide treatment of a crop ecosystem frequently produces a spectacular decline in the pest population, this result might be temporary. If the destruction of the target pest is accompanied by the destruction of other phytophagous species and most of the target's natural enemies, there can actually be a rapid rise in a pest population called **pest resurgence.** A variety of factors can cause this kind of population rebound. In a simple case, the surviving pests outnumber their surviving natural enemies, and if they have a high level of fecundity they can increase very rapidly (Figure 15-4). The rapid resurgence of a pest can also be due to the fact that the survivors are sometimes more vigorous and have a higher reproductive potential than many of those that did not survive. After a marked reduction in both inter- and intraspecific competition the surviving population can explode.

In most agroecosystems there are numerous phytophagous species, most of which do not cause any significant damage because they do not normally occur in sufficient numbers. Nevertheless, they have the potential to become pests, and if pesticide treatment releases such a population from the regulatory factors that normally keep it in check, a new, **secondary pest outbreak** can develop (Figure 15-5). This type of situation developed very early in the use of DDT to control apple pests. The DDT killed a wide range of phytophagous insects as well as variety of parasites and predators,

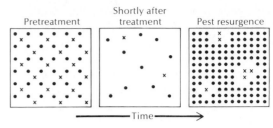

x = Predator-Parasite
• = Pest

FIGURE 15-4.
Diagramatic representation of pest resurgence. The three rectangles represent an agricultural plot immediately before, immediately after, and some time after treatment with an insecticide applied to control a pest (*represented by solid dots*). The immediate result is a significant reduction in the pest accompanied by an even greater reduction in the population of natural enemies (X). Because of the unfavorable ratio between the pest and its natural enemies, and because of the tendency for new pest individuals to invade the plot, there is a rapid resurgence of the pest. (Redrawn from Smith and Van den Bosch 1967.)

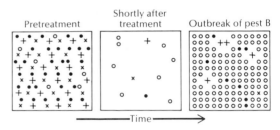

• = Pest A
o = Pest B
x = Predator A
+ = Predator B

FIGURE 15-5.
Diagramatic representation of a secondary pest outbreak. The three rectangles represent an agricultural plot immediately before, immediately after, and some time after treatment with an insecticide applied to control a pest (*solid dots*). The insecticide reduces the pest population but kills both parasites and predators (+ and ×) as well. A noninjurious population of a potential pest (O) is not affected by the insecticide. Released from predation and competition, the secondary species increases to a damaging level. (Redrawn from Smith and Van den Bosch 1967.)

but it was not lethal to spider mites. Without the competition and the natural enemies that normally regulated their populations, the spider mites often increased greatly in numbers and became pests of the orchards in their own right. Many of the most serious pest problems of cotton and alfalfa also originated in this way.

Finally, the favorable nature of most crop environments, combined with pest resurgence, secondary pest outbreaks, and more susceptible host plants, have too often created the need for repeated use of some pesticides. The intense selective pressure that this generates leads rapidly to the development of insecticide resistance in a large variety of insects. Once an insect is resistant to a chemical that previously controlled it, the insect becomes, in a sense, a new pest with an increased potential for causing economic damage. More will be said about pesticide resistance in Chapter 16.

MAJOR GROUPS OF PESTS

Almost any type of organism can be a pest in an isolated situation, but relatively few species cause economic damage throughout much of the world or on a more or less regular basis. Although some vertebrates, particularly rodents and birds, fall into this category, the major pests of agriculture and forestry are weeds, insects, mites and ticks, pathogens, nematodes, and, to a lesser extent, snails and slugs.

Weeds

Weeds are by far the most difficult pests to consider in a brief review because, as we stated earlier, any plant that grows where it is not wanted qualifies as a weed, and there is often little specificity between weeds and the crops they invade. The volume on weed control pre-

pared by the National Research Council (1969) contains a table listing almost two thousand species of weeds. The weed control manual published by Agri-Fieldman (1975) contains an index of about 370 weeds of importance in American agriculture, about 10 percent of which are serious pests to a variety of major crops. Which weeds are most important is a matter of opinion, but those listed in Table 15-1 would most certainly be considered as major pests of agricultural crops.

Weeds are undesirable for the following reasons: They compete with beneficial vegetation for space, nutrients, and water; they can create a fire hazard; some are toxic or cause allergic reactions in man and his animals; some are parasites on economically important plants; and many harbor injurious insects, plant pathogens, and rodents. Conversely, the same plants that cause crop losses can also reduce soil erosion, add organic material to the soil, fix nitrogen, and provide food and shelter for wildlife and other beneficial organisms.

Most weeds are plant species particularly well adapted to establishment and survival in disturbed areas that would be hostile environments for many plants. These species are adapted through their life histories, morphology, physiology, and reproduction to thrive in open habitats and to withstand frequent severe distrubances. Ecologically, most weeds are the pioneer species that modify harsh environments sufficiently to permit the entry of less tolerant species and thereby begin succession toward more stable communities. It is therefore not surprising that weed control is a difficult and costly procedure because it must counteract a strong, fundamental ecological tendency.

Agriculture is a disturbing force that repeatedly creates opportunities for aggressive plant species to invade disturbed sites. Such invaders are efficient competitors, and most of the losses they cause in crop ecosystems are the result of their capacity to grow rapidly

TABLE 15-1

Some major weed pests of agriculture in the United States and some of the principal crops in which each is a problem (arranged alphabetically by common name).

Common name	Scientific name	Principal crops affected
Bindweed	*Convolvulus arvensis*	Wheat, barley, oats, field corn
Canadian thistle	*Cirsium arvense*	Wheat, barley, oats, field corn, peas, beans
Chickweed	*Stellaria media*	Alfalfa, cole crops, potatoes, soybean, sugar beets
Cocklebur	*Xanthium* spp.	Cotton, wheat, barley, oats, sorghum, sweet corn
Crabgrass	*Digitaria* spp.	Beans, cole crops, cotton, corn, field crops, potatoes
Foxtail	*Setaria* spp.	Beans, cole crops, field corn, lettuce, soybeans
Goosegrass	*Eleusine indica*	Cole crops, cotton, peanuts, soybeans, tomatoes
Lamb's-quarters	*Chenopodium album*	Beans, carrots, celery, cotton, small grains, sugar beats
Mustard	*Brassica* spp.	Alfalfa, small grains, corn, sugar beets, rangeland
Nightshade	*Solanum* spp.	Beans, carrots, celery, small grains, onions, soybeans
Pigweed	*Amaranthus* spp.	Beans, carrots, cotton, peanuts, potatoes, orchards, pasture
Purslane	*Portulaca oleracea*	Cotton, lettuce, peanuts, potatoes, soybeans, sugar beets, rangeland
Ragweed	*Ambrosia* spp.	Cotton, small grains, field corn, soybeans, sugar beets, rangeland
Russian thistle	*Salsola kali*	Small grains, soybeans, sugar beets
Shepherd's purse	*Capsella bursa-pastoris*	Alfalfa, small grains, sugar beets, pasture, rangeland

and to use the water, nutrients, and sunlight that would otherwise support the growth of the crop (Figure 15-6). A relatively few species, such as dodder in alfalfa or witchweed in corn, soybean, and peanut, cause direct damage as parasites (Figure 15-7). Even fewer are actually toxic to humans or livestock.

Insects and Mites

Insects and mites are an enormous group of animals, roughly half of which are plant feeders. Thus it is not surprising that insects and mites form the most important group of agricultural pests. The number of such species that regularly attain pest status is difficult to determine but probably lies between 500 and 600 in American agriculture. Although some species are generalists, many are rather specific pests that attack a limited number of closely related plants. Insects and mites cause damage mainly by injuring, destroying, or consuming plant tissue, but they also have great impact as vectors of plant pathogens, especially viruses, and as consumers and foulers of stored products.

In evaluating injury caused by insects it is often useful to distinguish between **direct damage** and **indirect damage.** By direct damage we mean that the commodity of value is directly destroyed or downgraded. The consumption or spoilage of stored products by insects, the infestation of fruit or seed, the killing of trees, damage to lumber, or the death of a calf as a result of parasite attack are all examples of direct damage. Examples of indirect damage

FIGURE 15-6.
A heavy crop of weeds in a poorly attended plum orchard in north-central California. (Photograph by M. D. Atkins.)

FIGURE 15-7.
The parasitic pest, witchweed (*left*), emerging from the fringe of a corn plant's root system. Witchweed is an introduced pest of crops in the grass family and is currently on the United States quarantine list. (Photograph courtesy of the USDA.)

are the decline in crop yield because of damage to nonusable parts of the plant, the reduction in the incremental growth of forest trees because of defoliation, or a decline in milk production by dairy cattle because of irritation caused by parasites.

The most conspicuous damage to plants is generated when insects with chewing mouthparts feed directly on external parts such as the leaves or roots. The results of such damage can be quite spectacular on an individual plant (Figure 15-8) or over a large area (Figure 15-9). Other chewing insects might feed inside stems, fruits, seeds, and even between the surfaces of the leaves; this damage might be quite noticeable, but most internal damage is difficult to detect until after the serious injury has been done.

The second major form of plant damage is caused by insects with piercing-sucking mouthparts and by mites. Although it is not as conspicuous as damage caused by chewing insects, it can be just as significant because of the sap withdrawal and the creation of puncture wounds and groups of dead cells that may be grown over by scar tissue. Severely damaged foliage can lose much of its photosynthetic capacity, and photosynthesis can also be reduced by sooty molds that grow in the sticky exudate produced by sucking insects.

Other forms of plant damage are caused when insects inject toxins at the time of feeding and create wounds when they deposit their eggs. The toxins either kill the plant or cause deformation and galling, and the wounds that a variety of insects cause in laying eggs can

FIGURE 15-8.
Leaf damage to a grape vine being caused by the Japanese beetle. (Photograph courtesy of the USDA.)

FIGURE 15-9.
A corn field devastated by the European corn borer. (Photograph courtesy of the USDA.)

result in breakage or can create sites for invasion by pathogens.

All of these means can directly or indirectly damage crops, forests, and ornamental plantings. Most direct damage in agriculture is caused by insects that feed or oviposit on or in fruiting structures. A few well-known examples include the codling moth, the larvae of which feed in English walnuts and apples (Figure 15-10), and related species such as oriental fruit moth and cherry fruitworm, the larvae of which feed in other fruits. A great many insects, including the lygus bug, cotton bollworm (Figure 15-11), boll weevil, and pink bollworm, damage the squares (buds) and bolls of cotton. The destruction of leaf crops, such as lettuce and cabbage, by caterpillars of the imported cabbageworm and cabbage looper, and the damage to root crops caused by wire worms, are also classed as direct damage.

Sometimes a direct loss results from insect activities that affect the appearance of a product rather than a reduction in yield. Probably the best example is found in the citrus industry, where fruit marked by mites or scale insects is downgraded (Figure 15-12) even though the edible parts are unaffected—truly a reflection of our society's values. Some of the more important pests that cause direct damage in American agriculture are listed in Table 15-2.

Generally speaking, most indirect damage results from insect attack on vegetative parts (leaves, stems, roots), which reduces the yield of reproductive or storage parts (fruits, seeds, tubers) that are harvested. Most cereal pests are of this type, including a variety of aphids (which are juice feeders), the cereal leaf beetle and grasshoppers, (which chew on the leaves), wireworms and cutworms (which attack the roots), wheat stem maggot (which bores in the

FIGURE 15-10.
An apple sliced to show the damage caused to the fruit by the larva of the codling moth.
(Photograph courtesy of the USDA.)

FIGURE 15-11.
A cotton boll worm emerging from a damaged cotton boll in which it developed.
(Photograph courtesy of the USDA.)

FIGURE 15-12.
Oranges damaged only on the surface by a heavy infestation of rust mite. (Photograph by M. D. Atkins.)

stalk), and the wheat stem sawfly (which injures the stalk with oviposition punctures and by larval feeding.)

The European corn borer has become one of the most serious indirect agricultural pests in the United States, not only because of attacks on corn, but also because it feeds on over 200 other varieties of plants, including soybeans, sorghums, cereals, potatoes, and beans. Another important indirect pest, the Colorado potato beetle, reduces potato yield by devouring the leafy crown of the plant (Figure 15-13).

The order Homoptera contains a variety of important indirect pests that seriously affect the yield of many crops, particularly tree fruits. Probably the most important pest in this group is the San Jose scale, which was introduced into California about 1870. The San Jose scale does not usually cause direct damage to the fruit, but causes more permanent damage by killing twigs and branches. It is particularly destructive to apples, pears, peaches, sweet cherries, and citrus. Other members of this group include other scales, aphids, whiteflies, and mealybugs, many of which attack citrus. A few of the more important indirect pests of American agriculture are listed in Table 15-3.

Although heavy parasite infestations can kill livestock, most of the victims are usually young animals. Most insect-caused losses in the livestock industry are indirect and are the result of poor performance caused by annoyance or a weakened condition. Insects that attack livestock and poultry, like those that attack plants, vary from being quite host specific to being capable of attacking a variety of domesticated species. Members of the order Diptera, including the stable fly, horseflies, blackflies, the horn fly and the face fly, are the least specific pests of livestock. Livestock are sensitive to insect bites and to the feeding of parasitic larvae, and the annoyance and irri-

TABLE 15-2
Representative direct insect pests of agricultural crops (arranged alphabetically by family).

Common name	Order	Family	Scientific name	Crop affected
Boll weevil (I)	Coleoptera	Curculionidae	*Anthonomous grandis* Boheman	Bolls of cotton
Pink bollworm (I)	Lepidoptera	Gelechiidae	*Pectinophera gossypiella* (Saunders)	Bolls of cotton
Cabbage looper (N)	Lepidoptera	Geometridae	*Trichoplusia ni* (Hubner)	Lettuce and cruciferous crops
Lygus bug (N)	Hemiptera	Miridae	*Lygus hesperus* Knight	Legumes and cotton
Corn earworm	Lepidoptera	Noctuidae	*Heliothis zea* (Boddie)	Corn and vegetable crops
Tomato fruitworm				Tomatoes and tobacco
Cotton bollworm (N)				Bolls of cotton
Codling moth (I)	Lepidoptera	Olethreutidae	*Laspeyresia pomonella* (Linnaeus)	Apples and English walnuts
Oriental fruit moth (I)	Lepidoptera	Olethreutidae	*Grapholitha molesta* (Busck)	Stone fruits
Imported cabbageworm (I)	Lepidoptera	Pieridae	*Pieris rapae* (Linnaeus)	Cruciferous crops, vegetables
Japanese beetle (I)	Coleoptera	Scarabaeidae	*Popillia japonica* Newman	Tree fruits, truck crops, and turf
Cherry fruit fly (I)	Diptera	Tephritidae	*Rhagoletis cingulata* (Loew)	Sweet cherries
Mediterranean fruit fly (I)	Diptera	Tephritidae	*Ceratitis capitata* (Wiedemann)	All fruits including citrus
Oriental fruit fly (I)	Diptera	Tephritidae	*Dacus dorsalis* (Hendel)	Peach, apricot, plum, apple, pear, quince

(I) = Imported species.
(N) = Native species.

FIGURE 15-13.
The infamous Colorado potato beetle. (*Left*) An adult adjacent to a cluster of eggs on a potato leaf; (*right*) damage to young potato plants. (Photographs courtesy of the USDA.)

tation caused by biting flies can cause livestock to become restless, to curtail feeding, and even to stampede. Such abnormal activities cause weakness and weight loss.

Blackflies can be particularly irritating. In northern areas, where huge numbers of these flies emerge almost simultaneously nearly every year, some types of agricultural development, particularly the establishment of a cattle industry, have been curtailed. In northern Saskatchewan outbreaks of the blackfly *Simulium arcticum* kill cattle; an outbreak in 1946 is said to have killed 600 animals. *Simulium venustum*, which attacks both humans and domestic animals, can reduce the milk production of dairy cattle in northern areas by as much as 50 percent (Fredeen 1956). Similarly, the transmission by the tsetse fly of trypanosome parasites (causal agents of African sleeping sickness) from game animals to humans and livestock

has retarded settlement and the establishment of agriculture over much of central Africa.

The native screwworm (Figure 15-14), although now largely under control in the United States, has been a serious livestock pest. The damage is caused by the maggots of this species feeding on healthy tissue adjacent to minor wounds. During a severe outbreak in 1935, 1.2 million infected animals and 180,000 deaths were recorded. Before effective control was achieved, the screwworm caused an estimated $100 to $120 million in damage to the livestock industry in the southern United States.

The most important pests of poultry are biting lice of the order Mallophaga. Of nine species that attack chickens in this country, the chicken body louse, chicken head louse, and shaft louse are the most important. The feeding of heavy infestations of these ectoparasites causes severe irritation that produces

TABLE 15-3
Representative indirect pests of agricultural crops (arranged alphabetically by family).

Common name	Order	Family	Scientific name	Crop affected
Migratory grasshopper (N)	Orthoptera	Acrididae	*Melanoplus sanguinipes* (Fabricius)	Cereals and corn
Green bug (I)	Homoptera	Aphididae	*Schizaphis graminum* (Rondani)	Cereals, particularly wheat
Wheat stem sawfly (N)	Diptera	Cephidae	*Cephus cinctus* Norton	Wheat and other cereals
Cereal leaf beetle (I)	Coleoptera	Chrysomelidae	*Oulema melanopus* (Linnaeus)	Wheat and other cereals
Colorado potato beetle (N)	Coleoptera	Chrysomelidae	*Leptinotarsa decemlineata* (Say)	Potato, tomato, pepper, tobacco
Mexican bean beetle (N)	Coleoptera	Coccinellidae	*Epilachna varivestis* Mulsant	Beans
White fringed beetle (N)	Coleoptera	Curculionidae	*Graphognathus* spp.	Corn, peanuts, potato, cotton
San Jose scale (I)	Homoptera	Diaspidae	*Quadraspidiotus perniciosus* (Comstock)	Apple, pear, peach, cherry, citrus
Chinch bug (N)	Hemiptera	Lygaeidae	*Blissus leucopterus* (Say)	Wheat, corn, sorghums
Armyworm (N)	Lepidoptera	Noctuidae	*Pseudaletia unipuncta* (Haworth)	Cereals
Fall armyworm (N)	Lepidoptera	Noctuidae	*Spodoptera frugiperda* (Smith)	Cotton, corn, peanuts
European corn borer (I)	Lepidoptera	Pyralidae	*Ostrinia nubilalis* (Hubner)	Corn and field crops
Tobacco hornworm (N)	Lepidoptera	Sphingidae	*Manduca sexta* (Linnaeus)	Tobacco and tomato

(I) = Imported species.
(N) = Native species.

FIGURE 15-14.
A cow with a number of infected abscesses on its forequarter caused
by larvae of the screwworm feeding beneath the skin. The infestation
weakens the animal and devalues the hide. (Photograph by
T. Kamphuis, courtesy of the FAO.)

a failure to gain weight, reduces egg production, and sometimes kills young birds.

As mentioned earlier, a wide variety of insects feed on and despoil stored products. Unless precautions are taken the losses sustained in storage can equal or exceed those sustained by crops during cultivation. Various authors have claimed that in areas where little or no pest control is practiced, insects can destroy 50 percent of a crop before harvest and the other 50 percent after harvest! In the United States, where storage facilities are good and control methods are readily available, storage losses due to insects amount to $300 to $600 million annually. According to one estimate, 130 million people could have lived for a year on the grain consumed or spoiled throughout the world by insect pests in 1968 (Wilbur 1971). The most important storage pests are listed in Table 15-4.

Plant Pathogens

Plant pathogens are disease-causing microorganisms that derive their food and energy from plant hosts. The association of these organisms with plants often causes a physiologi-

TABLE 15-4
Important insect pests of stored products (arranged alphabetically by family).

Common name	Scientific name	Family
Lesser grain borer	*Rhyzopertha dominica* (F.)	Bostrichidae
Sawtoothed grain beetle	*Oryzaephilus surinamensis* (L.)	Cucujidac
Merchant grain beetle	*Oryzaephilus mercator* (Fauvel)	Cucujidae
Flat grain beetle	*Cryptolestes pusillus* (Schonherr)	Cucujidae
Rusty grain beetle	*Cryptolestes ferrugineus*	Cucujidae
Rice weevil	*Sitophilus oryzae* (L.)	Curculionidae
Maize weevil	*Sitophilus zeamaize* Motschulsky	Curculionidae
Granary weevil	*Sitophilus granarius* (L.)	Curculionidae
Khapra beetle and related species	*Trogoderma* spp.	Dermestidae
Angoumois grain moth	*Sitotroga cerealella* (Olivier)	Gelechiidae
Cadelle	*Tenebroides mauritanicus* (L.)	Ostomidae
Indian meal moth	*Plodia interpunctella* (Hubner)	Pyralidae
Mediterranean flour moth	*Anagasta kuehniella* (Zeller)	Pyralidae
Confused flour beetle	*Tribolium confusum* duVal	Tenebrionidae
Red flour beetle	*Tribolium castaneum* (Herbst)	Tenebrionidae

cal disturbance, or **disease,** that can be manifested in morphological changes, or **symptoms.** The study of plant diseases encompasses all forms of physiological disorders, including those caused by abiotic factors such as weather and chemicals. The similarity between the symptoms caused by abiotic factors and by pathogens is sometimes very great, but we will restrict our consideration to biotic causal agents, particularly **fungi, bacteria, viruses,** and **nematodes** (treated separately).

The fungi constitute the most diverse group of plant pathogens, and they infect all types of higher plants. Most agricultural crops are susceptible to fungal attack, and some, such as cotton, beans, and potatoes, can be infected by at least 30 different fungi. Fungi attack the roots, stems, leaves, flowers, fruits, and seeds of crop plants, thereby reducing seed germination, destroying seedlings, and reducing the vitality of older hosts. Apart from reducing yield during cultivation, fungal pathogens attack a variety of harvested products, such as fruits, tubers, and fiber, while in transit or storage.

Economically important pathogens appear in all the classes of fungi. Among the harmful Zygomycota is the fungus that causes club root of crucifers *Plasmodiophora brassicae;* a number of seed decays; downy mildews (*Peronospora* spp. and *Plasmopara* spp.) that damage the foliage of grapes, crucifers, and other crops; and a variety of important root-rotting fungi of the genus *Phytophthora.* The class Ascomycota is the largest group of fungi and includes the chestnut blight *Endothia parasitica* that wiped out the American chestnut, *Fusarium* wilt fungi, and the *Gibberella* species that attack a variety of cereals. The rusts, which have been especially damaging to wheat,

TABLE 15-5
Some important bacterial, fungal, and viral disease pests of crops grown in the United States (arranged alphabetically by common name).

Common name	Scientific name	Pathogen type	Crops affected
Apple scab	*Venturia inaequalis*	Fungus	Apples
Bacterial wilts	*Pseudomonas* spp.	Bacterium	Tomatoes
Black leg	*Erwinia atroseptica*	Bacterium	Potatoes
Brown rot	*Schlerotinia fructigena*	Fungus	Plums, cherries, peaches
Curly top		Virus[a]	Sugar beets, tomatoes
Damping off of seedlings	*Pythium* spp., *Rhizoctonia* spp.	Fungus	Most annual crops
Downy mildew	*Plasmopara viticola*	Fungus	Grapes
Fire blight	*Erwinia amylovora*	Bacterium	Pears, apples
Fusarium wilts	*Fusarium* spp.	Fungus	Cereals, vegetables, cotton, soybeans
Potato late blight	*Phytophthora infestans*	Fungus	Potatoes (also tomato-blight)
Rice blast	*Piricularia oryzae*	Fungus	Rice
Root rot	*Phytophthora megasperma*	Fungus	Soybeans, cauliflower
Southern corn blight	*Helminthosporium maydis*	Fungus	Corn
Stem rusts	*Puccinia graminis*	Fungus	Cereals
Tobacco mosaic		Virus[a]	Tobacco, tomatoes
Verticillium wilts	*Verticillium* spp.	Fungus	Cotton, tomatoes, some field crops
Yellow dwarf	*Avena sativa*	Fungus	Oats, barley

[a]Plant viruses are not usually designated by scientific names because they are more easily distinguished by the symptoms they cause than by morphological features. Many other viruses are important pests of a wide range of agricultural crops.

belong to the class Basidiomycota. Finally, the Fungi Imperfecti include a number of important crop pests including *Verticillium* wilt fungi and leaf-spot fungi that belong to several genera. Some examples of important fungal diseases of crops are presented in Table 15-5.

Relatively few major crop diseases are caused by bacteria, but some bacterial diseases have caused severe damage. About half the phytopathogenic bacteria belong to the genus *Pseudomonas*, species of which cause a variety of rots, wilts, and foliage blights on a wide range of crops (Table 15-5).

Almost all higher plants are infected by viruses, and a number of these organisms cause widespread damage to crops. The symptoms of virus infections vary widely in both their form and their severity. They can kill plants, but they more often reduce crop yield and quality. The viruses can be separated into two major groups: the **yellows,** which cause yellowing of the foliage, leaf curling, branching, and dwarfing (Figure 15-15); and the **mosaics,** which cause mottling and chlorotic spotting of the foliage (Figure 15-16).

One of the more interesting and important aspects of plant viruses is the variety of ways in which they are transmitted. All viruses that

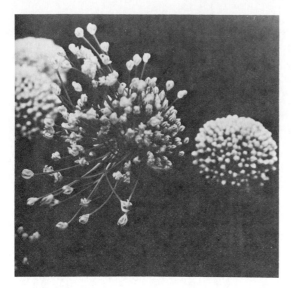

FIGURE 15-15.
An onion plant showing the symptoms of infection with a yellows virus. (Photograph courtesy of the USDA.)

a fairly long period of time, cause yellow diseases and are transmitted by leafhoppers.

A number of insect-vectored viruses infect both crop and noncrop plants, and so wild vegetation adjacent to agricultural land can be not only a reservoir from which the virus can be transmitted to the crop, but also a habitat for the virus while the fields are barren. Table 15-5 lists the more important plant virus disease problems.

Nematodes

Nematodes are small nonsegmented worms. Most nematodes are free-living, but about one thousand species attack green plants, and several hundred are pests of a variety of crops. Most plant parasitic nematodes belong to the order Tylenchida, and are characterized by specialized feeding stylets. These plant

spread systemically throughout their host can be transmitted by grafting. Vegetative propagation that uses parts from infected hosts will infect the new plants; this is important in the spread of virus diseases in fruit trees, potatoes, strawberries, and other crops that are propagated vegetatively. Other viruses are spread by planting infected seeds, but most are spread by biological vectors, including fungi, nematodes, mites, and particularly insects. Some viruses can be transmitted on the mouthparts of insects that move from infected to noninfected plants in the process of feeding. These viruses, called stylet-borne viruses, can only be transmitted for a short time; most are transmitted by aphids and cause a number of diseases referred to as mosaics. Other viruses are called circulative because they circulate within the body of the vector. This group is sometimes subdivided into propagative and nonpropagative, depending on whether they multiply within their vector. Most of these viruses, which can be transmitted over

FIGURE 15-16.
A potato leaf showing the symptoms of infection with potato virus A, commonly known as mild mosaic. (Photograph courtesy of the USDA.)

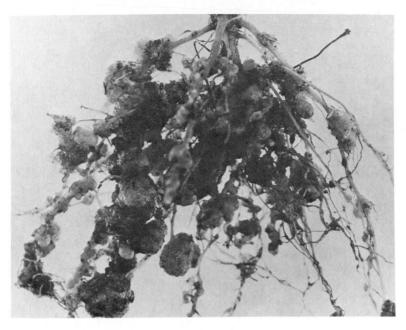

FIGURE 15-17.
Nematode galls, or root knots, on the roots of a soybean plant. (Photograph courtesy of
the USDA.)

parasites cause three basic kinds of damage: root-knot disease, necrotic foliar disease, and seed-gall disease.

Root-knot nematodes are the most important of these pests. They appear throughout the world and infect a variety of row crops such as tomato, tobacco, beans, and soybean. They can generate extensive economic loss, especially in areas with a long growing season conducive to the development of large nematode populations. The disease symptoms include the formation of root galls or tumefactions (Figure 15-17), and root damage that causes dwarfing and a decline in crop yield.

Birds and Mammals

Vertebrate pests are considerably less important than the classes of pests already discussed, but they can cause sporadic and local problems. Birds can be serious pests of crops such as berries and tree fruits but are probably most harmful to grain crops. The weaver finch, *Quelea quelea*, seriously damages wheat, sorghum, and millet in Africa; blackbirds cause serious damage to corn and other grains, particularly in Arkansas, Louisiana, and Mississippi, where they congregate during the autumn and winter; and starlings can be serious pests of livestock feedlots both by consuming grain and fouling the feed with their excrement. Starlings have also caused economic damage to grape crops in California.

Rodents are pests mainly of stored products, which they eat and spoil with their excrement; but small rodents also damage new orchard plantings by feeding on the bark just above the ground and can consume a fairly high percentage of newly planted seeds.

DETERMINING PEST STATUS

In the preceding paragraphs we have briefly reviewed the various origins of potentially harmful pests and some of the features of crop and forest management that favor the increase of pest populations. However, as indicated earlier, populations of organisms should not be suppressed simply because they have a history of causing damage or because an effective pesticide is available.

In forestry and agriculture, the object of pest control is to prevent losses and thereby to increase the profitability of the crop. There are, however, other considerations in public health that cannot be expressed in monetary terms, and related pest control decisions must be made in terms of less tangible criteria than monetary ones.

Generally, the accounting used in agricultural pest control is based primarily on the cost of control and the value of the potential benefits in the year of the control program. Long-term effects of control procedures are rarely considered, regardless of whether they are favorable or unfavorable. In reality, however, any expenditures made by government agencies to clean up pesticide pollution must be considered a cost of control that is paid by the public.

Pest species often appear in a crop at a population level that does not cause sufficient damage to generate a loss of income. Except for a few rather special cases, the mere presence of a potential pest is not reason enough to initiate control measures. Arbitrary decisions to use pesticides as a prophylactic treatment have led to unnecessary residue and resistance problems and are no longer acceptable. It is not always easy, however, to convince a grower who can see insects in his fields that control measures are not required. In principle, therefore, control should be undertaken only if the pest population reaches a density that will otherwise lead to economic loss. This population level is called the **economic threshold** and identifies a crucial

point in the control decision process. The term **economic injury level** is often used synonymously with economic threshold, but should be reserved to describe the level of pest population or damage that is of economic significance. These values obviously require that some relationship be established between the density of the pest and the damage it is likely to do to the crop yield. It is also clear that the population density must be determined by an appropriate sampling procedure.

The relationship between pest density and damage can vary considerably and is seldom a straight-line relationship. At low densities some pests act as thinning agents and can actually increase crop yield by reducing the competition between plants. Moreover, high pest populations experience intraspecific competition that can lower the impact of each individual on the crop. Between these extremes there may be a more direct relationship between the pest population and reduced yield, but even this can vary from one situation to another, depending, for example, on the stage of crop development. A population that can severely damage a crop in the early stages of development might have an insignificant effect on a mature crop near harvest. Similarly, the economic threshold might need to be adjusted to take into account certain events that occurred earlier in a crop's development; for example two different pests might be present at density levels that would be tolerable if they were completely independent of one another, but the damage done by one pest early in the development of a crop might make it impossible for the crop to withstand later damage done by the second pest.

Sometimes a large pest population develops in a crop or invades it after a critical stage in the crop's development. For instance, a large population of alfalfa caterpillars in a crop about to be mowed could be insignificant. Likewise, the invasion of a cotton crop by a potentially damaging population of pink bollworms or boll weevils might not significantly reduce the yield if the squares or bolls are developed beyond the stage of development that these insects prefer.

The situation can be even more complex if the potential problem involves insect transmission of a pathogen. A large vector population could be tolerated if the crop were a disease-resistant variety or if the pathogen reservoir were small or inaccessible to the vector. On the other hand, a susceptible crop could suffer severe damage from a much smaller vector population actively feeding and moving back and forth between the crop and an accessible pathogen reservoir. Consequently, in insect-vector disease situations, the establishment of an economic threshold requires information about the location and size of pathogen reservoirs as well as the vector population.

Although it is sometimes possible to tell from experience whether a given pest population attacking a particular crop at a particular time will reduce the crop yield, it is more difficult to determine what this will mean in lost income, which obviously affects any pest control decision. This may seem paradoxical in a world that is growing short of food, but it is nevertheless a factor in the economics of intensive agriculture. There must be a favorable cost-benefit ratio. The costs, which include the price of the pesticides and their application must be offset by at least an equivalent gain in income, and in fact the benefits of pest control should substantially exceed its cost. This means that the economic threshold must be closely related to the planned disposition of the crop. If, for example, the crop is grown for processing, as are some apples (for cider) and tomatoes (for catsup), more damage of certain kinds can be tolerated, and the reduced overall value of the crop limits the amount that can be spent on pest control. Sometimes the market conditions change because of changing patterns of supply and demand; it is indeed unfortunate when money is spent on pesticides to protect a crop for which there is no market. Such situations do develop but are often difficult to foresee at the time

when a control decision must be made. For example, in 1970 only about 76 percent of the 793,000-ton peach crop in California was purchased by canners; an estimated 79,400 tons were shaken from the trees before harvest and left to rot (Perelman and Shea 1972).

As pointed out by Woods (1974), the use of pesticides is subject to the law of diminishing returns. The initial pest control input might provide less extra yield than subsequent input, but a point is reached at which further control expenditures generate little or no increase in yield as the pest population is suppressed below the economic threshold. The concept of total pest control that once prevailed does not make economic sense even though some farmers still adhere to the "clean field" concept. Ideally, growers should be encouraged to curtail pest control at the point where the benefits to be expected from the increased yield drop below the cost of continued control.

Other considerations relate to whether the pest directly attacks the portion of the crop to be used or some other part in which damage reduces the yield of the desired part. In the first case, the greater the proportion of pests killed the greater the proportion of saleable yield; in the second case, lowering the level of control might not significantly decrease yield. For example, Wheatley and Coaker (1970) found that the percentage increase in the yield of cauliflower rose only slightly when the dosage of insecticide directed against cabbage root fly was increased from 3 milligrams per plant to 50 milligrams per plant, although the root damage itself was decreased from about 95 percent to 5 percent. Apparently, established cauliflower plants were able to tolerate considerable root damage and only low levels of pesticide application were necessary to protect the roots immediately following transplanting.

In forestry the economics of pest control is greatly affected by the time during the development of the stand when a potential pest problem develops. This also applies in some long-term agricultural crops, especially in the years before a harvestable crop develops. The annual increase in the value of a forest depends on incremental growth and is very low per acre. Consequently there is no annual income to offset management costs that include pest control. In the early stages of growth, it might be necessary to protect against some pests, such as shoot moths and weevils, in order to assure a growth-form that will lead to marketable trees later on; but some tree mortality may be beneficial as a form of natural thinning. During intermediate years, when growth is maximal, any losses in annual growth (due to defoliation, for example) might not be significant enough to warrant the expense of insect control. However, when the forest is in a harvestable condition it represents not only the value accumulated over a long time but the accumulated management costs as well. At this stage tree mortality represents a significant loss; the degree of damage that can be tolerated is therefore low, and a fairly expensive pest control program is justified.

Economic thresholds have not been established for very many crop pests, and those that have been established are usually set too low in order to be on the safe side; but this often leads to unnecessary control programs and unnecessary production costs for growers. However, the establishment of ideal economic thresholds is not a simple matter, as the foregoing discussion demonstrates. It is often necessary to evaluate many variables, including the ultimate disposition of the crop, the stage of crop development, crop variety, climatic conditions, prior investment, and market conditions.

Various workers have struggled with this problem and have derived models of varying complexity in an attempt to formulate an approach that would optimize the relationship between yield and pest control costs. A fairly simple example presented by Smith and van den Bosch (1967) illustrates that what might appear to be severe insect damage is not necessarily economic damage. Cotton has a limited capac-

ity to set bolls, and the squares that exceed that number drop whether or not they are punctured by lygus bugs or attacked by other square-feeding insects. Consequently the damage that the bugs do to the excess squares does not cause a reduction in yield. But some crop loss results when the insect damage reduces the boll load to a point below the plant's carrying capacity. A small reduction in the number of bolls below carrying capacity may not be economically significant, but control could be required to prevent further damage (Figure 15-18).

Ordish (1952) tabulated a cost-potential benefit for a series of crops that took into account a variety of situations. For example, he considered three kinds of apple orchards with different yield characteristics and determined the cost-benefit ratio for three levels of pest control (low, partial, and full) against severe, medium, and light attacks by fungal diseases and by arthropods. Headley (1972) developed a still more sophisticated approach aimed at establishing economic thresholds that permit optimization of net return from the control procedure. The following summary of Headley's approach incorporates the modifications presented by Woods (1974) but does not alter the essence of the original concept.

At a given time the grower of a particular crop operates according to a production function that relates input to output. Given a fixed amount of land and labor, capital applied in the best possible way generates a production curve similar to curve A in Figure 15-19. Assuming no pest attack, the most profitable relationship between cost and output occurs at the intersect of O_A and C_A. Beyond this point additional output is subject to the law of diminishing returns, and added units of capital input exceed the value of added units of output as indicated by the tangent of the curve at the above intersect.

If pest damage occurs after the full input of capital is committed, and no pest control is undertaken, a new production curve (B in Figure 15-19) is generated. Because the costs remain

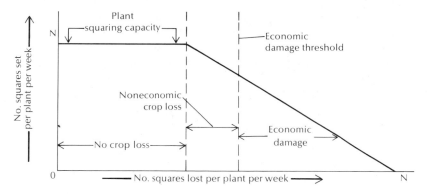

FIGURE 15-18.

The relationship between the maximum crop (plant squaring capacity) and the economic threshold for damage caused by insect attack. See text for additional details. (Redrawn from Smith and Van den Bosch 1967.)

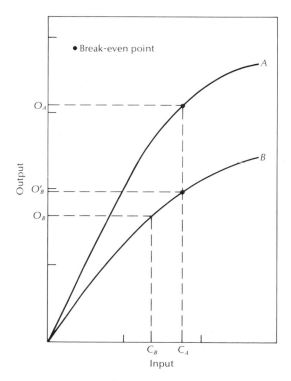

FIGURE 15-19.

Cost-output curves for an uninfested crop and a crop damaged by pests in the absence of pest control. (Modified from Headley 1972.)

the same but the output is reduced to O'_B the grower loses the difference between O_A and O'_B. Point O'_B lies on the new production curve beyond the profit maximization point represented by the intersect of O_B and C_B. If the grower has anticipated the pest attack he should reduce his capital input from C_A to C_B in order to make a profit of $(O_B - C_B)$, which is larger than $(O'_B - C_A)$. The input of capital equal to C_A without pest control actually increases the loss since the increased output between O_B and O'_B costs more than it is worth.

If the pest damage could be reduced partially by pest control, a third curve, lying between the other two, would be generated (Figure 15-20). The point of maximum profit on this curve, excluding the cost of pest control, is indicated by the intersect of O_C and C_C. The decision whether to engage in pest control can then be made on the basis of the control cost

relative to increased profit. With pest control, the maximum profit the grower could obtain would be $(O_C - C_C) - (\text{cost of pest control})$, and the extra profit he would gain above the maximum attainable without pest control would be $(O_C - C_C) - (O_B - C_B) - (\text{cost of pest control})$. If the cost of pest control is greater than the difference between $(O_C - C_C)$ and $(O_B - C_B)$ then it is not worthwhile. On the other hand, if the grower can change from production curve B to production curve C and increase his profit by doing so, the control action should be taken.

Headley's approach introduces the concept of economic efficiency, or the maximization of the difference between total benefits and total costs, as a means of avoiding the overestimation of pest losses and a corresponding overinvestment in pest control. This approach may require some modification in developing coun-

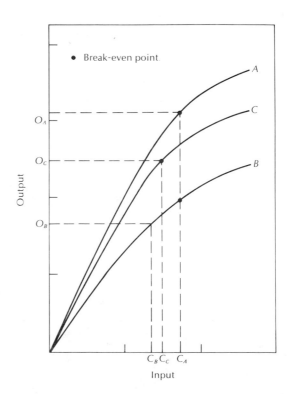

FIGURE 15-20.
Cost-output curves for a crop in which a pest population is partially controlled with pesticides. (Modified from Headley 1972.)

tries, where an increase in the production of food could be more important than the cost of each unit. Unfortunately, even in the developed nations the biological or economic data necessary to put Headley's method into operation is not readily available. Nevertheless, these ideas indicate that a more sophisticated approach to the concept of the economic threshold is possible and that such an approach could help to reduce the unnecessary use of pesticides. In the meantime we should apply the best of the simpler methods developed by Ordish and others and work at making economic thresholds realistic. Students interested in reading more on this topic should examine the excellent review by Stern (1973).

ESTIMATING PEST DAMAGE

There is no question that in the absence of pest control insect-caused damage can substantially affect the yield of some crops, but it is also true that some crops in certain areas suffer insignificant damage and require no pest control. The effects of indirect pest damage are certainly the most difficult to evaluate; some low levels of feeding can actually increase yield, whereas a high pest population can totally destroy a crop. Conversely, every unit of produce rendered unsaleable by direct pest damage represents a loss, and without control almost every unit can be infected. We have seen untended apple orchards, for example, in which almost every apple contained codling moth larvae. Likewise, grasshoppers can destroy almost every blade of grass over a wide area of rangeland.

Nobody really knows the monetary value of pest-caused damage, even when it is considered only in terms of material losses. If we attempt to consider, as well, the misery and suffering caused by starvation the problem becomes insoluble. Some good estimates of crop losses have been made in the United States and in other highly developed countries, but it is difficult to arrive at global estimates because of the variability in yields per acre and the amount spent on pest control, which in a sense is part of the loss when it is measured in monetary terms. Even when we confine our estimate to the economy of one nation, losses expressed in monetary terms can be misleading because of changes in commodity prices and the fact that the pest control industry, while adding to the cost of crop production, also contributes to the gross national product.

Expressing losses in terms of absolute yield or a percentage of yield is not an accurate portrayal because it does not permit the inclusion of any value losses that result from a decline in the quality of the produce as a result of inadequate pest control. Furthermore, losses cannot be expressed accurately as a reduction in the yield that would have been obtained in the absence of pests, because this is an unknown quantity.

Any attempt to determine losses during a single season is complicated by a variety of factors other than pests (e.g., weather) that make year-to-year comparisons difficult. On the other hand, if pest losses are based on an average for several seasons, they must take into account yield increases that may have accrued throughout the period from improved crop varieties, better pest control techniques, and increased uses of fertilizer and irrigation.

Ordish (1952) attempted to overcome some of these problems by introducing a unit he called the **untaken acre,** which represents the extra land that must be devoted to a crop to make up for the yield lost to pests. On an international scale, however, it is difficult to apply such a system because of the vast differences in the productivity of different agricultural systems.

One of the best attempts to estimate the agricultural losses due to pests was made by Cramer (1967), who surveyed the available data for the entire world and, using the prices paid to the farmer in each country, estimated losses amounting to between 70 and 90 thousand million U.S. dollars per year. He estimated that this amounts to about 55 percent of actual production, or 35 percent of potential production. The latter figure could be broken down to 13.8 percent attributable to insects, 11.6 percent to diseases, and 9.5 percent to weeds.

Cramer's figures, however, represent losses incurred up to harvest and do not include those that usually occur during transport and storage. Worldwide these losses may be as great as pre-harvest losses. A friend who was in India when a large amount of grain arrived from the United States, in an effort to relieve famine, claims that half the shipment was eaten by rats before it reached the people. In the Congo, insects in one year caused in-storage weight losses of 50 percent to sorghum, 20 percent to beans, and 15 percent to groundnuts (Woods 1974). Furthermore, it must be recognized that weight loss is not a perfect measure of stored product losses, because part of what remains is insect biomass, waste material, and badly damaged commodity.

SUMMARY

An organism only becomes a pest when human beings define it as one. The decision to call an organism a pest is a matter of judgment based on the level of conflict that develops between human objectives and the normal ecological role of the organism in question. The most important groups of pest organisms are unwanted plants (weeds) that compete with desired plants (crops or forage), herbivores (insects, mites, birds, and rodents) that are not exploited by man, and pathogenic organisms (bacteria, fungi, viruses, and nematodes) that cause diseases of crop, pasture, and range plants.

Pests can be either native or imported species that attack native and/or introduced cultivars or domesticated animals. They attain pest status for a variety of reasons ranging from the genetic makeup of the crop, through ecological changes associated with the cultivation of a crop, to more pronounced ecological disturbances caused by pest suppression. The damage caused by pest organisms varies widely in its extent but pests that directly damage the harvested commodity are usually more harmful than those that reduce yield by some form of indirect damage.

Estimates of agricultural losses caused by pests are difficult to make, and they vary widely according to the source of loss data and the way it is evaluated. The best estimates, however, indicate that pest losses before harvest amount to approximately one-half of the world's actual agricultural production, or about one-third of the potential production. The latter figure can be broken down and assigned almost equally to the insects and mites, plant pathogens, and weeds. Losses after harvest range from about 5 percent in the United States to 50 percent in some undeveloped nations.

Literature Cited

Agri-Fieldman. 1975. *Weed Control Manual* 31(2):30–46, la–22a.

Anderson, J. R. 1966. Recent developments in the control of some arthropods of public health and veterinary importance—muscoid flies. *Bull. Entomol. Soc. Amer.* 12:342–348.

Blair, W. F. 1964. An ecological context for the effects of chemical stress on population structure. *Bull. Entomol. Soc. Amer.* 10:225–228.

Cannon, W. N., and W. A. Connell. 1965. Populations of *Tetranychus atlanticus* McG. (Acarina: Tetranychidae) on soybean supplied with various levels of nitrogen, phosphorus and potassium. *Entomol. Exp. Appl.* 8:153–161.

Cook, W. C. 1967. *Life history, host plants and migrations of the best leafhopper in the western United States.* Tech. Bull. 1365. Washington, D.C.: USDA.

Cramer, H. H. 1967. Plant protection and world crop production. *Pflanzenschutz-Nachrichten, Bayer.* 20:1–524.

Elton, C. S. 1958. *The Ecology of Invasions by Animals and Plants.* London: Methuen and Co.

Fredeen, F. 1956. Blackflies (Diptera: Simuliidae) of the agricultural areas of Manitoba, Saskatchewan and Alberta. *Proc. Int. Cong. Entomol.* 10(3):819–923.

Graham, F., Jr. 1972. The war against the dreaded gypsies. *Audubon* 74(2):45–51.

Haynes, D. L. 1973. Population management of the cereal leaf beetle. In *Insects: Studies in Population Management.* P. W. Geier, L. R. Clark, D. J. Anderson, and H. A. Nix (eds.). pp. 232–240. Canberra: Ecological Society of Australia (Memoirs 1).

Headley, J. C. 1972. Economics of agricultural pest control. *Ann. Rev. Entomol.* 17:273–286.

LeRoux, E. J. 1954. Effects of various levels of nitrogen, phosphorus, and potassium in nutrient solution on the fecundity of the two spotted spider mite *T. bimaculatus* reared on cucumber. *Can. J. Agr. Sci.* 34:145–151.

MacArthur, R. 1960. On the relative abundance of species. *Amer. Nat.* 94:25–34.

Marchal, P. 1908. The utilization of auxiliary entomophagous insects in the struggle against insects injurious to agriculture. *Pop. Sci. Monthly* 72:352–419.

Michener, C. D. 1975. The Brazilian bee problem. *Ann. Rev. Entomol.* 20:399–416.

National Research Council. 1969. *Principles of Plant and Animal Pest Control*, vol. 3, *Insect-Pest Management and Control.* Washington, D.C.: National Academy of Sciences.
Disease Development and Control. Washington, D.C.: National Academy of Sciences.

Nene, Y. L. 1971. Plant protection: A key to maintaining present gains in food production. In *Some Issues Emerging from Recent Breakthroughs in Food Production.* K. L. Turk (ed.). pp. 365–377. Ithaca, N.Y.: State College of Agriculture.

Nye, P. H., and D. J. Greenland. 1960. *The soil under shifting cultivation.* Farnham Royal, England: Commonwealth Agricultural Bureaux.

Ordish, G. 1952. *Untaken Harvest.* London: Constable.

Perelman, M., and K. P. Shea. 1972. The big farm. *Environment* 14:10–15.

Pimental, D. 1961. Species diversity and insect outbreaks. *Ann. Entomol. Soc. Amer.* 54:76–86.

Pradhan, S. 1971. Revolution in pest control. *Pesticides* 5(8):11–17.

Rivnay, E. 1964. The inflence of man on insect ecology in arid zones. *Ann. Rev. Entomol.* 9:41–42.

Smith, R. F. 1971. *The impact of the Green Revolution on plant production in tropical and sub-tropical areas.* 1970 Founders Memorial Lecture. Presented to the Entomol. Soc. Amer. Nov. 29, Los Angeles.

Smith, R. F., and R. van den Bosch. 1967. Integrated control. In *Pest Control: Biological, Physical, and Selected Chemical Methods.* W. W. Kilgore and R. L. Doutt (eds.). pp. 295–340. New York: Academic Press.

Southwood, T. R. E. 1962. Migration of terrestrial arthropods in relation to habitat. *Biol. Rev.* 37:171–214.

Stern, V. M. 1973. Economic thresholds. *Ann. Rev. Entomol.* 18:259–280.

van Emden, H. F. 1966. Plant insect relationships and pest control. *World Rev. Pest Cont.* 5:115–123.

van Emden, H. G., and G. C. Williams. 1974. Insect stability and diversity in agroecosystems. *Ann. Rev. Entomol.* 19:455–475.

Watt, K. E. F. 1965. Community stability and the strategy of biological control. *Can. Entomol.* 97:877–895.

Wheatley, G. A., and T. H. Coaker. 1970. Pest control objectives in relation to changing practices in agricultural crop production. In *Technological Economics of Crop Protection and Pest Control,* Monograph No. 36, Society of Chemical Industry.

Wilbur, D. A. 1971. Stored grain insects. In *Fundamentals of Applied Entomology.* R. E. Pfadt (ed.). pp. 495–522. New York: Macmillan Co.

16

CHEMICAL PEST CONTROL

The use of chemicals to kill pests of all kinds has been a major factor in increasing the productivity of modern agriculture and the capacity of the world's food production systems to more or less keep pace with the demands of an ever-growing human population. In recent years much controversy has surrounded the use of pesticides, but it remains a vital part of crop production (Figure 16-1). There is no doubt that pesticides will continue to be the most important defense against pests that suddenly or unexpectedly break through the economic

FIGURE 16-1.
A tractor spray rig treating a large field of young lettuce in the Delta area of California's Central Valley. (Photograph courtesy of the USDA.)

threshold established for them; there is simply no other form of control that is as effective under such conditions.

Chemical pest control is a vast and complex subject. Like many of the subjects introduced in this book we can only touch upon some of the more important fundamentals and offer background information that will enable each reader to delve further into the subject.

HISTORY OF PESTICIDE USE AND REGULATION

No one knows when chemical pest control was first used, but there are references to insecticides in the writings of the Greeks, Romans, and Chinese, dating back to about 1200 B.C. Fumigation, to kill insect pests, accomplished by burning toxic plants, also dates back to about the same time in China. In Greece, Homer suggested using sulfur to avert pest problems, and by the first century A.D. both the Greeks and the Chinese were aware of the insecticidal properties of metals. For example, Shen Nung Pen Tshao Ching, in *Pharmacopoeia of the Heavenly Husbandman*, believed to have been written between 100 and 200 A.D., mentioned using mercury and arsenic to control the body louse and using a liliaceous plant to kill insect pests of man. Pig oil, lime, sulfur, copper, cinnabar (HgS), copper alum, and iron sulfate were all used for pest control in China before the tenth century (Konishi and Ito 1973).

Despite these early beginnings, the real era of chemical pest control is the twentieth century. The modern use of insecticides began in 1867, when Paris green was found to be effective against outbreaks of the Colorado potato beetle, *Leptinotarsa decemlineata*. At about the same time, powdered sulfur, lime sulfur, and a mixture of copper sulfate and lime called **Bordeaux mixture** gained widespread acceptance in England and France as fungicides especially useful for controlling diseases of grapes. These successes, particularly that of Paris green, stimulated experimentation with these and other compounds against a variety of plant pests. In the early years of the twentieth century, botanical insecticides and fluorine compounds were found to be effective against various pests.

The obvious value of pesticides to crop protection quickly led to the production and sale of a variety of ineffective remedies for a number of pest problems. This fraudulent activity led to the passage of the Federal Insecticide Act of 1910, which set standards for commercially sold insecticides and fungicides, prohibited the inclusion of phytotoxic materials in the formulations, and forbade the printing of false claims on pesticide labels.

Within a few years, the use of pesticides in American agriculture had become so common that people became concerned about residues in agricultural produce, particularly arsenic and lead on apples and pears. The result was the Federal Food, Drug, and Cosmetic Act of 1938, which established levels of tolerance for insecticide residues in agricultural products. The passage of this act was indeed well timed. It established basic protective regulations just before the spectacular increase in pesticide use that followed the discovery of the insecticidal properties of DDT in 1939 and the subsequent discoveries of pesticidal properties of other compounds such as 2,4–D (herbicide), TMTD (fungicide), and Warfarin (rodenticide).

The original Federal Food, Drug, and Cosmetic Act was improved in 1954 with the addition of the Miller Amendment, and again in 1960 with the addition of the Food Additives Amendment. The purpose of the law as it stands today is to maintain the purity and safety of the public food and drug supply by establishing tolerances for poisonous ingredients, food additives, and contaminants (including insect parts and excrement) in foods, and for pesticide residues in raw agricultural produce. Established pesticide tolerances set the maximum amount of the various chemicals that can remain in food crops, including meat, fat, milk, and eggs. The task of setting these tolerances and conducting the field surveillance necessary to enforce them originally fell to the Food and Drug Administration (FDA), but in 1970 the setting of tolerances was transferred to the Environmental Protection Agency (EPA).

In 1947 the Federal Insecticide, Fungicide, and Rodenticide Act (FIFRA) replaced the Federal Insecticide Act of 1910. This new law was extended to cover other agricultural chemicals, including molluscicides, nematicides, herbicides, defoliants, and disinfectants; it expanded labeling regulations; and it instituted a requirement that any pesticide marketed across state lines or used on produce entering interstate commerce be registered with the USDA. Before a product is registered, the manufacturer must provide scientific evidence that the product is effective for the purposes stated on the label and will not be injurious if used as directed.

The purpose of the 1947 FIFRA was to protect both the user of pesticides and the public from false claims and misuse. The label of every pesticide must show the name, brand, or trademark under which it is sold; the name and address of the manufacturer; the net contents; an ingredients statement; and a warning statement if one is necessary to prevent injury. Pesticide labels must also include directions for use that, if followed, will give the desired results and allow for compliance with established residue tolerances.

The FIFRA of 1947 was substantially amended in 1972 by the Federal Environment Pesticide Control Act (FEPCA). The amendment strengthens and expands the authority provided by the old law in that it

> extends federal registration and regulation to all pesticides including those distributed for use within a single state, requires proper application of pesticides to ensure greater protection of man

and the environment, prohibits use in a manner inconsistent with labeling instructions, authorizes classification of pesticides into "general use" and restricted use categores, specifies that federal standards for certification of applications of restricted use pesticides be established by October 21, 1974 as a basis for state programs for training and certifying pesticide applicators, and requires registration of plants manufacturing pesticides for interstate commerce and export.

SOME CHARACTERISTICS OF PESTICIDES

All pesticides have a number of characteristics in common, some of which are desirable and some undesirable. Ideally, pesticides should be highly effective against target organisms without harming beneficial or desired organisms. Unfortunately, most pesticides do not comply with this ideal. Many are broad-spectrum poisons and therefore injure beneficial species; some are toxic to crop plants, kill beneficial insects and microorganisms, taint the flavor of foods, or stain the desired product. Efforts are made, of course, to avoid as many of these undesirable features as possible, and a high agricultural priority should be for the development of chemicals with some degree of **specificity** (reduced spectrum of susceptible species). It has been difficult to obtain highly specific compounds partly because specificity narrows the number of uses and reduces marketability, but a growing interest in conserving beneficial species increases the demand for more specific chemicals, and manufacturers may find it profitable to produce them.

Because of the regulations governing residues, it is usually desirable that pesticides break down after they have been applied; however, they must be stable in storage and not be rendered inactive by other chemicals with which they are mixed to facilitate their application. Furthermore, it may be desirable for certain pesticides to remain active in the environment for several weeks in order to reduce the need for applying them at a critical time and to increase the time during which they are effective against the target pest. This is one of the criteria for selecting a pesticide as a crop approaches harvest, and the directions for use usually refer to the minimum time before harvest that each pesticide can be applied. The present trend in agriculture is to use pesticides, particularly insecticides, that have a low level of **persistence.**

The **mode of action** of pesticides, that is, the way they act upon organisms to cause death, has often proven difficult to ascertain. Although many can be classified generally as protoplasmic poisons (e.g., arsenic, copper, and sulfur), nerve poisons (e.g., organochlorines and organophosphates), respiratory inhibitors (e.g., oils, dithiocarbajate fungicides), or growth regulators (e.g., many organic herbicides), the precise ways in which they interfere with normal biochemical processes often are unknown.

Populations are made up of individuals that vary in a variety of ways, including their susceptibility to poisons, and so it is necessary to have a standard measure to determine the **toxicity** of pesticides, both for target and representative nontarget organisms. This is derived from a statistical method called **probit analysis,** which measures the response of a population sample, over a predetermined period of time, to various doses of the pesticide. If several fixed, geometrically increasing doses are applied to a population sample, the cumulative percentage killed increases with the dose level. Plotting cumulative percentage mortality for each dose against the logarithm of the dose generates a sigmoid curve similar to that shown in Figure 16-2. However, such a curve based on real data is difficult to plot, so the normal procedure is to convert the cumulative percentage mortality values to probit values using standard statistical tables. For convenience in calculation, 5 is added to each standard deviation; thus pro-

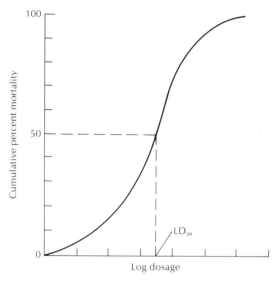

FIGURE 16-2.
A typical curve showing the relationship between cumulative insect mortality and the logarithm of insecticide dosage. The broken lines show the LD_{50} or log dosage required to kill 50 percent of the pest population.

bits for 0-, 5-, and 100-percent mortality are $-\infty$, 5.0, and $+\infty$, and if the data fit a normal distribution, probits 4.0 and 6.0 would represent 16- and 84- percent mortality, respectively.

When the data are plotted in this manner it is possible to fit a straight line to the data points as shown in Figure 16-3. Each probit thus represents a theoretical mortality that can be expected from the application of a particular dosage of pesticide. However, as one can predict from the shape of the sigmoid curve, the relationship between dosage and mortality is more accurate at 50 percent than at the extremes. Consequently, the standard measure of toxicity is the $\mathbf{LD_{50}}$, or the lethal dosage that will kill 50 percent of the population. The LD_{50} values of various pesticides are widely quoted for both pests and nontarget organisms (see, for example, Pimentel 1971). Because this value often varies according to the route of entry (oral or dermal) and sex of the organism,

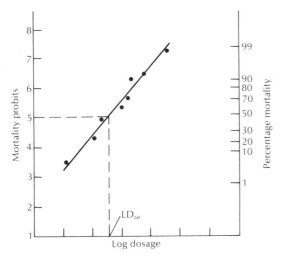

FIGURE 16-3.
A log dosage response line is generated when the percent mortality is transformed to units called probits (see text for details).

steep, accompanied each time by an increase in the LD_{50}.

It must be realized, of course, that well-designed experiments are essential to the determination of dosage-mortality relationships. Large samples, replicates, and controls must be used, and the controls must be handled the same as the treated samples in all respects except for the application of the pesticide being studied. Another important consideration is the possible synergistic action between the active pesticide ingredient and other chemicals used in its formation, and so the pure pesticide, not a formulation, should be used if the objective is to measure the toxicity of the pesticide by itself.

The fact that many pesticides are toxic to nontarget organisms makes it necessary that the labels carry relevant information. Even though there is no direct relationship between toxocity and the hazard because of differences in persistence, characteristics of particular formulations, and the dosage commonly used, it is wise to heed label notations that classify pesticides according to their acute oral and dermal toxicities (Table 16-1).

these are usually specified. An LD_{50} may be expressed either in terms of milligrams per individual or as milligrams per kilogram of body weight. The LC_{50}, a less useful lethal-inhalation- concentration value, is expressed as micrograms per liter ($\mu/1$).

Another important aspect of the probit transformation of dosage mortality data is that the use of the slope of the line can be used as a measure of the variance in a population's susceptibility to the pesticide. The shallower the slope of the line, the greater is the variation in the susceptibility to the toxicant of the individuals in the population. It would take a dose considerably higher than the LD_{50} to produce a reasonable level of mortality in such a population. A steep slope indicates homogeneity in the susceptibility of a population and a narrower range between nonlethal and lethal doses. Hoskins and Gordon (1956) suggested that changes in the slope of the line might indicate the development of resistance, particularly if the slope changed from steep to shallow and then back to

PESTICIDE FORMULATION

The active ingredient in a pesticide is often too toxic and too expensive to be used in a pure form and must therefore be mixed or formulated with other materials to facilitate dilution and the even distribution of the appropriate dosage of toxicant over the area treated. Although the formulation and application of pesticides depend heavily on the chemists and engineers who develop the compounds and design the spray equipment, they also involve considerable trial and error.

The objectives of **formulation** and **application** are to provide optimal coverage of the treated surface at the lowest possible cost per unit, with minimum crop damage and hazard to

TABLE 16-1
Toxicities of classes of pesticides, their LD_{50}s and LC_{50}s and required label statements.

Toxicity class	Oral LD_{50}	Dermal LD_{50}	Inhalation LC_{50}	Label statement required
Highly toxic	0–50 mg/kg	0–200 mg/kg	0–2000 $\mu g/l$	DANGER, POISON (carry skull and crossbones)
Moderately toxic	51–500 mg/kg	201–20,000 mg/kg	201–20,000 $\mu g/l$	WARNING
Slightly toxic	501–5000 mg/kg	1001–2000 mg/kg	none	CAUTION
Practically nontoxic	>5000 mg/kg	2001–20,000 mg/kg	none	no special wording

the applicator. The least expensive, safest, and most readily available diluent is water, but many pesticides are not water soluble and other means of dilution must be used. Materials other than diluents are used to act as carriers, spreaders, stickers, emulsifiers, and wetting agents. Although it is inappropriate here to go into details of formulation, application, and the calculation of concentrations, we will describe briefly the most common formulations.

Dusts

Dusts consist of a dry inert material, such as ground walnut shell or talc, mixed or impregnated with the toxic material so that the final concentration of toxicant in the mixture is less than 10 percent. Most dusts are fairly safe to handle and can be applied with relatively simple equipment, but precautions should be taken against inhalation, and if the formulation consists of a very small particle size, it can create a serious problem by drifting to other areas. Dusts are relatively inexpensive to purchase, but can be expensive to apply because they are used as manufactured (i.e., without dilution), thus requiring the handling of considerable bulk. Because of inhalation hazards and particulate pollution, dusts are gradually being swept off the market by the EPA.

Wettable Powders

Wettable powders consist of an inert dry material, a 25- to 75-percent concentration of the toxicant, and a dispersing or wetting agent. The wettable material is then diluted further by suspension in water. Reducing the amount of inert material in this way means the active ingredient can be handled in considerably less bulk than with dusts, and using water as a carrier reduces the overall cost. The main drawback to wettable powders is that they must be agitated constantly to prevent settling and more complex application equipment is needed than for dusting.

Granules

Granules are much like dusts except that the particle size of the inert material is larger and the toxicant concentration is much lower. Since the entire bulk must be transported and there is no further dilution, this kind of formulation is more inconvenient and costs more per unit of toxicant than other formulations. However, granules do not drift; they can be applied with simple equipment and can be directed easily into furrows or planting holes; and they penetrate through thick foliage when applied as a late soil treatment.

Emulsifiable Concentrates

Emulsifiable concentrates range from about 15- to 75-percent toxicant (by weight) combined with a solvent and an emulsifier. The concentrate is mixed with water to form an oil-in-water emulstion that is probably the most versatile of the pesticide formulations. Sprays of this type are effective contact poisons that spread and stick well on the surface treated, and they make good contact with the waxy cuticles of plants and with insects. The main problem with emulsifiable concentrates is that some of the solvents are phytotoxic.

Solutions

Solutions are simply toxicants dissolved in solvents. They have good spreading characteristics and can be applied with simple equipment because no agitation or mixing is required. Most pesticide compounds are insoluble in water, and so more expensive organic solvents must often be used, which increases phytotoxicity and hazard to the applicator. Nevertheless solvents can increase the effectiveness of the pesticide by increasing its penetration.

Fumigants

Fumigants are pesticides applied in a gaseous form and normally used in places like greenhouses and storage facilities, where the vapor can be confined, although soil fumigation is also a common practice. Most fumigants are held as a liquid under pressure and change to a gas when the pressure is released.

Seed Dressings

A variety of pesticides can be applied as a solution, dust, or slurry (thick suspension) as a protective dressing for seeds. The most common of these are insecticides and rodenticides used to prevent seed predation and fungicides used to prevent seed decay and the damping off of new seedlings. Seed dressings are economical and efficient because the only surface treated is that of the seeds themselves. These pesticides are generally safe, although some are highly toxic to birds. The treatment of seeds with mercurial compounds has been greatly curtailed because of the high concentrations of mercury that were showing up in the tissues of some game birds such as pheasants.

CLASSIFICATION OF PESTICIDES

There are several different ways that pesticides can be grouped. They can be classified according to the kind of pest that they control; for example, **insecticides** kill insects, **acaricides** kill mites, **herbicides** kill weeds, **fungicides** kill fungi, and so on. They can also be classified according to their chemical composition. Broadly speaking, there are **inorganics,** such as arsenic, copper, and mercury, and **organics** such as those derived from plants (**botanicals**) and those synthesized from organic or carbon-containing compounds. The **synthetic organic compounds** are then subdivided further on the basis of their general chemical nature; DDT, for example, belongs to the organochlorines. Other fairly broad classifications are based on the type of formulation in which the pesticide is applied (e.g., dust, spray, wettable powder) or on its basic mode of action (e.g., stomach poison, contact poisons, nerve poison, desiccant). Each system of classification has its own usefulness, but all require some degree of cross-referencing. For example, the organophosphate parathion might be formulated in a variety of ways and used as an insecticide, acaricide, or nematicide. Nevertheless, in our discussion we will group the major pesticides according to the types of pests they control. The discussion will, of necessity, be brief; but a number of rather extensive treatments that include most of the practical details are readily available.

Insecticides and Acaricides

It is convenient to study these two groups of pesticides together because of the similar ways in which phytophagous insects and mites cause damage, either directly or as vectors of plant pathogens. Both insects and mites are exposed (i.e., not covered or protected from a spray) sometime during their life history and feed either by chewing or sucking on plant or animal tissue. They can be controlled with contact, systemic, or stomach poisons or with fumigants, depending on their specific characteristics.

Stomach poisons are usually sprayed on the surface of the host on which the pest feeds, which means that coverage must be as complete as possible, and the application to growing plants must be repeated because coverage becomes diluted as the plant surface increases with growth. Surface applications of stomach poisons are less effective against tunneling species and those with piercing-sucking mouthparts because such insects frequently do not ingest a lethal dose of the chemical. A number of stomach poisons have some **systemic** qualities, however; that is, they can penetrate the host, in some way, and either the unmodified molecules or a toxic derivative is carried by the conducting system of the host to the feeding site of the pest. Some compounds of this kind can be applied to seeds or to the soil before planting and can provide protection for a short time after germination.

Contact poisons do not have to be ingested because they can penetrate the cuticle or gas exchange system of the target organism once contact has been made by the pest walking over a treated surface or by drops or dust particles falling directly on the pest during application. Such compounds are sometimes effective against all stages of pest development, including their eggs.

Fumigants are compounds that enter a vapor state under normal crop conditions or are applied as a gas; they enter the pest through its breathing apparatus. Fumigants are particularly useful for soil treatment or in confined spaces such as storage structures and greenhouses. Those applied to the soil can be formulated to provide for a fairly slow rate of release from various types of carrier compounds, thus spreading the treatment over a longer period of time.

INORGANIC COMPOUNDS   This group, which was used more widely in the past, includes calcium arsenate, lead arsenate, cryolite, and sulfur, most of which act as protoplasmic poisons following ingestion. They have been phased out of control programs because of such unfavorable characteristics as high mammalian toxicity, persistence as residues, and the fact that several target species have developed resistance.

BOTANICALS   The main insecticides and acaricides derived from plants are pyrethrum, nicotine, rotenone, and ryania, all of which are contact poisons. Only pyrethrum, obtained from a species of *Chrysanthemum*, is still used to any extent. It is a very ephemeral compound, being oxidized rapidly in the presence of sunlight, and so is seldom used outdoors. It has very low mammalian toxicity and is an effective nerve poison with a very rapid paralytic (knock-down) effect against insects. Both natural and synthetic pyrethroids are expensive to produce, however, and so tend to be used mainly in relatively high-cost household sprays.

The insecticidal qualities of nitotine have been known for a long time, but nicotine compounds have fallen into disfavor mainly because of their high mammalian toxicity. Nicotine acts mainly as a fumigant and its volatility is most often reduced by combining it into nicotine sulphate. It has low phytotoxicity, is very effective against insect larvae such as caterpillars, and has a low level of persistence in the environment.

Rotenone, formulated from the roots of *Derris*, is used to a limited extent to control both crop and livestock pests. It is now used primarily as a fish killer to rid bodies of water of unwanted species before restocking. Ryania is a rather specific contact and stomach poison of caterpillars, which is useful in integrated control programs where broad-spectrum pesticides might prove disruptive.

ORGANOCHLORINES   The organochlorines, of which DDT is by far the best known, were the first of the synthetic organic insecticides. Their great efficacy as both contact and stomach poisons and their inexpensiveness have made them the most widely used of all insecticides and acaricides.

Although it was first synthesized in 1874, DDT was not known to have insecticidal properties until 1939, when it was tested along with many other compounds in the search for chemicals that would provide woolens with extended protection against clothes moths. The use of DDT dust against the louse vectors of epidemic typhus in Naples in 1944 is credited with preventing an epidemic of disastrous proportions. After the war, DDT was used widely, almost as a panacea, to control insect pests of humans, livestock, forests, crops, and gardens. But the initial effectiveness and popularity of DDT were in a sense its undoing because overuse and misuse led to a series of ecological problems that have caused it to be banned in Sweden, Great Britain, Canada, and the United States.

The effectiveness of DDT led to the search for other man-made compounds for pest control, and the phenomenal proliferation of these substances since the mid-1940s is due in part to the wide variety of possibilities for making minor changes in their molecular structure. The DDT group is an excellent example of a group of similar compounds with different attributes as pesticides (Figure 16-4). DDT is a persistent, broad-spectrum insecticide (but not useful against mites) with a high affinity for milk fat and the fatty tissues of vertebrates. Methoxy-

chlor is also quite nonspecific as an insecticide, but has the virtue of a low affinity for milk fat and fatty tissue. TDE (also known as DDD) is a rather narrow-spectrum insecticide, and is particularly effective against caterpillars. Kelthane is a poor insecticide, but a very effective acaricide.

Other well-known organochlorines include benzene hexachloride (BHC) and the cyclodiene insecticides. Crude BHC tends to taint food products and is generally not used on crops except in the form of the gamma isomer, Lindane, which is relatively expensive. The cyclodienes particularly useful as soil insecticides.

Probably the best-known cyclodienes are aldrin and dieldrin, both of which have been used extensively against grasshoppers and locusts, soil pests, and cotton pests, but because of residue and resistance problems are now being replaced by organophosphates and carbamates. Other well-known representatives of this group include heptachlor, isodrin, endrin, toxaphene, and chlordane. Chlordane is particularly effective against ants and is used to prevent ants from interfering with biological control pro-

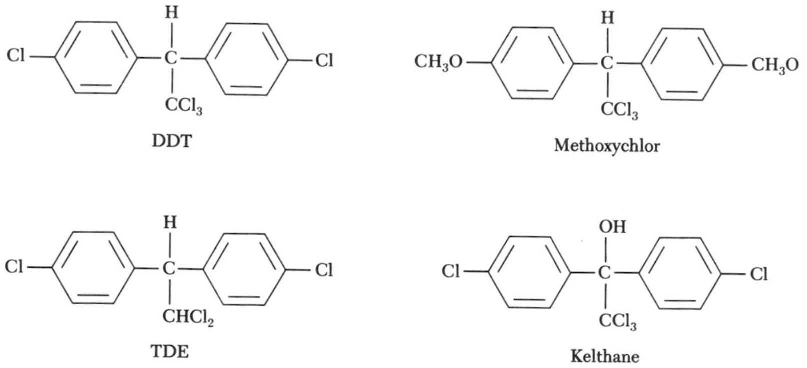

FIGURE 16-4.
The structural formulas of four closely related organochlorine compounds with substantially different uses. DDT and Methoxychlor are broad-spectrum insecticides; TDE is effective mainly against caterpillars; Kethane is an effective acaricide.

grams directed against a variety of plant sap feeders, such as whiteflies and scale insects, but recently has been banned in the United States.

ORGANOPHOSPHATES The organophosphates (OPs) form the largest and most versatile group of insecticides and acaricides. They can be used as contact poisons, stomach poisons, and fumigants and several are excellent plant systemics that provide a high level of protection against aphids, spider mites, and other plant juice feeders. A few are also useful as animal systemics for the control of livestock parasites.

OP compounds are related to nerve gases developed for military use, and they have a mode of action that irreversibly inhibits the enzyme cholinesterase, which is involved in the physiology of the nerve synapse. However, their range of mammalian toxicity is great. Parathion, for example, is extremely toxic to mammals and has fallen into disfavor, whereas malathion is safe enough to be sold widely for use by home gardeners.

The opportunities for synthesizing organophosphates are almost unlimited, and more than one hundred commercially produced compounds have already been introduced. The widespread interest in OP compounds stems from the fact that they are far less persistent in the environment than the organochlorines, and so far fewer target species have developed resistance to them.

A few of the better-known organophosphates and their uses are: diazinon, used to control livestock pests; fonofos, a persistent soil insecticide; and demeton, disulfoton, and dimethoate, which are highly effective systemics especially useful against sucking pests and insect vectors of plant viruses.

CARBAMATES The carbamates are mainly contact insecticides that, like the organophosphates, are cholinesterase inhibitors. Most have rather low mammalian toxicity and are readily degraded in the environment. Some have sys-

temic action, but only a few are used in this way. A few of the newer compounds have proved to be quite selective. Carbaryl and methomyl are probably the best known and most widely used carbamates. Carbaryl, although it is injurious to a variety of beneficial species, especially honeybees, has been useful against leaf-feeding caterpillars and is widely used in the control of fruit pests. Methomyl is used extensively to control caterpillars on vegetable crops. In addition to being effective against a wide range of insects, carbamates are also toxic to mites, snails, and slugs.

OILS Hydrocarbon oils have long been used to control both insects and weeds. As insecticides and acaricides, the less refined products are referred to as **dormant oils** because they are injurious to actively growing vegetation and must be used during the plant's dormant period. The more highly refined **summer oils** can be used on verdant vegetation, although phytotoxicity is the major limitation in the use of all oil sprays. Oils are particularly effective against the eggs and dormant stages of mites and scale insects because they interfere with the organism's gas exchange.

Oil sprays are gaining popularity because they are not toxic to mammals and are therefore not subject to residue restrictions. Furthermore, they are not as harmful to beneficial parasites and predators as the traditional insecticides and so are highly compatible with biological control methods. After more than 80 years of use there are no known cases of resistance development to oils.

Fungicides

There are basically two types of fungicides: One is used to kill exposed fungus mycelia; the second is used as a coating that prevents either spore germination or the penetration of the plant by the mycelia after spore germination. In recent years, a third group of promising compounds has been discovered with systemic qualities that might allow mycelia to be attacked within the plant host; until now this has not been possible. Fungicides are applied either as a foliar spray or dust, as a seed dressing, or as a soil treatment.

INORGANIC FUNGICIDES As indicated earlier, the use of sulfur as a pesticide has a long history, but it became a common means of controlling mildews on fruit trees and grapes during the early nineteenth century and various sulfur formulations are still used against fungal diseases of apples and pears. Sulfur is particularly useful against powdery mildews because their spores can germinate and infect a leaf in the absence of water. Sulfur sublimes and permeates the spores as a gas so that water is not needed for solubilization as it is in most fungicides. The more finely divided the sulfur the better are its fungicidal properties. Commercially produced lime sulfur, which consists mainly of calcium polysulfides that release elemental sulfur, is the form used most often. Sulfur is phytotoxic and so is being replaced by the organic fungicide dinocap, which is relatively nontoxic to plants.

Copper sulfate, like sulfur, has been used for a long time as a fungicide, but has gained popularity as a control agent for fungal diseases of grapes. Bordeaux mixture, developed in France in 1885 by Millardt, is still widely used; but the lime component of the mixture tends to abrade and block the nozzles of spray equipment, and less troublesome copper formulations have been developed, the most important being copper oxychloride. The main use of copper fungicides is directed against potato blights, downy mildew of grapes, *Phytophthera* diseases of cotton, rubber, and cacao, and seed-borne diseases of grains (Ordish and Mitchell 1967).

ORGANIC FUNGICIDES The original organic fungicides were all derivatives of dithiocarbamic acid used in the production of synthetic rubber. The first to be produced commercially was thiram, which has been used widely in treating seeds to prevent the development of seedling blight and other fungal diseases. Although it is still used, thiram has been replaced by a series of metallic salts of dithiocarbamic acid, such as zineb (zinc), maneb (manganese), and nabam (sodium). Although these earlier organic fungicides were effective against damping off, root rot, and various foliar blights, they were not very effective against powdery mildews. More recently, several organic compounds have been developed that provide good protection against these diseases, and some organophosphates have proved useful against rice blast and several powdery mildews. Further research on organophosphates for disease control could lead to the discovery of additional fungicides that can be used against soil-borne pathogens, such as *Verticillium* wilts, that cause a variety of vascular diseases. Two fairly new materials, vitavax and plantvax, show considerable promise against cereal smuts and rust diseases.

Herbicides

The chemical control of weeds, like that of arthropods and diseases, has become a complex aspect of agriculture important to both crop production and animal husbandry. As Woods (1974) clearly stated, ". . . weeds, unlike fungi and insect pests are not directly dependent upon their victims, but merely share the same habitat." Furthermore, weeds are closely related to crop and pasture plants, which often makes it difficult to kill one and not the other if they share the same habitat at the same time, and so, in addition to selecting the right chemicals it is important that they be applied in such a way that the desirable plants are not injured. This can be accomplished by selecting the herbicide that is appropriate to the nature of the crop and the weeds to be controlled and by adjusting the type and time of treatment. Herbicides are applied either to the soil or to the foliage of the weeds before the crop is planted (preplanting), after planting but before the crop germinates (preemergence), or after the crop is actively growing (postemergence). Depending on the herbicide, the weeds are killed by contact with either roots or the foliage, or by translocation of the herbicide after it enters through the roots or the leaves. The preplanting destruction of weeds with a broad-spectrum contact herbicide is clearly the most straightforward method. Preemergence treatment with a contact herbicide is also straightforward, but requires more critical timing.

With postemergence treatments, it becomes difficult to avoid herbicide contact with both the weeds and the crop, so destruction of the weeds depends on the choice of a selective material such as 2,4–D, which is toxic to dicotyledons but relatively nontoxic to monocotyledons. It is also possible to exploit differences in contact or in the retention of a toxic material on different forms of foliage. For example, broad-leaved weeds growing in cereals are not too difficult to control because the vertical, narrow leaves of the crop do not retain the herbicide as readily as the horizontal, spreading leaves of many annual weeds. Moreover, the sensitive young shoots of the weeds are fully exposed, whereas those of the cereal are ensheathed within older leaves.

INORGANIC HERBICIDES Inorganic compounds such as ashes and salt have been used to kill weeds for centuries, but it was not until Liebig developed the basic theory of plant nutrition in the nineteenth century that the search for more specific herbicides began. The first widely used inorganic plant killer was sodium arsenite, but successful weed control requires the application of 400 to 800 pounds of this substance per acre, and arsenical residues remain in the soil for many years. Other important inorganic herbicides are ammonium sulfamate, used in brush control; ammonium salts, used as foliage desiccants; various borates used in nonselective, long-term vegetation control; sodium chlorate, used as a soil sterilant; and sulfuric acid, which is applied in the form of brown oil of vitriol (BOV) against weeds of onions and cereals.

ORGANIC HERBICIDES Various petroleum oils have been used as contact herbicides for many years. Oils such as diesel oil that are not highly refined and that have a high proportion of unsaturated and aromatic compounds are useful for total vegetation control, whereas more highly refined oils can be used somewhat selectively.

The modern era of organic herbicides really began in 1932 with the discovery of 3,5-dinitro-o-cresol as an effective contact herbicide; its sodium salt, known as DNOC, became widely used, particularly in cereal weed control, despite its general biocidal properties. A few years after the introduction of DNOC, the herbicidal properties of 2,4–D were discovered. This chemical, an auxin (plant hormone)-type growth regulator with systematic action, can be used effectively in very small quantities, and it provides an inexpensive means of selectively controlling important annual and perennial weeds in cereals. This was a milestone in the development of organic herbicides, and it led to the search for additional compounds with similar properties.

Unlike the synthetic organic insecticides, the organic herbicides cannot be divided into a small number of families according to their chemical structure, nor can they be classified by their biochemical activity. Instead of simply mentioning the key characteristics of a few groups, therefore, we will have to just touch on some of the more important individual compounds.

The compound 2,4–D is a phenoxyaliphatic acid the derivatives of which form a major group of herbicides that include MCPA, 2,4,5–T, and 2,4–DES, to mention just a few. The hormonal or auxin-like effects of this group are still incompletely understood, but treated plants display cellular proliferation and reduced respiration. Like 2,4-D, MCPA is used as a post-emergence spray in cereal crops and grassland. 2,4,5-T is particularly useful against woody plants and is used widely in brush control projects; 2,4–DES is not an herbicide when applied but is converted to 2,4–D in the soil, so it is used as a soil treatment to kill weeds growing from seeds already present.

Two chlorinated aliphatic acids, TCA and dalapon, are widely used to control grasses. TCA is particularly effective against quackgrass when applied to the soil. Dalapon can also be applied to the soil and be absorbed through the roots, but it also acts systemically when applied to the foliage.

One of the better-known herbicides for the control of aquatic weeds is endothall, a contact herbicide also used in selective weed control in field crops. Endothall is readily decomposed into harmless chemicals by fish, soil microorganisms, and many nontarget plants.

Nematicides, Molluscicides, and Rodenticides

Nematodes, molluscs, and vertebrates, particularly rodents, are three additional groups of organisms that attain pest status and must be controlled, often with chemicals.

The most common method for controlling soil nematodes is a general sterilization that simultaneously eliminates bacteria, fungi, and weed seeds. The compounds used include methyl bromide and ethylene dibromide. Soil fumigation is expensive, however, and is not commonly practiced for field crops, although it is routine in nurseries and in some greenhouse crops. Some nematicides, particularly D-D mixture, are applied to field soils on a large scale, but the success of nematode control in the soil depends on a number of factors, including temperature, moisture, soil structure, and or-

ganic content (organic material tends to inactivate the fumigants).

Terrestrial snails and slugs are not major pests of agricultural crops, but they can cause sporadic local damage to the young seedlings of many crops, particularly of winter-sown wheat. The best-known chemical for use in crops and gardens is metaldehyde, which is applied in the form of treated bran.

Aquatic snails that are vectors of schistosomiasis are also an agricultural pest in that they are associated with waterways tapped for irrigation. The expansion of irrigation in Egypt, using the waters of the Nile, has increased the schistosomiasis problem in the agricultural areas where increasing numbers of people use the water for drinking, bathing, dumping excreta, and watering crops. Copper sulfate is the least expensive chemical for control of aquatic snails, but it is toxic to other aquatic organisms, including the crops to be watered. In the past decade an organic molluscicide, niclosamide, has been used with some success.

TRENDS IN PESTICIDE PRODUCTION AND USE

Data on the world production and use of pesticides is fragmentary and often difficult to interpret. Even in the United States it is difficult to obtain complete and up-to-date statistics on pesticides because some surveys that provide important information are conducted irregularly or at rather long intervals.

In 1974, the last year for which data are available, United States manufacturers produced 1417 million pounds of synthetic organic pesticides with a value of $1985 million. According to earlier estimates, the United States accounts for about 40 percent of the total world production, and so world consumption must be on the order of 3.5 billion pounds per year at a cost of more than $4.5 billion.

TABLE 16-2
Synthetic organic pesticide production and sales in the United States
from 1969 to 1974.

	Production		Sales (domestic and export)	
Year	(1000 pounds)	(1000 dollars)	(1000 pounds)	(1000 dollars)
1969	1,104,381	953,592	928,663	851,166
1970	1,034,075	1,058,389	880,914	870,314
1971	1,135,717	1,282,630	946,337	979,083
1972	1,157,698	1,344,832	1,021,565	1,091,708
1973	1,288,952	1,492,770	1,198,568	1,343,581
1974	1,417,158	1,984,794	1,365,214	1,815,433

Source: The Pesticide Review 1975. Agr. Stab. and Cons. Serv., USDA.

TABLE 16-3
Production ($\times 1000$ pounds) of pesticidal chemicals[a] by class in the
United States from 1969 to 1974.

Year	Insecticides[a]	Herbicides	Fungicides	Total[b]
1969	570,522	393,840	140,019	1,104,381
1970	490,132	404,241	139,702	1,034,075
1971	567,710	428,849	149,158	1,135,717
1972	563,575	451,618	142,505	1,157,698
1973	639,169	496,109	153,674	1,288,952
1974	650,209	604,288	162,661	1,417,158

Source: The Pesticide Review 1975. Agr. Stab. and Cons., USDA
[a]Includes small quantity of inorganic materials plus fumigants and rodenticides.
[b]Includes small amount of synthetic soil conditioners, rodenticides, and multipurpose fumigants.

The production and sales data presented in Table 16-2 show a steady increase from 1969 to 1974, a period during which a great deal was said about curtailing pesticide use. Table 16-3 shows the breakdown of pesticide production by class for the same period, and Table 16-4 presents the United States sales volume of pesticides by class for 1967, 1970, and 1973. These data show a small increase in production and sales of insecticides and fungicides, but a marked increase in the production and use of herbicides.

The last survey of the use of pesticides on farms in the United States was conducted in 1971 (USDA 1974). During that year about 833 million pounds of pesticides were used for pest control in the United States, of which 494 million pounds were used by farmers and the rest mainly by the forest industry, government agencies, and homeowners. The use of

TABLE 16-4
United States sales volume ($\times 1000$ pounds) and percent of total sales by class for 1967, 1970, and 1974.

Year	Insecticides[a]		Herbicides		Fungicides	
	Production	Percent	Production	Percent	Production	Percent
1967	503,796	44.8	439,965	38.3	177,886	15.9
1970	443,943	50.4	308,112	35.0	128,859	14.6
1974	691,865	50.7	528,469	38.7	144,880	10.6

[a]Includes a small quantity of synthetic soil conditioners, rodenticides, and multipurpose fumigants.
Source: The Pesticide Review 1975. Agr. Stab. and Cons. Serv., USDA.

household and garden pesticides continues to increase; the value of these products rose from $146 million in 1971 to $182 million in 1972. In addition to specific pesticides, farmers used an additional 112 million pounds of sulfur and 222 million pounds of petroleum for pest control.

Of the 565 million pounds of insecticides (including inorganics) produced in the United States in 1971, roughly 170 million pounds, or about 30 percent, were used by farmers (Figure 16-5). Of this amount, 154 million pounds were applied to crops and 15 million pounds used on livestock. In addition, farmers used 74 million pounds of petroleum for insect and mite control. Table 16-5 shows the quantities of major types of insecticide used by farmers for various purposes in 1971 and compares the 1971 total use with that for 1966. These data indicate a decline of about 22 percent in the use of organochlorines but respective increases of 77 percent and 93 percent in the use of organophosphates and carbamates. The approximate 540 percent increase in the use of petroleum is not only striking but environmentally encouraging because petroleum oils are less harmful to beneficial species than most organic insecticides. The large increase in the use of petroleum, mainly on citrus, apples, and other fruits, probably reflects the

development of integrated control programs for these crops.

The trend away from organichlorines has probably been the result primarily of the resistance to these compounds developed by a number of pests against which they were once highly effective. Most notably, the cotton bollworm and tobacco budworm have become resistant to DDT in many areas, and the corn rootworm is widely resistant to aldrin. Furthermore, DDT fell into general disfavor because of environmental contamination and was eventually banned by the EPA in January 1973.

In 1971, 73.3 million pounds or 47 percent of all the insecticides used by U.S. farmers was applied to cotton. Large quantities of insecticides were also used on corn and vegetables (Table 16-6).

In 1971, American farmers used almost 42 million pounds of fungicides and 112 million pounds of sulfur for disease control. About 95 percent of these materials were used on crops. Table 16-7 shows the quantities of major types of fungicides used by farmers for various purposes and compares the total farm use for 1966 and 1971. The comparison shows an increase of almost 70 percent over a five-year period.

In recent years the use of inorganic fungicides, particularly of the copper and zinc com-

FIGURE 16-5.
The aerial application of an insecticide to crops near Blythe, California. (Photograph by E. E. Hertzog, courtesy of the U.S. Bureau of Reclamation.)

TABLE 16-5
Comparison of the quantity ($\times 1000$ pounds) of different types of insecticides used by farmers in 1966 and 1971, and the pattern of use for 1971.

Type of insecticide	1966 Total	1971 Total	1971		
			Crops	Livestock	Other
Inorganics	5,784	3,232	3,042	189	1
Botanicals	204	213	69	144	
Organochlorines	89,239	69,873	61,876	7,627	370
Organophosphates	39,966	70,706	65,031	5,369	306
Carbamates	12,933	25,412	24,166	1,194	52
Other synthetics	798	334	72	261	1
Petroleum	11,419	73,950	60,721	13,126	103
Total (including petroleum)	160,343	243,720	214,977	27,910	833

Source: USDA. Farm use of pesticides in 1971.

438

TABLE 16-6
Farm use of insecticides, by crop, 1966 and 1971.[a]

	1966		1971	
Crop	Pounds of active ingredients[b]	Percentage of farm insecticides used	Pounds of active ingredients[b]	Percentage of farm insecticides used
	Million pounds	Percent	Million pounds	Percent
Cotton	64.9	47	73.3	47
Corn	23.6	17	25.5	17
Other field crops[c]	8.7	6	17.5	11
Vegetables	11.1	8	11.1	7
Fruits (not including apples and citrus)	6.6	5	6.3	4
Soybeans	3.2	2	5.6	4
Apples	8.5	6	4.8	3
Tobacco	3.8	3	4.0	3
Citrus	2.9	2	3.1	2
Hay and pasture	4.1	3	2.6	2
Other	0.2	—[d]	0.5	—[d]
All crops	137.6	100	154.3	100

[a]Does not include Alaska.
[b]Does not include petroleum.
[c]Includes wheat, sorghum, rice, peanuts, and sugar beets.
[d]Less than 0.5 percent.
Source: USDA. Farm use of pesticides in 1971.

TABLE 16-7
Comparison of the quantity ($\times 1000$ pounds) of different types of fungicides used by farmers in 1966 and 1971, and pattern of use for 1971.

	1966 Total	1971 Total	1971		
Type of fungicide			Crops	Livestock	Other
Inorganics (except sulfur)	7,567	15,987	15,857	98	32
Sulfur	57,101	112,453	112,093	358	2
Dithiocarbamates	15,146	13,042	11,932	3	1,107
Phthalimides	7,474	7,488	7,011	—	477
Other organics	3,017	5,210	4,755	413	42
Total	90,305	154,180	151,648	872	1,660

Source: USDA. Farm use of pesticides in 1971.

TABLE 16-8
Farm use of fungicides, by crop, 1966 and 1971.[a]

Crop	1966		1971	
	Pounds of active ingredients[b]	Percentage of farm fungicides used	Pounds of active ingredients[b]	Percentage of farm fungicides used
	Million pounds	Percent	Million pounds	Percent
Citrus	4.1	13	9.3	24
Apples	8.5	28	7.2	18
Vegetables	4.1	13	5.7	14
Peanuts	1.1	4	4.4	11
Irish potatoes	3.5	12	4.1	10
Deciduous fruit (excluding apples)	1.8	6	3.8	10
Other fruits and nuts	2.5	8	3.1	8
Other field crops[c]	4.5	15	1.7	4
Cotton	.4	1	.3	1
All crops	30.5	100	39.6	100

[a]Does not include Alaska.
[b]Does not include sulfur.
[c]Includes corn, sorghum, wheat, rice, soybeans, and sugar beets as well as other grains, other field crops, alfalfa, other hay and pasture, and nursery and greenhouse crops.
Source: USDA. Farm use of pecticides in 1971.

pounds, has increased substantially, but the use of mercury compounds has declined since they were banned for agricultural purposes. Of the organic fungicides, the dithiocarbamates (maneb, zineb, ferbam) and the phthalimide Captan make up almost 80 percent of the total used. The use of sulfur doubled between 1966 and 1971, maybe in part because of the trend to less noxious materials and the decline in the use of organochlorine insecticides with which lime sulfur was incompatible.

Fungicides are used primarily to control diseases in fruits, nuts, and vegetables. Protection of citrus fruits and apples alone accounts for 42 percent of the fungicides used by American farmers (Table 16-8).

The use of herbicides in U.S. agriculture has increased more rapidly than the use of any other

kind of pesticide; they have largely replaced mechanical weed removal in most major crops. Atrazine, which is particularly effective in the control of weeds in corn, is the most widely used of the many herbicides now available and accounts for about 25 percent of the total herbicide use. The use of the once popular 2,4–D declined markedly between 1966 and 1971.

Corn continues to receive more herbicides than any other crop, accounting for about 45 percent of the total herbicide use in 1971. Weeds in soybeans and cotton are also major herbicide targets (Table 16-9).

For any specific crop, the need for pesticide treatment varies according to the geographic area (Figure 16-6) in which the crop is grown, but in general the geographic pattern of pesticide use clearly reflects the major crops of the

TABLE 16-9
Farm use of herbicides, by crop, 1966 and 1971.[a]

Crop	1966		1971	
	Pounds of active ingredients[b]	Percentage of farm herbicides used	Pounds of active ingredients[b]	Percentage of farm herbicides used
	Million pounds	Percent	Million pounds	Percent
Corn	46.0	41	101.1	45
Soybeans	10.4	9	36.5	16
Cotton	6.5	6	19.6	9
Other field crops[c]	10.8	10	15.1	7
Wheat	8.2	7	11.6	5
Sorghum	4.0	4	11.5	5
Pasture and rangeland	10.5	9	8.3	4
Rice	2.8	2	8.0	3
Vegetables[d]	5.7	5	5.6	2
Peanuts	2.9	3	4.4	2
Fruits and nuts[e]	3.6	3	2.4	1
Summer fallow	.9	1	1.4	1
Nursery and greenhouse crops	.1	—[f]	.2	—[f]
All crops	112.4	100	225.7	100

[a]Does not include Alaska.
[b]Does not include petroleum.
[c]Includes tobacco, sugar beets, alfalfa, and other hay as well as other grains and other field crops.
[d]Includes potatoes as well as other vegetables.
[e]Includes apples and citrus as well as other deciduous fruits and other fruit and nut crops.
[f]Less than 0.5 percent.
Source: USDA. Farm use of pesticides in 1971.

various regions. The South traditionally accounts for a large share of the farm use of insecticides. In 1971, 40.4 million pounds (26 percent) were used in the southeast, 32.3 million pounds (21 percent) in the delta states, and 18.5 million pounds (12 percent) in the southern plains (USDA 1974). This high level of use reflects the fact that these are the main cotton growing regions and that among the other major crops grown in these regions are soybeans, peanuts, and tobacco, which also have a high demand for pesticides.

Herbicides are used most heavily in the corn belt, lake states, and northern plains, where most corn is grown; corn-growing regions also grow a substantial amount of soybeans, which can also require extensive use of herbicides.

The Southeast receives over one-third of the total applied fungicides, probably because the warm, humid climate is ideal for many fungi. In addition, there has been an increase in the use of dithiocarbamate summer sprays among citrus and vegetable growers. Other regions that account for a large percentage of

FIGURE 16-6.
The major agricultural regions of the United States. (Courtesy of the USDA.)

fungicide use include the apple- and nut-growing areas of the Northeast as well as the Pacific and Appalachian regions.

Despite the impression often created that this country is continuously saturated with pesticides, only about 5 percent of our agricultural land is treated with insecticides and about 12 percent with herbicides. If the land devoted to pasture and range is excluded, still only 12 percent of our crop acreage is treated with insecticides, 27 percent with herbicides, and 1.3 percent with fungicides (Pimentel 1973).

According to Pimentel (1973) annual losses in U.S. agriculture due to pests have increased steadily throughout this century. From 1951 to 1960 the annual losses due to insects, diseases, and weeds respectively were estimated at $3.8 billion (13 percent), $3.6 billion (12 percent) and $2.5 billion (8.5 percent), and despite the widespread use of pesticides agricultural losses totalled $9.9 billion or one-third of potential production. Pimentel calculated that in the absence of insecticides the annual damage

would increase to $12 billion, or almost 41 percent of production. He also estimated that the return on each dollar invested in pesticides of all types is about $2.82, a somewhat lower figure than previous estimates, which have been as high as $5.00. However, even the best cost-benefit ratios for pesticides are below those calculated for other control methods.

THE PESTICIDE CONTROVERSY

After the use of DDT against body lice averted a disastrous outbreak of epidemic typhus in Naples toward the end of World War II, many felt that a solution to the world's insect pest problems had been found. DDT, with its high insect toxicity and persistence and its low cost and ease of use, soon caused more conventional control methods to be replaced by the widespread use of synthetic organic chemicals. Although early warnings were voiced by prominent insect physiologists such as Sir Vincent Wigglesworth, pest-free crops became the ideal, and many agriculturalists felt that a panacea had been found. It was only a short time, however, before the unheeded warnings proved justified, as evidence appeared that DDT and related compounds had begun to lose their effectiveness. The insecticide industry responded with new materials, but these too lost some of their initial potency as time passed.

When the new organic pesticides first became popular, little was known about their mode of action or what happened to them after they were introduced into the environment. In fact, as Moore (1967) pointed out, it is regrettable that pesticides are often though of as substances applied to a pest population rather than to the ecosystem of which the pests are a part. The first indication that all was not well came in 1946 when resistance to DDT was encountered in several populations of the housefly in Sweden.

Resistance is defined by the WHO Expert Committee on Insecticides (1957) as "the development of an ability in a strain of insects to tolerate doses of toxicants which would prove lethal to the majority of individuals in a normal population of the same species." Resistance is the result of natural selection, not of the mutagenic action of the pesticides. Insects and mites that have gained resistance have done so because their populations contained genotypes that were preadapted for resistance. These were usually rare, probably because in insecticide-free environments they were less well suited for survival than their susceptible siblings. But when the population was subjected to insecticides, the resistant forms were better suited and thus survived and multiplied. When the selective pressure of the pesticide is a continuous component of the environment, the nonresistant forms are gradually eliminated and only the offspring of preadapted individuals survive. Clearly, then, resistance is more a result of the misuse of pesticides than of the pesticides themselves, be-

cause it is misuse that increases the intensity of selective pressure and allows the resistent strains to flourish.

When resistance to DDT was encountered, other cyclodienes, organophosphates, and carbamates were used instead, but many pests developed resistance to them as well. Pests that have developed what we commonly call multiple resistance have become increasingly difficult to control. Although some pests are not able to develop multiple resistance, and it is possible that some populations will undergo reversal once the selective pressure of pesticides is removed from their environment, it appears that the trend is against us. Over 200 species of arthropods are now resistant to one or more pesticides (Brown 1968; FAO 1969).

Resistance is not confined to insects and mites that have been treated with insecticides and acaricides. Bacteria are also known to have developed resistance to antibiotics, and rabbits in Australia have developed resistance to the pathogen that causes myxomitosis. Even chickweed, *Stellaria media,* has been known to become resistant to 2,4–D by the third generation (Woods 1974). Certainly resistance development is something that should be kept in mind whenever a new approach to pest control is devised. It is folly to assume arbitrarily that no behavioral changes will occur as a result of pheromone trapping (Atkins 1968), that insects cannot become resistant to compounds that mimic their hormones, or that sterile male releases will not lead to the development of parthenogenetic strains. These are all possibilities if we engage in the misuse of these control methods.

The recognition of other drawbacks to pesticides developed more slowly and first came to public attention after the publication of Carson's *Silent Spring* in 1962. Although, in our opinion, this book was somewhat more emotional than scientifically accurate, it did stimulate a closer examination of the environmental side effects of pesticides. The literature of the last few years is permeated with reports on the undesirable aspects of pesticides in the environment; some are exaggerations, but many are factual and therefore matters for real concern (Rudd 1964; Moore 1967; Slater 1967).

The problems of pest resurgence and secondary pest outbreaks that were discussed briefly in Chapter 15 have occurred as the result of pesticides killing beneficial species and disturbing the normal process of population regulation. Hazards to human health arise during the application of pesticides and from toxic residues in agricultural products. Moreover, crop productivity has sometimes declined because of pesticide misuse, as cotton yield did in the Canete Valley of Peru, and there is documented evidence that pesticides have entered food chains with severe consequences for animals on higher trophic levels (Figure 16-7).

However, this is not the place to review the details of problems that have arisen out of inadequate knowledge and an overzealous desire to increase food and fiber production and to stamp out the misery caused by diseases. These problems have been discussed many times before. What is more important is to recognize that properly used pesticides are still a highly effective means of reducing losses generated by pests. As Smith (1970) stated, "when pest populations approach economic levels, there is little other than pesticides we can use to avoid damage and which will have the desired immediate effect." Brown (1970) feels that pesticides will also be important, at least initially, in the protection of the new high-yielding varieties of grains, some of which have already been threatened by the appearance of new pests.

FIGURE 16-7.
The insecticide treatment of a New Hampshire orchard, showing the drift of spray into an adjacent aquatic ecosystem. (Photograph courtesy of the USDA, Soil Conservation Service.)

There are indeed alternative approaches to pest control, as we shall see in the following chapters. Unfortunately, one of the most serious consequences of relying so heavily on chemical pest control for two decades is that these alternative methods were not perfected sooner. To quote Smith (1970) again, " . . . it is irresponsible to advocate the total replacement of pesticides with sophisticated, but poorly tested alternative pest-management techniques. It is a disservice to society to discard the good crop protection currently available and to adopt in its place a glamorous new but untested methodology."

Clearly, we must adopt a balanced point of view, the aim of which is not to ban the use of pesticides, but rather to prevent their misuse. This must involve not only the elimination of unnecessary prophylactic treatments, but also the development of methods that will reduce the quantity and frequency of use. It is always easy to criticize in retrospect what was done in the absence of more adequate information, but it is more difficult to learn from mistakes and avoid repeating them. We realize now that many of the problems that developed could have been avoided simply by applying no more pesticide than was necessary; yet there are still growers who apply unnecessary pesticides, and the concept of weed-free farms that currently prevails in some sectors the way the goal of insect-free crops did in the 1950s could certainly lead to a similar misuse of herbicides.

Some of the procedures that could be necessary to maintain the effectiveness of pesticides in preserving adequate worldwide production of food and fiber will not be popular. At the risk of drawing considerable criticism from certain vested interests, we would like to mention a few controversial ideas:

1. *Require that pesticide use recommendations be separate from pesticide application:*

   Licensed consultants would make recommendations to growers in the form of prescriptions that would be "filled" by a pesticide purveyor or applicator much as a pharmacist fills a prescription written by a doctor. Severe penalties would be imposed for violations, and proof that a consultant received a kickback would result in a license suspension.

2. *Increase the labor involvement in pest control:*

   More workers would be involved in monitoring crop conditions and pest populations, and hand and mechanical weed removal or roguing would be substituted more often for chemical weed control and pathogen reservoir destruction.

3. *Introduce a system of taxes on pesticide use:*

   The use of certain kinds of broad-spectrum pesticides should be taxed and the revenue used as a *post facto* subsidy to defray research and development costs of companies that produce specific pesticides.

4. *Eliminate cosmetic standards for agricultural produce:*

   Removing cosmetic standards would eliminate the need to kill organisms that damage cosmetic quality but do not affect the nutritional or storage quality of the product, (e.g., organisms that cause skin blemishes on citrus fruits).

5. *Impose penalties for unharvested surpluses:*

   A system of tariffs should be developed to reduce the amount of agricultural produce that goes unharvested. Such surpluses should be stored or processed and distributed in the form of foreign aid as a means of offsetting world crop losses caused by pests.

6. *Eliminate the cultivation of certain luxury crops:*

Luxury crops that, when grown in certain areas, require heavy pesticide use, should no longer be cultivated until alternative pest control methods are established. For example, it could be advisable to stop new apple orchard plantings until coddling moth and apple maggot can be controlled with minimal use of pesticides.

7. *Increase government expenditures for the development of pest control methods:*

Increasing food production and reducing environmental contamination by pesticides should be high priority items in federal and state budgets. Considerably more support should be provided for pest control research and pest control advising by trained personnel.

8. *Expand educational programs:*

With financial support from all levels of government and grower cooperatives, universities should expand their training programs in all aspects of agricultural pest control.

9. *Improved safety conditions for pesticide applicators:*

The pesticide industry should be encouraged to develop new ways to reduce the hazards of handling and applying pesticides; one way would be to develop closed delivery systems.

SUMMARY

Although there are numerous ancient references to the use of poisons to control insect pests, chemical pest control belongs to the twentieth century. A wide variety of synthetic organic compounds capable of killing insects, plant pathogens, weeds, nematodes, and other less serious pests have been employed in agriculture over the past thirty years. These pesticides have contributed significantly to increased productivity in a wide range of agricultural systems; but the benefits of reduced crop damage and livestock losses have been gained at significant ecological cost. In response to the intense selective pressure of pesticide use many pests have evolved pesticide resistance. Secondary pest problems have arisen after the chemical destruction of the natural enemies of species that were relatively innocuous until their populations were able to increase. Large pesticide residues have accumulated in forests and agricultural ecosystems, and the undesirable side effects of some of these chemicals have appeared far from where they were applied.

Despite the problems mentioned above, pesticides remain a vital component of modern agriculture; they provide the only economical response to pest problems that develop suddenly, and without them the world's agricultural harvest could only be maintained at present levels by the use of more costly or more labor-intensive techniques of control. The challenge, therefore, is to find the best way to use our pesticide technology to maintain the high productivity of intensive agriculture while reducing the undesirable side effects that we have experienced in the past. At all levels we must strive for use without misuse.

Literature Cited

Atkins, M. D. 1968. Scolytid pheromones—ready or not. *Can. Ent.* 100: 1115–1117.

Brown, A. W. A. 1968. Insecticide resistance comes of age. *Bull. Ent. Soc. Ameri.* 14:3–9.

Brown, L. 1970. *Seeds of Change.* New York: Praeger Publishers.

Carson, 1962. *Silent Spring.* Boston: Houghton Mifflin.

FAO 1969. *Fifth Session of the FAO Working Party of Experts on Resistance of Pests to Pesticides.* September 1968. Rome.

Hoskins, W. M., and H. T. Gordon. 1956. Arthropod resistance to chemicals. *Ann. Rev. Entomol.* 1:89–122.

Konishi, M. and Y. Ito. 1973. Early entomology in East Asia. In *History of Entomology.* F. Smith, T. E. Mittler, and C. N. Smith (eds.). pp. 1–20. Palo Alto, Ca.: Annual Reviews, Inc.

Moore, N. W. 1967. A synopis of the pesticide problem. *Adv. Ecol. Res.* 4:75–130.

Ordish, G., and J. F. Mitchell. 1967. World fungicide usage. In *An Advanced Treatise.* D. C. Torgeson (ed.). pp. 39–62. New York: Academic Press.

Pimentel, D. 1971. *Ecological Effects of Pesticides on Non-Target Species.* Washington, D.C.: Executive Office of the President, Office of Science and Technology.

———. 1973. Extent of pesticide use, food supply and pollution. *N.Y. Entomol. Soc.* 81:13–33.

Rudd, R. L. 1964. *Pesticides and the Living Landscape.* London: Faber and Faber.

Slater, W. (ed.). 1967. A discussion on pesticides: benefits and dangers. *Proc. Roy. Soc. London* (B). 167:88–163.

Smith, R. F. 1970. Pesticides: Their use and limitations in pest management. In *Concepts of Pest Management.* R. L. Rabb and F. E. Guthrie, (eds.). pp. 103–113. Raleigh, N.C.: North Carolina State University.

USDA. 1974. *Farmers' Use of Pesticides in 1971.* Agr. Econ. Rep. No. 252. Econ. Res. Serv. 56 pp.

USDA 1976. *The Pesticide Review 1975.* Agr. Stab. and Cons. Serv. 60 pp.

Woods, A. 1974. *Pest Control: A Survey.* New York: John Wiley and Sons.

17

ALTERNATIVES TO CONVENTIONAL CHEMICAL PEST CONTROL

All forms of agricultural pests can be suppressed by a variety of special chemical or nonchemical methods, some of which are as old as agriculture itself and are time-tested as practical and ecologically sound ways to increase crop yields. Others are new and in some instances have not yet advanced beyond the experimental or conceptual stage. Although chemical pesticides have become the major means of pest control since the introduction of DDT and will certainly continue to be important for some time to come, their use is not the most satisfactory approach from an ecological point of view. Problems of pesticide pollution, resistance to pesticides, and secondary pest outbreaks have rekindled the interest in alternative control tactics that that can be used alone or in combination to provide effective and ecologically sound pest control.

Nonchemical pest control has been the subject of a number of books. It involves a variety of methodologies that draw from and integrate a wide variety of scientific and technological disciplines and each different tactic has a number of ecological and economic implications. Consequently, we make no apology for not treating the subject completely. Instead we will survey a number of pest control options so that each reader can determine for himself whether it is necessary to rely totally on the use of pesticides. At the same time we wish to point out that each alternative tactic has some limitations and that much existing and imaginative work is left to be done. We shall therefore identify those tactics that have been demonstrated to be effective, or to have the potential, either alone or in combination, to solve a variety of pest problems.

We have organized this review of pest control tactics largely according to personal preference. Although biological control in the broadest sense includes host resistance and autocidal methods as well as the manipulation of parasites, predators, and pathogens, each of these approaches is distinctly different and will be treated separately. This chapter is concerned with cultural control, legislative or regulatory control, special chemical methods, autocidal control, and some miscellaneous physical and mechanical methods. Biological control in the narrower sense is discussed in Chapter 18, host resistance in Chapter 19.

CULTURAL CONTROL

The term **cultural control,** as it is used here, refers to the manipulation of the crop (plant or animal) environment in order to make it less favorable to pest species and thereby to slow their rate of increase or suppress their effect. These manipulations make use of existing environmental factors and do not bring in other agents of mortality such as pesticides or additional natural enemies, so we will not discuss the cultivation of resistant varieties of plants or animals here. The extent of environmental manipulation can range from rather common cultivation procedures, such as tillage, to more radical environmental changes, such as the conversion of a natural community to a site for crop culture. This approach includes what is sometimes called **ecological control.**

Cultural control follows two basic approaches and includes some of the oldest and simplest methods of pest control, such as hand weeding and the removal of diseased plants (roguing). The two approaches can be classified as: (1) the avoidance of pest increase and related damage, and (2) the destruction or eradication of an existing pest condition. In some instances the techniques of cultural control are now required by law and form the basis of **legislative control,** which we will consider briefly a little later. However, an important aspect of cultural control is that many of the methods employed are concerned primarily with the growth of the crop and require little or no modification when used as pest control tactics.

Tillage

Tillage is the mechanical loosening or breaking up of the soil that prepares it for planting, improves water penetration and conservation, or destroys pests in the soil through exposure or mechanical injury. The development of tilling methods was a major step in the transition from food gathering to permanent agriculture (Cole and Matthews 1938) and has been an important factor in increasing crop production because it is a well-established technique for weed control. Autumn cultivation is extremely effective against some weeds, such as chickweed and wild oats, but the timing of effective cultivation depends on the germination cycle of the weeds involved, and no generalization can be made. Because most weeds are pioneer species well adapted to invade severely disturbed sites, seedbed preparation is often followed by a rapid and substantial weed invasion. In some circumstances, as in the establishment of a perennial forage crop, a second tilling before planting but after weeds have emerged can produce good results.

Tilling the soil at the appropriate time can be an effective way to control some insect pests by destroying the life cycle stages that occur in the soil or in crop residues. Fall plowing, for example, exposes grasshopper egg pods to desiccation and kills the pupae of the grape berry moth, *Polychrosis viteana*. Deep plowing after harvest buries infected plant parts and stubble and destroys the larvae of wheat-stem sawfly, corn earworm, and European corn borer.

Tillage is not always advantageous, however, and can actually aggravate some pest problems. If the soil surface tends to form a crust, for example, allowing this crust to remain intact can inhibit weed germination and prevent the penetration of some soil inhabiting pests such as the pale western cutworm, *Porosagrotis orthogonia*. Haynes (1973) found that conventional tillage had no direct effect on cereal leaf beetle populations, but that it destroyed over 97 percent of the parasite population.

Yarwood (1968) suggests that there are probably cases in which tillage reduces crop diseases, but he states that most experimental evidence indicates that it favors disease. For example, Boon (1967) reported that the incidence of the root disease of wheat caused by *Ophiobolus* increases on land that receives increased tillage. In some cereals, **direct drilling** (planting without seedbed preparation) seems to be more beneficial than conventional tillage in that it saves energy, reduces evaporation and erosion, and reduces the populations of several pests (see Chapters 14 and 24).

Crop Rotation, Mixed Cropping, and Fallowing

Growing the same crop in the same place for an unbroken sequence of years often encourages the build-up of pest populations. This is particularly true for pests such as nematodes and plant pathogens that infest the root zone because they tend to persist from crop to crop. Because crops are often rotated to conserve the soil and protect its fertility, this practice can provide a bonus in pest reduction; but when it is used mainly as a pest control measure, the crops rotated must be botanically unrelated. For example, there are a number of pests that damage almost all members of the Cruciferae, so that rotating crops belonging to this group might not reduce the pest problem. This technique is little use, of course, against pests with a wide range of hosts, as is the case for some nematodes (e.g., beet eelworm) and insects (e.g., cabbage looper).

Although some pests can be controlled effectively by crop rotation, the white-fringed beetle complex, all of which are unable to fly, have a higher reproductive rate when they feed on legumes such as peanuts and soybean than when they feed on grasses, including corn. Consequently the larvae do not damage grasses to the same extent that they damage legumes, and the populations of these pests decline during cultivation of a grass crop. Alternating corn with a legume crop can therefore result in excellent control. Rotating of potatoes with alfalfa simi-

larly reduces wireworm damage, and rotating oats and corn reduces corn rootworm damage. The winter grain mite can be controlled by planting small grains in the same field no more than two years in succession (National Research Council 1969).

Crop rotation is also effective against some, but not all, plant diseases. Outstanding results have been obtained in reducing *Ophiobolus* disease in wheat by using a wheat-oats rotation; and alternating potatoes with soybean and cotton with peas effectively reduce the incidence of potato scab and cotton root rot, respectively (National Research Council 1968).

Pasture rotation can help greatly to control livestock pests, particularly those like the cattle tick that have one primary host, as opposed to those like the lone star tick that attack a wide range of hosts. For example, keeping pastures free of cattle for a few months starves the cattle tick population, whereas pastures must be clear of all animals for several years in order to control the lone star tick.

Keeping pastures free of animals is similar to allowing crop land to lie fallow, a technique that is often used to conserve water and to improve fertility by allowing a period for nitrogen buildup. It can also be useful in pest control, however. **Fallowing** is sometimes carried out specifically for weed control; it is usually used against perennial weeds and is particularly effective when the fallow fields are tilled periodically. Fallowing, like resting pasture land, can reduce the population of insect and disease pests by depriving them of suitable hosts.

Mixed cropping or **interplanting** can be used as an alternative to crop rotation. Growing a mixture of crops, which is the practice in some forms of subsistance agriculture, can reduce pest damage by reducing the food supply for specific pests and by increasing the abundance of natural enemies. In mechanized agriculture this method has usually been limited to interplanting two species. Probably the best results of this method are achieved by interplanting strips of alfalfa in cotton fields for the control of lygus bugs (Stern 1969), which prefer alfalfa to cotton and will concentrate in the strips of alfalfa, leaving the cotton undamaged. The alfalfa can be harvested later as a forage crop.

Timing of Planting and Harvest

Organisms normally evolve life cycles that synchronize their development with the availability of suitable environmental conditions, including the appropriate developmental stage of their hosts, so that their survival is assured. Since many of these natural relationships have carried over to crop ecosystems many agricultural pests attack crops in a specific stage of development. It is sometimes possible to alter the timing of crop development by modifying normal cultural practice and thus to effect a substantial reduction in damage.

A rather simple way to alter the timing of crop development is to change the planting time, sometimes by delaying the planting date, other times by advancing it. Hessian fly adults usually live for only a few days after they emerge in the fall, and during this brief period they mate and lay their eggs on young winter wheat plants. If planting is delayed so that most of the flies have died before the wheat emerges, damaging infestations can be avoided. On the other hand, planting corn early instead of late reduces infestations of the southwestern corn borer because the moths tend to lay fewer eggs on more mature plants; planting corn early in northern areas reduces damage by the corn earworm populations that overwinter in the south and move northward each year. In situations where crops are invaded from elsewhere rather than attacked by a resident population, some crop protection can be obtained by delaying the planting of a crop until major pest migrations are over. Damage to sugar beets by curly top

virus can be diminished by planting the crop after the spring migration of the beet leafhopper, a vector for the disease.

Temperature is important in the development of many plant diseases, and by adjusting the time of planting the conditions that tend to favor a disease can sometimes to avoided. For example, if cool-weather crops such as peas are planted late, when the weather is warm, both the seeds and seedlings might be heavily damaged by fungi (National Research Council 1968).

Sometimes it is possible to reduce pest populations or pest damage by adjusting harvest time. Early harvest of the first two cuttings of alfalfa is useful in controlling alfalfa weevil larvae, which suffer from a shortage of suitable food and from the more severe physical environment in mowed fields. Wheat losses are reduced when wheat is harvested before attacks by the wheat-stem sawfly cause the plants to bend over.

Strip Harvesting and Trap Crops

At present, the technique of harvesting only part of a crop at one time (strip harvesting) is practical only for forage crops that are harvested several times each season. Nevertheless, this technique might also prove useful in the management of rapidly growing field crops that lend themselves to cultivation in strips of different ages. This method seems to have been most successful in reducing lygus bug damage in alfalfa. The alfalfa is harvested in alternate strips so that when one strip is cut the strips to each side are half grown (Figure 17-1). This technique works in two ways: It stabilizes the environment for both the lygus bugs and their natural enemies; and many of the lygus eggs that are laid in uncut strips by adults that have moved into them are removed before they hatch with the next harvest (Stern, van den Bosch, and Leigh 1964).

FIGURE 17-1.
Strip harvesting of alfalfa in California as a method of reducing lygus bug populations.
(Photograph courtesy of Vern E. Stern.)

Trap crops are small plantings of a highly favored host and are used to divert pests from the main crop or to draw them into a small area where they can be controlled without disrupting beneficial species in the main crop. For example, although lygus bugs attack cotton, they prefer alfalfa, and so strips of alfalfa planted within cotton fields attract these insects and reduce injury to the cotton plants. In Hawaii, corn, which is very attractive to melon flies, is planted around fields of melons and squash and is periodically treated with an insecticide. This technique is not widely used as a primary method of insect control, but it can be combined effectively with other methods. Trap crops can also provide effective control against plant pathogens; for example, creeping legumes planted between rows of rubber trees serve as a trap crop for several root pathogens. Interplantings of trap crops are also effective in reducing nematode infections in cruciferous crops and peaches.

Sanitation

Sanitation practices that are aimed at reducing pest populations by removing both suitable breeding material and the sites in which pests pass dormant periods have broad application, but they must be considered case by case since they can also have detrimental effects on populations of natural enemies.

Many weed problems can be diminished by cleaning farm machinery to reduce the transport of weed seeds and by using planting material that is certified as weed free in accordance with established standards. The reduction of weed populations along field margins, road sides, and irrigation ditches can substantially diminish the production of weed seeds and the invasion of crop acreage.

The impact of a number of plant diseases can also be lessened by sanitation. Removing certain wild plants and crop-plant volunteers from field margins can help to eliminate alter-

FIGURE 17-2.
An accumulation of postharvest sugarcane debris. Removing the residual plant material
would eliminate much of the refuge used by the sugarcane borer between crops.

nate hosts and refuges for vectors of plant pathogens and can destroy reservoirs of plant viruses that are transmitted by these vectors. Removing wild celery from the margins of water courses, for example, cuts down the incidence of aster yellows disease. Disease-free planting material is also necessary, especially when propagation is from vegetative parts such as seed potatoes. The destruction of residual plant material by burning is useful in preventing the development of a disease inoculum. Burning cotton after harvest has been used to control *Verticillium* wilt, and stubble burning is used in Oregon to prevent blind seed disease in ryegrass grown for seed. Disease inocula in surface soil can be buried by deep plowing.

Other sanitation methods, such as the destruction of prunings, for example, can control insect pests; this method is effective against the peach borer, *Conopia exitiosa,* and the lesser peach borer, *Synanthedon pictipes.* Collecting and using dropped fruit, or else destroying it, reduces populations of some important direct pests such as the plum curculeo, *Conotrachelus nenuphar,* the codling moth, and the apple maggot, *Rhagoletis pomenella.* Cutting and destroying cotton, corn, and sugarcane plants after harvest is an important measure in the control of the pink bollworm, the European corn borer and sugarcane borer, *Diatraea saccharalis* (Figure 17-2) (National Research Council 1969). In some instances cleaning field margins of wild vegetation eliminates refuges that crop pests can use during the absence of the crop, and it reduces pests that require alternate plant hosts.

Although most sanitation measures are directed at a specific pest and can therefore be carried out to produce mainly beneficial results, the practice of cleaning field margins requires careful scrutiny because it destroys vegetation that can provide certain benefits. Field margins lend a measure of diversity to areas of monoculture and can help to stabilize relationships between pests and their natural enemies. In the adult stage, many entomophagous parasites require pollen and nectar that might not be available if the field margins are bare. The presence of certain flowering plants can actually attract beneficial species.

Water and Fertilizer Management

The effects on pest problems of adding or withholding water and fertilizers from crop ecosystems can vary. The literature on the effects that different nutrients have on the fecundity of plant-feeding insects is extensive. Certain nutrients that adversely affect pest populations can be used to achieve some control; in other cases the growth response of some plants to water and nutrients helps them to overcome early pest damage. Yarwood (1970), however, found cases in which disease increased because of fertilization, and fertilizers can also stimulate weed growth. We have already mentioned examples of the effect that fertilizer has on insect pests (see Chapter 15).

The benefits of water management, like those of fertilizer, can be quite variable: In some cases irrigation helps to reduce pests, in others it is best to withhold water. Generally, moist soil conditions favor weeds and plant diseases, but both flooding and drying can be useful in controlling different soil insects. The method of applying irrigation water can actually be more significant than the amount used. Overhead sprinkling tends to reduce some orchard pests and is a quite effective means of reducing the populations of spider mites that are often protected by an accumulation of dust on the leaves. On the other hand, the higher humidity associated with sprinkling can favor the buildup of citrus rust mite. Overhead irrigation also encourages some wet-weather diseases such as halo blight in beans, black rot in melons, and angular leaf spot in cotton.

LEGISLATIVE OR REGULATORY CONTROL

Legislative or regulatory control is characterized by special laws or government regulations implemented to prevent foreign pests from entering or becoming established in a given area or to eliminate or contain those that have already become established. Many such regulations exist to prevent the spread of agricultural pests by commerce that involves certain agricultural products. Restricting the movement of infected plant and animal material is referred to as **quarantine,** and its purpose is to exclude potential pests or to prevent the spread of established pests that already have a well-defined but limited distribution. Action of this kind cannot be left up to individuals or private concerns and therefore requires specific regulations that can be enforced under the law. In the United States, the Plant Quarantine Act of 1912 gave the USDA the authority and responsibility for enforcing and implementing programs designed to eradicate or control introduced pests. State agencies were made responsible for enforcing regulations to prevent the spread of pests within their geographic boundaries.

Under the provisions of the United States Quarantine Act, restricted plants cannot be imported unless: (1) a special permit has been granted and obtained before shipment and all necessary information has been provided; (2) the shipment can be inspected; (3) the importer accepts the responsibility for meeting any special requirements such as fumigation; (4) the importer agrees that hazardous materials can be seized, treated, or disposed of; and (5) the products must remain in the custody of the Customs Department until released by a plant inspection officer.

Under the Quarantine Act there are three kinds of action that can be undertaken with public support when a pest in inadvertently introduced in a quarantined area:

1. **Eradication.** It might be possible to entirely eliminate the target species from the quarantined area. This type of action usually follows the discovery of introduction if the infested area is small and can be well defined through an intensive survey.

Both the Japanese beetle and the oriental fruit fly, for example, were successfully eradicated following their recent discovery in San Diego County, California.

2. **Containment.** If eradication is not feasible because of the area infested or because no suitable control method is known, efforts are made to limit the spread of the pest. The gypsy moth is currently subject to containment quarantine. The Secretary of Agriculture, after holding public hearings and providing adequate notice, can quarantine any portion of the United States; a number of pests are currently under federal interstate quarantine (Table 17-1).

3. **Suppression.** The quarantine act provides special assistance in controlling periodic pest outbreaks that are too large for a single agency or landowner to undertake. Large-scale outbreaks of grasshoppers or forest pests are controlled under this provision.

Most major countries now have quarantine laws and regulations that guard against the international spread of pests. Cooperation among many countries in the enforcement of quarantines was brought about by an FAO conference and agreement reached in 1951. One aspect of this cooperation is the clearance or certification of plant material or animals destined for export to foreign countries.

There are other laws besides the quarantine laws that are specifically directed at weed control and plant disease protection. As we have mentioned, the use of planting stock and seeds that are free of disease and relatively free of weeds is a sound cultural control procedure. This is aided by regulations that require propagative material to be inspected and certified as disease free. Federal and state seed laws regulate the importation of seeds into the country as well as the interstate transportation of seeds. For example, the Federal Seed Act of 1933 states that the labels on seeds entering interstate commerce must specify the percentage of pure seed of the designated crop, the percentage of other crop seeds that might be present, the percentage of weed seeds, and the names of noxious weed seeds present and their levels of occurrence.

AUTOCIDAL METHODS

Autocidal control can be defined as the use of an organism or one of its characteristics to destroy other members of the same species or to reduce its population by diminishing its reproductive capacity. The principal approaches involve sterilization and genetic manipulation.

Sterile-Male Technique

In the sterile-male technique male members of a species that have been sterilized with radiation or chemicals are released so that the females they mate with lay infertile eggs. The theory of this approach to pest control was developed by E. F. Knipling more than a decade before it was established as a practical means

TABLE 17-1
Pests subject to interstate quarantine within the United States in 1976.

Gypsy moth	*Porthetria diapar*
Brown-tail moth	*Nygmia phaeorrhoea*
Japanese beetle	*Popillia japonica*
Pink bollworm	*Pectinophora gossypiella*
Mexican fruit fly	*Anastrepha ludens*
White-fringed beetle	*Graphognathus* spp.
Imported fire ant	*Solenopsis saevissima*
Black stem rust	*Puccina graminis*
Witchweed	*Striga asiatica*

of reducing a pest population. Calculations made by Knipling (1955) indicated that the dosage level of an insecticide must remain the same in each generation to provide the same proportionate mortality. As the density of the pest population declines, fewer and fewer individual insects are killed with the same amount of chemical. From a cost benefit standpoint then, chemical control is subject to the law of diminishing returns. On the other hand, releasing the same number of sterile males in each pest generation would increase the ratio of sterile to nonsterile matings and would therefore become increasingly effective as the native population declines. It is obvious from the simple mathematical models prepared by Knipling that for this technique to succeed, it must be both practical and feasible to swamp the wild population with sterile males and thus create a ratio of sterile to fertile matings that would start a downward trend in the population. Once this has been achieved, fewer sterile males can be released each generation.

The initial project was directed against the screwworm fly, *Cochliomyia hominovorax*, a serious pest of cattle in the southern United States and most of tropical North and South America. Knipling noted that female screwworm flies mate only once, and he recognized this as a desirable attribute in that individual females would not receive both fertile and sterile sperm. Furthermore, the fly survived over the winter at fairly low population densities in the climatically favorable tropical and subtropical portions of its range. If the program could be initiated before the populations increased and moved northward in the spring, control seemed feasible since a manageable number of sterilized males could be used. Indeed it was feasible. The first totally successful field eradication occurred on the island of Curaçao off the north coast of Venezuela in 1954, only 13 weeks after the release program was completed. Later, over a span of 18 months in 1958 and 1959, more than two billion flies

were reared, sterilized with cobalt-60 radiation, and released in the southeastern United States (Figure 17-3). The result was the eradication of the screwworm from this region (Knipling 1960; Baumhover 1966). Sterile males are now released annually along the United States-Mexico border to reduce cattle losses caused by immigrant flies from Mexico, and a more extensive program is underway in Mexico with the object to eliminate the screwworm from North America.

Before the sterile-male program went into effect, estimated annual losses to the American cattle industry because of the screwworm were between $100 million and $140 million despite the use of pesticides and judicious animal husbandry to control the pest. These losses have been virtually eliminated after an expenditure of about $10 million for the sterile-male program.

The success of the sterile-male technique in controlling the screwworm stimulated considerable interest in the approach as a means of solving other kinds of pest problems. However, there are a number of requirements that must be met before successful control can be anticipated:

1. **Reproductive Biology.** If the female of the species mates only once, the program is more likely to succeed than if the female mates with several males. Monogamy is not essential, but if polyandry occurs, sterile males must produce active sperm that is competitive with normal sperm and results in the production of a high proportion of infertile eggs.

2. **Sex separation.** Unless the sexes can be separated by some efficient means, sterile males might be wasted in matings with released sterile females, which would reduce the program's effectiveness.

3. **Mass rearing.** It must be possible to rear and sterilize large numbers of individuals

FIGURE 17-3.
Loading boxes of sterilized male screwworm flies on an aircraft in preparation of aerial release over Texas. (Photograph courtesy of the USDA.)

in a short time in order to release enough individuals to swamp the wild population.

4. **Vigor.** The reared individuals must be at least as vigorous as members of the wild population, and after sterilization they must be able to compete with wild individuals for mates.

5. **Biological relationship.** If large numbers of individuals are to be released they should not have any serious economic impact on the agroecosystem. For example, males of the screwworm, unlike many biting flies, do not feed on blood and so are relatively harmless; this could not be said for a leaf-feeding beetle, the release of which would increase crop damage.

6. **Population characteristics.** The success of the campaign against the screwworm is predicated on the assumption that the wild population can be numerically swamped. If a large wild population exists,

it might be necessary to reduce it substantially with chemicals or other techniques before releasing the sterile males.

Several other control programs have been initiated as a practical means of pest suppression and a number are in fairly advanced states of development. Most notable are the California programs designed to eliminate Mexican fruit flies that enter from Mexico, and to prevent the spread of the pink bollworm from the Imperial Valley into the San Joaquin Valley. A table prepared by Knipling (1972) summarizes the recent status of the sterile-male technique (Table 17-2).

Sterilizing members of a naturally occurring population provides an alternative to programs that involve mass rearing, sterilization, and release, and it has some important differences. Wild individuals must somehow be brought in contact with a **chemosterilant** to replace the mass sterilization accomplished by radiation in a laboratory. This can be done by combining

TABLE 17-2
Objectives and developmental status of programs for use of the sterile-male technique.

Insect	Proposed manner of use	Status of research and development
Screwworm	For suppressing populations on regional basis	In practical use
Mexican fruit fly	For preventing establishment of incipient infestations	In practical use
Pink bollworm	For preventing establishment of incipient infestations and to eliminate low-level established populations	In practical use; additional improvements and pilot testing required.
Oriental and Mediterranean fruit flies, melon fly	To eliminate low-level populations and to prevent establishment of incipient populations	Effectiveness demonstrated in small island tests. Large pilot tests required.
Codling moth	To maintain suppression of populations achieved by cultural and chemical means	Effectiveness demonstrated in small orchard tests. Small pilot test underway.
Boll weevil	To eliminate low-level populations after prior suppression by chemical, cultural, and other means	Pilot tests planned
Bollworm and budworm	For area suppression of low-level populations	Pilot tests required
Cabbage looper	For area suppression of low-level populations	Pilot tests required
Fall armyworm	For area suppression of low-level populations	Pilot tests required
Tobacco hornworm	For area suppression after prior suppression by cultural means	Pilot test required
Gypsy moth	For preventing spread and to eliminate incipient infestations	Pilot test required
Mosquitoes (important vector species)	To maintain suppression after prior suppression by sanitary and chemical means	One pilot test underway, others required
Tsetse flies	To eliminate low populations after prior suppression by chemicals and brush clearing	Pilot test planned
Horn fly	To eliminate low populations after animal spraying	Pilot test required

Source: From Knipling, 1972.

the chemosterilant with an attractant or arrestant distributed in the population's habitat. Although sterilized individuals will not leave fertile offspring and will therefore contribute to the reduction of the population, they will persist in the population for their normal lifespan, and so will also contribute to crop losses on the one hand, but serve as hosts for natural enemies on the other. Once perfected, this technique would be preferable to the release technique for species in which released individuals could initially increase crop damage.

The most promising chemosterilants take effect when ingested, although some are effective simply through tarsal contact. In either case, the chemical must be presented with a bait, attractant, or arrestant in order to insure contact with a large portion of the wild population. Because the powerful chemosterilants are not specific, their preparation requires special precautions to minimize the danger to beneficial species and to man.

Genetic Techniques

Reducing pest populations by genetic means would involve some alteration of their genetic makeup that would cause genetic sterility, a reduction in fecundity, a reduction in vigor, or an inability to complete development in a normal manner. Manipulating the genetics of host species so that they can resist pest attacks is treated in Chapter 19.

The high genetic variability possessed by many insect species is often demonstrated in the existence of numerous morphological, physiological, and behavioral races or strains. This high level of population variability is an effective buffer against environmental changes in that it allows populations to adapt to new ecological situations. The fact that insects have high reproductive potentials and short lifespans provides the means by which the numerical distribution of variants in a population can change rapidly. One of the best examples of how this genetic plasticity can function is the rapid development of insecticide resistance by insect pests. Many entomologists conducting rearing programs under controlled conditions have also witnessed changes in some aspect of the biology or behavior of their colony.

Although the technique of genetic control is not yet at a stage refined enough for practical application, it has considerable potential. Some genetic manipulations that could be useful in pest control are: (1) creation of **cytoplasmic incompatibility,** (2) **hybrid sterility,** (3) introduction of **lethal** or **harmful genes,** and (4) **developmental alteration.**

In some insects there are strains that, when cross mated, produce some sterile offspring. The sterility results when the sperm enters the egg and stimulates meiosis but does not fuse with the egg nucleus. Most of the eggs fail to hatch, but a small proportion develop parthenogenetically to produce females (see Laven 1967 for more details). According to Woods (1974) this principle of cytoplasmic incompatibility has been used successfully in Burma to eradicate a population of a mosquito, *Culex pipiens fatigans,* in an ecologically isolated area near Rangoon.

Hybrid sterility is simply the production of infertile hybrid offspring as a result of a cross between related species. The offspring are viable and can even display hybrid vigor, thus competing with normal individuals for resources. Sometimes the females are normal and the males sterile, but the overall effect is to reduce the population. It might be possible to use this method of control by rearing and releasing a nonpest species into the habitat of a related pest species, or by bringing together a pair of allopatric pest species (i.e., species having distinctly separate ranges) that will mate but leave sterile progeny. This has been suggested as a way to reduce populations of the tsetse flies *Glossina swynnertoni* and *G. morsitans*.

Although introducing lethal or harmful genes into a population seems to contradict the process of natural selection, some workers have suggested that it might be possible to introduce genes that would be deleterious only at specific times or under particular circumstances and that could be propagated in a wild population when these conditions did not prevail. Whitten (1971) suggested that it would be desirable if a conditional lethal allele could be produced and be maintained in a homozygous condition. Hybridization with wild individuals would produce sterility, whereas the homozygotes would be destroyed only under certain conditions. For example, if the homozygote were susceptible to an insecticide to which the wild population was resistant, both populations could eventually be eliminated. As an alternative, Werhahn and Klassen (1971) suggested that destructive genes could be linked with those that confer insecticide resistance, so that the genetically manipulated individuals could be released at the same time that an insecticide is applied; this would favor the carriers of the lethal gene.

In some insects whose range spans several distinctly different climatic zones, strains have evolved with life cycle patterns that enable them to exploit their particular portion of the range. For example, populations in areas with severe winters might enter an obligatory dormancy (diapause) and therefore have only one generation per year. Populations in a warmer part of the range might have no dormant period and so might complete several generations per year. If these developmental characteristics are controlled genetically, it may be possible to reduce populations by mixing genotypes. Such a manipulation has been suggested as a way to control the European corn borer in the United States and the field crickets *Teleogryllus commodus* and *T. oceanicus* in Australia (Woods 1974). The genetic induction of diapause in warm area populations would reduce the number of generations per year, and the introduction of genes to prevent diapause in cool area populations could increase winter mortality.

SPECIAL CHEMICAL METHODS

A number of pest control methods that are particularly useful against insects involve the use of chemicals but are different from standard pesticides either in the way they are employed or in their modes of action. These include **feeding deterrents, antimetabolites, repellents, attractants** (including pheromones), and hormones and growth regulators.

Feeding Deterrents

Feeding deterrents are used to eliminate damage by keeping a pest from feeding rather than by reducing its population, although strong deterrents can induce starvation among species with a narrow range of hosts. The only such substance tested on a large scale is 4-(dimethyltriazcno) acetanilide. Treating foliage with this substance produced fair to excellent results against a variety of insect herbivores, including cabbage pests, the boll weevil, and cucumber beetles. However, it did not prevent damage caused by insects that feed from inside the plant tissue as the corn earworm and codling moth do (see Wright 1963). Combining the deterrent with an effective penetrant might prove useful.

A number of chemicals extracted from plants have been shown in the laboratory to be strong feeding deterrents for some important pests, such as the European corn borer and Colorado potato beetle (Smissen et al 1957; Maxwell et al 1965), but these chemicals have yet to be proven in field trials.

Feeding deterrents offer several advantages over conventional pesticides because they affect only the pest species. In addition, the fact that the pests are usually not killed means that they persist as hosts for their natural enemies. The main disadvantages of deterrents at present are that they are effective only against surface feeding pests, and that because the pests are not killed they can disperse to untreated areas. Furthermore, the host plants continue to grow after treatment and so provide a source of untreated food on which the pest population could concentrate and cause severe feeding damage.

Antimetabolites

Antimetabolites are compounds that resemble essential nutrients but inhibit the utilization of metabolites. When they are ingested with food, the insect displays the symptoms of a dietary deficiency that can result in death. The potential of such compounds as a practical control method is yet to be demonstrated and will probably be restricted to use against species with a limited range of foods or limited access to foods.

Repellents

Repellents are compounds that cause organisms to move away from the source. Insects usually respond to repellents from a relatively short distance, and they tend to avoid treated surfaces instead of leaving the area. Repellents are therefore more successful in protecting humans and domestic animals against irritating and disease vectoring species than in protecting crops.

A long lasting repellent can protect domestic animals from biting flies and can reduce the annoyance and loss of blood that can slow weight gain and reduce milk production; they can also help to reduce insect transmitted pathogens. However, of the hundreds of compounds that have been tested for use on domestic animals only a few are recommended, and these have limited value, mainly because they are effective for only a day or so.

Although there are many naturally occurring and synthetic compounds that repel plant-feeding insects, no practical methodology has been developed for their use in protecting crops. Repellents act mainly to prevent pests from invading a crop, not to reduce the pest population. Repellents would therefore probably have to cover a very large area to be effective. In special cases, such as a forest, where one part of a crop is more susceptible to pests than another, repellents could be used to disperse pests searching for susceptible hosts and could thereby increase mortality during dispersal.

Attractants

Chemical messengers are of considerable importance in regulating insect behavior. Odors given off by host plants and animals and by other sources of food are used by insects to locate feeding and breeding sites. Moreover, insects themselves produce a variety of highly specific compounds (**pheromones**) that are used in various kinds of communication including species recognition, aggregation for mating and host colonization, defense, sex attraction, and mating stimulation (Shorey 1973). The ability of these powerful chemicals to induce behavior modifications has long been a source of interest to entomologists, but only in recent years has chemical technology advanced to the point where small quantities of such complex substances could be extracted, identified, and synthesized.

In a strict sense, attractants are stimuli that draw organisms to their source and are therefore important components of insect communication systems. Because they bring individuals to a particular location, there are a variety of ways that they can be employed in insect control. First, because control decisions are frequently based on pest population levels, attractants can be used to monitor population levels. Because many insect attractants are specific, insect surveys that are directed against a limited number of pest species can be made quite efficient and are particularly useful for detecting the entry of pests into quarantined areas.

One of the largest survey programs uses disparlure, a synthetic female sex pheromone of the gypsy moth *Porthetria dispar*. Approximately 50,000 small cardboard traps (Figure 17-4), combining a disparlure-saturated wick with a sticky coating on the inside surface of the cardboard, are placed throughout the infested area of New England to monitor the flying population of male moths (Jacobson 1965). Attractant-baited traps are also used as in an early warning system to detect a number of potentially serious pests including the Mediterranean fruit fly in Florida, the Mexican fruit fly *Anestrepha ludens* along the U.S.-Mexico border, and the Japanese beetle and oriental fruit fly in California. When the traps

are set out properly in a systematic grid, such a detection program not only assists in the discovery of new invasions but also pinpoints the location so that an eradication program can be concentrated in a particular portion of the grid.

Some attractants, particularly those that stimulate population aggregation, can be used for pest control. Some pests can be drawn away from valuable or inaccessible resources and attracted to specific areas where they can be eliminated. For example, natural and synthetic bark beetle pheromones can be used to attract large numbers of beetles to trap trees or logs where they are allowed to reproduce before the infested host material is removed and/or destroyed.

Attractants can also be used to draw pests into traps as a direct means of population reduction. This requires an efficient trap design as well as an effective attractant that is released gradually over a period of time. The simplest arrangement involves a sticky surface on which the attracted insects become entangled (Figure 17-5). The efficacy of such approaches in suppressing pest populations is sometimes difficult to evaluate because the number of individuals trapped may be unrelated to the local population before the release of the attractant. Placing attractant throughout an area could change the behavior of dispersing individuals and actually draw additional individuals into the area.

Several other kinds of chemicals, which act as arrestants or feeding, mating, and ovipositional stimulants, can be included under the general heading of attractants. Some of these compounds can be used in combination with insecticides, sterilants, or pathogens to improve the efficacy of the control agent. The oriental fruit fly was eradicated from the island of Rota in the western Pacific by distributing five-centimeter cardboard squares treated with an attractant and an insecticide. The attractant was so effective against the males that 15 biweekly treatments with the cards reduced the

FIGURE 17-4.
Pheromone trap of the type used to survey gypsy moth populations. A pheromone-saturated wick is suspended inside the trap, which has a sticky inner surface. (Photograph courtesy of the USDA.)

FIGURE 17-5.
Large sticky trap baited with vials of attractant. Traps of
this type were used in a pilot project involving the use of
attractants to reduce bark beetle populations. (Photograph
courtesy of G. P. Pitman, B.T.I.)

sporozoan infection among corn earworms
and boll weevils respectively (National Research Council 1969). Thuron Industries has
developed a poison bait containing a fly attractant Muscemone, which is the first product with
an attractant component to be registered by
the EPA for insect control.

One of the basic problems in using pheromones and other behavior modifiers to control
insects results from the extent to which individuals in pest populations vary behaviorally.
Individuals respond to stimuli according to
their response threshold, which frequently
varies according to physiological conditions
and can change with time. For example, many
insects go through a dispersal phase during
which they are less responsive to hosts and
mates than they are after the dispersal drive
has been satisfied. The behavioral makeup of
a population can change according to how
much dispersal has occurred or can vary according to conditions during development.
Some populations might therefore be more
responsive to attractants than others, and
control programs using these substances could
produce variable results (See Atkins 1966).

male population to a level at which most females were not mated and consequently left
no fertile offspring.

As we indicated earlier, many chemosterilants are too dangerous for use by general
broadcasting and so must be combined with
baits or arrestants that attract only particular
species. Attractants can also be used to increase
the effectiveness of pathogens by drawing
pest individuals into the part of the environment where the pathogen has a better chance
to survive and infect the pest species. Laboratory tests using this principle have demonstrated an increased incidence of virus and

Insect Growth Regulators

In insects as in all organisms, the physiological
processes related to reproduction, growth,
molting, and metamorphosis are regulated by
hormones. Our understanding of the role of hormones in the developmental physiology of insects is far from complete, but improved
biochemical methods have produced major advances in recent years. There are three basic
types of hormones known at this time: the **brain
hormone,** which appears to mediate the production and interaction of the other hormones;
ecdysones or **molting hormones** principally involved with the periodic casting off of the cuticle; and the **juvenile hormone** that governs the
rate of maturation including the development

and maturation of the gonads. The chemical structure of the ecdysones and juvenile hormone have been determined, and juvenile hormone and alpha-ecdysone have been synthesized. In addition, numerous synthetic compounds (**juvenoids**) have been discovered that exhibit the major biological activities of juvenile hormone.

The principal function of juvenile hormone in insect development is to determine the result of each molt by suppressing maturation until the appropriate time. Treating an insect with juvenile hormone at intermediate stages of development can inhibit normal development and actually prevent maturation. Juvenoids have gonadotropic effects as well and can be used experimentally to disrupt dormancy, normal ovarian development, and the distribution of fat reserves between somatic and reproductive tissues. This has created considerable interest in the use of juvenile hormone mimics for pest control. Many of the juvenoids seem to affect fairly specific insect groups, unlike the natural juvenile hormone, which affects all insects; this specificity creates some interesting possibilities. Specially designed compounds could be used to affect adversely the development of selected pests without harming beneficial species.

Many problems will have to be solved through research before these compounds can be evaluated fully under field conditions, but progress has been rapid. Several products have already been tested in the field against specific pests with promising results. A California based company, Zoecon, has developed a compound distributed under the trade name Altosid, which has been registered by the EPA for the control of flood water mosquitoes (see Koslucher 1973). This was the first insect growth regulator the use of which was approved by a regulatory agency. According to company reports, Altosid fed to chickens and cattle has also proved effective in controlling houseflies and hornflies that breed in manure.

More recently a second Zoecon product, an insect growth regulator called Enstar-5E, has been approved for use in the control of aphids and whiteflies in greenhouses.

Another new group of compounds was recently reported in an undated Technical Information Bulletin of the Thompson-Hayward Chemical Company. Although they are not actually hormone mimics, the two 1-(benzoyl)-3-(phenyl)-urea compounds are believed to interfere with the formation of normal insect cuticle. The newly formed cuticle of treated larvae apparently cannot withstand the increased turgor and/or muscle movement that is necessary for molting, and so treated larvae survive until ecdysis, at which time they die from molting failure. Thompson-Hayward now produced a product commonly called diflurbenzuron under the trade name Dimilin, which has proved useful against larvae of the gypsy moth and range caterpillar *Hemileuca oliviae*.

MISCELLANEOUS METHODS

There are a number of physical and mechanical methods of pest control that involve the use of elements such as **electromagnetic radiation, heat, barriers,** and **trapping devices.** These are generally less important than the control methods already discussed, but some are effective in specific situations.

Low-energy radiation, for example, can be used as a means of increasing temperature, which appears to be a satisfactory means of controlling pests in stored products. The electrical properties of the pests are sufficiently different from the matrix that the pest tissue is heated more than the grain, and this either repels or kills the pest without the hazard of fire.

Heat produced by more conventional methods has been used to control insects and mites in stored commodities and in wood products. Steam or dry heat can be used for delousing and soil sterilization, and hot water treatment of planting stock can effectively control plant diseases. Phillips and McCain (1973) demonstrated that 90 seconds in water at 50 to 52°C was sufficient to control a rust on infected geranium cuttings. Hot water treatment is also effective against virus-infected fruit rootstock and a variety of nematodes.

Passive barriers of many kinds are standard forms of protection against a wide variety of pests. Examples include netting to protect fruits against birds, fences to protect livestock against predators, and various kinds of bands to protect young trees against rodents and insects that attack the root collar. Sticky bands are often used to curtail the movement of pests that crawl up and down the trunks of orchard trees.

Various mechanical methods are employed in pest control, the most noteworthy of which are the types of machinery used for weed control. Not too many years ago a horse-drawn device known as a hopperdozer was used in the United States to scoop up large numbers of grasshoppers that frequently infested the fields.

SUMMARY

A variety of nonchemical approaches to pest control, along with several new and sophisticated chemical methods, in many cases provide practical alternatives to traditional pesticides. Some of these alternatives, particularly those grouped as cultural techniques, have been known and practiced for many years. Most of them fit well into the normal activities of crop culture and require a minimal extra expenditure of time or energy, but some of these cultural techniques unfortunately are overlooked and omitted from modern systems of intensive crop production. In addition, there are a number of pest control alternatives that are yet to be proven widely applicable. These newer approaches include the use of genetic manipulations, pheromones, and insect growth regulators. Still other approaches, such as the sterile-male technique, although fairly new, seem to have quite wide applicability.

The significance of these various pest control tactics is that they offer a broader base for choosing between the use of pesticides and those strategies, such as biological control (see Chapter 18), with which most poisons are incompatible. There is no doubt that the use of pesticides will remain a most important weapon whenever a quick response is needed to a pest that has unexpectedly surpassed its economic threshold. There are many situations, however, in which we can reduce our reliance on pesticides by using alternative approaches that provide the needed level of control; and cases where pests have developed genetic resistance to all pesticides, these alternatives constitute our only hope for the immediate future.

Literature Cited

Atkins, M. D. 1966. Laboratory studies on the behaviour of the Douglas-fir beetle, *Dendroctonus pseudotsugae*, Hopk. *Can Entomol.* 98:953–991.

Baumhover, A. H. 1966. Eradication of the screwworm fly—an agent of myiasis. *J. Amer. Med. Assoc.* 196:240–248.

Boon, W. R. 1967. The quaternary salts of bipyindyl—a new agricultural tool. *Endeavor* 26:27–33.

Cole, J. S., and O. R. Matthews, 1938. Tillage. In *Yearbook of Agriculture*, Washington, D. C.: USDA.

Haynes, D. L. 1973. Population management of the cereal leaf beetle. In *Insects: Studies in Population Management*. P. W. Geier, L. R. Clark, D. J. Anderson, and H. A. Nix (eds.). pp. 232–240. Canberra: Ecological Society of Australia (Memoirs 1).

Jacobson, M. 1965. *Insect Sex Attractants*. New York: Interscience Publishers.

Knipling, E. F. 1955. Possibilities of insect control or eradication through the use of sexually sterile males. *J. Econ. Entomol.* 48:459–462.

——. 1960. The eradication of the screwworm fly. *Sci. Amer.* 203(4):54–61.

——. 1972. Sterilization and other genetic techniques. In *Pest Control Strategies for the Future*. pp. 272–287. Washington, D. C.: National Academy of Sciences.

Koslucher, D. 1973. Altosid: A new method of mosquito control. In *Proc. Louisiana Mosquito Control Assoc.* New Orleans.

Laven, H. 1967. Formal genetics of *Culex pipiens*. In *Genetics of Insect Vectors of Disease*. J. W. Wright and R. Pal (eds.). pp. 17–66. Amsterdam: Elsevier.

Maxwell, F. G., W. L. Parrott, J. N. Jenkins, and H. N. Lafever. 1965. A boll weevil feeding deterrent from the calyx of an alternate host, *Hibiscus syriacus*. *J. Econ. Entomol.* 58: 985–988.

National Research Council. 1968. *Principles of Plant and Animal Pest Control*. Vol. 1: *Plant-Disease Development and Control*. Washington, D. C.: National Academy of Sciences.

——. 1969. *Principles of Plant and Animal Pest Control*. Vol. 3: *Insect-Pest Management and Control*. Washington, D.C.: National Academy of Sciences.

Phillips, D. J., and A. H. McCain. 1973. Hot water therapy for geranium rust control. *Phytopathol.* 63:273–275.

Shorey, H. H. 1973. Behavioral responses to insect pheromones. *Ann. Rev. Entomol.* 18:349–380.

Smissen, E. E., J. P. Lapidus, and S. D. Beck. 1957. Corn plant resistant factor. *J. Org. Chem.* 22:220.

Stern, V. M. 1969. Interplanting alfalfa in cotton to control lygus bugs and other insect pests. *Proc. Tall Timbers Conf.* 1:55–69.

Stern, V. M., R. van den Bosch, and T. F. Leigh. 1964. Strip cutting alfalfa for lygus bug control. *Calif. Agric.* 18:406.

Werhahn, C. F., and W. Klassen. 1971. Insect control methods involving the release of a relatively few laboratory-bred insects. *Can. Entomol.* 103:1387–1396.

Whitten, M. J. 1971. Insect control by genetic manipulation of natural populations. *Science* 171:682–684.

Woods, A. 1974. *Pest Control: A Survey*. New York: John Wiley and Sons.

Wright, D. P., Jr. 1963. Antifeeding compounds for insect control. *Adv. Chem. Ser.* 41:56–63.

Yarwood, C. E. 1968. Tillage and plant diseases. *BioSci.* 18:27–30.

——. 1970. Man-made plant diseases. *Science* 168:218–220.

18

BIOLOGICAL CONTROL

Biological control means different things to different people. In the broadest sense it is the natural phenomenon of population regulation created by the interaction of the biotic components of an ecosystem. Viewed in this way biological control encompasses all biologically causes of mortality, including competition between individuals of the same and different species, the effects of host defenses and host resistance, and the direct or indirect results of attack by organisms belonging to higher trophic levels. More applied biological control involves the manipulation, conservation, and augmentation of specific kinds of organisms in order to regulate populations of undesirable species and thus to prevent or reduce their negative impacts on human well-being.

It is in this narrower sense that we will discuss biological control in this chapter, but we will leave our discussion of host plant resistance until Chapter 19 and we will not discuss autocidal mechanisms such as the induced lethal mutations mentioned briefly in Chapter 17. However, we will consider both viruses and bacteria whose uses are somewhat atypical, including the use of the proteinaceous parasporal bodies formed by some bacteria, even though the latter resemble insecticides in their mode of action.

In general we will follow the definition of Paul DeBach, one of the foremost experts in the field, who said that biological control is "the action of parasites, predators and pathogens in maintaining another organism's density at a lower density than would occur in their absence." We will therefore consider herbivores as parasites and predators of plants in the biological control of weeds.

THE ECOLOGICAL BASIS

All the organisms that make up natural ecosystems interact with each other and with their physical environment in such a way that each species achieves an average population density or characteristic level of abundance. Its numbers tend to fluctuate around this level in response to other events in the system that affect the relationship between the birth and death rates. This is just as true for plants as it is for the herbivores that feed on them and for predators and parasites of the herbivores. When conditions are favorable for a particular species, its numbers usually increase (or in the case of long-lived plants there is above-average vegetative growth). Sometimes favorable conditions are the result of better-than-average weather, the duration of which is entirely independent of biological activity within the ecosystem. When such weather conditions continue for a long period certain organisms grow unusually quickly or reach an abnormal abundance. This sets in motion other phenomena, such as increased competition or increased attack by natural enemies, that gradually slow the rate of increase. Eventually these interactions might reverse the trend of population change of the affected species.

This regulation in the population densities of organisms is often called natural control, and over the long term it tends to maintain populations at their characteristic levels of abundance so that the energy flow through the major trophic components of the ecosystem is more or less stabilized.

The intensity of some environmental factors that act on populations in the regulatory process is not affected by the size of the populations, whereas the intensity of other factors varies in relation to population density. Factors of the first type are referred to as **density independent factors** and those of the second type are called **density dependent factors.**

Most living organisms interact with each other in a density dependent way, and most relationships could be described as reciprocal in that the population density of one species tends to affect and be affected by the population densities of all the species with which it interacts. Thus when a host plant increases in abundance, the herbivores that feed upon it may also increase; but simultaneously the parasites of the herbivore will proliferate and eventually bring the herbivore population down again, and so on.

In natural communities the interplay of density independent and density dependent factors keeps most populations from attaining extraordinarily high densities. In agroecosystems and in some unstable natural systems the environment often favors certain kinds of organisms over others, and some of the regulatory influences may be absent, thereby leading to pest outbreaks. This is particularly true in ephemeral crop environments, because there is never time for well-forged relationships between different trophic levels to develop. However, even in more permanent crop situations that involve native species the characteristic population density of one or more organisms might be higher than desirable from a management standpoint. Furthermore, even the most stable crop ecosystems are challenged from time to time by new invaders. When an exotic species is introduced into a region where it has no natural enemies, it can reproduce actively but still maintain a low mortality rate, which leads to population outbreaks. The goal of applied biological control is to improve the biotic relationships in these situations in order to maintain undesirable species at population levels that are not injurious. Each of these situations obviously presents different difficulties and control opportunities.

The biological control of native pests with native natural enemies reveals limitations imposed by long-term relationships. We have to assume, for example, that the species have coevolved in such a way that the natural enemies do not overexploit their prey. Thus their capacity for suppressing the prey population equilibrium might be severely limited. At the same time some crop environments favor an increase in pest populations while impeding the regulatory efficiency of its natural enemies. In the case of native pests it may be necessary to establish a new relationship either by improving the crop environment for natural enemies or by introducing new enemies, such as species that attack relatives of the pest in other areas. According to ecological theory, it would be easier to implement biological control against exotic species that have not coevolved closely with their associates than against native species. This has

FIGURE 18-1.
A caterpillar carrying numerous cocoons of a parasitic wasp. The parasite larvae have finished their feeding inside the caterpillar, which usually dies shortly after they emerge. (Photograph by M. D. Atkins.)

indeed been the case; most successful cases of biological control have involved the control of introduced pests with natural enemies imported from the pest's native area.

METHODOLOGY

The methodology of biological control is a complex subject to which several authors have devoted entire books (DeBach 1964, 1974; Huffaker 1971; van den Bosch and Messenger 1973). Biological control is very much an entomological subject because nearly all its major areas of activity involve insects (Figure 18-1). These areas include the control of insects and mites with insects and other animals, the control of insects and mites with pathogens, the control of weeds with insects, and the control of plant diseases through the biological control of insect vectors of plant pathogens. In the next few pages we shall present some of the fundamental aspects of biological control by using successful examples as a basis for discussion.

Insect Control with Parasites and Predators

THE COTTONY CUSHION SCALE   The biological control of the cottony cushion scale, *Icerya purchasi*, is a success story that has been told many times, but it bears repeating because of its historical as well as its biological significance. The cottony cushion scale was unintentionally introduced into California from Australia in the 1860s and soon posed a serious threat to the state's citrus industry. C. V. Riley, the Chief of the Federal Division of Entomology, felt that importing natural enemies from Australia might provide a solution. Albert Koebele, in the guise of a government representative to the 1888 International Exposition at Melbourne, made an unofficial search for parasites and predators. In the meantime, near Adelaide, an Australian entomologist had found a small fly, *Cryptochaetum iceryae*, that was a parasite of the pest, and he sent a small shipment of these parasites to California. The live flies were released near San Francisco in early 1888.

While Koebele was in Australia, he discovered a small lady-beetle, *Rodalia cardinalis*, commonly called the vedalia beetle, that preyed on cottony cushion scale females. Koebele shipped 129 live vedalia beetles to Los Angeles between November 1888 and January 1889. These and two later shipments that arrived in February and March of 1889 were released on screened trees at three locations in the Los Angeles area. By June of 1889 over 10,000 offspring of the original 129 beetles had been sent to other parts of California, and within months of the release of the vedalia beetle, citrus trees were cleared of the cottony cushion scale. At the same time, *Cryptochaetum* became established and spread, probably from the San Francisco area, to the entire coastal region of California.

From a few years after the introduction of the two control agents from Australia until the system was recently disrupted by pesticides, the cottony cushion scale never attained pest status. Most of the credit for this success has been given to the predaceous vedalia beetle, but DeBach (1974) feels that the results would have been equally spectacular if *Cryptochaetum* alone had been introduced.

The successful control of *Icerya purchasi* in California was followed by similar successes in other citrus growing areas, and it clearly provided the impetus for an expansion of biological control. But in addition to this example's historical significance, it introduces one of the interesting questions in biological control: Do predators or parasites make the best biological control agents?

In more highly evolved parasitic relationships, the host is not usually killed, but the goal of biological control is the death of prey or host individuals. Although we normally distinguish between predation and parasitism, they are, in a sense, only different forms of the same relationship; and among insects we often find that parasites kill their hosts.

Generally **predators** are free-living forms in both the immature and adult stages, and they often attack the same species of prey throughout their entire life history. A fair number of species, however, are only predaceous either as immature forms or as adults. Most predators search out and consume a number of prey individuals in the course of their life. A lady-beetle, for example, might consume several hundred aphids during its development (Figure 18-2). This is due partly to the fact that it is active both in the immature stage and as an adult, and partly because it is larger than its prey. There is really no general statement that can be made about prey specificity, as it varies widely from species to species; but many predators will attack a variety of prey species concentrating on the one that is most abundant.

FIGURE 18-2.
A predaceous lady-beetle feeding on a group of aphids. (Photograph by Max Badgely.)

On the other hand, entomophagous **insect parasites** are usually free-living as adults and sometimes as first instar larvae. The rest of their life cycle is spent on or in the body of their host. In most species the adults either do not feed at all or else feed on pollen and nectar. Finding a host is usually the function of the female, and a single host, once invaded, provides sufficient food for the development of one or more immature parasites. Although there are exceptions, most parasites are rather host specific and consequently have life cycles that are well synchronized with that of the host on which they depend. Host specific parasites must also be able to search efficiently. It was first assumed that female parasites located the habitat of their host and then searched randomly for hosts, but at least some species have been found to leave trail odors that keep them from searching the same area twice (Price 1970).

Predators may appear to be better biological control agents than parasites simply because they consume more prey per individual, but this alone does not necessarily result in effective prey regulation over a range of population densities. If a predator is unable to find prey when they are present in low numbers, or if it switches to a more abundant prey species, its efficacy is reduced. Some predators, such as *Rodalia cardenalis*, are effective control agents because they are specific, they have a rapid rate of increase relative to that of their prey, and they prey upon sedentary and gregarious species. All insect predators do not display these attributes, and most workers now feel that host specificity, synchronized life cycles, and the superior searching ability that enables parasites to locate hosts at low densities, are attributes that tend to make them better biological control agents than predators.

It would be foolhardy, however, to exclude predators from the search for the natural enemies of pests, because a number have been highly effective and, as in the control of cottony cushion scale, they can work in conjunction with a parasite. Furthermore, not all parasites are good control agents. Some are too specific and are unable to survive the absence of their host by attacking an alternate host. Those that require flowers as a source of adult food for egg maturation may disperse away from the host population; but this can be corrected through vegetation management if the requirements of the parasite are understood.

THE ORIENTAL FRUIT FLY   The oriental fruit fly, *Dacus dorsalis*, is an Asiatic species that entered Hawaii near the end of World War II. It became a serious pest of many kinds of ripe fruits, and it posed a threat in the warm fruit-growing areas of the continental United States. Entomologists were sent to a number of tropical areas to search for parasites and predators of the fruit fly. Among the variety of natural enemies shipped to Hawaii in 1947 and 1948 were three species of parasitic wasps of the genus *Opius*. One of these, *Opius longicaudatus*, increased rapidly but lost its dominance to *O. vandenboschi*, which had been released at the same time. Later *O. vandenboschi* was replaced by *O. oophilus* (Clausen et al. 1965), which has since provided effective control of the oriental fruit fly.

This program, apart from being another successful case of biological control, illustrates the fact that several parasites of the same pest can be released without diminishing the overall level of control achieved. Although the three wasp species were competing for the same host, the one with superior qualities displaced the others and became dominant. *O. vandenboschi*, for instance, derived its advantage by attacking first instar larvae of the fruit fly and thereby inhibiting the eggs and larvae of *O. longicaudatus*, which favors older fruit fly larvae. Likewise, *O. oophilus*, which oviposits in the eggs of the host, were already present as larvae in hosts suitable for attack by *O. vandenboschi* (van den Bosch and Messenger 1973).

A similar sequence of displacements occurred

after the release over a number of years of parasites imported to control California red scale in California. As a result of competition among the parasites in different climatic zones, different species have become dominant, not because of the host stage they attack, but because they are best suited to different climates. At the present time the parasitic wasp, *Aphytis lingnanensis,* is dominant in coastal areas, where it has replaced *A. chrysomphali.* Similarly, *A. melinus* later eliminated *A. lingnanensis* from the warmer interior areas and remains dominant there. Along the coast the parasite *Prospaltella perniciosi* complements *A. lingnanensis,* while still another parasite, *Comperiella bifasciata,* complements *A. melinus* in the interior. Both the complementary parasites, although they are not dominant, have been able to survive because they attack different host stages than do the *Aphytis* species (DeBach 1974).

Although in the case of the oriental fruit fly the succession of parasites quickly led to the establishment of control by a single dominant species, the California red scale example illustrates the fact that successful control can take a great deal of work and persistance. The first natural enemies of red scale were imported in 1889, but a stable situation did not develop until after the introduction of *A. melinus* in 1957.

THE WINTER MOTH The geometrid moth, *Operophtera brumata,* or winter moth as it is commonly called, is a defoliator of hardwood forests and ornamental trees. The moth was introduced to eastern Canada from Europe during the 1930s. As a result of the gradual spread of the moth and considerable damage to oaks, a biological control program was initiated in 1954 with the introduction of six natural enemies from Europe. Two of these, a tachinid fly, *Cyzenis albicans,* and an ichneumon wasp, *Agrypon flaveolatum,* became established and brought the pest under control in about six years. There was no displacement in this case because the two species that brought about control were complementary. *C. albicans* is apparently very effective at high host densities, whereas the superior searching ability of *A. flaveolatum* makes it effective at low host densities. DeBach (1974) suggests that this characteristic of *A. flaveolatum* may have been reflected in the original parasite collections made during local winter moth outbreaks in parts of Europe where *C. albicans* was the predominant parasite. This suggests that the most common control agent may not be the best one. The average population of the pest would obviously be higher if it were not invaded at low densities by *A. flaveolatum.*

The example of the winter moth project is also useful in countering some of the criticisms that have been leveled against biological control. Many of the early successes were achieved on islands or in areas with island characteristics (small native floras and faunas that offer little resistance to invading or introduced species) and in places with favorable climates. Many workers thus came to believe that biological control would not be successful in continental areas, but examples like the Canadian winter moth project have shown this assumption to be incorrect.

Many of the early successes were against sedentary, gregarious insects such as scales and mealybugs, and the current record also shows more success against pests of this type than against more active pests like caterpillars, aphids, and beetles. Certainly the scales and mealybugs have many desirable attributes as targets for biological control, but we must also realize that many biological control programs were patterned after the successful program initiated against the cottony cushion scale. More recent successes, such as the control of the winter moth and the walnut aphid in California (van den Bosch et al. 1964), indicate that with conscientious effort there may be no such limitations to biological control in the sense that both active and sedentary species can be controlled.

A number of critics have suggested that in many cases only circumstantial evidence exists that introduced natural enemies have actually caused pest populations to decline, and these same critics argue that unless data are collected to prove such action, biological control should not be credited with pest reductions. This may be a legitimate criticism in that some introduced pest species tend to increase and spread rapidly at first but then to stabilize at lower densities. This could confuse the interpretation of a control program initiated during the pest's stabilization phase. However, the winter moth project was extremely well analyzed in this regard and life tables were used to develop a population model that clearly showed that parasitism was the key factor in controlling the moth (Embree 1971).

Insect Control with Pathogens

Disease symptoms in insects were recorded by the ancient Chinese and Greeks, but the experimental demonstration of microorganisms as causal agents did not occur until the nineteenth century. Interestingly, the first successful investigations involved the study of diseases in two beneficial insects, the silkworm and the honeybee. In 1835 an Italian, Agostino Bassi, discovered that a fungus, *Beauveria bassiana*, caused the white muscardine disease of silkworms. This was followed by Pasteur's famous discovery in 1870 of a microsporidian as the cause of pebrine and flacherie diseases of silkworms, and the recognition by Cheshire and Cheyne in 1885 that *Bacillus alvei* was the cause of European foulbrood of honeybees. These discoveries led to several early attempts to use microorganisms to control insect pests. Now six major groups and over 1200 species of microorganisms, most of which are pathogenic, have been found to be associated with

insects. We will consider briefly the three most important groups: **bacteria, fungi,** and **viruses.**

BACTERIA   The Japanese beetle, *Popillia japonica*, is a serious imported pest in parts of the eastern United States, where the adults feed on the foliage and fruits of more than 250 plant species, and the larvae damage the roots of grass, nursery stock, and vegetables. The discovery and application of the spore-forming bacterium, *Bacillus popilliae*, as a means of controlling the larvae of the Japanese beetle in the soil (White and Dutky 1940) provided the first encouragement for the use of bacteria in insect control. Now *B. papilliae* and *B. lentimorbus*, which cause the **Type A** and **Type B milky diseases** of the Japanese beetle (Figure 18-3), are mass produced and are sold as a spore dust for injection into the soil. Not only do

FIGURE 18-3.
Japanese beetle larvae: normal specimen on the left, specimen on the right showing white color symptomatic of infection with milky disease bacteria. (Photograph courtesy of the USDA.)

infected larvae that die in the soil contaminate other larvae feeding in the vicinity, but the *Bacillus* spores persist in the soil to infect larvae in later generations. Larval populations can thus be substantially reduced, but their stabilization at satisfactorily low levels takes several years to achieve, so some damage has to be tolerated after the disease preparation is applied to the soil.

Bacillus popilliae and *B. lentimorbus* are obligate parasites, which makes it impossible to propagate them on artificial media. Because their spores must be obtained by culturing the bacteria in host larvae, the dust preparation is expensive. A related spore-forming bacteria, *Bacillus thuringiensis*, is a facultative pathogen that infects a variety of insects, including lepidopterans, bettles, and flies. This bacterium can be cultured on artificial media and is therefore quite economical to produce.

Bacillus thuringiensis produces proteinaceous crystals as well as spores, and these crystals dissolve in the gut of a number of plant-feeding insects and quickly cause gut paralysis and cessation of feeding. The gut subsequently ruptures, which permits the bacterial spores to enter the insect's tissues, where they germinate and reproduce. Commercial preparations of *B. thuringiensis* containing both spores and crystals are registered with the USDA for use on a variety of crops (Table 18-1) as a **biological insecticide.** The cost of this material is competitive with that of chemical insecticides.

Bacillus thuringiensis has several desirable attributes as a means of controlling a variety of foliage-feeding insect pests. The paralytic action of ingested crystals causes a rapid cessation of feeding and thus a rapid reduction in crop damage; this is an improvement over the slower action of the milky diseases. Unfortunately, however, the spores do not seem to persist in the crop environment and so do not provide continuous control from generation to generation. Nevertheless, once a pest population reaches its economic threshold, inexpensive preparations can be applied as needed, much like regular insecticides. The fairly specific ability of *B. thuringiensis* to kill a few groups of foliage feeders without harming beneficial species is of great value in management programs designed to conserve natural enemies.

There have been a few reports that the repeated use of *B. thuringiensis* on the same land creates a residue of spores and crystals that are harmful to earthworms, but no evidence of parasite or predator mortality has been found. Mahr and Atkins (1976) fed infected caterpillars to several parasites and predators and found no indications that the bacteria caused any ill effects in the beneficial species. However, in an agroecosystem treated with *B. thuringiensis*, parasite and predator populations would decline in response to the mortality of their food species. In this respect *B. thuringiensis* sprays are much like insecticides.

FUNGI Fungi that attack insects belong to all of the four major taxonomic groups of true fungi, but only a few genera are frequently associated with insect disease outbreaks. Those most commonly used in insect control are *Beauvaria bassiana* (white muscardine disease) and *Metarrhizium anisopliae* (green muscardine disease), both of which are fungi imperfecti. A natural collapse of a chinch bug (*Blissus leucopterus*) population caused by *B. bassiana* prompted the first attempt to use a fungus to control an agricultural pest (Roberts and Yendol 1971). Although the induced infection of the chinch bug with its fungus did not result in satisfactory control, the project did provide the following basic insight into the problems of using fungi in microbial control.

Most entomogenous fungi are internal pathogens. The infective unit is usually a spore that germinates on the surface of the host's integument. A special rootlike structure then penetrates the integument by enzyme action and mechanical force, and once the host tissue is invaded, the fungus can normally complete

TABLE 18-1
Some registered uses for *B. thuringiensis* products in the United States.

Pest	Crop
VEGETABLE AND FIELD CROPS	
Alfalfa caterpillar (*Colias eurytheme*)	alfalfa
Artiohoke plume moth (*Platyptilia carduidactyla*)	artichokes
Bollworm (*Heliothis zea*)	cotton
Cabbage looper (*Trichoplusia ni*)	beans, broccoli, cabbage, cauliflower, celery, collards, cotton, cucumbers, kale, lettuce, melons, potatoes, spinach, tobacco
Diamondback moth (*Plutella maculipennis*)	cabbage
European corn borer (*Ostrinia nubilalis*)	sweet corn
Imported cabbageworm (*Pieris rapae*)	broccoli, cabbage, cauliflower, collards, kale
Tobacco budworm (*Heliothis virescens*)	tobacco
Tobacco hornworm (*Manduca sexta*)	tobacco
Tomato hornworm (*Manduca quinquemaculata*)	tomatoes
FRUIT CROPS	
Fruit-tree leaf roller (*Archips argyrospilus*)	oranges
Orange dog (*Papilio cresphontes*)	oranges
Grape leaf folder (*Desmia funeralis*)	grapes
FORESTS, SHADE TREES, ORNAMENTALS	
California oakworm (*Phryganidia californica*)	
Fall webworm (*Hyphantria cunea*)	
Fall cankerworm (*Alsophila pometaria*)	
Great Basin tent caterpillar (*Malacosoma fragile*)	
Gypsy moth (*Lymantria* [*Porthetria*] *dispar*)	
Linden looper (*Erannis tiliaria*)	
Salt marsh caterpillar (*Estigmene acrea*)	
Spring cankerworm (*Paleacrita vernata*)	
Winter moth (*Operophtera brumata*)	

Source: From information supplied, in part, through the kindness of International Minerals and Chemical Corp., Libertyville, Illinois and Nutrilite Products Inc., Buena Park, California. Compiled by L. A. Falcon 1971.

its life cycle. After the integument is successfully penetrated, mycelium usually spread throughout the host's body until the infected insect is almost filled with fungus. Fruiting bodies then erupt through the integument and produce spores that are released into the external environment. The survival and germination of the spores is therefore critical to the development of an epidemic.

Facultative fungi such as *Beauvaria* and *Metarrhizium* can be cultured on artificial media, which facilitates the production of spore preparations that can be used in biological control. Fungi, like most biological control agents, can be used for either persistent or short-term control.° A fungus can be introduced into an area where it becomes established and kills the host year after year, as in the control of the Japanese beetle with milky disease; or fungal spore preparations can be used as microbial insecticides similar to those prepared with *Bacillus thuringiensis.*

Relatively few attempts have been made to introduce insect-attacking fungi into new regions. Instead of importing foreign species, most projects have redistributed indigenous fungi or used those that were introduced along with introduced pests. The best example of attempts to establish new fungal pathogens in disease-free areas is the introduction of *Coelomomyces* into disease-free populations of the mosquito *Aedes polynesiensis.* The fungus became established, but its impact on the population of adult mosquitoes was inconclusive (Laird 1967).

On the other hand, repeated applications of fungal spores as microbial insecticides has been demonstrated (Müeller-Kögler 1965) as an effective means of achieving short-term reductions of pest populations. The major limiting factor in initiating fungal disease in insect populations is the microclimate, the effects of which can inhibit spore survival and germination. The optimal temperature range for the growth of insect-infecting fungi is fairly narrow, and relatively high humidity is needed by most fungi to germinate and successfully penetrate their host. Spores are also known to be killed by sunlight. Consequently, the application of a spore preparation must coincide with both the presence of susceptible hosts and the existence of suitable environmental conditions. Success can usually be obtained, therefore, by

applying the spores on a warm evening following rain or irrigation.

Roberts and Yendol (1971) outlined a number of areas in which further research is needed to improve the use of fungi in insect control. These include: (1) elucidating aspects of disease induction, including chemical constituents of the insect cuticle, that inhibit or enhance spore germination and penetration; (2) developing fungal strains suitable to microbial control; (3) developing methods for maintaining virulence in mass cultures; (4) standardizing microbial preparations; (5) further elucidating environmental effects on the pathogens; and (6) determining the effects of combinations of several fungi, of fungi with other pathogens, and of fungi with sublethal doses of insecticides.

VIRUSES The majority of insect pathogenic viruses are different from most viruses in that the virus particles or **virions** are enclosed in proteinaceous capsules or membranes. Such viruses are called **inclusion viruses** as opposed to noninclusion viruses in which the virions are free within the cells of the host.

There are two main types of inclusion viruses, the **polyhedroses** and the **granuloses.** The former consist of a number of virus particles embedded in a polyhedral-shaped protein matrix, and the latter consist of single virus particles enclosed in individual protein coats or capsules.

The polyhedroses are divided into nuclear and cytoplasmic forms, based on the site of virion multiplication within the cells of the host. **Nuclear polyhedroses,** most of which have been found in caterpillars and the larvae of sawflies, affect the epidermis, fat body, and blood cells. Virus particles infect the host through the mouth or cuticle or are passed from one generation to the next on or within the egg. Once nuclear polyhedrosis viruses start to multiply in the host tissue, the larvae become sluggish and often crawl to high branch tips where they die hanging from the prologs at the tip of their abdomen. At this stage the

---

°Even parasites and predators can be used to inundate an area as a means of reducing a pest population temporarily without concern for the permanent establishment of the control agent. For example, backswimmers (Hemiptera: Notonectidae) can be introduced into temporary ponds to reduce mosquito populations.

larval skin becomes fragile and ruptures easily, releasing a shower of virus particles onto the foliage below. **Cytoplasmic polyhedroses** cause diseases of caterpillars and the larvae of a few lacewings (Neuroptera). They invade primarily the cells of the alimentary canal, and infected larvae often become white and swollen. The virus is spread when polyhedral bodies are regurgitated or passed out with the feces.

Granulosis viruses are restricted to the larvae and pupae of Lepidoptera, in which the fat body is the main tissue attacked. The virions first multiply in the nuclei but later continue to replicate in the cytoplasm. The disease eventually kills the insect, leaving it hanging as a fragile sack of virus similar to that which is created by nuclear polyhedroses infection.

There are a few noninclusion viruses that attack specific insects, but, with the exception of a limited number that might prove useful in mosquito control, most attention has been given to the inclusion viruses. The fact that virus particles encapsulated in a protein matrix maintain their ability to infect for many years means that inclusion viruses can be stored as concentrated preparations for later application with conventional pesticide spray equipment.

The production of large quantities of virus is often difficult because they must usually be obtained from infected hosts, which requires an extensive rearing program. Although a supply of virus can be accumulated over a period of time, virulence is often lost in storage. Recently propagated virus is therefore most effective. Field trials involving virus applications have produced variable results, but the technique nevertheless promises to be an effective means of insect pest control.

A number of Canadian scientists have shown that a nuclear polyhedroses virus is highly effective against a variety of forest sawflies, and their work is particularly encouraging in that the virus has persisted in the environment and provides continuous regulation of the pest in some areas. One of the problems with viruses, however, is that there are periods when they seem to have little effect on pest populations. Naturally occurring virus may remain latent in a pest population for several generations, and then initiate epizootics when the pest population comes under some form of stress. Precisely what causes viruses to behave in this manner is poorly understood, and it is likely that more information on this subject will improve the effectiveness of viral pest control. Generally, short-term control can be gained by frequent applications of virus preparations so that there is an active inoculum in the pest environment for an extended period.

Several nuclear polyhedrosis viruses are being produced on a large scale for possible use against a variety of pests, including cotton bollworm, tobacco budworm, corn earworm, cabbage looper, forest tent caterpillar, and alfalfa caterpillar. However, virus preparations, like insecticides and bacterial preparations, can only be used experimentally in the United States unless they are properly registered. So far the progress in this area has been rather slow, due at least in part to registration restrictions imposed by the Environmental Protection Agency.

BIOLOGICAL CONTROL OF WEEDS

The control of weeds by various biological agents can be an effective alternative to cultural and chemical techniques, especially in inaccessible areas or on land with low agricultural value. However, because weeds are often useful plants elsewhere or are quite closely related to economically important species, using biological agents to reduce their numbers can have serious side effects on valuable plant species. For this reason there has been some reluctance to exploit the method fully, and there is no question that it must be employed with utmost care. Nevertheless, there have been some spectacular successes.

Plants form the foundation of all food chains, and of course there are many organisms that feed upon them. In natural communities, however, a balance has evolved between native organisms that produce and those that consume, and so the biological control of weeds, as of insects, has usually been employed against alien species or species that have escaped natural regulation. Potentially at least, weeds can be controlled by a variety of vertebrates, snails, insects, mites, parasitic plants, and plant pathogens including bacteria, fungi, viruses, and nematodes. So far the greatest success has been achieved using phytophagous insects.

The control of Klamath weed or St. Johns-wort, *Hypericum perforatum*, by an insect herbivore provides one of the best examples of biological weed control. A native of Europe and Asia, Klamath weed has invaded a number of subhumid areas of the world, where it displaces desirable forage plants and interferes with the raising of livestock because of its poisonous nature. The weed was first discovered in northern California in the early 1900s and quickly spread through a considerable area of valuable rangeland. Although it could be controlled by chemical and cultural methods, these were generally too costly over large and inaccessible areas. A similar occurrence in Australia led to the search for insects that feed on the weed in parts of its natural range.

Before an insect herbivore can be introduced into a new area for weed control it must be determined beyond a doubt that that organism is specific to the weed species and will not switch its feeding habits to desirable plants. Insects vary considerably in their host plant specificity, and the degree to which a species is tied to a particular host cannot be determined by observing it in its natural situation. An insect used for weed control should be monophagous or should at least restrict its feeding to a few related plants of no economic importance. To determine safely whether this is true requires rather exhaustive testing under experimental conditions.

The search in Europe for potentially useful natural enemies of Klamath weed led to the discovery of three species of leaf-feeding beetles (Chrysomelidae) that, after thorough testing, were shipped to Australia. In 1945 the beetles were imported from Australia to California for colonization, and within three years two of the beetles (Chrysolina spp.) had become established and were causing extensive defoliation of the weed in the areas where they had been released. The beetles both spread naturally and were distributed artificially to other areas of California infested by the weed. By 1956 Klamath weed in California was effectively suppressed (Figure 18-4).

This sort of widespread devastation of a host plant by an introduced insect often raises fears that the insects will eventually respond to the lack of food by switching to another host. However, if the specificity has been tested properly before introduction, there is little danger of this because the reciprocal relationship between food supply and population causes the herbivore population to decline along with that of the host. Some Klamath weed still exists in shaded locations where *Chrysolina* does not do particularly well, but this small reservoir of food maintains a low population of the insect. As the weed expands into sunnier sites the insect population increases and suppresses it, and good control of the weed still persists.

Outside of California, Klamath weed control by the leaf beetles has been less successful. This seems to be a feature of biological control. Although there have been a number of complete successes, there have also been a number of partial successes and of course some failures. Climate is always an important consideration, but numerous kinds of biological interactions also influence the success of a species after

478

FIGURE 18-4.
Steps in the biological control of Klamath weed in an area of California: (A) rangeland heavily invaded by Klamath weed; (B) same location one year after the release of the leaf-feeding beetle *Chrysolina;* (C) same area restocked with forage plants. (Photographs (A) and (B) by J. K. Holloway; photograph (C) by J. Hamai.)

its introduction into a new, complex community. A review of biotic interference with insects imported for weed control (Goeden and Louda 1976) indicates that this is a very important aspect of weed control by insects.

A list compiled by the National Research Council (1968) shows that 25 serious weed pests have been controlled by insects to various degrees in a variety of insular and continental areas. The most effective insects in weed control belong to the orders Lepidoptera, Hymenoptera, Coleoptera, Hemiptera, and Diptera. Those that attack aboveground vegetative parts generally seem to be more effective than those that attack roots and seeds, although complete control of puncture vine *Tribulus terrestris* on the island of Kauai was achieved with a seed-feeding weevil.

Pathogenic organisms have not yet been seriously tested as a means of controlling weeds, although their considerable potential against pest species is suggested by the fact that introduced diseases such as chestnut blight, white pine blister rust, and Dutch elm disease have completely destroyed several desirable plants in certain areas, and in Queensland the pamakani weed, *Eupatorium adenophorum,* was controlled by the inadvertent introduction of a fungus. Like insects plant pathogens must be subjected to vigorous screening to determine their host specificity before they are released for weed control. The complex interplay of factors such as climate, conditions needed for infection, size and location of the inoculum, and the availability of susceptible hosts can make diseases unreliable. However, it might be possible to supplement "wild" pathogens from laboratory maintained cultures and thereby achieve more predictable control. The use of plant pathogens vectored by insects could be a successful approach, because the insects could act both as effective dispersal agents and as agents of infection.

The most recent summary of the status of projects on the biological control of weeds in the United States and Canada (Goeden et al 1974) lists 78 species of weeds that are the subjects of biological control by about 110 species of insects and a few plant pathogens. Projects involving a total of 34 kinds of weeds have progressed to the stage at which natural enemies have become established, and the effect of the introduced insect herbivores on the weed population has been evaluated for 19 weed species. Table 18-2 lists the six weeds against which biological control has been completed, together with the biological control agent involved in each instance.

BIOLOGICAL CONTROL OF PLANT PATHOGENS

If we exclude the use of resistant plant varieties, the field of biological control of plant pathogens is still in its infancy. Nevertheless, the first book on the subject has been published (Baker and Cook 1974), and the potential in terms of scientific knowledge and practical applications is extensive.

Crop losses due to disease are determined by the interactions of the physical environment, the pathogen, **antagonists** of the pathogen, and the susceptibility of the crop plant. Baker and Cook (1974) developed the series of diagrams presented in Figure 18-5 to illustrate the interplay of these factors. On this basis those authors define the biological control of plant pathogens as " . . . the reduction of inoculum density or disease-producing activities of a pathogen or parasite in its active or dormant state, by one or more organisms, accomplished naturally or through manipulation of the environment, host, or antagonist, or by mass introduction of one or more antagonists." This is a broader definition than the one used at the beginning of this chapter, but this seems to be a necessity because of the importance of the

TABLE 18-2
Successful projects on the biological control of weeds by insects in the United States and Canada.

Target weed	Biological control agent	Family	Order
Canadian thistle (Cirsium arvense)	Altica carduorum	Chrysomelidae	Coleoptera
Pamakani (Eupatorium adenophorum)	Procecidochares utilis	Tephritidae	Diptera
St. Johnswort or Klamath weed (Hypericum perforatum)	Chrysolina spp. Agrillus hyperici Zeuxidiplosis giardi	Chrysomelidae Buprestidae Cecidomyidae	Coleoptera Coleoptera Diptera
Prickly pear (Opuntia litloralis)	Cactoblastis cactorum	Pyralidae	Lepidoptera
Prickly pear (Opuntia oricola)	Dactylopius opuntiae Chelinidea spp.	Coccidae Coreidae	Homoptera Hemiptera
Tansy ragwort (Senecio jacobaea)	Hylemya seneciella	Arctiidae	Lepidoptera

soil environment as both the habitat of many pathogens and the medium in which the plants grow.

Only those aspects of control that involve the manipulation of antagonists form a direct parallel with the other types of biological control that we have discussed. Baker and Cook divide antagonism into three kinds of activity: **antibiosis** and **lysis, competition,** and **parasitism** and **predation.**

Antibiosis involves the inhibition, and lysis the destruction or dissolution, of the cells of one organism by a metabolic product of another. Bacteria can be inhibited by antibiotics produced by other bacteria, by actinomycetes, or by fungi; for example, a strain of *Bascillus subtilis* has been reported to reduce the incidence of cotton wilt and fire blight in pears.

Competition is generated when two or more pathogens rely on the same substrate for requisites (usually nutrients) the supply of which is not sufficient for both. For example, the incidence of coniferous root disease, which is caused by *Fomes annosus*, an organism that is extremely sensitive to competition, can be reduced by the mass introduction of *Peniophora gigantea* as an antagonist.

A number of microorganisms are known to be parasites and predators of pathogens, but their usefulness in biological control under field conditions remains to be deomonstrated. Many small organisms, including protozoans, nematodes, mites, and collembolla, ingest spores and vegetative tissues of pathogenic organisms and undoubtedly play some regulatory role, but their importance in this regard is not known.

POPULATION REPLACEMENT

A somewhat different form of biological control that is potentially useful against insects and pathogens involves the replacement of a harmful population with an innocuous or less harmful one. The methodology may involve the gradual replacement of a pest by an introduced species, or the eradication of a pest by some other means of control, after which a more desirable species is introduced to fill the void and prevent the reestablishment of the original pest.

The concept of **competitive displacement** is based on Gaus's principle, which states that

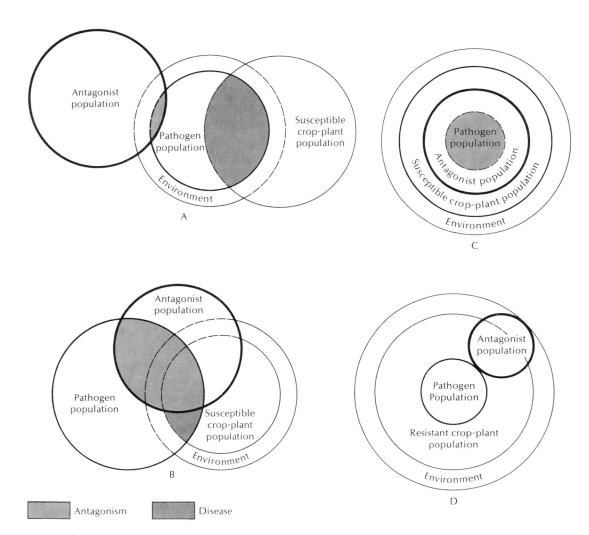

Antagonism   Disease

FIGURE 18-5.
Schematic diagram of interactions among groups of major factors in plant disease: (A) Severe disease loss. Susceptible crop moderately well adapted to the environment; pathogen well adapted; antagonists not well adapted and
ineffective. Exemplified by the fusarium wilt diseases in acid sandy soils. (B) Slight disease loss. Susceptible crop
well adapted to environment; pathogen poorly adapted; antagonists moderately adapted and quite effective.
Exemplified by the fusarium wilt diseases in alkaline clay soils. (C) No disease loss; biological control. Susceptible
crop, antagonists, and pathogen well adapted to environment. Antagonists have suppressed the pathogen.
Exemplified by *Phytophthora cinnamomi* root rot of avocado in Queensland. (D) No disease loss; resistance. Resistant crop, antagonists, and pathogen well adapted to environment. Host resistance prevents disease. Exemplified
by fusarium wilts in any soil when the crop carries monogenic resistance. (From *Biological Control of Plant Pathogens* by K. F. Baker and R. J. Cook. W. H. Freeman and Company. Copyright © 1974.)

ecological homologues, or species with the same ecological requirements, cannot coexist over an extended period of time. Eventually, the species with the highest rate of increase will survive. This does not imply that the species must have identical requirements, but they must actively compete for some life requisites such as food, nesting sites, or oviposition sites. There are a number of examples of species replacement that have occurred more or less naturally; that is, the displacement was natural but it occurred because man interfered by introducing a new species. The introduced cabbage butterfly, *Pieris rapae*, has unfortunately replaced the native *P. oleracea* throughout much of North America, but in this instance the successful species is the more troublesome.

Several possibilities for pest control exist. It might be possible, for example, to replace vectors of plant, animal, or human pathogens with species that are nonvectors or species that attack other plants or animals and minimize the nuisance they create. Several examples of this have occurred accidentally: The antimalarial campaign in Sardinia reduced the population of the vector *Anopheles labranchiae* until it was replaced by the normally rare nonvector *A. hispaniola*. The chemical control of *A. funestus* in East Africa resulted in its replacement by *A. rivulorum*, which attacks cattle instead of man. The use of a latent form of virus has been suggested as a possible means of preventing the infection of plants by a more serious virus.

GENERAL CONSIDERATIONS

Biological control presents the opportunity to suppress many pests of forestry and agriculture with a minimum of ecological disruption. It can be particularly effective as a colonization technique that imports the natural enemies of pest species that have become established in new geographic areas. Once a biological control agent has become established it becomes a part of the forest or crop ecosystem and contributes to the continuous, self-perpetuating regulation of the pest species. Unlike chemical control it often eliminates or markedly reduces the need for future human intervention. Since biological control agents are often fairly specific they have little or no adverse effect on beneficial species, and there is little danger of pest resurgence or secondary pest outbreaks as a result of their activity.

The development costs of most biological controls may be paid largely from public funds, but in many cases these are less than the research, development and production costs of an insecticide. Moreover, the cost is sometimes very small compared to the savings that accrue once the program has been implemented. A striking example is provided by the citrophilus mealybug project, which is estimated to have cost less than $2600 when it was implemented in 1928, and to have resulted in annual savings that amount to between one and two million dollars with no additional input of funds (Compere and Smith 1932). There are of course some costly failures, but no control tactic is a guarantee against losses.

The fact that, once established, biological control provides ongoing damage suppression with little or no additional cost makes it a highly desirable means of protecting crop values that accrue slowly, as in forestry, or low value resources such as grazing land. For example, the chemical control of Rhodesgrass scale, *Antonina graminis* (actually a mealybug), was economically unfeasible even though in Texas this species seriously damaged quality rangeland. The suppression of the pest to subeconomic levels through the importation from India of a parasite *Neodusmetia sangwani* provided a highly satisfactory solution.

Because biological control often generates substantial savings by freeing growers from the annual expense of chemical control, it is surprising that more money for research and development is not available. Perhaps growers should now be required to pay a small part of the savings they derive from biological control into a fund to support further research.

Not all agroecosystems are suitable for the establishment of natural enemies for ongoing pest regulation. As Southwood and Way (1970) pointed out, the best opportunities for mimicking nature occur in complex, more permanent ecosystems such as forests, orchards, plantations, vineyards, and other systems characterized by high species diversity, temporal continuity, reasonable climatic stability, and a minimal degree of isolation. Conversely, the use of colonized natural enemies is more difficult and less effective in short-term monocultures that are more or less isolated from other similar systems. Baker and Cook (1974) suggest that, at least for the biological control of plant pathogens, the opportunity in monocultures is increased by large-scale agriculture because the expenses of preparing the system for biological control can be borne more easily by large groups than by small growers.

However, biological control methods that involve the use of mass reared parasites and predators for temporary pest suppression by inundation can be used even in small, short-lived agroecosystems. Insect pathogens, particularly *Bacillus thuringiensis*, are also useful in protecting crops that do not lend themselves to natural enemy colonization. This approach encompasses all of the desirable attributes of biological control except self-perpetuation.

Nevertheless, there has been resistance to the use of biological control on the part of some growers who have grown accustomed to temporarily pest-free fields following the use of conventional chemical control. They find it difficult to accept the fact that biological control rarely attempts to eradicate pests. Merely suppressing the pest population below its eco-nomic threshold usually means that there is some evidence of pest presence, which some growers seem unable to tolerate. Furthermore, permanent, self-perpetuating biological control cannot be achieved overnight, and during the establishment period the project must not be disrupted by the use of broad-spectrum pesticides. This means that some damage usually has to be tolerated for a year or so and that the growers must be educated to the fact that the long-term results are worth waiting for. The use of biological insecticides during this period can be of considerable value and can also be a very persuasive tool.

SUMMARY

Biological control in the broad sense encompasses any population suppression that is a result of biologically induced mortality or of biologically reduced reproduction. In the applied sense it involves the use of one organism to regulate the population of another; it therefore includes the autocidal control methods discussed in Chapter 17, and the host plant resistance discussed in Chapter 19.

Insects play a major role in applied biological control. Insects attack not only each other, but also other kinds of pests, such as mites, snails and slugs, and weeds. In addition, insect pests are the hosts of a wide variety of pathogens, particularly bacteria, fungi, and viruses, that can be used to control them. In fact, there are only two potentially significant forms of biological pest control applicable to agriculture that do not involve insects. These are the use of plant pathogens to suppress unwanted vegetation and the use of antagonistic pathogens against plant pathogens of agricultural importance. Neither of these approaches has been developed to the point of practical application.

Biological control is an ecologically sound approach to pest suppression, because once it is established it is permanent, nondisruptive, and often self-perpetuating. Despite these strengths, however, biological control has been criticized and its proponents have had considerable difficulty in piloting the approach to widespread acceptance among farmers. Recent pressures to reduce the reliance of agriculture on pesticides has rekindled biological control research, and its future seems to be appropriately bright.

Literature Cited

Baker, K. F., and R. J. Cook. 1974. *Biological Control of Plant Pathogens.* San Francisco: W. H. Freeman and Company.

Burges, H. D., and N. W. Hussey. 1971. *Microbial Control of Insects and Mites.* New York: Academic Press.

Calusen, C. P., D. W. Clancy, and Q. C. Chock. 1965. *Biological control of the oriental fruit fly* (Dacus dorsalis *Hendel*) *and other fruit flies in Hawaii.* Tech. Bull. 1322. Washington, D.C.: USDA.

Compere, H., and H. S. Smith. 1932. The control of the citrophilus mealybug, *Pseudococcus gahani*, by Australian parasites. *Higardia* 6:585–618.

DeBach, P. 1964. *Biological Control of Insect Pests and Weeds.* New York: Reinhold Publishing Corp.

———. 1974. *Biological Control by Natural Enemies.* New York: Cambridge University Press.

Embree, D. G. 1971. The biological control of the winter moth in eastern Canada by introduced parasites. In *Biological Control.* C. G. Huffaker (ed.). pp. 217–268. New York: Plenum Press.

Falcon, L. A. 1971. Use of bacteria for microbial control. In *Microbial Control of Insects and Mites.* H. D. Burges and N. W. Hussey (eds.). pp. 67–96. New York: Academic Press.

Goeden, R. D., L. A. Andres, T. E. Freeman, P. Harris, R. L. Pienkowski, and C. R. Walker. 1974. Present status of projects on the biological control of weeds with insects and plant pathogens in the United States and Canada. *Weed Sci.* 22:490–495.

Goeden, R. D., and S. M. Louda. 1976. Biotic interference with insects imported for weed control. *Ann. Rev. Entomol.* 21:325–342.

Huffaker, C. B., ed. 1971. *Biological Control.* New York: Plenum Press.

Larid, M. 1967. A coral island experiment: a new approach to mosquito control. *Chron. Wld. Hlth. Org.* 21:18–26.

Mahr, D. L., and M. D. Atkins. 1976. Effect of *Bacillus thuringiensis* infected hosts on some entomophagous insects. (Unpublished.)

Müller-Kögler, E. 1965. *Pilzkrankheiten bei Insekten.* Berlin: Paul Parey.

National Research Council. 1968. *Principles of Plant and Animal Pest Control*, vol. 2: *Weed Control.* Washington, D. C.: National Academy of Sciences.

Price, P. W. 1970. Trail odors: Recognition by insects parasitic on cocoons. *Science* 170:546–547.

Roberts, D. W., and W. G. Yendol. 1971. Use of fungi for microbial control of insects. In *Microbial Control of Insects and Mites.* H. D. Burges and N. W. Hussey (eds.). pp. 125–149. New York: Academic Press.

Wouthwood, T. R. E., and M. J. Way. 1970. Ecological background to pest management. In *Concepts of Pest Management*. R. L. Rabb and F. E. Guthrie (eds.). pp. 6–29. Raleigh: North Carolina State University.

van den Bosch, R., and P. S. Messenger. 1973. *Biological Control*. New York: Intext Education Publishers.

van den Bosch, R., E. I. Schlinger, J. C. Hall, and B. Puttler. 1964. Studies on succession, distribution, and phenology of imported parasites of *Therioaphis trifolii* (Monell) in Southern California. *Ecology* 45:602–621.

White, R. T., and S. R. Dutky. 1940. Effect of the introduction of milky diseases on populations of Japanese beetle larvae. *J. Econ. Entomol.* 33:306–309.

19

HOST RESISTANCE

The susceptibility or resistance of an organism to damage caused by some other organism represents the current stage of the coevolution of interacting species. Resistant hosts occur naturally, and it would be surprising to learn that early man did not select at least some of his cultivars on the basis of their apparent freedom from damage by herbivores or infection by pathogens. Several very early authors, including the early Greek botanist Theophrastus, noted that some plants were less susceptible to insects and disease than others. Early in the history of agriculture, the selection of resistant forms was probably made on the basis of their observed condition and backed by little or no understanding of the causes of resistance. This is frequently the case today, in fact, and we are still far from completely understanding resistance mechanisms.

The importance of host resistance in modern agriculture dates back to the discovery in 1782 that Underhill wheat was resistant to attack by the Hessian fly. Other early examples include the discoveries that Winter Majetin apples were resistant to the woolly apple aphid; potatoes resisted late blight; and several varieties of wheat resisted rust.

Around 1850, when European grape varieties were highly susceptible to the grape phylloxera (Homoptera: Phylloxeridae), some American grape varieties were found to be resistant to this pest. American rootstock was therefore transported to Europe as a grafting base for the prized European varieties, but although this provided a solution to the phylloxera problem, it introduced another pest to the European continent. The American grapes were also resistant to downy mildew, *Plasmopara viticola*, which was not present in Europe until it was accidentally introduced on imported American rootstock. The European grapes were highly susceptible and a near catastrophe resulted. Fortunately, resistance to the disease was successfully incorporated into the European grapes by hybridization, and Bordeaux mixture was discovered as a means of chemical control of phylloxera.

The twentieth century has seen constant progress in the development of disease and insect resistant plants and animals. In recent years the search for ecologically sound alternatives to chemical control has stimulated a growing interest in breeding programs that are directed at pest resistance. Reviews by Painter

(1958), Beck (1965), Walker (1965), Day (1966), Maxwell (1972), and Gallun et al. (1975), to mention a few, provide ample evidence of both the complexity and the usefulness of this approach to pest control.

Plant resistance has been defined by many entomologists and plant pathologists in different ways, but one definition that seems applicable to both insect and disease resistance was provided by Painter (1951), who defined plant resistance as " . . . the relative amount of heritable qualities possessed by a plant which influences the ultimate degree of damage done. . . . In practical agriculture it represents the ability of a certain variety to produce a larger crop of good quality than do ordinary varieties at the same level of insect (pathogen) population."

MECHANISMS OF PLANT RESISTANCE

The mechanisms of resistance are variously grouped. Painter categorized the mechanisms of resistance to insects as: **Preference** and **nonpreference, antibiosis,** and **tolerance.** Disease resistance, however, does not fall clearly into these groups; most disease resistance appears to be antibiotic in nature, although there are some situations in which hosts are infected but seem to be tolerant.

Preference and Nonpreference

Preference and nonpreference are determined by host characteristics that attract or discourage a particular pest. Because this involves a response by the pest to particular host characteristics, it can be excluded as a mechanism of disease resistance. Insects and mites, however, do display a preference for certain foods, kinds of shelter, and oviposition sites, and recent research has demonstrated clearly that many such pests have stable, inherent responses to specific, genetically controlled characteristics of their hosts. The host characteristics to which arthropods respond are highly variable and include color, general conformation, surface and internal structures, and a number of chemical constituents perceived as odors or flavors.

Some insects will neither feed nor oviposit on a nonpreferred host. Others will oviposit on nonpreferred hosts, but their offspring will not feed there. Many, however, will entirely avoid nonpreferred hosts only when preferred hosts are available (Figure 19-1) and will feed on nonpreferred hosts when there is no alternative. After hatching on a nonpreferred host, the young larvae might feed very lightly and then stop. This is what happens when Colorado potato beetle larvae hatch on wild members of the potato family and when the European corn borer hatches on some strains of corn. In situations like these it is difficult to determine why the larvae stop feeding. Certain chemicals might produce a taste nonpreference but they also might actually act as toxic feeding deterrents. Under such circumstances nonpreference probably overlaps antibiosis.

Nonpreference for certain host varieties also occurs among insects and ticks that attack livestock. What appears to be nonpreference in this instance can actually be the result of avoidance behavior displayed by the host; but pest preference can be influenced by characteristics such as the chemical composition of sebaceous gland secretions and other features of the skin or coat. For example, zebu cattle (Figure 19-2) appear to be nonpreferred hosts for some ticks, horseflies, and mosquitoes because of their color and hide characteristics.

Antibiosis

Antibiosis is the result of characteristics of the host organism that adversely affect a pest organism's survival, developmental vigor, or reproductive potential. Resistance mechanisms

FIGURE 19-1.
Remains of a crop of corn after invasion by grasshoppers. Note that the sorghum in the background, a less preferred host plant, was not damaged. (Photograph courtesy of the USDA.)

FIGURE 19-2.
Light-colored cattle, such as these zebu grazing in Panama, not only are well suited to tropical climates but also are less preferred by some arthropod pests. (Photograph courtesy of the FAO.)

that operate against both insects and pathogens are included in this category.

In insects antibiosis results in premature death, lower fecundity, protracted development and associated mortality caused by longer exposure to adverse conditions, abnormal behavior, and reduced vigor and dispersal capacity. The mechanisms that create these effects include the presence of toxins, the presence of growth regulators or reproductive inhibitors, the absence of essential nutrients, and in some instances nutrient imbalance. Many host plants also display abnormal growth responses such as scar tissue formation, galling, and the development of secondary periderm that can also kill or injure attacking insects or can establish a feeding barrier.

Similar mechanisms are involved in the resistance of plants to microorganisms. Thick cell walls or a thick cuticle provides an effective mechanical barrier to penetration and subsequent infection. Many plants produce phenolic compounds and other chemicals that are toxic to certain microorganisms, and some synthesize fungitoxic compounds in response to attempted invasion. Baker and Cook (1974) point out that most host resistance to pathogens involves a response of the entire host-cell protoplast, which results in the exclusion or inhibition of the pathogen. It is clearly a dynamic rather than a passive process.

Tolerance

Tolerance is a form of resistance characterized by the host's ability to withstand or repair the effects of infection or infestation of a pest at a level that would cause economically significant damage to a susceptible host. In plants that are fed upon by chewing insects, resistance may result from an ability to bud adventitiously or simply to replace parts more rapidly than they are destroyed. Some varieties of corn, for example, can replace roots faster than they are destroyed by rootworms, *Diabrotica* spp (Figure 19-3), and therefore are resistant to attack by these insects (National Research Council 1969). The resistance of some varieties of wheat to the wheat stem sawfly is generally attributed to their solid stems, which deter both oviposition and the development of the larvae. Furthermore, the damage that occurs does not result in reduced yields of grain because the stems are not weakened enough to cause lodging (bending over) and related losses during harvest.

Some plants can tolerate invasion by pathogens; although infection takes place, the plants do not seem to suffer ill effects. The Huxley variety of strawberries, for example, can tolerate virus infection, whereas other varieties are completely destroyed by the same virus.

One of the problems that can arise from the development of commercial crop varieties that resist pests through tolerance is that this does nothing to reduce the pest population. In response to nonpreference or antibiosis the pest population declines because of emigration, reduced reproduction, or death, and so susceptible varieties of the crop grown in the same area can also benefit. Host tolerance, however, can simply provide the pest with a suitable breeding site, and a large pest population or inoculum can develop as a result. Tolerant hosts would thus serve as a reservoir from which susceptible varieties could be attacked and severely damaged.

THE GENETICS OF RESISTANCE

The inheritance of resistance to a specific pest is controlled by the same fundamental genetic mechanisms that determine the transfer of other traits. Resistance is often described in terms of the number of pests or pest **biotypes** against which it is effective, or according to the number of genes that govern it. Resistance to a sin-

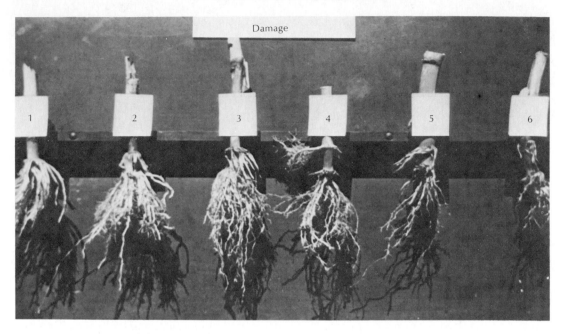

FIGURE 19-3.
Different levels of root damage caused by rootworms are displayed by these six varieties
of corn. (Photograph courtesy of Iowa State University.)

gle race or biotype of a pest is called **specific resistance,** whereas resistance to a variety of biotypes is called **nonspecific resistance.** If the resistance is controlled by a single gene it is known as **monogenic,** and if several genes are involved it is known as **polygenic.**

Specific resistance, also called vertical resistance, is usually monogenic. Although the exact nature of plant resistance is not known, geneticists have learned a great deal from studying plant-pest interactions in which a single gene for resistance in the plant is countered by a single gene for virulence in certain races of the pest.

Nonspecific resistance, also called horizontal resistance, is usually controlled by several genes and is usually effective to at least some degree against all races of the pest. Some geneticists consider that nonspecific resistance is simply what might be called general resistance; others feel that many plants have some level of resistance to a number of pathogens and that this is governed by genes that regulate ordinary physiological processes, not by special genes for resistance that have persisted because of the selective pressure of pathogens alone (van der Plank 1968). This type of more general resistance is thought to have developed in the long course of plant specialization. Baker and Cook (1974) state that " . . . the greater the [number of] gentic factors that condition resistance, the longer the evolutionary development of that resistance and the greater the spectrum of effectiveness. . . ." They conclude, therefore, that polygenic resistance indicates antiquity, whereas monogenic resistance reflects a recent response to a new selective pressure. Polygenic resistance probably begins as monogenic resistance that becomes more general with the accumulation of additional influential genes that broaden its effectiveness. However, we cannot say that polygenic resistance is in all cases an ancient characteristic. The potato, for example, has evolved a polygenic resistance to late blight since the famine of 1840.

When resistance is conditioned by a single gene it can be either completely dominant or recessive. For example, the resistance of IR8 rice to green leaf hopper and the resistance of other varieties to rice blast are cases in which single dominant genes have been identified. On the other hand, some varieties of peas that resist powdery mildew and corn varities that resist corn borer adults have single recessive genes for resistance (Sifuentes and Painter 1964). Some plants possess two or more genes that act independently to condition resistance to two or more different pests, and some cereals have as many as six independently acting dominant genes that condition resistance to different races of rust (Williams 1975).

Polygenic resistance, which occurs in a number of crop plants, is more complicated and less completely understood than monogenic resistance. The protection provided by two or more interacting genes is the sum of the ranges of resistance governed by each gene, and the key genes can be either dominant or recessive, just as they can in monogenic resistance. Some strains of tobacco that are resistant to bacterial wilt have a collection of recessive genes, whereas the polygenic resistance of corn to European corn borer is dominant (Penny and Dicke 1966).

In single species of plants the genetics of the resistance to different pests can be highly variable. Pathak (1969) analyzed the inheritance patterns for resistance in rice to three different insect pests. Resistance to the striped borer, *Chilo suppressalis*, is polygenic; resistance to the plant hopper *Nilaparavata lugens* is created by both dominant and recessive interacting genes; and resistance to the leafhopper *Nephotettix impicticeps* is the result of a single dominant gene.

Additional evidence suggests that some microorganisms act as antagonists to pathogens by stimulating resistance in the host plant. Littlefield (1969) succeeded in inducing resistance in flax to a virulent strain of the rust fungus *Melampsora lini* by inoculating the leaves with an avirulent strain. Some metabolites produced by avirulent microorganisms seem to be able to induce physiological processes controlled by specific genes, so that a cell environment is created that is detrimental to certain parasitic organisms.

THE IMPORTANCE OF INSECT BIOTYPES

Genetic diversity exists not only in plants but in insects and pathogens as well. Consequently, when an insect population is subjected to the selective pressure of host resistance, variants of the insect that survive are those that are least affected by the resistance. These survivors might interbreed to form populations of a new biotype to which the previously resistant plant is susceptible.

In both insects (such as aphids) and pathogens that reproduce without sexual recombination (and therefore without diluting favorable genes) and that produce several generations per year, new biotypes can predominate in a population in a very short time. For example, the parthenogenic greenbug (an aphid that attacks grains) is reported to have three biotypes that differ in their ability to attack different varieties of certain cereals. Until recently, two varieties of wheat grown widely in the United States were resistant to normal field populations of the greenbug, but a second biotype has appeared that can kill both the previously resistant varieties. A third biotype has recently developed that is able both to feed and to reproduce on Piper sudangrass, a completely new host, whereas the other two biotypes cannot (Wood 1961; Wood et al. 1969).

It is difficult to determine how long a newly developed resistant variety will remain resistant. Generally, monogenic resistance is less persistent than polygenic resistance because, when the resistance depends on a gene-for-gene relationship between the host plant and the pest, it can be overcome by a single gene mutation in the pest species. For example, a resistant strain of alfalfa was developed in order to reduce the damage caused by the spotted alfalfa aphid *Therioaphis maculata*. After only five years, however, a new biotype of the aphid appeared that was able to attack the previously resistant alfalfa strain. On the other hand, polygenic resistance rarely breaks down. Several varieties of tobacco with polygenic resistance to bacterial wilt have been planted in wilt-infected soil for twenty years with no evidence that new virulent strains of the bacteria have developed that are capable of infecting the tobacco.

Gallun, Starks, and Guthrie (1975) present an interesting example involving eight biotypes of Hessian fly. Table 19-1 lists the biotypes of Hessian fly and the wheat genes that provide resistance to each biotype. The wheat genes that confer the resistance are dominant and are specific for comparable recessive genes for

TABLE 19-1
Genes in wheat that condition resistance to races of Hessian fly.

Hessian fly biotypes	Genes in wheat
GP	$H_1, H_2, H_3, H_5, H_6, H_7, H_8$
A	H_3, H_5, H_6
B	H_5, H_6
C	H_3, H_5
D	H_5
E	$H_1, H_2, H_5, H_6, H_7, H_8$
F	$H_1, H_2, H_3, H_5, H_7, H_8$
G	H_1, H_2, H_5, H_7, H_8

virulence in the Hessian fly. A wheat variety remains resistant, therefore, as long as the insect does not have a set of genes for virulence that can counter all of the opposing wheat genes. However, a wheat plant can have any number of the genes for resistance and any combination can provide resistance as long as the insect does not have the specific genes for virulence to overcome that resistance. In other words, for every gene that provides resistance in a plant, there may be a comparable gene in the insect that can overcome that resistance. However, in the Hessian fly the fact that the genes for virulence are recessive and those for avirulence are dominant led to the suggestion that this might be used as a means of genetic control. If an avirulent strain of the pest were released in areas populated by the virulent strain, the progeny of matings between the biotypes would be avirulent and would perish in the presence of a resistant variety of wheat.

PRACTICAL CONSIDERATIONS

As long as a host remains resistant to attack by a pest, its effect on the pest is specific, cumulative, and persistent and will usually tend to reduce the pest population. These features make host resistance a desirable alternative to pesticides, which are usually nonspecific, noncumulative, and temporary. Furthermore, the use of resistant crop varieties does not harm either beneficial species or the environment. As we have mentioned, total resistance is not necessary; any level of resistance that will maintain the pest population at a level below the established economic threshold meets the objective of pest control. Those pest individuals that remain in the crop ecosystem are themselves hosts for natural control agents that help to stabilize the pest population further.

Resistant varieties can be obtained by: (1) selecting resistant individuals from heterogenous populations that are exposed to pest species, (2) crossing varieties that carry resistance with varieties that display other desirable characteristics, (3) hybridizing resistant wild species with susceptible cultivars and then selectively breeding the resistant progeny, (4) backcrossing to the original parent to maintain desirable characteristics after first crossing different varieties or species, (5) budding and grafting to maintain a resistant stock or scion, and (6) subjecting wild host plants to radiation or other mutagenic agents (National Research Council 1968).

The development of resistant varieties can be time consuming and of course carries no guarantee of success. Monogenic resistance is the easiest to obtain, as one might expect. Polygenic resistance is difficult to obtain and is complicated by the fact that the greater number of genetic adjustments required increases the likelihood that changes will be induced in associated genes that govern other characteristics, some of which may be undesirable. Moreover, inducing resistance to one pest can inadvertently cause increased susceptibility to others. For example, progeny from a cross of Richland and Victoria oats were resistant to stem and crown rust and to loose and covered smut, but were highly susceptible to Victoria blight.

Once a resistant variety has been developed, the benefits can be substantial even if the resistance breaks down several years later. The cost of this form of pest control to the grower is generally low, perhaps no more than a reasonable increase in the cost of seed, and the benefits can be great. The cost of more conventional forms of pest control are eliminated and increased yields often result. Unfortunately, because crop losses caused by a specific pest are difficult to evaluate, it is often difficult to determine the exact savings attributable to resistance. In addition to direct benefits for the resistant variety there may be side benefits such as an overall reduction in the pest population and a related decline in the damage to susceptible varieties grown in the same area.

It is estimated that the programs designed to develop crop varieties resistant to the Hessian fly, wheat stem sawfly, and spotted alfalfa aphid have reduced losses by $308 million per year in the United States. On the basis of research and development costs of $9.3 million, it is calculated that the benefit-to-cost ratio of these programs for a ten-year period would exceed 300:1 (Luginbill 1969). The cultivation of crop varieties that are resistant to pathogens has been estimated to provide a monetary gain to American farmers of more than $1 billion per year (National Research Council 1968).

SUMMARY

The resistance of organisms to species that attack them is a natural occurrence. All organisms display some degree of general resistance that has developed in the long course of their evolution, but varieties periodically appear that are resistant to specific enemies. Resistance conditioned by two or more genes (polygenic) is in general thought to be older and more advanced than resistance conditioned by a single gene (monogenic). Polygenic resistance is demonstrably less likely to break down.

The genes that condition resistance to a particular enemy often counteract corresponding genes for virulence in the pest species. Consequently, there is a dynamic genetic relationship between hosts and their parasites that results from the selective pressure of virulent parasites on susceptible hosts and that of resistant hosts on avirulent parasites.

Resistant varieties of plants and animals can be selected from wild populations, developed by crossing, or produced by other forms of genetic manipulation such as induced mutation. The development of genetically resistant varieties is an ecologically sound alternative to chemical pest control, and the use of resistant varieties is widespread in American agriculture and generates major economic and productivity benefits. Although the benefits to be derived from host resistance seem unlimited, it must be realized that the dynamic nature of host-parasite relations makes resistance unstable. We must therefore be constantly on the lookout for a breakdown in resistance, and the development of resistant varieties must be an ongoing process.

Literature Cited

Baker, K. F., and R. J. Cook. 1974. *Biological Control of Plant Pathogens.* San Francisco: W. H. Freeman and Company.

Beck, S. D. 1965. Resistance of plants to insects. *Ann. Rev. Entomol.* 10:207–232.

Day, P. R. 1966. Recent developments in the genetics of the host-parasite system. *Ann. Rev. Phytopathol.* 4:245–268.

Gallun, R. L., K. J. Starks, and W. D. Guthrie. 1975. Plant resistance to insects attacking cereals. *Ann. Rev. Entomol.* 20:337–357.

Littlefield, L. J. 1969. Flax rust resistance induced by prior inoculation with an avirulent race of *Melampsora lini. Phytophol.* 59:1323–1328.

Luginbill, P., Jr. 1969. *Developing Resistant Plants—the Ideal Method of Controlling Insects.* Prod. Res. Rep. 111. Washington, D.C.: USDA.

Maxwell, F. G. 1972. Host plant resistance to insects—nutritional and pest management relationships. In *Insect and Mite Nutrition.* J. G. Rodriguez (ed.). pp. 599–609. Amsterdam: North-Holland.

National Research Council. 1968. *Principles of Plant and Animal Pest Control,* vol. 1: *Plant-Disease Development and Control.* Washington, D.C.: National Academy of Sciences.

———. 1969. *Principles of Plant and Animal Pest Control,* vol. 3: *Insect-Pest Management and Control.* Washington, D.C.: National Academy of Sciences.

Painter, R. H. 1951. *Insect Resistance in Crop Plants.* New York: Macmillan Publishing Co.

———. 1958. Resistance of plants to insects. *Ann Rev. Entomol.* 3:267–290.

Pathak, M. D. 1969. Stemborer and leafhopper-planthopper resistance in rice varieties. *Entomol. Exp. Appl.* 12:789–800.

Penny, L. H., and F. F. Dicke. 1966. Inheritance of resistance in corn to leaf feeding of the European corn borer. *Agron. J.* 48:200–203.

Sifuentes, J. A., and R. H. Painter. 1964. Inheritance of resistance to western corn rootworm adults in field corn. *J. Econ. Entomol.* 57:475–477.

Van der Plank, J. E. 1968. *Disease Resistance in Plants.* New York: Academic Press.

Walker, J. C. 1965. Disease resistance in the vegetable crops. *Bot. Rev.* 31:331–380.

Williams, P. J. 1975. Genetics of resistance in plants. *Genetics* 79:409–419.

Wood, E. A., Jr. 1961. Biological studies of a new greenbug biotype. *J. Econ. Entomol.* 54:1171–1173.

Wood, E. A., Jr., H. L. Chada, and P. N. Saxton. 1969. *Reaction of Small Grains and Grain Sorghum to Three Greenbug Biotypes.* Okla. State Univ. Agr. Res. Prog. Rep. 618.

20

INTEGRATED PEST MANAGEMENT

In the foregoing chapters we have presented some of the important methods of pest control and some of the major principles underlying them. Some of the methods are time-tested, whereas others have great potential but are not yet proven. Each of these approaches to pest control either has played or will play an important role in pest suppression; each can be used alone and is subject to failure if it is abused or used improperly.

For the most part, these methods have been used in an ad hoc manner, frequently with great success. Sometimes, however, less than the desired results have been achieved, and new pest problems have arisen far too often out of trial-and-error attempts to eliminate old pests. Chemical control has clearly produced some of the greatest benefits on the one hand and some of the most serious problems on the other. These problems—in the form of ecological disruption, pest resurgence, secondary pest outbreaks, and the induction of pesticide resistance discussed in Chapter 16—have stimulated a reexamination of pest control methodology and particularly of the practice of using any one tactic separately and to the exclusion of others.

The development of pesticide resistant strains and the appearance of biotypes that overcome host resistance provide ample evidence of the genetic plasticity of pest organisms. Pest control tactics create strong selective pressure to which target organisms are bound to respond, and experience has shown that no single form of pest control can be expected to provide a permanent solution. New biotypes that do not respond to attractants, that avoid male sterilization by becoming parthenogenic, that become resistant to primary natural enemies, or that adjust to certain cultural practices loom as distinct possibilities. We should never again make the mistake of assuming that pest resistance will not develop. Over the past few decades we have grown to realize that we must replace empirical control methods with approaches that are based on sound ecological and evolutionary principles. **Integrated pest management** is one such approach and has the potential to provide the solution we are seeking.

The literature contains many definitions of integrated pest management. The early definitions stressed a blend of biological and chemical control, and this is still a popular view. Some of the more recent definitions, however, such as the one proposed by an FAO panel in 1967, stress that an integrated control strategy is a combination of all suitable techniques. The FAO experts defined integrated pest management as:

. . . a pest management system that in the context of the associated environment and the population dynamics of the pest species, utilizes all suitable techniques and methods in as compatible a manner as possible and maintains the pest populations at levels below those causing economic injury.

Taking into account the need for systems based on sound ecological principles, an alternative definition would be:

The combination of as many suitable control methods as practical into an ecologically harmonized system designed to maintain pest populations at levels below those which would cause economically significant losses to agriculture and forestry or would endanger human health.

This definition does not exclude the use of pesticides, but it implies that the control methods used must be reasonably compatible if a harmonized system is to be the result. The value of chemicals in pest control is clearly established and recognized, but it is also clear that the use of chemicals should be substantially reduced whenever possible. Furthermore, the multifaceted nature of integrated pest management reduces the intensity with which any single method acts upon a pest population, so that there is less selective pressure leading to the development of new biotypes that are difficult to control.

The practice of integrated pest management is not new. Before modern synthetic pesticides offered growers the hope of reducing pest populations to near zero at very little cost, a combination of methods was used to reduce crop losses. In many instances, pest control was an art, one that was unfortunately pushed aside by the proliferation of pesticides. Farmers who developed their own forms of integrated pest control did so largely by trial and error without fully understanding the underlying ecological principles. It is rather interesting that many of these farm-grown programs involved cultural practices that made the crop environments less favorable to pests and more favorable to their natural enemies.

Because biological control was a well established alternative to chemical control it is not surprising that the first conceptualization of integrated pest management involved mainly a combination of these two approaches. Some programs still lean heavily toward biological control carefully augmented by the use of pesticides; but as the concept has evolved, various cultural methods, plant resistance, pheromones, and autocidal methods have all been considered part of integrated control programs. Chemical control is an important component in some programs, but is not used at all in others.

The terms integrated control and integrated pest management are variously applied, but the concept has become so widely used and publicized that there seems no practical value in placing unnecessary restrictions on their meaning. For example, there is no need to distinguish between pest suppression and eradication when either is achieved by a multiplicity of control methods.

Some workers consider eradication to be incompatible with the integrated concept of pest control because the total elimination of a species reduces diversity, but species designated for eradication are often introduced species. If they can be eliminated most efficiently from an ecosystem in which they do not belong by combining several control methods, an integrated approach should be used.

In order to maximize the efficacy of an integrated pest management program it is important to consider the whole crop or forest eco-

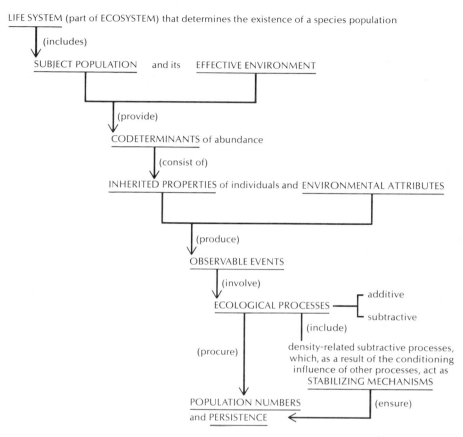

FIGURE 20-1.
Diagram of the components of a life system. (Redrawn from Clark et al. 1967.)

system in which the pests exist. It is necessary to recognize that several pests of varying importance might be present and to evaluate each pest's population dynamics. The multiplicity of factors that influence a species population density has been described as its **life system** (Figure 20-1), but usually such systems are imperfectly known. Consequently, although integrated pest management programs are based on a broad consideration of ecological relationships, they still tend to be rather empirical.

The successful application of integrated control certainly depends on an understanding of the agroecosystem and the biological relationships of the pests within it. Southwood and Way (1970) discussed the important characteristics of forest and crop ecosystems that influence pest population dynamics and that form a basis for selecting suitable control methods. Two of the important considerations are the size and delimitation of the ecosystem to be treated. As we have mentioned, large monocultures form large targets for dispersing pests and natural enemies, but they are also less affected than more diverse systems by the composition of adjacent ecosystems or by the natural occupants of field margins, roadways, and other peripheral areas. In such cases the flora and fauna of adjoining areas influence the ecology of the marginal portions of the crop considerably more than they influence the more isolated central portions. Conversely, a small crop area would be heavily influenced by adjacent ecosystems. Any integrated pest management program must take these features into account because accessible wild areas form refuges for both pests and beneficial organisms when the crop ecosystem is disrupted. Depending on the pest problem and the choice of control tactics, it might be necessary to decide whether the control program should be extended into marginal ecosystems.

Crop permanence is another feature that influences the selection of pest control methods. The organisms associated with long-lived crops, such as those grown in vineyards, orchards, or groves, have less need to disperse in and out, and so major pests of these crops are frequently less inclined to move about than those of annual crops. Such agroecosystems tend to become more diverse and stable with time and to provide ideal situations for the application of biological control. The relative permanence of the system permits the implementation of management procedures that conserve and augment natural enemies; for example, parasites, predators, and pathogens can be introduced and can become permanently established in these ecosystems. On the other hand, the permanence of a crop might also encourage the establishment of some pests such as root pathogens.

Ideally, when an integrated control program is formulated, the entire pest complex should be considered at once. Entomologists, plant pathologists, and weed control experts are generally adopting this approach to the control of pest insects, pathogens, and weeds, but there are still only a few cases where all types of pests are considered together. This should be the ultimate goal of any program designed for a specific crop ecosystem, even though it can complicate the situation considerably.

Plant pathologists often attempt to control a variety of diseases simultaneously, and most forms of weed control are not particularly specific. Generally speaking, however, entomologists seem to have led the development of a formalized, multiple-pest approach to integrated control.

To help organize the development of integrated control programs for entire pest complexes, Chant (1964) categorized insect pests according to four classes: Class I pests are those that appear perennially and that normally attain population levels high enough to cause economic damage unless some control measure is implemented. These **key pests** usually dominate the development of a control program; when chemical control is used alone, they are the principal targets. Class II pests are species that usually are controlled adequately by natural enemies, but that occasionally escape natural regulation and reach population levels at which economic damage results. Class III pests are those that normally do not cause significant damage, even though they might occur in large numbers. However, changes within the agroecosystem, perhaps from the use of a pesticide, can elevate these secondary pests to primary pest status. Class IV pests are nonresident species that might invade a crop ecosystem and then cause economic damage. The invasion of Class IV pests is often unpredictable, but when it occurs, direct chemical control that disrupts the ongoing integrated control program can become necessary.

Integrated pest management programs, like chemical control programs, are usually based on the most important pests. Consequently, the Class I pests are those for which we must develop well defined economic thresholds and whose population dynamics we must understand reasonably well. The need for more ecologically sound approaches to pest control is immediate, however, and some changes in a control strategy could be desirable even before long-term studies are carried out. Consequently, more often than not, integrated control programs are initiated as pilot studies based on common sense and a fundamental understanding of ecosystem function. In many respects this is a practical approach, because changes in control practices tend to generate changes in the agroecosystem, and subsequent adjustment in the integrated pest management program can be related to these responses. Thus the agroecosystem and the control program evolve together, and well-integrated programs actually go beyond pest management to total crop ecosystem management.

Pest problems arise either because the equilbrium population density of a species lies above the established economic threshold, or because the population density of a species regularly or occasionally rises above the economic threshold. The aim of an integrated pest management program therefore depends on the nature of the population suppression that is needed. In the first case, for example, it might be necessary to reduce the overall favorability of the environment in order to reduce the level of the pest's normal population equilibrium; in the latter cases the goal of the control program is to dampen population oscillations. Even when a control program seems to be working well, however, a population can rise unexpectedly above its economic threshold as a result of the temporary relaxation of some constraining factor or of an unpredictable influx of individuals from elsewhere. Under such circumstances it might be necessary to react quickly with an insecticide. If an integrated control program has been developed over an extended period of time to achieve a satisfactory state, the degree to which the use of a pesticide can disrupt the program must be considered; it may be better to absorb the current crop loss than to sacrifice the future of the integrated program.

If the use of chemicals appears to provide the best solution to a temporary problem, or if it is necessary as a component of integrated pest management, there are a number of ways to minimize the disruption of normal functioning of the crop ecosystem. When a chemical application is necessary, a highly specific pesticide provides the least disruption, but a compound with a high level of specificity is seldom available. Steps must therefore be taken to apply the available broader-spectrum compounds in such a way as to narrow their specificity and avoid diminishing the efficacy of the other control tactics in the program. Barlett (1964) discussed several ways to reduce the impact of pesticides on natural enemies:

1. Preservation of natural enemies outside treated area. Although it is best to conserve the natural enemies within the crop area, this is not always possible. The ultimate decision depends on the dispersal and searching characteristics of parasites or predators. For example, *Tetrastichus julis* is a good gregarious parasite of the cereal leaf beetle, but it has poor powers of dispersal and cannot keep pace with shifting populations of the pest (Haynes 1973). Such a parasite could fit nicely into an integrated pest management program that involved insecticide use because it can be protected, moved, and manipulated in a way that is compatible with the application of insecticides.

   Natural enemy populations can be maintained in environments adjacent to the crop or in parts of the crop that do not need to be sprayed, and the natural vegetation in field margins, roadsides, and irrigation ditches can provide valuable refuges. A crop can also be spot treated. DeBach and Landi (1959) applied a nonspecific oil spray to alternate pairs of citrus trees at six-month intervals to reduce purple scale populations without eliminating their natural enemies. In Israel, strips of trees were sprayed with an attractant to concentrate adult Mediterranean fruit flies where they can be treated with malathion without causing widespread mortality to scale parasites formerly killed by blanket applications of insecticide (Harpaz and Rosen 1971). An effective but less desirable alternative (because it is riskier, more expensive, and more time-consuming) is to culture natural enemies in insectaries and then release them in the crop once the danger of the pesticide has subsided.

2. Selectivity based on differences in pest susceptibility to insecticides. Adult parasites are generally more susceptible to insecticides than those at other stages of the life cycle. The nymphs or larvae of many herbivores, however, are often more susceptible than the adults. Thus it might be possible to arrange the timing of spray applications to coincide with the presence of susceptible stages of the pest and less susceptible stages of their natural enemies.

3. Selectivity based on mode of feeding. Since parasites and predators of pests do not feed on the vegetative parts of plants, plant systemic poisons are preferable to contact poisons in integrated pest management. The pest ingests the poison with its food, while its natural enemies are protected. However, natural enemies can be killed if they feed on hosts that have accumulated the toxic material.

4. Selectivity based on life cycles and habitats. It could be possible to apply a pesticide when the pest is in a stage of development not under attack by its more important natural enemies or when it is in a portion of the environment not frequented by the beneficial species.

5. Selectivity based on characteristics, formulation, or application of the pesticide. If a pesticide must be chosen from several nonspecific compounds, a nonpersistent one would be preferred in order to reduce the risk of exposing natural enemies to it. Because the goal of an integrated approach is to suppress a pest below the economic threshold rather than to kill more than 90 percent, it is often possible to reduce the dosage and thus to reduce the danger to beneficial species. However, some parasites are so susceptible to poisons that they will be killed by dosage levels that permit the pest to survive, thereby aggravating the pest problem.

Although little is known about the effect that the formulation of pesticides has on beneficial species, some obvious possibilities exist. For example, granular formulations, soil fumigants, or seed dressings applied against pests in the soil environment would be generally less harmful to parasites and predators active above the soil.

Some groups of insecticides, such as petroleum oils commonly used against phytophagous mites and scale insects, are generally nontoxic to parasites and predators. Biological insecticides such as *Bacillus thuringiensis* or special virus preparations are particularly useful in integrated pest management because they are specific to herbivorous species.

There is one important aspect common to the use of any insecticide in integrated pest management: Regardless of the care taken to reduce the direct impact of poisons on natural enemies, the desirable species will be affected by the reduction in the number of hosts or prey available. The balance between an adequate or inadequate number of the pest species can be rather critical, and if a pesticide application seriously depletes the food resource available to

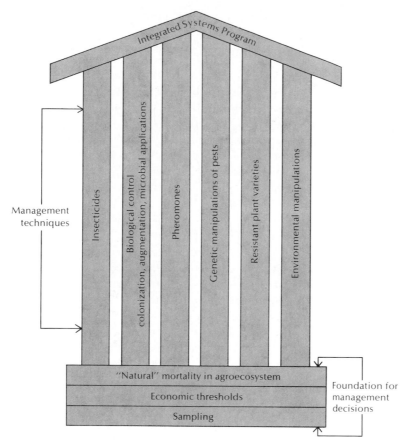

FIGURE 20-2.
Diagram comparing the development of a sound integrated pest management system to the construction of a house. (Redrawn from Gonzalez 1970.)

parasites and predators they can die out or leave the crop environment.

Although existing pesticides can be used in a manner that is fairly compatible with other control methods there is still a great need for new chemical compounds that are more specific, safer, less persistent, and biodegradable. It is unfortunate that, given the vast increase in the technological capability of the chemical industry over the past few decades, more desirable pesticides have not been developed. But the turning point could be approaching as the potential of some of the new control methods becomes more fully realized.

Gonzalez (1970) likened developing an integrated pest management program to building a house. The foundation is formed by population sampling, establishing economic thresholds, and understanding natural mortality in agroecosystems. The wall studs are the variety of control methods available, such as insecticides, biological control, host resistance, and autocidal methods, and the roof is an integrated pest management program (Figure 20-2). How well the roof holds up depends on the strength of the foundation and the number of wall studs that support it.

EXAMPLES OF INTEGRATED PEST MANAGEMENT

The formalization of integrated pest management as an approach to pest suppression in most crops is rather recent. The immediate need to implement controls that are less reliant on pesticides has usually made it impossible to spend the time and effort really necessary for a good understanding of the population dynamics of the pest species. Current integrated pest management programs are therefore being developed step by step. None can be considered to be in anything like a final form, and the cases given below as examples should be viewed only as stages in a dynamic developmental process. The purpose of the examples is to illustrate the ways in which various tactics can be harmonized into a more ecologically sound approach to pest control.

Cotton Pest Control in the Canete Valley, Peru

This example clearly shows the detrimental impact of empirical pest control with broad-spectrum insecticides and the recovery that can result from the implementation of an ecologically sound program (Doutt and Smith, 1971). The Canete Valley is an irrigated coastal valley isolated from a number of similar agricultural areas by rugged topography and arid uplands with native vegetation. In the 1920s the farmers of the valley switched their emphasis from sugarcane to cotton. Before the development of the broad-spectrum organochlorines, six prominent pests including the bollworm *Heliothis virescens*, the boll weevil *Anthonomus vestitus*, and the leafworm, *Anomis texana* were kept in check with arsenical and nicotine insecticides and by manual removal. From 1943 to 1948 the cotton yield in the valley ranged from 466 to 591 kilograms per hectare. In 1949 it dropped to 366 kilograms per hectare because of outbreaks of an aphid and *Heliothis*. In the years that followed, organochlorine insecticides became available and the farmers relied heavily on the application of DDT, BHC, and Toxaphene; cotton yields almost doubled as a result.

The farmers became convinced that their yields were related directly to the amount of insecticide applied and began to blanket the fields with aerial applications of these materials. Within a few years, the pests developed resistance to the organochlorine pesticides. Organophosphates such as parathion were substituted for the compounds used previously, and the interval between insecticide applications was shortened. Beneficial species were almost totally destroyed, and there was a rapid resurgence of the old pests. In addition, previously innocuous species rose to pest status, and, finally, most of the pests developed resistance to all available insecticides, and the isolated nature of the valley prevented the entry of natural enemies from elsewhere. As a result, by 1956 the valley's cotton yield dropped to 332 kilograms per hectare despite the continued massive use of insecticides.

The farmers then turned to their own research organization for assistance and a variety of new regulations were imposed. The use of synthetic organic insecticides was prohibited unless special dispensation was granted. Several cultural methods were used, including uniform planting dates and fallow periods. More acreage was devoted to other crops, such as corn and potatoes, and flax was introduced; this increased the ecological diversity and provided alternative hosts for parasites and predators. Natural enemies were brought in from other valleys and from abroad, and arsenicals and nicotine sulfate were reinstated as the main insecticides. As a result, the major pest problems subsided and the secondary pests reverted to their previously innocuous status. In the following year cotton yields increased to 526 kilograms per hectare and for a number of years thereafter remained at the highest levels in the history of the valley.

Cotton Pest Control in the United States

Cotton seems to be plagued by a variety of pests regardless of where it is grown, but the pests that are most important seem to be markedly different in different areas. The key pests in Texas are the boll weevil and the cotton fleahopper *Psallus seriatus*; in the Imperial Valley of California they are usually the pink bollworm and the cotton leaf perforator *Bucculatrix thurberiella*; and in the San Joaquin Valley of California they are the lygus bugs. In all of these areas the traditional approach to pest control has been the repeated use of chemicals. In fact in 1971, the last year for which data are available (USDA 1974), 47 percent of the insecticides used by American farmers were applied to cotton. Most of the major cotton pests have now developed resistance to one or more of the insecticides registered for use on the crop, and in situations where insecticides are still effective they often cause secondary pest problems. Thus in the cotton-growing areas of the United States the situation is similar to that which developed in the Canete Valley, and a widespread effort has therefore been made to devise integrated pest management programs.

According to Adkisson (1972), chemical control of the boll weevil and cotton fleahopper in Texas disrupts natural enemy populations and releases the bollworm and tobacco budworm from their regulation. Early attempts to control these key pests without disrupting the parasites and predators of the *Heliothis* species involved early termination of the crop with defoliants and desiccants, treating the boll weevil with late season insecticides to reduce the diapausing population, and plowing plant residues under immediately after harvest.

The present program involves the same cultural and chemical practices, as well as an attempt to avoid treating the fleahopper whenever possible plus the use of cotton varieties that can be forced to early maturation by withholding water and fertilizer. This early maturation reduces the amount of host food material available to the key pests some time before they are ready to enter dormancy, which in turn reduces the number of pests that overwinter. However, if cool moist weather delays the the maturation of the crop, pest populations can increase, and if outbreaks that demand the use of chemicals occur, the integrated pest management program is disrupted.

New short-season cotton varieties are now being developed that can be grown in dry areas without large doses of fertilizer, water, and pesticides. These varieties are designed to be resistant to some pests and they should fruit early enough and over a short enough time that they can escape damage by others (Walker and Niles 1971). Additional studies are under way to develop still other tactics to be integrated into the control program. These include the use of insect pathogens, pheromones, sterile males, and the development of more selective insecticides.

In the San Joaquin Valley of California, the most important pest of cotton has been the lygus bug, *Lygus hesperus*. Three lepidoterans, bollworm, cabbage looper, and beet army worm, also become pests from time to time, particularly when their natural regulation is disrupted by insecticides directed against the lygus bug. The present integrated pest management program, outlined by van den Bosch and Messenger (1973), requires a more careful evaluation of the lygus bug population and a reduction of insecticidal treatment. Previously, sprays were used as a prophylaxis or whenever the lygus bug population reached the economic threshold. Now insecticide treatment is recommended only when the population is found to

exceed the economic threshold on two consecutive sampling dates, three to five days apart,
during the period of flowering and budding. In
some areas lygus bug populations are substantially reduced by cultural practices instead of
insecticides. The strip harvesting of alfalfa, the
favored host of *Lygus*, prevents the large-scale
movement of the bugs into cotton after mow-
ing, and interplanting strips of alfalfa as a trap
crop in cotton fields also reduces the lygus population on the cotton plants and improves natural control. However, as is so often the case
in integrated pest management, additional
changes in the pest management strategy may
be required as a result of the pink bollworm invasion of the valley in 1975.

Orchard Pests

An orchard crop environment is quite different
from that of cotton. Tree fruits are a long-term
crop and therefore provide a relatively stable
environment in which natural controls can
flourish if they are not disrupted by chemical
sprays. Most of the key pests attack the fruits
directly, however, and can cause serious losses
even at relatively low populations. Fruits are
high-value crops whose marketability declines
rapidly if the fruit is not of the highest quality,
and so the economic threshold for pests that do
direct damage can be very low. Moreover, fruit
trees are plagued by a variety of diseases that
can require numerous applications of sprays.
Any attempt to develop an integrated program
is therefore complicated by the effect of these
sprays on the resident parasites and predators.

Like cotton, tree fruits grown in different regions are often plagued by different pests, but
the integrated pest management programs are
nevertheless similar for most orchards. They all
involve efforts to reduce the use of insecticides
against the key pests and thus to reduce secondary pest problems (see Hoyt and Burts 1974).

In apple orchards, traditional pests such as
codling moth and apple scab have determined
the spray schedules, and various secondary pest
problems have developed, particularly those involving phytophagous mites. In Nova Scotia,
codling moth populations are usually low and
can be controlled satisfactorily with a relatively specific botanical insecticide, ryania,
which allows populations of native parasites
and predators to build up and in some orchards
eliminates the need for sprays against codling
moth. The apple maggot is the key pest, however, and since it is essentially free from attack
by predators it must be controlled chemically.
Lead arsenate is fairly specific for the apple
maggot, but it might be banned as a pollutant,
which would make if more difficult to control
the pest within the integrated program. The
recent discovery of a pheromone that deters
oviposition by apple maggot females could alleviate the problem in the future (Prokopy 1972).

The integrated management of peach pests
in California (Hoyt and Caltagirone 1971) is
particularly interesting because it closely parallels the theoretical program suggested by
Chant (1964) as a framework for his discussion
of the strategy and tactics of insect control. In
most California peach orchards (Figure 20-3)
the oriental fruit moth and the peach twig borer
are pests almost every year. Both of these insects attack both fruits and twigs and cause
economic damage at low population levels.
Neither pest is controlled satisfactorily by natural enemies, and so insecticides have traditionally been used against them on a regular
schedule; these are clearly the key or Class I
pests in Chant's classification. The two-spotted
spider mite, European red mite, and San Jose
scale are the Class II pests. In the absence of
disruption due to pesticides, these pests are normally held in check by natural enemies, but
when they escape natural regulation they must
be controlled chemically. The two mite pests
can be particularly difficult to deal with be-

FIGURE 20-3.
Typical California peach orchard showing indications of intensive management in the
form of pruning, weed removal, and cultivation. (Photograph by J. C. Dahilig, courtesy of
the U.S. Bureau of Reclamation.)

cause of the rapidity with which they develop resistance to acaricides. The peach silver mite is a typical Class III pest; it is often present in very large numbers but apparently does not cause much economic damage. Nonetheless, some growers spray for it. There are a few Class IV pests, such as the western peach tree borer, that become important from time to time and require special applications of insecticides.

Several considerations are important in designing an integrated program to control a pest complex such as the foregoing one. The predaceous mite *Metaseiulus occidentalis* effectively controls the phytophagous mites as long as its population can be maintained at high levels. This requires reducing the use of pesticides and preserving the peach silver mite as a food source to support an expanding population of *Meta-*

seiulus. The disruption caused by most broad-spectrum insecticides can be reduced by partially controlling the peach twig borer, San Jose scale, and phytophagous mites with winter sprays, such as petroleum oil, that are less harmful to beneficial species. Early-season spraying for the oriental fruit moth is eliminated in order to avoid disrupting the spring population expansion of beneficial species. Chemical control of the peach silver mite is no longer recommended because an adequate number of them must survive as prey for a large population of *Metaseiulus*. This in turn reduces the need to apply acaricides against the two-spotted spider mite and European red mite.

So far it seems that the developing integrated pest management program is producing satisfactory results, but incorporating additional tactics into the program could generate even more benefits; for example, removing twigs or fallen fruits infested with oriental fruit moth would be advantageous if it were economically feasible. The oriental fruit moth also has a pheromone which might be useful in control. Traps baited with the pheromone could be used to reduce the population of adult moths, or a system that combined the pheromone with chemosterilents could reduce the number of fertile moths.

EVALUATION AND PROSPECT

According to Wildbolz and Meier (1973), the development of integrated pest management programs in affluent countries like the United States is motivated by the biological and environmental consequences of the continuous use and misuse of pesticides. However, in some crops, such as cotton, where pests have developed multiple resistance to previously useful pesticides, no alternative is left but to search for new approaches to pest suppression. In other managed ecosystems, such as grassland and forests, the economic thresholds for most pests do not warrant the relatively high per acre cost of repeated pesticide use. In these systems, methods of pest control that are self-perpetuating and less expensive are more desirable and easier to establish given the temporal and spatial continuity of these systems.

The economic advantages of mass producing uniformly high quality produce, and the fact that consumers have become accustomed to these advantages, have made modern mechanized agriculture a standard cultural practice. This approach to food production and marketing is not tolerant of pest damage, and it has led to the development of highly simplified, unstable agroecosystems. These features hardly favor the development of integrated pest management, which in any case is particularly difficult to achieve for short-term crops.

Implementing integrated pest management is also complicated by some resistance among growers, of whom many have become accustomed to the idea of pest-free fields and to relying on pesticides that produce more than 90-percent pest mortality. Furthermore, the implementation of integrated pest management is more complex than that of chemical control, and so growers feel they lose independence when they must seek outside assistance in their pest control programs. It also sometimes takes several years before full benefits are obtained from integrated pest management, and growers who fear a loss of income from pest damage sometimes revert to more familiar methods that disrupt the program during this period. The extent of grower participation is another problem. Sometimes it is difficult to persuade all the growers in an area to cooperate, and if one or two decide to rely on a previously established spray schedule, integrated pest management becomes, difficult or impossible. The best success is achieved when grower cooperatives or a government agency such as Israel's Citrus Board regulates the pest control program over a large area.

Despite the above difficulties, the developed countries have the research and development capability to provide the information and technology necessary to implement sound integrated pest management programs, but widespread adoption of the integrated approach will not come about easily. The separation that has existed between agriculture and ecology will have to be overcome; consumers will have to be reeducated; governments will have to make more funds available for research; and full cooperation between growers and scientists will have to be accomplished.

In developing nations the problems are somewhat different. Many of the agricultural countries that have not yet realized their full potential are in tropical areas, where pest outbreaks are often sporadic and thus discourage thorough investigation, When these outbreaks occur they are usually combated with pesticides and little is learned about their cause or prevention. This, combined with a tendency for administrators to equate crop protection with chemical control and with the prestige gained by outside agencies from spectacular short-term successes, creates heavy reliance on pesticides (Wood 1973). This is especially unfortunate because the tropical environment seems to favor integrated pest management, and such a program would be highly advantageous in economies that can ill afford to import chemicals for which other control methods can be substituted. However, these countries lack the scientific expertise to implement integrated pest management and their training facilities are inadequate. Regional governments are often too unstable to give any assurance that a program can be brought to fruition, especially if other projects are more politically useful to them. Yet these agricultural areas will play an important role in providing the world's population with adequate food and fiber, and unless a conscientious effort is made to support the development of integrated pest management programs in agriculturally emerging nations, the increasing reliance on pesticides will make the task considerably more difficult as time passes.

Integrated pest control seems to have advanced most rapidly as an approach to insect pest suppression. Unfortunately, we have not yet reached the stage in which many workers think in terms of controlling the whole pest complex, including plant diseases and weeds. This is understandable given the degree to which our training programs encourage specialization; but the changeover to broader curricula in agriculture, ecology, entomology, pathology, and other related areas has begun. This is an important development because agroecosystems cannot be evaluated or managed in terms of single classes of pests. For example,

clean field margins might be a useful means of removing a source of weed seeds or eliminating the reservoir of a plant virus, but, as mentioned earlier, the wild vegetation of field margins and roadsides can also provide the pollen and nectar required by parasites and can create zones of diversity in which relationships between pests and their natural enemies can flourish. As long as such situations are viewed only in terms of one kind of pest, any decisions to alter them can easily damage the overall crop management program.

Considering the entire pest complex will of course further complicate an already complicated situation, but it seems to be the only logical avenue to follow. Fortunately, computer science and modeling techniques are sufficiently advanced to be useful in sorting out the complexity of these systems. Computers can deal with the multiplicity of factors that must be considered and can store large quantities of data to provide a historical perspective. Developing models is a means of testing the effects of various management strategies, and computer simulation permits us to explore many more strategies than we could in field experimentation. However, models must not be developed in a vacuum; there must be substantial input from entomologists, pathologists, and population ecologists who work in the field, and at present there are too few pest control specialists to provide the input to modeling projects and to carry the results of computerized gaming back to the crop systems for field testing.

We have come a long way, but we still have a long way to go. A strong commitment to research will add new and improved techniques for pest control. In the near future we can expect major advances in pheromones, chemosterilants, hormones, autocidal control methods, and the biological control of plant pathogens; but we must not be blinded by any successes we achieve, and we must always bear in mind that overemphasizing a single mortality agent invites a compensatory response from nature. The success of integrated pest management will depend heavily on the open-mindedness of everyone involved.

SUMMARY

Integrated pest management minimizes pest damage through the combined use of several compatible control tactics. A major motivation for such an approach is to avoid a variety of problems that frequently result from the extensive use of ecologically disruptive pesticides. Integrated pest management does not exclude the use of pesticides, but it does require that whenever possible they be selected and used to enhance rather than to impede other control tactics.

Integrated pest management is not new, but in recent years it has evolved from an art relying heavily on trial and error to a technology relying on scientific research and analysis. Well-conceived integrated pest management programs that have replaced a total reliance on pesticides have markedly reduced pest resurgence and secondary pest problems and have often increased crop yield.

Despite some success, however, integrated pest management still has a long way to go, and its path will be strewn with obstacles not easily overcome. In developed agricultural countries with the scientific and technological expertise necessary to develop the programs, implementation will be hampered by the need to educate farmers to tolerate low levels of pest damage and to convince consumers that cosmetic imperfections in produce are unimportant. The obstacles in developing agricultural countries may be even more difficult to overcome. Many of these countries lie in the tropics, where the periodic nature of many pest problems, combined with a lack of research, has led to a strong reliance on pesticides. The social and political pressure to increase food production dramatically has in many instances placed short-term gains ahead of sound long-term development policies.

The ultimate success of integrated pest management will depend on our ability to adopt a holistic view of agroecosystems. We will have to treat the entire pest complex—insects, pathogens, and weeds—together, so that the suppression of one group of pests does not disrupt the effort to suppress another.

Literature Cited

Adkisson, P. L. 1972. Integrated control of insect pests of cotton. *Proc. Tall Timbers Conf. on Ecol. Animal Control by Habitat Manag.* 4:175–189.

Bartlett, B. R. 1964. Integration of chemical and biological control. In *Biological Control of Insect Pests and Weeds.* P. De Bach (ed.). pp. 489–511. New York: Reinhold Pub. Corp.

Chant, D. A. 1964. Strategy and tactics of insect control. *Can. Entomol.* 96:182–201.

De Bach, P., and J. Landi. 1959. Integrated chemical, biological control by strip treatment. *Calif. Citrograph* 44:324–352.

Doutt, R. L., and R. F. Smith. 1971. The pesticide syndrome—diagnosis and suggested prophylaxis. In *Biological Control.* C. B. Huffaker (ed.). pp. 3–15. New York: Plenum Press.

Gonzalez, D. 1970. Sampling as a basis for management strategies. *Proc. Tall Timbers Conf. on Ecol. Animal Control by Habitat Manag.* 2:83–101.

Harpaz, I. and D. Rosen. 1971. Development of integrated control programs for crop pests in Israel. In *Biological Control.* C. B. Huffaker (ed.). pp. 458–468. New York: Plenum Press.

Haynes, D. L. 1973. Population management of the cereal leaf beetle. In *Insects: Studies in Population Management*. P. W. Geier, L. R. Clark, D. J. Anderson, and H. A. Nix (ed.). pp. 232–240. Canberra: Ecological Society of Australia (Memoirs 1).

Hoyt, S. C., and L. E. Caltagirone. 1971. The developing programs of integrated control of pests of apples in Washington and peaches in California. In *Biological Control*. C. B. Huffaker (ed.). pp. 395–421. New York: Plenum Press.

Hoyt, S. C., and E. C. Burts. 1974. Integrated control of fruit pests. *Ann. Rev. Entomol.* 19: 231–252.

Prokopy, R. J. 1972. Evidence for a marking pheromone deterring repeated oviposition in apple maggot flies. *Environ. Entomol.* 1:326–332.

Southwood, T. R. E., and M. J. Way. 1970. Ecological background to pest management. In *Concepts of Pest Management*. R. L. Rabb and F. E. Guthrie (eds.). pp. 6–29. Raleigh, N. C.: North Carolina State University.

USDA. 1974. *Farmer's Use of Pesticides in 1971*. Agro. Econ. Rep. No. 252., Econ. Res. Ser.

van den Bosch, R., and P. S. Messenger. 1973. *Biological Control*. New York: Intext Education Publishers.

Walker, J. K., Jr. and G. A. Niles. 1971. *Population Dynamics of the Boll Weevil and Modified Cotton Types*. Texas Agr. Exp. Sta. Bull. 1109.

Wildbolz, T. and W. Meier. 1973. Integrated control: critical assessment of case histories in affluent economies. In *Insects: Studies in Population Management*. P. W. Geier, L. R. Clark, D. J. Anderson, and H. A. Nix (eds.). pp. 221–231. Canberra: Ecological Society of Australia (Memoirs 1).

Wood, B. J. 1973. Integrated control: critical assessment of case histories in developing economies. In Insects: Studies in Population Management and Control. P. W. Geier, L. R. Clark, D. J. Anderson, and H. A. Nix (eds.). pp. 196–220. Canberra: Ecological Society of Australia (Memoirs 1).

Part Three

AGRICULTURE
AND THE FUTURE

21

GENETIC VULNERABILITY AND
GERM PLASM RESOURCES

Ecological and evolutionary processes are inseparably related in the dynamics of plant and animal populations. One observer has likened this relationship to that of an "ecological theater and the evolutionary play." Because this interconnection is as strong in agroecosystems as it is in natural ecosystems, we can see that agricultural genetics—that is, the study of the genetics and evolution of the member species of agricultural systems—is strongly complementary to the field of agricultural ecology (Baer 1977).

In examining agricultural genetics we shall focus on inherited patterns of resistance to pest and disease organisms and on the potential to improve, through selective breeding, the productivity of crops and domestic animals. In Chapter 19 we explored certain aspects of breeding for host resistance as a general tool for controlling pests and pathogens. In this chapter we shall consider the nature of **germ plasm resources** (the genetic variability potentially usable by domestic plant and animal breeders) and the extent to which the vulnerability of domestic species to pests and disease is a result of plant and animal breeding strate-

gies. In Chapter 22 we shall examine in specific terms the potential for breeding for improved productivity and performance.

THE GERM PLASM RESOURCE

The local populations of most widely distributed plant and animal species are, to a significant degree, genetically adapted to the specific environmental conditions in which they exist. This variability, termed **ecogeographic variation,** affects morphological, physiological, and (in animals) even behavioral characteristics. Where variability occurs in conspicuous morphological characteristics and on a geographical scale it is often manifested in subspecific taxonomy; that is, in the recognition of named subspecies or varieties. However, much of this variation, especially that which involves primarily physiological characteristics, is not recognized taxonomically.

Ecogeographic variation was first recognized by the plant ecologist Turesson (1922) in populations of various plant species that occupied the distinctive, severe environment of coastal sea cliffs. Ecogeographic variability, at first thought to be an uncommon and specialized phenomenon, was shown, through a series of classic studies in the late 1930s and 1940s, to be widespread and important in plants (see, for example, Clausen, Keck, and Hiesey 1940; Olmsted 1944). Among a number of species of perennial range grasses of the Great Plains, for example, these studies demonstrated significant regional differences in morphology and vegetative growth patterns as well as genetically based differences in photoperiodically controlled activities such as flowering. Some of the most detailed and extensive studies were carried out by McMillan (1959) on a number of these species. He was able to compare plants of the same species from populations that ranged from southern Canada to northeastern Mexico. These studies showed that genetic adaptations of local populations to local environments was one of the important patterns of adaptation in the grassland ecosystem (Figure 21-1). These adaptations permitted remarkably constant species composition and general vegetational structure to be maintained over a wide geographical range of temperature, rainfall, and photoperiod conditions.

Similar patterns of variability have subsequently been recognized in animal populations (Mayr 1963). In fact, zoologists seem to have shown a stronger tendency than botanists to recognize ecogeographic morphological variation through subspecific taxonomy. Animal physiologists, on the other hand, tend to refer to such patterns as **physiological races** when the characteristics involved are primarily physiological; they do not give such races taxonomic recognition, even though they are the result of the same phenomenon that creates similar variations in plant populations.

More recently, it has been found that significant patterns of genetic variation in some species occur along habitat gradients in quite localized situations. Moreover, it has been shown that many such patterns have originated by strong selection within historical time. Antonovics (1970) and a number of his students have shown, for example, that, in a number of plant species, populations with different degrees of tolerance to toxic materials in mine tailings have evolved locally within recent historical time. Likewise, where different-colored rock and soil substrates exist close to each other, populations of animals such as grasshoppers that characteristically rest on the bare ground may show highly localized, genetically based color patterns that match the substrate color (Cox and Cox 1974).

Leopold (1966), in discussing patterns by which animals adapt to habitat change, has shown that genetic plasticity—that is, the ability of populations to evolve adjustments to local conditions—is especially important to species in early stages of ecological succession. In fact, it is their overall plasticity, which is in part genetic, that enables them both to occupy such environments and to tolerate intensive human exploitation. Habitats of early successional systems, which are the product of disturbance, tend to be more variable in time and space than the habitats of climax systems. The ability of a successional species to adjust to this variation, coupled with a high reproductive potential, is thus essential to its survival. It is this kind of species, of course, that man has incorporated into agricultural systems; and in fact this has been responsible to a great extent for the fact that agroecosystems possess patterns of structure and function similar to those of successional communities in nature. It is therefore reasonable to expect that genetic plasticity and rapid evolutionary response should characterize species in agricultural systems.

These patterns of ecogeographic variation are part of the variability that is translated to the interspecific level by processes of **speciation;** that is, by the establishment of reproductive isolation between populations that were formerly capable of gene interchange. Specia-

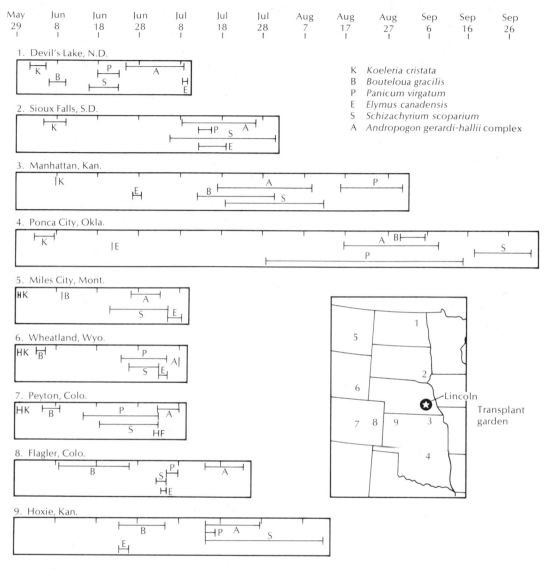

FIGURE 21-1.
The dates on which clones of various perennial bunchgrasses first flowered when transplanted from different parts of the great plains to an experimental garden in Lincoln, Nebraska. (Modified from Macmillan 1959.)

tion is less absolute than we might suppose; for example, in Chapter 4 we discussed the significance of hybridization in the origin of domestic crops and animals.

Thus the germ plasm resource consists not only of the variation among individuals in a single population, but also of variations in other local populations and in the populations of related species that occupy various environments and that come in contact with different biotic associates (Figures 21-2, 21-3). As we shall see, the genotype of a species is adjusted not only to the physical environment, but to the biotic environment as well.

COEVOLUTION AND COUNTERADAPTATION

Coevolution is the continuing process by which two or more species affect each other's evolution and adaptation. In natural communities such coadjustments take a variety of forms, all of which are related in some way to the use of food, nutrients, or other resources by the species involved. Seed plants, for example, exhibit a host of specialized relationships that vary in specificity and affect pollination and seed dispersal. Food chain interactions, including plant-herbivore, predator-prey, and parasite-host relationships, inevitably lead to evolutionary change in the participating species. Competition creates mechanisms of differentiation and antagonism between species. Biologically, these adjustments improve their survival rate by delimiting a dependable resource supply. Mutually beneficial symbiotic relationships, such as those between nitrogen-fixing bacteria and their legume hosts, are perhaps more common than we have yet realized. These types of relationships all exist in nature with bewildering diversity and greatly varying degrees of specificity.

In a sense it is these relationships that organize biotic communities. Because living organisms increase or decrease in abundance and vary their activities in other ways, they exert changing degrees of control or regulation on the other species with which they interact. The structure of a biotic community—the types, abundance, and biomass of organisms, together with their spatial distribution—is in large measure the product of such interactions. It is clear that, when the organization of a community is the product of interaction among highly co-evolved species, the resulting structure and function of the system will differ greatly from those of a system that consists of a random set of species with no coevolution.

Coevolutionary processes tend, in a general fashion, to lead toward a state of mutualism, that is, toward conditions that are most favorable to the survival and population stability of all the interacting species. Let us illustrate this pattern by two general examples, one involving plant-herbivore interactions, the other a disease agent and its host. When an herbivore overexploits available plant food, strong evolutionary pressure is exerted on the plant species involved. Environmental change can also take place that is unfavorable to the herbivore, such as habitat deterioration or the invasion of other plant species. Thus in the exploited plant populations characteristics will be favored that tend to protect plants from the most intense exploitation; spines, chemical substances, and protective growth forms are examples. In the herbivore population, behavioral characteristics, such as territoriality, that delay or prevent plant overexploitation will be favored. In time, a harmonious relationship might develop, with the herbivores grazing with moderate intensity, which can stimulate the recycling of nutrients through the system and promote a higher level of primary productivity. Actually, this beneficial function might be fulfilled more often than we have recognized, even in cases that appear superficially to be examples of destructive overexploitation (Mattson and Addy 1975).

Disease agents and their hosts tend to interact in a similar manner. For a disease organism, killing the host is equivalent to killing itself. The longer the host survives, the greater is the probability that the infecting agent will be able to pass to a new host and perpetuate its genotype. Introducing a highly virulent agent to an unresistant host population will thus create two strong evolutionary tendencies: a decrease in the virulence of the agent and an increase

FIGURE 21-2.
A genetic conservation specialist of the FAO collecting samples from a wheat field in Greece. From 200 to 500 heads are randomly selected from a field such as this and are placed in collections designed to preserve germ plasm resources. (Photograph by F. Botts, courtesy of the FAO, Rome.)

FIGURE 21-3.
Wheat spikes from a collection made in a wheat field in Afganistan. Note the variations in the characteristics of the stalk, spike, seeds, and awns. (Photograph by E. Bennett, courtesy of the FAO, Rome.)

in the resistance of the host. This logical pattern has been verified in several situations involving both human and animal diseases. Perhaps the best documented and most dramatic example is the introduction of **myxomatosis** virus to Australia as a biological control for the introduced European rabbit. The virus, a minor disease agent in South American rabbits, at first proved to be highly virulent for the European animal; but after a short time the virulence of field strains of the virus declined rapidly, and the resistance of the rabbit population increased appreciably (Fenner and Ratcliffe 1965). An important factor in the decline of viral virulence was the fact that transmission from rabbit to rabbit was largely by mosquito vectors. Less virulent strains, because they lengthened the period of time during which mosquitos could feed on an infected animal, were obviously favored by this fact. Needless to say, the rabbit was not eliminated, although populations were reduced somewhat.

Counteradaptation is closely related to co-evolution. We can define it as the total set of adjustments that members of a biotic community make to any single species in that community (Ricklefs and Cox 1972). The number and intensity of these adjustments determine the role and importance of each species in a community. This implies that species of destructive pests or disease agents are forms recently introduced to the biotic system in question, for which coevolved mechanisms of counteradaptation are absent or weakly developed. Herbivorous insects, for example, tend to be pests only when they have been recently introduced to a region or when the artificiality of the system has excluded counteradaptive mechanisms. Disease agents or parasites also tend to be virulent only when they have recently come into evolutionary association with the affected hosts, regardless of whether the hosts are plants, domestic animals, or humans.

Coevolutionary relationships have considerable significance for agricultural strategy.

In almost all cases, pest or disease organisms are derived from natural systems in which their significance is minor because strong counter-adaptation mechanisms have evolved. These genetically based mechanisms represent a resource that can be manipulated to incorporate counteradaptive control into agricultural systems. The frequency with which pest and disease species create problems in agricultural systems emphasizes the fact that these systems are new, that their structure is artificial, and that they lack the coevolved relationships that create stable relationships in natural systems (Levins 1974). Thus many of our agricultural systems possess what we term genetic vulnerability. We shall now examine more specifically the conditions that create genetic vulnerability in agroecosystems.

GENETIC VULNERABILITY

Genetic vulnerability is the susceptibility of important crop or domestic animal varieties to destructive attacks by recently introduced pest, parasite, or disease agents; it is caused by the genetic uniformity of the host species. We have already seen that outbreaks of such agents occur for various reasons, but our concern at the moment centers on the extent to which genotypic uniformity of the host species encourages such outbreaks or intensifies their seriousness.

There are four major characteristics of the relationship between uniformity and outbreaks of destructive agents (Adams, Ellingboe, and Rossman 1971):

1. A host variety with a uniform pattern of genetic resistance to a pathogen must have become widely distributed throughout a particular region.
2. A virulent race of the pathogen must exist at low frequency within the region or must appear suddenly by evolutionary change or dispersal.

3. Environmental conditions must favor the growth, multiplication, and dispersal of the pathogen within the region.
4. The host variety must provide a highly favorable substrate for the growth and reproduction of the pathogen.

All four of these conditions must obviously be satisfied for a destructive outbreak to occur; but the first is the one for which man is most directly responsible. Genetic uniformity is favored in domestic plants and animals in several possible ways. The selection of varieties with desirable inherited characteristics is inherent in modern agriculture, and extreme uniformity arises from this and from the way in which these varieties reproduce or are propagated. Vegetative propagation, as it is carried out with potatoes and various tree fruits, produces genetically identical individuals. In addition, self-fertilization systems of reproduction in plants generally promote genetic homozygosity.

Added to these basic biological relationships that favor uniformity is the intense, standardizing selection employed in modern plant and animal breeding. Varieties are bred for many uniform characteristics for convenience in increasingly mechanized production, processing, and marketing. Genetic characteristics are also introduced to assist in the mass production of seed material, and these too contribute to uniformity in genetic constitution.

Let us now examine some examples of genetic vulnerability in species of crops and domestic animals and attempt to evaluate their significance.

GENETIC VULNERABILITY AND PLANT EPIPHYTOTICS

The Irish potato famine is perhaps the best known and most dramatic example of the dangers intrinsic in genetic vulnerability. After it was introduced into Ireland early in the seventeenth century, the potato quickly became the principal carbohydrate food plant because of its adaptability to the damp, cool climate. By the mid-1800s, most of the arable land in Ireland was planted to potatoes, predominantly a variety known as "lumpers" that was highly susceptible to the disease called late blight and caused by the fungus *Phytopthora infestans*. This disease agent was unknown in Europe before this period. In 1845, however, a major outbreak of late blight occurred, and in the following year and again in 1848 major late blight outbreaks occurred not only in Ireland but throughout much of Europe. Destruction of the potato crop triggered famine in 1845 and 1846, and at least one million people died of starvation or starvation-related diseases. Between 1846 and 1852 an additional 1.5 million people emigrated from Ireland to North America.

This classic example illustrates several of the points mentioned earlier. Genetic uniformity in the crop resulted from the widespread reliance on a single, vegetatively propagated potato variety. The sudden appearance of a virulent pathogen, apparently by natural dispersal from someplace in the New World (possibly Mexico, where the fungus disease is native), created an immediate threat of an epiphytotic. The timing of the outbreak, however, was apparently determined by the occurrence of cool, damp conditions particularly favorable to growth of the blight fungus.

This disease continues to be a serious problem in growing potatoes. The genetic relationship between the crop and its pathogen is complex (Vanderplank 1968), and blight resistance in potato varieties developed for this quality tends to be overcome quickly by new strains of the blight fungus. At least 11 genes for resistance are known in the potato (Day 1974), but reliable genetic resistance to late blight has nevertheless not been established. Fungicidal control is relied upon in many localities, although the blight fungus has shown resistance to fungicides.

About 23 percent of the potatoes grown in the United States are Russet Burbanks, a variety that is highly susceptible to late blight. Its successful cultivation clearly depends on the continued adequacy of fungicidal control.

A more recent example of genetic vulnerability is provided by the Victoria oat blight epiphytotic of 1946 and 1947. This case is a good illustration of how pathogens from an unexpected source can take advantage of a vulnerable crop. In 1942, a new oat variety, Victoria, was introduced to the midwestern United States. This variety was resistant to all known races of the crown rust, a serious oat pathogen, as well as to stem rust and two important smut diseases. It was therefore a successful and highly desirable variety. By 1945, about 97 percent of the oat crop acreage in the Midwest was planted to Victoria and related varieties. In 1946 and 1947, however, the Victoria variety was attacked by a different pathogen, the blight fungus *Helminthosporium victoriae*, so named because it was first discovered during this epiphytotic. It was subsequently discovered that this fungus was a minor parasite of several species of wild grasses and was widely distributed throughout the region. Susceptibility of the Victoria strain to *H. victoriae* was found to be the result of pleiotropic or secondary effects of the gene that conferred resistance to crown rust. The oat blight attacked seedlings of the crop plant, killing them through the production of a highly specific toxin to which plants that were homozygous or heterozygous for the dominant allele for crown rust resistance were sensitive, but to which those homozygous for the recessive allele were resistant. Destruction of the oat crop was nearly complete, and the Victoria variety was completely abandoned (Day 1974).

In the United States oats are a grain of secondary importance. It wasn't until an epiphytotic affected corn, a major grain crop, that the scientific community focused on the seriousness of the genetic vulnerability problem. The southern corn leaf blight epiphytotic of 1970

and 1971 accomplished this. To understand this important event, however, we must go back to the early years of the twentieth century and trace the development of modern corn seed production.

The technique of cross breeding several inbred lines of corn to produce hybrid seed for planting was discovered in 1917 by D. F. Jones, then a graduate student (NAS 1972). The increase in production gained from the use of hybrid seed was great, and it led gradually to an entirely new technology in corn farming. By the mid-1940s essentially all the corn acreage in the United States was planted with hybrid seed.

Hybrid seed production, however, was not easy. The ears of one parental strain had to be pollinated by pollen from a second strain, which meant that in seed production fields the plants of the female parent, which were to produce the final seed, had to be detasseled by hand to prevent self-fertilization or fertilization by plants of the same line. Nevertheless, hybrid corn was worth the effort.

In the early 1950s, however, an interesting mutation was discovered in Texas—a cytoplasmically inherited factor for pollen sterility. This gene, termed the **Texas cytoplasmic male sterile** or **Tcms gene,** was immediately recognized as being useful for hybrid seed production. Once it was bred into the female parental line, it would eliminate the need to detassel these plants, since their pollen would then be sterile. Because it was a cytoplasmic characteristic, however, it would produce the pollen infertility trait in the seed. Fortunately, a second gene, called the **restorer gene,** was identified that restored fertility. This gene, bred into the male line of the final cross, led to the production of seed that would once again produce plants with fertile pollen. In this way, a genetic system was developed to reduce the need for hand labor in detasseling during the production of seed corn (Ullstrup 1972).

D. F. Jones, the pioneer in corn hybridization, also participated in much of this work, and pointed out the possibility that genetic uniformity of the general type encouraged in corn by such breeding programs might make the crop vulnerable to some virulent new parasite. Use of the Tcms and restorer gene systems became widespread, however, and by 1970 about 85 percent of the hybrid seed corn used in the United States carried these factors.

As early as 1961, a hint of trouble appeared. Two workers in the Philippine Islands reported that plants carrying the Tcms factor were highly susceptible to the blight fungus *Helminthosporium maydis*, a species long known to be a minor or occasional pathogen of corn. This possibility was then examined experimentally in the United States, but no evidence for high susceptibility was found using the local *H. maydis* strains. The Philippine problem was consequently thought to be caused by a factor unique to that area (NAS 1972).

FIGURE 21-5.
Damage to ears of corn by the agent of the southern corn leaf blight, *Helminthosporium maydis*, during the 1970 epiphytotic. (Photograph courtesy of the USDA.)

FIGURE 21-4.
Leaf damage caused by the agent of the southern corn leaf blight, *Helminthosporium maydis*, during the 1970 epiphytotic. (Photograph courtesy of the USDA.)

In 1969, small outbreaks of the southern corn leaf blight occurred in the midwestern United States (Ullstrup 1972). In 1970, a severe epiphytotic developed, apparently involving a novel strain of the blight fungus *H. maydis* (Figures 21-4, 21-5). Losses in 1970 amounted to about 15 percent of the entire United States corn crop, with 50 to 100 per cent losses occurring in some areas of southern Illinois and Indiana. Official estimates placed the value of these losses at nearly one billion dollars.

Seed producers and plant breeders responded rapidly, but much seed corn available for the 1971 planting still carried the Tcms factor. Fortunately, 1971 weather conditions over much of the corn belt did not favor growth of the blight fungus, and so the overall impact was minor.

The case involving corn demonstrates the rapidity with which new genetic races of pathogens can arise and attack an important crop. Most of the corn affected was destined for animal feeding, and supplies of other grains such as sorghum were available to fill the need created by corn underproduction. Had the grain been rice or wheat destined for human consumption in an area such as southeast Asia, what was a serious but strictly economic problem might instead have caused a catastrophic famine.

These three examples clearly indicate that the extensive monocultures of genetically similar individuals fostered by intensive agriculture invite evolutionary responses by pests and pathogens. Moreover, increased international trade involving food materials, and the greater mobility of society in general, mean that newly evolved agents of this type are more and more likely to be dispersed to new areas.

GENETIC VULNERABILITY OF MAJOR CROPS

Genetic vulnerability in any crop changes constantly as new varieties are developed, new systems of resistance are employed in breeding, and changes take place in planting acreage and varietal use. Specific statements about particular crops thus become outdated quickly. Nevertheless, an examination of crop vulnerability patterns may reveal something about the general level of vulnerability in modern agriculture.

A survey by the National Academy of Sciences in 1972 revealed that, because of their genetic homogeneity, a number of crop species in the United States were highly vulnerable to pests and diseases. Table 21-1 summarizes general information about the number of varieties used for several important crop species in the United States. These data show that although several varieties exist for most species, a few varieties, generally less than five, dominate most of the crop acreage.

We should note that this emphasis on limited varietal diversity results from a number of factors. Uniformity is useful to the farmer and the processor because it facilitates the handling of the crop or product by labor-efficient mechanized techniques. Uniformity is also promoted in marketing techniques that tend to overemphasize appearance.

Uniformity within each variety is also encouraged by government regulations that pertain to varietal uniformity or that grant patent protection to varieties with demonstrably consistent uniformity (NAS 1972). Obviously, such laws and policies can be counterproductive to the goal of reducing genetic vulnerability.

In terms of total planted acreage, wheat is the world's most important crop. The existing germ plasm resources for cultivated wheats are extraordinarily diverse. Wheat, one of the earliest domesticates, has been transported widely, and literally tens of thousands of local varieties exist. Moreover, interspecific gene transfer is quite useful; the wheat group includes a large number of wild and cultivated forms between which hybridization is readily carried out. Many of these forms carry distinctive genetic resistance patterns to insects and disease as well as patterns of tolerance to drought and unfavorable temperature (Kuckuck 1970).

The genetic diversity of wheat has seriously eroded, however. Intensified cultivation and grazing are reducing the areas occupied by populations of wild wheat relatives, and local or "field" varieties can be replaced almost overnight by new mass-produced strains promoted for their high-yield potential. Intensive breeding programs for wheats are also using some of the same techniques employed for corn; for example, hybrid seed wheat, produced by crossing inbred lines and employing genetic systems of male sterility and fertility restoration, is being used more and more (NAS 1972).

Quisenberry and Reitz (1967) describe some 49 major diseases of wheat caused by viruses, bacteria, fungi, and nematodes. Resistance to most of these has been found in the form of

TABLE 21-1
Varietal dominance of crop acreage for various major crops in the United States in 1969.

Crop	Hectares (10⁶)	Value ($ × 10⁶)	Total varieties	Major varieties	
				Number	Acreage (%)
Corn	26.8	5,200	197	6	71
Soybean	17.2	2,500	62	6	56
Wheat	17.9	1,800	269	9	50
Cotton	4.5	1,200	50	3	53
Millet	0.8	?	3	3	100
Rice	0.7	449	14	4	65
Bean (dry)	0.6	143	25	2	60
Peanut	0.6	312	15	9	95
Potato	0.6	616	82	4	72
Sugar beet	0.6	367	16	2	42
Peas	0.2	80	50	2	96
Bean (snap)	0.1	99	70	3	76
Sweet potato	0.05	63	48	1	69

Source: National Academy of Sciences 1972.

specific genetic factors, several of them derived from wild relatives of cultivated forms. Such resistance is temporary (Johnson 1961), however, and is often overcome by evolutionary change in the pathogen. In the northwestern United States, the average "lifetime" of a new variety is about five years (NAS 1972). Intensive wheat cultivation is thus an ongoing genetic battle between breeders and evolving pathogens.

These facts raise an important issue: The use of uniform varieties with high-yielding potential depends on an efficient technological support system of continuous breeding activity that usually exists in developed countries but not in emerging nations. The distribution of new mass-produced high-yielding varieties must therefore be coupled with an ability to update continuously the systems of genetic resistance in wheat varieties. At present this system seems to be inadequate. Stewart et al. (1972) have found that wheat varieties introduced to the Near East are highly susceptible to a number of regional wheat diseases and possess vulnerable single-gene systems of resistance to others.

Similar vulnerability patterns exist for the other major grain crops of the world—rice, corn, sorghum, and others—although the evolutionary response of pathogens to new systems of genetic resistance does not seem to be as rapid in these grains as in wheat. However, breeding programs for all these crops increasingly utilize genetic systems for dwarfing (short stems), male sterility, and fertility restoration. The dwarfing gene, for example, is the same in all the new high-yielding rice varieties that are widely distributed throughout tropical and subtropical areas. As we have seen, these systems, although unrelated to pathogen resistance in their intent, can inadvertently confer susceptibility to pathogens. Still other genetic systems, widely used by breeders, are responsible for making varieties of some of these grains insensitive to photoperiod variations.

Among other major crops, the soybean, which is one of the most important potential sources of plant protein, shows a high degree of genetic vulnerability (NAS 1972). Soybeans are susceptible to 14 viral, bacterial, and fungal diseases,

4 nematode diseases, and at least 10 important insect pests. A few soybean varieties dominate the production acreage (Table 21-1), and genetically based resistance to any of these diseases can be traced to a single genetic source. Attempts to achieve major soybean yield increases by intensifying breeding efforts, are likely also to involve the use of special genetic mechanisms like those involved in grain breeding.

For crop plants in general, genetic resistance to disease is high and is increasing. Fortunately, much of the germ plasm variability for these species still exists, and this constitutes a resource on which we are becoming more and more dependent.

GENETIC VULNERABILITY IN DOMESTIC ANIMALS

Infectious and parasitic diseases of domestic animals constitute one of the most serious problems in animal agriculture; for example, in swine and poultry production in the United States, losses from disease amount to about 20 percent and 15 percent, respectively (Schwabe 1974). There is abundant evidence, however, that major, genetically based variability exists with respect to disease resistance in all domestic animals (Hutt 1958, 1965). Nevertheless, breeding for genetic resistance to disease has not been emphasized, nor has genetic vulnerability been recognized as a significant problem in animal agriculture.

There are many reasons for this. The larger domestic animals have long generation times, and so breeding programs must be of comparable duration. In addition, disease resistance in most animals appears to be a polygenic characteristic, and selection combined with progeny testing must usually be carried out over several generations to realize major improvements (Hutt 1965). Furthermore, a number of alternative approaches to disease control and prevention exist for animal species that are not available for plants. Quarantine, sanitation, vaccination, and treatment with modern drugs have all been effective in dealing with animal disease problems.

We suggest, however, that the importance attached to disease eradication programs and to preventing the introduction of foreign diseases to the United States is a direct measure of the genetic vulnerability of American livestock. This vulnerability is recognized by veterinary scientists (Schwabe 1974); the increase in both legal and illegal international movements of animals, animal products, semen, and animal handlers makes the threat of introductions of exotic diseases greater than ever. Fortunately, more is being learned about the genetics of disease resistance, and it is clear that this approach will rapidly become more important in animal agriculture.

Perhaps the most comprehensive studies of genetically based resistance to disease have been carried out for poultry, where mortality losses are great and production techniques favor disease transmission. Genetic resistance to several bacterial and viral diseases is known in chickens. **Pullorum disease,** caused by the bacterium *Salmonella pullorum,* can be eliminated by culling individuals that react positively to an agglutination test involving the bacterial antigen (Hutt 1965). The most serious disease of chickens at present is the avian leucosis disease complex. **Leucosis** is a cancerlike disease complex apparently induced by certain viruses (Figure 21-6). One form of the disease, caused by a myxovirus, is termed **lymphoid leucosis** since it involves tumors of the lymphoid system. A second form, known as **Marek's disease,** is triggered by a Herpes virus and involves the nervous system. For both of these diseases, resistant and susceptible lines can be produced by selection (Figure 21-7) (Hutt 1965; Morris et al. 1970). Other studies have shown that these viruses are able to infect only certain chicken cell types in tissue culture, and that the resistant and susceptible characteristics of these cells show a simple pattern of Mendelian inheritance (Nordskog 1974).

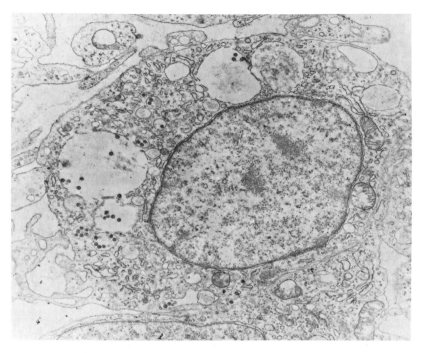

FIGURE 21-6.
Electron micrograph of a nine-day chick embryo, showing lymphoid leucosis virus (large black spots) infecting the limb buds of the developing embryo. The large circular object in the center of the photograph is the nucleus of one limb bud cell. (Photograph courtesy of the USDA.)

Similar observations have been made on the viral disease of sheep known as **scrapie** (Figure 21-8), and in this case too, resistant and susceptible flocks have been established by selection (Nussbaum et al. 1975). For the larger domestic animals, including sheep and cattle, one of the approaches that seems to advance research in genetic resistance is the detection of characteristics correlated with resistance. Such characteristics then permit selection for resistance to be carried out without exposing entire animal groups to the actual disease agent.

There are a number of interesting examples of genetic susceptibility and resistance in cattle. Hereford cattle, for example, have long been known to show an unusually high frequency of carcinoma of the eyeball and eyelid, and a number of studies have shown conclusively that individuals with pigmented, rather than white,

eyelids are less susceptible to this disease than others. Individuals with extensive pigmentation appear to be nearly immune to cancer of the eye (Hutt 1965). This is an example of a situation in which breeding for a particular trait, an all-white face, has produced genetic vulnerability to disease.

There is also evidence of differential patterns of genetic resistance to infectious disease in cattle. **Mastitis,** a troublesome udder disease of dairy cattle caused by *Streptococcus* bacteria, is much more frequent in the offspring of susceptible cows than in those of resistant cows. Zebu cattle, derived from *Bos indicus*, have several well-known physiological characteristics that have helped them adapt to tropical climates. In addition, some of these zebu strains show patterns of genetic disease resistance that could be of considerable value; for example,

526

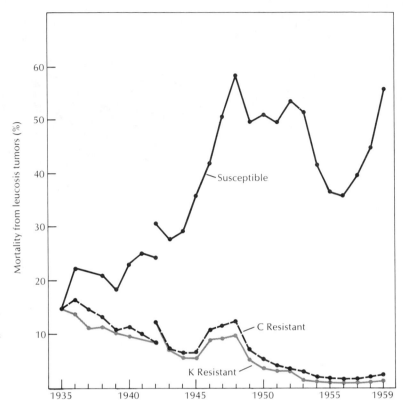

FIGURE 21-7.
Effect of selection for resistance and susceptibility on the mortality rate from tumors triggered by viruses of the leucosis complex. One line of chickens was selected for resistance, two (C and K) for susceptibility. (Modified from Hutt 1965, copyright 1965 by Pergamon Press Ltd. Reprinted by permission of author and publisher.)

FIGURE 21-8.
Scrapie is an infectious viral disease that attacks the nervous system of sheep. The name describes a conspicuous symptom: the animal scrapes off large patches of wool by rubbing against objects to relieve intense itching. (Photograph courtesy of the USDA.)

zebu cattle show higher resistance to tick parasitism, which can seriously inhibit weight gains by European breeds, and a greater tolerance of intestinal parasitic worms (Rendel 1974).

One of the most intriguing possibilities for cattle breeding is to establish resistance to the protozoan disease **trypanosomiasis** in Africa. This disease, widely and patchily distributed in tropical Africa (Figure 21-9), renders much potential rangeland uninhabitable by cattle. The current approach to this problem is to eradicate the tsetse fly vector of the disease. Since trypanosomiasis is also a disease of humans, this approach should obviously not be abandoned; but certain native African cattle breeds (Figure 21-10) tolerate trypanosome infections, and it may well be possible to combine this genetic tolerance in cattle with some form

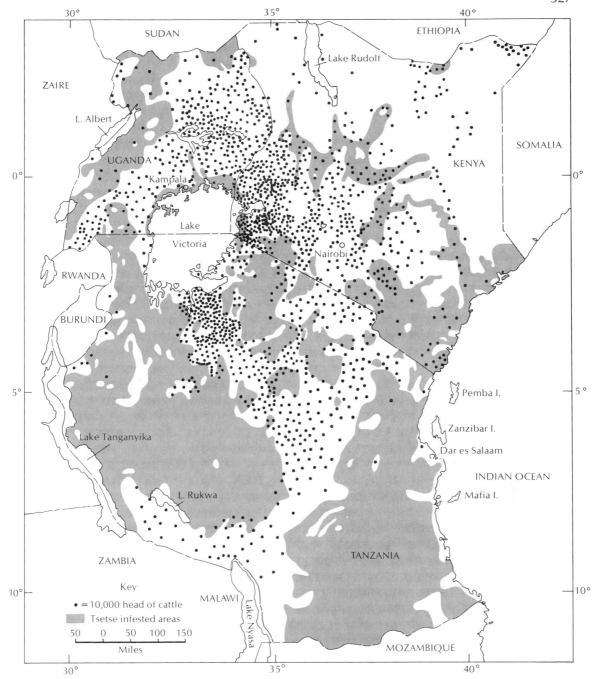

FIGURE 21-9.
Cattle populations and areas of tsetse fly infestation in East Africa. Each dot represents 10,000 cattle. (Based on the FAO East Africa Livestock Survey. From J. R. Peberdy, Rangeland, in *East Africa: Its Peoples and Resources.* Copyright © 1972 by Oxford University Press, London.)

FIGURE 21-10.
Bull of the N'dama breed of cattle from West Africa. This
breed is well known for its tolerance of trypanosomiasis.
(Photograph courtesy of the FAO, Rome.)

of protection for humans that will permit cattle
production even if it proves impossible to eradicate the disease and vector.

Thus we can see that the importance of genetic variability within and between breeds of
domestic animals is gradually being recognized
and documented. Poultry and livestock are
very vulnerable to disease, especially in the
developed countries where production is carried out in facilities such as broiler factories
and feed lots where large numbers of animals
are in close contact. These facts are certain to
lead to greater appreciation of germ plasm resources in domestic animals.

PROTECTING GERM
PLASM RESOURCES

Germ plasm resources for agriculture are diverse and include the gene pools not only of existing major varieties of domesticates but also
of primitive cultivars, wild relatives, and other
members of natural communities (Figure 21-
11). All these forms are potential sources of

genetic material ranging from individual alleles
to groups of genes integrated into coadapted
gene complexes. The importance of genetic
conservation increases with the increased complexity of the genetic systems involved, however. As a source of new alleles for genes already
present in major varieties, artificial techniques
such as irradiation that increase mutation rates
are valuable but probably inadequate. Furthermore, since the great majority of radiation-
induced mutations are recessive, finding and
transferring dominant alleles by traditional
breeding techniques is often easier and less expensive (Brock 1971). New types of genes cannot be created in the laboratory, however, and
so primitive cultivars and wild relatives are of
major importance as sources of new gene material. Wild relatives of crop species have already
contributed to many breeding programs by
providing genes for resistance to pathogens,
improved adaptation to climatic stress, higher-
quality yield, increased crossability, male sterility, and other characteristics (Harlan 1976).

As agriculture moves into more difficult and
marginal environments for organized food production, the need for complex, multigenic
systems will grow. Such systems, termed coadapted gene complexes, consist of genes for
different characteristics in linked and genetically buffered arrangements that produce adaptive phenotypes of coordinated expressions. In
order to effect major breeding changes that
create varieties adapted to deserts, the humid
tropics, and saline soils, for example, existing
crops and domestic animals must undergo many
modifications that together must create a total
adaptive package.

Protecting these different types of germ
plasm resources will require a variety of effective conservation systems, and both artificial
and natural germ plasm banks have important roles.

The concept of formal genetic banks has
evolved gradually, and implementation of the

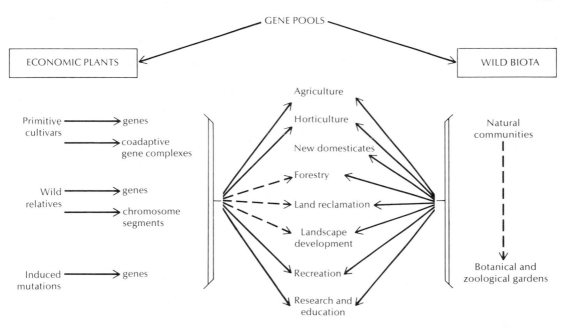

FIGURE 21-11.
Sources of new species and new forms of genetic variability for applied plant sciences. (From Frankel 1974.)

concept has unfortunately been piecemeal. In 1972, however, the United Nations Conference on the Human Environment, held in Stockholm, endorsed the idea of an international system of genetic resource banks (Frankel 1974), which suggests that major improvements in existing systems may soon take place.

Genetic banks fall into two groups: conservation centers and working collections. Conservation centers are responsible primarily for storing and preserving seeds or other materials for long periods of time and for providing materials to the research centers actively engaged in breeding activities. Working collections are those held by organizations that are engaged in active breeding programs or that serve as major suppliers of varietal material to breeders. Obviously, many intermediate situations exist.

In the United States, the main genetic conservation center is the National Seed Storage Laboratory in Fort Collins, Colorado (Creech and Reitz 1971). This facility is operated by the United States Department of Agriculture (USDA) and maintains some 78,000 seed lines in a series of cold storage rooms (Figures 21-12, 12-13). Included are world collections of wheat, oats, barley, sorghum, soybeans, tobacco, cotton, safflower, and sesame, as well as more localized collections of many other crop species. The stored materials are tested for viability every five years, and when viability of the seed deteriorates, they are increased by propagation in contract with growers in appropriate areas. The technology of storage permits such materials to be kept safely and relatively inexpensively for decades, and it appears that prospective advances in storage technology will soon extend this capability to centuries (Frankel 1974).

The most extensive working collection of crop plants in the United States is the USDA

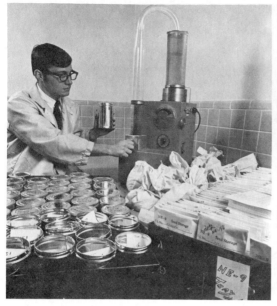

FIGURE 21-12.
The National Seed Storage Laboratory in Fort Collins, Colorado is the primary storage center for germ plasm of economically important plants in the United States. Here, a technician is cleaning seed of foreign matter before placing the seed in storage containers. (Photograph courtesy of the USDA Agricultural Research Service.)

FIGURE 21-13.
This man is checking the continuous temperature recorder that monitors seed storage rooms at the National Seed Storage Laboratory. These rooms are maintained at a temperature of 4.4°C (40°F) and a relative humidity of 32 percent. (Photograph courtesy of the USDA Agricultural Research Service.)

working collections of small grains, based at the Agricultural Research Service Plant Industry Station in Beltsville, Maryland. This collection has mainly wheat, oats, and barley, but includes important collections for other grains such as corn, sorghum, and rice. Figure 21-14 diagrams the way the collection functions (Creech and Reitz 1971). The reserve storage location in this system is the National Seed Storage Laboratory in Fort Collins; otherwise the Beltsville collection is independent, and the stocks included are used in active breeding programs.

The USDA also maintains a network of federal and cooperative federal-state **Plant Introduction Stations,** which hold important collections of regionally important crop species. Three federal Plant Introduction Stations—at

Glenn Dale, Maryland; Savannah, Georgia; and Miami, Florida—now exist. Cooperative centers are located at Geneva, New York; Ames, Iowa; Experiment, Georgia; and Pullman, Washington. In addition, an Interregional Potato Introduction Station is maintained at Sturgeon Bay, Wisconsin, at which some 2000 potato clones are propagated (Rowe 1969). From 1950 to 1967 more than 22,000 samples were distributed by this station alone to researchers working with potato breeding and related topics. Still other collections are maintained by other federal and state experiment stations and by universities.

A number of important conservation centers and working collections exist elsewhere in the world. Seed storage facilities are maintained in

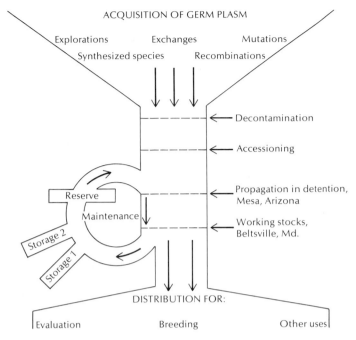

FIGURE 21-14.
Outline of the major activities involved in acquiring and maintaining germ plasm in the USDA small grains collection. (From Creech and Reitz 1971.)

Japan, Turkey, Italy, Germany, the Soviet Union, and the Philippines. More limited, within-country collections are being developed in several other locations, including India, Poland, Bulgaria, and Yugoslavia (Creech and Reitz 1971). For certain crops, especially corn and rice, organized international networks of storage and propagation centers have developed. For corn, a comprehensive collection of indigenous varieties in the New World was undertaken from 1951 to 1961. The USDA, the Rockefeller Foundation, and the government of Brazil have cooperatively established collection and seed storage centers for corn in Mexico, Colombia, Brazil, and the state of Maryland. More than 12,000 varieties were initially included in these collections. Since then, collections from various parts of Europe have been made and placed in storage in Italy (NAS 1972).

More than 30,000 collections of cultivated varieties of rice, together with over 900 collections representing 38 wild forms relatives, are held at the International Rice Research Institute in the Philippines (Creech and Reitz 1971; NAS 1972). Other major rice collections are held in Japan, India, and the United States.

Even with improved facilities and organization, however, artificial genetic banks will serve only part of our genetic conservation needs. Such banks will probably be the major means of preserving primitive cultivars and local varieties of domesticated species; preserving them by closing off the agricultural systems involved —that is, by isolating major regions and their human populations from the modifying impacts of modern technology—is economically and socially impossible (Frankel 1974). However, the preservation of wild relatives and other wild species not presently regarded as economically important can be accomplished only by establishing natural ecosystem preserves. Zoos and botanical gardens are inadequate for this purpose because they cannot accommodate the normal range of diversity and variability that is essential for the continued evolution of wild species.

Ecologists and population biologists have only recently begun to explore strategies appropriate for preserving major elements of natural plant and animal diversity. Many of the initial ideas on this subject have come from the recent wave of interest in island biogeography (Diamond 1975; Sullivan and Schaffer 1975). It is now apparent, for example, that the number of species that exists in a region bounded by strong habitat or geographic barriers represents a balance within the area between rates of addition (by speciation and immigration) and the rate of extinction (Figure 21-15). For a given area, the rate of immigration of new species declines as the number of species in the area increases, since the greater the number of species already present, the less likely it is that any arrival will represent a new species. The absolute rate of immigration under given circumstances obviously depends on the size of the area, the degree of its isolation from source areas, and other factors. Extinction tends to increase as the number of species in the region increases, since the greater the number of species, the lower the populations of each must be and the greater the chance that competition and predation will eliminate some of the species. The chance of extinction is, of course, greater in small areas, which support fewer individuals and in which accidental factors are more likely to destroy a total species population.

A number of examples are now available in which species diversity in a region or an environmental preserve has declined because of the size of the area (Diamond 1975). Thus in order to create a comprehensive system of

reserves designed to perpetuate wild plants and animals and to preserve their gene pools, we must consider not only the locations of reserve units, but also their size, number, and connections between them (Figure 21-16). The bigger a reserve (other factors being equal) the more species can be maintained. In a network of reserves, the largest units could be adequate to maintain populations of the larger carnivores, encompassing several hundred to

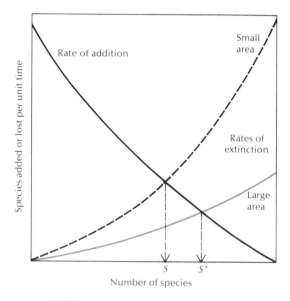

FIGURE 21-15.

The number of species in an area is the result of a balance between the rate at which species are added (by speciation and immigration) and the rate at which they become extinct. When few species are present in an area, the rate at which new species are added by immigration is high. As the number present increases, the rate of immigration of new species drops off because many of the individuals arriving by dispersal are already represented in the biota. For simple numerical reasons, the rate of extinction increases as the size of the biota increases (the more species present, the more can become extinct). Moreover, the more intense biotic interactions in complex biotas tend to increase the probability of extinction. Extinctions are more likely to occur in small than in large areas (if the total population of species is proportional to area of habitat), and so the equilibrium number of species is greater in large areas (S^1) than in small (S).

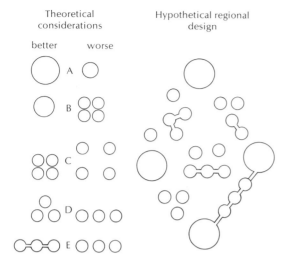

Theoretical considerations

better   worse

Hypothetical regional design

FIGURE 21-16.
Considerations of the size and arrangement of preserve units are important to the long-term preservation of biotic diversity. In theoretical terms, larger preserves should be more effective than smaller ones (A); single units more effective than several smaller units of the same total size (B); and units that are closely grouped or connected by corridors more effective than other less proximate arrangements (C, D, E). An overall regional design of preserves of various sizes, in which small preserves act as stepping stones or corridors for dispersal between larger preserve units, may well be the most effective approach to preservation of the wild biota. See text for further discussion. (Modified from Diamond 1975.)

a few thousand square kilometers (Sullivan and Shaffer 1975). If a single large area is set aside, this space can support more species than a set of smaller, separate preserves of the same habitat totalling the same area. Similarly, as Figure 21-16 suggests, more species survive when separate preserves are close together, because there are greater rates of migration between units, which reestablishes populations that have died out in individual areas. Corridors between units also favor reinvasion of species into units where accidental extinctions have occurred.

Any network of preserves must also be designed to include as many of the world's species as possible. Such a network would logically

consist of a series of first-order major reserves in the world's major biotic and biogeographic regions, combined with carefully located smaller reserves that serve as distributional stepping stones and offer insurance against total extinction of species (Sullivan and Shaffer 1975). Obviously, even if such a system were established, man would have to assist in the maintenance of species populations by habitat manipulation, deliberate reintroductions of species, and artificial diversification of genetic variability. Without such a comprehensive system, the chances of preserving the wild biota of the planet are minimal.

REDUCING GENETIC VULNERABILITY

The idea of reducing genetic vulnerability implies improving both our knowledge of evolutionary trends in pathogens and our ability to combat such pathogens by breeding programs and other techniques (Stewart, Hafiz, and Hak 1972). Improvements are needed in the system of detecting and classifying new races of pathogens and in the distribution of information regarding the virulence of such races for crop varieties with particular resistance genes or gene complexes. An organized international system of this type might, for example, have prevented the 1970 southern corn leaf blight epiphytotic.

For crop plants, comprehensive inventories of germ plasm sources of resistance to various pathogens are needed, together with information on obtaining seed material from genetic banks or field populations. The need to make rapid responses in breeding for resistance is growing, and immediate access to information is critical. Efforts should of course be expanded to increase the inventory of resistance genes by field collection and by inducing mutations in the laboratory.

In breeding programs, broadening the genetic basis of resistance in widely distributed, high-yielding varieties is a clear necessity. Moreover, efforts should be made to incorporate new and broader resistance systems into local varieties that are already adapted to local climates, soils, and cultivation techniques. Abandoning these local varieties is not justified; their preservation is of value, and one way to encourage their preservation is to strengthen their systems of resistance.

Finally, cultivation strategy can be employed to reduce genetic vulnerability by diversifying agroecosystems in terms of crops and varieties so that all the eggs are not in one basket. This is especially important for regions that depend greatly on internal production to meet food needs. Diversity of crop species should be encouraged, for example, and the interplanting of multiple varieties, with different resistance systems, may be advisable. These approaches, along with the use of polygenic resistance systems as described in Chapter 19, can greatly reduce genetic vulnerability.

SUMMARY

Almost all widely distributed plant and animal species show patterns of genetic adaptation to local factors of the habitat and regional climate, and the member species of natural communities show evidence of coevolution; that is, of having influenced each other's evolutionary development and adaptation. Variations in genetic adaptation to physical and biotic factors represent the germ plasm resources of a species. For crop and domestic animal species, germ plasm resources include the gene pools of domestic varieties plus those of wild relatives with which they can be crossed artificially.

Genetic vulnerability develops because crop ecosystems have no evolutionary steady-state; crops and associated species have not achieved coevolutionary adjustment to each other. Vulnerability to pests and pathogens is high when the mechanism that confers resistance at a given moment can be overcome by small genetic changes in a pest or pathogen, or when a host's freedom from attack can be breached by dispersal of a new pest or pathogen into a region. Genetic vulnerability has led to epiphytotics such as the Irish potato famine, caused by the dispersal of a fungal pathogen into Europe from North America; the Victoria oat blight, in which a susceptible variety of oats was attacked by a fungus associated with other grass species; and the southern corn leaf blight, caused by a sudden evolutionary change in a fungus to which corn had previously been resistant.

The genetic vulnerability of major crops is high in the industrialized nations, where a few varieties with high genetic homogeneity tend to dominate the planted acreage. Genetic vulnerability also exists for domestic animals, although it has received little attention; but the increasing dangers of the accidental dispersal of animal pathogens make it likely that more efforts will be made to breed animals for genetic resistance.

Germ plasm resources are now beginning to be lost at a serious rate. Although comprehensive, organized gene banks have not yet been created, systems of genetic banks for plants have begun to develop. These banks consist of conservation centers, mostly seed storage facilities, and working collections of varieties held by organizations closely associated with active breeding programs.

In addition to artificial gene banks, a worldwide system of carefully designed environmental preserves is essential to the protection of wild species for which no economic value may yet be known. Both artificial and natural genetic banks are essential to the strategy of reducing genetic vulnerability by diversifying the crop species in agroecosystems, the varietal composition within species, and the resistance mechanisms within varieties.

Literature Cited

Adams, M. W., A. H. Ellingboe, and E. C. Rossman. 1971. Biological uniformity and disease epidemics. *BioSci.* 21:1067–1070.

Antonovics, J. 1970. Evolution in closely adjacent plant populations. VII. Clinal pattern at a mine boundary. *Heredity* 25:349–362.

Baer, A. S. 1977. *Genetic Perspective.* Philadelphia: W. B. Saunders Co.

Brock, R. D. 1971. The role of induced mutations in plant improvement. *Rad. Bot.* 11:181–196.

Clausen, J., D. D. Keck, and W. M. Hiesey. 1940. *Experimental studies on the nature of species. I. Effects of varied environments on western North American plants.* Carnegie Inst. Wash. Pub. 520. Washington, D.C.

Cox, G. W., and D. G. Cox. 1974. Substrate color matching in the grasshopper *Circotettix rabula* (Orthoptera: Acrididae). *Great Basin Nat.* 34:60–70.

Creech, J. L., and L. P. Reitz. 1971. Plant germ plasm now and for tomorrow. *Adv. in Agron.* 23:1–49.

Day, P. R. 1974. *Genetics of Host-Parasite Interaction.* San Francisco: W. H. Freeman and Company.

Diamond, J. M. 1975. The island dilemma: lessons of modern biogeographic studies for the design of natural preserves. *Biol. Cons.* 7:129–146.

Fenner, F., and F. N. Ratcliffe. 1965. *Myxomatosis.* London: Cambridge University Press.

Frankel, O. H. 1974. Genetic conservation: our evolutionary responsibility. *Genetics* 78:53–65.

Harlan, J. R. 1976. Genetic resources in wild relatives of crops. *Crop Sci.* 16:329–332.

Hutt, F. B. 1958. Genetic resistance to disease in domestic animals. Ithaca, N.Y.: Cornell University Press.

——. 1965. The utilization of genetic resistance to disease in domestic animals. Genetics today. *Proc. XI Intern. Cong. Genetics, The Hague, 1963.* pp. 775–782. New York: Pergamon Press.

Johnson, T. 1961. Man-guided evolution in plant rusts. *Science* 133:357–361.

Kuckuck, H. 1970. Primitive wheats. In *Genetic Resources in Plants—Their Exploration and Conservation.* IBP Handbook No. 11. O. H. Frankel and E. Bennett (eds.). pp. 249–266. Philadelphia: F. A. Davis Co.

Leopold, A. S. 1966. Adaptability of animals to habitat change. In *Future Environments of North America.* F. F. Darling and J. P. Milton (eds.). pp. 66–75. Garden City, N.Y.: Natural History Press.

Levins, R. 1974. Genetics and hunger. *Genetics* 78:67–76.

Mattson, W. J., and N. D. Addy. 1975. Phytophagous insects as regulators of forest primary production. *Science* 190:515–522.

Mayr, E. 1963. *Animal Species and Evolution.* Cambridge: Belknap Press, Harvard University.

McMillan, C. 1959. The role of ecotypic variation in the distribution of the central grassland of North America. *Ecol. Monog.* 29:285–308.

Morris, J. R., A. E. Ferguson, and F. N. Jerome. 1970. Genetic resistance and susceptibility to Marek's disease. *Can. J. Animal Sci.* 50:69–81.

National Academy of Sciences. 1972. *Genetic Vulnerability of Major Crops.* Washington, D.C.: National Academy of Sciences.

Nordskog, A. W. 1974. Breeding for eggs and poultry meats. In *Animal Agriculture.* H. H. Cole and M. Ronning (eds.). pp. 319–333. San Francisco: W. H. Freeman and Company.

Nussbaum, R. E. 1975. The establishment of sheep flocks of predictable susceptibility to experimental scrapie. *Res. in Vet. Sci.* 18:49–58.

Olmsted, C. E. 1944. Growth and development of range grasses. IV. Photo-periodic responses in twelve geographical strains of side-oats grama. *Bot. Gazette* 106:460–474.

Quisenberry, K. S., and L. P. Reitz (eds.). 1967. *Wheat and wheat improvement.* Agron. Monog. 13. Madison, Wisc.: Amer. Soc. Agron.

Rendel, J. 1974. The role of breeding and genetics in animal production improvement in the developing countries. *Genetics* 78:563–575.

Ricklefs, R. E., and G. W. Cox. 1972. Taxon cycles in the West Indian avifauna. *Amer. Nat.* 106:195–219.

Rowe, P. R. 1969. Nature, distribution, and use of diversity in the tuberbearing *Solanum* species. *Econ. Bot.* 23:330–338.

Schwabe, C. W. 1974. Management and disease. In *Animal Agriculture.* H. H. Cole and M. Ronning (eds.). pp. 637–654. San Francisco: W. H. Freeman and Company.

Stewart, D. M., A. Hafiz, and T. A. Hak. 1972. Disease epiphytotic threats to high-yielding and local wheats in the Near East. *FAO Plant Protection Bull.* 20:50–57.

Sullivan, A. L., and M. L. Shaffer. 1975. Biogeography of the megazoo. *Science* 189:13–17.

Turesson, G. 1922. The genotypical response of plant species to habitat. *Heriditas* 3:211–350.

Ullstrup, A. J. 1972. The impacts of the southern corn leaf blight epidemics of 1970–1971. *Ann. Rev. Phytopath.* 10:37–50.

Vanderplank, J. E. 1968. *Disease Resistance in Plants.* New York: Academic Press.

22

BREEDING FOR IMPROVED PERFORMANCE

In this chapter we shall examine some of the major possibilities for improving crop and domestic animal production through breeding. We shall see that these improvements depend on the genetic resources that we have just examined and that it is important for domestic species to be genetically well adjusted to their agricultural environment.

During recent agricultural history, the idea of improved performance, or increased efficiency, has been interpreted in many and changing ways. Mechanized monoculture, as it is practiced in American grain farming, has been described as highly efficient in terms of crop yield per farm worker. But if mechanized monoculture is viewed in terms of yield per hectare, or per dollar or calorie of fossil fuel invested, it would be seen as less efficient than many other cropping systems. Ecologists, who are interested in the rate at which nutrients are incorporated into new organic matter, view efficiency in still different terms. Performance and efficiency are thus changeable concepts that express the output of useful foods and fibers from agricultural ecosystems in terms

of inputs of economically important materials and effort.

In the agricultural systems of the developed, industrialized nations, improvements in performance have come to be virtually synonymous with increases in yield per hectare, which have been achieved through various technological improvements. Some of these sources of improved performance are recurring, others are nonrecurring (Brown 1967). Recurring sources are those that, like fertilization and soil management, can be improved more or less continuously. Improvements in the composition and application of fertilizers, in their formulation for controlled release, and in many other factors can gradually be made, and these can be combined with plant breeding techniques that increase the crop species ability to utilize the fertilizer element. Nonrecurring sources of improvement, on the other hand, are those that, like herbicides, generally make a striking one-time contribution to yield increase. Once weed competition is largely eliminated, for example, no further benefit can be gained from an herbicide. Obviously, the line between

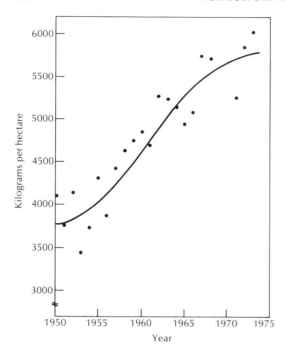

FIGURE 22-1.
Rice yields in Japan from 1950 through 1970 (three-year sliding average). The dashed line indicates the S-shaped trend of the plotted data points.

One interesting and important principle that applies to such improvements is that they may be more difficult to achieve in the centers of origin or evolutionary diversification of crop and domestic animal species (Jennings 1974), where coevolutionary processes have closely integrated the species with other components of the living environment, which makes yield-limiting factors most diverse and complex. Many of the most spectacular yield-per-hectare breakthroughs have occurred when crop and animal species have been moved to new regions where they are freed from such constraints; this phenomenon clearly parallels the outbreaks of introduced pest species (see Chapter 15).

Despite the many interpretations of efficiency and performance, there is now a strong tendency toward a clearer and more unified view of the relationship between the two, which is similar to the ecological concept of yield per unit of productive work and materials invested. In the past, the application of improvement techniques has been guided by short-term economic considerations, so that what represented an improvement under one set of conditions easily became the opposite when costs of fuels or fertilizers changed. In our discussions of improved performance, therefore, we shall emphasize the need to increase yield per unit of energy or other invested resources, and, in most cases, we shall consider that an improvement has been made when inputs from nonrenewable sources are replaced with equally effective inputs from renewable sources.

CROP IMPROVEMENT

The goal of breeding for improved performance is to maximize the economic yield of a crop per unit of invested resources. In manipulating crop plant physiology, this goal comprises two aspects: increasing net primary production per unit area of cropland; and promoting the maximum accumulation of this production in

recurring and nonrecurring inputs is not sharp, but some sources of improved performance are most often recurring and others nonrecurring.

As more and more inputs of productive chemicals and work are applied to a crop, the yield increases. This "yield per acre takeoff" tends to follow a characteristic pattern, the familiar S-shaped curve of biological growth processes. As initial inputs are applied, yields start to rise slowly (Figure 22-1); and as major nonrecurring and recurring sources of improved performance are discovered and applied the yield per hectare rises rapidly. Then, as nonrecurring sources are exhausted and as improvements in recurring sources become more difficult to achieve, improvement levels off and approaches what might be termed a "biological ceiling" (Wilson 1973).

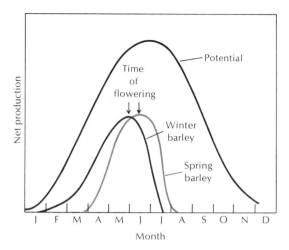

FIGURE 22-2.
The observed rate of dry matter accumulation by winter and spring barley in England relative to the potential defined by seasonal weather conditions. (Modified from Ivins 1973.)

the desired harvestable parts (National Research Council 1975). These objectives must be pursued in relation to several other constraints, including breeding for disease and pest resistance, and compatibility with the technology available for planting, cultivating, and harvesting.

The seasonal pattern of the performance of a crop species can be illustrated by a curve similar to that shown in Figure 22-2, in which the potential curve of dry matter production represents that achieved by the plant members of natural communities under regional regimes of temperature and photoperiod where conditions of moisture and nutrients are favorable. In other words, this potential represents that which is demonstrated in systems of coevolved species adjusted to local conditions. A specific crop, such as barley, clearly realizes only a portion of this productive potential; at any given time, its rate of net production is somewhat below that shown on the potential curve. In addition, it does not make maximum use of the entire season of favorable conditions. In

England, to which this particular curve pertains, barley completes its development in July, and for several weeks of favorable weather thereafter it realizes no significant amount of net production.

With these facts in mind, we can see that examining the life cycles of crops in relation to environmental potential will offer fairly straightforward predictions of how better advantage can be taken of this potential. In barley and other grains, one breeding strategy is to extend the period between anthesis, or flowering, and senescence, or temination of metabolic activity. Extending this period would increase the amount of photosynthetically-produced material available for storage in the seed. In potatoes, which initiate tuber formation early in this growth process, the breeding strategy is to maximize the capacity of the leaf canopy to take advantage of potential solar energy by increasing the leaf area index and preventing its senescence while conditions are still favorable. In sugar beets, which develop a dense canopy and maintain it through the end of the growing season, the obvious strategy is to promote early growth and maturation of the canopy (Figure 22-3).

These discrepancies between crop life cycle and environmental potential are obviously the result of transporting successful domesticates into environments quite different from those to which they are native. Although it is obvious that plant breeding can considerably modify life cycles to suit new environmental regimes, it seems unlikely that a single species can be modified to achieve the performance shown by the diverse plant species that exist in natural communities, in which the physiology of each species is specialized for a limited portion of the seasonal spectrum. During this period, the species grows efficiently, and then in other seasons it becomes dormant or dies and survives as seed. The integration of many species, each adapted to particular portions of the seasonal spectrum, is thus a vehicle for the reali-

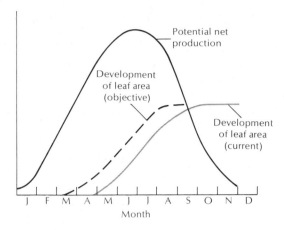

FIGURE 22-3.
Once sugar beets develop a mature canopy, the foliage produces photosynthate for storage in the root. This storage continues until the end of the growing season. Thus the breeding goal for this crop is to achieve earlier development of a mature canopy. (Modified from Ivins 1973.)

zation of maximum environmental potential.

Few efforts have been made in crop breeding to duplicate this community pattern, but the potential exists (Trenbath 1974). Most species combinations that have been examined experimentally produce combined yields whose values are intermediate compared to those of the member species grown by themselves. A few cases are known, however, in which combined yields were greater than those of each species in monoculture, and several patterns of ecological complementarity might be expected to create such "over-yields". This topic obviously deserves more thorough investigation.

PHOTOSYNTHETIC AND RESPIRATORY SYSTEMS

Studies of photosynthesis and respiration in plants native to different environments have recently revealed a number of basic differences that could have major significance to agricultural plant breeding. The most important of these concern photorespiration and its relationship to photosynthesis.

Photosynthesis has two stages: a set of light reactions followed by a set of dark reactions that take place during or shortly after light periods but that do not require light energy. In the light reactions, solar energy received by chlorophyll and associated pigments energizes one of the electrons associated with the chlorophyll molecule (Figure 22-4). Two somewhat different photosystems, based on chlorophyll "light traps," are actually present: photosystem I (cyclic) and photosystem II (noncyclic). Both of these photosystems produce the high-energy compounds ATP and NADP·H_2. These reactions take place in the chloroplast.

In most higher plants, the dark reactions proceed as shown also in Figure 22-4. The high-energy compounds formed during the light reactions are used to assimilate CO_2 into reduced organic compounds. The sequence of reactions is cyclic and is called the **Calvin Cycle** after its discoverer, Melvin Calvin, a University of California scientist who received the Nobel Prize in 1961 for his studies of photosynthesis. It involves an initial reaction in which CO_2 combines with the 5-carbon molecule ribulose diphosphate to form a complex that immediately splits into two 3-carbon molecules known as phosphoglyceric acid. Since the initial compound in which the newly-fixed CO_2 appears is a 3-carbon molecule, this pattern is termed **C_3 photosynthesis.**

Higher plants also possess two systems that cause respiratory release of CO_2: dark respiration and photorespiration. **Dark respiration,** somewhat misnamed because it can occur either in the dark or the light, is the general pattern of oxidative breakdown of organic materials carried out in the mitochondria. This breakdown, of course, is accompanied by the production of high-energy molecules, such as ATP, that are required for various maintenance and

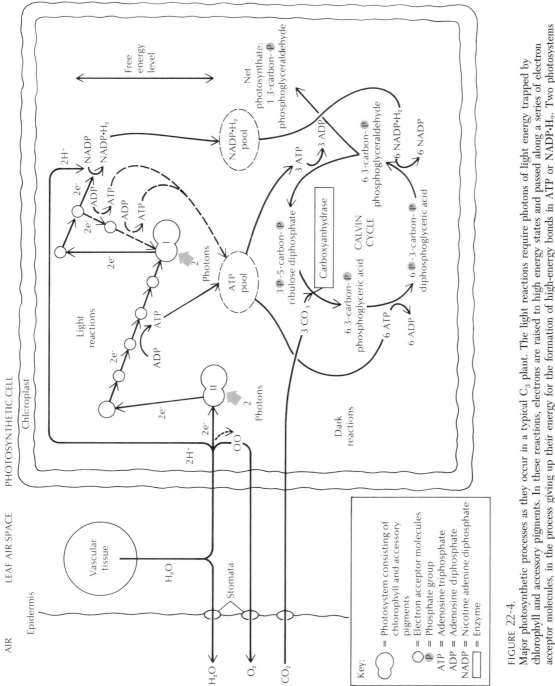

FIGURE 22-4.

Major photosynthetic processes as they occur in a typical C_3 plant. The light reactions require photons of light energy trapped by chlorophyll and accessory pigments. In these reactions, electrons are raised to high energy states and passed along a series of electron acceptor molecules, in the process giving up their energy for the formation of high-energy bonds in ATP or NADP·H₂. Two photosystems exist that differ somewhat in the structure and organization of various pigments. Photosystem I typically operates in the manner indicated by the solid lines, producing NADP·H₂, but may also function as indicated by the dashed lines to produce ATP. The dark reactions can actually occur either during light periods, since the energy for their operation is produced in the light reactions, or during dark periods, as long as adequate supplies of ATP and NADP·H₂ are available in the chloroplasts. Note that CO_2 is incorporated into a 3-carbon compound under the influence of the enzyme carboxyanhydrase.

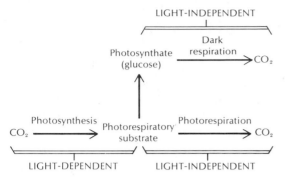

FIGURE 22-5.
Schematic representation of the relationships among photosynthesis, dark respiration, and photorespiration. (Modified from Chollet and Ogren 1975.)

growth processes in the cells and tissues involved. In daylight, dark respiration may be partially or even greatly reduced (Chollet and Ogren 1975), but the evidence on this point is still inconclusive.

Photorespiration is a distinct process that is closely linked to photosynthesis (Figure 22-5) and that causes the release of CO_2 in addition to that produced simultaneously by dark respiration. Although the reactions of photorespiration are light independent, the process involves early derivatives of photosynthetic reactions, and so it occurs only during the daytime and the very earliest portion of the dark period. The tentative pathways of photorespiration are shown in Figure 22-6. In the chloroplast, a portion of the ribulose diphosphate is apparently broken down to 2-carbon glycolate fragments that pass into structures known as peroxisomes, located in close proximity to both chloroplasts and mitochondria, where they are oxidized and aminated to form glycine. Then glycine molecules combine to form the amino acid serine in the mitochondria and in doing so produce CO_2. This is thought to be the main source of the CO_2 produced through photorespiration. Although photorespiration is difficult to separate completely

from all the other processes that influence CO_2 gain and loss, estimates suggest that photorespiration occurs at about one-sixth the rate of net photosynthesis under normal conditions (Chollet and Ogren 1975).

Recently, however, a quite different system of photosynthesis and associated photorespiration has been discovered in certain plants, including a number of major crop species. In this system, termed **C_4 photosynthesis**, CO_2 is initially combined with the 3-carbon compound phosphoenolpyruvate (PEP) under the influence of the enzyme PEP carboxylase. Thus a 4-carbon compound is initially formed (giving this system its identifying name), which then passes from the mesophyll cells in which it was formed into a specialized cell layer immediately surrounding the vascular bundles of the leaf. Here, the CO_2 is liberated and enters into the normal Calvin Cycle (Figure 22-7). Plants with the C_4 system thus have a distinctive morphology characterized by a cylindrical arrangement of mesophyll and sheath cells about the vascular bundles (Figure 22-8).

The C_4 system is both a biochemical and morphological addition to the normal pattern of leaf structure and photosynthesis. Correlated with this complication of the photosynthetic system in C_4 plants, however, is the virtual absence of photorespiration. These two features are combined in some members of at least ten plant families, including both monocots and dicots, and several hundred C_4 species are now known (Bjorkman and Berry 1973). Curiously, in all the families, and in at least 11 genera, both C_3 and C_4 species may occur in the same taxon. In general, the species with C_4 characteristics are inhabitants of hot, arid, or saline environments. Three major species of crop plants—corn, sorghum, and sugarcane—are C_4 plants (Figure 22-9).

Apparently, the C_4 system of photosynthesis increases net photosynthesis and reduces water loss by transpiration under situations of high photosynthetic potential. Oxygen stimulates

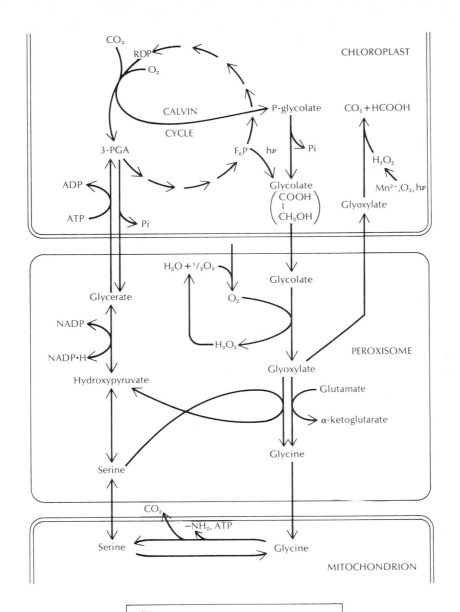

FIGURE 22-6.
Tentative outline of the biochemical pathways of photorespiration. See text for explanation. (Modified from Chollet and Ogren 1975.)

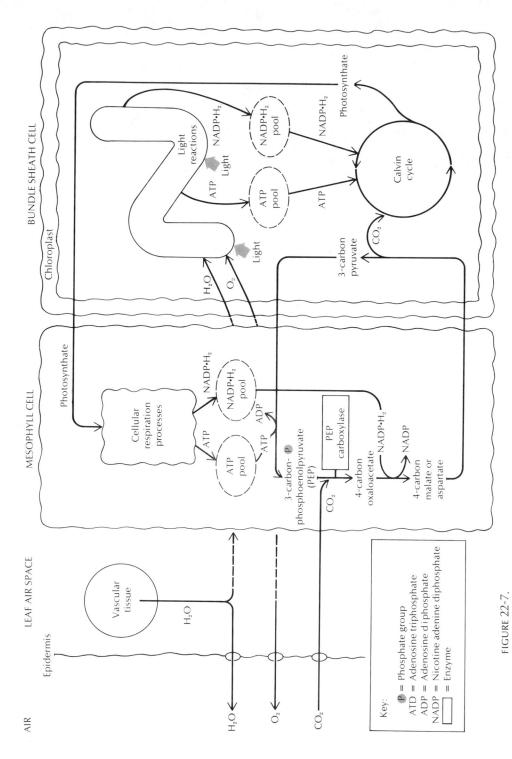

FIGURE 22-7.

Distinctive features of C_4 photosynthesis. In the mesophyll cells, CO_2 is combined with phosphoenolpyruvate (PEP) under the influence of the enzyme PEP carboxylase to form a 4-carbon compound, oxaloacetate. This substance is converted to malate or aspartate and passes into the chloroplast-containing cells of the bundle sheath, where CO_2 is released and a 3-carbon pyruvate molecule reformed. Pyruvate returns to the mesophyll cells, while the CO_2 enters the Calvin Cycle. This sequence, which requires the use of high-energy molecules from cellular respiration, apparently functions as a highly efficient CO_2 uptake system that minimizes the need for stomatal gas exchange and the concomitant loss of water by transpiration. Contrast this figure with Figure 22-4.

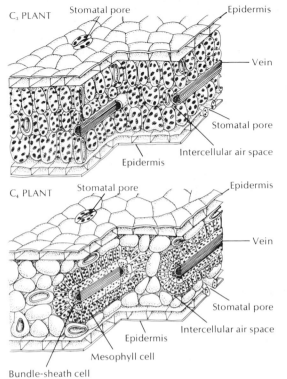

C₃ PLANT — Stomatal pore, Epidermis, Vein, Stomatal pore, Intercellular air space, Epidermis

C₄ PLANT — Stomatal pore, Epidermis, Vein, Stomatal pore, Intercellular air space, Epidermis, Mesophyll cell, Bundle-sheath cell

FIGURE 22-8.
Distinctive structural differences between the leaf structure of C_3 and C_4 plants are illustrated for the C_3 species *Atriplex patula* (*upper*) and the C_4 species *A. rosea* (*lower*). In the C_4 leaf the chloroplast-containing cells are grouped in concentric sheathlike layers around the vascular bundles. (From "High-Efficiency Photosynthesis" by Olle Björkman and Joseph Berry. Copyright © 1973 by Scientific American, Inc. All rights reserved.)

FIGURE 22-9.
Sorghum, together with corn, sugarcane, and a number of other crop plants of tropical and subtropical origin, is a C_4 photosynthetic species. The capacity of C_4 species for net photosynthesis in general exceeds that of C_3 species. The taller weedy grasses in this field are Johnson grass, a member of the same genus as sorghum and also a C_4 species. (Photograph by Bob Kral, courtesy of the USDA Soil Conservation Service.)

photorespiration, and increases in temperature differentially increase the rate of photorespiration relative to that of photosynthesis. However, the enzyme PEP carboxylase that initially assimilates CO_2 in C_4 photosynthesis is able to withdraw CO_2 to virtually zero concentration in the air space within the leaf; this level is much lower than that to which CO_2 can be reduced by the enzyme that links CO_2 to ribulose diphosphate. This same C_4 enzyme probably serves as an effective scavenger of any CO_2 that is released by dark respiration and photorespiration in the same tissues, which means that less gas exchange is necessary to provide the CO_2 needs of C_4 plants and that less transpirational water loss will accompany such exchange.

As a result, net photosynthetic production is higher by a factor of 2 or more in C_4 plants than in C_3 plants (Table 22-1). This fact has provided strong incentive for breeding programs with C_4 species. First of all, it is evident that some minor crop species, such as the grain amaranths of Central and South America, that

TABLE 22-1
Typical rates of net photosynthesis for C_3 and C_4 plants under high light intensities and warm temperatures (25°–30°C).

Species	Net photosynthesis (mg CO_2/dm²/hr)
C_3 species	
Spinach	16
Tobacco	16–21
Wheat	17–31
Rice	12–30
Bean	12–17
C_4 species	
Maize	46–63
Sugarcane	42–49
Sorghum	55
Bermuda grass	35–43
Pigweed	58
(*Amaranthus edulis*)	

Source: Data from Zelitch 1971.

have the C_4 characteristic, deserve much more serious attention. Second, the fact that the characteristic is widely distributed among plants and often varies in occurrence among the members of the same genera suggests that breeding and hybridization programs might be able to transfer the system into species now lacking it (Bjorkman and Berry 1973). Other workers currently doubt that such an introduction will become common, but they emphasize the possibilities of selecting for lower levels of photorespiration in C_3 crop species and of developing techniques to inhibit photorespiration in other ways (Chollet and Ogren 1975).

Still another photosynthetic system that is closely related to C_4 photosynthesis is called **Crassulacean Acid Metabolism** or CAM (Ting 1973; Evans 1975). This system, first observed in plants of the family Crassulaceae, occurs in several families of succulent plants, including crop species such as pineapple (Figure 22-10), sisal and maguey, and prickly pear cactus. This system functions very similarly to C_4 photosynthesis in that the initial product of CO_2 assimilation is a 4-carbon compound (Figure 22-11). However, this fixation occurs almost entirely at night, when the stomata of the plant are open and gas exchange occurs. At this time, atmospheric humidity is also highest, and the transpirational loss of water is less than it would be during daytime exchange. During the daytime the stomata close, and CO_2 is released from the 4-carbon storage compound to enter the Calvin Cycle. CAM photosynthesis is thus clearly adaptive in arid environments, and many of the succulent plants that show it are characteristic of such regions. The potential of CAM plants as crop species should also be explored more fully (Evans 1975).

These recent developments in plant physiology provide excellent examples of the long-range value of basic research in areas peripheral to traditional agricultural science. The esoteric projects that led to the understanding of these systems of respiration and

FIGURE 22-10.
Pineapple, seen here on the island of Maui, Hawaii, is one of the few CAM photosynthesis species that is an important food crop. (Photograph by Arnold Nowotny, courtesy of the USDA Soil Conservation Service.)

photosynthesis, many of them carried out on noneconomic species, may have been viewed by some as a waste of time and effort. But now we can see how they might well contribute to the development of new crops and agricultural systems of particular value in some of the most challenging environments. The physiological characteristics involved, furthermore, are already proving of productive value in other regions, such as the corn fields of the midwestern United States.

NITROGEN-FIXING SYMBIOSES

Increasing attention is being paid to the encouragement of biological nitrogen fixation in crop ecosystems. As we noted in Chapter 13, a number of important discoveries have recently been made about the basic genetics and biochemistry of nitrogen fixation. These findings suggest a number of profitable directions for plant breeding, which are quite different for legume species than for nonlegumes.

The most immediate opportunities for breeding exist for legumes, most cultivated species of which possess symbiotic associations with *Rhizobium* bacteria in root nodules. For legumes, the objective is to increase both the output of fixed nitrogen by the root symbionts and its subsequent use by the host or by other plants. There seem to be several possible ways to achieve this.

Nitrogen fixation, first of all, is an energy demanding process. Approximately 24 ATP molecules are required to fix one molecule of N_2 (Hardy and Havelka 1975). Minchin and Pate (1973) have estimated that about 32 per-

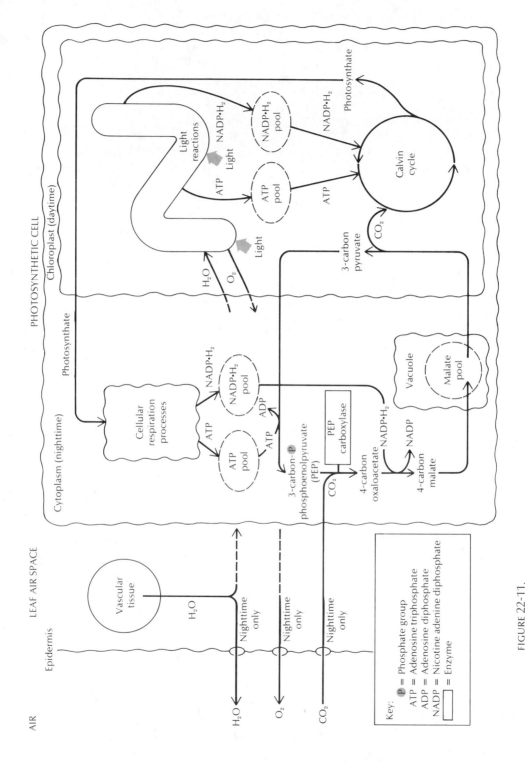

FIGURE 22-11.

Crassulacean Acid Metabolism (CAM) is a form of photosynthesis similar to C4 photosynthesis. In CAM, the initial fixation of CO_2 in a 4-carbon compound occurs at night, separated in time rather than space from other photosynthetic reactions. This allows the plant to keep the stomata closed during daylight hours, when humidities are very low, and thus to minimize water loss without severely impairing photosynthetic activity. Contrast with Figures 22-4 and 22-7.

cent of the carbon fixed in photosynthesis in peas is transported to the root nodules, where about 45 percent of this quantity is converted into nitrogen-containing compounds and 55 percent is respired to provide the energy needed for fixation. This cost, in all probability, accounts for the lack of nitrogen-fixing capability in most higher plants (Shanmugam and Valentine 1975); and it results in a direct relationship between the rates of photosynthesis and nitrogen fixation in legumes. The supply of photosynthate to the nodules is therefore the major limiting factor in the rates of fixation that occur under otherwise favorable conditions (Evans 1975). Thus breeding strategies to increase photosynthesis in legumes, or to reduce losses through respiration, should simultaneously increase biological nitrogen fixation.

A second approach with legumes involves improving the match between legume and *Rhizobium* strains. The symbiosis established in the field depends on four basic factors: the ability of *Rhizobium* strains in the soil to compete with each other and with other microorganisms; the capability of these strains to infect the legume hosts present; their capacity for inducing nodulation once they are established; and their ability to fix nitrogen once nodules have developed. All these processes involve genetically based variability (Evans 1975). The genotype of the *Rhizobium* bacterium influences all four processes; the genotype of the host influences the last three. One major line of research has therefore become centered on ways of specifying the host-bacterium relationship that will produce the most fixed nitrogen. One recent breakthrough in this area is the discovery that a group of proteinaceous compounds known as **lectins** serve as agents that bind specific *Rhizobium* cells to cells of legume roots and that this mechanism is capable of great specificity (Bohlool and Schmidt 1974).

A third approach is based on the fact that high levels of NH_4^+, the end product of fixation,

seem to inhibit the production of nitrogenase enzyme. This repression is particularly strong in some important legumes, such as soybeans, and it appears to be one of the major obstacles to yield improvements in this species. Symbiotic nitrogen fixation accounts for only about 25 to 40 percent of the total nitrogen required by soybeans (Evans 1975). This repression appears to be subject to modification by breeding, however.

Establishment of nitrogen-fixing symbioses with other kinds of crop plants, such as grains, is considered by some workers to be a long-range possibility. We have already discussed the reported association of certain bacteria of the genus *Spirillum* with the roots of tropical grasses and corn (Chapter 13). These associations are not nodule-inducing, and, because oxygen irreversibly inactivates nitrogenase, it is difficult to see how high rates of fixation could occur when good aeration conditions also prevail. Nevertheless, breeding activities might succeed in improving the capacity of this relationship.

The establishment of symbioses of *Rhizobium* or other bacteria with crop species carries with it the energy cost of nitrogen conversion. Thus the total yield of a grain might suffer somewhat, although the yield per unit of input cost might be higher due to the reduced need for synthetic nitrogen fertilizers. Recent studies of the nitrogen content of certain tropical forage grasses inoculated with *Spirillum* cultures suggest that nitrogen fixation by this bacterium might contribute significantly to increased total productivity, particularly in combination with intermediate levels of nitrogen fertilizer use (Smith et al. 1976). It could well turn out, however, that the most satisfactory approach is not to create new symbioses, but to develop legumes that can be used as winter cover crops or can be interplanted so as to carry out nitrogen fixation separate from the grain crop itself. Engineering completely new symbiotic relationships is obviously not an imminent possibility.

IMPROVING PROTEIN QUALITY

An increasingly important aim of crop breeding programs is to improve specific nutritional and chemical characteristics of harvestable crop materials (Simmonds 1973). To some degree, this aim can be achieved in other ways than breeding; for example, during the processing stage, harvested foods can receive special treatment and supplementation with materials (such as added protein or vitamins) from other sources. Breeding techniques, however, generally offer the least energy-intensive solution to problems of quality.

At present, most major-crop breeding programs are designed particularly to increase the protein content and improve the protein quality in grains and legumes (CIMMYT 1975; Milner 1975). About 70 percent of the dietary protein consumed by human beings comes from basic food grains (Harpstead 1971). As competition for other, more concentrated food proteins increases, overall reliance on this source is likely to grow as well. Protein quality relates to the amino acid content and balance of the protein present. Of the 18 amino acids required by humans, adults must obtain 10 through diet and children 11. These are termed essential amino acids because the human body is unable to synthesize them. Of the essential amino acids, grain protein is most often deficient in lysine and tryptophan, the former of which is more critical.

Grain and seed crops differ greatly in their content of protein, oils, and carbohydrates. Much of this variation can be expressed graphically by plotting the gram weight of seed produced per gram of photosynthetic glucose against the quantity of nitrogen, contained only in proteins, per gram of glucose used in seed production (Figure 22-12). Differences in the position of species on this graph reflect seed composition, because one gram of glucose produced initially in photosynthesis can be converted into about 0.83 grams of carbohy-

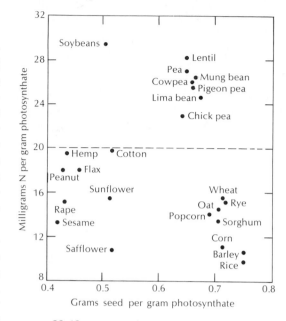

FIGURE 22-12.
Milligrams of nitrogen required and grams of seed produced per gram of photosynthetically-produced glucose for various crop species. The greater the nitrogen requirement, the higher in general is the protein content of the seed. The lower the amount of seed produced per gram of glucose, the higher in general is the oil content of the seed. The dashed line indicates the nitrogen requirement that corresponds to the maximum rate of nitrogen uptake from the soil; plants above this line must also transfer quantities of nitrogen to the seed from other plant tissues during the period of seed maturation. (Modified from Sinclair and De Wit 1975. Copyright © 1975 by the American Association for the Advancement of Science.)

drate, 0.40 grams of protein, or 0.33 grams of fat stored in the seed (Sinclair and de Wit 1975). Thus the world's staple grain crops, which are relatively low in both protein and fats, are clustered in the lower right quadrant of the figure. The major oil seeds, low also in protein, lie in the lower left quadrant. The upper right quadrant contains the principal protein-rich legumes used directly for human consumption. These legume foods are often paired with grains in the basic diets of human populations that depend on subsistence agriculture. The sole species in the upper left quadrant is the soybean, which contains high percentages of protein (40 to 45 percent) and appreciable amounts of oil (18 to 20 percent) in the seed (Dovring 1974).

The differences among these crops are obviously important to plant breeders since changing the composition of the seed clearly generates some degree of trade-off involving total yield (i.e., gross weight of harvested material per hectare). Furthermore, Sinclair and de Wit (1975) show that there is a basic difference in the ability of different plants to meet the nitrogen needs of seed formation. Assuming that healthy plants can take up about 5 kg/ha/day of nitrogen through the root system (including fixation in legumes) and can produce about 250 kg/ha/day of photosynthate, the species above the dashed line in the figure would not be able to meet their seed nitrogen needs entirely from such uptake. For them, the extra nitrogen must be withdrawn from vegetative tissues, which would lead to the senescence of these tissues as their protein content is depleted. Thus breeding for increased seed protein in species that already have high protein levels could well reduce total yields both by changing seed composition and by accelerating foliage senescence.

The above relationships, even in grains, generate what may be termed a **protein-yield threshold** (Gomez and De Datta 1975), which is simply a point at which selection for increased protein content is coupled with a decrease in total yield per acre. Because yield per acre is of such overriding importance to the farmer, this threshold becomes very important to the acceptance of a new variety.

Many breeding programs are directed toward increasing the protein content and quality of major grain species. Two somewhat different strategies are employed: the improvement of existing crop species; and the creation of new crop types.

Of the major grains, rice has the lowest protein percentage, typically 6 to 7 percent. Corn, barley, and wheat have about 11 to 14 percent protein. Protein content of grains, however, is strongly influenced by environmental conditions, and it varies widely. Selection for increased protein levels in certain of the high-yielding rice varieties appears to have the potential to raise protein content to 8 to 10 percent before reaching the protein-yield threshold (Figure 22-13). Since rice protein has a desirable amino acid balance, this is a significant improvement in quality.

In corn, the principal effort has been to improve protein quality by increasing the amount of available lysine and tryptophan. Although corn has a protein level of 10 to 11

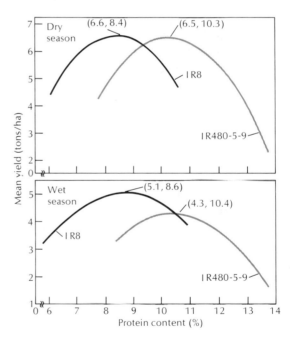

FIGURE 22-13.
Yields, in relation to protein content, of two irrigated rice varieties during wet and dry seasons in the Philippine Islands. Data are from 1968–1972 tests at the International Rice Research Institute. These relationships indicate that a protein-yield threshold exists, above which further increase in protein content can only be achieved at the expense of reduced yields per unit area. (Modified from Gomez and De Datta 1975.)

percent, about 50 percent of the protein present is in the form of zein, a substance that is not readily digested by nonruminants. The available protein fraction includes about 2 percent lysine and only a negligible amount of tryptophan. The poor quality and small amount of available protein in corn are major factors in protein deficiency malnutrition, or kwashiorkor, in regions where the population depends heavily on corn.

In the early 1960s, two mutant characteristics, termed opaque and floury, were discovered as factors that influence the lysine content of the corn kernel. These mutants reduce the production of zein and increase the remaining available fraction of protein. The term **opaque** refers to the fact that the endosperm of the kernel looks dull and faded instead of bright and translucent as it is in most corn. The **floury** mutant has a similar appearance. The allele **opaque-2** has been most widely used in corn breeding and has been found to produce about a 60 percent increase in the quantity of lysine present (Harpstead 1971), raising it from about 2 grams per 100 grams protein to about 3.4 grams per 100 grams protein.

The opaque-2 allele has a number of major shortcomings, however. The grain yield by weight is about 10 to 15 percent lower than that of traditional varieties, mainly because the endosperm of the kernel has a floury, loosely packed arrangement of starch grains. The moisture content is also somewhat higher, making post-harvest drying more critical, and susceptibility to certain insect pests and seed rots is somewhat higher. Finally, when used for human food, this variety tends to mill to a fine flour that does not behave in food preparation in the same way that traditional corn does; this characteristic requires cultural modification by the human populations using the grain. Nevertheless, current studies of the opaque-2 gene system suggest that some of the yield and kernel characteristics could be

restored to the "normal" condition by contined breeding activities. High-lysine corn is currently receiving considerable interest in the United States because of its higher value as livestock feed.

Grain sorghum also has a serious lysine inadequacy in its protein spectrum. The improvement of sorghum has great potential significance, because it is one of the most suitable grains for the dry, rocky soils of many of the semiarid regions of the world (Shapley 1973). In Africa, nearly 300 million persons rely on sorghum as their principal grain food. Genes for improved lysine levels in sorghum have recently been found in Ethiopia, but this genetic system is likely to have certain weaknesses like those of the opaque-2 gene for corn. Incorporating such a genetic system into the many local sorghum varieties, and maintaining it under local conditions of seed propagation, will both be extremely difficult in areas where sorghum is an important human food.

Still another crop in which high lysine varieties have been sought is barley (Munck et al. 1971). By screening material from the USDA Small Grains Collection, which contains some 2500 collected samples, a variety known as Hiproly was found to have both total protein and lysine levels well above those found in standard varieties. The barley collection involved also came from Ethiopia, one of the world regions that has an extraordinary diversity of small grain varieties, a fact that underlines the importance of genetic diversity and genetic banks (Chapter 21).

In the case of wheat, many qualitative characteristics are of major importance, several of which are related to this grain's major role in bread making. As we shall see below, one of the most important developments in this crop has been its use in an intergeneric cross with rye, designed to create a new grain species that combines the best characteristics of both grains.

HYBRIDIZATION AND CROP IMPROVEMENT

As a technique of crop breeding, the term **hybridization** has two somewhat different meanings. In the commercial sense, it refers to the process of producing seed material by crossing inbred lines of different varieties of a single species. To the plant breeder searching for quantum improvements in crop characteristics, it can also mean crossing species that are much more distantly related, in an attempt to obtain dramatic new combinations of characteristics.

The production of hybrids in the commercial sense involves two steps: the development of inbred or homozygous populations of different varieties that have desirable properties and favorable "combining" ability; and the subsequent crossing of these lines to give hybrid seed. Favorable combining ability simply means that the cross results in plants showing **hybrid vigor** or **heterosis,** a condition in which the expression of important characteristics, such as yield, is greater in the hybrid than in either parent. Heterosis is the result of combining, in the F_1 hybrid, contrasting but complementary dominant characteristics that relate to germination, seedling vigor, photosynthesis, dry matter accumulation, flowering, seed set, and other important processes (Sinha and Khanna 1975). Homozygous, or inbred, lines can be obtained in various ways, depending on the reproductive system of the plant. Plants that are self-pollinating are automatically inbred. For those that are open-pollinated, but that have flowers containing both stamens and pistil, simply enclosing the flower in a bag excludes all pollen except that from stamens of the flower itself. Still other techniques, including those of tissue culture, might be needed in other cases.

Once a number of inbred lines are obtained, various crosses can be made. Not all resulting hybrids will be of superior performance, of course. It is necessary to screen the various possible crosses and to select those with desirable patterns of hybrid vigor. Once the crosses that yield desirable hybrids are found, techniques for mass production of hybrid seed must be developed. The production of hybrid seed in this manner is now a standard procedure for many crop species, including grains, forage crops, vegetables, fruits, and fiber plants.

Corn furnishes the classic example of yield improvement through the development of hybrid seed. The history of hybrid corn is a long one (Mangelsdorf 1951); the basic idea of hybrid seed production in corn was first proposed in 1909. The feasibility of growing hybrid corn on a large scale was not established until 1918, however, when it was discovered that a double-step hybridization, involving four inbred lines, could provide simultaneously both a high level of hybrid vigor and large quantities of seed (Figure 22-14). Initially, since corn is open-pollinated by wind-borne pollen, the female lines in each cross had to be detasseled by hand to prevent self-pollination. With the discovery of cytoplasmic male sterile genes (see Chapter 21), the seed production process was facilitated, as shown in Figure 22-15. Although this last system led to some unfortunate problems with leaf blight susceptibility, it has otherwise proved a very useful improvement in corn seed production.

Hybrid seed corn was first introduced in the 1920s. During the 1930s and 1940s virtually all of the United States corn crop shifted to reliance upon hybrid seed. The advantage of hybrid seed is about 20 to 30 percent over the yields of the varieties used in creating the hybrid. Thus the increase in yield per hectare from 54 bushels in the early 1930s to over 198 bushels (average) in 1970 is only partially due to the introduction of hybrid seed.

Despite the early development of theory and practice of hybrid seed production in corn,

554

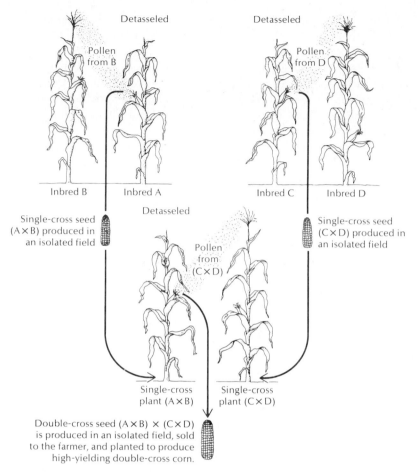

Detasseled   Detasseled

Pollen from B

Pollen from D

Inbred B   Inbred A   Inbred C   Inbred D

Single-cross seed (A×B) produced in an isolated field

Detasseled

Pollen from (C×D)

Single-cross seed (C×D) produced in an isolated field

Single-cross plant (A×B)   Single-cross plant (C×D)

Double-cross seed (A×B) × (C×D) is produced in an isolated field, sold to the farmer, and planted to produce high-yielding double-cross corn.

FIGURE 22-14.
The detasseling system of hybrid corn production. This procedure requires the manual removal of tassels (male inflorescences) from the intended female parents of the cross so that cross-pollination will occur. (From *Plant Science*, second edition, by Jules Janick et al. W. H. Freeman and Company. Copyright © 1974.)

the development of comparable systems for other major grain crops has been slow. Corn was perhaps an ideal grain for the development of such a system, since the male and female inflorescences are separate, large, and easy to handle. For other small grains, male and female flower parts occur together in tiny florets, and finding ways to prevent self-pollina-

tion and to promote cross-pollination have been major limiting factors.

Recent discoveries of male sterility systems that compare to those used in modern corn breeding have been made for both barley (Wiebe and Ramage 1971) and wheat (Livers and Heyne 1968). Hybrid barley is still in the development stage, but hybrid wheat is now

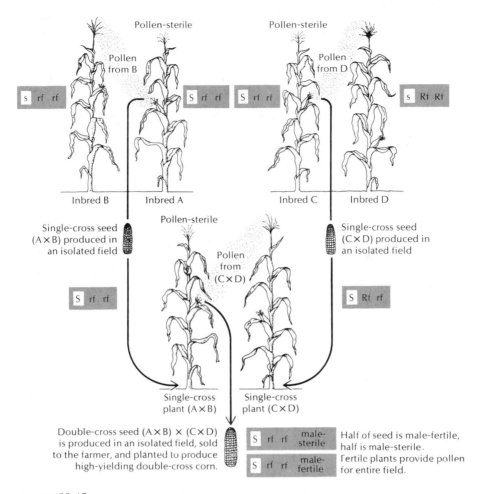

FIGURE 22-15.
The cytoplasmic male-sterile system of hybrid corn production. This system utilizes a cytoplasmic gene for sterility (S = dominant sterile, s = recessive fertile) and fertility restoring genes (Rf = dominant fertile, rf = recessive sterile. (From *Plant Science*, second edition, by Jules Janick et al. W. H. Freeman and Company. Copyright © 1974.)

available commercially. The varieties for which hybrid seed has been developed are hard red winter wheats, where the hybrid advantage is an increase in production of about 32 percent (Table 22-2). Hybrid seed was first released to growers in 1972, and several varieties suited for areas of the Great Plains from Texas to Montana are just now becoming available in quantity. The cost of hybrid seed is roughly four times that of regular wheat (Deterling 1975).

The second approach to hybridization is what can be termed **wide hybridization,** or the crossing of distantly related species. The most intensive program directed at developing a new crop species through wide hybridization

TABLE 22-2
Yields of hybrid wheats compared with those of parental varieties in tests over a four-year period in Kansas.

Average yield, hybrids (bu/hectare)	92.2
Average yield, parental varieties	69.9
Difference (bu/hectare)	22.3
Percent hybrid advantage	32
Yield of best hybrid (bu/hectare)	105.8
Yield of best parental variety (bu/hectare)	80.6
Difference (bu/hectare)	25.2
Percent hybrid advantage	31

Source: Modified from Livers and Heyne 1968.

involves **triticale,** a grain created by the hybridization of forms of wheat, *Triticum* spp., with forms of rye, *Secale* spp. This case illustrates many of the techniques required for wide hybridization, as well as the opportunities and problems involved.

Wheat and rye possess several characteristics that would seem desirable in combination. Rye is quite winter-hardy, does well in poor soils, and has a protein fraction rich in lysine, although its total protein content is low. Wheat has relatively high overall protein levels, high total grain yield, and many desirable food qualities. Thus considerable opportunity for crop improvement appears to exist.

Hybrids between members of the two genera have been known for over 100 years (Zillinsky 1974) but until the 1930s little interest existed in these forms, and little progress in triticale breeding was achieved. With the advent of the colchicine technique of chromosome doubling, however, the ability to create fertile hybrids of these two plants suddenly appeared.

As noted earlier (Chapter 4), wheats have chromosome numbers that are multiples of 7; the durum, or macaroni, wheats have haploid chromosome numbers of 14, the bread wheats haploid numbers of 21. Members of the rye genus *Secale* possess haploid chromosome numbers of 7. The cells of a hybrid embryo between wheat and rye will thus possess either 21 or 28 chromosomes, depending on what type of wheat parent it has. Generally, this hybrid will be infertile, because when meiosis occurs in this individual, pairing between chromosomes cannot occur. However, if the early embryo or seedling of the hybrid is treated with colchicine, which disrupts the spindle in cells undergoing mitosis, the chromosome number per cell may be doubled. In this doubled condition, with 42 or 56 chromosomes, pairing at meiosis can occur, and fertility is thus restored to the hybrid. The colchicine technique has made it possible to create hybrids and to establish their fertility at will (Figure 22-16).

In the 1950s, intensive breeding programs with triticale began in several locations, including the University of Manitoba and the Center for Wheat and Maize Improvement in Mexico (Hulse and Spurgeon 1974). These activities followed the general crossing strategy outlined in Figure 22-17. The 42-chromosome triticales, created from crosses of durum wheats and rye, proved to be the most satisfactory for further breeding work. Over a period of 10 to 15 years this breeding program gradually perfected triticale varieties that gave yields comparable to those of wheat in at least some situations. In 1970 the first commercial varieties were placed on the general market.

At least a portion of the expectations regarding triticale appear to have been realized. The species does well under certain environmental conditions, including high elevations and cooler regions. The lysine content of triticale is, in fact, higher than that of wheat, although lower than that of rye (Hulse and Spurgeon 1974), and total protein content is equal to or slightly above that of wheat. It appears now that the basic genetic format of triticale has been established and that straightforward breeding programs will be able to produce continuous

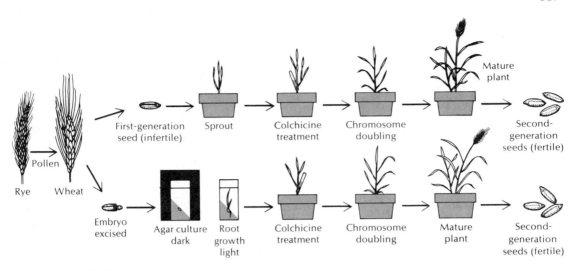

FIGURE 22-16.
The technique of using colchicine to produce fertile triticale plants from hybridizations of wheats ($n = 14$ or 21) and rye ($n = 7$). The seeds and seedlings of the hybrids, which contain one set of chromosomes from each parent ($14 + 7$ or $21 + 7$) produce sterile adults because chromosome pairing and subsequent meiotic divisions cannot occur normally. Colchicine treatment of young seedlings leads to chromosome doubling due to interference with the formation of mitotic spindles in cell division. Adult tissues can therefore have 42 or 56 chromosomes ($14 + 14 + 7 + 7$ or $21 + 21 + 7 + 7$) in sets that can pair and allow meiosis to occur, thus permitting fertile seeds to be produced. (Modified from "Triticale" by Joseph H. Hulse and David Spurgeon. Copyright © 1974 by Scientific American, Inc. All rights reserved.)

improvements in yields, nutritional quality, resistance to grain pests and diseases, and other important features.

We can see from the foregoing that the success of this intergeneric cross rests in large measure on the fact that cross-fertilization and the development of hybrid individuals occur and that colchicine treatment can restore fertility to these hybrids. The former is not accomplished as easily with all plants; basic barriers to crossing exist between many varieties and species (Bates and Deyoe 1973). These can be grouped in two categories: gametic incompatibility and hybrid breakdown. **Gametic incompatibility** includes any mechanism that prevents pollen grains from germinating on the stigma, growing through the style tissue, or fertilizing the egg nuclei. Mechanisms that operate in this fashion appear to be similar to

the immune response in animals—for example, the biochemical incompatibility that interferes with the success of organ transplants. **Hybrid breakdown** involves the inability of the hybrid genetic system that results from fertilization to guide development. At present, one of the major areas of research interest is the effort to overcome gametic incompatibility and thus, it is hoped, to make wide hybridization possible in new situations.

DOMESTIC ANIMAL IMPROVEMENT

The modification of production systems for domestic animals is likely to occur rapidly in the near future in both the developed and developing countries. The economics of intensive production technologies will force

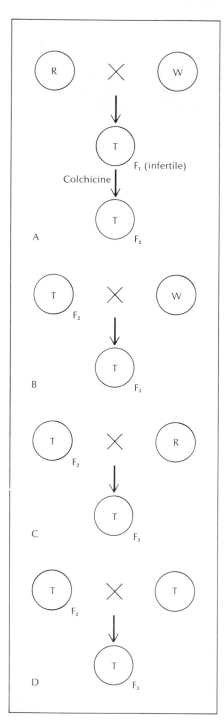

FIGURE 22-17.
Breeding of triticale (T) begins (A) with a cross between wheat (W) and rye (R). The first-generation triticale plant (F_1) is made fertile with colchicine treatment, and the resulting seeds become the second generation (F_2). This generation and subsequent ones can be backcrossed to either parent or crossed with entirely different varieties of wheat or rye (B, C). Triticale-triticale crosses are also made (D). A primary triticale is one that is obtained from a wheat and rye cross or from crossing two triticales, each obtained from the same wheat and rye forms. A secondary triticale is the result of crossing different varieties of triticale or of crossing a triticale with a wheat or a rye that was not a parental stock. (From "Triticale" by Joseph H. Hulse and David Spurgeon. Copyright © 1974 by Scientific American, Inc. All rights reserved.)

change in the industrialized nations; the severe protein deficiencies in developing areas (see Chapter 1) will at the least create the need for improvement.

Like plant breeding, animal breeding represents the genetic adjustment of a species to the physical, biological, technological, economic, and social features of the production environment (Donald 1973). Climate and the existence of pests and diseases, for example, influence the breeding strategies to be undertaken. But the nature of the livestock production system must also be considered. With cattle, for example, four production systems can be distinguished: pastoralism, permanent diversified farming, intensive beef ranching, intensive dairying. The breeding strategy must also be geared to the system in which the animals will be used (Rendel 1974). For cattle, it is clear that most efforts have been directed toward improving the last two of the above systems.

In animal production systems, certain features are of major importance to efficiency that are relatively minor in annual crops. With animals, the costs of maintaining basic breeding stock are major. This means that efficiency must be considered for entire **production units** instead of on the number of individuals that are actually harvested for meat or that yield other useful products. The smallest production unit in livestock systems consists of a sire, a dam, and their progeny (Cartwright 1974). Furthermore, an evaluation of efficiency must take into account the total costs and yields during the entire production cycle, including the full cost of livestock feed. Otherwise, an inefficiency that is merely shifted from one portion of the production system to another might seem to be an efficiency gained. Improved feed conversion and increased growth rates in livestock "factory farming" systems in large measure reflect the use of more refined and digestible feed formulations, which means that the less digestible components are eliminated and the subsequent efficiency of their use not considered. Selection for rapid growth rates in feeder cattle, which improves efficiency at the feed lot, also creates larger individuals, and thus greater maintenance costs, in the adult breeding herds.

Keeping these points in mind, breeding programs could nevertheless increase efficiency in several general areas: reproduction; productive longevity; growth rate; feed conversion rate; and disease resistance (Wilson 1973). For each of these categories we may recognize a sort of "biological ceiling" toward which improvements can reasonably aspire.

Reproductive efficiency simply means getting the most out of the animals maintained as breeding stock. The more new animals, or other products, that can be obtained from each individual that is supported for breeding purposes, the greater is the biological efficiency of the system. Artificial insemination has been one way of improving reproductive efficiency by reducing the number of males that must be supported for reproduction alone. This practice, now standard with cattle, is likely to be extended to other livestock. Female reproductive efficiencies might also be improved (Table 22-3). Increase in number of young per litter seems possible for all livestock, and an increase in number of litters per year seems possible for sheep and swine. In the near future, major increases in reproductive efficiency could be gained for these latter two species, and at least a slight improvement achieved for cattle. Eventually, however, twinning might make

TABLE 22-3
Current standards, biological ceilings, and economically probable levels achievable by 1985 for reproductivity of major livestock species.

	Cow	Ewe	Sow	Hen
Current standard				
(1) Offspring per litter	1.0	1.5	9	—
(2) Offspring per year	0.9	1.5	18	220
Biological ceiling				
(3) Offspring per litter	2.0	5.0	20	—
(4) Offspring per year	2.2	10.0	44	365
(2) As percentage of (4)	41	15	41	60
Improvement estimate—1985				
(5) Offspring per litter	1.2	3.0	14	—
(6) Offspring per year	1.2	4.5	31	300
(6) As percentage of (4)	55	45	70	82

Source: Modified from Wilson 1973.

TABLE 22-4
Current standards, biological ceilings, and economically probable levels achievable by 1985 for growth rates (live weight gain in grams per day) of selected livestock species.

	Broiler	Lamb	Steer	Hog
(1) Current standard	32	114	1140	636
(2) Biological ceiling	46	456	2500	912
(3) Improvement estimate—1985	36	184	1590	773
(1) As percentage of (2)	70	25	46	70
(3) As percentage of (2)	78	40	64	85

Source: Modified from Wilson 1973.

greater improvements possible in beef cattle production.

Productive longevity refers simply to the length of the period of adult life during which the individual produces offspring, milk, eggs, or other products. The longer this period, the smaller are the average costs of supporting the individual's preproductive life. For example, in dairy cattle, if the current adult productive life of about six years could be increased to 10 years, considerably fewer heifers (young, non-milking animals) would need to be maintained through two years of age to replace the adults going out of production. In other animals, it might be possible to extend the productive life through earlier maturity.

Improved growth rates seem possible for most species, but particularly for lambs (Table 22-4). In livestock, however, increasing the growth rate of young animals can increase adult size and thus the cost of maintaining the adult breeding herd. This is particularly significant for cattle, in which these two characteristics are strongly correlated, and in which single births are the norm. A number of breeding strategies are possible to circumvent this dilemma, such as crossing large sires with females from a line characterized by relatively small size but high milk production. In theory, this should give calves with moderately high growth potential

from mothers well equipped to nourish their early growth.

The conversion of feed into animal tissue depends on several factors, including feed digestibility, body maintenance requirements (resting metabolism), and the activity of the animal beyond resting levels. The digestion processes of ruminants, such as cattle and sheep, differ strikingly from those of other domestic animals. A ruminant's stomach has several parts. In one compartment, the primary activity is fermentation of plant materials by bacteria. This assists with the digestion of this material, but also leads to the loss of about 8 percent of the total food intake in the form of methane gas (Byerly 1967). In addition, much of the food ingested is not absorbed, being passed out as feces. As a result, of the food they eat, ruminants digest and absorb from about 40 percent for materials such as wheat straw to 80 percent for corn. Other animals achieve about the same digestive efficiency for foods such as corn, but they do not digest roughage foods as well.

A portion of the food materials digested and absorbed is converted to new tissue, including flesh, milk, eggs, wool, or other materials used by man. The rest is used in various metabolic processes. The expenditures for basic maintenance, or resting metabolism, are roughly proportional to the weight of the animal raised to

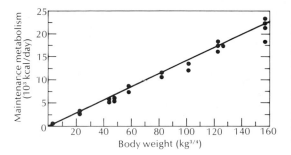

FIGURE 22-18.
Maintenance metabolism of domestic animals ranging in size from chickens to beef cattle is roughly proportional to the body weight raised to the three-quarters power. Data are from several sources for chickens, sheep, swine, dairy cattle, and beef cattle. (Modified from Byerly 1967. Copyright © 1967 by the American Association for the Advancement of Science.)

the ¾ power (Figure 22-18). Additional metabolic expenditures are exacted when animals must move around actively to find food, generate metabolic heat to keep warm, or do work. All of these processes reduce feed conversion efficiency.

The feed conversion efficiencies for various livestock species and production activites are given in Table 22-5. These efficiencies represent those observed under the best practice (Wilson 1973) and of course do not reflect efficiency for overall production units. Many apparent improvements in these efficiencies have come out of improvements in the nutritive value of the food used, rather than in the animal's ability to process food materials. For chickens, however, a marked improvement in feed conversion efficiency has occurred in recent years (Figure 22-19); but the extent to which feed conversion efficiency can be improved in other species is somewhat uncertain.

As in crop improvement, one of the increasingly important techniques in animal improvement is crossbreeding inbred lines. For characteristics with a strong degree of genetic determination—such as milk butterfat levels in dairy cattle—little benefit results from crossbreeding; hybrids tend to possess intermediate values. For cattle, crossbred lines show about 5 percent heterosis, or hybrid vigor, in characteristics such as total milk yield and body

TABLE 22 5
Current efficiencies of crude protein conversion achieved under good practice for specific livestock classes.

Livestock class	Annual production level	Protein consumed (kg)	Protein produced (kg)	Conversion efficiency (%)
Cow	6800 l milk per individual	586	222	38
Hen	300 eggs per hen	7.3	2.3	31
Broiler	1.8 kg gross carcass wt	3.6	1.1	31
Fish	carp in 0.4-ha warm pond	3400	680	20
Rabbit	4 litters of 10	48	8	17
Hog	2¼ litters of 12	840	123	15
Ewe	2 litters of 3	125	11	9
Steer	295 kg gross wt increase	568	34	6

Source: Modified from Wilson 1973.

562

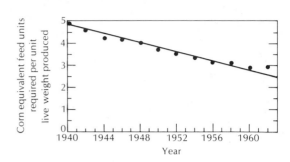

FIGURE 22-19.
Trend in food conversion efficiency in broiler production from 1940 through 1967. Data are expressed in corn equivalent feed units per unit weight of broiler produced. Note that the improvement in this relationship appears to have slowed toward the end of this period. (From Byerly 1967. Copyright © 1967 by the American Association for the Advancement of Science.)

size, and even greater degrees of heterosis in viability and fertility (Donald 1973). Characteristics in which heterosis is appreciable tend to be those that are less rigidly determined by genetic mechanisms and that are therefore more variable even in inbred lines.

Incorporating crossbreeding in livestock production systems obviously means that two or more pure lines must be maintained to provide the parents for crossbreeding. Thus the advantages of heterosis must generally be substantial to make the system profitable for cattle and other livestock with low reproductive potential.

APPLIED PRODUCTION EFFICIENCY

Plant and animal breeding activities have sometimes been pursued without close attention to the needs and constraints of most farmers. Crop yields and the efficiency with which feed materials are converted into animal products can be raised to record levels by applying enough technology. On demonstration plots soil structure, water and nutrient levels, and pest control can be adjusted to create the best possible conditions for a crop variety selected for its high response to such conditions. In intensive animal "feeding factories" environmental conditions and diet can be controlled closely, with minimal extra metabolic demands placed on individuals. It is rarely possible to achieve such control on the average farm, however, and varieties of crops and domestic animals that do well under optimal conditions might perform below average under average farm conditions.

Breeders have recently placed more and more emphasis on the responsiveness of crops and livestock to purchased inputs and to energy-intensive technology. This has been true not only for species intended for use in the production systems of industrialized nations, but also for those promoted in underdeveloped areas. The dwarf rice and wheat varieties that combine rapid maturation with high yields are examples of this approach. Because they are short-stemmed plants, they require more effective protection against weed competitors. Their higher yield and rapid growth depend on high nutrient levels and thus on the increased use of chemical fertilizers. Early maturity, to be taken advantage of, requires quick harvesting by machine and perhaps supplementary drying of the grain. Where these inputs can be supplied, there are advantages to be gained.

Now, however, it is clear that breeding activities in both developed and developing areas must attempt to maximize net productivity, the difference between output and input. Furthermore, emphasis must be placed on using inputs to one production sector that compete as little as possible with those required by other sectors. Efficiency and breeding improvement must be considered more and more critically in relation to the entire food production systems that serve regional human populations.

SUMMARY

Improved performance, in ecological terms, is an increase in the yield of crops or animal products per unit of energy or resources invested in production. The sources of increased yield are either recurring and subject to continuous improvement or nonrecurring, giving a one-time increase.

Crop improvement basically involves increasing both the net production of the crop and the fraction of this production that is stored in harvestable parts. One important strategy is to increase the rate of crop production under given conditions; a second, to improve the adjustment of crop life cycles to seasonal conditions. The existence of modified systems of photosynthesis and respiration in the plants of hot and arid environments offers a potential for developing new crops with more efficient patterns of net photosynthesis and possibly for incorporating improvements in the photosynthetic and respiratory systems of existing species. Studies of symbiotic nitrogen-fixing microorganisms suggest that the function of these symbionts might be enhanced and symbioses established with nonlegume species. The nutritional quality of basic foods, particularly the total protein and lysine content, is also subject to potential improvement.

Hybridization of inbred crop lines, giving offspring with hybrid vigor, or heterosis, is one major technique that it might be possible to extend to new crops. Wide hybridization, or the crossing of more distantly related species, offers the potential of creating new crop forms with radically new combinations of characteristics.

Improvements in domestic animal production must be attempted and evaluated in terms of the performance of total production units rather than that of selected individuals. The efficiency of such units can be increased by improving reproductivity, longevity, feed conversion rate, growth rate, and resistance to disease. Crossbreeding inbred lines to gain hybrid vigor is an important technique, as it is in crop breeding.

In time, increased efficiency must come to be viewed according to the way in which energy and other resources are utilized in entire crop and animal production systems, with careful consideration being given to yields per unit input of energy and other productive resources.

Literature Cited

Bates, L. S., and C. W. Deyoe. 1973. Wide hybridization and cereal improvement. *Econ. Bot.* 27:401–412.

Bjorkman, O., and J. Berry. 1973. High-efficiency photosynthesis. *Sci. Amer.* 229(4):80–93.

Bohlool, B. B., and E. L. Schmidt. 1974. Lectins: A possible basis for specificity in the *Rhizobium*-legume root nodule symbiosis. *Science* 185:269–271.

Brown, L. R. 1967. The world outlook for conventional agriculture. *Science* 158:604–611.

Brown, W. L. 1975. Worldwide seed industry experience with opaque-2 maize. In *High-Quality Protein Maize.* Stroudsburg, Pa.: Dowden, Hutchinson and Ross. pp. 256–264.

Byerly, T. C. 1967. Efficiency of feed conversion. *Science* 157:890–895.

Cartwright, T. C. 1974. Net effects of genetic variability on beef production systems. *Genetics* 78:541–561.

Chollet, R., and W. L. Ogren. 1975. Regulation of photorespiration in C_3 and C_4 species. *Bot. Rev.* 41:137–179.

CIMMYT-Purdue. 1975. *High-Quality Protein Maize.* Stroudsburg, Pa.: Dowden, Hutchinson and Ross.

Deterling, D. 1975. Hybrid wheat: What it can mean to you. *Prog. Farmer* 90(9):34, 50.

Donald, H. P. 1973. Animal breeding: Contributions to the efficiency of livestock production. *Phil. Trans. Roy Soc. London.* (B) 267:131–144.

Dovring, F. 1974. Soybeans. *Sci. Amer.* 230(2):14–21.

Dudley, J. W., D. E. Alexander, and R. J. Lambert. 1975. Genetic improvement of modified protein maize. In *High-Quality Protein Maize.* pp. 120–135. Stroudsburg, Pa.: Dowden, Hutchinson and Ross.

Evans, H. J. 1975. *Enhancing Biological Nitrogen Fixation.* Washington, D.C.: National Science Foundation, Div. Biol. Med. Sci.

Evans, L. T. 1975. *Crop Physiology—Some Case Histories.* London: Cambridge University Press.

Gomez, K. A., and S. K. De Datta. 1975. Influences of environment on protein content of rice. *Agron. J.* 67:565–568.

Hardy, R. W. F., and U. D. Havelka. 1975. Nitrogen fixation research: A key to world food? *Science* 188:633–643.

Harpstead, D. D. 1971. High-lysine corn. *Sci. Amer.* 225(2):34–42.

Hulse, J. H., and D. Spurgeon. 1974. Triticale. *Sci. Amer.* 231(2):72–80.

Ivins, J. D. 1973. Crop physiology and nutrition. *Phil. Trans. Roy. Soc. London* (B)267:81–91.

Jennings, P. R. 1974. Rice breeding and world food production. *Science* 186:1085–1088.

Livers, R. W., and E. G. Heyne. 1968. Hybrid vigor in hard red winter wheat. In *Proceedings of the Third International Wheat Genetics Symposium.* K. W. Finlay and K. W. Shepherd (eds.) pp. 431–438. New York: Plenum Press.

Mangelsdorf, P. C. 1951. Hybrid corn. *Sci. Amer.* 185(2):39–47.

Milner, M. (ed.) 1975. *Nutritional Improvement of Food Legumes by Breeding.* New York: Wiley-Interscience.

Minchin, F. R., and J. S. Pate. 1973. The carbon balance of a legume and the functional economy of its root nodules. *J. Expt. Bot.* 24:259–271.

Munck, L., K. E. Karlsson, and A. Hagberg. 1971. Selection and characterization of a high-protein, high-lysine variety from the world barley collection. In *Proceedings of Second International Barley Genetics Symposium.* A. Nilan (ed.). pp. 544–558. Pullman: Washington State University Press.

National Research Council. 1975. *Enhancement of Food Production for the United States.* Washington, D.C.: National Academy of Sciences.

Rendel, J. 1974. The role of breeding and genetics in animal production improvement in the developing countries. *Genetics* 78:563–575.

Shanmugam, K. T., and R. C. Valentine. 1975. Molecular biology of nitrogen fixation. *Science* 187:919–924.

Shapley, D. 1973. Sorghum: "Miracle" grain for the world protein shortage? *Science* 182:147–148.

Simmonds, N. W. 1973. Plant breeding. *Phil. Trans. Roy. Soc. London* (B) 267:145–156.

Sinclair, T. R., and C. T. de Wit. 1975. Photosynthate and nitrogen requirements for seed production by various crops. *Science* 189:565–567.

Sinha, S. K., and R. Khanna. 1975. Physiological, biochemical, and genetic basis of heterosis. *Adv. in Agron.* 27:123–174.

Singh, J., and V. L. Asnani. 1975. Present status and future prospects for breeding for better protein quality in maize through opaque-2. In *High-Quality Protein Maize.* pp. 86-101. Stroudsburg, Pa.: Dowden, Hutchinson and Ross.

Smith, R. L., J. H. Bouton, S. C. Schank, K. H. Quesenberry, M. E. Tylor, J. R. Milam, M. H. Gaskins, and R. C. Littell. 1976. Nitrogen fixation in grasses innoculated with *Spirillum lipoferum. Science* 193:1003–1005.

Ting, I. P. 1973. Comparative carbon metabolism in plants. *What's New in Plant Phys.* 5(3):1–4.

Trenbath, B. R. 1974. Biomass productivity of mixtures. *Adv. in Agron.* 26:177–210.

Weibe, G. A., and R. T. Ramage. 1971. Hybrid barley. In *Proceedings of second international barley genetics symposium.* R. A. Nilan (ed.). pp. 287–291. Pullman: Washington State University Press.

Wilson, P. N. 1973. Livestock physiology and nutrition. *Phil. Trans. Roy. Soc. London* (B) 267:101–112.

Zelitch, I. 1971. *Photosynthesis, Photorespiration, and Plant Productivity.* New York: Academic Press.

Zillinsky, F. J. 1974. The development of triticale. *Adv. in Agron.* 26:315–348.

23

FOOD HARVEST AND PRODUCTION
IN AQUATIC SYSTEMS

Human food production activities form an interconnected network of relationships with all the world's major ecosystems. Aquatic ecosystems, both freshwater and marine, are no exception. Fisheries harvests have become an important source of protein for both humans and domestic animals. So enmeshed are these harvests in the total food production system that the sudden decline in the production of anchoveta off the Peruvian coast reverberates through soybean and livestock production in the American Midwest and even influences the basic plant food composition of the Japanese diet. In examining agricultural ecology, we cannot avoid considering the role and potential of these systems as sources of human food.

Moreover, because the potential for expanding arable lands suitable for conventional agriculture is very limited, considerable interest has arisen in attempting deliberately to farm or husband aquatic systems. Many aquatic systems—such as marshes, ponds, and mangrove swamps—that can be converted for traditional agriculture only at great expense are among the most naturally productive ecosystems that exist. Furthermore, some land areas that might not be reclaimable as farmland could be con-

verted at reasonable cost into aquatic systems. Aquatic agriculture thus represents a technological frontier whose potential is not yet completely known.

Marine ecosystems cover some 361 million square kilometers, nearly 71 percent of the total surface area of the earth. Freshwater systems, both lakes and streams, cover about 2 million square kilometers, which is about 0.4 percent of the total surface of the earth, or 1.4 percent of the continental surface. When we consider that the arable areas of the continents amount to only 10 to 11 percent of their area and nevertheless supply most of the food for the world's population, the development of these fresh water systems is clearly worth considering.

Aquatic ecosystems share a number of basic structural and functional characteristics with terrestrial ecosystems, but they also differ in certain important ways (Odum 1962). Both kinds of ecosystems possess the same major groups of organisms, plant producers, animal consumers, and microorganism decomposers, although the species and their patterns of adaptation to their habitat are obviously different. Both kinds of ecosystem are limited in productivity by light energy and by the availability of particular nutrients such as phosphorus and nitrogen. Both have two strata: an upper stratum where green plants and their capture of light energy by photosynthesis predominate; and a lower stratum, the soil or deep water and sediments, in which secondary production and decomposition are the principal activities.

Aquatic systems are different from terrestrial systems in major ways, which are ultimately related to the density of the medium and to the frequent absence of surfaces to which both plants and other organisms can attach themselves. Along shorelines and in shallow waters attached plants can occur: encrusting algae; macrophytic algae such as the kelps and many others; and submerged, floating, or emergent higher plants. A few forms, such as the duckweeds and the tiny aquatic fern *Azolla*, live free-floating on the surface of quiet waters. Although all these can be important in particular situations, by far the greatest portion of the primary production in aquatic systems is carried out by microscopic or submicroscopic phytoplankton living in the surface waters. Because they live in water, a relatively dense medium, they do not need the extensive supporting tissues of terrestrial plants. Large size and extensive differentiation of specialized parts are also less of an advantage; in an unstable medium that lacks attachment surfaces such characteristics can actually be disadvantageous. Thus the plant biomass in open water systems is small and inconspicuous, but that which is present is highly active photosynthetically. Aquatic systems often have a plant biomass smaller than that of the animal consumers present (Figure 23-1) but capable of a higher net primary production per unit biomass than is typical in terrestrial systems.

As a result, aquatic food chains are different from terrestrial ones. Small producers are fed upon by small consumers, the zooplankton, which include the copepods, cladocerans, rotifers, and other small animals able to filter algal cells or colonies from the water. These are fed on in turn by slightly larger animals. Finally, still larger fish, aquatic mammals, and other large species appear in the food chain—species that are both existing and potential sources of food for human beings. In contrast to the terrestrial environment, where man generally harvests herbivores as food animals, the aquatic animals of harvestable nature are usually two or more links further along the food chain. As we shall see, this severely constrains the exploitation of aquatic productivity. Nevertheless, man's harvest of food from aquatic systems is quantitatively important. To appreciate the relative extent of this harvest, and to assess its future potential, we must first look more closely at the productivity of aquatic ecosystems.

PRODUCTIVITY IN AQUATIC ECOSYSTEMS

Not too long ago, the prevailing assumption was that the basic productivity of the oceans per unit area was about equal to that of the land. With the development of quantitative methods for measuring primary production, and with the steady accumulation of production measurements from different parts of the world's oceans, it is now clear that this is far from the case (Morris 1974; Ricker 1969; Ryther 1969). Marine ecosystems in general appear to be more severely limited by nutrient shortages than terrestrial ecosystems, and high levels of primary production occur only in situations in which special factors increase nutrient availability (Table 23-1).

The most productive marine systems, reefs and estuaries, have productivities that are roughly equal to those of the best terrestrial systems, such as humid tropical forests. These systems act as nutrient traps, however. Coral reefs, located in nutrient-poor tropical seas, capture nutrients from the water that flows across the reef through the feeding of reef animals upon pelagic organisms carried in the

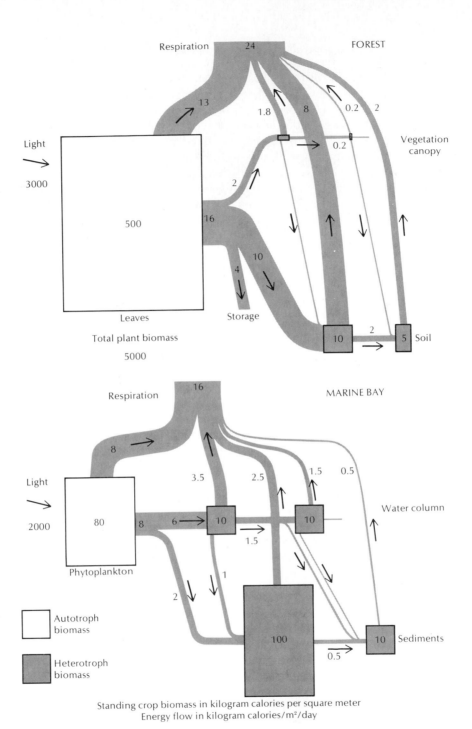

FIGURE 23-1.
Biomass and energy flow relationships for a temperate forest and a marine bay. Note that the aquatic ecosystem has a small biomass of producers (phytoplankton), but that net primary production per unit of this biomass is much higher than for the forest. (Modified from Odum 1962.)

TABLE 23-1
Area and primary production level for various marine and freshwater ecosystems.

	Area	Net primary production		Total for area	
	(10^6 km^2)	(g/m^2/yr)	(kcal/m^2/yr)	10^9 t	10^{15} kcal
Oceanic					
Reefs and estuaries	2.0	2000	9000	4.0	18.0
Upwellings	0.4	500	2500	0.2	1.0
Continental shelf	26.0	350	1600	9.3	42.6
Open ocean	332.0	125	600	41.5	199.2
Total	360.4	155	725	55.0	260.8
Fresh waters					
Swamp and marsh	2.0	2000	8400	4.0	16.8
Lake and stream	2.0	500	2300	1.0	4.6
Total	4.0	1250	5350	5.0	21.4
Continental land	145.0	670	2800	95.2	404.7

Source: Leith 1972.

water. Similarly, estuaries trap the nutrients that enter them in the flow of fresh water from the land (Odum 1961). In addition, tides create a mixing action that promotes the active circulation of the nutrients present and removes the waste products of metabolism and decomposition. In estuarine salt marsh areas, changing water levels also permit the existence of three major and distinctively different producer life-forms: rooted higher plants such as marsh grasses; benthic and epiphytic algae living as attached films on plant and substrate surfaces; and phytoplankton in the water that covers the marsh at high tide.

Upwelling areas (Figure 23-2) are also among the most productive marine environments. They are located mainly along the western margins of the continents and in the Antarctic, and they comprise less than 1 percent of the area of the oceans. In these areas, deep, cold water rises to the surface. These upwelling waters, rich in nutrients, support high levels of primary production and are the locations of the major offshore marine fisheries. The level of primary produc-

tion realized, however, is only equal to that of temperate zone grasslands (Table 23-1).

The waters overlying the continental shelves are next in productivity. These waters benefit from the tendencies of continental margins to stimulate upwelling, but they also receive the nutrients that are carried into the seas from the continents. The latter is probably the more important source of their overall productivity. Away from the continental shelves, in deep water, nutrients are depleted by the continual settling of dead organic matter and wastes into deep areas where light is unavailable for photosynthesis. These regions, which make up over 90 percent of the area of the oceans, genuinely deserve to be characterized as "biological deserts." The level of primary production in the open ocean lies between the values given by Leith (1972) for arctic tundra and desert scrub.

The values for net primary production in the ocean must be adjusted, however, because what man is really interested in is the productivity of harvestable animal species at some point several links along the food chain. Ryther (1969)

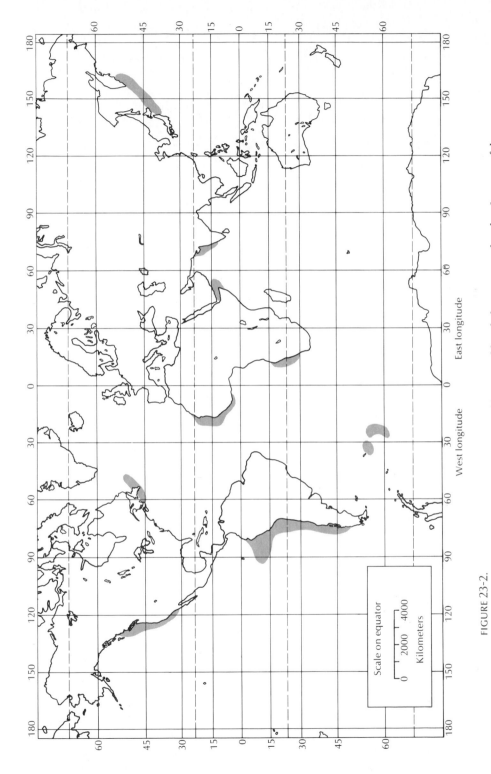

FIGURE 23-2.
Location of the major upwelling regions of the world oceans. Although they constitute less than 1 percent of the total area of the oceans, these regions are the locations of the most important marine fisheries.

TABLE 23-2
Estimates of the potential production of harvestable fisheries for the world oceans
by major productive regions.

	Reef-estuary	Upwelling	Coastal	Open sea
Percentage of ocean area. (a)	0.6	0.1	7.2	92.1
Net primary production in kcal/m^2/yr. (P)	9000	2500	1600	600
Efficiency of trophic level transfer. (e)	0.10	0.20	0.15	0.10
Trophic levels to point of harvestable production. (n)	3	1.5	3	5
Overall efficiency. (e^n)	10^{-3}	9×10^{-2}	3.4×10^{-3}	10×10^{-5}
Fish production in kcal/m^2/yr. $(P \cdot e^n)$	9.0	225.0	5.4	0.006
Total fish production in 10^6 tons/yr.° $(360.4 \cdot a \cdot P \cdot e^n)$	1.9	81.1	141.2	2.0
Percentage of fish production.	0.8	35.9	62.4	0.9

°Assuming 1 kcal = 1 gram of fish flesh.
Source: Modified from Ryther 1969.

pointed out that the length of the food chains leading to harvestable fish production varied for the major productive zones of the ocean. In upwelling areas, where the primary producers tend to be phytoplankton of relatively large size, and the harvestable fish tend to be species such as sardine and anchovy, only one or two links exist. In other words, harvestable fish species in these areas feed on a combination of phytoplankton and herbivorous zooplankton for the most part. In continental shelf waters, the average sizes of the producers are smaller, and food chains possess about three links before food energy reaches harvestable species. Finally, in open ocean, the plant producers are microscopic forms, and several small animal links are interposed in the food chain, in a sense magnifying food particles to a size that harvestable species can take. Here, food chains average about five links.

Each of these links demonstrates a certain efficiency in converting the stored energy of the tissues eaten into new tissue at its own level.

Ryther (1969) stated that this efficiency of conversion was directly related to productivity per unit area in open water areas, ranging from about 10 percent in the open sea zone to 20 percent in upwelling areas. Combining these estimates with those for area and primary production in the three zones he recognized, he was able to calculate the production of harvestable fish for the ocean as a whole. We repeat these calculations in Table 23-2, adding in those for reef and estuary areas, which Ryther did not separate out. The food chain lengths and conversion efficiencies in such areas are somewhat difficult to determine as averages. However, it seems likely that food chain length is intermediate and conversion efficiency low, since much of the primary production is processed through detritus food chains, and considerable energy is lost through decomposer activities.

In any case, these calculations suggest that the potential fisheries production of the seas is between 200 and 250 million tons per year, our figure of 226.2 million tons agreeing closely with Ryther's figure of 221.6 million tons. Man cannot, of course, harvest all of this production without seriously disrupting the normal function of the food chains involved. Ryther's estimate is that the long-term sustainable harvest might be about 100 million tons, assuming that this is taken in a careful manner. Clearly, if the technology emerges that will allow us to exploit species lower in the food chain than those we now harvest, the estimates of harvestable production will change. The interested student can explore this by modifying the calculations given in Table 23-2.

In assessing the potential productivity of freshwater systems, we should recognize first of all that such systems are in even closer association with land areas than coastal marine waters are. We would therefore expect them to be relatively fertile. Recent estimates given by Leith (1972) suggest that levels of primary production in swamp and marsh areas are comparable to those in reefs and estuaries, and that lake and stream systems are roughly as productive as upwelling zones in the ocean.

HUMAN EXPLOITATION AND IMPACT

Human exploitation of the productivity of the oceans corresponds more nearly to hunting and gathering in terrestrial environments than it does to herding and farming (Emery and Iselin 1967). This is true both for plant and animal materials. Sea "farming," or the cultivation of various seaweeds, is largely limited to Far Eastern nations such as Japan, Korea, and China, and probably involves the annual harvest of about 300,000 tons dry weight of algae for food (Bardach et al. 1972). The harvest of algae is accomplished by gathering a great variety of untended wild species that have both food and industrial uses (Levring et al. 1969). The total harvest of such materials is difficult to determine but probably amounts to about 400,000 tons annually, giving a total seaweed harvest of about

700,000 tons (Ricker 1969). It appears likely that both of the above can be increased several-fold as demand increases.

The total commercial fisheries harvest for the world, excluding whales and other marine mammals, was 69.8 million tons in 1974 (National Marine Fisheries Service 1976). Of this total, some 9.8 million tons were harvested from inland waters and 60.0 million from the oceans. In the early 1970s the total world production by aquaculture and mariculture constituted about 4.7 million tons of the total harvest (Pillay (1973), and of this 4.7 million tons, about 3.7 million consisted of freshwater fish and the rest of cultured marine organisms, largely oysters and mussels.

From these figures we can obtain some idea of the relative importance of aquatic and terrestrial food sources, as well as that of hunting and gathering compared to farming and herding (Table 23-3). Table 23-3 represents a rough updating of a table presented by Emery and Iselin (1967) and is modified to show the food coming from all aquatic environments rather than just the oceans. The data in this table demonstrate that food production and harvest in aquatic systems are still of very minor importance except for the commercial fisheries harvests from wild species, which equal nearly 10 percent of the production of animal tissue by terrestrial agriculture. It must be remembered that not all of this production ends up as human food; about 30 percent of the fisheries catch is converted into fish meal and used to feed livestock and pets (National Marine Fisheries Service 1976). Another portion of this gross tonnage is lost as inedible portions discarded during processing and preparation.

In addition to the harvests already mentioned, somewhere between 0.5 and 1.0 million tons of whales are harvested annually, a figure that is based on the quota of 37,300 animals of various species set for the 1974–1975 season (Table 23-4). Much of this tonnage is reduced to oil, but a significant fraction is eaten. In 1973,

TABLE 23-3
Relative importance of food harvest and production from aquatic and terrestrial systems by techniques of hunting and gathering as opposed to farming and herding.

	Annual production (tons $\times$ 10^6)			
	United States		World	
	Aquatic	Terrestrial	Aquatic	Terrestrial
Plant foods				
Gathering	0.2	2.5	0.4	124
Farming	0.0	286	0.3	2480
Animal foods				
Hunting	2.1	1.6	64.4	33.5
Herding	0.4	105	4.7	645

Source: Modified from Emory and Iselin 1967.

TABLE 23-4
Original and present estimated whale stocks, together with hunting quotas for 1974–1975.

Species	Estimated original stock	Estimated present stock	1974–1975 quota	Comments
Humpback	30,000+	4,000–9,000	Protected	Showing little recovery.
Gray	20,000	7,500–13,000	Protected	Western Pacific race extinct; Eastern Pacific race recovering.
Right (bowhead) (4 species)	300,000	4,000 or fewer	Protected	Recovering very slowly.
Blue	200,000	6,000–9,500	Protected	Recovering slowly.
Pygmy blue	?	6,000+		
Sperm	934,000	619,000	23,000	Quotas broken down by sex and ocean region.
Fin	443,000	97,000	1,300	Quota broken down by region.
Sei and bryde's	210,000	85,000	6,000	Intensive hunting begun in 1960 after fin whale decline.
Minke	150,000–300,000	150,000–300,000	7,000	Smallest of great whales; exploited little before 1970.

Source: Myers 1975.

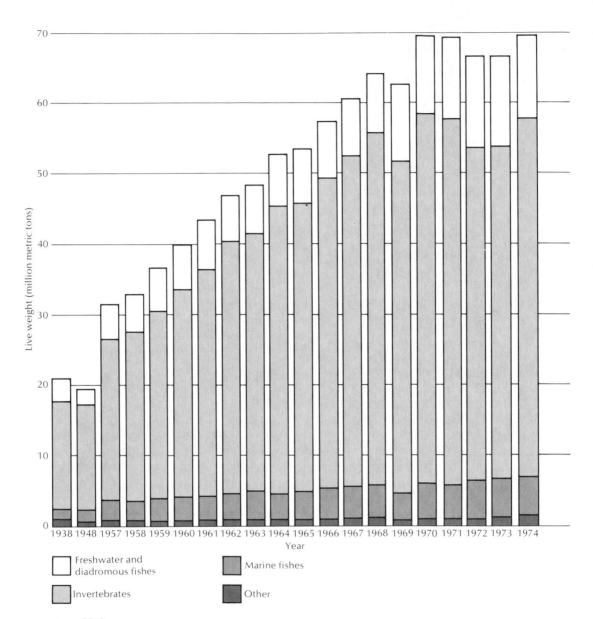

FIGURE 23-3.
World harvest of marine and freshwater fish, invertebrates, and other aquatic foods (including plants) for various years from 1938 through 1975. (Modified from "The Food Resources of the Ocean" by S. J. Holt. Copyright © 1969 by Scientific American, Inc. All rights reserved.)

FIGURE 23-4.
New technology has been a major factor in the growth of marine fisheries harvests. Here, a power block is being used to pull in a large purse seine. This device passes the entire seine through a mechanically operated, V-shaped pulley. The power block has not only reduced the labor involved in net recovery, but has permitted the use of nets 1000 meters or more in length and 100 meters in depth to encircle and trap large schools of fish. (Photograph courtesy of the FAO, Rome.)

122,700 tons of whale meat were consumed in Japan alone.

Still another harvest of some consequence is the recreational catch of freshwater and marine animals. In the United States, in 1970, this catch amounted to 717,000 tons (National Marine Fisheries Service 1976), which suggests that, for the world at large, the total recreational or noncommercial catch might equal 2 million tons or perhaps somewhat more.

Marine fisheries harvests have grown rapidly over recent decades, even more rapidly than the human population. In the century ending with 1950 the total world catch had increased more than tenfold (Holt 1969). After the end of World War II, which interrupted marine affairs, annual marine fisheries harvests in-

creased explosively from less than 20 million tons to well over 60 million tons in 1971 (Figure 23-3). During this postwar period, there was general optimism about the limits to which this growth could be carried. Several relatively low estimates were surpassed, and various workers made a number of new estimates ranging from about 100 million to nearly 1 billion tons per year (Ricker 1969).

Late in the 1960s, however, it became clear that much of this rapid growth was more the product of powerful technology applied to weakly exploited stocks than of the immense productive capacity of the oceans (Figure 23-4). Typically, a virgin stock comprises a high percentage of large, older fish. Each year a portion of these die and are replaced by reproduction and the growth of younger animals. In general, the reproductive rate of fish species is far more than adequate to provide for this replacement. When this stock first begins to be exploited, the catch, per boat, of high-quality fish is great, the fishery is highly profitable, and rapid expansion occurs. During this period, however, a great deal of the harvest comes from "fishing-up the accumulated stock" (Ricker 1969). Gradually, the accumulated stock of large, mature animals is depleted, and the catch per boat declines along with the average size of the fish taken. At this point, it might still be profitable for fishing to continue, even though catches are declining. But continued exploitation can eventually reduce the population to a point at which reproduction can no longer replace the total population, and the stock as well as the catch declines.

In 1949 an international conference sponsored by the United Nations, while recognizing that some stocks were already overfished, identified some 30 major marine fisheries stocks that appeared to be underfished at that time. By 1968, when the exploitation of these stocks was reviewed, it was concluded that for half of these, harvests had reached or exceeded the sustainable levels (Holt 1969). Thus it seems that the spectacular postwar growth of marine fisheries was greatly influenced by the tapping of previously unfished stocks.

Overexploitation of marine fisheries is a pattern that began as early as 1890 with the decline of plaice harvests in the North Sea. It has now extended to a number of ocean regions and to stocks as different as whales and sardines. For whales, the first major decline was that of the blue whale in about 1935 (Figure 23-5). Populations of several other whales have also been reduced to very low levels (Table 23-4), and it is clear that many decades will be required for the recovery of some of these species, assuming that recovery is still possible.

One of the most spectacular collapses was that of the California sardine harvest (Figure

FIGURE 23-5.
Whaling has reduced the stocks of the blue whale, the largest of the group, from an estimated 200,000 animals to less than 10,000. The species is now protected. (Photograph courtesy of the National Oceanic and Atmospheric Administration.)

23-6). This fishery rose from one of very minor importance in the early 1920s to one that harvested about 800,000 tons in 1936 (Murphy 1966). Following this period of exponential growth, the harvest fluctuated violently for nearly a decade and then crashed. At the time, the irregularities of the catch were attributed largely to changes in water temperature and other environmental factors, and little restraint was placed on harvests. Although the cause of this decline is still not entirely clear, it seems likely that it was either triggered or accelerated by overfishing, and that its continued decline was maintained by the expansion of populations of anchovy, a member of the same group of

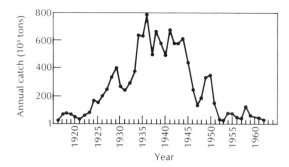

FIGURE 23-6.
Annual catch of sardines along the Pacific Coast of the United States from 1916 through 1962. (Modified from Murphy 1966.)

fishes and a presumed competitor (Gulland 1974).

In spite of this well-known example of the explosive rise and fall of a member of the anchovy group of fish, an even more serious case may have occurred in recent years (Idyll 1973a). In the mid-1950s a fishery developed off the coast of Peru that centered on the local anchovy species, known as the anchoveta, that occurs in one of the most productive upwelling zones of the world oceans (Figure 23-7). The anchoveta, a small fish with a life span of only about three years, is the single dominant fish species of this region and at times is present in incredible abundance. As we noted earlier, it feeds at a point low in the food chain and thus converts a large fraction of the primary production of this upwelling zone into harvestable animal tissue.

The Peruvian coastal region is located where cold, northward-flowing coastal and oceanic currents approach the equator and turn westward, driven by the strong tradewinds. The coastal current itself receives most of the upwelling water and therefore most of the nutrients that permit its high levels of production. Near the equator, these northward-flowing currents meet a southward-flowing countercurrent that consists of warm, nutrient-poor

water from the Bay of Panama. During the Southern Hemisphere summer, this countercurrent strengthens and tends to force its way further south. The anchovetas are largely confined to the coastal current, and when this current is narrowed and restricted to the immediate coast they become highly vulnerable to capture by purse seiners.

The anchoveta population, however, is severely affected by unusual patterns in these currents. In some years, the southward-flowing countercurrent can become strong enough to displace completely the cold coastal current along much of the Peruvian coast. This condition is called **El Niño** (the child) because it often occurs at Christmas time. When El Niño conditions develop, the anchovetas disperse

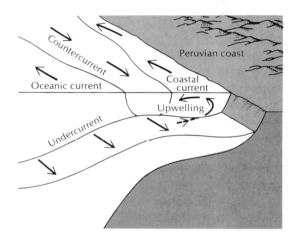

FIGURE 23-7.
In this diagram, two southward-flowing components of the Peru current are seen schematically in relation to the two northward-flowing components. The countercurrent (*dark color*), a surface or near-surface stream of water, intrudes between the northward-flowing, cold Coastal and Oceanic currents. Normally the countercurrent does not extend much south of the Equator, but when the wind that moves the northward-flowing currents falters or changes direction, its warm water pushes far to the south with disastrous biological consequences. Far deeper than all three currents is the second, far larger southward-flowing component, the undercurrent (*light color*). (From "The Anchovy Crisis" by C. P. Idyll. Copyright © 1973 by Scientific American, Inc. All rights reserved.)

FIGURE 23-8.
Anchoveta being pumped aboard a Peruvian purse seiner. (Photograph courtesy of the
FAO, Rome.)

and die, the sea bird populations that depend on them suffer catastrophic mortality, and the fisheries harvest drops sharply. The interval between these occurrences is quite irregular but averages about seven years. In a sense, the Peruvian anchoveta fishery has a serious natural problem built into it.

With the growth of intensive poultry and livestock feeding activities in North America and Europe, the anchoveta became a species of major importance, not as a human food, but as a source of fish meal for animal feeds (Figures 23-8, 23-9). Over a period of about 15 years, the Peruvian anchoveta fishery grew from less than one million tons per year to over 12 million tons (Figure 23-10). At this level, this single fishery constituted over 20 percent of the entire world fish catch, and, through the sale of fish meal, provided Peru with its most important source of foreign exchange.

In 1972, however, events took a disastrous turn. At a time when the annual level of exploitation had risen above 10 million tons, a figure considered by fishery scientists to represent a maximum sustained yield level, one of the most severe El Niño conditions in recent history occurred. Anchoveta catches in the spring of 1972 declined to nearly zero, and the catch for the entire year amounted to about 4.7 million tons. In 1973, instead of experiencing a recovery, the fishery harvest remained at a very low level. Total catches in 1973 were even lower than in 1972, equalling about 2.3 million tons. Since then, even though current patterns have returned to normal, the fishery remains well below the peak yields it achieved in the 1960s and early 1970s.

Again, the complete cause of this fishery's decline is still poorly understood, although it took place under the noses of a group of fishery

FIGURE 23-9.
A Peruvian fish meal plant. The bins in the foreground hold anchoveta before their processing by the grinding and drying machinery in the background. During the peak anchoveta season, fish meal plants operate at night (as in this photo) as well as during the day. (Photograph by R. Coral, courtesy of the FAO, Rome.)

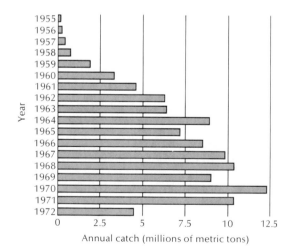

FIGURE 23-10.
Before its crash in 1972, the Peruvian anchoveta harvest had risen to the highest level for any single marine fishery. (From "The Anchovy Crisis" by C. P. Idyll. Copyright © 1973 by Scientific American, Inc. All rights reserved.)

scientists established specifically to study the anchoveta and to advise the Peruvian government on management of the fishery. It appears, however, that overexploitation, as well as natural events, played a significant role.

A few important marine fisheries remain to be tapped. Perhaps the most important is the krill fishery in the cold, upwelling areas of the Antarctic (Figure 23-11). **Krill** are not fish, but small crustaceans and other invertebrates that are dominant members of the marine zooplankton. The most important of these are a group known as the euphausid shrimp. Although these are spoken of as zooplankton, they are much larger than the organisms that make up the zooplankton in fresh waters. They

FIGURE 23-11.
Krill comprise a variety of small, pelagic, marine invertebrates including euphausid shrimp (the larger shrimplike forms), chaetognaths or arrow worms (the leaflike forms), polychaetes (the feathery worm near the top), and other organisms. The euphausid shrimp in this photo are about 2 to 3 centimeters in length. (Photograph courtesy of the National Oceanic and Atmospheric Administration.)

are also longer-lived, reaching ages of two or even three years. Feeding largely on phytoplankton and smaller zooplankton in Antarctic waters, these small animals essentially take the place of the anchovylike fishes. Krill, of course, form the principal food of the baleen whales of the Antarctic; any exploitation of krill would have to be closely examined for its impact on whale populations.

The total annual production of krill is still poorly known, but probably lies somewhere between 25 and 50 million tons (Ryther 1969). The sustainable yield of a krill fishery would necessarily be much less than this, probably less than half the annual production. Problems in the development of a krill fishery relate to the difficulties of harvesting such small animals and of processing the organisms obtained into material usable as food for man or livestock. It seems certain that these problems can be overcome, but it is also likely that the energy and cost required will be greater than for conventional fisheries.

These patterns of exploitation are only one aspect of man's impact on marine fisheries productivity. Human activities have greatly modified nutrient relationships in estuaries and in the coastal zone in general, and they have created pollution problems that are potentially dangerous to marine animals. Heavy metals, pesticides, petroleum compounds, and many other organic and inorganic substances are being introduced into the oceans in large quantities. Most of these have demonstrable toxic effects on marine organisms from phytoplankton to fish, and the serious possibility of multiplying interactions among some of these materials exists. Nevertheless, at the present time, man's impacts on nutrient relationships probably have their greatest effect on the productivity of fish and shellfish resources.

The most severely affected coastal marine environments are the estuaries and sounds along immediate coastlines (Figure 23-12). The filling and development of salt marshes (Figure 23-13)

FIGURE 23-12.
Estuaries and salt marshes, such as this area on the middle Atlantic Coast of the United States, are highly productive ecosystems that serve as spawning and nursery areas for important marine fishery species. (Photograph courtesy of the National Oceanic and Atmospheric Administration.)

FIGURE 23-13.
Industrial development on coastal marshlands is rapidly destroying and polluting these valuable ecosystems. (Photograph by Hope Alexander, courtesy of EPA-DOCUMERICA.)

and the dredging of water areas occupied by eelgrass beds have damaged or destroyed some of the most productive portions of the coastal aquatic system (Teal and Teal 1969; Thayer et al. 1975). These areas often provide the plant material, converted to detritus, that supports secondary production in the deeper water areas nearby. In addition, the modification of freshwater inflows to estuaries and sounds can change important features of the physical environment as well as reduce nutrient inputs. Along the Gulf Coast of the United States, for example, several major fisheries depend on the special characteristics of the coastal sounds— the sheltered waters on the landward sides of barrier islands. These waters are generally hyposaline and highly productive. They are permanent environments for certain important species such as oysters and blue crabs and are "nursery" areas in which other species, including white and brown shrimp, spend important developmental periods. Salinities below those of the open sea benefit these species by excluding predators, parasites, diseases, and competitors in a number of cases (Copeland 1966). The inflow of fresh water is essential to maintaining this condition. However, with the increased damming of rivers that enter these sounds, due to flood control and demands for urban, industrial, and agricultural water use, the inflow of fresh water is being seriously reduced.

Reduced inflows of fresh water to the coastal marine environment also mean reduced inputs of nutrients to these systems. Nowhere has this impact been demonstrated more strikingly than in the eastern Mediterranean following the damming of the Nile at Aswan. The Nile is the only river of any consequence that enters the eastern part of the Mediterranean, and periods of heavy river discharge have always produced major blooms of phytoplankton. With the closure of the Aswan Dam, however, all discharge of the river ended, apparently for good. Since that time the production of Egyptian

fisheries in the Mediterranean has steadily declined from an annual yield of 37,500 tons (1962) to 11,900 tons (1969). There is strong evidence that this is directly due to nutrient impoverishment of offshore waters (Ryder and Henderson 1975).

Freshwater fisheries, which make up slightly over 10 percent of the world's total fisheries harvest, have fared little better than marine fisheries. Here, however, essentially no important unexploited stocks remain; the trick is simply to improve our management of those that exist. The creation of reservoirs, of course, creates new freshwater areas, but, as we have seen, there is a trade-off between the gain in freshwater fish production and losses in coastal marine areas. For instance, in the case of the Aswan Dam, the newly created fisheries in Lake Nasser seem to have a potential annual yield of between 12,000 and 23,000 tons, depending on the final reservoir level achieved (Ryder and Henderson 1975). Both these values are below the observed decline in Mediterranean fisheries that followed the creation of the lake.

Major freshwater lake fisheries, such as those of the St. Lawrence Great Lakes, have suffered a variety of impacts, including overexploitation, pollution with toxic chemicals, eutrophication, and the introduction of alien species. **Eutrophication** is the increase in productivity of an aquatic system that results from an increased rate of nutrient inflow to the system. At first glance, such an increase may appear beneficial. When it is carried to an extreme, however, it may change the character of the lake system to the detriment of important fisheries species. Increased productivity in the surface waters, for example, may lead to an increase in the rate at which dead organic matter such as wastes and dead organisms is transferred into the bottom waters of stratified lakes. Decomposition of this material may consume all the oxygen in the bottom water, creating anaerobic conditions detrimental to fish. Changes in the

FIGURE 23-14.
This Lake Michigan trawler is bringing aboard a bag of alewives, a species used primarily in the production of fish meal for livestock feed preparation. (Photograph courtesy of the National Oceanic and Atmospheric Administration.)

plant and invertebrate species of the lake may also occur. In particular, these changes adversely affect species such as lake trout, whitefish, and other valuable food fish typical of clear, cold lakes with relatively low primary productivity. They favor, instead, a group of less valuable "trash" fish such as carp, perch, and catfish. Although total fish production might be higher in a eutropic lake than before, the type of fish it favors might be usable only in fish meal and not as human food (Figure 23-14). This is essentially what has happened in Lake Erie. Lake Erie is far from "dead"; it is more alive than ever before. Since 1900, however, the character of the fish community has changed from one dominated by valuable food fish to one dominated by low-value species.

The introduction of foreign fish species is also a serious problem in fresh waters. In Lake Michigan, the introduction of smelt and alewife has apparently contributed to the decline of some of the formerly important whitefish species. Recently, the introduction of Coho salmon as a predator of the less desirable species has partially mitigated the loss of these whitefish and created an important new fishery. The sea lamprey, entering the lake via the Welland Canal around Niagara Falls, was directly involved in the decline of the lake trout fishery in the lake. This species, a parasite, was a direct cause of trout mortality.

River fisheries, especially those for anadromous fish such as salmon, have also been abused. In the Columbia River, for example, the com-

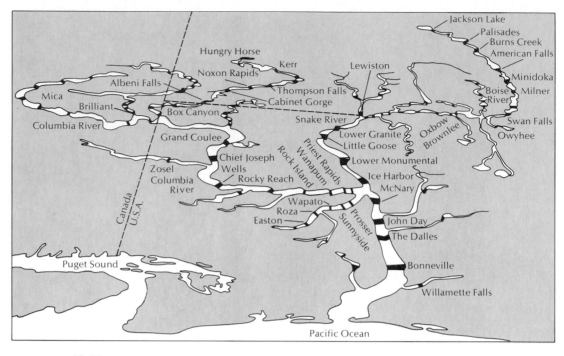

FIGURE 23-15.
Major dams on the Columbia River system of the northwestern United States and neighboring parts of Canada. (Modified from Trefethen 1972.)

mercial harvest of salmon and other fish in 1911 was some 22,400 tons. Since then, the river has been transformed by more than 50 dams, some of them among the largest in the world (Figure 23-15). Although special efforts have been made to provide for fish migration (Figure 23-16), it has nevertheless been inhibited, and certain features of the river environment have been detrimentally modified as well. In recent years, the annual commercial fish harvest from the river has been about 6800 tons, which is less than one-third its 1911 yield (Trefethen 1972).

All in all, it can be seen that, instead of continuing its spectacular growth toward projected annual levels of 100 or more million tons, world fisheries production seems to have plateaued in the range between 60 and 70 million tons (Figure 23-3). Both marine and freshwater fisheries are beset by a number of very serious human impacts whose correction will require major efforts. The new stocks available to be exploited are limited and difficult to capture and utilize. Moreover, the costs of boats, equipment, fuels, and labor have increased, making profitable fishing more difficult to conduct. This is reflected in the fact that, despite the decrease in total fish harvest in the past few years, the total dockside value of the catch has continued to grow. In other words, the cost per unit weight of fish is increasing rapidly. Clearly, world fisheries represent an important source of protein and must be managed as carefully and efficiently as possible. There is little hope, however, that the oceans will provide vast quantities of cheap protein to alleviate the problems of hunger and malnutrition throughout the hungry areas of the world.

FIGURE 23-16.
Fish ladder at Ice Harbor Dam on the Snake River, a branch of the Columbia River system. This ladder is designed to permit salmon to migrate upstream to their spawning grounds. (Photograph by J. D. Roderick, courtesy of the U.S. Bureau of Reclamation.)

AQUATIC AGRICULTURE: FRESHWATER AQUACULTURE AND MARICULTURE

Aquaculture is the husbandry of aquatic plant and animal species in fresh or salt waters. **Mariculture** specifically designates this activity in marine waters. Freshwater aquaculture and mariculture are widely believed to hold promise of major contributions to human food production. As we have seen, however, their present contributions are quite small (Table 23-3). Nevertheless, many different systems of aquatic agriculture have been developed, and at least some of these are quite productive. In basic design they vary from open systems in which the species husbanded are in unconfined seminatural environments, to closed systems in which they are carefully shielded from the surrounding environment. The latter systems resemble the glass-house vegetable farms or broiler factories of land-based agriculture. A few systems of aquatic agriculture involve plants, including certain marine algae, water chestnuts, and a few other species; but most are concerned with animal production.

In general, several characteristics combine to make a species suitable for aquaculture (Bardach et al. 1972). First, the species should have a high reproductive potential, and its breeding activities should be subject to human control. In other words, it should be possible to obtain large numbers of eggs or young at chosen or at least predetermined times. Second, the species should have an uncomplicated and hardy larval stage. Some aquatic animals pass through complicated sequences of larval stages, each of which might show specialized environmental and nutritional requirements. Certain marine algae similarly show life cycles involving alternation of generations that complicates the husbandry process. Third, the organism must have a high tolerance for crowding. Most aquacultural systems are monocultures, and high densities of individuals are required to give profitable yields per unit area. Finally, of course, the species must tolerate and accept the water conditions and food materials that are available. While almost any condition of purity or chemical composition can be created artificially, the cost of maintaining special conditions quickly becomes prohibitive; the same is true for foods.

Let us examine several representative systems of freshwater aquaculture and mariculture. Once we have done this, we can attempt to evaluate the future potential of these systems of aquatic agriculture.

Perhaps the most important system of aquatic plant farming is that for the production of nori. **Nori** is the name of the basic food product made from the harvested portions of various species of red algae of the genus *Porphyra*. Nori is produced mainly in Japan but also to a limited extent in Korea, Bangladesh, and a few other areas. The red algae of this genus have a life cycle in which two different plant forms alternate; one reproduces sexually and the other reproduces by spores. The sexually-reproducing phase is the one that is harvested for food. The plants of this stage form thin leaf-like blades 10 to 15 centimeters in length. Special racks are used to grow the algae; these are made of bamboo poles between which nets innoculated with *Porphyra* spores are hung. Blades of the length mentioned develop during a single tidal cycle of about 15 days and may be harvested several times before the nets need to be "reseeded." The spore-producing stage has quite a different appearance and is an encrusting plant that grows on shell material. This stage is now generally cultured by a few regional laboratories. Individual farmers take their nets to these centers and pay a fee to immerse them long enough for the spores to attach. The nets are then transferred to the "farm" areas in sheltered marine waters. There are about 70,000 nori growers in Japan alone, and in recent years annual production has been about 120,000 tons. At a density of about 100 nets per hectare, the yield of nori is about 750 kilograms per hectare, distributed over a 6-to-8-month period (Bardach et al. 1972).

Nori, which is used largely as a condiment, is only a small fraction of the Japanese diet. It is highly nutritious, however, consisting of between 30 and 40 percent protein in its dried form. Thus it represents a food whose use is limited largely by technological and cultural factors.

In addition to the red *Porphyra* algae, brown algae of the genera *Undaria* and *Laminaria* are farmed in the Far East. Annual production of the former may be about 60,000 tons and of the latter, 100,000 tons. Along the coast of China, where *Laminaria* farming is conducted,

the goal seems to be to increase production to an annual level of one million tons. From figures such as these, it is clear that the farming of a number of marine algae could increase by 1 or 2 orders of magnitude, given an acceptance of and demand for the food products obtained.

The oldest, and, today, probably the most quantitatively important, system of animal aquaculture is for carp, including the common carp *Cyprinus carpio* and a number of its relatives. Beginning in China, the common carp has been cultured for food and ornamental use for over 2000 years. This species is truly domesticated and is the only aquatic species for which breeding programs are practiced that are similar in sophistication to those used with livestock.

Carp have all the characteristics mentioned earlier as desirable in an aquacultural species; they are easily bred, tolerant of water conditions and crowding, and catholic in their food habits. Common carp farming is distributed widely among a number of tropical and subtropical countries. When eggs and sperm are mature, carp are easily induced to spawn by placing them in water slightly warmer than normal and by providing aquatic plants to which the eggs are attached. Other techniques may also be used including stripping the eggs and sperm by hand or injecting sexually mature fish with pituitary extracts. Eggs or newly hatched fry are then introduced into ponds in which they grow to maturity. At this point, carp farming varies greatly, depending on whether techniques of fertilization and supplementary feeding are used. Carp are heavy feeders, and when the composition of the supplementary feed is adjusted properly, feed conversion efficiency is high, equalling 20 to 30 percent or more. Yields per hectare range from less than 100 kilograms annually, in infertile ponds, to from 1000 to 5000 kilograms annually under intensive management.

In some areas, particularly in China, polyculture of several carp species is employed

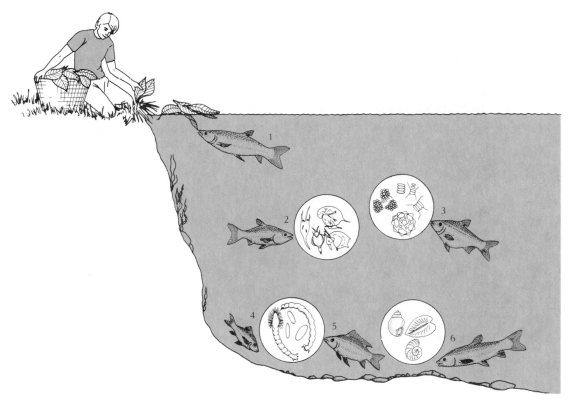

FIGURE 23-17.
Habitat and feeding niches of the principal species in Chinese carp culture: (1) grass carp (*Ctenopharyngodon idellus*) feeding on vegetable tops; (2) big head (*Aristichtys nobilis*) feeding on zooplankton in midwater; (3) silver carp (*Hypophthalmichtys molitrix*) feeding on phytoplankton in midwater; (4) mud carp (*Cirrhinus molltorella*) feeding on benthic animals and detritus, including grass carp feces; (5) common carp (*Cyprinus carpio*) feeding on benthic animals and detritus, including grass carp feces; (6) black carp (*Mylopharyngodon piceus*) feeding on mollusks. (From Bardach et al. 1972.)

(Figure 23-17). The species utilized have complementarity feeding habits, usually consuming different types of foods. Under favorable conditions these polycultures yield 7500 to 8000 kilograms per hectare.

In the mid-1960s, carp farming in countries other than China produced a combined total per year of about 210,000 tons of harvest; in China, at least 1.5 million tons of carp were harvested annually. This makes carp production the largest single aquacultural system. Furthermore, it suggests that if the rest of the world adopted fish farming on the scale that it is

practiced in China, more than 10 million tons of fish could be produced annually by this system alone.

In countries such as the United States, species other than carp are preferred for food. However, commercially profitable farming operations have been developed for two other types of fish; catfish and trout. Catfish farming is centered on the native North American channel catfish, although several other species are also used to some extent (Figures 23-18, 23-19). The farmer usually obtains the fish as fry from a hatchery and stocks them in ponds much the

FIGURE 23-18.
Channel catfish are raised in large ponds, such as this one one a farm in the Imperial Valley of California. Here, a pelletized feed containing at least 8 percent fish meal is being distributed. (Photograph by E. E. Hertzog, courtesy of the U.S. Bureau of Reclamation.)

FIGURE 23-19.
Harvesting channel catfish in the Imperial Valley, California. This farm, with 130 hectares of ponds, has the potential of producing 272,000 kilograms of fish annually. (Photograph by E. E. Hertzog, courtesy of the U.S. Bureau of Reclamation.)

way carp are stocked. Some interest has also developed in polycultures of one or more catfish together with largemouth bass or African tilapias. Feeding is more critical for catfish than for carp, and one apparent essential for good production is a minimum of 8 percent fish meal in the feed. Catfish thus become predators on anchovetas from the coastal marine waters of Peru! Proper farming operations give yields that average about 900 kilograms per hectare. About 30,000 tons are now produced annually in the United States (Bardach et al. 1972).

Trout are farmed in what is probably the most artificial and highly mechanized system of aquaculture. Trout farming, unlike several other fish production systems, is essentially a closed system. Eggs are obtained, incubated, and hatched under very carefully regulated conditions. The fry, fingerlings, and larger fish are maintained in ponds and raceways carefully designed to allow close control over water temperature and flow rate. Prepared foods are used, which contain 35 to 40 percent protein, at least 30 percent of which is in the form of fish meal, and a number of vitamin additives. In return, a high feed conversion efficiency is achieved, reflecting, of course, the careful preparation of the feed. Production of trout in the United States in 1969 amounted to 5000 tons. Denmark, with an annual production of 10,000 tons, and Japan, with 5100 tons, are also major trout producers.

Many other fish are being farmed in smaller quantities or are under study as potential species for aquaculture. Milkfish, *Chanos chanos*, are grown extensively in Java, the Philippines, and Taiwan. This fish, which has not yet been bred in captivity, is collected as fry and placed in ponds. It is tolerant of widely varying salinities. Various mullets, *Mugil* spp., are also cultured in brackish water areas. They too must be collected as fry, but considerable effort is being devoted to perfecting systems of spawning induction and of early rearing of mullets (Nash and Kuo 1975). Still another fish group,

the tilapias of Africa, offer considerable promise in aquaculture. One of the common features of these species is that they are plant or zooplankton feeders, and are thus able to convert a high percentage of natural primary production into harvestable fish tissue.

The oldest and still the most successful system of mariculture is that involving oysters. At least five species, one of the genus *Ostrea* and four of the genus *Crassostrea*, are now farmed commercially. Culture techniques vary greatly in detail, but the basic method involves inducing larval stages to settle on shell or other substrates and then placing them in areas that favor growth. Spawning is easy to induce in oysters, since it is triggered mainly by water temperature. Fertilized eggs hatch to form a planktonic larval stage that must grow and develop for about two weeks before they settle. Several systems have been developed to rear oyster larvae through this stage in closed systems. In many instances, however, settled larvae are simply collected in natural areas where wild populations are reproducing. In any case, a settling substrate, termed **cultch**, is provided. The most commonly used cultch is cleaned oyster or other shell. Once the larvae have settled, the cultch is transferred to carefully chosen bottom areas, termed oyster **beds**, or to rafts where the oysters grow suspended above the bottom (Figure 23-20). Bottom culture is practiced primarily in the United States, but most oyster growing in Japan employs rafts.

As we have already noted, hyposaline conditions are most favorable for oysters, partly because they discourage a number of important predators, parasites, and diseases. Where bottom culture is practiced, chemical treatments, including applications of quicklime and chlorinated hydrocarbon pesticides, are often used to eliminate starfish and predatory mollusks. These chemicals have undesirable aspects as pollutants, of course.

Yields of oysters vary, depending on the type of culture (bottom or raft) and the favorability

FIGURE 23-20.
Shells attached to lines suspended from rafts form the substrate upon which oysters grow
to maturity, as in this experimental operation at the National Marine Fisheries Service
Laboratory at Oxford, Maryland. (Photograph courtesy of the National Oceanic and Atmospheric Administration.)

of the environment in which they are grown.
Oysters feed by filtering materials from the
water that flows over and around them. They
are thus capable of harvesting the productivity
of an area much larger than that in which they
actually live. In natural oyster beds, where little
or no management is practiced, annual yields
may be about 10 to 100 kilograms per hectare.
In well-managed beds, however, this may be
increased to 100 to 500 kilograms, and in raft
culture, yields up to more than 20,000 kilograms of meat may be obtained. Total worldwide production is difficult to estimate but is
probably in the range of 700,000 to 800,000
tons annually.

From the foregoing examples, it can be seen
that freshwater aquaculture and mariculture
in open systems suffer from a variety of problems relating to contact between the cultured
populations and their surroundings. Through-
flows of water are likely to bring in a variety
of undesirable organisms and pollutants. These
throughflows also tend to carry away fertilizers,
foods, pesticides, and even the "seed organisms"
of the species being cultivated. This, together

with legal aspects of ownership and responsibility for damages, make mariculture in particular subject to many factors that are hard to
control or predict. Developing increasingly
closed systems of aquaculture and mariculture,
however, rapidly adds to the expense of production. Highly efficient and reliable production
systems are thus practical only for luxury food
species that command high market prices.

Ecologically, the opportunities for the development of aquacultural systems as important
contributors to human food production appear
greatest for open systems where most of the
work is done by nature. Specifically, the best
opportunities appear to be in husbanding
various species of brackish and freshwater fish
that are herbivores or first-level carnivores,
and in developing raft systems of shellfish culture for oysters, mussels, and other species that
are able to harvest production from extensive
areas of the coastal marine environment. An
important aspect of aquacultural policy is
creating an interest and demand for the type
of product that can be produced efficiently.

THERMAL AND
SEWAGE AQUACULTURE

A number of special opportunities exist for developing aquaculture systems that take advantage of what are now merely waste discharges from industrial or urban systems. Waste heat and treated sewage effluents are two such discharges that are now being investigated.

Thermal discharges are the result primarily of the use of water as a coolant in electrical power plants that use fossil or atomic fuels. Such discharges could be used to regulate water temperatures in open or closed aquaculture systems in temperate or higher latitude regions (Sylvester 1975). Warm water conditions are not automatically beneficial to all species, however, since they can increase metabolic energy dissipation, reduce the oxygen-carrying capacity of water, and favor disease organisms. Furthermore, the heated discharge water might also contain chemicals added to prevent fouling by organisms within the power plant cooling system, and these can also detrimentally affect aquaculture species.

One of the most ambitious proposals for the use of such thermal effluents is for heating an entire agriculture complex consisting of complementary production systems of vegetables, swine, chickens, and fish (Beall 1973). This system would be designed not only to take advantage of the heating potential, but also to allow maximum recycling of the wastes of one production component as inputs to others. Thus crop residues and swine wastes might be usable in part as food inputs to catfish production. Complete recycling of all wastes would be difficult to achieve, however, and a number of other problems remain to be solved. For example, heating the air space of greenhouses by means of the relatively small heat gradient available, without simultaneously creating undesirable humidity problems, presents certain difficulties. Nevertheless, some potential certainly exists for the utilization of waste heat.

More ambitious is the attempt by a group of workers at Wood's Hole Oceanographic Institution to develop a system that combines tertiary sewage waste treatment with mariculture (Dunstan and Tenore 1972; Ryther et al. 1972, 1975). The objective of this system is to utilize the organic matter and nutrient content of secondary sewage waste to support useful food production, and at the same time to remove these materials from the water so that it can be discharged into the environment without polluting it.

The system now being explored is an integrated one consisting of marine phytoplankton, filter-feeding bivalves, bottom-feeding fish and polychaetes, and other attached algae (Figures 23-21 to 23-23). In this system, secondary sewage wastes are mixed with sea water in a ratio of about 88 percent sea water to 12 percent waste water. This creates a salinity in which marine phytoplankton can still flourish. Marine phytoplankton grown in this water remove most of the nutrients from solution. The water in which they have grown is then fed into tanks containing oysters and clams. These species filter out the phytoplankton and organic detritus, consuming a portion and sedimenting the rest as pseudofeces (aggregated matter produced but not ingested by such mollusks). Pseudofeces and feces of the bivalves settle to the tank bottom, where they are fed on by polychaete worms. These in turn are periodically harvested by introducing bottom-feeding fish such as flounder. The water that leaves these tanks passes through cultures of other attached algae, from which it emerges essentially nutrient free.

The virtue of this system is that the costs of its operation are offset by the savings that result from the fact that wastewater treatment is accomplished simultaneously. Many problems still exist, however. The question exists, for example, whether food produced in this way can be used by man, given the possibility of contamination by viruses or other microorgan-

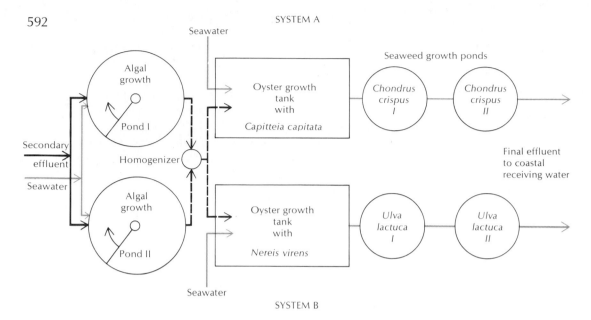

SYSTEM A

Seawater

Seaweed growth ponds

Algal growth

Pond I

Secondary effluent

Seawater

Homogenizer

Algal growth

Pond II

Oyster growth tank with

Capitteia capitata

Oyster growth tank with

Nereis virens

Chondrus crispus I

Chondrus crispus II

Ulva lactuca I

Ulva lactuca II

Final effluent to coastal receiving water

Seawater

SYSTEM B

FIGURE 23-21.
Diagram of experimental marine polyculture and tertiary waste treatment system at Woods Hole Oceanographic Institution. *Capittela* and *Nereis* are marine polychaete worms; *Chondrus* and *Ulva* are marine green algae. (From Ryther et al. 1975, by permission of Elsevier Scientific Publishing Company, Amsterdam.)

FIGURE 23-22.
Algal growth ponds (see Figure 23-21) in the experimental marine polyculture and tertiary waste treatment facility at Woods Hole Oceanographic Institution. (Photograph by G. W. Cox.)

FIGURE 23-23.
Oyster culture trays used in the experimental marine polyculture and tertiary waste treatment system at Woods Hole Oceanographic Institution. Trays with oysters are stacked in the adjacent concrete growth tanks, which receive a mixture of seawater and water from algal growth ponds (see Figure 23-22). (Photograph by G. W. Cox.)

isms. Another problem relates to the need to control temperature and other factors year-round to make the system a reliable one for tertiary waste treatment.

OUTLOOK FOR AQUATIC FARMING AND FOOD HARVEST

Human food production activities in aquatic systems are interrelated and interdependent with those in terrestrial environments. Nutrients that are diverted at one point in their flow from the continents to the oceans may increase the fertility of one system of food production, but at the cost of simultaneously impoverishing other systems. Primary production that is directed by man into one food chain

cannot also support the productivity of species in other food chains.

We must therefore recognize that if we convert coastal marshes to terrestrial agriculture or to intensive aquaculture of some sort, the gains in productivity by these systems will be to some extent at the expense of the productivity of the wild populations for which these are breeding or nursery areas. If we intensify harvests of protein at lower links in marine food chains, we must recognize that the productivity of other wild species that depend on these same species must also decline. Harvesting krill reduces the food supply for whales and other krill predators, which may also be valuable species for man.

The harvest of wild species in the oceans and fresh waters of the earth appears to be approaching its maximum sustainable limit. Clearly, improved management of fisheries exploitation might allow harvest levels to be increased somewhat, but an order of magnitude increase seems impossible. It does seem possible, however, that aquaculture and mariculture, at present of quite minor importance in overall quantitative terms, could expand by one or two orders of magnitude (Idyll 1973b). This estimate is based on the fact that very large areas of unused freshwater and coastal saline swamplands actually exist in various areas of the world, particularly in Southeast Asia, Africa, and Latin America. Presently, about 2.2 million hectares are utilized for finfish culture. Tens of millions of hectares of freshwater marshes, coastal lagoons and mangrove swamps, and irrigated rice paddies exist, and significant portions of these appear to be suitable for aquaculture (Pillay 1973). In spite of this, the basic limitations on productivity of marine ecosystems will keep their productivity well below that of the terrestrial environment, and the difficulties of harvesting this productivity mean that man's share is likely to remain much less than is true for the land.

However, aquatic systems do represent one of our most important potential sources of a particular food component—protein. Thus our efforts should be to manage and develop aquatic systems for their food protein potential.

SUMMARY

The ocean covers 71 percent of the earth's surface, fresh water ecosystems another 0.4 percent. Foods harvested or produced in these environments are integral parts of man's overall supply. However, although certain aquatic environments such as reefs, estuaries, and marshes have productivity levels that match those of the most productive terrestrial systems, the average productivity of aquatic systems is well below that of the continents. Over 95 percent of the production of harvestable fish occurs where nutrients are especially abundant; that is, in coastal and upwelling areas of the oceans, making up less than 10 percent of their area.

The sustainable yield of marine fisheries may be about 100 million tons annually. The current level of exploitation has now reached 65 to 70 million tons, but it seems to have plateaued. A number of major stocks now appear to be exploited at or beyond their maximum capacity, and there have been a number of spectacular recent fisheries collapses. The largest and most recent of these occurred in 1972 for the Peruvian anchoveta, a species that had risen to become more than 20 percent of the total world fish catch. A few marine stocks, krill in particular, remain as yet untapped. In fresh waters, few or no stocks remain unexploited, and serious damage has been done to many by overexploitation, pollution, eutrophication, and the introduction of foreign species. It appears unlikely that the fisheries harvests will exceed the 100 million ton estimate.

Freshwater aquaculture and mariculture have yet to approach the limits of their capability, however. To be suitable for such production, species must be easy to breed, possess relatively simple life cycles, and tolerate high crowding and the available water conditions and foods. A number of successful farming systems have been developed, such as those for nori, carp, catfish, trout, milkfish, and oysters. It seems likely that the production of animal tissue can be increased by one or two orders of magnitude by future intensification of such systems. Additional possibilities exist for developing aquacultural systems that take advantage of thermal and sewage inputs. All these systems should be viewed as contributors to food protein production, rather than as major sources of inexpensive basic foods for the growing human population.

Literature Cited

Bardach, J. E., J. H. Ryther, and W. O. McLarney. 1972. *Aquaculture—The Farming and Husbandry of Freshwater and Marine Organisms.* New York: Wiley-Interscience.

Beall, S. E. 1973. Conceptual design of a food complex using waste warm water for heating. *J. Env. Qual.* 2(2):207–215.

Copeland, B. J. 1966. Effects of decreased river flow on estuarine ecology. *J. Water Poll. Contr. Fed.* 38(11):1831–1839.

Dunstan, W. M., and K. R. Tenore. 1972. Intensive outdoor culture of marine phytoplankton enriched with treated sewage effluent. *Aquaculture* 1:181–192.

Emery, K. O., and C. O'D. Iselin. 1967. Human food from ocean and land. *Science* 157:1279–1281.

Gulland, J. A. 1974. *The management of marine fisheries.* Seattle: University of Washington Press.

Holt, S. J. 1969. The food resources of the ocean. *Sci. Amer.* 221(3):178–194.

Idyll, C. P. 1973a. The anchovy crisis. *Sci. Amer.* 228(6):22–29.

——. 1973b. Marine aquaculture: Problems and prospects. *J. Fish. Res. Bd. Canada* 30:2178–2183.

Leith, H. 1972. Modelling the primary productivity of the world. *Nature and Resources* 8(2):5–10.

Levring, T., H. A. Hoppe, and O. J. Schmid. 1969. *Marine Algae—A Survey of Research and Utilization.* Hamburg, Germany: Cram, De Gruyter and Co.

Morris, I. 1974. The limits to the productivity of the sea. *Sci. Prog.* 61:39–57.

Murphy, G. I. 1966. Population biology of the Pacific sardine (*Sardinops caerulea*). *Proc. Calif. Acad. Sci.* 34(1):1–84.

Myers, N. 1975. The whaling controversy. *Amer. Sci.* 63:448–455.

Nash, C. E., and C. Kuo. 1975. Hypotheses for problems impeding the mass propagation of grey mullet and other finfish. *Aquaculture* 5:119–133.

National Marine Fisheries Service. 1976. *Fisheries of the United States, 1975.* Current Fisheries Statistics No. 6900.

Odum, E. P. 1961. The role of tidal marshes in estuarine production. *The Conservationist* 15(6):12–15, 35.

——. 1962. Relationships between structure and function in the ecosystem. *Japanese J. Ecol.* 12(3):108–118.

Pillay, T. R. V. 1973. The role of aquaculture in fishery development and management. *J. Fish. Res. Bd. Canada* 30:2202–2217.

Ricker, W. E. 1969. Food from the sea. In *Resources and Man.* pp. 87–108. San Francisco: W. H. Freeman and Company.

Ryder, R. A., and H. F. Henderson. 1975. Estimates of potential fish yield for the Nasser Reservoir, Arab Republic of Egypt. *J. Fish. Res. Board Canada* 32(11):2137–2151.

Ryther, J. H. 1969. Photosynthesis and fish production in the sea. *Science* 166:72–76.

Ryther, J. H., W. M. Dunstan, K. R. Tenore, and J. E. Huguenin. 1972. Controlled eutrophication—Increasing food production from the sea by recycling human wastes. *BioSci.* 22(3):144–152.

Ryther, J. H., J. C. Goldman, C. E. Gifford, J. E. Huguenin, A. S. Wing, J. P. Clarner, L. D. Williams, and B. E. Lapointe. 1975. Physical models of integrated waste recycling—marine polyculture systems. *Aquaculture* 5:163–177.

Sylvester, J. R. 1975. Biological considerations on the use of thermal effluents for finfish aquaculture. *Aquaculture* 6:1–10.

Teal, J., and M. Teal. 1969. *Life and Death of a Salt Marsh.* New York: Ballantine Books.

Thayer, G. W., D. A. Wolfe, and R. B. Williams. 1975. The impact of man on seagrass systems. *Amer. Sci.* 63:288–296.

Trefethen, P. 1972. Man's impact on the Columbia River. In *River Ecology and Man.* R. T. Oglesby, C. A. Carlson, and J. A. McCann (eds.). pp. 77–98. New York: Academic Press.

24

ENERGY COSTS OF AGRICULTURE

The controlling factors of natural ecosystems, as we noted in Chapter 2, comprise the macroclimate, the geological context (including parent materials, topography, and structural features that affect groundwater), and the potential set of organisms available to occupy the area in question. Given these conditions, developmental processes are set in motion that lead to ecosystems with particular characteristics. The major features of these systems, which may be termed **dependent factors,** are those of soils, microclimate, vegetation, and animal and decomposer communities. The developmental processes that shape these features are those of ecological and evolutionary succession, both of which lead toward ecosystems of greater organizational complexity and closer adjustment to the controlling factors of the environment (Collier et al. 1973).

The developmental processes of these natural systems are intrinsic to them and are defined by the controlling factors of the natural environment. Thus when human beings attempt to modify natural systems in order to increase the yields of particular materials through agricultural activity, they must oppose certain of these developmental processes. Agroecosystems differ in structure from mature natural systems, often corresponding in general structure to systems in early stages of ecological succession. As a result, man must do work, and thus expend energy, to offset the processes that tend to bring about natural developmental change.

To a great extent, agroecosystems also lack the regulatory processes that natural ecosystems possess as a result of the evolutionary coadjustment of species. Thus, man must also assume the work involved in many of these regulatory processes.

As a result of all of this, agroecosystems are dependent upon two quite different **energy inputs:** ecological and cultural. **Ecological inputs** include incident solar radiation that fuels photosynthesis, controls environmental temperature, and creates patterns of atmospheric circulation and precipitation. **Cultural inputs** can be grouped into biological and industrial ones. Human and animal labor are the two most obvious forms of **biological energy input,** but the addition of organic matter, such as animal manures, is another. **Industrial energy inputs** are those derived, by means of modern technologies of electric power production, from fossil fuels, or from sources such as radioactive materials, geothermal sources, or flowing waters. Some sources of industrial energy, such as fossil fuels, are nonrenewable; others, such as hydroelectric energy, are renewable. It should also be noted that the energy inputs to agricultural activities are both immediate and indirect. A farmer plowing with a tractor, for example, is expending energy both in human labor and through fuel consumption. However, he is also utilizing a device, the tractor, that requires large and varied expenditures of energy to produce.

In general, we may suggest that the cultural energy requirement in agriculture is related to the degree of modification of natural ecosystem processes. The costs are small when man simply tries to increase the abundance of a native species in its natural system; for example, in the culture of "forest coffee" in Ethiopia. When man substitutes a desired species for one that is less easily utilized—for example, when bison are replaced by beef cattle on western rangelands— the cultural energy costs are higher, although still relatively low. Cultural energy costs rise steeply, however, when a complex natural system is replaced by a crop monoculture of a different life form, as in the midwestern United States where corn has replaced areas of deciduous forest. Finally, where such replacement also strives to increase the level of productivity (or fixation of solar energy) above that shown by the intact natural system, cultural energy costs may become very high indeed.

In recent years it has become clear that man has rapidly been exhausting some of the most easily harnessed forms of cultural energy—fossil fuels. At the same time, it has become obvious that intensive mechanized agriculture depends greatly on such energy inputs. In this chapter, therefore, we shall examine various agricultural systems, characterize their cultural energy costs, and discuss the implications of these input requirements for the future of modern agriculture.

CULTURAL ENERGY IN NONMECHANIZED AGRICULTURAL SYSTEMS

Analyses of cultural energy inputs as they relate to food energy outputs have been made for many agricultural systems in the past few years. These have employed a wide variety of techniques and have incorporated varying assumptions, so that precise comparisons of the results of different investigators are certainly not possible. To give the reader some idea of the relationships upon which these analyses are based, we have included a table of energy values and equivalents (Table 24-1). These are only approximate in many cases and are not always the values that particular workers used in their energy budgets.

Nonmechanized agricultural systems can be divided roughly into three groups: pastoralist systems; cropping systems that rely exclusively on human labor; and cropping systems that use draft animals.

To our knowledge, no detailed study has been conducted of the energetics of pure pastoralist systems of agriculture. From the data supplied by Brown (1971) for Africa south of the Sahara, however, we can derive approximate estimates (see Chapter 5). If we assume that about one-third of the total energy expended by two members of an average family group of 6.5 persons is devoted to herding and to related activities that yield milk, blood, and meat, the ratio of food energy yield to cultural energy investment is about 9.7. Actually, for such peoples, any division of this sort must be almost completely arbitrary because virtually all aspects of culture and life activity are closely involved with the animals on which they depend.

Odum (1967) has presented an approximate analysis of energy relations within a semipastoralist system (Figure 24-1). This analysis, which deals with the Dodo tribe in Uganda, shows that most of the food consumed is grain produced by a cultivation system that yields

TABLE 24-1
Energy values and equivalents (approximate).

Solar constant	2.00 kcal/cm²/min
Heat of combustion	
Coal	6,650 kcal/kg
Gasoline	36,225 kcal/gal
Human and animal labor	
Human labor	175 kcal/hr
Large draft animal labor	2,400 kcal/hr
Energy quality equivalents	
Coal and oil	1.0 fossil fuel kcal/heat kcal
Electric power	4.0 fossil fuel kcal/heat kcal
Water (elevated)	3.0 fossil fuel kcal/heat kcal
Wood	0.5 fossil fuel kcal/heat kcal
Gross plant production	0.05 fossil fuel kcal/heat kcal
Solar radiation	0.0005 fossil fuel kcal/heat kcal
Heat radiation from sunlight	0.0001 fossil fuel kcal/heat kcal
Agricultural inputs	
Fertilizers—nitrogen	17,600 kcal/kg
—phosphorus	3,190 kcal/kg
—potassium	2,200 kcal/kg
Pesticides	27,170 kcal/kg
Machinery	20,712 kcal/kg

more than five times the cultural energy invested. Grain production is coupled with the herding of cattle, however, in which the cultural energy investment exceeds the food energy yield. Animal foods, therefore, seem to play a special role, meeting specific nutritional needs and acting as a form of food storage for periods of crop failure. Overall, this system yields food energy at a ratio of about 2.9 times the cultural energy investment. In general, therefore, it appears that pastoralist and semi-pastoralist systems show a positive energy balance, but one that is not particularly large because these systems must harvest the low productivity of a large area and because they possess special security measures against seasons of poor production.

In Chapter 5 we discussed in some detail cropping systems that rely exclusively on human labor and systems that utilize draft animals. Here, we shall concentrate on general patterns emerging from studies of several systems of these two types. Studies of cropping systems that depend on human labor (Table 24-2, numbers 3-7) reveal several points that deserve emphasis. First, there is a high degree of variability in the absolute productivity of such systems, which is primarily a function of climate. Some systems, such as the mixed crop system in New Guinea and the cassava system in Zaire, achieve yields per hectare that are comparable to those of intensive mechanized farming systems. These systems require a human labor investment that is high by the standards of mechanized farming. However, the labor requirement per unit of food energy obtained is smaller than that in either pastoralist systems or cropping systems with less favorable environments than those in New Guinea or Zaire. In the drier areas of Mexico and the Sudan, for ex-

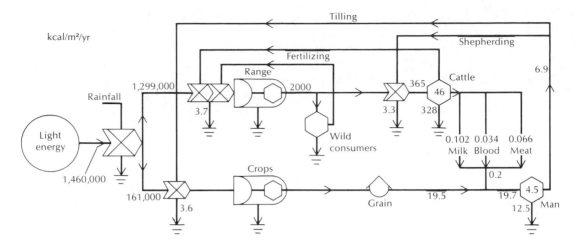

kcal/m²/yr

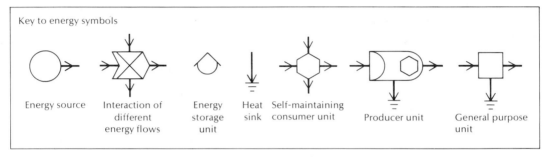

FIGURE 24-1.
Energy flow in a semipastoralist food supply system, as exemplified by the Dodo of Uganda. Food is derived principally from grain cultivation; but cattle (an average of 4.5 per person) are also tended, and these animals supply food in the form of milk, blood, and meat. The population density of the region occupied by the Dodo is about 27 people per square kilometer. (Modified from Odum 1967.)

ample, yields per hectare are lower, and the ratio of food calories obtained to cultural energy investment is lower. Nevertheless, for these cropping systems, in which the only energy investment from industrial sources is that involved in manufacturing simple tools such as a hoe, the energy balance is strongly positive, with ratios of output to input ranging from 10 to one to nearly 40 to one.

The use of draft animals generally corresponds to the change from shifting cultivation to permanent cultivation. Again, we can examine the results of several analyses of agroeco-system energetics (Table 24-2, numbers 8–13). These systems, like those that depend on human labor, also demonstrate high variability. However, it is clear that the use of draft animals does not result in levels of production per hectare greater than those achieved in systems that depend on human labor (Figure 24-2). Instead it appears that draft animals make it possible to maintain, on a permanent basis, yields that are comparable to those obtained by shifting cultivation, which harvests the accumulated fertility of lands that have been fallow for several to many years. Moreover, the addition of draft ani-

TABLE 24-2
Cultural energy inputs and food energy gain for various nonmechanized agricultural systems.

System	Foods	Kcal/m²/yr					Ratios		Reference
		Cultural input				Output	Food / Industrial	Food / Total	
		Human	Animal	Industrial	Total	Food			
1. Pastoralist, Africa	Milk, blood, meat	0.51	—	—	0.51	4.95	—	9.7	(Brown 1971)
2. Semipastoralist, Uganda	Milk, blood, meat, grains	6.9	—	—	6.9	19.7	—	2.9	(Odum 1967)
3. Shifting cultivation, New Guinea	Mixed crops	139	—	—	139	2278	—	16.4	(Rappaport 1971)
4. Shifting cultivation, Thailand	Rice	34.0	—	0.6	34.6	622	1037	18.0	(Hanks 1972)
5. Shifting cultivation, Mexico	Maize	65.92	—	1.65	67.57	684.3	415	10.1	(Pimentel 1974)
6. Shifting cultivation, Sudan	Sorghum	19.33	—	1.65	20.98	297	180	14.2	(Pimentel 1974)
7. Shifting cultivation, Zaire	Cassava	55.62	—	1.65	57.27	2145	1300	37.5	(Pimentel 1974)
8. Flood irrigation, Thailand	Rice	12.4	1.0	1.8	15.2	573	318	37.7	(Hanks 1972)
9. Paddy irrigation, Thailand	Rice	34.8	1.9	4.7	41.4	940	200	22.7	(Hanks 1972)
10. Permanent farming, Mexico	Maize	20.84	69.30	4.14	97.94[a]	331.23	80.0	3.38	(Pimentel 1974)
11. Permanent farming, India	Wheat	33.47	224.70	4.14	283.76[a]	270.93	65.4	0.95	(Pimentel 1974)
12. Permanent farming, Philippines	Rice	31.35	95.20	16.19	183.13[a]	600.40	37.1	3.28	(Pimentel 1974)
13. Permanent farming, Nigeria	Sorghum	6.31	255.50	4.14	272.22[a]	247.17	59.7	0.91	(Pimentel 1974)

[a]Totals include caloric value of seeds and other miscellaneous inputs.

Input	Quantity/ha	Kcal/ha
Human labor	383 man hours	208,448
Ox labor	198 animal hours	693,000
Equipment	41,400 kcal*	41,400
Seeds	10.4 kg	36,608
Total		979,456
Corn yield	941 kg	3,312,320
Kcal return / Kcal input		3.38

*Energy cost of manufacture of plow prorated over its useful life.

FIGURE 24-2.
Energy budget for corn farming in Mexico using oxen for plowing. (Data from Pimentel 1974; photograph by F. Botts, courtesy of the FAO, Rome.)

mal labor does not greatly reduce the need for human labor. Instead, the total investment of cultural energy is increased. In some cases, such as wheat farming in India and sorghum farming in Nigeria, this total investment exceeds the energy gain in the food obtained. Farming in these cases becomes an exercise in trading calories of one type (those consumed by draft animals) for calories that can be consumed by man.

Most of these systems incorporate very small inputs of industrial energy, however. Rice farming in the Philippine Islands uses small amounts of commercial fertilizers and insecticides, and

all systems use a steel plow that represents an indirect industrial energy cost. Nevertheless, nearly all the cultural energy invested is biological and is therefore from renewable sources. As we noted in Chapter 5, in our discussion of the energetics of Indian cattle (Odend'hal 1972), most of the food calories consumed by draft animals come from sources that cannot be used directly by man. In India, the primary cattle feeds are rice straw, banana trunks, and sugarcane leaves that would not otherwise enter the food chain leading to man. Thus, while the total energy balance of systems that use draft animals falls below that of systems that rely exclusively on human labor, the more significant relationship might be that between food calorie yield and industrial energy investment; the latter varies from about 30 to 300 for the systems examined. We should also note that the very favorable energy balance of the flood irrigation system for rice in Thailand is largely the result of a considerable "natural" input of work in flood irrigation and fertilization; that is, the energy that comes from the hydrologic cycle.

CULTURAL ENERGY IN MECHANIZED AGRICULTURAL SYSTEMS

With the mechanization of agriculture both the direct and indirect uses of industrial energy increase. It becomes important at this point to distinguish among the forms of energy that enter into the food production process. Different forms of energy might possess the same heat equivalent, but not at all the same capacity to be put to useful work. Forms of energy thus differ in quality, and this difference essentially reflects the degree of concentration of energy. For example, one kilocalorie of solar radiation has much less capacity to do useful work than a kilocalorie of energy in the form of electricity; in fact, it has approximately 2000 times less. The differences in the quality of energy forms

SUBSISTENCE FARMING

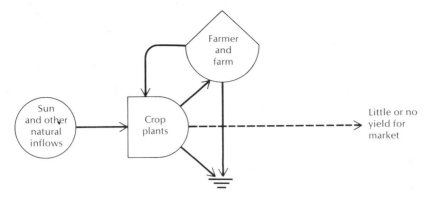

MECHANIZED FARMING

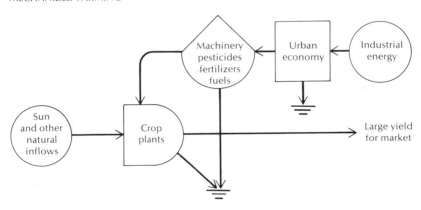

FIGURE 24-3.
Pattern of energy flow through subsistence and mechanized, market-oriented farming systems. Subsistence systems use only small inputs of industrial energy and yield little marketable produce. Mechanized systems use fossil fuels to produce large net yields for the marketplace. See Figure 24-1 for definition of symbols. (Modified from Odum and Odum 1976.)

are reflected in the cost of transforming one form to the other.

The fact that energy forms differ in quality is significant because energy flows must be analyzed, through complex processes, in energy units of the same quality. In Table 24-1, therefore, we have indicated quality equivalents of different energy forms.

Mechanized agricultural activities really amount to using high-quality energy forms to enhance the conversion of solar energy into stored food energy (Odum and Odum 1976). This is shown diagrammatically in Figure 24-3, which compares energy flows in subsistence and mechanized farming. In mechanized farming, high-quality forms of energy, principally fossil fuels, are used to manufacture agricultural machinery, fertilizers, pesticides, and other chemicals. On the farm, they are used to operate farm equipment, pump irrigation water, and

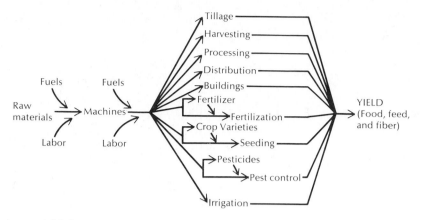

FIGURE 24-4.
The major direct and indirect inputs of cultural energy to mechanized crop production. (Modified from Heichel 1973.)

treat harvested materials. Energy is also required for other special tasks, such as the production of hybrid seed with special characteristics. All these inputs represent the cultural energy subsidy of the farming process (Figure 24-4).

The utility of such high-quality forms of energy is in their ability to interact with and to amplify flows based on lower-quality energy sources (Odum and Odum 1976). In other words, high-quality energy is most useful for its control capability, or its ability to do especially useful types of work. It is wasted if it is not amplified in the processes to which it is applied. Thus in looking at the energetics of mechanized agriculture, our objective is to see how high-quality energy, which we have termed **industrial energy,** is used, whether it is being used in the most desirable fashion, and how we might improve our use of it.

The important role of high-quality energy in mechanized agriculture was emphasized by a series of analyses carried out by David Pimentel and a number of his students and colleagues at Cornell University. The first of these focused on energy inputs to corn production in the United States over the period 1945 to 1970

(Pimentel et al. 1973). These analyses did not strictly convert energy inputs to energy units of the same quality, but most energy inputs represent energy of fossil fuel quality.

The years from 1945 to 1970 constitute the period of major transition of United States agriculture from a lightly mechanized to a heavily mechanized system. For corn, this was also a period of revolutionary yield increases, with the introduction of hybrid seed, intensive use of inorganic fertilizers, and modern pest control. Average yields per hectare rose from 84 bushels in 1945 to 200 bushels in 1970, an increase of 240 percent. As indicated in Table 24-3, these yields were the result of several trends in agricultural practice. The use of inorganic fertilizers increased more than tenfold, and significant use of insecticides and herbicides began. In fact, by 1970, about 17 percent of all insecticides and 41 percent of all herbicides used in American agriculture were applied to corn (Pimentel 1973). The use of farm machinery such as tractors, spraying equipment, and corn-pickers has increased over this period, and the horsepower of individual items of machinery also increased (Figure 24-5). Artificial drying to prevent the harvested grain from spoiling (Figure 24-6) has

TABLE 24-3
Average inputs and yields per acre for corn production in the United States for selected years from 1945 to 1970.

Inputs	1945	1950	1954	1959	1964	1970
Labor (hours)	23	18	17	14	11	9
Machinery (kcal $\times 10^3$)	180	250	300	350	420	420
Gasoline (gallons)	15	17	19	20	21	22
Nitrogen (pounds)	7	15	27	41	58	112
Phosphorus (pounds)	7	10	12	16	18	31
Potassium (pounds)	5	10	18	30	29	60
Seeds for planting (bushels)	0.17	0.20	0.25	0.30	0.33	0.33
Irrigation (kcal $\times 10^3$)	19	23	27	31	34	34
Insecticides (pounds)	0	0.10	0.30	0.70	1.00	1.00
Herbicides (pounds)	0	0.05	0.10	0.25	0.38	1.00
Drying (kcal $\times 10^3$)	10	30	60	100	120	120
Electricity (kcal $\times 10^3$)	32	54	100	140	203	310
Transportation (kcal $\times 10^3$)	20	30	45	60	70	70
Corn yields (bushels)	34	38	41	54	68	81

Source: Pimentel et al. Copyright © 1973 by the American Association for the Advancement of Science.

FIGURE 24-5.
Harvesting corn in the San Joaquin delta region of California. This combination picker and sheller harvests the mature ears and removes the grain from the cobs, leaving the latter in the field. (Photograph by A. G. D'Alessandro, courtesy of the U.S. Bureau of Reclamation.)

FIGURE 24-6.
A corn drier. The shelled corn is fed from the bin (*right*) into the drier (*center*). From the
drier it is carried through the pipe (*left*) into a storage area. (Photograph courtesy of the
USDA.)

become more important as it has become more
common to allow the crop to remain in the field
later into the autumn (this takes maximum advantage of late season production). Accompanying these trends has been an increase in the use
of fuels for machinery and in the use of electricity in farming activities. Furthermore, as the
use of machinery and purchased inputs in corn
production has increased, the energy costs of
transporting all these materials to the farm have
become considerable.

When these inputs are evaluated in terms of
energy value (Table 24-4), their total is large;
and it has increased by over 300 percent between 1945 and 1970. Except for human labor,
the energy used corresponds to energy of fossil
fuel quality. The energy output, as corn grain,
is also a relatively high-quality form of energy.
If we overlook the differences in energy quality

involved, this analysis shows that, for every
calorie of cultural energy invested in corn production in 1945, 3.7 calories were recovered in
grain; by 1970 this return had declined to 2.5
calories. Although this analysis shows a positive
energy balance for the production process, it
is clearly much less than the return on the investment of industrial energy in less intensively
mechanized systems. Furthermore, we should
recognize that the product obtained is almost
entirely an animal feed grain, so that before
corn calories end up as food they must be fed to
cattle and hogs, where other large losses and
energy costs are incurred.

Finally, we should note in regard to this example that one of the most dramatic trends related to this energy intensiveness is the decline
in human labor input to the farming process.
The labor required in 1945 was more than 2.5

TABLE 24-4
Energy value (kcal) of inputs and yield for corn production per acre in the United States from 1945 to 1970.

Input	1945	1950	1954	1959	1964	1970
Labor	12,500	9,800	9,300	7,600	6,000	4,900
Machinery	180,000	250,000	300,000	350,000	420,000	420,000
Gasoline	543,400	615,800	688,300	724,500	760,700	797,000
Nitrogen	58,800	126,000	226,800	344,400	487,200	940,800
Phosphorus	10,600	15,200	18,200	24,300	27,400	47,100
Potassium	5,200	10,500	50,400	60,400	68,000	68,000
Seeds for planting	34,000	40,400	18,900	36,500	30,400	63,000
Irrigation	19,000	23,000	27,000	31,000	34,000	34,000
Insecticides	0	1,100	3,300	7,700	11,000	11,000
Herbicides	0	600	1,100	2,800	4,200	11,000
Drying	10,000	30,000	60,000	100,000	120,000	120,000
Electricity	32,000	54,000	100,000	140,000	203,000	310,000
Transportation	20,000	30,000	45,000	60,000	70,000	70,000
Total inputs	925,500	1,206,400	1,548,300	1,889,200	2,241,900	2,896,800
Corn yield (output)	3,427,200	3,830,400	4,132,800	5,443,200	6,854,400	8,164,800
Kcal return/input kcal	3.70	3.18	2.67	2.88	3.06	2.82

Source: Pimentel et al. Copyright © 1973 by the American Association for the Advancement of Science.

times that employed in 1970. Large inputs of high-quality energy have thus been utilized to gain increased production per hectare with greatly reduced labor costs. The student can also determine from Table 24-4 that the incremental effect of additional energy inputs, both on corn yield and labor reduction, has declined during the period 1964 to 1970, compared to the period just before. In other words, a strong "diminishing returns" effect is appearing in the system.

The relationships seen in corn production are not unusual. Analyses of the efficiency of production processes for wheat, rice, and potatoes show that the return in food calories for these crops is even less than that for corn (Table 24-5). The energy costs of machinery and fuel, per hectare of wheat, are nearly as high as those for corn, but the total weight of harvested grain averages less than half that obtained from corn.

The energy cost of seed grain is also greater for wheat. For rice, irrigation costs are nearly equal to all other energy costs combined, and the artificial drying of rice requires more energy than drying other grains. Potatoes, which yield more calories per hectare than any of the grains, of course require greater amounts of all fertilizers per hectare.

Independent analyses for various crops in the United States have also been conducted by Gary Heichel, in studies conducted at the Connecticut Agricultural Experiment Station. His analyses give similar results for the basic crops discussed above, but they extend the available information to many other species, including fruits and vegetables (Heichel 1973, 1974, 1975). These analyses show that most vegetable crops and fruits yield food energy in amounts equal to or less than the cultural energy investment (Figure 24-7). A variety of factors contri-

TABLE 24-5
Energy inputs and return for mechanized farming of wheat, rice, and potatoes in the United States (per hectare).

Item	Wheat	Rice	Potatoes
Inputs			
Labor	6,531	16,328	32,655
Machinery	1,037,400	1,037,400	1,000,000
Fuel	1,339,800	2,153,250	1,971,420
Nitrogen	1,284,800	2,358,400	2,601,280
Phosphorus	54,230	—	818,235
Potassium	37,400	147,400	546,920
Seeds	552,750	813,120	269,500
Irrigation	—	6,545,880	—
Insecticides	26,620	135,520	135,520
Herbicides	—	135,520	135,520
Fungicides	—	—	135,520
Electricity	370,500	—	765,700
Drying	—	1,070,597	—
Transportation	86,450	172,900	250,000
Total	4,796,481	15,352,015	8,662,270
Yield	8,428,200	21,039,480	19,712,000
Yield/total inputs	1.76	1.37	2.28

Source: Pimentel. Copyright © 1974 by the American Association for the Advancement of Science.

bute to these high costs, but pesticide and irrigation costs are major in almost all cases. Over 40 percent of the fungicides used in the United States, for example, is applied to citrus and apples (Pimentel 1973). Here it must be borne in mind that these crops are not consumed simply for their energy content, but for other nutritional and esthetic reasons as well.

We should also note that some basic energy foods, sugar in particular, fall in a special category because of the major energy costs associated with processing. Sugar is produced primarily from sugar beets and sugarcane (Table 24-6). Sugar production from beets is more highly mechanized than that from cane, and production costs are therefore higher (Figures 24-8, 24-9). Refining requires 32.5 percent of the total cultural input for sugar beets and 44.4 percent of all cultural energy for sugarcane. Processing costs per unit of sugar obtained, however, are twice as great for sugar beets as for sugarcane (Figure 24-10). Overall, the production of sugar from beets yields only 0.8 food calories per calorie of cultural energy invested, while the yield is 2.5 for cane sugar. It is interesting to note that these energy-intensive systems of sugar production have evolved under a complex system of tariffs and import quotas that have protected United States producers from intense competition by maintaining basic sugar prices that are nearly twice the price of sugar in free international trade (Janick et al. 1974). Efficient use of energy in production has thus not been encouraged by competition in the market.

The extreme of industrial energy subsidizing of crop production is shown in the glasshouse industry (Sheard 1975). Glasshouses (greenhouses or hothouses) are being used increasingly in various countries for the production not only

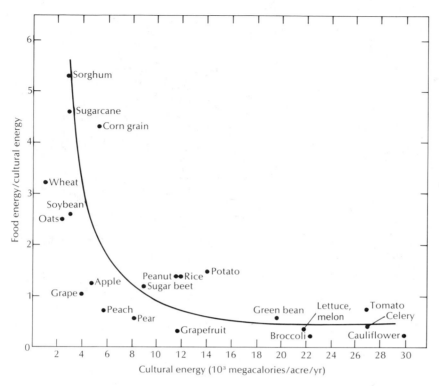

FIGURE 24-7.
Ratios of food energy output to cultural energy input for various crops. The horizontal axis shows the actual input of cultural energy per hectare. (Modified from Heichel 1975.)

TABLE 24-6
Comparison of energy inputs, sugar yields, and energy efficiency ratios for Hawaiian sugarcane and Californian sugar beets in 1970.

Item	Sugar beets	Sugarcane
Labor input (kcal $\times 10^3$/ha)	12.6	19.3
Other inputs (kcal $\times 10^3$/ha)	16,383.5	7,607.6
Processing (kcal $\times 10^3$/ha)	7,918.8 (32.5%)	6,100.9 (44.4%)
Total input (kcal $\times 10^3$/ha)	23,314.7	13,728.3
Sugar yield (kcal $\times 10^3$/ha)	19,913.1	34,957.9
Yield/energy input	0.8	2.5

Source: Heichel 1973.

FIGURE 24-8.
Harvesting sugar beets near Boise, Idaho. Mechanized operations are more extensive for
sugar beet than for sugarcane production, partly because soil preparation and planting
must be done each growing season for sugar beets. Sugarcane, a perennial, is replanted
only after a number of years. (Photograph by J. D. Roderick, courtesy of the U.S. Bureau
of Reclamation.)

FIGURE 24-9.
Harvesting sugarcane in Puerto Rico. After cutting, the cane will regrow from the perennial root system of the plant. Note that the cane planting has been burned with a light fire
that removes the dead lower leaves without burning the upper leaves or destroying the
stalks. (Photograph by R. A. Garcia, courtesy of the USDA Soil Conservation Service.)

FIGURE 24-10.
Processing sugar beets, as in this plant near Nampa, Idaho, requires about twice as much energy per unit of sugar produced as the processing of sugarcane. (Photograph by Glade Walker, courtesy of the U.S. Bureau of Reclamation.)

of horticultural crops, but also of high-value vegetables such as tomatoes, lettuce, cucumbers, asparagus, and brussels sprouts. Some fruits, including apples, are also grown in glasshouses. In Temperate Zone countries such as Britain and the Netherlands, large quantities of vegetables are grown in glasshouses, the cost of their production being less than that of shipping fresh produce from areas further south.

In Britain there are some 2580 hectares of glasshouses, most of which are heated to optimal growing temperatures for a considerable part of the year. In fact, about 35 percent of the total agricultural use of energy in Britain is consumed in heating glasshouses, an amount that corresponds to 800,000 tons of oil plus 100,000 tons of coal annually (Sheard 1975). Heating currently accounts for about 30 percent of the cost of crop production under glasshouses, and so this technology of production is very sensitive to fluctuations in fuel prices.

These studies show that high-quality energy is being used in agricultural crop production at a nearly break-even efficiency. Current practices do not use such high-quality energy to amplify total energy storage in agricultural production. As we suggested at the beginning of this section, such amplification should be encouraged.

THE ENERGETICS OF LIVESTOCK AND PROTEIN PRODUCTION

In the developed countries, much of the total yield of crop agriculture, produced by the technology described above, goes to feed livestock and is thus converted into meat, milk, eggs, and other animal products. Of the 27.1 million tons of grain, legume, and vegetable protein produced in the United States in 1975, for example, 24.6 million tons, or 91 percent, was fed to livestock. This creates a major loss of energy

TABLE 24-7
Energy cost of animal protein production (per hectare) in the United States.

| Animal product | Animal protein yield (kg) | Feed protein input (kg) | Feed energy input (10^3 kcal) | Fossil energy input (10^3 kcal) for the production of: | | | Labor (man-hours) | Kilocalorie ratio | |
				Feed	Feed and animal	Animal		Feed input/ protein output	Fossil energy input/ protein output
Milk	59	188	6,963	2,382	8,561	6,179	23	30	35.9
Eggs	182	672	14,406	6,070	9,560	3,490	174	20	13.1
Broilers	116	651	8,886	6,446	10,233	3,787	38	19	22.1
Catfish	51	484	5,007	2,180	7,068	4,888	55	25	34.6
Pork	65	689	17,021	6,774	9,212	2,438	28	65	35.4
Beef (feedlot)	51	786	24,952	7,129	15,845	8,716	31	122	77.7
Beef (rangeland)	2.2	33	1,420	0	89	89	1	164	10.1
Lamb (rangeland)	0.17	3	128	2	11	9	0.2	188	16.2

Source: Pimentel et al. Copyright © 1975 by the American Association for the Advancement of Science.

and protein through the food chain, so that, in mechanized agricultural systems, the cultural energy subsidy per calorie of animal food products becomes very large. For beef and pork production in the United States, about 0.2 and 0.6 calorie respectively, are obtained for each calorie of cultural energy invested (Heichel 1975).

With animal foods, however, the energy value of the food material is not as great a consideration as the protein content. We should therefore examine animal agriculture in terms of the energy efficiency of unit protein production, and in terms of the ratio of protein in animal food products to that in the feed consumed. The caloric and protein investments in animal food production must also be determined and must include both the inputs needed to maintain the "nonproductive" members of breeding herds and the inputs to animals whose production is actually harvested.

The industrial energy and labor inputs for various systems of livestock and dairy production in the United States have recently been

estimated by Pimentel and a number of his associates (1975). These analyses show that milk production is the most efficient system of converting feed protein (total protein in food consumed) into animal protein; but it is not the most efficient in terms of energy (Table 24-7). In the process of milk production, about 31 percent of the protein in feed is converted into milk protein, but about 36 calories of cultural energy are needed for each calorie of milk protein yielded. Of course milk also contains food energy in the form of butterfat.

According to Pimentel's analyses, poultry are among the most efficient producers of animal protein (Table 24-7). Egg production systems realize about a 27 percent protein conversion efficiency, not greatly below that for milk production, but they require an industrial energy subsidy of only about 13 calories per calorie of egg protein. Broiler production is efficient compared to other systems of meat protein production, showing a protein conversion ratio of nearly 18 percent and requiring only about

FIGURE 24-11.
Feedlot production of beef, shown in this 40,000-head feedlot near Billings, Montana,
requires almost 80 calories of cultural energy input per calorie of protein obtained.
(Photograph by E. C. Nielson, courtesy of the USDA Soil Conservation Service.)

22 calories of industrial energy investment per calorie of protein yielded.

Pimentel's data also show the dramatic contrast between the energy efficiency of beef production in feedlots and on rangeland. In both systems the efficiency of converting food protein into beef protein is about the same. The industrial energy required for feedlot production, however, is 77.7 calories per calorie of beef protein, which is 7.7 times that for rangeland beef (Figure 24-11). One of the principal reasons for the low efficiencies shown by beef production are the large inputs required to maintain the breeding herd.

Overall, in the United States, some 26.1 million tons of feed proteins are fed each year to livestock. From this come about 5.3 million tons of meat, milk, and eggs that Americans consume. The protein conversion ratio for the whole system is thus about 20 percent. Taking into account the fact that Americans annually consume about 54 kilograms of beef, 30 of pork, 23 of poultry, 6 of fish, 2 of lamb, 129 of milk products, and 36 of eggs, the average industrial energy subsidy for animal food protein is about 30 to 35 calories per food calorie.

Oceanic fisheries are frequently looked to as a source of inexpensive, high-quality protein. Data on the energy costs of trawlers (with freezer storage) that operate off the coast of England, however, indicate that the subsidy for fish at the dockside amounts to about 20 calories per calorie of fish (Leach 1973). For distant ocean fishing, this subsidy is obviously higher.

For the world as a whole, most of the protein in the human diet comes from plant foods (see Chapter 1). The very large subsidies of industrial energy for animal protein, and the increasing cost of such energy, mean that plant protein will probably become more important in the future. Estimates of the energy efficiency of plant protein production under mechanized agriculture vary widely (see Heichel 1975; Pimentel et al. 1975), but all show that the greatest efficiencies are achieved by legumes. Alfalfa, which has one of the highest nitrogen-fixing abilities of any major crop, is the most energy efficient crop (Figure 24-12). This plant is a forage, of course,

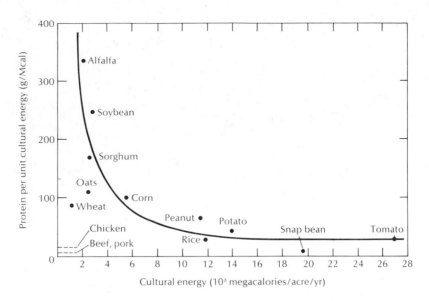

FIGURE 24-12.
Protein yield per unit of cultural energy input for various crop plants. (Modified from Heichel 1975.)

and the protein contained in it is unavailable to man without extensive processing either by livestock production or by expensive industrial technology. Soybean protein is the most efficiently produced nonanimal protein that can be processed easily for human consumption; the energy cost is about 2 to 3 industrial calories per protein food calorie. Hannon et al. (1976) have estimated that the total cost of preparing textured soybean protein (for use as a food additive) would be about 13 to 14 industrial calories per protein calorie, while processing into Unitex, a complete meat analogue, would be possible at a cost of about 15 industrial calories per protein calorie.

For the world as a whole, about 47 to 50 percent of dietary protein intake comes from grains. We should note here that corn and sorghum, two species with efficient photosynthetic systems (see Chapter 22), occupy positions of some importance. We have already discussed the efficiency of corn in food energy produc-

tion, noting that it was higher than that of wheat and rice. The digestible energy yield of sorghum under intensive cultivation is also high —5.3 calories per calorie of cultural energy invested (Heichel 1973). Since the yields per hectare of corn and sorghum are high, their protein content relatively high, and their cultural energy requirements only moderate, their protein production efficiency is also high, lying in the range of 3 to 5 cultural calories per calorie of food protein.

Thus, within intensive agricultural systems, we find a wide range in the efficiency with which food protein is produced. For some of the currently most preferred animal protein products, including milk, pork, and feedlot beef, values lie in the range of 30 to 80 calories of input energy per protein calorie. For efficient small-animal production systems and systems for producing meat analogues from soybean protein, values lie in the range of 10 to 20 input calories per food calorie. Finally, efficient grain production systems yield one calorie of protein for every 3 to 5 calories of cultural energy invested.

IRRIGATION, FERTILIZATION, AND ENERGY NEEDS

Irrigation is a major technique of mechanized agriculture, and one of the main goals for future agricultural development is to increase the percentage of arable land irrigated from assured water sources. The energy cost of irrigation accomplished by the pumping of deep wells or by transporting water long distances is considerable, however. Likewise, the cost of water supplied through gravity flow systems usually includes, in reality, large energy costs for the construction of dams and canals that make such flow possible. It is evident from several studies that these costs are a major portion of the energy investment of modern agriculture. In Nebraska, where irrigated farming has become much more important in recent years, more than 9000 center-pivot watering systems (Figures 24-13, 24-14) are now in operation, each of which draws water from an average well depth of 180 feet at a rate of 900 gallons per hour (Splinter 1976). The fuel requirements for pumping water in these and other irrigation systems in that state constitute an estimated 43 percent of the total energy input to agriculture in that region.

Estimates of the average cost for the application of irrigation water are only approximate. Pimental et al. (1973) estimate 7.34×10^6 kcal/hectare-meter, based on studies in Nebraska. Calculations derived from data on fuel use for the center pivot systems (Splinter 1976) estimate 8.0×10^6 kcal/hectare-meter.

FIGURE 24-13.
A center-pivot irrigation system in operation near Amarillo, Texas. A pump at the center of the system draws water from a well and creates the necessary pressure in the sprinkler system. Wheeled supports at intervals along the main pipe permit the system to pivot in a circle. The system is moved by small gasoline or electric motors. (Photograph by F. S. Witte, courtesy of the USDA.)

FIGURE 24-14.
Closeup view of a center-pivot sprinkler irrigation system for pasture land in Washington. This system is propelled by electric motors. (Photograph by Ron York, courtesy of the U.S. Bureau of Reclamation.)

Considering that these represent only the immediate fuel costs of these systems, and not the energy costs of constructing them, it is clear that total energy costs must be greater, with 8.1×10^6 kcal/hectare-meter being our minimum estimate. This value may not differ greatly from the total energy cost for other irrigation systems, when the energy costs of constructing dams, canals, and other distribution systems are fully considered. In the United States, current water use for irrigation is about 16.4 million hectare-meters (see Chapter 12). Thus, a rough estimate of the energy cost of irrigation in the United States is about 133×10^{12} kcal/year. This makes irrigation one of the most important energy users in the agricultural sector (see discussion of national energy budgets below.)

Fertilizers are also energy-expensive inputs. Nitrogen fertilizers are manufactured largely from natural gas or from petroleum feedstocks by a process that requires high temperatures and pressures (see Chapter 12). The energy cost of this process is about 17,600 kilocalories per kilogram of nitrogen (Pimentel 1974). Phosphorus and potassium fertilizers are obtained by mining and processing mineral deposits rich in these elements. The energy costs are about 3190 kilocalories per kilogram for phosphorus and 2200 kilocalories per kilogram for potassium (Pimentel 1974).

The use of fertilizers is absolutely essential to the high productivity of mechanized agriculture. In the example of midwestern corn farming discussed earlier, total fertilizer use increased more than tenfold from 1945 to 1970,

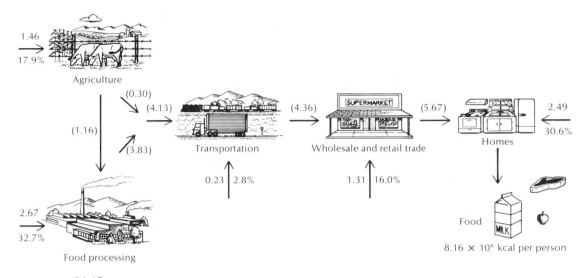

FIGURE 24-15.
Cultural energy inputs to various components of the food supply system as a whole in the United States. (Modified from Hirst 1974.)

and the use of nitrogen fertilizers alone increased sixteenfold (Table 24-3). By 1970 these fertilizer inputs accounted for more than 36 percent of the total energy costs of corn production (Table 24-4).

Using these figures, and considering the use of nitrogen, phosphorus, and potassium fertilizers in 1972 and 1973, Pimentel (1974) has estimated that the total energy cost for fertilizers in North America is about 166.2×10^{12} kcal/year. About 90 percent of this, or 150×10^{12} kcal, applies to the United States. These estimates, of course, apply simply to production and not to transportation or application.

It is important to emphasize the dependence of irrigation and fertilizer technology on fossil fuels, and the degree to which changes in the price of fossil fuels will be reflected in the costs of these practices. Furthermore, it is necessary to note the dependence of modern mechanized crop production systems upon heavy inputs of both water and fertilizers.

NATIONAL AGRICULTURAL ENERGY BUDGETS

Data like those described above may be combined to estimate the total energy input to food production as well as to "downstream" portions of the food supply system: processing, marketing, and preparation. For the United States, two independent comprehensive studies have been made to date. Hirst (1974) calculated total energy costs for the United States food supply system for 1963 (Figure 24-15). These calculations show that the actual production of food consumes a smaller fraction of energy than other portions of the system. The percentage of energy expended in production was 17.9, while 32.7 percent was used for processing and 30.6 percent for preparation in the home. Overall, Hirst estimated that for each calorie of food energy produced, 1.14 calories of industrial energy and 0.01 calory of human energy were expended. For the food supply system as a whole,

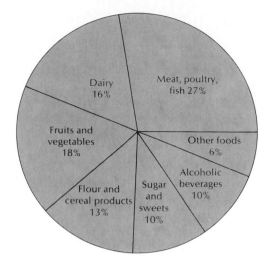

FIGURE 24-16.
Cultural energy input to major food classes in the food
supply system of the United States. (From Hirst 1974.
Copyright © 1974 by the American Association for the
Advancement of Science.)

this means that for each calorie of food actually
consumed, about 6.4 calories of energy are expended. It is also interesting to note that, when
energy costs are broken down by major food
groups, 43 percent are associated with dairy
and meat production and some 10 percent with
the production of alcoholic beverages (Figure 24-16).

Steinhart and Steinhart (1974) have also compiled estimates of energy use for the entire food
system for various years from 1940 through
1970 (Table 24-8). Their analysis also indicates
that "on farm" energy costs are only a fraction
of the total input to the food supply system—
about 24 percent in 1970. It seems, however,
that the Steinharts' analysis has seriously under-
estimated the energy costs of irrigation and of
fertilizer production. Data presented by Pimentel (1974) suggest that they also underestimate
fuel use. Furthermore, in relating the energy
investment to food production, no account
seems to have been taken of the large quan-

tities—40 million tons in 1970—of food grains
exported.

Given these shortcomings, we have constructed an alternative energy budget estimate
for the United States in 1970 (Table 24-9). This
set of estimates suggests that slightly more
than two calories are invested per calorie obtained for all agricultural production, and
slightly over three calories are invested per
calorie of food produced for consumption
within the United States. Subtracting estimated
production costs for export grain, and combining the resulting internal food production
costs with the Steinharts' estimates of "processing" costs and of "commercial and home"
costs, we estimate that the total investment in
the United States food supply system is about
9.8 calories of cultural energy per calorie of
food actually consumed here.

Summaries of a similar nature have been developed for agriculture in the United Kingdom
(Blaxter 1975) and in Israel (Stanhill 1975). In
the United Kingdom, about 43 percent of the
total food consumed is imported. The energy
costs of producing the remaining 57 percent
average about 1.60 calories of cultural energy
per calorie of food leaving the farm. This value
is somewhat lower than that obtained above
for the United States. In large measure this
probably reflects the lower production and
consumption of meat per capita in the United
Kingdom, together with smaller energy expenditures for irrigation. The latter is of negligible
importance at present.

In Israel, where almost all fossil fuels must be
imported and where irrigation is a necessity in
virtually all areas, the energy picture is more
critical than in the United States (Stanhill 1975).
Israel ranks sixth in the world in the percentage—41.5 percent—of cultivated land under
irrigation (USDA 1974). The major energy
flows through the Israeli food supply system are
shown in Figure 24-17, in which it can be seen
that about 37 percent of the food for the human

TABLE 24-8
Energy use in the United States food supply system (kcal $\times 10^{12}$ per year) over the period 1940–1970.

Component	1940	1947	1950	1954	1958	1960	1964	1968	1970
On farm									
Fuel (direct use)	70.0	136.0	158.0	172.8	179.0	188.0	213.9	226.0	232.0
Electricity	0.7	32.0	32.9	40.0	44.0	46.1	50.0	57.3	63.8
Fertilizer	12.4	19.5	24.0	30.6	32.2	41.0	60.0	87.0	94.0
Agricultural steel	1.6	2.0	2.7	2.5	2.0	1.7	2.5	2.4	2.0
Farm machinery	9.0	34.7	30.0	29.5	50.2	52.0	60.0	75.0	80.0
Tractors	12.8	25.0	30.8	23.6	16.4	11.8	20.0	20.5	19.3
Irrigation	18.0	22.8	25.0	29.6	32.5	33.3	34.1	34.8	35.0
Subtotal	124.5	272.0	303.4	328.6	356.3	373.9	440.5	503.0	526.1
Processing industry									
Food processing industry	147.0	177.5	192.0	211.5	212.6	224.0	249.0	295.0	308.0
Food processing machinery	0.7	5.7	5.0	4.9	4.9	5.0	6.0	6.0	6.0
Paper packaging	8.5	14.8	17.0	20.0	26.0	28.0	31.0	35.7	38.0
Glass containers	14.0	25.7	26.0	27.0	30.2	31.0	34.0	41.9	47.0
Steel cans and aluminum	38.0	55.8	62.0	73.7	85.4	86.0	91.0	112.2	122.0
Transport (fuel)	49.6	86.1	102.0	122.3	140.2	153.3	184.0	226.6	246.9
Trucks and trailers (manufacture)	28.0	42.0	49.5	47.0	43.0	44.2	61.0	70.2	74.0
Subtotal	285.8	407.6	453.5	506.4	542.3	571.5	656.0	787.6	841.9
Commercial and home									
Commercial refrigeration and cooking	121.0	141.0	150.0	161.0	176.0	186.2	209.0	241.0	263.0
Refrigeration machinery (home and commercial)	10.0	24.0	25.0	27.5	29.4	32.0	40.0	56.0	61.0
Home refrigeration and cooking	144.2	184.0	202.3	228.0	257.0	276.6	345.0	433.9	480.0
Subtotal	275.2	349.0	377.3	416.5	462.4	494.8	594.0	730.9	804.0
Grand total	685.5	1028.6	1134.2	1251.5	1361.0	1440.2	1690.5	2021.5	2172.0

Source: From *Energy: Use and Role in Human Affairs* by Carole E. Steinhart and John S. Steinhart. Copyright © 1974 by Wadsworth Publishing Company, Inc., Belmont, California 94002. Reprinted by permission of the publisher, Duxbury Press.

TABLE 24-9

Approximate energy budget for United States agriculture in 1970 (based on several sources).

Cultural energy inputs	
Irrigation	133×10^{12} kcal
Fertilizers	150×10^{12} kcal
Pesticides	12×10^{12} kcal
Fuel (other than for irrigation)	326×10^{12} kcal
Machinery	101×10^{12} kcal
Electricity	107×10^{12} kcal
Total	829×10^{12} kcal
Food energy output	
Consumption (200×10^6 people at 1.095×10^6 kcal/ind/yr)	219×10^{12} kcal
Wastage of produced food (20% of above)	24×10^{12} kcal
Grain export (40×10^6 tons at 4×10^6 kcal/ton)	160×10^{12} kcal
Total	403×10^{12} kcal
Ratios	
Food consumed	
Input/output	3.08
Food exported	
Input/output	0.50
Total production	
Input/output	2.06

population and 43 percent of livestock feeds are imported, as are nearly all fossil fuels and agricultural machinery.

Energy costs for irrigation are the largest component of Israel's agricultural energy budget. Water is taken primarily from Lake Kinneret (Sea of Galilee) and lifted some 200 meters to be introduced into the distribution system that serves much of the country. Nearly 38 percent of the total energy invested for agriculture is expended in connection with irrigation.

Israeli agriculture, comparable in approach to that of mechanized agriculture in California, thus emerges as one of the most fossil fuel-dependent systems in the world. If we count imported animal feeds at their actual caloric value, the cost of the food produced for human consumption within Israel is about 3.59 calories per food calorie, or more than twice that for the United Kingdom. Under these circumstances, it might be thought that Israeli agriculture would emphasize the greatest degree of self-sufficiency possible in human food production. Actually, nearly the opposite is true. Their dependence on imported machinery and fossil fuels also means that they must generate large amounts of foreign exchange to pay for these materials.

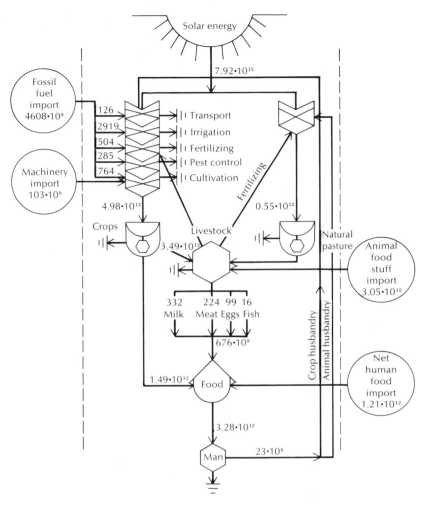

FIGURE 24-17.
Energy inputs to the national food supply system of Israel. Values are in kilocalories per year. (From Stanhill 1975.)

Much of the production in the country therefore consists of materials of high monetary value but relatively low energy content, such as citrus fruits, ornamental flowers, and cotton. These crops, given water and nutrients, do well in Israel's subtropical climate, and they find a ready market in Europe. Ornamental flowers are cut in the late afternoon, placed on cargo jets in the evening, and by morning are offered for sale in major European cities. Other crops similar to those just mentioned, such as bananas, are being grown more and more in Israel.

Thus the agricultural systems of many industrial nations, some of which are considered models of agriculture for the future, are highly dependent on a reliable supply of inexpensive fossil fuels. The application of modern fossil fuel-based technology to farming in Israel, for example, has created a good share of that country's so-called "modern miracle." Unfortunately, such success also creates vulnerability. A number of industrialized nations have made such heavy committments to energy-intensive agriculture that their production systems are at the mercy of supply and price manipulations by suppliers of fossil fuels. Clearly, much closer attention to these problems must be given by politicians and economists, as well as by agricultural scientists.

IMPROVING ENERGY EFFICIENCY IN MECHANIZED AGRICULTURE

It goes almost without saying that increased efficiency must be sought in agricultural systems. This will be a major consideration not only in this chapter, but also in the next, Chapter 25. Here we shall consider only some of the immediately applicable approaches that are evident from comparing existing systems and practices for energy efficiency. We should also note that major improvements in energy efficiency might also be made in portions of the food supply system "downstream" from the farm, including processing, distribution, and marketing, as well as home storage and preparation. As we have noted, these portions of the American food supply system consume almost two-thirds of the total energy expended in food related activity.

Two of the main characteristics of energy intensive agriculture are its high production per unit area and its minimum human labor input, and so some of the most obvious approaches to the problem of reducing energy costs relate to these two factors. Shifting production to less intensive systems spread over larger areas than are currently used offers the possibility of energy reduction. This is particularly true for livestock production, where increased dependence upon range-fed beef could greatly reduce the energy costs per unit of meat production. A complete shift of this sort, for cattle, swine, and poultry, would not be possible, however, without markedly reducing the total amount of animal protein produced; the estimated reduction in the United States would be from the current six million tons per year to only two million tons (Pimentel et al. 1975).

Directly substituting human labor for the use of machinery might also reduce energy use. Herbicides, for example, can be applied to agricultural land with tractor and sprayer rigs or with hand sprayers. Equally efficient application can be achieved either way, but about one-sixtieth as much energy is used to make the application with human labor (Pimentel et al. 1973). Here, of course, the problem is the cost of labor, which at current rates is about four times as great for application by hand. Thus, in the context of modern agriculture, the practical consideration is not to reintroduce unskilled manual labor in such a fashion, but to use increased amounts of skilled labor in a way that both reduces energy costs and either reduces or maintains the dollar costs of production.

As Spedding and Walsingham (1976) have argued, the principal value of energy analyses lies in comparing the efficiencies of alternative systems of producing particular desired outputs. An example is a comparison of corn production by alternative systems of tillage and fertilization (Heichel 1975). In this comparison (Table 24-10), conventional procedures included broadcast fertilization, liming, disking, plowing, planting, weed and insect control operations, and various harvesting operations. Total cultural energy input was 7.225×10^6 kcal/ha (about 2.5 times the average energy input to corn production in the United States!). Minimum tillage for corn or soybeans involves the use of a single implement that replaces separate disking, plowing, planting, and weed control operations with a single operation that simultaneously plants, fertilizes, and applies herbicide in untilled soil (Figure 24-18). This reduces the fuel use during preharvesting operations by about 80 percent, and it also somewhat reduces the indirect energy investment required in machinery. The need for

TABLE 24-10
Comparison of energy inputs to corn production by systems of
conventional tillage and fertilization, minimum tillage, and manure
fertilization.

Kcal × 10³/ha Item	Conventional tillage Inorganic fertilizers	Minimum tillage Inorganic fertilizers	Conventional tillage Manure fertilization
Fertilizers	3199	3199	—
Lime	49	49	49
Pesticides	109	178	109
Machinery	1806	1504	1934
Production fuel use	324	130	2523
Harvesting fuel use	1739	1739	1739
Total	7226	6799	6354

Source: Heichel 1975.

FIGURE 24-18.
Zero tillage equipment such as this, prepared for use in Illinois, permits simultaneous
planting and fertilizer application without previous soil preparation. Here, soybeans are
being planted in wheat stubble. The two larger white containers on the planting rig carry
dry fertilizer. Ahead of them, barely visible under the left end of the planting rig, are implements that cut furrows for the soybean seed, which drops into the furrows from the
four cylindrical containers. The wheels at the rear of the rig close the furrow over the
seed. (Photograph by William Schaller, courtesy of the USDA Soil Conservation Service.)

chemical herbicides is increased, however, by the loss of tillage operations that reduce weed growth. In all, the reduction in energy requirements of the total growing and harvesting process is about 6 percent.

Fertilization with manure, at a rate of 37 tons per hectare with conventional tillage, also reduces energy input requirements under certain conditions (Table 24-10). The reduction in this case is achieved by eliminating the cost of producing inorganic fertilizers. This saving is partially offset by the considerable increase in fuel costs associated with loading and spreading manure on the field. If manure is locally available, its use for fertilizer reduces total crop energy requirement by about 12 percent. However, if the manure must be hauled a distance of more than about three miles, the operation becomes uneconomical as an energy saving activity. We should note, however, that organic fertilizers have important functions other than the release of nutrients for crop growth (see Chapters 10 and 12). In any case, it is clear that alternative production systems could have significantly different energy requirements and that these deserve analysis to determine their potential contribution to energy conservation.

The foregoing analyses highlight an important question: Can the energy intensiveness of agriculture be reduced without diminishing the profitability of farming operations? This question has been examined recently by a group of scientists at the Center for the Biology of Natural Systems at Washington University in St. Louis (Lockeretz et al. 1975). This group compared the profitability and energy intensiveness of farming activities in 32 midwestern farms, 16 of which were "organic" farms that did not use highly processed inorganic fertilizers and pesticides, and 16 of which were conventional farms that did use these materials routinely. The farms in these two groups were as closely matched in size, soils, and location as possible. All used the grains and forages produced for feeding cattle and hogs for meat production.

The results of this comparison were striking. Yields per harvested acre were quite similar for the two types of farms for corn, wheat, oats, and soybeans. For matched pairs of farms (selected as identical in size, soils, and location), the yields of grains averaged less for organic farms, although this difference was not statistically significant. Because the organic farms relied on crop rotation systems involving temporary cultivation of grass and other noncommercial plants, the average area of harvested cropland was much less than that of conventional farms, 55.9 percent for the former, 77.5 percent for the latter (harvested acres relative to total acres). Thus the total income from harvested crops was much higher for the conventional farms, and gross production was correspondingly greater.

Organic farms incurred lower production costs because of their nonuse of pesticides and more expensive fertilizers. They relied on manures and rock phosphate to provide adequate nitrogen, phosphorus, and potassium. As a result, their costs were low enough to offset almost exactly their lower level of gross income. Considering all hectares that were ever in cropland at some stage in the rotational cycle, net profit per hectare was $331 for organic farms and $326 for conventional farms.

When the results of this comparison were examined for energy intensiveness per unit of harvested food material, an even greater difference appeared. The energy cost per dollar of harvested crop was 1.71×10^3 kcal for organic farms, 4.63×10^3 kcal for conventional farms, an almost threefold difference. The efficiency gained in energy use was therefore much greater than the difference shown in net profit.

Of course, the total output of food from organic farms was less, so that to realize the same level of production by organic methods on a regional basis, more acres would have to be brought into the farming cycle. Farming would

FIGURE 24-19.
Amish farmer plowing with a team of mules in central Pennsylvania. (Photograph courtesy of Richard Brown, University Park, Pennsylvania.)

have to become more "extensive" to give the same total output. Nevertheless, the savings in energy are great enough that, even if this extension occurred, total energy intensiveness of regional farming would be markedly reduced.

Johnson et al. (1976) have taken a second approach to the question of whether productivity per hectare can be maintained at a high level with less energy intensive technology. These workers compared the energy efficiency of production on Amish farms with that on selected non-Amish farms of similar nature. In central Pennsylvania, where dairy farms were compared, the study examined both Old Order Amish, who use several modern technologies such as modern refrigeration, and Nebraska Amish (i.e., a sect from Nebraska, now in Pennsylvania) who use only an old tractor for belt power (Figure 24-19). Non-Amish dairy farmers in this area invested about 1.81 calories of cultural energy per calorie of food energy at the farm gate (as milk). For Old Order Amish, this subsidy was lower, about 0.99 calories per output calorie, but production per hectare was slightly greater (4 percent) than on non-Amish farms. Nebraska Amish had an even lower energy subsidy, 0.66 calories per output calorie, but their production level was 47 percent lower than that of non-Amish farms. This study demonstrates that for some farming operations energy intensiveness can be significantly reduced without reducing yield, but it also shows that this is true only up to a point.

There are, of course, a great many facets to the agricultural energy question. For example, if energy intensiveness were reduced in a major region, how would the energy saved be utilized? If it were saved for use in farming in the same region at a later time, or if it were diverted to an area where it could be applied to upgrade farm production where need is great, the saving would be beneficial. If it were used for trivial nonagricultural purposes, however, the result could be detrimental.

ENERGY, AGRICULTURE, AND THE FUTURE

The realities of the dependence of intensive, mechanized agriculture on industrial energy are important for economic planning and development strategy. Although it is certainly true that agricultural energy use is only a small fraction of total use, equalling, for example, only 12 to 15 percent for the total food supply system in the United States (Pimentel et al. 1975), the changing relations between the supply and price of energy will be a major determinant of future patterns of agriculture.

Modernization of world agriculture is frequently considered as synonymous with the adoption of mechanized or "green revolution" technologies similar to those employed in the United States. However, several simple analyses indicate that it is already impossible to feed the current world population (slightly over four billion) at the United States level of nutrition with such a technology. Pimentel et al. (1975) have calculated, for example, that the annual energy input to American agriculture, for the total food supply system, is 1250 liters of gasoline per capita. If such inputs were employed to feed the entire world population, and if known petroleum reserves were the source of the energy involved, these reserves would be exhausted in only 13 years. This calculation of course is based on an improbable situation, but it serves to emphasize the position in which the human population exists with respect to an intensification of agricultural production based on fossil fuels.

As the human population grows and if the quality of diet is to improve, production per unit of agricultural land must be increased. Studies of protein yields per hectare from various systems show that the yields bear a close relation to energy subsidy (Figure 24-20). Stated simply, greater yields are obtained only with proportionately greater increases in energy subsidies (Slesser 1973). Other studies show that our ability to meet these increasing needs for energy-related inputs is becoming more and more limited. The natural gas supply in the United States is dwindling so that major impacts on nitrogen fertilizer supplies and prices are almost certain to develop within ten years unless significant new technologies are introduced (Chancellor and Goss 1976).

It seems certain that an increasing percentage of the world's energy use must in the future be directed into food production. Similarly, it seems certain that the agricultural systems of many of the industrialized nations will feel the severe stress of increasing energy costs. An agricultural policy that emphasizes efficiency in energy use and the ability to adjust to changes in its price and availability will be essential in meeting this challenge.

SUMMARY

The essence of agriculture is the application of cultural energy in order to direct biological production into channels where it can be utilized more easily by man. The greater the modification of natural processes that man attempts in this process, the greater is the energy cost he must bear. Nonmechanized systems of pastoralism and shifting cultivation use cultural energy in the form of human labor and realize returns that vary from 5 to nearly 40 calories of food energy per calorie of cultural energy invested. Permanent farming systems in which draft animals are used increase the cultural energy investment, yet still retain a very high efficiency in relation to the investment of industrial or fossil fuel energy.

In mechanized farming, however, industrial energy is invested in large quantities, enabling high levels of production to be achieved with minimum investments of human labor. For most grains, such as corn, wheat, and rice, these

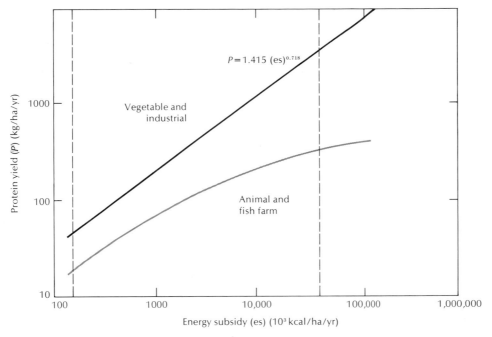

$P = 1.415 \ (es)^{0.718}$

Vegetable and
industrial

Animal and
fish farm

Protein yield (P) (kg/ha/yr)

1000

100

10

100   1000   10,000   100,000   1,000,000

Energy subsidy (es) (10^3 kcal/ha/yr)

FIGURE 24-20.
Protein yields per hectare per year as they relate to cultural energy subsidy. (Modified
from Slesser 1973.)

systems obtain only 1 to 3 calories of food energy per calorie of cultural energy. For many fruits and vegetables, and for all forms of animal protein, the return is less than the investment. The energy costs of protein production vary greatly, however. Plant protein in grains can be produced for about 3 to 5 calories per calorie of protein; concentrated plant protein analogues of meat for 10 to 20 calories; and milk, pork, and feedlot beef for 30 to 80 calories.

Irrigation, fertilizer production, and fuels for machinery are the major energy costs of mechanized agriculture. Analyses of total energy budgets for national farming systems reveal that, on the whole, food production alone requires from about 1.5 to about 3.5 calories of cultural energy per food calorie obtained, varying with climate and type of food products emphasized.

The heavy dependence of mechanized agriculture on fossil fuels makes the nations that depend on such systems vulnerable to price and supply manipulations. It also creates the need for careful analysis of alternative practices that might maintain high productivity at lower costs of fossil fuels.

Literature Cited

Blaxter, K. L. 1975. The energetics of British agriculture. *J. Sci. Fd. Agric.* 26:1055–1064.

Brown, L. H. 1971. The biology of pastoral man as a factor in conservation. *Biol. Cons.* 3:93–100.

Chancellor, W. J., and J. R. Goss. 1976. Balancing energy and food production, 1975–2000. *Science* 192:213–218.

Collier, B. D., G. W. Cox, A. W. Johnson, and P. C. Miller. 1973. *Dynamic Ecology.* Englewood Cliffs, N.J.: Prentice-Hall.

Hanks, L. M. 1972. *Rice and Man.* Chicago: Aldine-Atherton.

Hannon, B. M., C. Harrington, and K. Kirkpatrick. 1976. *The dollar, energy, and employment costs of protein consumption.* Presented at AAAS Annual Meeting, Boston, 23 Feb.

Heichel, G. H. 1973. *Comparative Efficiency of Energy Use in Crop Production.* Conn. Agr. Exp. St. Bull. No. 739.

——. 1974. Energy needs and food yields. *Technol. Rev.* 76:3–9.

——. 1975. *Energy use and crop production.* Presented at AAAS Annual Meeting, New York City, 30 Jan.

Hirst, E. 1974. Food-related energy requirements. *Science* 184:134–138.

Janick, J., R. W. Schery, F. W. Woods, and V. W. Ruttan. 1974. *Plant Science,* 2nd ed. San Francisco: W. H. Freeman and Company.

Johnson, W. A., V. Stoltzfus, and P. Craumer. 1976. Energy flow in Amish agriculture. (Unpublished manuscript.)

Leach, G. 1973. In *The Man-Food Equation.* A. Burne (ed.). New York: Academic Press.

Lockeretz, W., R. Klepper, B. Commoner, M. Gertler, S. Fast, D. O'Leary, and R. Blobaum. 1975. *A Comparison of the Production, Economic Returns, and Energy Intensiveness of Corn Belt Farms That Do and Do Not Use Inorganic Fertilizers and Pesticides.* St. Louis: Ctr. Biol. Nat. Syst., Washington University.

Odend'hal, S. 1972. Energetics of Indian cattle in their environment. *Human Ecol.* 1:3–22.

Odum, H. T. 1967. Energetics of world food production. In *The World Food Problem,* vol. 2. pp. 55–94. Washington D.C.: President's Science Advisory Committee.

Odum, H. T., and E. C. Odum. 1976. *Energy Basis for Man and Nature.* New York: McGraw-Hill.

Pimentel, D. 1973. Extent of pesticide use, food supply, and pollution. *Proc. N. Y. Ent. Soc.* 81:13–33.

——. 1974. *Energy Use in World Food Production.* Report 74–1, Dept. Ent. and Sec. Ecol. Syst. Ithaca, N.Y.: Cornell University.

Pimentel, D. L. E. Hurd, A. C. Bellotti, M. J. Forster, I. N. Oka, O. D. Sholes, and R. J. Whitman. 1973. Food production and the energy crisis. *Science* 182:443–449.

Pimentel, D., W. Dritschilo, J. Krummel, and J. Kutzman. 1975. Energy and land constraints in food protein production. *Science* 190:754–761.

Rappaport, R. A. 1971. The flow of energy in an agricultural society. *Sci. Amer.* 224(3):117–132.

Sheard, G. F. 1975. Energy utilization in the glasshouse industry. *J. Sci. Fd. Agric.* 26:1071–1076.

Slesser, M. 1973. Energy subsidy as a criterion in food policy planning. *J. Sci. Fd. Agric.* 24: 1193–1207.

Spedding, C. R. W. and J. M. Walsingham 1976. The production and use of energy in agriculture. *J. Agric. Econ.* 27:19–30.

Splinter, W. E. 1976. Center-pivot irrigation. *Sci. Amer.* 234(6):90–99.

Stanhill, G. 1975. Energy and agriculture: a national case study. In *Heat and Mass Transfer in the Biosphere*, part I. *Transfer Processes in the Plant Environment.* D. A. de Vries and N. H. Afgan (eds.). pp. 513–533. Washington, D.C.: Hemisphere Publishing Corporation.

Steinhart, J. S., and C. E. Steinhart. 1974. Energy use in the U. S. food system. *Science* 184: 307–316.

USDA. 1974. *The World Food Situation and Prospects to 1985.* Foreign Agricultural Economic Report No. 98. Washington, D.C.

25

THE FUTURE OF AGRICULTURE IN THE DEVELOPED WORLD

The developed nations, with about one-fourth of the world's population, possess almost half the arable land of the earth. The success with which this land is used will be critical for the entire human population over the coming decades. In this chapter we shall review the advantages and constraints of intensive agriculture as it is practiced in the developed nations and attempt to define ecologically sound ways to increase the productivity and efficiency of the systems involved.

Agriculture in the developed nations has both strengths and weaknesses. These countries, located almost entirely in the north and south temperate regions of the earth, contain most of the cropland that is best suited to continuous cultivation. In particular, the soils that developed under temperate forests and grasslands, many of them young glacial soils, have demonstrated a capacity to remain productive for several centuries under a variety of techniques of cultivation, fertilization, and irrigation. Although production is greatly reduced on most of this land during the winter, it must be remembered that much useful ecological work nevertheless goes on during this season. Microbial activity important to fertility and favorable soil structure occurs, soil moisture is recharged, and insect and weed populations are reduced.

Agriculture in the developed nations is also associated with a scientific and technological establishment of remarkable capability. This establishment, which has grown up in close association with mechanized agriculture, is capable of responding rapidly to many problems encountered by the farmer. For example, seed producers responded quickly to the southern corn leaf blight epiphytotic of 1970 (see Chapter 21), and American cotton farmers were able to cope with pest problems that had eliminated cotton farming in nearby regions of Mexico. This research and development ability makes feasible in developed nations certain approaches to future techniques that are effectively closed to many nations of the developing world.

We should also note that agriculture in the developed world is highly efficient in its use of labor because the labor force is skilled and has at its disposal powerful machinery to do the work of plowing, planting, treating, and harvesting. This means in turn that, given the required inputs of fuels, seeds, fertilizers, water, and other materials, major increases in total production can be obtained without a major increase in the size of the labor force.

These advantages are also coupled with constraints, however. The dependence of intensive mechanized farming on energy and purchased inputs makes this system vulnerable to supply and price manipulations. The degree of this vulnerability was made apparent by the petroleum embargo implemented by the OPEC nations in 1973. In general it means that the price of food materials produced in the developed nations is tied to the costs of many input resources, some of which are nonrenewable and limited. Moreover, because the labor force is skilled and efficient and because farm labor is expensive and the supply of potential labor quite limited in most rural areas, agriculture in the developed nations is constrained in its capacity to become more labor intensive.

Another constraint lies in the fact that the infrastructural commitment to mechanized farming is high. At all levels in the food supply system, from the farm up, individuals and corporations have made massive, long-term investments in machinery and facilities geared to mechanized farming. This investment extends well into the urban economy, where machinery, farm chemicals, and other inputs are manufactured. Thus, for every farm worker, there are two or more additional farm-support workers (Pimentel et al. 1973) who are tied very closely to the system of mechanized agriculture. As a result, agriculture in the developed nations can change only gradually, the alternative being severe disruption of both rural and urban economies.

Finally, because intensive mechanized agriculture is more a business than a way of life, the tendency to cut costs by externalizing them is very strong indeed. External costs, or costs that must be borne by society at large, are therefore significant in agriculture in the developed nations. Examples of such external costs include the effects of pesticide pollution, of fertilizer and salt discharges into natural waters, and of siltation from erosion on agricultural lands. For the most part, many of these costs went unrecognized until recent years. Now, however, their recognition and integration into the economies of food production systems is becoming increasingly important.

Although the future of agriculture in the developed nations is difficult to predict in detail, we can indicate several trends that will certainly occur and identify adjustments that, on ecological grounds, will become more and more important. Because the developed nations supply the basic food commodities in international trade, and because problems of food inadequacy will not be solved quickly in the developing world, the production of foods for export will become increasingly important, at least for some time. As this sector of the economy grows, an increased fraction of fuels and other productive resources is likely to be drawn into food production activity. Intensification of production is therefore likely to continue. At the same time, however, the increasing costs of energy and materials will create greater emphasis on the efficient use of these materials, a trend which will probably grow rapidly.

The future concerns of agriculture in the developed world will therefore center on four major areas: (1) increasing and safeguarding the supplies of productive inputs to agriculture, (2) improving the efficiency with which these inputs are used in food production, (3) developing new sources of basic food materials, and (4) coping with the external costs that tend to appear in intensive mechanized agriculture.

PRODUCTIVE RESOURCE SUPPLIES

In the recent past agricultural development has emphasized major land and water development projects that have high energy costs and require large, continuous inputs to realize

TABLE 25-1
Major patterns of cropland use in the United States from 1949 through 1974.

Cropland use	Millions of hectares						
	1949	1954	1959	1964	1969	1972	1974
Harvested	142	137	128	118	116	117	130
Crop failure	4	5	4	2	2	2	4
Cultivated summer fallow	11	11	13	15	17	15	11
Total for crops	157	153	145	136	135	135	146
Idle cropland	9	8	13	21	21	19	11
Total cropland	166	161	158	157	155	155	156

Source: Modified from Anderson et al. 1975.

high production levels. Dams, interbasin water transport projects, and other large-scale reclamation projects are examples. Although it is certain that some projects of this type will still be undertaken, we believe that the most economical approaches to increasing the supplies of productive resources for agriculture will be local and more modest in scale. We predict that these developments will center on four resource categories: land, energy, water, and fertilizers.

Land must be considered as the basic resource of agriculture, and the protection and efficient use of prime agricultural land a major goal of agricultural policy. In the United States little attention has been given to identifying and preserving productive farm land. Since 1949 the total amount of cropland has fluctuated between 155 and 166 million hectares (Table 25-1), with little indication of consistent decrease. However, this cropland area is dynamic in nature; the annual loss of about 607,000 hectares from agriculture to urban development is being compensated for by land brought into production by irrigation, drainage, and clearing (Anderson et al. 1975; Isberg 1973). Much of the land converted to urban use is first-class agricultural land, since many cities are located where they are because of proximity to productive agricultural land (Figure 25-1). In the

Santa Clara Valley of California, for example, the growth of San Jose and other cities between 1950 and 1970 consumed 57,000 hectares of first-class agricultural land (Isberg 1973).

It must also be recognized that intensive agriculture in the developed nations depends on particular land and soil conditions for high realized productivity. In Wisconsin, analysis of land use trends over the post-settlement history of that state showed that, since 1934, during the period of growth of mechanized farming, crop cultivation has been concentrated increasingly on soils with the most favorable moisture characteristics (Auclair 1976). Much Wisconsin land with less favorable conditions has been abandoned as farmland during this same period. In essence, where cultural practices have removed limitations once set by low fertility and other factors, moisture then becomes the critical factor. This trend will continue because of the increasing costs of inputs required for production through mechanized farming, and it is obviously important for the sake of agricultural efficiency that the best lands for mechanized farming be protected. Thus, in addition to developing new farmland where reasonable opportunities exist, we can predict a growing emphasis upon methods of protecting first-class farmland from other types of development.

FIGURE 25-1.
Sprawling residential subdivisions, such as these developments east of Dallas, Texas, annually remove about 607,000 hectares of cropland from production in the United States. (Photograph by Hugo Bryan, courtesy of the USDA Soil Conservation Service.)

Near cities, much of the loss of farmland to urban uses occurs because of two factors: a higher value of the land for development, and higher taxes based on this potential value. One approach to land preservation therefore is **differential assessment.** Under an arrangement of this sort, farmland is valued for tax purposes in accordance with its actual use rather than its development potential. In receiving this tax break, the landowner can agree to pay a penalty if he sells to a developer, or perhaps forfeit his right to make such a sale for a certain period of time. Programs of this type are now operating in California and New York. In California, local governments are empowered to create agricultural preserves in which landowners can contract for farming-based assessments for ten-year periods, forfeiting in the process their rights to other development. In 1973–1974, some 12.6 million acres, or about 35 percent of the state's farmland, was in such

status (Anderson et al. 1975). One of the difficulties of such a program is that individuals holding land parcels most in danger of development are least likely to participate.

Stronger systems are being considered in New Jersey (Blueprint Commission on the Future of New Jersey Agriculture 1973), Connecticut (Connecticut Governor's Task Force 1974), and a number of other states. These systems propose that local municipalities designate agricultural preserves that contain major fractions of the remaining high-class farmland of the area, and that a state administration agency purchase the development rights for these lands. In New Jersey, the goal of this program is to preserve 0.4 million hectares of farmland, three-fourths of which would be of high quality. In Connecticut, the goal is to preserve some 131,000 of the remaining 202,000 hectares of agricultural land in the state.

Similar ideas have been put into practice in some other countries, including Canada and West Germany (Reilly 1976). In Germany a strong zoning system that designates development lands and nondevelopment lands combined with a favorable property and estate tax structure favor the preservation of agricultural land. Furthermore, the purchase of farmland is limited to persons with farming experience or education, which reduces the acquisition of farmland for purely speculative purposes.

Energy is the second most critical resource for mechanized agriculture. Energy use in agriculture can only be considered within the framework of national energy policy, since agricultural energy requirements, although critical, are only a small fraction of the country's total energy needs. We should note here that a number of the major energy development projects now under way, including Alaskan oil development and the North Sea oil development, are marginal in their net energy characteristics. In other words, the energy invested to obtain these resources is a large fraction of the resource energy obtained, the incentive being simply the rising price of petroleum.

We can identify several areas that need emphasis with regard to the supply of energy for agriculture. First, because liquid fuels are critically important to the operation of farm equipment, we believe that programs to conserve such fuels in nonagricultural sectors of the economy are justified. The greatest uses of petroleum are in transportation and space heating, and major savings could be achieved here that would preserve fuels for more essential uses in agriculture. This effort is justified on economic grounds because of the rising importance of agricultural produce as export from the developed countries.

Second, we believe that efforts should be intensified to develop alternative fuel systems for farm vehicles. These efforts should explore liquid fuel systems based on fossil fuels such as coal and on other materials, such as methane, and should consider nonliquid fuel systems as well. The possibility of practical, on-farm production of fuels from wastes and crop residues should also be carefully assessed.

Finally, systems of wind and solar energy might be employed more easily in agriculture than in any other sector of the economy, and their potential to supply energy for pumping water, drying harvested materials, heating and cooling, and generating electricity must be exploited effectively (Figure 25-2). Thermal effluents from power plants and certain other industrial operations also offer limited opportunities for agricultural use. We shall discuss these later in this chapter.

The possibilities proposed for increasing the supply of water for agricultural use exemplify the dilemma of mechanized agriculture. Demand for additional irrigation water in various regions and river basins in the United States has stimulated interest in massive interbasin transfer systems. In addition to projects already in operation, such as the California Water Project, two major interbasin transfer systems have been seriously proposed: the Texas Water Plan and the North American Water and Power Alliance. The former proposes to transport water westward from the Mississippi River to high plains areas of Texas and Oklahoma; the latter to transfer water from the Yukon River in northern Canada and Alaska southward through the Columbia River Basin into the southwestern United States (Figure 25-3). The costs of the energy required to pump water through such systems are considerable. In California, approximately 38,880 kilowatt-hours of electricity per hectare-meter of water are required for the total pumping lift of 1154 meters over the entire length of the California Water Project (Figure 25-4). Present estimates are that, when existing contracts for this power terminate in 1983 and are renegotiated, the

FIGURE 25-2.
An experimental solar grain-drying apparatus. The solar collector at the left feeds hot air into the small building where heat is stored in a rock-filled bin. Air is then blown through the rock bed and into the metal grain-drying bin at right. The solar collector is movable and can be adapted to other heating functions. (Photograph courtesy of Charles C. Smith, Solar Energy Applications Laboratory, Colorado State University, Fort Collins.)

price per kilowatt-hour will increase tenfold (Charles F. Cooper, personal communication 1976). We must therefore question seriously the practicality of additional projects of this type.

Precipitation augmentation and **desalination** are two other approaches widely believed to have major potential for increasing water supplies. Cloud seeding to increase precipitation has, in fact, been characterized as an emerging technology. Two somewhat different objectives have been pursued in cloud seeding studies (Cooper 1973): to increase precipitation over agricultural areas deficient in moisture, and to increase precipitation over mountain regions, where the added moisture can be stored in reservoirs or in snowpack for later distribution via irrigation in downstream areas. In either case, it has become clear that seeding operations can only increase precipitation from cloud systems from which rainfall is imminent or in progress. Seeding orographic storm systems (see Chapter 7) appears to have the potential to increase precipitation by 10 to 15 percent under favorable conditions (Sax et al. 1975). Some potential does exist, then, to increase winter precipitation in areas such as the Rocky Mountains or the Sierra Nevada, and to make use of the added water through irrigation (Figure 25-5). Seeding convectional storm systems, the thunderstorms that bring much of the rainfall over the Great Plains during the summer, has been less successful. In either case the success and predictability of augmentation efforts are likely to remain highly variable, and little possibility exists for using these techniques to relieve droughts. Furthermore, because the effects of cloud seeding extend over a wide area and affect individuals who do not desire extra rainfall as well as those who do, programs of this type must be designed and carried out with the greatest caution (Cooper, Cox, and Johnson 1974). Augmenting precipitation thus appears to have only limited and local potential.

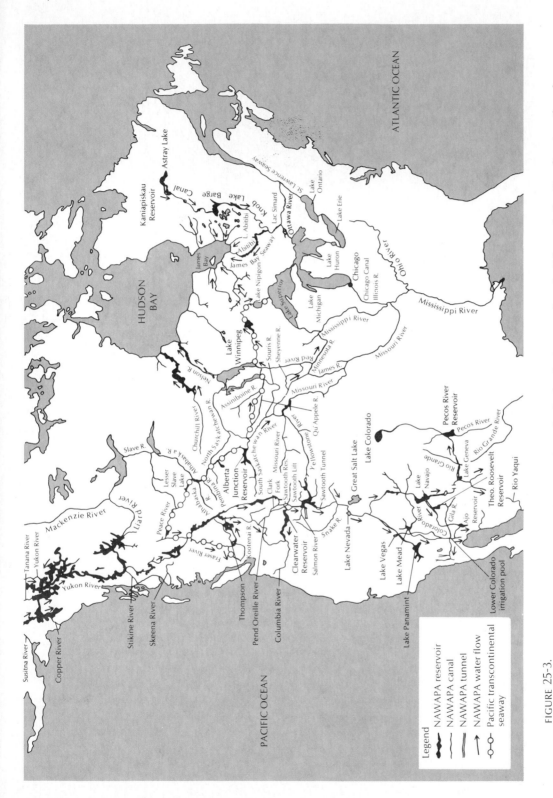

FIGURE 25-3.
The North American Water and Power Alliance (NAWAPA) is a plan for distributing water from the Yukon and other Canadian rivers through the central and western United States and a portion of northern Mexico. The water distribution system is combined with a series of hydroelectric power plants and an interocean seaway. The cost of such a system would be well over $100 billion, and 30 years or more would be required for its construction. (From H. B. Hawkes, in Guy-Harold Smith, *Conservation of Natural Resources.* Copyright © 1965 by John Wiley and Sons, New York.)

Legend

- NAWAPA reservoir
- NAWAPA canal
- NAWAPA tunnel
- NAWAPA water flow
- Pacific transcontinental seaway

FIGURE 25-4.
The Dos Amigos Pumping Plant of the California Water Project. This station lifts water as it moves south along the western edge of the San Joaquin Valley; it is one of several that provide a combined lift of 1154 meters. (Photograph by G.W. Cox.)

FIGURE 25-5.
One technique of seeding orographic storms to increase precipitation is to use ground generators in which silver iodide is burned, releasing billions of small silver iodide crystals that act as nuclei for the condensation of moisture droplets. Here a rancher is lighting such a generator as part of a U.S. Bureau of Reclamation experimental program for snowpack augmentation in the mountains of southern Colorado. (Photograph by W. L. Rusho, courtesy of the U.S. Bureau of Reclamation.)

Large-scale systems of desalination have also been viewed as an ultimate solution to water shortages in coastal deserts and in regions with brackish water supplies. Desalinated sea water is now being used to produce vegetable crops in small-scale greenhouse systems (Figure 25-6) in several locations, including the United States, Mexico, the Soviet Union, and Abu Dhabi (Fontes 1973; Hodges and Hodge 1971; Jensen and Teran 1971). Because these systems are enclosed, they suffer only very small losses of water by evaporation. They are expensive to construct and operate for other reasons, however, and are competitive in production costs only for high-value crops that would otherwise have to be shipped long distances (Figure 25-7). It has been argued that large-scale desalination systems, particularly those combined with nuclear power plants in agroindustrial complexes, will eventually produce desalted water cheaply enough for open irrigation of basic crops such as wheat, corn, and rice (Young 1970). Recently, this hope has been questioned by the Panel on Promising Technologies in Arid-Land Water Development of the National Academy of Sciences (NAS 1974). The costs of plant construction, operation and maintenance, waste brine disposal, and water distribution make such developments unlikely in the near future. Instead, small-scale operations, using solar distillation for the most part, will probably continue to be developed for local production of specialty crops. More promising for coastal desert areas are systems of salt or brackish water irrigation of crops selected for high salinity tolerance (see Chapter 26).

We suggest that the more profitable approaches to increasing water supplies involve reducing runoff, reducing storage and transport losses, and reclaiming waste water. These approaches can be implemented on-site and with minimum cost.

Collecting runoff and concentrating it directly on areas intended for cultivation is termed **water harvesting.** In desert regions, it has been suggested that water could be col-

FIGURE 25-6.
These greenhouses, located in Abu Dhabi on the Arabian Peninsula, utilize water produced by solar distillation to irrigate vegetable crops. (Photograph courtesy of Carl N. Hodges, Environmental Research Laboratory, University of Arizona, Tucson.)

FIGURE 25-7.
Inside the greenhouses shown in Figure 25-6, high-value crops such as tomatoes are grown in sandy soil and irrigated with desalinated water fortified with essential fertilizer nutrients. (Photograph courtesy of Carl N. Hodges, Environmental Research Laboratory, University of Arizona, Tucson.)

lected from large areas of otherwise nonproductive watershed that have been modified to increase runoff (see Chapter 26). Large-scale conversion of desert areas to barren, lifeless water catchments does not, however, represent wise use of the desert environment, considering its value for wildlife and recreation. Furthermore, such extreme modification is likely to have profound second-order effects relating to climate, groundwater systems, and flood probability. The development of limited catchments by such modification might, however, permit more extensive use of desert rangelands by grazing animals.

Despite its drawbacks, however, it is possible that the application of water harvesting techniques in areas of dry-land farming could significantly increase crop production over

wide areas. In Texas, experiments have shown that the construction of bench terraces (Figure 25-8) can harvest runoff from sloping areas between terraces and concentrate it on bench areas to eliminate runoff losses (Jones and Hauser 1975). In addition to controlling erosion, these terraces led to sorghum yields 43 percent greater than those obtained on similar unterraced land with normal runoff.

Effective increases in the amount of water available for irrigation can also be achieved by reducing seepage and evaporation losses from reservoirs and irrigation ditches. Seepage losses can be reduced by constructing lined or covered distribution systems (Figure 25-9), and by sealing the bottoms of reservoirs (NAS 1974). The technology for achieving such reductions is relatively well established, and

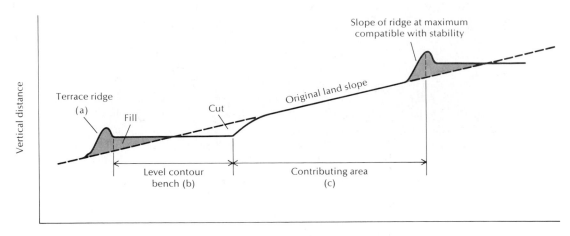

FIGURE 25-8.
Bench terraces may serve not only to prevent soil erosion, but also to capture and retain surface water. The level portion of the terrace (B) acts as a catchment for water from the sloping area (C). The terrace ridge impounds this water until it is absorbed. (From Jones and Hauser 1975.)

FIGURE 25-9.
This flow of 4 cubic feet per second was completely lost by seepage before the ditch was lined with concrete. The location is near Vermillion, Utah. (Photograph by Carl Wilker, courtesy of the USDA Soil Conservation Service.)

considerations relate mainly to cost. In areas such as the Imperial Valley of California, where evaporation losses from open water surfaces are almost three meters annually, investment in reducing evaporation is well justified. Major problems exist in reducing evaporation from open water surfaces, however. The application of chemicals, such as certain alcohols, that form monomolecular surface films has theoretical potential, but also involves many practical problems. Films that reduce evaporation but permit solar radiation to enter the water result in heating of the water mass, which creates various problems, including a tendency for increased evaporation when and where the protective film is disturbed. Other coverings that reflect or reradiate incident radiation might be practical on small bodies of water but would be difficult to use on the surfaces of large reservoirs, where wave action may be severe.

Perhaps the greatest opportunity for increasing the supply of water for agriculture in the vicinity of large cities is to reuse municipal waste water. A recent report notes that some 338 municipalities in the United States practice some sort of waste water reuse (Schmidt and Ross 1975). The great majority of these projects are located in Texas and California. Of the total volume of water reused, about 58 percent is for irrigation of crops, pastures, and landscaped areas. In almost all cases, crop irrigation involves plants such as tree fruits or corn, in which the applied effluent does not come into close contact with the harvested material. Use of waste water in this manner in effect eliminates the need for tertiary wastewater treatment, since nutrients are removed by biological uptake by the crops or plantings involved.

Supplies of fertilizers for intensive agriculture are subject to similar considerations. Phosphate fertilizers are now obtained by mining deposits of phosphate rock, which are in limited supply (Institute of Ecology 1972). The fact that man's use of phosphorus resources accelerates the flow of this element into the oceans and disperses fertilizer phosphorus as a fixed mineral in soils (see Chapters 10 and 12) has serious implications for the long-term future of man. Although new deposits of phosphate rock might be found and exploited, the costs of such recovery will certainly increase. At present, our mining and use of phosphorus are inefficient and careless. Mining operations now recover only about 40 to 60 percent of the phosphate in the ore removed. At the same time, it is predicted that the high-grade phosphate ores in Florida will become exhausted in about 30 years (NAS 1975b). Because this element is indispensable to agriculture, several reforms regarding the use of these resources should be implemented. These include curtailing the use of phosphates in major nonessential materials such as detergents; improving the recovery of phosphate in mining operations; and protecting tailings of low-grade ore that are stripped to allow access to high-grade ores, since these are equivalent to the ores that will have to be exploited in the near future.

Major adjustments must be made in the technology of manufacturing nitrogen fertilizers. The supply of natural gas, the most common hydrogen source for ammonia production in the United States, is dwindling so that half the total supply will be exhausted by 1980, even if use decreases in proportion to remaining supplies (Chancellor and Goss 1976). Cost and supply relationships will thus force us to use new sources of hydrogen, such as hydrolysis of water or processing of other fossil fuels. These will have the potential to produce large quantities of fertilizer, but production costs will also be much higher.

The increasing costs of inorganic fertilizers will stimulate the use of alternatives, including animal manures and urban wastes. We believe that this is not only an inevitable trend, but one that will prove beneficial. A strong case can be made for the conclusion that exclusive reliance on inorganic chemical fertilizers has in many instances created a vicious circle. Failure to maintain adequate levels of organic matter in many soils has led to a reduction in their nutrient-holding capacity and has increased their bulk density and erodibility. The ability to counteract these changes over the short run by abundant use of highly available fertilizers and more powerful cultivation equipment has permitted continuous cultivation of soils under conditions that lead to progressive degradation. To a great extent this trend has been fostered by commercial interests that have actively publicized the short-term effectiveness of marketable supplies and equipment in coping with fertility problems. As a result, however, the farmer is often placed in a position where his use of these effective production inputs increases his dependence upon them and his quantitative need for them.

Nevertheless, the farmer must make his decisions on a fairly short-term basis, and the economics of using organic materials are often unfavorable. Such materials are bulky and expensive both to transport and to apply. Furthermore, little has been done to develop more efficient systems for distributing and applying these materials, since this has not emerged as a commercially attractive area. Instead, organic residues have been treated as having only negative value, that is, the costs of their disposal. It is clear that programs for utilizing organic wastes in agriculture should count reduced disposal cost, as well as the direct agronomic value of these materials, as economic benefits.

Animal manures have proved useful agents for maintaining soil fertility, although, as several studies have shown, manures and inorganic fertilizers in combination provide even better performance (see Chapter 14). Manures from different livestock species differ somewhat in nutrient content (Table 25-2) but contain large amounts of the basic fertilizer nutrients, nitrogen, phosphorus, and potassium. These nutrients are in organic form and are released gradually by microbial action. Roughly half the quantities present are released during the first year after application (McCalla 1974), which must be taken into account in planning application levels. For nutrients to be largely retained within the soil system, manures should be plowed or disked into the soil shortly after application. Otherwise nutrients can be lost through volatilization of nitrogen and through surface runoff carrying soluble nutrients. From the data in Table 25-2 it can be seen that applying 25–50 tons of manure per hectare provides a level of fertilization adequate for most crops. As we noted earlier (Chapter 24), this

TABLE 25-2
Characteristics of various animal manures.

| Animal | Moisture (%) | Kg/ton manure | | | | | | | | |
		N	P	K	S	Ca	Fe	Mg	Volatile solids	Fat
Dairy cattle	79	5.6	1.0	5.0	0.5	2.8	0.04	1.1	166	3.5
Finishing cattle	80	7.0	2.0	4.5	1.8	1.2	0.04	1.0	198	3.5
Hogs	75	5.0	1.4	3.8	1.4	5.7	0.28	0.8	200	4.5
Horses	60	6.9	1.0	6.0	0.7	7.8	0.14	1.4	193	3.0
Sheep	65	14.0	2.1	10.0	0.9	5.8	0.16	1.8	284	7.0

Source: Modified from McCalla 1974.

TABLE 25-3
Yields of leaf protein for various species under various
harvesting conditions.

Species	Period (days)	Protein extracted (%)	Protein yield (kg/ha/day)
Medicago sativa (alfalfa)	50	57	3.76
	26 r[a]	57	8.65
	26 r	57	8.42
	26 r	57	8.88
	26 r	57	7.69
	26 r	57	9.19
Amaranthus paniculatus	46	52	4.52
	44 r	52	8.52
Pennisetum sp.	55	27	1.58
	25 r	27	2.00
Tithonia sp.	40	72	8.55

[a]Harvest of material regrown after earlier cutting.
Source: Data from Crewther 1976.

level of application requires a fair amount of fuel but can reduce the total energy intensiveness of the farming operation, providing the source of manure is nearby.

Urban wastes are also a source of potential fertilizer and soil conditioner materials (Carlson and Menzies 1971). These wastes fall into three categories: municipal garbage, treated sewage and sludge, and wastes from food processing plants. Municipal refuse consists of about 50 percent paper products, 20 percent food wastes, and 20 percent nondegradable materials. The nutrient value of this material is very low, and its best potential is as a soil conditioner. Various systems of composting with municipal refuse have been developed and utilized, especially in Europe. In the Netherlands, about 17 percent of all municipal refuse is made into compost and sold for various uses (Table 25-3). Much of it is used in specialty agriculture, including hotbed vegetable farming where the compost, through its decomposition, provides heat to warm the soil (Satriana 1974). Because this use is concentrated in specialty areas, the composting process has had to be somewhat refined. If more of the compost were used on field crops, less expensive composting procedures would be adequate. It should be noted that the sale of compost in European systems does not pay the cost of production; the operation is subsidized by local governments as a means of solid waste disposal. Other countries, including France, Germany, Italy, and Switzerland, also have significant municipal waste composting systems.

The potential of composting municipal waste has not been exploited fully in any region, however. In the United States, for example, such material could find an important role in the reclamation of eroded or strip-mined land. Composted materials could also serve as mulches in situations where they might inhibit weed growth, retain soil moisture, or prevent erosion.

In contrast to municipal garbage, sewage and sludge from sewage treatment plants possess fertilizer value comparable to that of animal manures. The disadvantages of these materials relate to the fact that, since they originate in urban rather than rural areas, they must usually be transported greater distances; moreover, considerations of human pathogens and materials toxic to crops are of major importance. Experiments conducted from 1963 to 1970 in the sprinkler application of municipal sewage effluent to various crop systems in Pennsylvania demonstrated that the dual ends of crop production and disposal can be served by this technique (Sopper and Kardos 1973). Applying effluent rates of 2.5 to 5.0 centimeters per week gave yields of various grains and forages equal to or better than those on control plots that received heavy applications of inorganic fertilizers. One of the species that converted the nutrients in effluent most efficiently into harvestable plant material was reed canary grass, a perennial forage species. The perennial nature of this species meant that early in the growing season a good standing crop already existed to take up nutrients, whereas corn, for example, had only a small standing crop of seedlings present (Figure 25-10).

The development of extensive cropland systems that receive effluent is now being tested (Bauer and Matsche 1973). In Muskegon County, Michigan, a system that draws effluent from a city of 160,000 people is now in operation. This system includes 2428 hectares of irrigated land designed to receive 2.3 meters of effluent per year. Because the effective irrigation period is only 30 weeks during spring, summer, and fall, the system also incorporates storage lagoons that allow the storage levels to fluctuate considerably. A similar but much larger system serving the Chicago Metropolitan Area has been proposed. If it is developed, this system will include several hundred kilometers of waste water transmission tunnels,

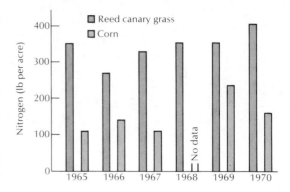

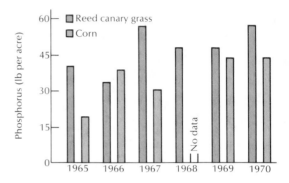

FIGURE 25-10.
Quantities of nitrogen and phosphorus in harvested forages under experimental irrigation with municipal sewage effluent in Pennsylvania. (From W. E. Sopper and L. T. Kardos, *Recycling Treated Municipal Wastewater and Sludge Through Forest and Cropland.* Copyright © 1973 by The Pennsylvania State University Press, University Park, Pa.)

194 square kilometers of lagoons, and 1813 square kilometers of irrigated area (Bauer and Matsche 1973).

A somewhat different approach to utilizing waste materials in agricultural production involves thermal discharges (Beall 1973; Oswald 1973). Several integrated approaches have been suggested for using heated water from the cooling systems of large power plants. One suggests that such thermal effluent be used to heat sewage effluent in an anaerobic system that would produce methane gas for fuel (Figure 25-11). The remaining effluent would then

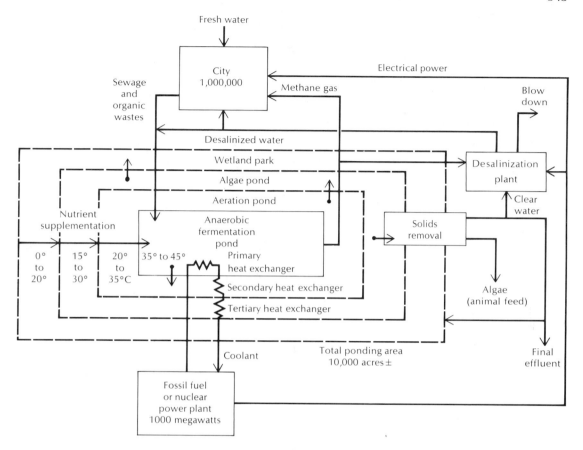

FIGURE 25-11.
In this hypothetical integrated system, municipal sewage is first treated by anaerobic fermentation, which leads to the production of methane. The effluent then passes to an algal growth pond. Harvested algae and other solids are used as animal feed, and water from the algal ponds may be reclaimed for other uses. The various ponds also serve as a source of water for cooling an associated power plant. (Reproduced with modification from Oswald, *Journal of Environmental Quality*, vol. 2, 1973, by permission of the American Society of Agronomy, Crop Science Society of America, and Soil Science Society of America.)

be aerated and used as fertilizer for the production of algae, which would then be harvested for animal feed. A second system (Beall 1973) envisions the use of thermal effluent to heat greenhouse systems in which complementary systems of plant and animal production would be established (Figure 25-12). In this "foodplex" system, most plant residues and animal wastes would be processed for use as inputs to other components of the production system. The technology of recycling animal wastes, particularly cattle, swine, and chicken manures, is now under careful study (Bhattacharya and Taylor 1975). The energy and protein contents of these materials are considerable, and various tests indicate that processed wastes can be recycled successfully when these materials constitute 5 to 40 percent of feed composition; this varies considerably with the animal and situation, however. Although

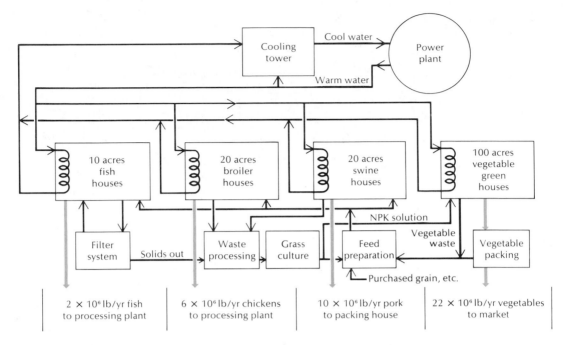

FIGURE 25-12.

A hypothetical integrated system in which thermal effluent from a large power plant is used to heat a food production complex in which crops and domestic animals are raised. Note that wastes from the various production units are processed and recycled as feed components or plant fertilizers. (Reproduced from Beall, *Journal of Environmental Quality*, vol. 2, 1973, by permission of the American Society of Agronomy, Crop Science Society of America, and Soil Science Society of America.)

complete recycling could almost certainly not be achieved, a great deal of it could be reused in a system involving vegetables, swine, broilers, catfish, and outdoor forage production areas. In both of the integrated systems described above, the agricultural operations would partially or completely cool the water heated by use in the power plant cooling system.

In the foregoing examples, many of the specific suggestions are perhaps optimistic or speculative. However, they do demonstrate the fact that much of what we now regard as wastes would be useful in food production. Major efforts should certainly be made to devise ways to make these wastes useful to agricultural systems in the developed nations.

EFFICIENCY OF RESOURCE USE

The key strategy for increasing the efficiency of agricultural production in the developed countries is to maximize net production. Considerable publicity has been given to increasing yields per hectare, or maximizing *gross* production. This has been done both by agricultural administrators, who see these yields as the measure of progress in meeting food deficiencies, and by agrobusiness interests, who are well aware that high yields are achieved with the aid of heavy inputs of purchased materials. The best interests of the farmer, and of society in the long run, are served by maximizing the difference between the value of the harvest and the cost of its production. In this light, the responsibilities of agricultural administrators are to see that resources and produce are valued appropriately for the long-term interests of the nation, that adequate supplies of productive resources are available, and that the production achieved is effectively distributed and utilized.

We shall identify several approaches to increasing the efficiency of resource use in modern agriculture. The first is to adjust water and nutrient supplies more closely to crop demands during the growing cycle. The technology of irrigation and fertilization, in the past, has meant that large quantities of water and fertilizers have had to be applied at certain times to ensure that needs are met over long portions of the crop cycle. These applications lead to large losses by runoff, leaching, or volatilization at some times, and to inadequate supplies at other times. It has been suggested, for example, that over half the irrigation water received at the farm level is wasted through overirrigation. A variety of irrigation systems including both drip and sprinkler systems, now exist as alternatives to past approaches. These systems can provide frequent irrigation and fertilization, and can also apply certain other agricultural chemicals, such as fungicides (NAS 1974; Rawlins and Raats 1975). Systems of this sort are expensive to install, but functional and economic improvements can be expected in them.

A second approach is that to minimize the demand of crop systems for costly inputs, especially water. Water is lost from crop systems through evaporation from the soil surface and transpiration from leaves. The use of various mulches can reduce soil evaporation losses (NAS 1974), and rates of transpiration might also be reduced if these are above the levels adequate to supply nutrients to aboveground portions of plants. Transpiration serves other functions, of course, including the regulation of leaf temperatures. A number of studies (e.g., Doraiswamy and Rosenberg 1974) have shown, however, that dusting crop canopies with kaolinite, a light, reflectant clay, reduces both leaf temperature and transpiration (see Chapter 8). The demand of crops for nutrients might also be modified, particularly by breeding programs that modify the distribution of assimilated materials between vegetative and food storage organs.

The demand for water might also be better adjusted to the supply if water supply forecasts can be made more accurate. Accurate forecasts would enable irrigation farmers to select crops and adjust acreages in a manner appropriate to the supplies of water to be expected to be available for the growing season (Moore and Armstrong 1976).

Greater efficiency in the use of farm machinery could also be achieved. During recent years there has been a strong trend toward increasing size and power in tractors and other machinery. This has reduced some labor costs, but at some point this advantage is offset by increased fuel costs (Pimentel et al. 1973). It may be possible to reduce costs through greater efforts to scale farm equipment to the job in question and to operate equipment at efficient speeds. In addition, developing more systems for sharing the use of specialized equipment could reduce costs related to equipment purchase and maintenance. Minimum tillage systems (see Chapter 24) also offer the potential to reduce energy inputs to certain crop systems (Triplett and Van Doren 1977). Here, of course, reduced fuel costs (Figure 25-13) must be balanced against the costs and problems associated with greater reliance on selective chemical herbicides.

The reintroduction of ecosystem controls to agroecosystems offers still another broad area of emphasis. We can cite several examples of this approach. Increasing the role of biological nitrogen fixation by incorporating

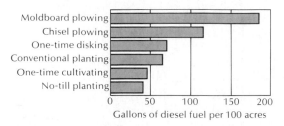

Gallons of diesel fuel per 100 acres

FIGURE 25-13.
Comparison of fuel costs for preparation and planting by
conventional and no-tillage systems, assuming use of a 100-
horsepower tractor. The fuel costs of conventional tillage
are those of plowing (with either moldboard or chisel
plow), disking, planting, and cultivating. Those of no-
tillage include that of the single planting operation. (From
"Agriculture Without Tillage" by Glover B. Triplett, Jr.,
and David M. Van Doren, Jr. Copyright © 1976 by
Scientific American, Inc. All rights reserved.)

legumes in crop systems or rotations and by
encouraging symbiotic associations between
nitrogen-fixing organisms and nonlegumes is
equivalent to reestablishing natural controls
on the nitrogen cycle (Figure 25-14). Increasing
the use of biological control and designing
crop systems that increase the potential for
biological control, is a second area of emphasis.
Grazing a combination of species of range animals might permit both a fuller exploitation of
range production and the control of nonpreferred species that tend to proliferate under
grazing by a single species (NAS 1975a). In
Wyoming, for example, range stocking rates for
combinations of sheep and pronghorn antelope
are 50 to 60 percent higher than for either species alone, reflecting the fact that food pref-
erences of these two species virtually do not
overlap (Figure 25-15).

Because one of the major trends in agriculture in the developed countries has been to
substitute fossil fuel energy for human labor,
it seems inevitable that as energy costs rise,
the reverse of this trend will follow. With
unemployment a critical problem throughout
the developed world, it seems logical that this
opportunity should be recognized and en-

couraged. We believe, however, that manpower
must be reintroduced to agriculture as skilled
rather than unskilled labor. In essence, we
are suggesting that trained individuals, through
their ability to make specific adjustments to
immediate and local conditions, can reduce
the cost of inputs and raise the value of production to an increasingly effective degree.

This opportunity is illustrated by the rapidity
with which independent pest management
consultation has grown as a profession (Hall
et al. 1975) as pest problems have escalated
(due to agroecosystem simplification, secondary
pest outbreaks, and evolution of resistance to
pesticide chemicals) and as the costs of chemicals and treatment have increased. In the San
Joaquin Valley of California, pest management
consultants, who operate on a per-hectare
contract with farmers, are able to provide
higher yields of citrus and cotton per hectare
at lower costs of treatment per hectare, and
the difference more than covers the cost of
the service.

We believe that similar professional opportunities are developing in several other aspects
of agroecosystem management, and that consultants could provide many more services,
including: (1) design of fields to minimize erosion and maximize water and nutrient retention;
(2) comprehensive fertility management of
cropland and range soils; (3) design and management of irrigation systems; (4) agrochemical
application, as distinct from prescription and
sales; (5) custom crop harvesting; and (6) range
condition and stocking advising. Several of
these professions already exist to a significant
extent, of course. The potential for developing
such professions needs to be evaluated further,
however, and programs to encourage their
development should be considered by state and
federal governments.

Finally, we suggest that as international
grain markets expand more emphasis will be
placed on rangeland systems of livestock production and on shortened finishing periods

FIGURE 25-14.
Vetch, a legume, is interplanted with grain sorghum in this field near Plainview, Texas. The vetch not only serves as a cover crop to reduce wind erosion, but adds a significant quantity of biologically fixed nitrogen to the soil. (Photograph by John McConnell, courtesy of the USDA Soil Conservation Service.)

FIGURE 25-15.
Rangelands, such as this area in Wyoming, can be stocked with a mixture of sheep and pronghorn antelope at densities 50 to 60 percent higher than those possible for either species alone. (Photograph courtesy of the USDA Soil Conservation Service.)

FIGURE 25-16.
Rangelands like this one in Santa Cruz County, Arizona, are highly susceptible to degradation by overgrazing. Scattered through this range, which is otherwise in good condition, are occasional plants of prickly pear, cholla, yucca, and mesquite. These species tend to increase rapidly when semidesert rangelands are overgrazed. (Photograph courtesy of the USDA Soil Conservation Service.)

for grass-fed animals in feed lots. To meet this need, more careful and intensive management of rangelands will be necessary, including greater attention to rangeland fertilization and condition. Under present conditions, a complete shift to livestock and animal protein production on rangeland in the United States would reduce total animal protein production by about two-thirds (Pimentel et al. 1975). If such a change were encouraged, therefore, strong incentives would exist to expand the area of rangeland and to intensify production practices. Such intensification will probably occur, and it must be pursued with care because when rangeland is overstocked, much of it is subject to degradation by erosion and invasion by undesirable plants (Figure 25-16).

The key strategy in the use of resources in agriculture in the developed world is thus to make production more dependent on renewable resources and to maximize the net return of production per unit of resources invested (Wittwer 1975).

NEW SOURCES OF FOOD PRODUCTION

A variety of systems have been suggested for the production of unconventional foods, primarily protein. When considered realistically, many of these are no more "unconventional" than margarine, beer, or shrimp. Of the proposed sources of new foods, we suggest three that deserve serious attention: leaf protein, single-cell protein, and insect protein.

The possibility of use of **leaf proteins** for human and livestock food has been investigated for over 30 years by N. W. Pirie (1942, 1975). Yields of leaf protein from plants grown under favorable conditions exceed those of seed protein taken from high-yielding legumes (Table 25-3). Of course plant species vary greatly in the quantity and type of protein present and in the rate of protein production. Furthermore, there are major differences in the extractability of total protein present in leaf material (Crewther 1976). In some species, enzymes or other chemicals released when leaves are crushed cause the precipitation of proteins onto the fibrous matter of the leaves, which prevents the proteins from being recovered in the liquid expelled. The various leaf components, however, can be put to use in several ways (Figure 25-17). The liquid from crushed leaves can be coagulated, to separate the proteins, which can then be refined. The remaining fluid still contains organic and inorganic nutrients, and can be used as a medium for growing microorganisms (see below). Finally, the fibrous material from which the protein-containing liquid was extracted still contains certain quantities of protein and other organic substances, and can be used as fodder for ruminant animals.

Ideally, the goal of leaf protein technology should be to obtain such protein from sources that do not compete with other agricultural production systems; for example, from the leaves of woody plants, sugarcane, or aquatic weeds. The most successful sources to date, however, are crop plants or plants grown under crop conditions. In addition, the greatest potential of leaf protein so far seems to be as an alternative to fish meal or soybean meal in poultry and livestock diets. Several commercial operations now exist for producing such feed materials in the United States (Crewther 1976).

Single-cell protein is obtained from microorganisms such as bacteria, yeasts, and algae that have been cultured on some medium (Loosli 1974; MacLennan 1976). Bacteria and yeasts are about 40 to 80 percent protein by dry weight and are capable of extremely rapid growth. Furthermore, microorganisms exist that are capable of growing on virtually all organic materials, various petroleum sub-

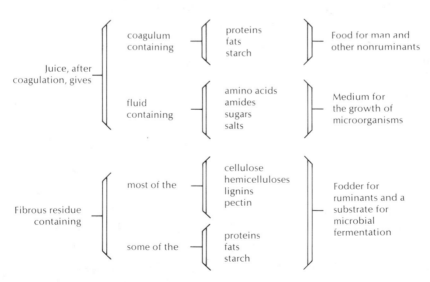

FIGURE 25-17.
Products derived by fractionation of fresh leaf material. (Modified from Pirie © CRC Press, Inc., 1976. Used by permission of the CRC Press, Inc., Cleveland, Ohio.)

stances, and certain inorganic media. Various yeasts, including brewers yeast from brewing operations and torula yeast from the fermentation of cellulose, have been used in animal feeds for many years (Loosli 1974). More recently, the petroleum industry has explored the use of straight-chain paraffins, which comprise about 10 percent of crude oil, as a substrate for yeast growth. Because protein from these sources may carry traces of certain carcinogenic hydrocarbons, it is likely that this will remain a source of protein for animal feeding. Several plants designed to produce protein in this fashion are in operation or under construction in the United Kingdom, France, Italy, Japan, the USSR, the United States, and a number of other countries. The future of this technology, as of all activities that depend on petroleum, is obviously limited and uncertain (MacLennan 1976).

In Australia, investigators at Sydney University are exploring a system of single-cell protein production based on starch fermentation (Figure 25-18). This system would use starch derived from a high-yielding but protein-poor crop such as cassava to produce a protein-rich derivitive (MacLennan 1976). Because starch is an abundant and renewable resource, free of carcinogens, this technique seems to have greater potential for human food production than systems based on petroleum. The yeast produced in this system could be used in relatively unrefined form for animal feeds, or could be refined further for use as human foods or food additives (Figure 25-19). Still other methods of culturing microorganisms on sewage

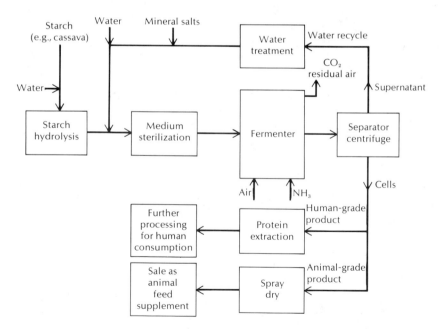

FIGURE 25-18.
In a system now being studied at the University of Sydney, Australia, starch from cassava or other sources is fermented by yeasts. The resulting yeast cells are then extracted or treated to obtain protein for human foods or protein-rich livestock feed supplements. (Modified from MacLennan 1976.)

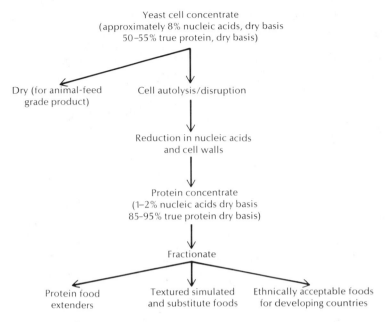

Yeast cell concentrate
(approximately 8% nucleic acids, dry basis
50–55% true protein, dry basis)

Dry (for animal-feed
grade product)

Cell autolysis/disruption

Reduction in nucleic acids
and cell walls

Protein concentrate
(1–2% nucleic acids dry basis
85–95% true protein dry basis)

Fractionate

Protein food
extenders

Textured simulated
and substitute foods

Ethnically acceptable foods
for developing countries

FIGURE 25-19.
Processing of yeast cell concentrates for human use requires fractionation to reduce the content of nucleic acids (for which the maximum daily adult intake should not exceed about 2 grams) and to increase the concentration of protein. The resulting concentrate may be used to produce meat substitutes and other foods for human consumption. (Modified from MacLennan 1976.)

and other organic wastes have been proposed that, if proven practical, would both provide animal feed and assist in waste disposal (Loosli 1974).

Insect protein is a source that has been almost completely overlooked (DeFoliart 1975). House fly pupae are 60 to 65 percent protein by dry weight, termites 36 to 46 percent, and locusts (grasshoppers) 50 to 75 percent. Scattered human cultures make use of insect foods, but major opportunities seem to exist for developing them as protein-rich components of domestic animal feeds. Poultry manure, as well as other organic wastes, may be used as a substrate for growing certain species. In other situations, large quantities of locusts may be collected; a record 274 tons were collected in Utah in the early 1900s when a bounty of 60 cents a bushel was offered (DeFoliart 1975). The diversity and abundance of insects, reflected in their importance as pests in all kinds of crop systems, suggests that man might benefit by seeking better ways to exploit them directly.

AGROECOSYSTEMS AND ECONOMICS

The development of agricultural economies tends to follow a general sequence—a sort of economic transition—that can be outlined in three stages (Headley 1972). In its early stages, as agriculture moves out of the subsistence phase, primary concerns are with organizing systems of trade and marketing and with evolving a political system compatible with these activities. In the second stage, as industrialization brings powerful resources to bear upon the production process, problems of supply and distribution of foods rise to prominence, and efforts tend to concentrate on continuously increasing production to meet overall needs. There is a third stage, however. At this point, both population densities and human exploitation of resources have reached high levels, so that further efforts to meet problems through increased production are often coupled with diseconomies equal to or greater than the productive gains. The appropriate strategies for this stage are to emphasize efficiency, reduce the detrimental side effects of intensified production, and adjust human populations to accord with environmental carrying capacity. The developed world is now in transition from the second to the third of these stages, yet most of the approaches advocated for dealing with food problems are still those appropriate to the second stage.

Diseconomies may be of two types: **external** and **internal.** External diseconomies, or **externalities,** relate to general features of the production environment that are essential to the production process, but for which no actual monetary values are taken into account. A general example is the air environment; clean air is essential to agricultural production and is "used" freely by many kinds of agricultural and nonagricultural enterprises. Some uses, however, can degrade air quality so that it interferes seriously with other uses. In the Los Angeles Basin in California, for example, urban air pollution has led to about a 50 percent reduction in citrus productivity (Thompson et al. 1970).

Certain other externalities are less obvious. The biotic organization of both natural and man-dominated ecosystems is a useful resource in that it exerts certain controls over the populations of organisms in those systems. All farmers rely on ecological organization; it is referred to as "natural" biological control. Yet this natural control can be weakened or destroyed by extensive use of chemical pesticides. This externality has obvious economic costs, just as air pollution does.

Another "common property resource" of agriculture is the set of prevailing weather conditions, which can be modified deliberately or inadvertently by urban air pollution that changes precipitation patterns and the frequency of hail (Huff and Changnon 1972; Changnon 1976) (Figure 25-20). Still another is the ability of the environment to accept wastes and outflows from agricultural land: water, eroded soil, fertilizers, pesticides, and other substances. Finally, we suggest that natural mechanisms of fertility regeneration, like natural biological control, are ecological resources that are significant but that can be damaged by careless management.

Internal diseconomies may exist because of the immediate values attached to various resources, goods, and services. In attaching value to various goods, the market does not distinguish among renewable resources, nonrenewable resources, manufactured products, or human services; all are valued on the same scale. We now realize that certain nonrenewable resources have been undervalued and as a result, consumed at a faster than optimal rate. Natural gas, an important fuel and raw material, has been consumed for such trivial purposes as heating swimming pools. The diseconomy in this case results from the fact that agriculture will soon be forced to turn to more expensive sources of raw materials for the production of nitrogen fertilizers.

To illustrate the fact that these sorts of diseconomies tend to increase as production intensifies, let us consider the example of continuous corn cultivation in the midwestern United States (Headley 1972). Shifting to continuous corn from rotation systems that included legumes and small grains has weakened

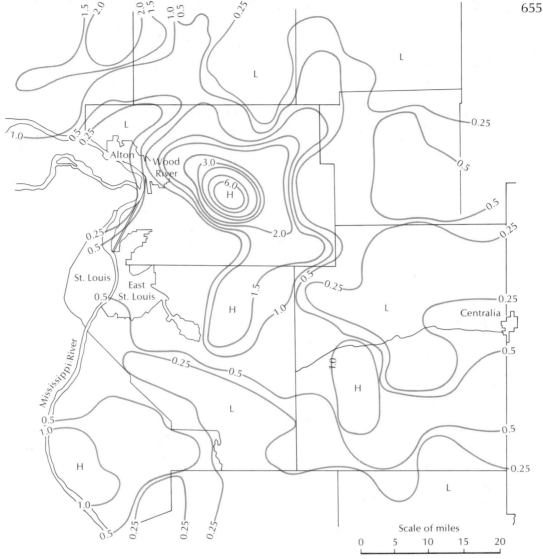

FIGURE 25-20.
Pattern of increased hail damage to agricultural crops downwind from the St. Louis, Missouri, metropolitan area. Isopleths indicate ratios of hail losses to insurance liability. (Modified from Changnon 1976.)

natural processes that relate to soil structure and nitrogen availability. As a result, the use of nitrogen fertilizers increased; this was made possible in part because of the unrealistically low cost of natural gas for fertilizer production. Without crop rotation, problems of pest insects, weeds, and diseases specific to corn proliferate and thus require intensified treatments. The evolutionary responses of these organisms, accelerated by continuous cultivation, require more intense research to develop more effective controls for them. With deteriorating soil structure and increasingly aggressive pests, the influences of unfavorable weather are magnified. These problems are not simply acts of God but the consequences of diseconomies incurred in the intensification of corn production. They must be recognized and taken into account.

Internalizing costs and adjusting them to optimal levels for the long-term best interests of society are enormous tasks. It will be increasingly necessary to assign certain rights to particular groups and to limit the rights of others. Soil conservancy laws (see Chapter 14) are one example of this necessary discrimination. In effect, these laws state that the public holds the rights to the condition and water quality of streams, and that the rights of land-owners to allow excessive erosional discharge into streams are limited.

Pest control practice has become one of the most active topics of definition and limitation of rights, and this will continue actively into the near future. At present, businesses have considerable freedom to advertise and recommend pesticides to the farmer; the chemicals involved must, of course, be approved for the uses recommended. Similarly, farmers have considerable freedom to select, determine the intensity of application, and choose the timing and frequency of application of various chemicals. Increasingly, however, these rights are being limited and defined, and it appears likely that chemical pest control will evolve into a prescription operation in which the right to determine type, quantity, method, and timing of pesticide use will be assigned to pest management specialists independent of both the supplier and the farmer.

SUMMARY

The future of agriculture in the developed countries is favored by an abundant supply of good land, a capable scientific and tech-nological establishment, and efficient use of skilled labor made possible by mechanization. It is constrained in a related manner by the high cost of labor, the infrastructural commitment to mechanization, and the strong tendency to externalize costs in mechanized farming.

Significant opportunities exist for increasing or safeguarding the supplies of land, water, energy, and fertilizers for use in mechanized agriculture. In general, however, the best opportunities do not appear to lie in developing new land areas, transporting water between major basins, and tapping new sources of non-renewable fuel and fertilizer resources. Instead they lie in the utilization of locally available resources, including materials now regarded as wastes. The preservation and efficient use of prime farmland is a major priority, together with increased reliance upon renewable sources of energy and other inputs. Water supplies may effectively be increased by systems that capture and retain rainfall on farmland itself, and by reducing storage and transport losses in irrigation systems. Increased utilization of manures and urban wastes offer one possibility for increasing fertilizer supplies.

The second emphasis of mechanized agriculture in the future will be on increasing the efficiency with which resources are used, which can be accomplished by adjusting input to crop needs, reducing unnecessary losses from crop systems, increasing the efficiency of equipment use, increasing the reliance of production upon ecosystem processes, and increasing the use of skilled manpower in farming.

New sources of food protein will be brought into production, including leaf protein, single-cell protein, and possibly even protein from more novel sources, such as insects.

The recognition of external diseconomies in agriculture will be a major area of concern. This will lead to the assignment of more rights to certain groups and will limit the rights of others.

Literature Cited

Anderson, W. D., G. C. Gustafson, and R. F. Boxley. 1975. Perspectives on agricultural land policy. *J. Soil and Water Cons.* 30:36–43.

Auclair, A. N. 1976. Ecological factors in the development of intensive-management ecosystems in the midwestern United States. *Ecology* 57:431–444.

Bauer, W. J., and D. E. Matsche. 1973. Large wastewater irrigation systems: Muskegon County, Michigan, and Chicago Metropolitan Region. In *Recycling Treated Municipal Wastewater and Sludge Through Forest and Cropland.* W. E. Sopper and L. T. Kardos (eds.). pp. 345–363. University Park, Pa.: Pennsylvania State University Press.

Beall, S. E. 1973. Conceptual design of a food complex using waste warm water for heating. *J. Environ. Qual.* 2:207–215.

Bhattacharya, A. N., and J. C. Taylor. 1975. Recycling animal waste as a feedstuff: A review. *J. Anim. Sci.* 41:1438–1457.

Blueprint Commission on the Future of New Jersey Agriculture. 1973. The report of the commission. Trenton, N.J.

Carlson, C. W., and J. D. Menzies. 1971. Utilization of urban wastes in crop production. *BioSci.* 21:561–564.

Chancellor, W. J., and J. R. Goss. 1976. Balancing energy and food production, 1975–2000. *Science* 192:213–218.

Changnon, S. A., Jr. 1976. Inadvertent weather modification. *Water Res. Bull.* 12:695–718.

Connecticut Governor's Task Force. 1974. Report of the task force for the preservation of agricultural land. Hartford, Conn.

Cooper, C. F. 1973. Ecological opportunities and problems of weather and climate modification. In *Modifying the Weather: A Social Assessment.* Western Geographical Series, vol. 9. W. R. D. Sewell (ed.). Victoria, B.C., Canada: University of Victoria.

Cooper, C. F., G. W. Cox, and A. W. Johnson. 1974. *Investigations Recommended for Assessing the Environmental Impact of Snow Augmentation in the Sierra Nevada, California.* Center for Regional Environmental Studies, San Diego State University.

Crewther, W. G. 1976. Leaf proteins for human consumption. *Search* 7:140–145.

DeFoliart, G. R. 1975. Insects as a source of protein. *Bull. Ent. Soc. Amer.* 21:161–163.

Doraiswamy, P. C., and N. J. Rosenberg. 1974. Reflectant induced modification of soybean canopy radiation balance. I. Preliminary tests with a kaolinite reflectant. *Agron. J.* 66: 224–228.

Fontes, M. R. 1973. Controlled-environment horticulture in the Arabian Desert at Abu Dhabi. *HortSci.* 8:13–16.

Hall, D. C., R. B. Norgaard, and P. K. True. 1975. The performance of independent pest management consultants in San Joaquin cotton and citrus. *Calif. Agr.* 29(10):12–14.

Headley, J. C. 1972. Agricultural productivity, technology, and environmental quality. *Amer. J. Agr. Econ.* 54:749–756.

Hodges, C. N., and C. O. Hodge. 1971. An integrated system for providing power, water, and food for desert coasts. *HortSci.* 6:30–33.

Huff, F. A., and S. A. Changnon, Jr. 1972. Evaluation of potential effects of weather modification on agriculture in Illinois. *J. Appl. Meteor.* 11:376–384.

Institute of Ecology. 1972. Cycles of elements. In *Man and the Living Environment.* pp. 41–89. Madison: University of Wisconsin Press.

Isberg, G. 1973. Controlling growth in the urban fringe. *J. Soil and Water Cons.* 28:155–161.

Jensen, M. H., and M. A. Teran R. 1971. Use of controlled environment for vegetable production in desert regions of the world. *HortSci.* 6:33–36.

Jones, O. R., and V. L. Hauser. 1975. Runoff utilization for grain production. In *Proc. Water Harvesting Symposium.* pp. 277–283. Phoenix, Arizona, March 26–28. Agr. Res. Serv. USDA (ARS W–22).

Loosli, J. K. 1974. New sources of protein for human and animal feeding. *BioSci.* 24:26–31.

MacLennan, D. G. 1976. Single-cell protein from starch: A new concept in protein production. *Search* 7:155–161.

McCalla, T. M. 1974. Use of animal wates as a soil amendment. *J. Soil and Water Cons.* 29: 213–216.

Moore, J. L., and J. M. Armstrong. 1976. The use of linear programming techniques for estimating the benefits from increased accuracy of water supply forecasts. *Water Resources Research* 12:629–639.

National Academy of Sciences. 1974. *More Water for Arid Lands.* Washington, D.C.: National Academy of Sciences.

———. 1975a. *Agricultural Production Efficiency.* Washington, D.C.: National Academy of Sciences.

———. 1975b. *Enhancement of Food Production in the United States.* Washington, D.C.: National Academy of Sciences.

Oswald, W. J. 1973. Ecological management of thermal discharges. *J. Environ. Qual.* 2:203–207.

Pimentel, D., L. E. Hurd, A. C. Bellotti, M. J. Forster, I. N. Oka, O. D. Sholes, and R. J. Whitman. 1973. Food production and the energy crisis. *Science* 182:443–449.

Pimentel, D., W. Dritschilo, J. Krummel, and J. Kutzman. 1975. Energy and land constraints in food protein production. *Science* 190:754–761.

Pirie, N. W. 1942. Green leaves as a source of proteins and other nutrients. *Nature* 149:251.

———. 1975. Leaf protein: A beneficiary of tribulation. *Nature* 253:239–241.

———. 1976. Production and use of unconventional sources of food. In *Man, Food, and Nutrition.* M. Rechcigl, Jr. (ed.). pp. 189–202. Cleveland: CRC Press.

Rawlins, S. L., and P. A. C. Raats. 1975. Prospects for high-frequency irrigation. *Science* 188: 604–610.

Reilly, W. K. 1976. Thoughts on the second German miracle. Washington, D.C.: *Conservation Foundation Newsletter.* August.

Satriana, M. J. 1974. *Large Scale Composting.* Park Ridge, N.J.: Noyes Data Corp.

Sax, R. I., S. A. Changnon, L. O. Grant, W. F. Hitschfeld, P. V. Hobbs, A. M. Kahan, and J. Simpson. 1975. Weather modification. Where are we now and where should we be going? An editorial overview. *J. Appl. Meteor.* 14:652–672.

Schmidt, C. J., and D. E. Ross. 1975. *Cost-effectiveness Analysis of Municipal Wastewater Reuse.* Washington, D.C.: Environmental Protection Agency (WPD–4–76–01).

Sopper, W. E., and L. T. Kardos. 1973. Vegetation responses to irrigation with treated municipal wastewater. In *Recycling Treated Municipal Wastewater and Sludge Through Forest and Cropland.* W. E. Sopper and L. T. Kardos (eds.). pp. 271–294. University Park, Pa.: Pennsylvania State University Press.

Thompson, C. R., O. C. Taylor, and B. L. Richards. 1970. Effects of air pollutants on L. A. Basin citrus. *Citrograph* 55:165–166, 190–192.

Triplett, G. B., Jr. and D. M. Van Doren, Jr. 1977. Agriculture without tillage. *Sci. Amer.* 236(1):28–33.

Wittwer, S. H. 1975. Food production: Technology and the resource base. *Science* 188:579–584.

Young, G. 1970. Dry lands and desalted water. *Science* 167:339–343.

26

THE FUTURE OF AGRICULTURE
IN THE DEVELOPING WORLD

Three-fourths of the world's people live in the developing nations. Although these nations do not have identical characteristics, they share certain basic features that relate to the balance between human populations and their food supply systems. In these nations lie the centers of world population growth as well as the world centers of illiteracy, poverty, and malnutrition. Despite these last three drawbacks, events of recent years have supported the hypothesis that world influence and power tend to shift toward the world's demographic centers (Dorn 1962). It appears inevitable that the future of the human race as a whole will be determined to a great extent by our success in dealing with the imbalances between population and food in the developing nations.

The natural environments of the developing world are diversified. Spanning most of the tropics and subtropics, the developing world has more diverse climates, soils, and biotas than those of the developed world. However, many of the world's problem climates and soils also lie in the developing world, and the affected areas must solve the difficult problems pre-sented by the sustained human use of lateritic soils of humid tropical forest regions and of lands that border the world's subtropical deserts.

Both the renewable and nonrenewable resources available for agricultural development are generally more limited in the developing areas of the world. The goal of agricultural development in these areas is thus to devise sustained-yield agricultural systems geared to the carrying capacity of a small resource base (Janzen 1973). To achieve this end, accommodating changes will be necessary in economic, social, and political institutions, as well as in agricultural technology. The farm economies of many developing nations are locked in institutional rigidity: Feudal land tenure systems crowd most of the rural population onto areas of poor land. Lending and marketing systems tend to exploit small producers. Although they are not universal, such institutional deficiencies prevail in too many places.

Before we attempt to suggest lines of future emphasis for agriculture in the developing nations, we shall examine the nature and success of past approaches. In doing so we shall identify the favorable and unfavorable conditions for future development, as they exist now, and evaluate the consequences of continuing to emphasize development along these lines.

EXISTING PATTERNS OF AGRICULTURAL DEVELOPMENT

The existing agricultural economies of most of the developing nations evolved under the influences, first, of colonialism, and second, of finished-product-for-raw-material trade patterns established under colonialism and intensified after the industrial revolution. During the period of colonial expansion, new social, economic, and political conditions were imposed on indigenous societies that were engaged, for the most part, in subsistence agriculture. The political boundaries established during this process were often arbitrary and unrelated to natural ecological or ethnic patterns. Arbitrary boundaries of this sort still survive in many areas, particularly in northeastern South America, the West Indies, and Africa.

Most of the indigenous societies affected by colonialism were ecologically stabilized societies (Wilkinson 1974). Human populations were stabilized at relatively low densities under the influence of several cultural mechanisms and natural causes of mortality. The subsistence food production systems that supported these populations were complex and diversified (see Chapter 5) but technologically simple.

Colonialism established a hybrid social and economic structure that still survives in most underdeveloped countries (Janzen 1973). One portion of the national system in these countries has developed in response to the economic interests of the colonizing nation. This sector typically emphasizes the production of basic raw materials such as mineral ores and raw fuels, or basic, unprocessed food commodities (**cash crops**) such as coffee, tea, cacao, sugar,

raw fruits, oils and oil seeds, and plant fiber crops (Figure 26-1). The export of these materials provides foreign exchange with which the colonial nation purchases finished products and other imports of generally higher value than that of the exports. Once established, such trade patterns are difficult to discontinue, and so, even after decolonization, the nations involved remain tied to the developed, industrialized nations by these old patterns of trade.

The second portion of the national system in colonial areas is the continuation of the indigenous culture and system of food production. These areas of society however, are exposed to many unbalanced influences that disrupt their ecological stability. Natural mortality and cultural practices that serve to reduce population growth rates are modified with no accompanying decline in birth rates. Indigenous crop and dietary complexes are modified by the introduction of new and unfamiliar food plants that promote nutritional imbalances when not carefully incorporated into food culture. An awareness of the advantages of modern material culture is disseminated but is not combined with adequate educational and economic preparation for participation in such a system.

The consequence of these relatively independent patterns of development is a dualistic agricultural economy (Lofchie 1975). One portion of this economy, the cash-crop sector, tends to attract technological innovation and investment. Modern equipment and purchased inputs are employed much as they are in the developed nations. Road and rail systems, irrigation projects, and other major developments are planned mainly to facilitate the production of cash crops for export. The other sector of the agricultural economy, that engaged in production of food for internal consumption, remains virtually unmodified. In large areas of Africa, for example, subsistence farming remains virtually the same as in its

FIGURE 26-1.
A henequen (*Agave henequen*) plantation in El Salvador. Henequen is grown for fiber, which is obtained from the leaves (note the plants in the left foreground from which the leaves have been cut). This large plantation, occupying the fertile bottomland of the valley, exemplifies the dilemma created by cash-crop farming, which produces materials for export, in regions where malnutrition is a serious problem. (Photograph by J. E. Avila, courtesy of the USDA.)

precolonial days (Figure 26-2); even the ox-drawn plow is a rarity in many areas (Figure 26-3).

The dangers of dualistic economies of this type are well illustrated by a number of African nations that have suffered recent drought conditions. The agricultural economies of several of these countries are still active in the cash crop sector despite near famines caused by failure of the subsistence food supply systems. In Mali, for example, production of corn for local consumption declined by more than one-third between 1969 and 1971; production of cottonseed, groundnuts, and rice (for export) all reached record levels during the same period (Lofchie 1975). The cash-crop sector thus shows greater resistance to the effects of unfavorable weather, at least as long as the required inputs of water, fertilizers, and other components of modern cash-crop production are still available.

This imbalance within the agricultural sector received little attention from national governments and international agencies during the period following World War II. Instead, emphasis was placed on developing the cash crop sector and on rapid industrialization. Economic planners, in effect, encouraged many nations to create shortcuts in the development process by undergoing rapid industrialization without previously or simultaneously forming a strong agricultural sector (Schaefer, May, and McLellan 1970). Influential groups such as the U.S. Agency for International Development supported this approach. International aid,

FIGURE 26-2.
In many parts of Africa, subsistence farming remains little changed from precolonial times. These men in rural Ethiopia are threshing sorghum. (Photograph courtesy of Dr. Solomon Bekure.)

FIGURE 26-3.
Throughout much of Africa, soil is still tilled by human labor. These farmers in Upper Volta are preparing their land for planting. (Photograph by James Pickerell, courtesy of the World Bank.)

FIGURE 26-4.
The Volta River dam at Akosokombo created a lake 8482 square kilometers in area and displaced 85,000 rural farming people. The principal purpose of this dam was to generate hydroelectric power for industrial use, primarily for the refining of aluminum ore, much of which was shipped into Ghana from South America. (Photograph by Peyton Johnson, courtesy of the FAO, Rome.)

whether supplied by the United States, the Soviet Union, or other developed nations, was channeled primarily into large, highly visible projects such as dams, industrial plants, refineries, and highway projects (Figure 26-4).

The general consequence of this pattern of unbalanced development was the increased likelihood of failure on the part of the food producing sector of the economies of developing nations. Pressed by continued population growth, relegated to the less favorable lands by cash-crop developments, and ignored as a sector that deserves study and investment, food production has declined in productive capacity relative to other portions of the national economy. Even when food supplies are adequate, inadequate distribution and marketing systems lead to shortages in some areas and spoilage of surpluses in others. These facts are often buried by the increases achieved in cash-crop production, but their existence is attested to by the postwar dependence of most developing nations on the import of basic foodstuffs.

At this point, we should carefully assess the status of the major determinants of agricultural development as they exist in the developing nations. These determinants fall into four major groups: labor; energy and materials resources; climate and soils resources; and biotic resources.

In most developing countries, urbanization of the population is still at an earlier stage than it is in the developed countries. Thus in rural agricultural areas the potential labor supply is still large. Its cost remains low, however, a condition that must eventually be improved if development is to be successful. These characteristics, which are advantages to development, are combined with certain disadvantages: the facts that rural laborers are largely unskilled; they are often tied to subsistence farming so that their continuous availability is limited; and they may be unresponsive to the normal incentives of a labor-for-wages economy.

Most developing nations are disadvantaged with regard to certain energy and raw materials resources. Although the Middle East and parts of Latin America hold a major portion of the world's petroleum reserves, the use of these reserves is dominated by the developed countries, which have evolved strong economies that depend greatly on petroleum. Not only can these nations afford to pay higher petroleum costs than can the developing nations, the potentially catastrophic worldwide impact of their economic collapse means that they will probably continue for some time to receive most of the world's production. Further development of hydroelectric projects in the developing nations is also limited, since many of the areas in which such development is still possible are far away from the areas where energy is needed. These limited energy resources in turn limit the potential of the developing nations to utilize ore and other mineral resources. Thus it seems likely that many of these countries will continue to export unrefined raw materials and to import finished goods.

On the positive side, however, most developing nations lie in the tropics and subtropics, and so energy needs for space heating and similar purposes are lower and the potential for harnessing solar and wind energy seems greater than in temperate areas. Some developing nations have reserves of major raw materials that are critically important to the developed world; this must also be counted as an advantage and one that is already being capitalized upon in several instances.

As a whole, the developing portion of the world has highly diversified climates and soils. The Andean countries of South America, for example, possess a greater variety of climates and soils than most developed regions of comparable size. Similar conditions exist in parts of Africa and Southeast Asia. These climates and soils provide suitable environments for virtually all of the major temperate zone crops and domestic animals, as well as for all of the tropical species that cannot be grown in most of the developed world. Unfortunately, however, most of the world's low latitude deserts and virtually all of the areas of tropical laterites and lateritic soils are also included in the developing world. Individual developing nations also vary widely in the diversity of conditions in their national domains.

Finally, important characteristics relating to biotic resources must be recognized. The more humid areas of the tropics and subtropics contain the greatest portion of the world's biota. These organisms, both plants and animals, include a very large number of species with proven economic value, as well as many species of unrecognized potential. The possibilities for designing new crops, animal domesticates, and productive agroecosystems is therefore greater in these areas than elsewhere. For similar reasons, the species pool from which agricultural pests may emerge is also very large. The fact that in these diverse tropical systems cold seasons do not decimate pest populations also means that biological mechanisms and spatial isolation of host individuals are likely to be more important here than in temperate areas as controls of potential pests (Janzen 1973). The dangers of crop monocultures may therefore be greater.

It is in this context that future lines of developmental emphasis must operate in the developing world. Obviously, these conditions do not define a single narrow pathway for development. They do, however, call into question two major areas of past emphasis: cash-crop cultivation and intensive mechanized agriculture as it is currently practiced in the developed countries.

First, cash-crop production clearly has a role in the future of the developing countries. It is a form of renewable productivity, given proper production practice, and for a number of foods and other products it represents a more efficient system than other alternatives for the use of land, energy, and other resources. It must be recognized, however, that the populations of the developed countries are growing much more slowly than those of the developing world and may virtually stabilize in the near future. Economic growth in the developed world, expressed in buying power, is also likely to slow or to stabilize. Thus the market for cash crops cannot continue to expand at a rate that exceeds or even equals the population rate of growth in the developing world (Norse 1976).

Second, the intensification and expansion of mechanized agriculture faces several very serious limitations. Most energy and equipment together with many specialized inputs such as pesticides and nitrogen fertilizers, must be imported by most developing countries. In many of the arable areas of the tropics and subtropics, moreover, the quantity of inputs of various materials per unit of output appears to be greater than in the Temperate Zone. As we have indicated in a number of earlier chapters, the transfer of Temperate Zone technology to the tropics is ecologically risky and often unsuccessful. In fact, this approach to development in the developing world has been characterized as a new form of economic colonialism in which the developed nations cultivate the dependence of the developing nations on expensive imported materials and equipment (Feder 1974).

The intensification of mechanized agriculture is furthermore linked to increased urbanization as rural labor is displaced and to increased emphasis upon economic development of urban areas. It is clear, however, that in the developing world unlimited growth of urban economies like those of Europe, Japan, and the United States, all manufacturing goods for internal consumption and for export, is not possible. Limited energy and materials resources, and the partitioning of world markets for exports, both con-

strain such growth. Thus contiued emphasis upon industrialization and agricultural intensification through mechanization may lead developing nations into a dead-end commitment from which retreat may be very difficult.

With these comments in mind, let us examine the developmental pathways that are fairly accessible to the developing nations.

TRADITIONAL APPROACHES TO INCREASING FOOD PRODUCTION

Four major pathways exist for increasing the production of foods by agricultural systems in general (Norse 1976):

1. Expanding, improving, or rehabilitating the overall area of cultivated land.
2. Modifying the cropping pattern to replace nonfood or low-yield crops with high-yielding food species.
3. Increasing outputs per unit area of cropland by using improved seed, irrigation and fertilization, and pest and disease control.
4. Shifting from intensive feeding of livestock to extensive rangeland production of meat, thus making land on which feed grain is now grown available for human food production.

The last of these four possibilities is one that must be carried out in the developed countries, where most intensive livestock production is centered. For example, in 1968 the developed nations used about 350 million tons of animal feed grains, enough to feed perhaps 1.4 billion persons. Such a substitution will obviously not occur, however, unless the developing nations can stimulate it by paying competitive prices and shipping costs. Nevertheless, if the economies of developing nations improve, international trade offers considerable capacity for adjustment.

Earlier we noted that only 45 percent of the potentially arable land of the world is now under cultivation, and that large fractions of the unused land are in South America and Africa (see Chapter 1), but everywhere the remaining potentially arable land is generally less fertile and more costly to bring into production than that already in use. A major portion of this land cannot be cultivated without irrigation, and in many cases there is no likely source of irrigation water. Nevertheless, at the recent United Nations World Food Conference in Rome (1974), the development of 34 million hectares of new land in Africa and 85 million in South America between 1974 and 1985 was proposed. The respective estimated costs of bringing these lands into cultivation were $44 and $150 per hectare.

It is likely that these estimates are unrealistically low (Norse 1976). In fact, the costs of clearing land now generally exceed $200 per hectare, and in tropical rain forest areas they may be several thousand dollars per hectare. The cost of new irrigation developments is now between $1000 and $1500 per hectare and is rising at a rate of about $27 per year (Figure 26-5). Historically, the tendency has been to underestimate the final costs of such developments.

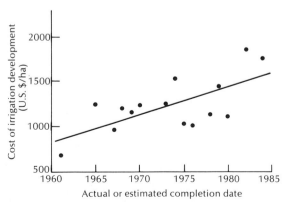

FIGURE 26-5.
The trend in costs per hectare (1968 U.S. dollars) of new irrigation developments. (Modified from Norse 1976.)

On the other hand, although much irrigated land now under cultivation in the developing countries suffers from poor or unreliable water supplies, or is deteriorating because of waterlogging, salination, or alkalinity buildup, rehabilitating or improving such lands can generally be carried out with smaller investments. Examples of such projects include the reclamation of saline and saline-alkali soils and the construction of concrete-lined canals, improved water storage and distribution systems, and tile drainage systems. It thus appears that the greatest returns are to be gained by improving currently cultivated lands instead of by developing new arable lands.

The substitution of staple crops such as grains, tubers, and legumes for nonfood species such as tobacco, or for cash crops such as sugarcane, is still another way to increase internal food production in developing areas. Of course, such changes may require adjustments in other portions of the national economy in order to maintain balance-of-trade relationships. Furthermore, not all such substitutions are desirable, since some cash crops, such as cacao, coffee, and tree crops, not only yield valuable materials but also protect soils that might otherwise by subject to severe erosion or to deterioration in fertility.

Improving both the quality and quantity of productive inputs to food crop ecosystems is still another area in which there is great opportunity to increase food production. In many areas, for example, fertilizers are applied primarily to nonfood and cash crops such as rubber and tea. In Latin America, it has been estimated that only about 10 percent of all fertilizer used is applied to basic food crops (Vega 1971). In most developing countries the use of nitrogen fertilizers on grain crops falls in the lower part of the effective application range, and common application rates are 20 to 40 kg/ha (Norse 1976). Empirically determined curves that reflect the yield responses of these grains to fertilizer input indicate that major yield increases

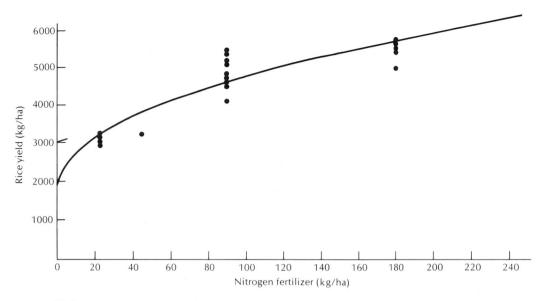

FIGURE 26-6.
Rice yield (kg/ha) as it relates to the use of nitrogen fertilizer (kg/ha) in paddy rice production in humid tropical regions. (Modified from Norse 1976.)

are possible through additional fertilizer use (Figure 26-6). Major problems exist, of course, in making additional fertilizers available at a reasonable cost. The use of nitrogen fertilizers produced from fossil fuels will probably be an increasingly expensive approach to meeting this need. Moreover, as the cost of these fertilizers increases, they will probably be used to produce the commodities that contribute most to the favorable balance of international trade.

Nevertheless, efforts designed to develop and supply fertilizers in other ways offer considerable potential for increasing the yields of food crops. In this connection, it should be noted that China has pursued a variety of technological approaches to fertilizer production (Sprague 1975). Human and animal wastes and crop residues are composted to produce organic fertilizers on a greater scale than in any other country, and many small factories are developed for the production of nitrogen fertilizers from lignite, a low-grade fossil fuel.

Since about 1965 the introduction of new, high-yielding grain varieties, particularly of wheat and rice, has been a highly publicized approach to improving food production in the tropics and subtropics. The varieties involved, designed especially for irrigated farming, have very high yield potentials when grown under optimal conditions. Furthermore, under optimal conditions, they show a strong fertilizer response, partly because their short stature and strong stems protect them from losses due to lodging, which becomes more serious with tall varieties as the heads grow heavier (Figure 26-7). These varieties have been well accepted. By 1972-1973 high-yielding rice varieties were being planted on 70 percent of the irrigated paddy rice lands of the Philippines (Atkinson and Kunkel 1976), and about 52 percent of the wheat acreage and 25 percent of the rice acreage in India were in high-yielding varieties (Dalrymple 1975).

The success of these varieties, and their con-

FIGURE 26-7.
Mexican agronomists examine one of the high-yielding dwarf varieties of wheat
developed at the Centro Internacional de Mejoramiento de Mais y Trigo (CIMMYT) near
Mexico City. Dwarf varieties are much less susceptible to lodging. (Photograph courtesy
of the FAO, Rome.)

tribution to actual production increases, is not easy to determine, however, although careful analysis of rice yields in the Philippines suggests that the gains are modest. From 1968 to 1973 the yield differential for high-yielding varieties was about 14 percent in irrigated paddy fields and 4 percent in rainfed lowland paddies; taking the areas of these croplands into account, the average differential was 11 percent (Atkinson and Kunkel 1976). The realized yields were also considerably below those obtained under experimental conditions.

Analysis of the yields of high-yielding and other varieties of rice in the Philippines in relation to other variables, including fertilizer and pesticide use, revealed several interesting points. High-yielding varieties were three times as responsive to fertilizer inputs as other vari-

eties. However, other (traditional) varieties were about twice as responsive to herbicide and pesticide inputs. Overall, these and other variables accounted for only about 17 percent of yield variation in high-yielding varieties and 23 percent in other varieties, indicating that a great deal remains to be learned about the factors that influence rice yields under farm conditions (Atkinson and Kunkel 1976).

Because the scale of pest damage to tropical and subtropical crop systems is potentially greater than to other systems (Janzen 1973), increased investment in pest and disease control is certainly warranted. However, the weaknesses and self-defeating tendencies of chemical control methods appear to be greater in these environments. Thus greater attention must be paid to alternative pest and disease control approaches, such as genetic resistance, cultural control, and biological control in its broadest sense; these are the areas of greatest potential for yield improvement.

A much neglected aspect of the pest problem relates to losses of harvested materials during storage, transportation, and processing. Storing grain under improper conditions of temperature and dryness can lead to major losses through germination, respiration of the grain, and the activity of microorganisms. Inadequately protected grain, as well as other harvested food materials, may also be attacked by various insects and rodents. It is commonly suggested that, worldwide, some 10 percent of all grain harvested is lost by pest feeding or fouling (with urine, feces, and dead carcasses), but in many areas of the developing world these losses are as high as 25 to 50 percent (Barrass 1974).

In India, where grain supplies for cities are usually stored temporarily in **godowns** (small warehouses), which are often in poor repair, losses to bandicoot rats (*Bandicota bengalensis*) and other rodents can be quite large (Frantz 1976). These rats live not in the godowns, but in outside burrows that they leave to enter the godowns in search of food after dark. Each rat consumes about 30 grams of grain daily. In Calcutta, Frantz (1976) estimated that in a godown area covering slightly over 2 hectares there were at least 5000 adult and 10,000 total bandicoots. Assuming that nonadults consumed roughly half the quantity of food eaten by adults, a minimum annual grain loss by consumption alone would exceed 82 tons in this one area!

Here, we should note that the technology for reducing many of these losses is simple and well understood. In the case of rodent problems in Calcutta godowns, it is obvious that the two approaches called for are, first, to block the animals' access to the godowns by modifying and repairing the godowns themselves and, second, to eliminate the sites favorable to burrow con-struction. Neither of these activities has to be energy intensive or difficult. In similar fashion, improvements in the design, cleanliness, and monitoring of stored grains, combined with fumigation and other procedures, can reduce losses of stored products to insect pests.

The scale of stored food losses in the developing countries is great enough that an investment in improved control of stored food pests is probably the most effective method now available for increasing the quantity of available food.

INNOVATIVE APPROACHES TO INCREASING FOOD PRODUCTION

Agriculture in the developing nations is not irreversibly committed to a particular pattern of intensification. Although mechanized monocultural techniques have been introduced in many areas, the fraction of total arable land involved is still small. This is an advantage because it allows for varied alternatives to be examined and perhaps to be implemented in diverse ways. In the developed countries, on the other hand, the total investment, by both farming and industry, in the mechanized monocultural approach is so great that other approaches can only gradually be emphasized.

The richness of the biotic resources in the tropics and subtropics, combined in many situations with the ecological constraints on sustained cultivation, demand that innovative approaches to plant and animal food production be explored seriously. Several innovative approaches have been recognized: mixed cropping, forest farming, game cropping or farming, runoff farming, and saltwater farming. All of these approaches share the feature that the design of the productive system can be adjusted to the general nature of the environment involved, instead of the environment having to be transformed to suit the crop or animal species.

Mixed Cropping

In many areas of the developing world, **mixed cropping,** or growing more than one crop species in a single farm plot, is standard practice (Norman 1974). Several variations on this approach are used (Figure 26-8). **Interplanting,** for example, is the practice of planting two or more species with different patterns of seasonal activity, so that while one species is reaching maturity a second is undergoing early growth to mature at a later time. **Intercropping,** on the other hand, is the cultivation of two or more species whose growth and maturation tend to be synchronous. Mixed cropping is usually practiced in areas of highly seasonal rainfall where no irrigation systems have been developed, and is also typically a subsistence farming practice.

Mixed cropping has several benefits that are only now being recognized and evaluated. The yield of many mixed crop systems is greater than that of monocultures, particularly when the species involved possess complementary morphological or physiological capabilities. In an area of Nigeria studied by Norman (1974),

for example, 24 crop species were cultivated, and 156 intercropping and interplanting combinations of up to six species were noted. Of the total acreage in the area studied, 83.4 percent was farmed by some sort of mixed cropping system. In the mixed cropping systems studied in detail by Norman, the gross return per hectare from crop production was about 60 percent higher for mixed than for single-crop plantings, not because of higher output per unit of labor, but simply because of a greater gross yield (Table 26-1). At the International Institute of Tropical Agriculture in Ibadan, Nigeria, controlled studies of inter-cropped cowpeas and corn have also shown that yields are quite significantly higher in mixtures (Table 26-2; Greenland 1975).

A second apparent advantage of mixed cropping is that it minimizes the risk of production failure (Norman 1974). Under indigenous practice in Nigeria, the probability of obtaining a higher gross yield per unit of input is greater in mixed crop systems than in single-crop systems. Although the reasons for this are still poorly understood, the reduction of damage by pests and disease organisms in

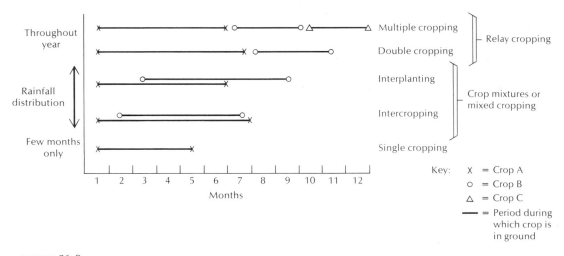

FIGURE 26-8.
Terminology describing multiple species cropping on the same area. Where three or more crops are grown in mixture, elements of both interplanting and intercropping may exist. (Modified from Norman 1974.)

TABLE 26-1
Labor inputs, crop yield, and gross and net returns for single- and mixed-cropping practices in Nigeria.

Variable	Crop mixtures				
	Single crop	Two crops	Three crops	Four crops	All mixtures
Labor input (hours/hectare/year)	362.2	582.2	556.7	669.9	586.4
Crop yields (kg/hectare/year)					
Millet	—	370	359	365	366
Sorghum	786	726	536	374	644
Groundnuts	587	437	389	429	412
Cowpeas	—	98	138	139	132
Sweet potatoes	—	—	194	—	194
Cotton	213	215	161	—	189
Gross return (sh/hectare/year)[a]	379.6	594.5	567.8	842.4	613.6
Gross return on labor (sh/hour)[a]	1.3	1.1	1.0	1.1	1.1
Net return on costed labor (sh/hectare/year)[a]	183.1	285.4	260.2	456.2	296.0

[a]One Nigerian shilling = $0.076
Source: Norman 1974.

TABLE 26-2
Yields of cowpeas and intercropped cowpeas and maize under different tillage systems at the International Institute of Tropical Agriculture, Ibadan, Nigeria.

Tillage method	Yield (kg/ha)[a]			
	Single cropping	Intercropping		
	Cowpea	Cowpea	Maize	Total
Plowed and ridged	1185	665	1705	2370
Plowed, flat bed	1274	725	1675	2300
Strip tillage	1538	1022	2337	3359
Zero tillage	1649	941	2809	3750

[a]All plots received 30 kg N/ha, 30 kg P/ha, and 30 kg K/ha.
Source: Greenland 1975.

mixed plantings appears to be a significant factor (Greenland 1975).

Studies of mixed cropping sytems and of ways to improve them further have only begun. Such practices seem likely to provide better soil protection, more efficient use of nutrients and solar energy, and a higher level of biological pest and disease control than monocultures. It is possible that in some systems labor requirement per unit yield may be reduced because of a decline in problems from weed growth.

It is inaccurate to suggest that such mixed cropping is completely incompatible with mechanization. Mechanized techniques of soil preparation can be used, and so can mechanized cultivation equipment if different species are interplanted in rows or other regulated fashion. It is likely, however, that mechanization of such systems can only proceed to an intermediate scale.

Forest Farming

Forest farming, or agrisilviculture, is yet a more radical and less explored approach to tropical agriculture (Douglas 1973). Temperate Zone agriculture depends mainly on annual herbaceous plants, particularly grains, as producers of staple foods. Vegecultural systems that originated in the tropics, in contrast, emphasize root staples produced by perennials, at least some of which are woody or semiwoody. The goal of forest farming is to develop a multispecies system of woody plants that simulate in composition the diverse structure of tropical forests or woodlands and yield a variety of food and non-food materials. Such a system would theoretically retain certain ecological advantages related to soil protection, retention of nutrients, and minimization of pest outbreaks. The harvestable products from such a system would include not only those traditionally obtained from woody plants—lumber, resins and other industrial chemicals, fibers, fruits, and

nuts—but also staple carbohydrate foods and animal forages. The last two categories are essential components of a complete agricultural system and at the same time represent the two categories toward which little effort has been directed in forest crop development.

A number of leguminous trees, such as members of the genus *Prosopis* (mesquites or algarobas), produce large quantities of beanlike fruiting pods. The seeds of certain species can be processed into a flour or meal and used as food for either humans or domestic animals. The whole pods of many species can be fed directly to livestock. The carob, *Ceratonia siliqua,* is another widely grown species whose seeds are suitable for being processed into meal or flour. Under favorable conditions, algarobas and carobs yield 50 tons of pods per hectare, making them competitive with other crop species. A large number of other legume and nonlegume trees also have the potential to be developed as basic carbohydrate foods or fodder. Domestic animals, such as the goat, and potential domesticates, including various African antelopes, are species that might be able to utilize tree forage foods.

The potential for the development of forest farming extends from the humid tropics into the drier tropical environments, where the natural vegetation is woodland or thorn scrub. The necessary impetus for the development of such systems, however, is research into the production, processing, and utilization of staple foods and forages that are central to agriculture. Significant results in these areas would go far to stimulate interest in diversified forest farming systems.

Game Cropping and Farming

In animal agriculture, similar possibilities exist for the husbandry of multispecies systems in several tropical areas. Nowhere is this possibility greater than in Africa, where a diversified fauna of large game species still exists in many locations. The species all differ in habitat, body size, and feeding behavior. Their habitats are distributed along a moisture gradient from humid forest to desert edge. Some species, such as the eland and the oryx, are closely adjusted in physiology and feeding behavior to drier areas where the availability of free water is limited (Taylor 1969). Many species are browsers, and in areas of woodland or thorn scrub various species coexist whose browsing patterns are complementary. For example, giraffes feed at the highest levels; gerenuks feed at intermediate heights reached by their long necks and their habit of standing erect on their hind legs; impalas feed at lower levels; and diminutive species such as the dik-dik feed near the ground (Crawford and Crawford 1974). In open savanna areas, other groups of species utilize the range production in a sequentially diversified manner (see Figure 2-15) that is correlated with seasonal patterns of migration (Beall 1971). This intensive utilization apparently stimulates range productivity by stimulating nutrient recycling in these grasslands (McNaughton 1976). The native fauna thus seems to utilize the natural primary production very intensively but in a manner to which the primary producers are well adjusted. Moreover, these native animals are resistant to trypanosome parasites, which are a serious disease problem for introduced livestock.

The potential of **game cropping,** the sustained yield harvesting of animals from wild populations of mixed species, has been variously estimated. Estimates of the animal biomass supported by some of the African parks and game refuges range from 1 to about 30 tons per square kilometer (Table 26-3; Bourliere and Hadley 1970), depending largely on the wetness of the region in question. These biomass values are appreciably higher than those of cattle that could be maintained on comparable rangeland, and the harvestable yields of meat also seem to be higher. For example, directly comparing cattle and game production on different parts of a 56,658 hectare ranch in Rhodesia indicated that the yield of game meat, about 430 kg/ha/yr, was about 26 percent higher than that of beef. Furthermore, the lower "production" costs for game, essentially costs of harvesting, meant that the net profit from game cropping was six times that of cattle ranching (Dasmann 1964). So far

TABLE 26-3
Standing crop biomass of ungulates in some African parks and game preserves.

Locality	Habitat	Number of species	Biomass (tons/km²)
Tarangire Game Reserve, Tanzania	Open *Acacia* savanna	14	1.1
Kafue National Park, Zambia	Tree savanna	19	1.3
Kagera National Park, Rwanda	*Acacia* savanna	12	3.3
Nairobi National Park, Kenya	Open savanna	17	5.7
Serengeti National Park, Tanzania	Open and *Acacia* savannas	15	6.3
Queen Elizabeth National Park, Uganda	Open savanna and thickets	11	12
Queen Elizabeth National Park, Uganda	Same, overgrazed	11	27.8–31.5
Albert National Park, Zaire	Same, overgrazed	11	23.6–24.8

Source: Bourliere and Hadley 1970.

game cropping has developed only to a moderate extent in Rhodesia, South Africa, Uganda, and other scattered areas of Africa (Pollock 1969). Although the potential for meat production by such techniques has sometimes been overrated, it nevertheless appears that rational management of the native wildlife in Africa could go far to alleviate protein deficiencies in that area of the world and could perhaps make the continent one of the world's major meat suppliers (Crawford and Crawford 1974).

Game farming, or the deliberate husbandry of game species, also has considerable potential in the tropics. Essentially this amounts to the domestication of new species with characteristics adapted to tropical environments. The eland, for example, which is native to Africa, produces more milk than cattle in the same range areas, and its milk has higher fat and protein levels than cow's milk. The eland is an excellent meat animal, too; as with most game species, the percentage of the eland's body weight that comprises fat is small, 5 percent or less, compared to about 35 percent for a steer in good condition. An additional advantage is that the digestive tracts of game species constitutes a smaller fraction of overall body weight than do those of domestic cattle. Thus there are several good reasons for seriously considering the agricultural potential of various game species.

Runoff Farming

Large areas of the developing world have arid climates or saline soils. Innovative approaches to food production are needed in these environments, too. One such effort, which is being made in the Negev Desert of Israel (Evenari et al. 1971, 1975), began with the reconstruction of the sytems of **runoff farming** used by the Nabataeans some two thousand years ago (Chapter 5). Avdat Farm, the most extensive of the reconstructed sites, was put into production in the winter of 1960–1961 (Figure 26-9). The reconstructed farm plots were irrigated from two watersheds. One area of 1.2 hectares was supplied from a small watershed of 30.7 hectares (25:1 ratio of watershed to farm plot), and a second plot of 1.6 hectares was supplied from a 345 hectares watershed (215:1 ratio) from which the runoff per unit area was less. Over the years since this experiment was begun, it has shown that excellent yields of grains, vegetables, forages, and fruits of a wide variety of species can be obtained. Yields vary considerably from year to year, of course, since the quantity and timing of the winter rains are highly unpredictable. In 1966, a relatively favorable year, respective yields of wheat and barley were 4.4 and 4.8 tons/ha, while peas yielded 5.6 tons/ha of seed. Several forage crops, including alfalfa and a number of forage grasses also yield well now, and fruits such as figs, pomegranates, grapes, peaches, apricots, almonds, and loganberries have been grown successfully (Figure 26-10).

One new approach at Avdat Farm is the construction of **microcatchments** ("negarin") that vary from 16 to 1000 m^2 in size and that supply water to a single tree or vine (Figure 26-11). This idea was developed as a result of the observation that the actual runoff from a watershed was inversely related to its size, the greatest percentage runoff being from small watersheds. Several fruit species, including pomegranates, carobs, and olives, have survived and grown well in these microcatchments. In addition, a saltbush, *Atriplex halimus*, a salt-resistant desert forage plant, performed very well in small catchments (31.2 m^2 in area), indicating that a density of 320 plants per hectare of this species could be grown in areas with an average of 90 millimeters of annual rainfall, a significant improvement in productivity of a native forage species.

Subsequently, Evenari and his colleagues (undated) have established a third farm, Wadi Mashash, to test the economic practicality of

FIGURE 26-9.
Avdat Farm in the Negev Desert, south of Be'er Sheva, Israel. The slender dark lines crossing the slopes near the left center margin of the photo are ancient stone conduits that directed water to particular portions of the farm as it existed in Nabatean time. (Photograph by G. W. Cox.)

FIGURE 26-10.
Almond (*right foreground*) and peach (*left rear*) trees at Avdat Farm. The ruins of the ancient city of Avdat can be seen on top of the hill in the background. In the lower right is a stone dike with a floodgate to control the flow of water from one section of the farm to another. (Photograph by G. W. Cox.)

FIGURE 26-11.
A microcatchment at Avdat Farm. This catchment, 250 square meters in area, supplies water to a single grapevine and is delimited by low dirt ridges (the hat rests on the near border ridge; the person standing near the vine is at the right rear border ridge). The area immediately around the plant is deepened to receive and hold the runoff as it infiltrates. (Photograph by G. W. Cox.)

runoff farming. This farm, with a total area of about 2000 hectares, was begun in 1971, with the construction of 3000 microcatchments for almond, pistachio, and olive trees. Other plantings of saltbush and drought-tolerant legumes were made to improve grazing conditions for sheep. An experimental flock of about 300 ewes and 10 rams was created. One of the objectives of this part of the project was to determine whether runoff harvesting techniques could be utilized by the 35,000 or so Bedouins who graze sheep and goats over much of the Negev Desert.

Interest in runoff farming has developed in several other regions, including Botswana, Afganistan, India, Australia, and other areas bordering the Sahara Desert in Africa. The world distribution of loess soils in semiarid regions suggests that the techniques utilized in the Negev may be applied to many other areas (Evenari et al. undated).

Saltwater Farming

A second innovative approach in desert coastal regions, or in areas with soil and water salinity problems, is saltwater agriculture. The use of seawater or brackish water for crop irrigation requires, first of all, excellent drainage and good soil aeration to prevent salt buildup and to permit efficiency of root metabolism in nutrient uptake and salt exclusion (Boyco 1967). Sandy soils are therefore a virtual necessity. One of the important phenomena in such soils is the formation of subterranean dew. This moisture, which condenses in the air space of the soil, is in only partial contact with soil solutes and thus constitutes a source of low-salinity moisture for plants. Experiments in Israel have also shown that certain useful plants such as barley, and various grasses, rushes, and desert succulents, can be grown experimentally under irrigation with full seawater, and that other plants, such as sugar beets, can be grown in low-salinity seawater. Still other studies show that careful adjustment of fertilizer levels may increase the tolerance of sensitive crops to brackish irrigation water (Raikovitch 1973). It is thus apparent that several profitable lines are open to research in the utilization of arid and saline lands for agricultural production.

OVERALL EMPHASIS FOR AGRICULTURAL DEVELOPMENT

Several areas of general emphasis emerge from our consideration of the problems and opportunities of agriculture in the developing world. First, major attention must be devoted to adjusting agricultural systems to the ecological realities of tropical and subtropical areas. At the present time, the overwhelming emphasis is on modifying the conditions of these environments to suit preselected species that in most cases have been perfected as important crops in temperate regions. The costs in energy and materials needed to intensify and expand this current approach in the developing countries are greater than the costs of carrying out such activity in the developed countries, and the geographical distribution and limited reserves of the required resources are unfavorable. The valid developmental alternative seems to be to design productive systems that retain a maximum of natural ecological organization and function.

Second, labor intensiveness of the agricultural sector should be maintained or increased. Reducing the labor requirement per hectare by substituting machinery and fuels for manpower is not a valid developmental goal. Rather, the objective should be to use manpower to the same or a greater extent and in a more efficient and skilled manner, so that yield per hectare and yield per man-hour are increased. The increasing displacement of rural labor can only magnify the economic problems of urban areas within the developing world. As for the agricultural sector, the limitations of energy and materials constrain the potential of expanding urban, consumer-product, exporting economies. For these nations to be economically stable over the long run, they must establish a strong sector whose productivity is based on renewable resources. For most developing nations this is the only realistic approach to agriculture.

The development of the rural sector of the economy, the sector that is generally ignored or given inadequate investment as described earlier, thus demands increased attention. This need has been recognized recently, and the response of a number of international agencies, particularly the World Bank, is evident (Lele 1975; World Bank 1975). The percentage of lending by the World Bank for agricultural development has risen from about 6 percent in 1948 to 24 percent in 1973–1974. However, total lending by the World Bank has increased enormously in recent years, meaning that the greatest absolute increase in lending has been in nonagricultural sectors of developing economies (Figure 26-12). The need for further expansion of aid in the agricultural sector is evident from the World Bank's own estimate that 85 percent of all absolute poverty in developing areas is in rural areas (World Bank 1975).

The World Bank has recently reviewed rural development projects now operating in Africa under the sponsorship of various groups (Lele 1975). Many of the existing projects are concerned with export crops and products such as coffee, cacao, tea, groundnuts, cotton, tobacco, and beef (Figure 26-13). Although these projects contribute to the ability of the countries involved to purchase other goods on international markets, emphasizing cash crops alone does little to solve the immense problems of malnutrition in Africa. Efforts in two African nations, however, deserve special mention.

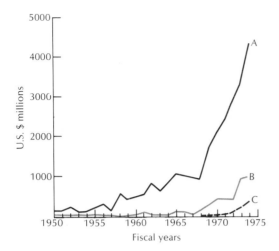

FIGURE 26-12.
World Bank lending for (A) all purposes, (B) agriculture, and (C) rural development from 1950 to 1974. (From World Bank, *The Assault on World Poverty.* Copyright © 1975 by The Johns Hopkins University Press, Baltimore.)

participation in the national economy has been moderately successful. Between 4 and 6 million people are believed to have been organized into ujamaa villages. A second objective, fostering communal farming activity by members of the ujamaa community, has been largely unsuccessful. Nevertheless, it appears that this program, supported almost entirely by the Tanzanian government, has been one of the

In Ethiopia, several major programs, financed by several groups including the Swedish International Development Authority, the United Kingdom, United Nations agencies, and the Ethiopian government, have undertaken broad assistance in rural development. These programs have focused on dairying activities and on the production of basic food crops such as barley, wheat, corn, and tef (an endemic grain crop of major importance). Emphasis has been on the incremental improvement of seed, fertilizer use, and farming practices, the last by encouraging the development and use of labor-intensive farm implements. In Tanzania, one of the most controversial projects of rural development is the **ujamaa village** program. The term ujamaa means "familyhood" in Swahili and refers to the goal of creating village units that constitute or function as extended family groups. Socialist in idealogy, this program was initiated in 1967. The objective of drawing dispersed rural populations into small village units that can benefit from cooperative

FIGURE 26-13.
These workers are picking cotton in the Shire Valley of Malawi. This area was included in the Shire Valley Development Project supported by the World Bank. The project provided expanded extension services, soil conservation, access roads, marketing and storage facilities, and access by farmers to credit for the purchase of seeds, fertilizers, and equipment. (Photograph by James Pickerell, courtesy of the World Bank.)

most innovative and distinctively African of the rural development projects on the continent.

Given rural development as a major priority, one of the major developmental emphases that must also be recognized is reliance on intermediate, or appropriate, technology (Schumacher 1973). The concept of **intermediate technology** is applicable to many areas of development and refers simply to the employment of methods and equipment that are cheap enough to be available to most people and are suitable for small scale application and compatible with the human need for satisfaction and enjoyment of work. For example, instead of promoting the use of modern tractors, which have a high work capacity but are expensive to buy, operate, and maintain, intermediate technology emphasizes animal power or smaller tillage machinery that is cheaper, easier to adapt to, and less likely to displace rural labor (Figures 26-14, 26-15). Intermediate technology is particularly well suited to the development of techniques for reducing losses of stored foods to pests (Figures 26-16, 26-17).

Farming operations may, in fact, be separated into two groups: technology-centered activities, in which farming operations are determined by the available technology; and farm-centered activities, in which the constraints of the farm situation determine the technology that can be applied (Uchendu 1975). Agricultural assistance to developing areas has frequently offered highly advanced technology to farm-centered operations. What is generally needed instead is technology that is:

1. independent of scale and incorporable on small farms

FIGURE 26-14.
Tie-ridging is an African system of field preparation that reduces erosion by creating a series of ridges, on which the crops are planted, with periodic cross connections that form dams to hold water until it infiltrates. These ridges and cross-ties were made by hand. See Figure 26-15. (Photograph courtesy of the FAO, Rome.)

FIGURE 26-15.
A tie-ridger developed by the Intermediate Technology Development Group and designed to be pulled by a team of oxen. The front blade and flanges throw the soil into ridges on either side. The rear blade accumulates soil which at intervals forms a cross-tie when the rear blade is lifted. (Photograph courtesy of the Intermediate Technology Development Group, London.)

FIGURE 26-16.
In many parts of Africa, grain is stored in small grass-roofed, wattle and daub granaries that do not effectively exclude rodents and other pests. These granaries are in the Ethiopian highlands. (Photograph by G. W. Cox.)

FIGURE 26-17.
A grain storage silo designed by the Intermediate Technology Development Group. This silo was constructed mainly out of local materials and entirely by local labor. (Photograph courtesy of the Intermediate Technology Development Group, London.)

2. adaptable to local manufacture and repair
3. compatible with other such practices and equipment
4. compatible with labor-intensive farming

Recently, for example, studies have been carried out in Uganda to assess the value and acceptability of ox-cultivation technology as an alternative to existing hoe-cultivation practices (Dima and Amann 1975; Okai 1975). Making even as small a shift as this in farming practice actually requires quite a number of adjustments. Pasture areas must be set aside for oxen. Farmers must be trained in the use and care of oxen and must be provided the opportunity to purchase oxen, ox-drawn equipment, and crop inputs appropriate to the farming to be undertaken. Lastly, of course, they must be assured of a market for their production.

An optimum farm plan and loan package was developed from farm case studies in the Busoga District of Uganda and from analyses of the optimal crop combination for profitable initiation of ox-cultivation farming (Dima and Amann 1975). This package (Table 26-4) provides for the acquisition of oxen and equipment together with the costs of purchased inputs for the first growing season. It is assumed that input costs for subsequent seasons, as well as for loan repayment, will come from profits on the expanded production and sale from the farm. All told, for ox cultivation to be introduced to a 5.2-hectare farm, a loan of about 1120 Ugandan shillings, or about $160.00, is projected. Farm case studies also suggested that in most cases conversion to ox cultivation was profitable. Increases of 75 to 100 percent in income were realized by three-fourths of the farmers involved.

TABLE 26-4

Optimum loan package for the introduction of ox-cultivation technology to small farms in the Busoga District of Uganda.

Item		Cost (Ugandan sh.)[a]
One ox team (shared)		500.00
Plow frame, plow, weeders and planters (shared)		200.00
First year inputs		
Bananas	1.2 hectares	—
Maize	0.4 hectares (fertilizers + seed)	55.00
Groundnuts	1.2 hectares (fertilizers + seed)	135.00
Finger millet	0.4 hectares (seed)	5.00
Soybeans	0.7 hectares (fertilizers + seed)	67.00
Cotton	1.3 hectares (insecticides)	30.00
Additional cash credit		125.00
Total	5.2 hectares	1117.00

[a]One Ugandan shilling = $0.14.
Source: Dima and Amann 1975.

POPULATION AND ECONOMIC DEVELOPMENT POLICY

We have concentrated on the supply aspects of the food problem in the developing world, but demand is at least equally important (Norse 1976). There are two major problems that relate to demand; a rapid increase in the numbers of people needing food; and the inability of many of these individuals to buy food at prices that will stimulate production. Resolving the food-population dilemma requires attention to both these problems.

In several countries, such as the United States, birth rates have declined naturally as economic development has occurred and as societal values have changed. In some developing countries—Costa Rica, for example—the lower birth rates in recent years have also supported the idea that economic and educational development can lead automatically to slower population growth. However, it is naive to expect that attention to economic development alone can create such a change throughout the developing world. The momen-

tum of population growth in many parts of Asia, Africa, and Latin America is enormous, and the economic gap between existing conditions and those at which birth rates are likely to decline is too great to allow such an approach to succeed. Formal policies must therefore be adopted in many areas to reduce population growth.

Programs to reduce population growth cannot be limited to clinical approaches that simply distribute information and contraceptives. These services must be combined instead with incentive programs that are carefully designed to influence individuals, couples, social groups, and whole communities (Kangas 1970). The objective must be to encourage reduced birth rates without unfairly penalizing the children who are born. Logically, incentives must be an integrated set of values that extend throughout society and complement each other (Figure 26-18).

In dealing with the inadequacy of purchasing power, the overall pattern of economic development must be considered. In particular, informed policy must be formulated pertaining

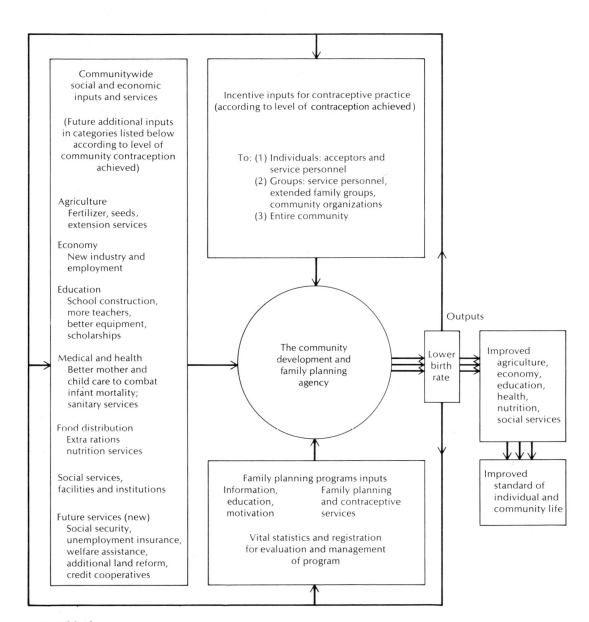

FIGURE 26-18.
Schematic example of the integration of individual, family, group, and community incentives designed to reduce community birth rates. (Modified from Kangas 1970. Copyright © 1970 by the American Association for the Advancement of Science.)

to the relative development of urban and rural sectors of the national economy. The factors that have permitted the development of successful, highly urbanized societies in Europe and the United States are first, high labor productivity in agriculture, industry, and manufacturing; and second, rapid and inexpensive transportation. All of these depend heavily on inexpensive petroleum fuels and feedstocks. The era of cheap petroleum is rapidly coming to an end, however, and limitations on other nonrenewable resources are also appearing. As the raw materials that support production in highly urbanized economies become more and more limiting to such production, the supply of labor and therefore its value will become less limiting. Under these circumstances it is doubtful that purchasing power in the urban sector can be increased enough to stimulate greatly increased agricultural production in the rural sector. Thus resource considerations indicate that it is unlikely that urbanization can proceed successfully to the point already achieved in developed countries, where 80 to 90 percent of the population live in cities. In many parts of the developing world urbanization has probably already gone beyond the point of a stable rural-urban balance.

The direction that development takes must very likely be shifted toward a more agrarian society in which a large fraction of the population would remain engaged in the production of basic foods, fibers, and other renewable products. This economic group, which would meet its own food needs, must be balanced with an urban group in trade relationships that would maintain the appropriate balance of rural and urban populations.

Coupled with stabilization of population size, this would reestablish ecologically stable societies (Cox 1978). But these societies would be stabilized in a new way instead of being adjusted by superstition, barbarism, and the rigorous challenges of a unbuffered natural environment to the productive capacity of an unsophisticated technology. The developmental goal must be to use intelligent and humane means to adjust human populations to the capacity of the environment under sophisticated technology, but the new relationship must be equally stable.

SUMMARY

The agricultural economies of the developing nations have evolved out of ecologically stabilized conditions that existed in indigenous societies occupying many of the world's tropical and subtropical regions. In most developing countries, however, the influences of colonialism and international trade have led to serious imbalances between urban and rural development and between cash-crop production and domestic food-crop production.

The pathways of future development in these nations must be predicated on the particular conditions of existing labor supply, energy and material resources, climate and soils, and biota. These conditions create both limitations and opportunities. Abundant rural labor, diversified climates, soils, and biota, and limited supplies of cheap energy are dominating considerations.

Of the traditional approaches to increasing productivity, the ones that offer greatest potential are to invest in improving or rehabilitating existing cultivated lands, and to improve the quality and quantity of productive inputs such as fertilizers and improved seed, particularly seed of varieties bred for pest and disease resistance. Investment in improved protection for harvested materials during storage and transportation also offers potential benefits.

A variety of innovative approaches that represent the adjustment of agricultural techniques to environmental characteristics also deserve attention and research; these include mixed cropping systems, forest farming, game cropping and farming, and, in desert and coastal regions, runoff farming and saltwater irrigation.

Both traditional and innovative approaches assume several general developmental emphases, including increased investment in rural development by national and international groups, emphasis on labor-intensive approaches or techniques that do not displace labor, and utilization of intermediate technology. The goal of development must be to reestablish an ecologically stabilized society at new levels of both human development and technological interaction with the natural environment.

Literature Cited

Atkinson, L. J., and D. E. Kunkel. 1976. *High-yielding Varieties of Rice in the Philippines.* Foreign Agr. Econ. Rept. No. 113. Washington, D.C.: USDA.

Barrass, R. 1974. *Biology: Food and People.* New York: St. Martin's Press.

Bell, R. H. V. 1971. A grazing ecosystem in the Serengeti. *Sci. Amer.* 225(1):86–93.

Bourliere, F., and M. Hadley. 1970. The ecology of tropical savannas. *Ann. Rev. Ecol. Syst.* 1:125–152.

Boyko, H. 1967. Salt-water agriculture. *Sci. Amer.* 216(3):89–96.

Cox, G. W. 1978. The ecology of famine: an overview. *Ecology of Food and Nutrition* 6:207–220.

Crawford, S. M., and M. A. Crawford. 1974. An examination of systems of management of wild and domestic animals based on the African ecosystems. In *Animal Agriculture.* H. H. Cole and M. Ronning (eds.). pp. 218–234. San Francisco: W. H. Freeman and Company.

Dalrymple, D. G. 1975. *Measuring the Green Revolution: The Impact of Research on Wheat and Rice Production.* Foreign Agr. Econ. Rept. No. 106, USDA.

Dasmann, F. R. 1964. *African Game Ranching.* New York: Macmillan Co.

Dima, S. A. F., and V. F. Amann. 1975. Small holder farm development through intermediate technology. *E. Afr. J. Rural Dev.* 8(112):215–245.

Dorn, H. F. 1962. World population growth: an international dilemma. *Science* 135:283–290.

Douglas, J. S. 1973. Forest-farming: an ecological approach to increase nature's food productivity. *Impact of Sci. on Soc.* 23:117–132.

Evenari, M., L. Shanan, and N. Tadmor. 1971. *The Negev.* Cambridge, Mass.: Harvard University Press.

———. 1975. The agricultural potential of loess and loess-like soils in arid and semi-arid zones. In *Land Evaluation in Arid and Semi-arid Zones of Latin America.* pp. 489–507. Rome: Arti Gratiche e. Cossidente.

Evenari, M., L. Shanan, N. Tadmore, U. Nessler, A. Rogel, O. Schenk. Undated. *Fields and Pastures in Deserts. A Low Cost Method for Agriculture in Semi-arid Lands.* Darmstadt: Eduard Roether Verlag.

Feder, E. 1974. Notes on the new penetration of the agricultures of developing countries by industrial nations. *Boletin de Estudios Latinoamericanas y del Caribe* (Amsterdam) 16: 67–74.

Foster, G. M. 1962. *Traditional Cultures: and the Impact of Technological Change.* New York: Harper and Row.

Frantz, S. C. 1976. Rats in the granary. *Nat. His.* 85(2):10–21.

Greenland, D. J. 1975. Bringing the Green Revolution to the shifting cultivator. *Science* 190: 841–844.

Janzen, D. H. 1973. Tropical agroecosystems. *Science* 182:1212–1219.

Kangas, L. W. 1970. Integrated incentives for fertility control. *Science* 169:1278–1283.

Lele, U. 1975. *The design of rural development.* Baltimore: Johns Hopkins University Press.

Lofchie, M. F. 1975. Political and economic origins of African hunger. *J. Mod. Afr. Stud.* 13 (4):551–567.

McNaughton, S. J. 1976. Serengeti migratory wildebeeste: facilitation of energy flow by grazing. *Science* 191:92–94.

Norman, D. W. 1974. Rationalising mixed cropping under indigenous conditions: the example of northern Nigeria. *J. Dev. Studies* 11(1):3–21.

Norse, D. 1976. Development strategies and the world food problem. *J. Agr. Econ.* 27:137–157.

Okai, M. 1975. The development of ox cultivation practices in Uganda. *E. Afr. J. Rural Dev.* 8:191–214.

Pollock, N. C. 1969. Some observations on game ranching in southern Africa. *Biol. Cons.* 2(1): 18–24.

Ravikovitch, S. 1973. Effects of brackish irrigation water and fertilizers on millet and corn. *Exp. Agr.* 9(2):181–188.

Schaefer, A. E., J. M. May, and D. L. McLellan 1970. Nutrition and technical assistance. In *Malnutrition is a Problem of Ecology.* P. Gyorgy and O. L. Kline (eds.). Bibliotheca Nutritio et Dieta, No. 14, pp. 101–109. Basel: S. Karger.

Schumacher, E. F. 1973. *Small Is Beautiful.* London: Blond and Briggs.

Sprague, G. F. 1975. Agriculture in China. *Science* 188:549–555.

Taylor, C. R. 1969. The eland and the oryx. *Sci. Amer.* 220 (1):88–96.

Uchendu, V. C. 1975. The role of intermediate technology in East African agricultural development. *E. Afr. J. Rural Dev.* 8:182–190.

Vega, J. 1971. Fertility, the soil condition which is most limiting to agriculture in Latin America. In *Systematic Land and Water Resource Appraisal. Proceedings of the FAO/ UNDP Latin America, Mexico, November.*

Wilkinson, R. G. 1974. Reproductive constraints in ecologically stabilized societies. *In Population and Its Problems: A Plain Man's Guide.* H. B. Parry (ed.). pp. 294–299. Oxford: Clarendon Press.

World Bank. 1975. *The Assault on World Poverty.* Baltimore: Johns Hopkins University Press.

27

INTERNATIONAL AGRICULTURAL POLICY

The food production systems upon which different segments of the world's population depend have become increasingly interrelated. Political and economic ties have created a complex, yet imperfect, set of trade channels for basic food commodities, channels upon which many nations have grown dangerously dependent. Systems of virtually instantaneous mass communication transmit news and influence public opinion within much of the world population. It is literally true that major successes or failures of food production in local areas have worldwide effects. The success or failure of North American grain crops has a growing influence on the price of bread in many other parts of the world. The decline of an important marine fishery, such as the Peruvian anchoveta fishery, affects the price of soy products for human consumption in areas as distant as Japan. Famine in areas such as the Sahel of Africa or Bangladesh draws food aid from various parts of the developed and developing worlds.

At an even more basic level, the biosphere, that portion of the earth on and within which life exists, is another system of closely interrelated components. Terrestrial, freshwater, and marine ecosystems are bound together by global flows of energy and cycles of matter. Terrestrial climates are influenced by the location and circulation systems of nearby oceans. The fertility and productivity of fresh and marine water areas are strongly determined by the rate at which these areas receive critical nutrients from the land. Here again, local events can exert global effects. Volcanic action in a single location can introduce quantities of ash into the upper atmosphere, and this can modify solar radiation and climate over latitudinal belts around the globe.

These systems of human food production and global flows of energy and matter are of course cross-linked. Human activities, including those concerned with agriculture, presently affect global patterns of energy exchange and the cycling of nearly all elements (Hutchinson 1970). In turn, the processes of the biosphere set limits upon terrestrial agriculture and aquatic food production by determining patterns of oceanic circulation, global climate, and soil development. The consequence of these cross linkages is that the strategy of agriculture employed in one region does influence both the economics and ecology of food production in other regions. Thus we may conclude (1) that agricultural policy at the international level is essential and (2) that such policy must have a sound ecological basis.

We must also recognize that this policy must accommodate the wide range of relationships that exist between human populations and their productive environments. To emphasize this point we may distinguish three major groups of nations according to their population densities and food production. The first group includes several large developed countries, such as the United States, the USSR, Canada, and Australia, that possess relatively low population densities and relatively low levels of food production per unit area of land, but high levels of food production per capita. Members of the second group are the smaller developed countries, such as Japan, England, the Netherlands, and Israel, that possess higher population densities and high levels of production per unit area of land, but lower levels of food production per capita. The first group includes several of the major food exporting nations of the world; the second includes countries that have considerable food production capacity combined with economies strong in other areas, but that depend to some degree on net importation of basic food commodities. The third group of nations, however, in a sense combines the least favorable aspects of the first two groups: high population densities, low levels of food production per unit of land area, and low levels of food production per capita. These are the developing nations with the most severe problems of food supply; for example, India, Indonesia, Egypt, and various nations of Latin America.

Obviously, many nations do not fit neatly into these classifications, and in fact their utility is to emphasize the range of difference among nations rather than to classify them in logical fashion. Agricultural policy at the international level must be adapted to this full range of conditions of population density, production per capita, and production per unit area.

In discussing international agricultural policy, we shall concentrate on the major questions on which international coordination of activity and effort must center. Each of these questions implies a goal that is more or less clearly defined in its nature, but each is more immediately concerned with the actions that must be taken to move in the general direction of that goal. In no case is the course of appropriate action clearly mapped in considerable detail. Neither is it likely for the course of appropriate action to be identical for all nations. The following questions of international policy are central to the problem of mankind's survival:

1. *What strategy should be followed in developing and allocating of energy resources for use in agriculture?*

Before the industrial revolution, human populations were dependent upon traditional systems of agriculture for which most of the needed cultural energy was supplied by human and animal labor. The capacity of these systems varied greatly but averaged about 1100 kilograms of staple carbohydrate food per hectare (Chancellor and Goss 1976). Subsequently, the relationship between populations and food production has deviated from this balance in two directions: One segment of the human population has mobilized nonrenewable sources of energy to greatly increase production per hectare and more than meet their own needs; a second has expanded in size far beyond the capacity of traditional production methods to meet needs from the available land.

The solution to our growing dependence upon the most rapidly exhaustible energy resources requires many adjustments, the most crucial of which relate to the input of energy to the production process. Many existing and potential sources of energy are available to man (Figure 27-1), and it is apparent that the development and allocation of energy resources must involve them all. We should note that these

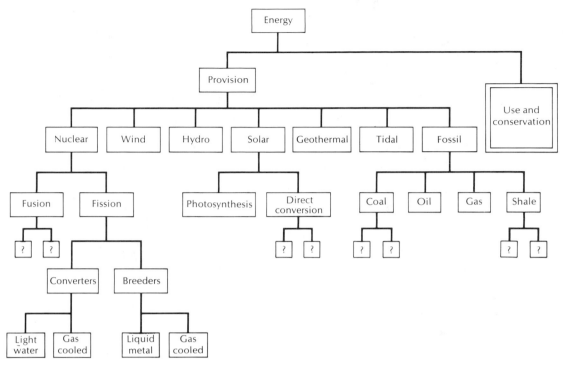

FIGURE 27-1.
A variety of important energy resources are available to man. The potential of these resources has not undergone a balanced exploration, however. In particular, until recently, a very large fraction of United States government research effort concentrated on a single system, the liquid metal fast breeder reactor. (Modified from Rose 1974.)

energy sources fall into two major categories: **exhaustible resources,** including fossil and nuclear fuels; and **flow resources** such as wind, water, the sun, and related sources that are not exhausted by their use. Quantities of fuel available for certain fusion power sources that may or may not eventually be developed are effectively unlimited, but until these are demonstrated practical we shall consider nuclear energy an exhaustible resource. Geothermal energy, although theoretically exhaustible, appears virtually unlimited compared to the potential for exploiting it; we have therefore considered it a flow resource (Figure 27-2).

Past energy policies have had several major weaknesses. First, they have strongly emphasized the development of technologies that increase the rate at which exhaustible resources are consumed (Randall 1975). Examples of this approach are efforts to find ways of substituting less scarce, but also exhaustible, fossil fuels—such as coal and oil shale—for scarce fuels such as petroleum and natural gas (Figure 27-3). Second, they have emphasized the development of a very limited number of long-range technologies at the expense of other potentially valuable lines of effort (Rose 1974). United States government funding of the liquid metal fast

FIGURE 27-2.
Geothermal power plant at Cerro Prieto, Baja California, Mexico. The potential of geothermal power production, and the technology for exploiting this potential, remain largely unexplored. (Photograph by E. E. Hertzog, courtesy of the U.S. Bureau of Reclamation.)

FIGURE 27-3.
Large coal reserves exist in parts of the western United States. Many of these deposits can be mined by stripping, as in this operation near Billings, Montana. The impacts of such operations upon the long-term productivity of the land must be weighed carefully against their benefits. (Photograph by Boyd Norton courtesy of EPA-DOCUMERICA.)

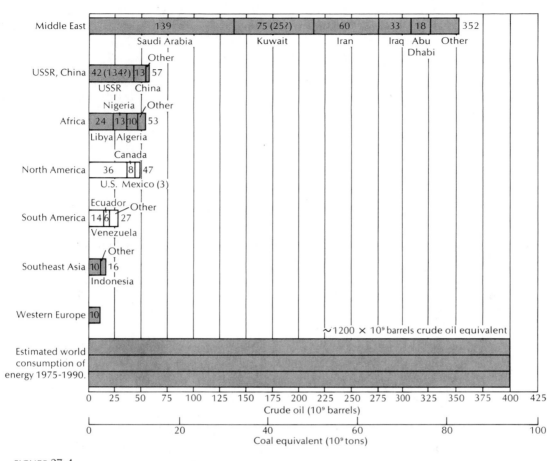

FIGURE 27-4.

Proved world reserves of crude oil total 562 billion barrels; their geographical distribution is shown here. More than half the proved reserve is concentrated in five Middle East nations. The USSR and China together possess about 10 percent of the total. The entire Western Hemisphere has 13 percent. The estimate for the United States includes 10 billion barrels from the North Slope of Alaska. The National Petroleum Council estimated that, at the end of 1970, some 385 billion barrels of oil remained to be discovered on United States territory or immediately offshore. This is believed to represent about one-half of all oil that is ultimately discoverable. For the world as a whole, the National Petroleum Council estimates that proved reserves can be doubled in the next 15 years. For this reason the projected world total consumption of energy between 1975 and 1990 (some 1200 billion barrels) is not as alarming as it might appear. In fact, natural crude oil will probably be supplying at least 60 percent of the world's total energy demand even in 1990. (From "Energy Policy in the U.S." by David J. Rose. Copyright © 1974 by Scientific American, Inc. All rights reserved.)

breeder reactor is an example of this emphasis. Third, flow resources in general have been ignored. In particular, these include solar and wind energy. Fourth, the potential to increase energy conservation and the efficiency of energy use has received only token attention.

In some cases the dangers of accelerating the exploitation of exhaustible resources are recognized; in other instances overoptimism prevails. Proved reserves of crude oil are quite limited relative to the rate of world energy consumption (Figure 27-4). World coal reserves are

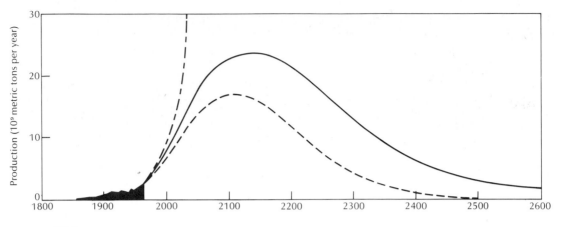

FIGURE 27-5.
Cycle of world coal production is plotted on the basis of estimated supplies and rates of production. The top curve reflects an estimate of 7.6×10^{12} metric tons as the initial supply of minable coal; the bottom curve reflects an estimate of 4.3×10^{12} metric tons. The curve that rises to the top of the graph shows the trend if production were to continue rising at the present rate of 3.56 percent per year. The amount of coal mined and burned in the century beginning in 1870 is shown by the black area at left. (From "The Energy Resources of the Earth" by M. King Hubbert. Copyright © 1971 by Scientific American, Inc. All rights reserved.)

much larger, but the probable pattern of exploitation (Figure 27-5) is such that the importance of even this resource is likely to decline between the years 2100 and 2200 (Hubbert 1971). Global reserves of uranium oxide, which is required for existing fission reactors and future breeder reactors, are also limited (Figure 27-6). Moreover, recent analyses of the rate at which minable reserves or uranium oxide are discovered in the United States suggest that past estimates may be as much as three and one-third times too high (Lieberman 1976). This implies that most of our uranium resources may be exhausted before satisfactory breeder reactor technology can be developed and implemented on a wide scale.

We therefore suggest three main areas of emphasis for agricultural energy policy: (1) the conservation of exhaustible energy resources in other sectors of the economy, which would preserve these resources for critical roles in agriculture; (2) reduction in the dependence of agriculture production upon the most limited exhaustible sources of energy; and (3) development of flow resources for more extensive use in agriculture.

Economic growth, particularly in the industrialized nations, has emphasized the production and consumption of material goods and services at rates increasing faster than the rate of population growth in order to achieve higher standards of living (Irwin and Penn 1975; Randall 1975). The enormous current scale of these activities in developed nations consumes most of their total energy. Although the benefits of such energy use are in some cases great, it is also clear that there is great potential to reduce energy use in this sector and to conserve it for use in more essential areas such as food production.

Agriculture is currently too dependent on the two most critical forms of exhaustible energy: natural gas and petroleum. Because natural gas and naphtha (from crude oil refining) are basic feedstocks for the production of most ammonia fertilizers, the production of nitrogen fertilizers is one of the most fossil fuel-intensive activities

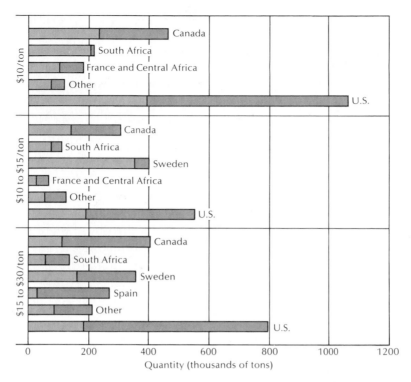

FIGURE 27-6.
World reserves of uranium oxide (U_3O_8). The basal portion of each bar represents reasonably certain supplies, and the outer part represents estimated additional supplies. (From "The Energy Resources of the Earth" by M. King Hubbert. Copyright © 1971 by Scientific American, Inc. All rights reserved.)

associated with intensive mechanized farming. The development of alternative ways of providing abundant nitrogen to crops is an immediate priority. Considerable potential for such development lies in the area of biological fixation (accomplished by interplanting legumes with nonlegume crops) and in improving alternative industrial processes of ammonia production, such as the splitting of water molecules to provide hydrogen for combining with nitrogen to yield ammonia. This process is currently very costly in energy.

The development of fuels and energy systems based on solar, wind, and biomass resources also needs emphasis. One goal of such efforts should be to develop power units for farming equip-

ment that can be recharged or refueled from on-farm energy sources such as windmills, solar converters, or methane generators.

2. To what extent should we encourage the development of food supply systems that require high inputs of materials and energy?

In most industrialized nations, agriculture has developed according to the rules of free enterprise. When energy and materials resources were both abundant and inexpensive, labor costs quickly became a critical constraint on profitability. Substituting cheap energy and materials for labor was thus favored to the point where the dollar gain from such additional input equalled the cost of the input itself. These favorable cost relationships have prevailed so long that agriculture in the developed world now has a strong infrastructural commitment to this approach. At all economic levels, from the farm through farm-related business to the organization of society as a whole, the long-term investment in energy-intensive food production systems is high. Indeed, it is believed by many to represent the only economically viable system compatible with a high standard of living.

This approach to agriculture has been exported during recent years under the name of the Green Revolution. The high-yielding varieties of rice and wheat upon which the Green Revolution was based were bred specifically for their responsiveness to energy and materials inputs, particularly to fertilizers and pesticides. The adoption of these crop varieties with the complete package of inputs for which they are adjusted requires a commitment to the style of agriculture practiced in the industrialized nations. Because resource availability and the industrial potential of most developing nations are low, these countries must purchase the necessary machinery, pesticides, and other manufactured inputs from the developed nations, a relationship that can create a serious form of economic imperialism (Feder 1974). Furthermore, in many of the developing countries, the prospects for providing the quantities of required fertilizers, especially nitrogen and phosphorus, are not good. To achieve an annual economic growth rate of 2.1 percent per capita, coupled with an annual population growth rate of 2.4 percent, it would be necessary to increase fertilizer use in the developing nations from 14.8 million tons in 1970 to 48 million tons in 1985, according to USDA estimates (Allen 1976). The escalating costs of petroleum and the limited number of high-quality phosphate rock deposits (see Chapter 12) will make such increases very costly. The

price of rice paid to the producer in many areas of Southeast Asia, despite the need for more food, already stands in a highly unfavorable ratio to the cost of fertilizer, a relation that correlates closely with the average rice production achieved per hectare (Table 27-1).

Clearly, the efficient use of energy and materials inputs in agriculture is as important a goal as increasing the availability of productive inputs. Several areas of opportunity exist for improving the efficiency with which fossil fuels and other exhaustible resources are used: closer adjustment of machine power to job requirements; increased sharing of specialized equipment; replacement of machine use by other techniques; a shift from chemical control to increased reliance on genetic resistance and biological control; a reversion to extensive grazing instead of intensive feeding for livestock production; and the substitution of skilled or unskilled labor for operations that now consume large quantities of exhaustible energy resources.

Increasing the efficiency of energy and materials uses in food production implies the need to measure total inputs carefully and to relate them to the actual yield obtained. Analyses of this sort for energy efficiency (see Chapter 24) indicate that agriculture in industrialized nations consumes one to several calories of energy for every food calorie yielded. The importance of these kinds of energy input-output relations has led to the suggestion that **net energy analysis,** the analysis of input-output relations in energy units of the same quality, offers a clearer guide to public policy than does monetary cost analysis (Gilliand 1975). It is certainly true that such analyses deserve to be carried out and examined carefully. However, the assumption that energy considerations alone, weighing energy units from different sources equally, override all other resource considerations is unrealistic. Energy policy in agriculture must be meshed

TABLE 27-1
Rice and fertilizer prices and rice yields in various Asian
countries in 1970.

Country	A Price for rice paid to producers (U.S. $/kg)	B Fertilizer price paid by producers (U.S. $/kg)	Price ratio (A/B)	Rice yield (tons/ha)
Japan	0.307	0.215	1.43	5.64
South Korea	0.184	0.191	0.96	4.55
Taiwan	0.117	0.262	0.45	4.16
Malaysia	0.088	0.203	0.44	2.72
Ceylon	0.113	0.158	0.72	2.64
Indonesia	0.045	0.152	0.30	2.14
Thailand	0.045	0.143	0.32	1.97
Philippines	0.070	0.173	0.41	1.72
Burma	0.031	0.251	0.12	1.70

Source: Allen 1975.

with policy that relates to many other resources and social considerations (Huettner 1976; Spedding and Walsingham 1976).

3. *What emphasis should be placed on the roles of skilled and unskilled labor in agriculture?*

The inevitable increase in costs of energy and materials inputs for intensive agriculture will lead inevitably to a readjustment in the role of manpower in food production. The rate and extent of this readjustment and the particular use made of manpower in agriculture are matters of importance to agricultural planners (Figure 27-7). In the developed nations, the idea of any such readjustment is often dismissed out of hand on the assumption that increased labor intensiveness means a return to the most tedious physical drudgery of farming. For this reason, we wish to distinguish between skilled and unskilled labor and to emphasize that the

appropriate role of labor intensification may be strikingly different for developing and developed countries.

In developing countries, where rural labor supplies are relatively large and inexpensive (see Chapter 26), the goal must be to utilize the labor force without displacing it. Because most of its members are unskilled, this must be done in a technologically uncomplicated fashion. This is the objective of intermediate technology (Schumacher 1973). Even in regions where agriculture is carried out largely by human and animal labor, bottlenecks to the production process may exist because of labor shortages. For example, enough animal and manpower may be available to plant and harvest a larger area than can effectively be weeded by hand during the growing period. Introducing a simple and inexpensive device to increase the capability of a man to weed crops is therefore an obvious need. The strategy of intermediate technology is thus to identify

FIGURE 27-7.
Certain types of farming activities remain relatively labor intensive, even in the
developed countries. Whether a greater degree of automation is desirable, even in
agricultural systems of the developed world, is highly questionable. (Photograph by E. E.
Hertzog, courtesy of the U.S. Bureau of Reclamation.)

such bottlenecks and to alleviate them by developing simple, inexpensive, and energy efficient equipment.

In areas of intensive mechanized agriculture, increases in energy and materials costs tend to favor the reintroduction of labor to farming, but with greater emphasis upon skilled labor. This is already beginning to happen in many areas of the United States with the emergence of several private enterprises dealing in pest management and other advisory services. The success of these services rests on the concept that increases in net profits may be higher than the cost of hiring individuals with special expertise in evaluating pest control, fertilizer, and other costly inputs.

In the San Joaquin Valley of California, cotton and citrus growers have been hiring integrated pest management services for nearly 20 years. Typically, these services are provided by individuals who have specialized university-level training in entomology and pest management and who monitor pest populations and recommend specific control actions.

The success of this approach has recently been evaluated (Hall, Norgaard, and True 1975). Dollar yields (gross) and pesticide treatment costs were compared for farms that used and did not use pest management services in 1970 and 1971. It was found that cotton farmers who used these services spent about 59 percent less per acre for insecticides and also averaged $22.40 more in gross yield of cotton. Citrus growers reduced their insecticide costs by 51 percent and obtained average yields $12.95 greater in value per acre. The net financial gain to the farmer was therefore the value of the increased yield plus the savings on insecticide expense. Because the services of pest management consultants cost only $2.50 per acre, clear advantages of this system are demonstrated.

One objective of agricultural policy should therefore be to identify areas in which skilled services can reduce the need for expensive inputs to production, as well as to promote training programs for such individuals.

4. What changes in rural economic and development policy must be implemented to achieve the appropriate level of labor participation in agriculture?

The increase in the costs of materials and energy inputs to agricultural production, which in theory makes alternative labor-intensive approaches more competitive, may not be great enough to stimulate optimal participation of labor in food production. Some of the chronic problems in developing countries at present are inadequate food production, soaring urban unemployment, and agricultural input costs close to those that prevail in developed countries. As we have noted, the rate of urbanization in developing countries is greater than in the developed nations (see Chapter 26). The rapid growth of cities in the developing world is a product of three major forces: higher material standards of living in cities due in part to the developmental emphasis on industrialization and urban development; unavailability of land to large segments of the rural population; and high rates of population growth (Turner 1976). Under these circumstances, higher energy and materials costs to farmers may lead to reduced production instead of to increased reliance on labor.

At a somewhat more sophisticated level, the same is true of the developed nations, in which the population is already highly urbanized. Urban residents tend not to perceive rural life, particularly on farms, as attractive. Thus despite economic difficulties in the large cities of the United States and various European countries, large numbers of chronically unemployed persons stay in the cities and press for welfare and other aid programs designed to resolve the problem there.

It is clear in both cases that programs must be developed to encourage and assist the reentry of labor, both skilled and unskilled, into the rural segment of the economy. Two major kinds of effort can be distinguished here: First, a change in the cultural and developmental emphases that relate to rural and urban life; and second, implementation of land reform systems that enable individuals to actively reenter farming.

In the developing nations, serious and sustained attention to basic rural development programs is long overdue. Since World War II, the emphasis in these nations has been on developing the urban-industrial and cash-crop sectors of the economy, with little attention to improving the infrastructure of basic food production activities (Lofchie 1975). The need for rural development programs of various sorts, including improved technology, agricultural extension services, marketing systems, and agricultural credit, is obvious (Lele 1975). Land reform is also needed in many countries. There is strong evidence that food production per hectare is greater on smaller farm units; labor intensiveness on such farm units is also higher (Table 27-2). In Colombia, for example, studies have shown that labor input to farming is about 2.7 man-years per hectare of farm plots 0.5 hectare or less in size, and that this figure declines steadily to a value of 0.17 man-years per hectare on ranches 500 to 1000 hectares in size (World Bank 1975). There is therefore considerable reason to believe that well-designed land reform programs can both increase production and absorb labor.

An additional consideration, however, is the fact that smaller farm units generally produce a smaller marketable surplus than do larger farms; increased labor intensiveness also means that there are more mouths to be fed on the farm! Despite this relationship, the total

TABLE 27-2

Ratios of output per hectare and output per worker on small subfamily farms (numerator) to outputs on large multifamily farms (denominator) for various Latin American countries.

Country		Output per hectare	Output per worker
Argentina	1960	8.20	0.21
Brazil	1950	8.80	0.14
Chile	1955	8.20	0.23
Colombia	1960	14.30	0.10
Ecuador	1954	2.80	—
Guatemala	1950	3.90	0.14

Source: World Bank 1975.

market output from an area that has undergone land reform may increase through higher production per hectare, even though the marketable surplus per hectare is a smaller percentage of the total production. In any case, the goal of increasing the number of people in active working relations to food production is achieved through land reform. The World Bank has supported carefully planned land reform programs in recent years (World Bank 1975).

In the developed nations, government policies designed to protect family farms provide improved social and cultural services to rural areas and encourage the skilled individuals to move into small communities where their services can be utilized. Deliberate efforts should be made to promote the advantages of rural living and to offer economic redevelopment programs, such as job training and housing development, that are more strongly competitive with those provided in urban areas.

5. *What emphasis should be placed upon expanding the area of agricultural land as opposed to intensifying production on land already being used for agriculture?*

When we focus on the supply side of the food-population question, we recognize many ways in which production can be increased. In fact, many economists assume that levels of production per hectare can be increased considerably through the same techniques that have been used in the past few decades in the developed nations. Some warning signs have appeared in this regard, however. One of the most important indexes of production as it relates to effort is yield related to the quantity of fertilizer nutrients required per hectare; this index has shown a strong tendency to level off as input levels continue to increase (Figure 27-8). The alternative way of increasing food supply, which has been important through much of human history, is to expand the area of agricultural land.

Here we are faced with conflicting considerations, each with a strong ecological basis. Should the additional productive inputs that become available to us, such as fertilizers and fuels, be applied to production on new agri-

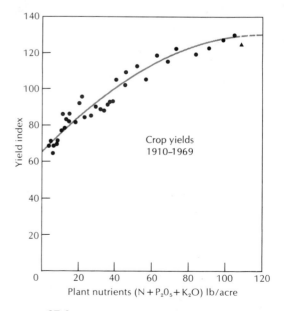

FIGURE 27-8.
Fertilizer use and crop yields in the United States from 1910 through 1969. (From NAS 1975.)

cultural lands, where their effectiveness has not yet approached the point of diminishing returns? Or should these inputs be applied to the most productive existing lands, to avoid possible damage to lands that are very likely marginal for agriculture and ecologically fragile? In effect, should we strive to maximize the efficiency with which we use productive inputs, or minimize the long-term degradation of marginal land areas?

Both considerations are obviously important. The policy problem thus becomes one of evaluating production efficiency, assessing the quality of potential new land areas, and weighing developmental efforts appropriately in these two directions. The appropriate balance is likely to differ widely in different parts of the developed and developing worlds.

In the United States, several attempts have been made to predict the effects of various trends and policies on the agricultural land area and the food production capacity of the nation through the year 2000 (Carr and Culver 1972; Heady et al. 1972). These projections have assumed various patterns of population and economic growth, as well as various constraints upon agricultural technology, such as restricted fertilizer and pesticide use. The projections themselves are the product of complex computer models and are "scenarios" or illustrative possibilities, for future conditions, rather than strong predictions.

Carr and Culver (1972) present five such scenarios (Table 27-3), all of which, incidentally, were prepared before the Arab oil embargo in 1974. Scenario IV suggests that with a slow rate of population growth and no restriction upon fertilizer and pesticide use, the food needs of an additional 62 million people in the year 2000 could be met by less than a 5 percent increase in the area of harvested cropland. Most of the additional food production, which is projected to increase by 49 percent over the 1970 level, would therefore be due to increased production per hectare.

At more rapid rates of population growth, or with restrictions on fertilizer and pesticide use, some expansion of the harvested cropland is projected. In scenario II, rapid population growth, coupled with such restrictions, leads to an increase of over 27 percent in harvested cropland, largely at the expense of pastures and rangelands. Total food production increases more than in scenario IV; but the increase is less for livestock than for crops, and projected food prices, especially for meat, are higher than for scenario IV. With the restriction on agrochemical use in scenario II, it may be noted that the labor input in agriculture remains much higher than for scenario IV.

Heady et al. (1972) have conducted similar analyses in which they examined nine scenarios projecting population levels of 280 to 325 million and various constraints imposed by irrigation water cost, restricted insecticide use, removal of land from use, and technology trend. These analyses suggest that a great deal of adjustment is possible by means of regional shifts in crop production patterns in response to economic pressures. As a result these workers project that most conditions envisioned by 2000 can be responded to with only minor adjustments in total cropland area.

We believe that it is very difficult in these studies to estimate the quantities and costs of the added inputs required to achieve increased production per unit land area. Projections based on past trends are inadequate for this purpose. Furthermore, the effectiveness of additional inputs begins to decline at some point. In Chapter 6 we examined the trend of energy input and corn production in the midwestern United States. This example suggests that in some areas of the developed world mechanized production has reached the point of rapidly diminishing return on additional inputs, a point that seems to lie at a level of net primary production close to that typical of natural ecosystems of the same region. Elevating agroecosystem productivity

TABLE 27-3
Scenarios of United States agriculture in 2000 A.D.

	1970	Year 2000 scenarios				
		I	II	III	IV	V
Population (10^6)	204	321	321	321	266	266
GNP ($\$ \times 10^9$)	831	2991	2991	2414	2774	2239
Technology (U or R)[a]		U	R	R	U	R
Land (10^6 acres)						
Cropland	440	442	471	468	410	422
Harvested	344	391	438	436	359	390
Unharvested	96	31	33	32	51	32
Pasture and range	636	617	580	580	613	617
Other	29	27	27	27	27	27
Total	1105	1066	1078	1075	1050	1066
Inputs						
Nitrogen (10^6 tons)	6.3	14.8	9.4	9.2	12.9	8.2
Pesticides (10^6 lbs)	410	662	137	134	608	122
Labor (10^9 man-hours)	6.5	2.5	4.0	3.9	2.3	3.5
Production (value for 1967 = 100)						
Crops	101	175	165	164	157	149
Livestock	104	166	150	154	139	129
Overall	102	171	161	159	149	139
Prices (value for 1967 = 100)						
Crops	91	101	117	113	85	93
Livestock	107	109	123	113	97	100
Overall	100	105	120	113	92	97

[a]U = unrestricted growth in the use of fertilizers, pesticides, and other agrochemicals;
R = severe restriction on the use of agrochemicals.
Source: Carr and Culver 1972.

to higher levels may be justified only under conditions that allow the increase to be achieved in a highly efficient manner; for example, by irrigation or fertilization in cases where these practices alleviate critical limitations at small energy cost. As a result, in the developed world, the need to use resources efficiently may be a strong force favoring the expansion of agricultural land area.

In most areas of the developing world, on the other hand, production per unit area is well below that which could be realized with reasonable inputs and sound husbandry. Moreover, most land with good potential for permanent agriculture is already in use. The expansion of cultivation in these regions is, in fact, creating serious dangers to fragile environments, particularly those of mountain regions (Eckholm 1976) and tropical forests (Farnworth and Golley 1973). Efforts to expand the areas of land in agriculture in the developing nations are nevertheless strongly supported by national and international agencies. We believe these efforts should be evaluated much more carefully for their long-term ecological dangers, and that more effort should be devoted to increasing production on existing lands.

6. *What strategies should be employed to adjust food production and supply to variable or changing climatic conditions?*

Long-term planning of agricultural development by national and international agencies has grossly neglected the effects of major patterns of year-to-year variability and the possibility of secular climatic change (Allen 1976). Most climatologists agree that, within recent historical time, there have been significant shifts in average conditions of temperature and precipitation and that some of these changes have persisted for years, decades, or centuries (Figure 27-9). Although the degree of change may be quite small, some of these changes have strikingly influenced human activity and occupation in regions of marginal climate (Bryson 1974a). There is general agreement that, most recently, mean temperature, at least for the Northern Hemisphere, increased gradually from the late 1800s or early 1900s through the early 1940s and then declined progressively and somewhat more rapidly (Bryson 1974b; Lamb 1974). Opinions differ considerably about the role that man may have played in these trends by accelerating CO_2 release through fossil fuel burning and by introducing particulates into the atmosphere by various processes.

Opinions also vary widely over whether the most recent trend—declining global temperature—signals the return to a cooler mean climate that is likely to persist for some time. Bryson (1973) and Lamb (1974) believe that this may be the case and that the specific climatic changes involved will include cooler, wetter, and shorter summers in the higher latitudes and more severe droughts in the Sahel of Africa and in areas that depend on summer monsoon rains.

Perhaps the most pessimistic assessment of the situation is that prepared recently by the U.S. Central Intelligence Agency (1974). This report, based heavily on Bryson's views, concludes that a climatic shift to conditions like those of the 1880s is likely, and that these changed conditions will probably persist for at least 70 years. The report further explores the possible effects of such change on agriculture in a number of critical areas. Drought conditions, capable of inducing extensive famines unless countered by massive grain imports, are foreseen once every four years in India and once every five years in China. Serious reductions in grain production capacity are also predicted for northern Europe, Canada, and the Kazakhstan region of the Soviet Union.

Whether or not secular shifts do occur on such a scale, it seems clear that the world has enjoyed better than average weather conditions for the past several decades. It is also apparent that unfavorable weather can very greatly influence modern systems of grain production. In 1975, for example, grain production fell 76.6 million tons—more than 35 percent—below expectation in the Soviet Union due to unfavorable weather (*Time* 1975). These relationships obviously deserve serious attention by organizations engaged in developing world food policies (Figure 27-10).

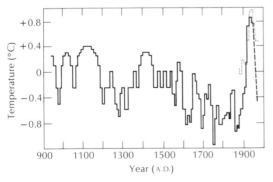

FIGURE 27-9.
Variation in the mean annual temperature of Iceland by decades over the past one thousand years. The dashed line indicates the rate of temperature decline in the 1961–1971 period; the dotted line is the variation in mean temperature for the Northern Hemisphere as a whole. (From Bryson 1975.)

FIGURE 27-10.
In spite of storage facilities like these grain elevators at Hutchinson, Kansas, carryover stocks of world grain now equal less than 10 percent of annual grain consumption, a value that corresponds roughly to the difference in production between a favorable year and an unfavorable year. (Photograph courtesy of the USDA.)

Two major areas are indicated in which research should be intensified. First, a greater understanding of climatic trends and their possible influence on human activities is needed. Second, strategies for adjusting agriculture to changing climates must be explored. Some agricultural economists have generally assumed that such adjustments will occur automatically as a result of market relationships and the development of new technology. Although considerable potential for adjustment does exist in areas where shifts in temperature or precipitation occur in ranges that do not come near the limiting extremes of these factors, this is not necessarily the case in areas already on the margins of drought or extreme cold. Furthermore, the investment that farmers and supporting business and industry have in the technology related to

certain crops may limit the rate of adjustments. Thus the degree of adjustment possible in the face of climatic deterioration needs to be determined, and specific plans must be formulated to enable that adjustment to be made quickly and efficiently. Provision must also be made for climatic impacts that may render significant land areas unproductive, both in terms of the use of the land itself and in compensating for lost production.

7. *What means should be employed to avoid and combat the evolutionary responses of agricultural pests and disease organisms?*

A special committee of the National Academy of Sciences of the United States has recently issued a report warning that the effectiveness of chemical pest control techniques may soon break down (Carter 1976). Several developments make this a serious possibility. First, genetic resistance to pesticide effects has been found in over 200 species of insects and other pests, including virtually all the major pests of agriculture. Second, an increasing number of pest problems are being recognized as resulting from the fact that pesticides applied to control one species can disturb the natural biological mechanisms that control other species. Third, the use of chemical pesticides, which are tending to become more exotic and toxic, is increasingly being constrained by regulations concerning residues on harvested materials, environmental side effects, and occupational health and safety. In Chapter 16 we noted many of these points and observed that, when they were first brought into use, synthetic insecticides were highly effective in crop protection. More recently, however, the losses due to insects have increased despite the increasing use of insecticide. This is certainly a reflection of several factors, including lower crop tolerances for insect damage; but declining insecticide effectiveness is one important cause.

Similar patterns of evolutionary response can be expected in response to herbicide use, a danger that has not yet been generally recognized. Herbicides now represent the class of pesticides whose use is increasing most rapidly. Moreover, the effectiveness of herbicides in reducing losses from weed competition is impressive. In their basic biological characteristics, however, weeds are similar to pest insects in having high reproductive potential, rapid growth, and high dispersal potential; counter-evolutionary response is only a matter of time. But because weeds are physiologically more similar to crop plants than insects are, solving the problems of herbicide resistance without causing physiological damage to the crop itself may be even more difficult than overcoming insecticide resistance.

The dangers of microbial diseases—bacterial, fungal, and viral—are also likely to increase as world agriculture expands and intensifies. In human disease it has long been recognized that particular cultures are characterized by their patterns of disease. In the modern world, for example, man is on the verge of eliminating certain of the most feared diseases of the past, such as leprosy and smallpox. With the development of large cities and rapid air transportation, however, he has set the stage for epidemics of other diseases, such as viral influenza, that are highly mutable and can be dispersed suddenly into human populations that lack immunity to the strains that cause them. Similarly, the development on several continents of extensive monocultures of a few basic crops, such as wheat, sets the stage for similar patterns of crop disease outbreaks. New varieties of disease agents are more likely to originate in scattered, semi-isolated host populations. As international trade increases, so does the likelihood that new strains will enter areas where crops lack resistance. This is essentially what occurred in the case of the 1970 corn blight epiphytotic in the United States (see Chapter 21). The same is true for viral and bacterial diseases of domestic animals, whose vulnerability to the sudden appearance of a number of existing diseases is already high in areas such as the United States.

These problems present severe difficulties for international agricultural policy, since some of the most straightforward approaches to their treatment impede the freer and more rapid movement of basic food products. Legislative control, exercised through expanded systems of inspection and quarantine, will nevertheless be required to an increasing degree.

Intensified research into pest and disease control systems less easily overcome by pest species is another priority. The approaches that seem to be most promising (Carter 1976) include breeding for pest and disease resistance, developing specific bacterial and viral control agents, designing other specific chemical agents useful in control, using sterile male techniques, and relying more on integrated control systems (see Chapter 20).

8. What balance should be struck between domestic food production and reliance on imported food?

Individual nations must strive to meet internal food needs by some combination of domestic production and importation of basic foods. In the latter case, they must rely on areas of the national economy that gain foreign exchange and on international market relationships that permit the import of needed food quantities at reasonable prices. During the early 1970s, many of the developing nations found themselves in the uncomfortable position of having inadequate internal production patterns and of having to operate in very erratic international trade situations. The questions highlighted by this situation concerned the degree of self-sufficiency that should be encouraged in basic food production, the nature of international systems of trade stabilization that might be implemented, and the strength and long-range reliability of segments of the national economy that support net food importation patterns. We shall begin with the second of these questions.

Stabilizing international food trade by establishing internationally administered **food banks** or food funds has become the object of considerable interest. The typical argument for food banks is that they could smooth out uncontrollable year-to-year variations in world food production, mainly grains. The need for food banks cannot be due to this factor alone, however (Johnson 1975). The pattern of variation in total world grain production over the past 25 years has been such that it would be economical, in terms of the relation between storage costs and grain price, to hold reserves of over 10 million tons in only 1 out of 20 years (with average world production equalling about 1.2 billion tons).

The critical consideration relates to the price of grain in international trade, and specifically to efforts by various nations and national consortia to stabilize the price of grain within their own regions. This price stabilization would require either supply or demand relationships to be very elastic, that is, to be subject to broad adjustments via economic relationships. A change in demand requires a change in a populations eating habits; for example, the consumption of more or less grain-fed beef. In most developing countries, the opportunity to make this sort of adjustment in "luxury consumption" does not exist. Supply can be made elastic instead either through manipulating exports and imports or through a food commodity storage system. Such systems tend to stabilize prices within the area they serve.

Much of the instability in the price of grain in international trade during recent years has been the result of attempts by developed nations to stabilize grain prices internally (Johnson 1975). All the variability in production and in supply-demand interaction has thus been thrown into the international market. Furthermore, as long-term trade relationships are established among various developed nations, the remaining variability affects primarily the developing nations.

If it were possible to make international trade in grains freer, much of the irregularity in world supply and price would be eliminated. This is not likely to happen, however, because of the strong concern of the United States and groups such as the European Economic Community for internal stabilization. What course of action remains, particularly for the developing nations? Greater emphasis on internal self-sufficiency of production is one possibility. This, of course, would ultimately reduce both the potential export market for developed nations and perhaps their freedom in using export-import mechanisms to stabilize food prices internally. A second possibility is a food reserve or food bank that would store grains when production is high and prices are low and would

FIGURE 27-11.
In the 1970s, exports of grain by the United States increased sharply, and agricultural sales abroad contributed significantly to a favorable balance of payments in international trade. (Photograph courtesy of the USDA.)

release grain at reasonable prices during years of shortage. In principle, it seems that this approach could be implemented more rapidly than could major improvements in internal self-sufficiency.

There is considerable evidence that food importing nations would, in fact, benefit from such a system (Johnson 1975). As grain prices have increased, demand appears to have become less elastic. In recent years, this has worked to the advantage of grain exporting nations, enabling them to improve their balance of trade relationships with grain importing countries (Figure 27-11). This has obviously encouraged favor for food bank systems among importing nations and dampened enthusiasm among grain exporting countries.

Nevertheless, the degree of reliance upon food banks, however effective, must be carefully assessed. In cases where much of a nation's foreign exchange is derived from the sale of

nonrenewable resources, such as Chile's sale of copper, greater reliance upon internal self-sufficiency of basic food production would seem to be the only realistic course of action. In other areas, where sustained production of important commodities such as lumber, sugar, coffee, and other items can be achieved with efficiencies that are high compared to those possible in grain exporting countries, greater reliance on international trade seems practical. Here, the liberalization of nonagricultural trade arrangements may help developing countries to improve their capacity to trade in food markets.

9. What policies will be followed to avoid calorie/protein and other nutritional trade-offs of detrimental nature?

A strong case can be made for the argument that the introduction of corn and cassava to

Africa from the New World has contributed significantly to widespread protein-calorie malnutrition in Africa. These two crops, both of which have high yields of basic carbohydrate food material, are seriously deficient in protein and essential amino acids. Because of their high yield potential, however, their introduction to Africa was in some sense an earlier Green Revolution. To a lesser degree, the accomplishments of the most recent Green Revolution, which involve high-yield varieties of wheat and rice, carry the same side effect. In areas where these high-yielding varieties have been heavily adopted, one consequence has been a decline in the acreage devoted to legumes (Aylward and Jul 1975). Thus it is clear that certain kinds of agricultural development programs, especially those designed by agricultural scientists and economists alone, may contain hidden, detrimental nutritional trade-offs.

In matters of nutritional policy, one of the most severe problems is that integration of the relevant disciplines is rarely achieved. Economists, for example, concentrate on food commodities that are highly quantifiable and that move in channels of trade; they also tend to encourage approaches that are theoretically cost-efficient, but not always practical (Dwyer and Mayer 1975). Physicians and professional nutritionists often show tunnel vision, as well; the solutions to nutritional problems are frequently viewed only within the narrow professional area involved. One of the major objectives of health organizations must therefore be to develop a comprehensive approach that can influence various components of the food supply system that affect nutrition.

A second need is an ecological approach to nutrition that accepts conditions as they exist in areas where subsistence agriculture is the rule, and that employs appropriate and indigenous approaches to righting nutritional imbalances. It is now apparent that, although problems of nutritional protein inadequacy exist, no "protein gap" exists that cannot be closed except by revolutionary new approaches (Aylward and Jul 1975). The protein needs of individuals are now better understood, and are generally less than they once were thought to be (see Chapter 1). These needs are generally met when adequate caloric intake is realized, except under conditions in which this intake is composed largely of a protein-deficient carbohydrate food, or when illness, pregnancy, or lactation place extra protein demands upon the individual.

Diversification is one important approach to achieving an adequate intake of protein and other nutritional factors. No ideal diet, with respect to protein, really exists, but many suitable diets are possible. These are made possible by mixing the intake of various foods so that the protein or specific amino acids lacking in one food can be obtained from other foods (Aylward and Jul 1975). Likewise, increased efforts to encourage the consumption of fruits, vegetables, meat from small animals such as rabbits or fowl, and other locally produced food items, are likely to be more successful than large-scale programs that involve unfamiliar food materials and new techniques of food preparation. These possibilities are often ignored because they lie outside the limits of organized agriculture that are recognized by economists and government planners.

Furthermore, it is apparent that education can play an important role in the area of nutrition, especially in developing countries (Dwyer and Mayer 1975). Here, much of the population is rapidly changing from cultural patterns in which traditional knowledge was generally an adequate nutritional guideline to new patterns for which they will require a new formal understanding of nutrition. Changes in income alone do not guarantee adequate nutrition in many such situations.

10. *In resolving population and food imbalances, what emphasis should be placed upon programs of population regulation?*

It is clear to virtually all who examine the question of world population growth that all such growth must eventually cease. It is also clear that if population stabilization is not achieved by design and in a humanitarian fashion it will very likely be effected by war, pestilence, and famine. The level at which world population could be maintained, with a reasonable material standard of living, is not agreed on, however. De Wit (1967), assuming no limitations of water or minerals in agricultural production, calculated in simplistic fashion that populations ranging from 79 to 175 billion might be supported. Revelle (1974) suggested that applying agricultural technology like that used for Iowa corn to 2.5 billion hectares of the earth's surface (the 1.4 billion now cultivated plus another 1.1 billion!) could provide a high-quality diet for 38 to 48 billion people. As we have seen, the energy and materials intensiveness of this approach are not likely to be extended worldwide, and it is doubtful that they can even be maintained for long in areas where they are currently practiced. Others (e.g., Brown 1974) suggest that a steady-state population must be limited to 6 billion, or roughly one and one-half the times the present world population—about the population predicted for the year 2000 at current growth rates. Still others have suggested that the human population has already exceeded the carrying capacity of the planet.

In our view, this carrying capacity lies in the lower portion of the range indicated above, probably close to that suggested by Brown. Leveling off the growth of world population therefore requires immediate action. Even if we can successfully apply appropriate means for reducing birth rates, the youthful populations that already exist will create continued growth long after self-replacement birth rates are achieved. What, then, are the appropriate means that we must quickly apply?

Strongly divergent views have recently emerged regarding the course of action and particularly the role of assistance by developed nations. One extreme, the **lifeboat ethic,** has been offered by Hardin (1974). Essentially, this view states that economic and food assistance to many of the poor nations that still possess high population growth rates will only lead to more reproduction and faster growth rates. Therefore, until these nations demonstrate forceful action to reduce birth rates, assistance is not only useless, but suicidal; letting too many people into the lifeboat leads to deaths not only of those in the water, but also of those that were in the lifeboat.

Others, including Murdoch and Oaten (1975), argue that the lifeboat ethic is not only inhumanitarian but unrealistic; nations cannot be likened to lifeboats in their potential ability to repel influences from the world outside. It is also apparent that economic, educational, social, and medical factors of many sorts influence the number of children that people have (Figure 27-12). Furthermore, fairly significant declines in birth rates have been recorded in recent years in a number of the more advanced developing nations (Table 27-4); this suggests that the influences of economic development may not be unique to the already industrialized nations in encouraging smaller family size. Murdoch and Oaten (1975) suggest that the common denominator of the nations in which birth rates have declined is the fact that the economic improvement that did occur was spread broadly throughout the population. In contrast, in countries such as Mexico, Brazil, and Venezuela, where a great deal of development has also occurred, the benefits have been shared very unequally and their populations have continued to increase.

Nevertheless, the rates of population growth in the countries listed in Table 27-4 are still very high. These nations, as well as most of the remaining countries of the developing world, now have family planning or birth control programs of some sort. In some nations, such as India, despite the fact that family plan-

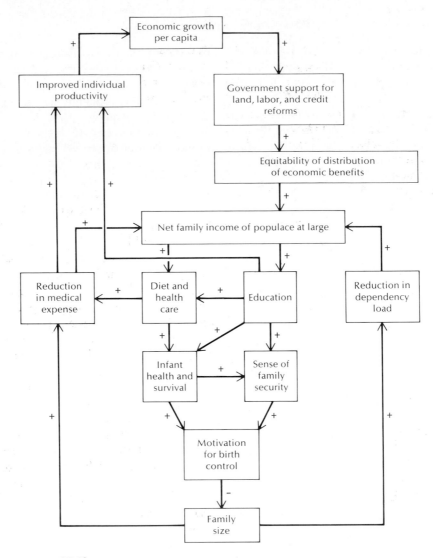

FIGURE 27-12.
The equitability with which economic benefits are distributed throughout the population as a whole may be an important factor in mediating the relationship between economic development and population growth. This diagram suggests some of the ways in which equitability of distribution of such benefits may influence family income and well-being and so favor reduces birth rates. Direct relationships are indicated by + signs, inverse relationships by − signs.

ning programs have existed for many years, birth rates have slowed very little (Figure 27-13). It is only reasonable to conclude that both agricultural development and population control programs must be pursued with greater vigor than ever.

SUMMARY

The fact that the world is an interdependent whole means that agricultural policy must be global in scale. Moreover, the fact that this whole consists of distinct national units with

TABLE 27-4
Declining birth rates and per capita income in selected developing countries.

| Country | Period | Crude birth rate (Births/1000/year) | | 1973 per capita income |
		Average annual decline	1972 rate	
Barbados	1960–69	1.5	22	$570
Taiwan	1955–71	1.2	24	390
Tunisia	1966–71	1.8	35	250
Mauritius	1961–71	1.5	25	240
Hong Kong	1960–72	1.4	19	970
Singapore	1955–72	1.2	23	920
Costa Rica	1963–72	1.5	32	560
South Korea	1960–70	1.2	29	250
Egypt	1966–70	1.7	37	210
Chile	1963–70	1.2	25	720
China			30	160
Cuba			27	530
Sri Lanka			30	110

Source: Murdoch and Oaten 1975.

FIGURE 27-13.
Signs encouraging family planning are familiar sights in India, but progress toward significant reduction of birth rates has been slow. (Photograph by H. Null, courtesy of the FAO, Rome.)

very distinctive problems and potentials of their own means that such policy must be flexible and differentiated. The fact that human food production is a major activity within the biosphere—an ecological enterprise—means that such policy must have a strong ecological basis.

Agricultural policy for the future cannot now be rigidly specified. Certain broad limitations to potential activity exist: the limits of land areas, the influential factors of climate, the potential reserves of nonrenewable resources, and the impossibility of infinite growth in a limited system. The questions of policy must be stated as questions, however, since they correspond to the immediate, unsolved problems with which governments and international agencies must work from day to day and year to year.

These questions fall into several major groups, the first of which concerns the energy and materials resources that must be relied upon by agriculture in the future: the nature of these resources, their quantitative availability, and their cost. Tied inseparably to these considerations is the question of the role of human labor, as well as that of the rural-urban design of future societies.

A second group of questions is more strictly ecological and concerns methods of dealing with climate, the evolutionary responses of man's competitors for food, and the intrinsic factors in nature that tend to limit the productivity of regional environments. These questions relate to those of resources and the efficiency with which man uses them.

A third group of questions is still broader. This group concerns international trade, food reserves, human nutrition, and direct approaches to population limitation. The questions posed in this discussion cannot be deferred until those of politics and human rights are resolved; their solution is a major part of man's adjustment to life on his planetary spaceship.

Literature Cited

Allen, G. 1975. *Agricultural Policies in the Shadow of Malthus.* Lloyd's Bank Review, No. 117, pp. 14–31.

——. 1976. Some aspects of planning world food supplies. *J. Agr. Econ.* 27:97–119.

Aylward, F., and M. Jul. 1975. *Protein and Nutrition Policy in Low-Income Countries.* New York: John Wiley and Sons.

Brown, L. R. 1974. *In the Human Interest.* New York: W. W. Norton and Co.

Bryson, R. A. 1973. Drought in Sahelia: Who or what is to blame? *Ecologist* 3(10):366–371.

——. 1974a. *World Climate and Food Systems.* III. *The Lessons of Climate History.* Univ. Wisc. Inst. Env. Stud., Rept. 27.

——. 1974b. A perspective on climatic change. *Science* 184: 753–760.

——. 1975. The lessons of climatic history. *Env. Cons.* 2:163–170.

Carr, A. B., and D. W. Culver. 1972. *Agriculture, Population, and the Environment.* Commission on Population Growth and the American Future, Research Report No. 3. pp. 183–195. Washington, D. C.: U. S. Government Printing Office.

Carter, L. J. 1976. Pest control: NAS panel warns of possible technological breakdown. *Science* 191:836–837.

Central Intelligence Agency. 1974. *A Study of Climatological Research as it Pertains to Intelligence Problems.* Washington, D.C.: Office of Research and Development, CIA.

Chancellor, W. J., and J. R. Goss. 1976. Balancing energy and food production, 1975–2000. *Science* 192:213–218.

de Wit, C. T. 1967. Photosynthesis: Its relationship to overpopulation. In *Harvesting the Sun*. A. San Petro, F. A. Greer, and T. J. Army (eds.). pp. 315–320. New York: Academic Press.

Dwyer, J. T., and J. Mayer. 1975. Beyond economics and nutrition: The complex basis of food policy. *Science* 188:566–570.

Eckholm, E. P. 1976. *Losing Ground*. New York: W. W. Norton and Co.

Farnworth, E. G., and F. B. Golley (eds.). 1973. *Fragile Ecosystems*. New York: Springer-Verlag.

Feder, E. 1974. Notes on the new penetration of the agricultures of developing countries by industrial nations. *Boletin de Estudios Latinoamericanas y del Caribe* (Amsterdam) 16:67–74.

Gilliand, M. W. 1975. Energy analysis and public policy. *Science* 189:1051–1056.

Hall, D. C., R. B. Norgaard, and P. K. True. 1975. The performance of independent pest management consultants in San Joaquin cotton and citrus. *Calif. Agr.* 29(10):12–14.

Hardin, G. 1974. Living on a lifeboat. *BioSci.* 24(10):561–568.

Heady, E. O., H. C. Madsen, K. J. Nicol, and S. H. Hargrove. 1972. *Agricultural and Water Policies and the Environment*. CARD Report 40T, Ames, Iowa: Iowa State University.

Hubbert, M. K. 1971. The energy resources of the earth. *Sci. Amer.* 224:(3):60–70.

Huettner, D. A. 1976. Net energy analysis: An economic assessment. *Science* 192:101–104.

Hutchinson, G. E. 1970. The biosphere. In *The Biosphere*. pp. 2–11. San Francisco: W. H. Freeman and Company.

Irwin, G. D., and J. B. Penn. 1975. Energy, government policies, and the structure of the food and fiber system. *Amer. J. Agr. Econ.* 57(5):829–835.

Johnson, D. G. 1974. World agriculture, commodity price, and price variability. *Amer. J. Agr. Econ.* 57(5):823–828.

Lamb, H. H. 1974. Reconstructing the climatic patterns of the historical past. *Endeavour* 33:40–47.

Leith, H. 1972. Modeling the primary productivity of the world. *Nature and Resources* 8(2):5-10.

Lele, U. 1975. *The Design of Rural Development*. Baltimore: Johns Hopkins University Press.

Lieberman, M. A. 1976. United States uranium resources—an analysis of historical data. *Science* 192:431–436.

Lofchie, M. F. 1975. Political and economic origins of African hunger. *J. Mod. Afr. Stud.* 13(4):551–567.

Murdoch, W. W., and A. Oaten. 1975. Population and food: Metaphors and the reality. *BioSci.* 25(9):561–567.

Pimentel, D., L. E. Hurd, A. C. Bellotti, M. J. Forster, I. N. Oka, O. D. Scholes, and R. J. Whitman. 1973. Food production and the energy crisis. *Science* 182:443–449.

Randall, A. 1975. Growth, resources, and environment: Some conceptual issues. *Amer. J. Agr. Econ.* 57(5):803–809.

Revelle, R. 1974. Food and population. *Sci. Amer.* 231(3):160–170.

Rose, D. J. 1974. Energy policy in the U.S. *Sci. Amer.* 230(1):20–29.

Schumacher, E. F. 1973. *Small Is Beautiful*. London: Blond and Briggs Ltd.

Spedding, C. R. W., and J. M. Walsingham. 1976. The production and use of energy in agriculture. *J. Agr. Econ.* 27:19–30.

Time, Inc. 1975. Reaping a bad harvest. *Time* 106(24):35.

Turner, F. C. 1976. The rush to the cities in Latin America. *Science* 192:955–962.

World Bank. 1975. *The Assault on World Poverty*. Baltimore: Johns Hopkins University Press.

INDEX

Boldface numbers indicate pages on which topic is defined. *Italicized* numbers refer to figures; references to tables are indicated by the letter *t* following a number.